Lectures on Differential Geometry

GRADUATE STUDIES IN MATHEMATICS 245

Lectures on Differential Geometry

Bennett Chow
Yutze Chow

AMS
AMERICAN MATHEMATICAL SOCIETY
Providence, Rhode Island

2020 *Mathematics Subject Classification.* Primary 53-01, 53C18, 53C20, 53C21, 53C22, 53C40, 53E20, 53E10, 57R18, 58-01.

For additional information and updates on this book, visit
www.ams.org/bookpages/gsm-245

Library of Congress Cataloging-in-Publication Data

Names: Chow, Bennett, author. | Chow, Yutze, author.
Title: Lectures on differential geometry / Bennett Chow, Yutze Chow.
Description: Providence, Rhode Island : American Mathematical Society, [2024] | Series: Graduate studies in mathematics, 1065-7339 ; volume 245 | Includes bibliographical references and index.
Identifiers: LCCN 2024014161 | ISBN 9781470477677 (hardcover) | ISBN 9781470478049 (paperback) | 9781470478032 (ebook)
Subjects: LCSH: Manifolds (Mathematics) | Riemannian geometry. | Submanifolds. | Riemannian manifolds. | Geodesics (Mathematics) | Geometry, Differential. | AMS: Differential geometry – Instructional exposition (textbooks, tutorial papers, etc.). | Differential geometry – Global differential geometry – Global Riemannian geometry, including pinching. | Differential geometry – Global differential geometry – Methods of Riemannian geometry, including PDE methods; curvature restrictions. | Differential geometry – Global differential geometry – Geodesics. | Differential geometry – Global differential geometry – Global submanifolds. | Manifolds and cell complexes – Differential topology – Topology and geometry of orbifolds. | Global analysis, analysis on manifolds – Instructional exposition (textbooks, tutorial papers, etc.).
Classification: LCC QA613 .C486 2024 | DDC 516.3/6–dc23/eng/20240412
LC record available at https://lccn.loc.gov/2024014161

10 9 8 7 6 5 4 3 2 1   29 28 27 26 25 24

We dedicate this book to Wanlin Chow, the wife of the senior author and the mother of the junior author. Additionally, Bennett would like to thank his wife, Jingwei Xia, for her patience, encouragement, and support.

Contents

Preface

Manifolds are essential objects of study in both differential geometry and algebraic topology. In algebraic topology, we define algebraic structures — through homology and homotopy in particular — using the topological structures of the manifolds. In differential geometry, we consider geometric-analytic structures defined on smooth manifolds in ways that are compatible with the smooth structures of manifolds. Notably, these geometric structures are not unique and are in fact quite flexible.

Of the many possible questions, it is natural and pertinent to ask under what conditions there exist canonical geometric structures on manifolds. More generally, one may ask if all smooth closed manifolds admit decompositions into pieces that admit canonical geometric structures. In dimension 3, this question was posed by Thurston, who made substantial progress and finally answered affirmatively by the Hamilton–Perelman proof of the geometrization conjecture. The existence of geometric decompositions has topological consequences, and in particular, the Poincaré conjecture follows. The proof employed the analysis of the Ricci flow, which is a central equation in geometric analysis. In Kähler geometry, the analytic minimal model program asks a similar question. Regarding canonical geometric structures *in* manifolds, minimal submanifolds are important objects of study. Other important canonical geometric objects of study include Einstein metrics; Yang–Mills, self-dual, and Seiberg–Witten connections (which are in the setting of bundles); constant scalar curvature metrics; harmonic maps; and solutions to the Einstein field equations in general relativity.

The goal of this book is to provide an introduction to differential geometry with a focus on Riemannian geometry at the level of a beginning graduate course, where the capstone is an introduction to modern geometric analysis. Motivations for the study of differential geometry come from many sources. For example, in theoretical physics, Einstein's theory of general relativity is based on Lorentzian geometry, while gauge theories and string theory are discussed in the language of differential geometry.

Our goals for readers include the following objectives: to ease the transition from an undergraduate-level to a graduate-level course in differential geometry; to introduce fundamental concepts in differential geometry that are beneficial for applications to fields beyond mathematics such as physics, engineering, biology, and computer science; and to equip them with essential material for further study in geometric analysis. While our treatment of the basics of differential geometry is not intended to be extensive or comprehensive, we aim to provide a clear and engaging pathway toward the study of related or more advanced topics.

Differential geometry is connected to a wide range of subjects in mathematics. As such, we recall basic concepts in topology, real analysis, and abstract algebra to make the book more self-contained, although some prior exposure to these subjects on the reader's part is ideal. Since our treatment is superficial in this respect, we provide some references at the undergraduate level for readers interested in learning more about the background material in these subjects. At a minimum, the reader is assumed to have at least a background in vector calculus and linear algebra. Readers should also have a good understanding of what a mathematical proof is, at the level of an introductory upper-division undergraduate course.

The idea of this book is that differential geometry, as well as geometric analysis, is just supercharged vector calculus. As we will see, the vector calculus aspect is familiar ground, whereas the supercharging aspect is actually quite approachable. In particular, familiarity with partial differential equations is not a necessary prerequisite for the geometric analysis part of this book. In fact, for the most part, one only needs real analysis at the level of Rudin's classic text *Principles of Mathematical Analysis* [**Rud76**], informally known as 'Baby Rudin', and predominantly the key a priori estimates stem from the first and second derivative tests as well as integration by parts in calculus.

We strive for this book to be readable, although at certain points the book is not wholly self-contained in that it assumes certain standard results whose statements are relatively easy to understand even if their proofs are beyond the scope of this book. The hope and expectation, at the cost of such minor inconveniences, are that new topics are opened up for the reader. For those results we do not prove in the book, we provide references that contain examples and accessible details.

Some of the topics we cover are as follows.

Part I on the Geometry of Submanifolds of Euclidean Space. Chapters 1 and 2: Submanifolds in Euclidean space; we study this notion with an eye toward its generalization to differentiable manifolds in Part II. Chapter 3: Tangent, cotangent, normal, and tensor spaces and bundles. Chapter 4: The local extrinsic and intrinsic geometry of submanifolds in Euclidean space. Chapter 5: We discuss some classical theorems on the global extrinsic geometry of submanifolds.

Part II on Differential Topology and Riemannian Geometry. Chapter 6: We introduce abstract smooth manifolds. Chapter 7: We introduce some basic concepts and results in Riemannian geometry, with an emphasis on comparison geometry. Chapter 8: We introduce differential forms and Cartan's method of moving frames. Chapter 9: Using the set-up of the previous chapter, we prove the Gauss–Bonnet formula for surfaces. Chapter 10: We discuss fiber, vector, and principal bundles and Lie groups. Using the unit tangent bundle, we discuss the Chern–Gauss–Bonnet formula.

Part III on Elliptic and Parabolic Equations in Geometric Analysis. Chapter 11: We discuss Bochner formulas, the Bishop–Gromov volume comparison theorem, and linear elliptic and parabolic equations in Riemannian geometry. Chapter 12: We study the use of nonlinear elliptic equations in differential geometry, with the focus on minimal surfaces. Chapter 13: We begin our study of geometric evolution equations, which are parabolic equations in differential geometry. Here, geometric objects evolve according to their curvature and become more uniform. In this chapter our focus is on the curve shortening flow, where embedded plane curves shrink to 'round' points. Chapter 14: We consider the uniformization of surfaces by heat flow techniques. More specifically, we use Hamilton's Ricci flow on surfaces to prove that arbitrary metrics evolve to constant curvature metrics.

Acknowledgments

We would like to express our gratitude to all of the students and colleagues that have assisted us in the preparation of this book. We are greatly indebted to all of the reviewers of the book; their suggestions, without exception, have greatly improved the book. We are grateful to the editors of the AMS Graduate Studies in Mathematics series for all of their help. Special thanks to AMS publisher, Sergei Gelfand, and AMS book acquisitions editor Ina Mette. We are grateful to Ina for her constant splendid work and help as our book editor. We would like to thank Auden Boatman, associate editor Christine Thivierge, and the production editors for their help with the book. Special thanks to Arlene O'Sean for her magnificent copyediting. We would like to thank Peng Lu for comments and suggestions that have improved the book. We would like to thank Chuu-Lian Terng, Dick Palais, and The 3D-XplorMath Consortium for permission to include figures from `https://virtualmathmuseum.org/` in this book. We are indebted to Peter Chow for helping make this book possible. We are also grateful to him for all the help that he has given us.

Bennett Chow

Professor of Mathematics, University of California, San Diego

Yutze Chow

Professor Emeritus of Physics, University of Wisconsin-Milwaukee

Suggested Topics for Courses

The book contains material that can be covered in a year-long graduate course. For a 3-quarter course, selections from Parts I, II, and III may correspond to the fall, winter, and spring quarters, respectively. For a 2-semester course, selections from Chapters 1–8 and Chapters 9–14 may be used as a basis for the fall and spring semesters, respectively. Depending on the level and scope of the course, topics may be skipped at the discretion of the instructor.

Individually, each part of the book may be used as a quarter-long or semester-long course at the appropriate undergraduate or graduate level. For example, Part I may serve as a 1-quarter or 1-semester course for advanced undergraduates with a solid background in point-set or metric space topology, linear algebra, and multivariable calculus. Parts I and II may be used for a 1-semester or 2-quarter introductory course in Riemannian geometry. For students who have already taken a course in smooth manifolds at the graduate level, Part II may be covered in 1 quarter or 1 semester. A follow-up course may consist of Chapters 10 and 11 and selected parts of Chapters 12–14. Part III may also provide a reference for an independent reading course at the graduate level.

Detailed descriptions of the chapters are as follows.

Chapter 1 introduces the idea of submanifolds and gives examples, which include spheres, graphs, and surfaces of revolution. The intention of this chapter is to develop intuition for manifolds, which are the main objects of study in differential geometry.

Chapter 2 gives the rigorous definition of a submanifold of a Euclidean space. Our discussion brings in notions from vector calculus in all dimensions, topology, continuity, differentiability, the implicit function theorem, parametrized submanifolds, immersions, embeddings, and local graphs.

Chapter 3 discusses various concepts within the context of Euclidean submanifolds, including tangent spaces and bundles, normal spaces and bundles, cotangent spaces and bundles, tensors, and tensor products. These objects are essential in furthering our understanding of the geometry of submanifolds.

Chapter 4 discusses the local geometry of submanifolds of Euclidean space. This includes the first and second fundamental forms, the Weingarten map, length and distance, the mean and Gauss curvatures, the Levi-Civita connection, Riemann curvature tensor, Gauss and Codazzi equations, Gauss's Theorema Egregium, and the fundamental theorem of hypersurface theory. We also discuss plane and space curves.

Chapter 5 discusses the global geometry of submanifolds. We look at totally umbilical hypersurfaces, convex hypersurfaces, the Gauss map, the support function, the Laplacian, the Hessian, hypersurface integral identities, and the rigidity of convex surfaces.

Chapter 6 is an introduction to abstract smooth manifolds, including manifolds with boundary, tangent and cotangent spaces, smooth maps, derivatives, diffeomorphisms, tangent and cotangent bundles, vector fields, the Lie bracket, tensors, push-forwards, pull-backs, the Lie derivative, Killing vector fields, and the topological and smooth categories of manifolds.

Chapter 7 discusses the basics of Riemannian geometry, including metrics, length, distance, isometries, connections, parallel transport, geodesics, the Riemann, Ricci and scalar curvatures, the Bianchi identities, the first and second variation of arc length, the exponential map and its properties, Jacobi fields and their properties, conjugate points, the index form and lemma, the cut locus, the injectivity radius, holonomy, and constant curvature spaces. We also discuss the curvature tensor decomposition, the Weyl tensor, and locally conformally flat manifolds. Key results include the Hopf–Rinow theorem, the Gauss lemma, Myers's theorem, the Rauch theorem, and the Cartan–Hadamard theorem.

Chapter 8 discusses moving frames and the Cartan structure equations, which form a useful method of calculation in Riemannian geometry. We also discuss orientations, differential forms, exterior differentiation, integration, and the fundamentally important Stokes's theorem, which is the generalization to manifolds of the fundamental theorem of calculus. We sketch the idea of the proof of Stokes's theorem and refer the reader to other books for its full proof. We also discuss conformally equivalent metrics and hyperbolic 2-space and we state the uniformization theorem for surfaces, which is proved in Chapter 14. Finally, we discuss submanifolds of Riemannian manifolds using moving frames.

Chapter 9 discusses one of the most iconic results in differential geometry, the Gauss–Bonnet formula for surfaces and 2-dimensional orbifolds. Our discussion relies on moving frames. We also develop aspects of intersection number, degree of a map, and Morse theory, and we state the Poincaré–Hopf theorem.

Chapter 10 puts into the modern abstract context some Riemannian geometry concepts. We consider the Liouville theorem for the geodesic flow on the unit tangent bundle, vector bundles, Lie groups, fiber bundles, principal bundles, connections on bundles, index theorems on 4-manifolds (statements only), and Einstein manifolds. The highlight of this chapter is the Chern–Gauss–Bonnet formula.

Chapter 11 discusses basic concepts regarding elliptic equations in Riemannian geometry. The fundamental Bochner–Weitzenböck formula plays a key role by bringing the Ricci tensor to the fore. We discuss eigenvalues and eigenfunctions of the Laplacian, the Lichnerowicz eigenvalue theorem, the geometry of geodesic spheres, the Riccati equation, the Laplacian comparison theorem, the Bishop–Gromov volume comparison theorem, the Cheeger–Gromoll splitting theorem, Cheng's diameter rigidity theorem, Yau's Liouville theorem for harmonic functions, the Li and Yau eigenvalue estimate, the heat equation, and Hodge theory.

Chapter 12 discusses minimal surfaces, the monotonicity and mean value formulas, minimal graphs, the Bernstein theorem, the Simons identity, curvature estimates, and stable minimal surfaces in 3-manifolds with positive scalar curvature. We also state the Plateau problem and give an introduction to harmonic maps.

Chapter 13 discusses the isoperimetric inequality, geometric flows of plane curves, namely the inverse curve shortening flow and an introduction to singularity analysis for the curve shortening flow, including the Gage and Hamilton theorem for convex embedded curves and the Grayson theorem for general embedded curves. In the context of the curve shortening flow, we discuss methods such as Gage's monotonicity formula, Hamilton's entropy and Harnack estimates, Huisken's monotonicity formula and chord-arc estimate, and blow-up techniques for singularity analysis.

Chapter 14 discusses Hamilton's work on the Ricci flow on surfaces and in particular a proof of the differential geometric version of the uniformization of complex analysis. We also consider 2-dimensional orbifolds. General ideas which translate to Ricci flows in general dimensions include short-time existence, the pointwise maximum principle estimates, higher derivatives of curvature estimates, integral monotonicity formulas, differential Harnack estimates, entropy estimates, noncollapsing, and Ricci solitons. An appendix on the Kazdan–Warner identity is also included.

Notation and Symbols

$\odot$	Kulkarni–Nomizu product
$A \setminus B$	set difference between A and B
$\alpha \smile \beta$	concatenation of paths α and β
b	Bianchi map
$B_r(p)$	open geodesic ball of radius r centered at p
$B(V,W)$	second fundamental form of a submanifold of any codimension
C^k	k times continuously differentiable
C^∞	infinitely differentiable
$\mathrm{C}\,M$	bundle of curvature tensors
cn_K	modulated cosine function
d	distance or exterior derivative
$d(x,y)$	distance between x and y
d_p	distance to p
D_p	injectivity domain of $\exp_p$
$\mathcal{D}_p$	injectivity range of $\exp_p$
ds	arc-length parameter
$D_V W$	Euclidean covariant derivative
δ	codifferential
Δ	Laplacian, a.k.a. Laplace operator
Δ_d	Hodge Laplacian
$\det$	determinant

diam	diameter
dA	volume form of an oriented 2-dimensional Riemannian manifold; a.k.a. area form
$d\mu$	volume form of an oriented Riemannian manifold
div	divergence
e	Euler's number
$\{e_i\}_{i=1}^n$	orthonormal frame or orthonormal frame field
$\mathbb{E}^n$	Euclidean space ($\mathbb{R}^n$ with the Euclidean metric $g_{\mathbb{E}}$)
$\exp_p$	exponential map at p
F	a parametrization of a hypersurface
g	Riemannian metric
$g_{\mathbb{E}}$	Euclidean metric
g_{S^n}	standard metric on S^n (the unit n-sphere)
$[g]$	pointwise conformal class of g
G	Gauss map
G^n	graph that is an n-dimensional hypersurface
$\mathrm{Gr}(k,n)$	Grassmannian of k-dimensional vector subspaces of $\mathbb{R}^n$
$\mathrm{Gr}(2,T_xM)$	Grassmannian of 2-planes in T_xM
Γ_{ij}^k	Christoffel symbols
γ_V	constant-speed geodesic with initial velocity V
H	mean curvature
$\mathbf{H}^p(M^n)$	space of harmonic p-forms on M^n
$*$	Hodge star operator
H_K	mean curvature in the simply connected space form of sectional curvature K
$\mathfrak{hol}_p$	holonomy Lie algebra
Hol_p	holonomy group at p
Hol_p^0	restricted holonomy group at p
I	first fundamental form
II	second fundamental form
$\langle\cdot,\cdot\rangle$	inner product; when applied to tangent vectors; this is the same as the metric g
J	Jacobian of the exponential map
J	Jacobi field
K	Gauss curvature of a hypersurface
κ	geodesic curvature of a path

$\langle\langle\cdot,\cdot\rangle\rangle_{L^2}$	L^2-inner product of tensors
$\Vert\cdot\Vert_{L^2}$	L^2-norm of a tensor
$\lambda_1(\Delta)$	first eigenvalue of the Laplacian
$\Lambda^p T^* M$	the vector bundle of p-forms on M^n
$\Lambda^2 M$	the vector bundle of $\Lambda^2 T^* M$
L	length or Weingarten map
$[\cdot,\cdot]$	Lie bracket
$\mathcal{L}$	Lie derivative
Loop_p	set of piecewise smooth loops based at p
M^n	smooth manifold of dimension n
(M^n, g)	Riemannian manifold of dimension n
∇	covariant derivative or gradient
∇^Γ	covariant derivative along the map Γ
∇^2	Hessian operator
N	unit normal to a hypersurface
$\vert\cdot\vert$	norm of a tensor (e.g., tangent vector) on a manifold
$\Vert\cdot\Vert$	norm of a vector in Euclidean space
$\{\omega^j\}_{j=1}^n$	orthonormal coframe or orthonormal coframe field
ω_n	volume of the unit Euclidean n-ball
ω_i^j	connection 1-form
Ω_i^j	curvature 2-form
$\Omega^p(M^n)$	space of p-forms (sections of $\Lambda^p T^* M$)
$\Omega_p M$	domain of the exponential map $\exp_p$
$\mathrm{O}(n)$	orthogonal group
$\mathrm{O}(T_pM, g_p)$	orthogonal group of (T_pM, g_p)
$P_\gamma^\Vert$	parallel transport along γ
π_1	fundamental group
$\mathbb{R}_+$	the set of strictly positive real numbers
R	scalar curvature
Ric	Ricci tensor
$\overset{\circ}{\mathrm{Ric}}$	trace-free Ricci tensor
Rm	Riemann curvature tensor or Riemann curvature operator
s	the arc-length parameter
S^n	the unit n-sphere
S^n/Γ	spherical space form

$S_r(p)$	geodesic sphere of radius r centered at p
S_pM	unit tangent space at p
S^2M	bundle of symmetric 2-tensors
S_0^2M	bundle of symmetric trace-free 2-tensors
σ	the support function of a hypersurface
sect	sectional curvature function
sn_K	modulated sine function
TM	tangent bundle of M^n
T_xM	tangent space of M^n at x
$\otimes$	tensor product
$\mathrm{trace}_{k,\ell}$	trace on the k-th and ℓ-th components of a tensor
$\mathfrak{U}$	open cover
$\mathrm{W}M$	bundle of Weyl curvature tensors
Wy	Weyl tensor
$\mathbf{x}$ or $\mathbf{y}$, $\cdots$	typically a point in $\mathbb{R}^n$ or $\mathbb{R}^{n+1}$

Part I

Geometry of Submanifolds of Euclidean Space

Chapter 1

Intuitive Introduction to Submanifolds in Euclidean Space

This book is about differential geometry, which is the study of the geometry of *manifolds*. In Part I of this book, we study the special class of manifolds that are defined as being contained in some higher-dimensional Euclidean space, which are called *submanifolds*. In Part II we begin the study of manifolds in general. In Part III we study analysis on manifolds and its geometric applications.

Regarding our everyday experience, submanifolds are everywhere. For example, the surface of a ping pong ball is a 2-dimensional manifold; so are the surfaces of doughnuts and Möbius strips. In this chapter, we discuss surfaces in an intuitive manner and through pictures help the reader visualize these spherical, toroidal, and twisted shapes, waiting until later to define them rigorously.

1.1. The idea of a submanifold as being locally Euclidean

What is a submanifold? In this chapter we will start with a more intuitive approach to answering this question:

> The idea of an n-dimensional submanifold is that it is a subset of Euclidean space $\mathbb{R}^N$ that is *locally Euclidean* of dimension n, where $N \geq n$.

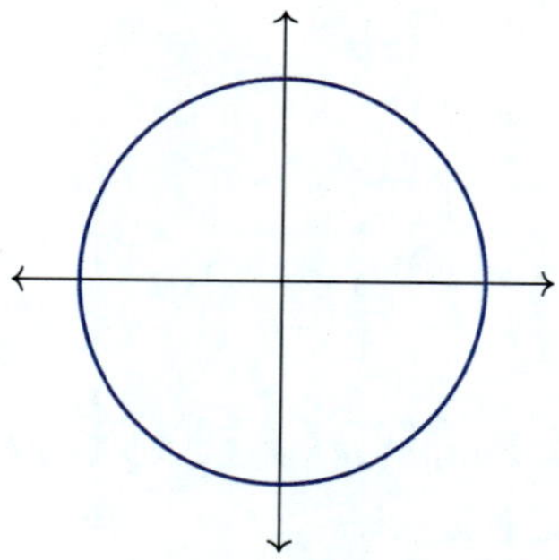

Figure 1.1.1. A **circle** is a 1-dimensional submanifold of $\mathbb{R}^2$.

Our present discussion of submanifolds will largely be based on examples and visualization. In the next chapter, we will introduce the rigorous definition of submanifold. However, in this chapter, we will freely use the term 'submanifold' in a less formal manner before subsequently presenting its definition.

As one of the simplest and yet important examples, a *circle* is a 1-dimensional submanifold of $\mathbb{R}^2$. See Figure 1.1.1.

Take a point on the circle and look at a *neighborhood* of it (informally, we say that we are looking *locally* at the circle). The intersection of the neighborhood and the circle is an arc. If we straighten out an arc, we get an interval, which is locally what 1-dimensional Euclidean space $\mathbb{R}$ looks like. See Figures 1.1.2–1.1.4.

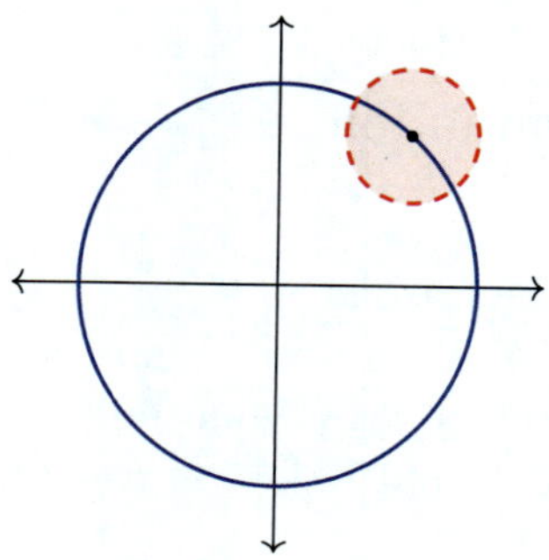

Figure 1.1.2. A neighborhood (the pink disk) in $\mathbb{R}^2$ of a point on a circle.

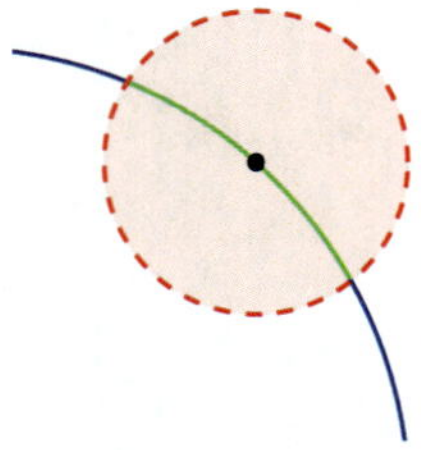

Figure 1.1.3. The green arc represents the intersection of the circle and the neighborhood.

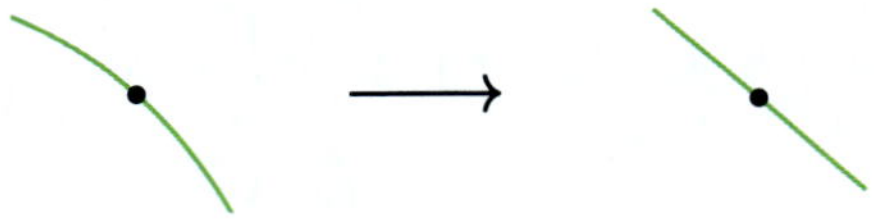

Figure 1.1.4. The green arc in Figure 1.1.3 can be smoothly straightened out to an interval. In this sense, the circle is locally Euclidean.

1.2. Euclidean and vector inner product spaces

The study of the *geometry* of submanifolds begins with considering Euclidean space as an inner product space. The Euclidean inner product yields the notions of *distance*, *length*, and *angle*, and hence geometry.[1]

Euclidean space $\mathbb{R}^n$ is defined to be the n-fold Cartesian product of $\mathbb{R}$ with itself. We may consider elements of $\mathbb{R}^n$ as either **points** or **vectors**. We use lowercase letters to denote elements considered as points and uppercase letters for elements considered as vectors. We may write a point $\mathbf{x} \in \mathbb{R}^n$ as an n-tuple

$$\mathbf{x} = (x^1, x^2, \ldots, x^n), \tag{1.1}$$

where $x^i \in \mathbb{R}$ for each $1 \leq i \leq n$. Note that the components x^i of $\mathbf{x}$ have *upper* indices; this will be our convention. We denote the **origin** in $\mathbb{R}^n$ by

$$\mathbf{0} = (0, 0, \ldots, 0). \tag{1.2}$$

We denote vectors similarly.

[1]In dimension 2, we are thus considering the geometry of Euclid. The notions of length and angle lead to the concepts of similarity and congruence of triangles and the definitions of special types of triangles such as isosceles and equilateral triangles. However, the notion of inner product is a much more modern development, introduced around the 18th or 19th century.

1.2.1. Vector spaces. It is natural to generalize the algebraic operations of addition and multiplication from $\mathbb{R}$ to $\mathbb{R}^n$. Euclidean space $\mathbb{R}^n$ is an example of a *vector space*. We can add vectors $\mathbf{X} = (X^1, X^2, \ldots, X^n)$ and $\mathbf{Y} = (Y^1, Y^2, \ldots, Y^n)$:

$$\mathbf{X} + \mathbf{Y} = (X^1 + Y^1, X^2 + Y^2, \ldots, X^n + Y^n), \tag{1.3}$$

and we have scalar multiplication:

$$a\mathbf{X} = (aX^1, aX^2, \ldots, aX^n). \tag{1.4}$$

Let $\mathbf{X}, \mathbf{Y}, \mathbf{Z} \in \mathbb{R}^n$ and $a, b \in \mathbb{R}$. The vector space properties for addition are

(Associative law)	$(\mathbf{X} + \mathbf{Y}) + \mathbf{Z} = \mathbf{X} + (\mathbf{Y} + \mathbf{Z}),$
(Commutative law)	$\mathbf{Y} + \mathbf{X} = \mathbf{X} + \mathbf{Y},$
(Additive identity)	$\mathbf{X} + \mathbf{0} = \mathbf{0} + \mathbf{X} = \mathbf{X},$
(Inverse)	$\mathbf{X} + (-\mathbf{X}) = (-\mathbf{X}) + \mathbf{X} = \mathbf{0}.$

The vector space properties for scalar multiplication are

(Multiplicative compatibility)	$a(b\mathbf{X}) = (ab)\mathbf{X},$
(Multiplicative identity)	$1\mathbf{X} = \mathbf{X},$
(Vector addition distributivity)	$a(\mathbf{X} + \mathbf{Y}) = a\mathbf{X} + a\mathbf{Y},$
(Scalar addition distributivity)	$(a + b)\mathbf{X} = a\mathbf{X} + b\mathbf{X}.$

With the introduction of the Euclidean inner product, the study of Euclidean geometry and trigonometry becomes more algebraic.

In general, when a set $\mathbb{V}$ with a binary operation $+$, called **addition**, and a field $\mathbb{F}$ (such as $\mathbb{R}$) with an action (a.k.a. external binary operation) $\cdot : \mathbb{F} \times \mathbb{V} \to \mathbb{V}$, called **scalar multiplication**, satisfies these axioms for all $a, b \in \mathbb{F}$ and $\mathbf{X}, \mathbf{Y}, \mathbf{Z} \in \mathbb{V}$, some $\mathbf{0} \in \mathbb{V}$, and $1 \in \mathbb{F}$ the field identity,[2] then we say that $\mathbb{V}$ is a **vector space over** $\mathbb{F}$. In this book, in the context of real differential geometry, we will always take $\mathbb{F} = \mathbb{R}$. In complex differential geometry, which has relations with algebraic geometry, one takes $\mathbb{F} = \mathbb{C}$, the field of complex numbers.

Let $\mathbb{V}$ be a real vector space. The dual space $\mathbb{V}^*$ is the space of **linear functions** from $\mathbb{V}$ to $\mathbb{R}$. We denote this as

$$\mathbb{V}^* := \mathrm{Lin}(\mathbb{V}, \mathbb{R}). \tag{1.5}$$

Another notation for the right-hand side is $\mathrm{Hom}(\mathbb{V}, \mathbb{R})$, where Hom is for *homomorphism*.

[2] We denote the additive identity of $\mathbb{V}$ by $\mathbf{0}$.

1.2.2. Inner product vector spaces. We use the notations $\cdot$ or $\langle\, ,\rangle$ to denote the **Euclidean inner product** defined by

$$\mathbf{X}\cdot\mathbf{Y} := \langle \mathbf{X},\mathbf{Y}\rangle := X^1Y^1 + X^2Y^2 + \cdots + X^nY^n. \tag{1.6}$$

Let $\mathbf{X},\mathbf{Y},\mathbf{Z}\in\mathbb{R}^n$ and $a,b\in\mathbb{R}$. The inner product properties are

(Symmetry) $\mathbf{Y}\cdot\mathbf{X} = \mathbf{X}\cdot\mathbf{Y}$,

(Bilinearity) $(a\mathbf{X}+b\mathbf{Y})\cdot\mathbf{Z} = a(\mathbf{X}\cdot\mathbf{Z}) + b(\mathbf{Y}\cdot\mathbf{Z})$,

(Positive-definite) $\mathbf{W}\cdot\mathbf{W} > 0 \quad$ for $\mathbf{W}\in\mathbb{R}^n\setminus\{\mathbf{0}\}$.

The second property is linearity in the first argument, but since $\cdot$ is symmetric, it implies *bilinearity*; namely, we also have

$$\mathbf{Z}\cdot(a\mathbf{X}+b\mathbf{Y}) = a(\mathbf{Z}\cdot\mathbf{X}) + b(\mathbf{Z}\cdot\mathbf{Y}).$$

In general, when a vector space $\mathbb{V}$ over a field $\mathbb{F}$ with a map $\cdot : \mathbb{V}\times\mathbb{V}\to\mathbb{F}$ satisfies these axioms for all $a,b\in\mathbb{F}$, $\mathbf{X},\mathbf{Y},\mathbf{Z}\in\mathbb{V}$, and $\mathbf{W}\in\mathbb{V}\setminus\{\mathbf{0}\}$, then we say that $\cdot$ is an **inner product** on $\mathbb{V}$ and that $(\mathbb{V},\cdot)$ is an **inner product space**. In Exercise 1.1 you are asked to give the precise definition of an inner product (vector) space.

1.2.3. General inner products on $\mathbb{R}^n$. The notion of Euclidean inner product extends to general inner products on vector spaces; for simplicity, we consider the special case where the vector spaces are Euclidean spaces. Let

$$A : \mathbb{R}^n\times\mathbb{R}^n\to\mathbb{R} \tag{1.7}$$

define an inner product on $\mathbb{R}^n$. That is, we are taking $\mathbb{V}=\mathbb{R}^n$ in the definition of an inner product space, but we are not necessarily taking $\cdot$ to be the Euclidean inner product. By definition, A is a symmetric, positive-definite bilinear function; i.e.,

$$A(c\mathbf{V}_1+\mathbf{V}_2,\mathbf{W}) = cA(\mathbf{V}_1,\mathbf{W}) + A(\mathbf{V}_2,\mathbf{W}), \tag{1.8a}$$

$$A(\mathbf{V},c\mathbf{W}_1+\mathbf{W}_2) = cA(\mathbf{V},\mathbf{W}_1) + A(\mathbf{V},\mathbf{W}_2), \tag{1.8b}$$

$$A(\mathbf{V},\mathbf{W}) = A(\mathbf{W},\mathbf{V}), \tag{1.8c}$$

$$A(\mathbf{V},\mathbf{V}) > 0 \quad \text{if } \mathbf{V}\neq\mathbf{0}. \tag{1.8d}$$

Note that the second equation (1.8b) follows from the first (1.8a) and the third (1.8c). Given A, we have correspondingly the symmetric positive-definite matrix

$$a := (a_{ij})_{i,j=1}^n \tag{1.9}$$

defined by

$$a_{ij} := A(\mathbf{e}_i,\mathbf{e}_j), \tag{1.10}$$

where $\{\mathbf{e}_1, \ldots, \mathbf{e}_n\}$ denotes the standard Euclidean basis of $\mathbb{R}^n$; that is, $\mathbf{e}_i := (0, \ldots, 0, 1, 0, \ldots, 0)$, where the 1 is in the i-th slot. We have (see Exercise 1.2)

$$A(\mathbf{V}, \mathbf{W}) = \sum_{i,j=1}^{n} a_{ij} V^i W^j. \tag{1.11}$$

Conversely, given a symmetric positive-definite matrix

$$a = (a_{ij})_{i,j=1}^{n},$$

the bilinear function $A : \mathbb{R}^n \times \mathbb{R}^n \to \mathbb{R}$ defined by (1.11) is the unique inner product satisfying

$$A(\mathbf{e}_i, \mathbf{e}_j) = a_{ij}.$$

So we have a one-to-one correspondence between inner products on $\mathbb{R}^n$ and symmetric positive-definite $n \times n$ matrices.

The Euclidean inner product is just one inner product on $\mathbb{R}^n$, namely the one corresponding to the $n \times n$ identity matrix I_n:

$$\mathbf{e}_i \cdot \mathbf{e}_j = \delta_{ij} = I_n(\mathbf{e}_i, \mathbf{e}_j),$$

where δ_{ij} is the Kronecker delta function defined by

$$\delta_{ij} := \begin{cases} 1 & \text{if } i = j, \\ 0 & \text{if } i \neq j. \end{cases} \tag{1.12}$$

1.2.4. Euclidean distances and angles. The Euclidean inner product determines distances and angles and hence the geometry of Euclidean space. This principle is fundamental to the study of differential geometry. The **length** of a vector $\mathbf{X} \in \mathbb{R}^n$ is defined by

$$\|\mathbf{X}\| := \sqrt{\mathbf{X} \cdot \mathbf{X}} = \sqrt{(X^1)^2 + (X^2)^2 + \cdots + (X^n)^2}. \tag{1.13}$$

The Euclidean distance $d_{\mathbb{E}}$ between two points $\mathbf{x}$ and $\mathbf{y}$ is defined by

$$d_{\mathbb{E}}(\mathbf{x}, \mathbf{y}) := \|\mathbf{x} - \mathbf{y}\| := \|\mathbf{X} - \mathbf{Y}\| \tag{1.14}$$

where $\mathbf{X}$ and $\mathbf{Y}$ are $\mathbf{x}$ and $\mathbf{y}$ considered as vectors, respectively.[3]

We have for all $\mathbf{x}, \mathbf{y}, \mathbf{z} \in \mathbb{R}^n$:

(1) $d_{\mathbb{E}}(\mathbf{x}, \mathbf{y}) = d_{\mathbb{E}}(\mathbf{y}, \mathbf{x})$.

(2) $d_{\mathbb{E}}(\mathbf{x}, \mathbf{y}) \geq 0$ with $d_{\mathbb{E}}(\mathbf{x}, \mathbf{y}) = 0$ if and only if $\mathbf{x} = \mathbf{y}$.

(3) (Triangle inequality)

$$d_{\mathbb{E}}(\mathbf{x}, \mathbf{z}) \leq d_{\mathbb{E}}(\mathbf{x}, \mathbf{y}) + d_{\mathbb{E}}(\mathbf{y}, \mathbf{z}). \tag{1.15}$$

[3]In the subsequent discussions, we will often not distinguish between points and vectors, treating them as equivalent.

The proof of this is left as Exercise 1.4. The reason why (1.15) is called the 'triangle inequality' is that it implies that the sum of the lengths of any two sides of a triangle is greater than or equal to the length of the remaining side.

Definition 1.1. In general, when a set $\mathfrak{M}$ with a function $d : \mathfrak{M} \times \mathfrak{M} \to \mathbb{R}$ satisfies these axioms for all $\mathbf{x}, \mathbf{y}, \mathbf{z} \in \mathfrak{M}$, we say that $(\mathfrak{M}, d)$ is a **metric space** (see Exercise 1.3).

The **Cauchy–Schwarz inequality** says that for all vectors $\mathbf{V}, \mathbf{W} \in \mathbb{R}^n$,

$$|\mathbf{V} \cdot \mathbf{W}| \leq \|\mathbf{V}\| \, \|\mathbf{W}\|. \tag{1.16}$$

For the proof, see Exercise 1.5. Moreover, the Cauchy–Schwarz inequality is used to prove the triangle inequality (1.15).

We have that $\mathbf{X} \cdot \mathbf{Y} = 0$ if and only if $\mathbf{X}$ and $\mathbf{Y}$ are perpendicular (orthogonal) to each other.

The concept of perpendicularity is a special case of the more general relationship between vectors, which can be expressed through the angle between them. The **angle** $\angle(\mathbf{X}, \mathbf{Y})$ between two nonzero vectors $\mathbf{X}$ and $\mathbf{Y}$ in $\mathbb{R}^n$ is

$$\angle(\mathbf{X}, \mathbf{Y}) := \cos^{-1} \frac{\mathbf{X} \cdot \mathbf{Y}}{\|\mathbf{X}\| \, \|\mathbf{Y}\|}. \tag{1.17}$$

For the proof of this, see Exercise 1.6. Figure 1.2.1 shows the idea of the proof, which involves the vector projection map.

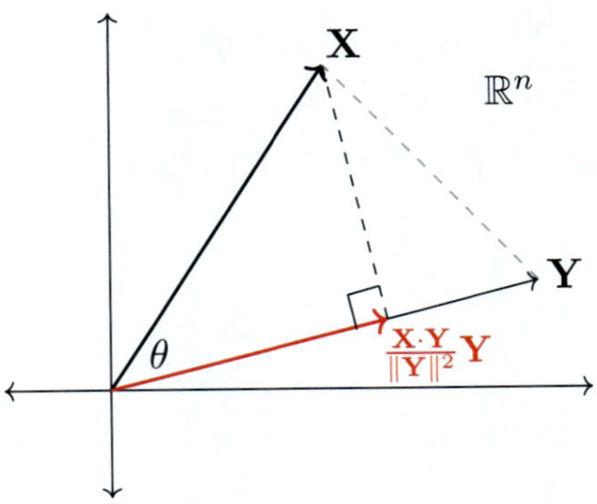

Figure 1.2.1. The vectors $\mathbf{X}$ and $\mathbf{Y}$, making an angle $\theta = \angle(\mathbf{X}, \mathbf{Y})$, and the vector projection of $\mathbf{X}$ onto $\mathbf{Y}$.

1.3. Introduction to submanifolds via examples

As we stated at the beginning of this chapter, *submanifolds*, loosely speaking, are locally Euclidean spaces inside higher-dimensional Euclidean spaces. Before we give the formal definition of submanifolds at the end of the next chapter, we discuss several examples.

1.3.1. Hyperplanes.

1.3.1. Hyperplanes. Hyperplanes are examples of 'linear' submanifolds. As such, we will at first understand them algebraically, via the Euclidean inner product.

1.3.1.1. *Coordinate hyperplanes.* Recall that a **bijection** is an injective and surjective map. Recall that in general a map $f : X \to Y$ is a bijection if and only if it has an inverse $g : Y \to X$. That is, $g \circ f = \mathrm{id}_X$ and $f \circ g = \mathrm{id}_Y$. Often, a convenient way to show that a function is a bijection is to show that it has an inverse.

Define the j**-th coordinate hyperplane**

$$\mathbb{R}^n_j := \{\mathbf{X} \in \mathbb{R}^{n+1} \colon X^j = 0\} \tag{1.18}$$

for $1 \leq j \leq n+1$; see Figure 1.3.1. As a vector subspace of $\mathbb{R}^{n+1}$, $\mathbb{R}^n_j$ is an n-dimensional vector space over $\mathbb{R}$. We have the bijective linear map (vector space isomorphism) $\rho_j : \mathbb{R}^n_j \to \mathbb{R}^n$ defined by

$$\begin{aligned}\rho_j : (X^1, \dots, X^{j-1}, 0, X^{j+1}, \dots, X^{n+1}) \\ \mapsto (X^1, \dots, X^{j-1}, X^{j+1}, \dots, X^{n+1}).\end{aligned} \tag{1.19}$$

This map in effect identifies $\mathbb{R}^n_j$ with $\mathbb{R}^n$. Note that the inverse map $\rho_j^{-1} : \mathbb{R}^n \to \mathbb{R}^n_j$ is also a linear map. Thus, $\mathbb{R}^n_j$ is Euclidean, both globally and locally, making it a submanifold of $\mathbb{R}^{n+1}$.

1.3.1.2. *Affine hyperplanes.* An **affine hyperplane** in $\mathbb{R}^{n+1}$ is a subset of the form

$$\mathbb{A} = \{\mathbf{X} \in \mathbb{R}^{n+1} \colon \mathbf{X} \cdot \mathbf{N} = a\}, \tag{1.20}$$

where $a \in \mathbb{R}$ and $\mathbf{N}$ is a fixed nonzero vector, called a **normal vector** to $\mathbb{A}$. See Figure 1.3.2.

If $a = 0$, then $\mathbb{A}$ passes through the origin $\mathbf{0}$ and we call $\mathbb{A}$ a **linear hyperplane**.

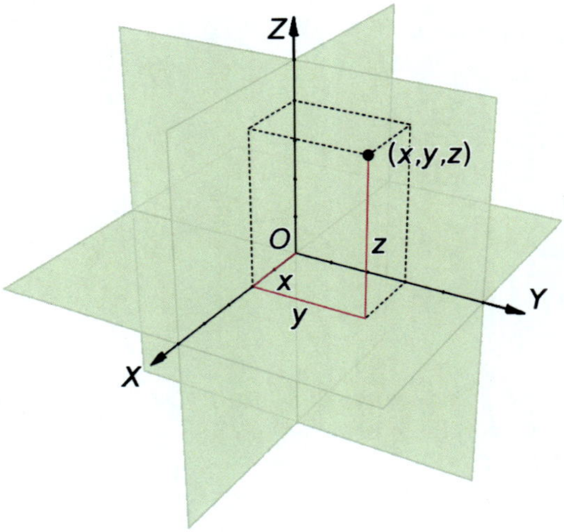

Figure 1.3.1. The coordinate hyperplanes of $\mathbb{R}^3$. Credit: By Jorge Stolfi – Own work, Public Domain, `https://commons.wikimedia.org/w/index.php?curid=6692547`

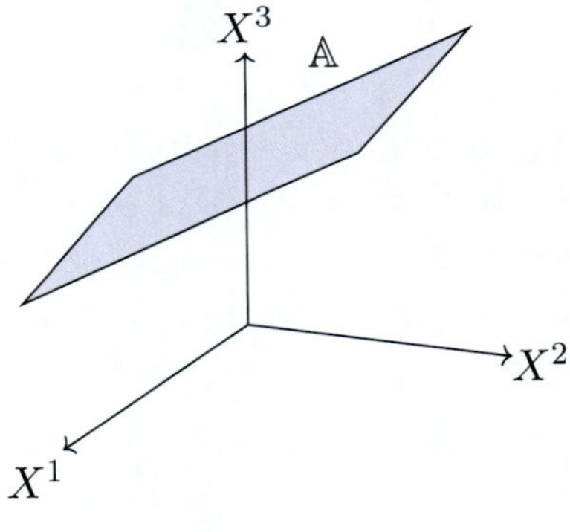

Figure 1.3.2. An affine hyperplane $\mathbb{A}$ in $\mathbb{R}^3$.

Example 1.2. *Any affine hyperplane is a submanifold of dimension n. In fact, it is* globally *Euclidean, not just* locally *Euclidean. We can see this as follows. Since the normal vector $\mathbf{N} = (N^1, \ldots, N^{n+1}) \neq \mathbf{0}$, there exists $1 \leq i \leq n+1$ such that $N^i \neq 0$. Then the map $\pi_i : \mathbb{A} \to \mathbb{R}^n_i$ defined by*

$$\pi_i(\mathbf{X}) = (X^1, \ldots, X^{i-1}, 0, X^{i+1}, \ldots, X^{n+1}) \tag{1.21}$$

is a bijection; see Figure 1.3.3 *and Exercise* 1.7.

To the affine hyperplane $\mathbb{A}$ we can associate the parallel linear subspace

$$\mathbb{L}_{\mathbb{A}} := \{\mathbf{X} \in \mathbb{R}^{n+1} \colon \mathbf{X} \cdot \mathbf{N} = 0\}. \tag{1.22}$$

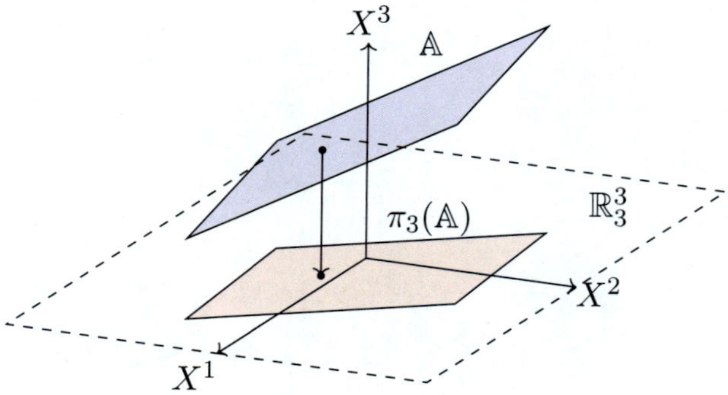

Figure 1.3.3. An affine hyperplane $\mathbb{A}$ in $\mathbb{R}^3$ and its projection under π_3 onto the third coordinate plane $\mathbb{R}^3_3$.

We can identify $\mathbb{A}$ and $\mathbb{L}_{\mathbb{A}}$ by the map $\phi_{\mathbb{A}} : \mathbb{L}_{\mathbb{A}} \to \mathbb{A}$ defined by (see Exercise 1.8)

$$\phi_{\mathbb{A}}(\mathbf{X}) := \mathbf{X} + a\frac{\mathbf{N}}{|\mathbf{N}|^2}. \tag{1.23}$$

In general, we say that two affine hyperplanes are **parallel** if their respective nonzero normal vectors are scalar multiples of one another. The relation of two affine hyperplanes being parallel is an equivalence relation.

To recap, affine hyperplanes $\mathbb{A}$ are examples of submanifolds of $\mathbb{R}^{n+1}$, as they can be made Euclidean by the identifications $\pi_i \circ \phi_{\mathbb{A}}$. Since the maps are defined on the *whole* spaces, we say that affine hyperplanes are *globally* Euclidean.

1.3.2. Spheres.

1.3.2. Spheres. Generalizing round circles to all dimensions, we have round spheres. A sphere is the set of points of constant distance to a fixed point in a Euclidean space. Let $\mathbf{c}_0 \in \mathbb{R}^{n+1}$ and $r > 0$. The n-dimensional **sphere** of radius r and center $\mathbf{c}_0$ is

$$S^n(r) := \{\mathbf{x} \in \mathbb{R}^{n+1} \colon \|\mathbf{x} - \mathbf{c}_0\| = r\}. \tag{1.24}$$

To keep track of the dimension in our terminology, we call $S^n(r)$ an n-**sphere** of radius r. See Figure 1.3.4.

Example 1.3. *Any n-sphere is an n-dimensional submanifold of $\mathbb{R}^{n+1}$.*

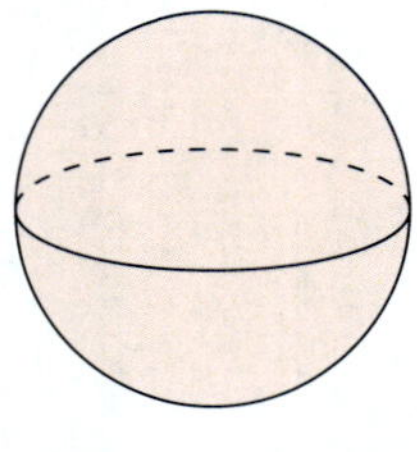

Figure 1.3.4. An n-sphere. This is the boundary sphere and does not include the interior of the $(n+1)$-dimensional ball that it bounds.

1.3.2.1. *Spheres are locally Euclidean.* Let

$$S^n := S^n(1) = \{\mathbf{x} \in \mathbb{R}^{n+1} \colon \|\mathbf{x}\| = 1\}. \tag{1.25}$$

That is, S^n is the unit n-sphere centered at the origin.

We now explicitly explore in what way spheres are locally Euclidean. Define $2(n+1)$ hemispheres of S^n by

$$S^n_{i,\pm} := \{\mathbf{x} \in S^n \colon \pm x^i > 0\}. \tag{1.26}$$

For example,

$$S^n_{n+1,-} = \{\mathbf{x} \in S^n \colon x^{n+1} < 0\}. \tag{1.27}$$

Let $S^n_{i,+}$ and $S^n_{i,-}$ be called the northern and southern i-th hemispheres, respectively. Taking $n = 2$, we have that a model for the southern hemisphere of the Earth is $S^2_{3,-}$.

Let $\mathbf{x} \in S^n$ be any point. Since $\|\mathbf{x}\| = 1 > 0$, there exists i such that $x^i \neq 0$. If $x^i > 0$, then $\mathbf{x} \in S^n_{i,+}$. Otherwise, if $x^i < 0$, then $\mathbf{x} \in S^n_{i,-}$. This shows that

$$S^n = \bigcup_{i=1}^{n+1} S^n_{i,\pm}. \tag{1.28}$$

That is, the collection $\{S^n_{i,\pm}\}_{i=1}^{n+1}$ of hemispherical subsets of the unit sphere S^n is a **covering** of S^n.

Consider the open unit n-dimensional ball in the i-th coordinate hyperplane $\mathbb{R}^n_i = \{\mathbf{x} \in \mathbb{R}^{n+1} \colon x^i = 0\}$ defined by

$$B_i := \{\mathbf{x} \in \mathbb{R}^n_i \colon \|\mathbf{x}\| < 1\}. \tag{1.29}$$

Each hemisphere $S^n_{i,\pm}$ is similar to (an open subset of) Euclidean space in the sense that we have maps, which we will call 'coordinate charts',

$$\phi_{i,\pm} : S^n_{i,\pm} \to B_i \tag{1.30}$$

defined by the coordinate projections

$$\phi_{i,\pm}(\mathbf{x}) = (x^1, \ldots, x^{i-1}, 0, x^{i+1}, \ldots, x^{n+1}). \tag{1.31}$$

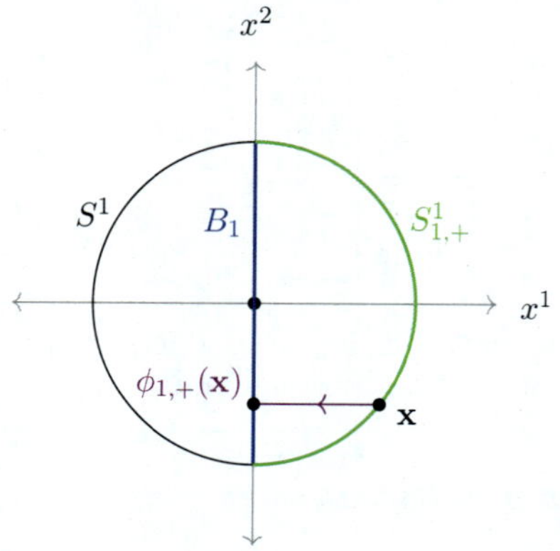

Figure 1.3.5. Visualizing the map $\phi_{1,+} : S^1_{1,+} \to B_1$ in the case $n = 1$.

See Figure 1.3.5. These coordinate charts are bijections and differentiable with injective derivatives; the precise properties of these maps will be explained in detail in the next chapter when we discuss 'embeddings'. In any case, in this sense, S^n is locally Euclidean. So spheres are submanifolds. See Exercise 1.9.

1.3.2.2. *Stereographic projection.* An alternative way to see that spheres are locally Euclidean is via stereographic projection. Let S^n be the unit n-sphere in $\mathbb{R}^{n+1}$. Let $\mathbf{N} := \mathbf{e}_{n+1} := (0, \dots, 0, 1)$ be the **north pole**. Let $\mathbb{L}$ be the hyperplane defined by

$$\mathbb{L} := \{\mathbf{y} \in \mathbb{R}^{n+1} : y^{n+1} := \mathbf{y} \cdot \mathbf{N} = -1\}. \tag{1.32}$$

Then $\mathbb{L}$ is tangent to S^n at the south pole $\mathbf{S} := -\mathbf{N} = (0, \dots, 0, -1)$. In Figure 1.3.7, which visualizes stereographic projection, $\mathbb{L}$ is the blue ocean the globe is sitting on.

Stereographic projection is the map G from $S^n \setminus \{\mathbf{N}\}$ to $\mathbb{L}$ which is defined to take a point $\mathbf{x}$ to the intersection of the hyperplane $\mathbb{L}$ and the unique line passing through $\mathbf{N}$ and $\mathbf{x}$. See Figure 1.3.6.

The stereographic projection map is equal to the map $G : S^n \setminus \{\mathbf{N}\} \to \mathbb{L}$ defined by

$$G(\mathbf{x}) = \frac{2\mathbf{x} - (1 + \mathbf{x} \cdot \mathbf{N})\,\mathbf{N}}{1 - \mathbf{x} \cdot \mathbf{N}} \tag{1.33}$$

for $\mathbf{x} \in S^n \setminus \{\mathbf{N}\}$. See Exercise 1.10; Exercise 1.11 provides a formula for the inverse of G.

In Figure 1.3.7, we have $n = 2$ and

$$G(x, y, z) = \left(\frac{2x}{1 - z}, \frac{2y}{1 - z}, -1\right). \tag{1.34}$$

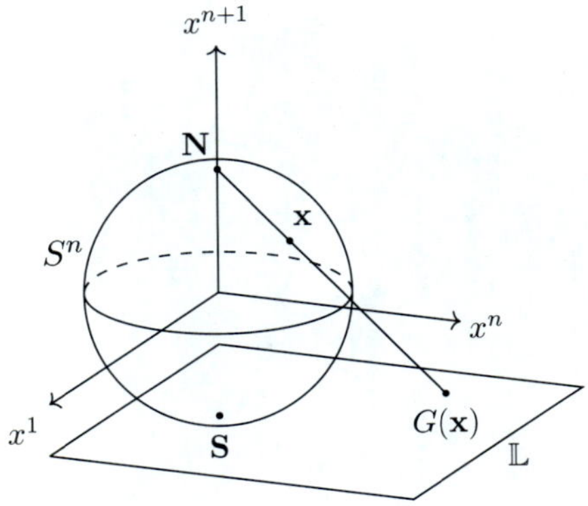

Figure 1.3.6. Picturing the stereographic projection map from the sphere minus the north pole $G : S^n \setminus \{\mathbf{N}\} \to \mathbb{L}$.

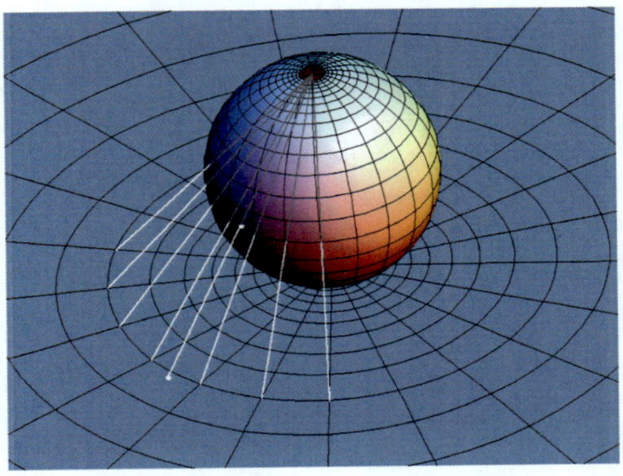

Figure 1.3.7. Stereographic projection. Credit: 3DXM Virtual Mathematics Museum by The 3D-XplorMath Consortium. © Copyright 3DXM Consortium. `https://3d-xplormath.org/`

Finally, we mention that spherical geometry has a long history, having been studied for thousands of years. It has been developed for centuries for various practical applications such as navigation and astronomy.

1.3.3. Graphs. Graphs are 'easy' examples of submanifolds. Moreover, they fully reflect the local nonlinearity of submanifolds and thus present an important class to study.

1.3.3.1. *Graphs of functions.* Let $\mathcal{U}$ be a nonempty open subset of $\mathbb{R}^n$.[4] Let $f : \mathcal{U} \to \mathbb{R}$. The **graph** of f is defined as

$$G^n := \{(\mathbf{x}, f(\mathbf{x})) : \mathbf{x} \in \mathcal{U}\}, \tag{1.35}$$

[4]See §2.1.1 below for the definition of *open subset* of $\mathbb{R}^n$. Since we are thinking intuitively for now, its precise definition is not necessary for this discussion.

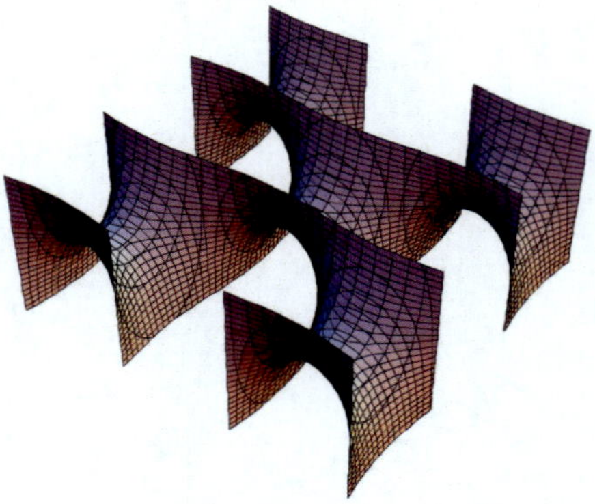

Figure 1.3.8. Scherk's surface. Credit: By Erminia Naccarato (Miriane) – Own work, CC BY-SA 3.0, `https://commons.wikimedia.org/w/index.php?curid=2240083`

which is a submanifold of $\mathbb{R}^{n+1}$. A graph is globally Euclidean in the sense that we have the bijective map $\phi : G^n \to \mathcal{U}$ defined by

$$\phi(\mathbf{x}, f(\mathbf{x})) := \mathbf{x}. \tag{1.36}$$

Indeed, its inverse $\phi^{-1} : \mathcal{U} \to G^n$ is given by

$$\phi^{-1}(\mathbf{x}) = (\mathbf{x}, f(\mathbf{x})). \tag{1.37}$$

Observe that each subset $S^n_{i,\pm}$ of S^n is the graph of a function.

Example 1.4 (Scherk [**Sch35**]). *Define* $f : \mathcal{D} \to \mathbb{R}$ *by*

$$f(x^1, x^2) := \ln \frac{\cos(x^2)}{\cos(x^1)}, \tag{1.38}$$

where $\mathcal{D}$ *is the largest subset of* $\mathbb{R}^2$ *for which the expression on the right-hand side is defined. The graph of* f *is called Scherk's first surface; see Figure* 1.3.8. *Note that the function* f *is smooth on its domain* $\mathcal{D}$, *which is the set of* $(x^1, x^2) \in \mathbb{R}^2$ *for which* $\cos(x^1)\cos(x^2) > 0$; *see Exercise* 1.12.

1.3.3.2. *Affine graphs.* In short, affine graphs are a natural generalization of graphs over Euclidean spaces, and we now proceed to describe them. Let $\mathbb{A}$ be an affine hyperplane. Let $\mathbf{N}$ be a unit vector normal to $\mathbb{A}$. Then

$$\mathbb{A} = \{\mathbf{x} \in \mathbb{R}^{n+1} : \mathbf{x} \cdot \mathbf{N} = c\} \tag{1.39}$$

for some $c \in \mathbb{R}$. The orthogonal projection map $\pi_{\mathbb{A}} : \mathbb{R}^{n+1} \to \mathbb{A}$ is defined by

$$\mathbf{y} - \pi_{\mathbb{A}}(\mathbf{y}) := a\mathbf{N} \tag{1.40}$$

for some $a \in \mathbb{R}$. Since $\pi_{\mathbb{A}}(\mathbf{y}) \in \mathbb{A}$ implies that $\pi_{\mathbb{A}}(\mathbf{y}) \cdot \mathbf{N} = c$, we obtain

$$\mathbf{y} \cdot \mathbf{N} - c = a. \tag{1.41}$$

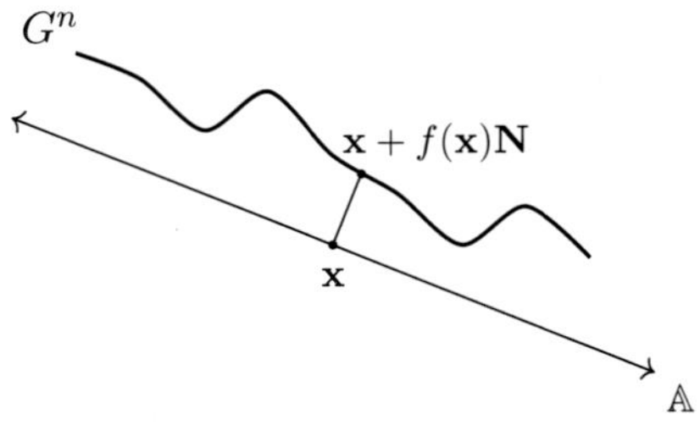

Figure 1.3.9. The affine graph G^n of a function $f : \mathbb{A} \to \mathbb{R}$.

Thus,

$$\pi_{\mathbb{A}}(\mathbf{y}) = \mathbf{y} - (\mathbf{y} \cdot \mathbf{N} - c)\mathbf{N}. \tag{1.42}$$

Let $f : \mathbb{A} \to \mathbb{R}$ be a function. The **affine graph** of f is

$$G^n := \{\mathbf{x} + f(\mathbf{x})\mathbf{N} \colon \mathbf{x} \in \mathbb{A}\}. \tag{1.43}$$

See Figure 1.3.9. The restriction map

$$\pi_{\mathbb{A}}|_{G^n} : G^n \to \mathbb{A} \tag{1.44}$$

is a bijection; see Exercise 1.13.

Given $\mathbf{V} \in \mathbb{R}^n$, denote

$$\mathbb{R}\mathbf{V} := \{t\mathbf{V} \colon t \in \mathbb{R}\}. \tag{1.45}$$

Note that $\mathbb{R}^n = \mathbb{L}_{\mathbb{A}} \oplus \mathbb{R}\mathbf{N}$, where $\mathbb{L}_{\mathbb{A}}$ is defined by (1.22). That is, for each $\mathbf{X} \in \mathbb{R}^n$ there exist unique vectors $\mathbf{Y} \in \mathbb{L}_{\mathbb{A}}$ and $\mathbf{Z} \in \mathbb{R}\mathbf{N}$ such that

$$\mathbf{X} = \mathbf{Y} + \mathbf{Z}. \tag{1.46}$$

Indeed, these unique vectors are given by

$$\mathbf{Y} = \mathbf{X} - (\mathbf{X} \cdot \mathbf{N})\mathbf{N} \quad \text{and} \quad \mathbf{Z} = (\mathbf{X} \cdot vN)\mathbf{N}. \tag{1.47}$$

See Exercise 1.14 for the case where the linear hyperplane $\mathbb{L}_{\mathbb{A}}$ is replaced by an affine hyperplane $\mathbb{A}$.

In the definition of an affine graph, we may replace $\mathbb{A}$ by a subset of $\mathbb{A}$ which is relatively open in $\mathbb{A}$.

Example 1.5. *Consider the case where the affine hyperplane $\mathbb{A}$ is the i-th coordinate hyperplane $\mathbb{R}^n_i$ in $\mathbb{R}^{n+1}$. Let $f : \mathbb{R}^n_i \to \mathbb{R}$. Choosing the unit normal to $\mathbb{R}^n_i$ as $\mathbf{e}_i = (0, \ldots, 0, 1, 0, \ldots, 0)$, where the 1 in the i-th slot, we have that the affine graph of f is*

$$G^n = \{\mathbf{x} + f(\mathbf{x})\mathbf{e}_i \colon \mathbf{x} \in \mathbb{R}^n_i\}. \tag{1.48}$$

Equivalently,

$$G^n = \{(x^1, \ldots, x^{i-1}, f(\mathbf{x}), x^{i+1}, \ldots, x^{n+1}) \colon \\ x^1, \ldots, x^{i-1}, x^{i+1}, \ldots, x^{n+1} \in \mathbb{R}\}. \tag{1.49}$$

For example, consider the unit 2-sphere S^2 in $\mathbb{R}^3$ and take $i = 3$. The equation for S^2 is $(x^1)^2 + (x^2)^2 + (x^3)^2 = 1$. Consider the northern third-hemisphere

$$S^2_{3,+} = \{\mathbf{x} \in S^2 \colon x^3 > 0\}. \tag{1.50}$$

We can express the hemisphere $S^2_{3,+}$ as the graph of a function as follows. Define $f : \mathbb{R}^2 \cong \mathbb{R}^3_3 \to \mathbb{R}$ by

$$f(x^1, x^2) := \sqrt{1 - (x^1)^2 - (x^2)^2}. \tag{1.51}$$

Then $S^2_{3,+}$ is the graph of f. For the general case of $S^n_{i,\pm}$, where $1 \leq i \leq n+1$, see Exercise 1.15.

1.3.3.3. *Radial graphs.* Compact hypersurfaces cannot be graphs over Euclidean spaces, but as will be evident, they can be radial graphs. In short, radial graphs are graphs over the unit sphere, which we now describe beginning with a familiar example.

Consider an ellipse E^1 in the plane, which is the set of all points $(x, y) \in \mathbb{R}^2$ satisfying the equation

$$\frac{x^2}{a^2} + \frac{y^2}{b^2} = 1, \tag{1.52}$$

where a, b are positive constants. This is a 1-dimensional submanifold of $\mathbb{R}^2$. One can represent the ellipse E^1 in polar coordinates. Namely,

$$E^1 = \{(r(\theta), \theta) \colon \theta \in [0, 2\pi]\}, \tag{1.53}$$

where the formula for the radial function is

$$r(\theta) := \frac{ab}{\sqrt{b^2 \cos^2(\theta) + a^2 \sin^2(\theta)}}. \tag{1.54}$$

See Figure 1.3.10.

Note that, instead of θ being an element of the closed interval $[0, 2\pi]$, we can think of θ as being an element of the **quotient set** $\mathbb{R}/2\pi\mathbb{Z}$. That is, consider the equivalence relation $\sim$ on $\mathbb{R}$ defined by $x \sim y$ if and only if $x - y$ is an integer multiple of 2π. Defining $2\pi\mathbb{Z} := \{2\pi k \colon k \in \mathbb{Z}\}$, we have $x \sim y$ if and only if $x - y \in 2\pi\mathbb{Z}$. For this reason we denote by $\mathbb{R}/2\pi\mathbb{Z}$ the quotient set $\mathbb{R}/\sim$ of equivalence classes $[\theta]$, where $\theta \in \mathbb{R}$.

Note that we can identify $\mathbb{R}/2\pi\mathbb{Z}$ with the unit circle S^1 by the map

$$[\theta] \mapsto (\cos(\theta), \sin(\theta)). \tag{1.55}$$

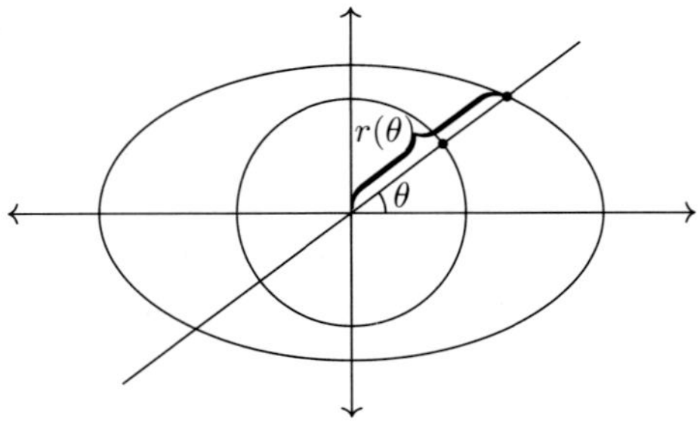

Figure 1.3.10. An ellipse E^1 in polar coordinates is given by $r(\theta)=\frac{ab}{\sqrt{b^2\cos^2(\theta)+a^2\sin^2(\theta)}}$.

That is, the element $[\theta] \in \mathbb{R}/2\pi\mathbb{Z}$ is identified with the point in S^1 with polar coordinates $(1,\theta)$, that is, radius 1 and angle θ; see Exercise 1.16. The considerations above form a special case of the notion of *quotient space* in topology (see Definition 2.5 below).

We can parametrize the ellipse by a map $F : S^1 \to E^1$ defined by

$$F(1,\theta) = (r(\theta),\theta) \tag{1.56}$$

for $[\theta] \in \mathbb{R}/2\pi\mathbb{Z}$ and where $r(\theta)$ is defined by (1.54). See Figure 1.3.10.

More generally, we can define a graph over the unit sphere S^n, where $n \geq 1$. Let $f : S^n \to \mathbb{R}_+$. The **radial graph** of f is

$$R^n_f := \{f(\mathbf{x})\mathbf{x} \colon \mathbf{x} \in S^n\}. \tag{1.57}$$

We also say that R^n_f is **starshaped**.

Define the **radial projection map** $\pi : R^n_f \to S^n$ by

$$\pi(f(\mathbf{x})\mathbf{x}) := \mathbf{x}; \tag{1.58}$$

see Figure 1.3.11. This map is well-defined; see Exercise 1.17.

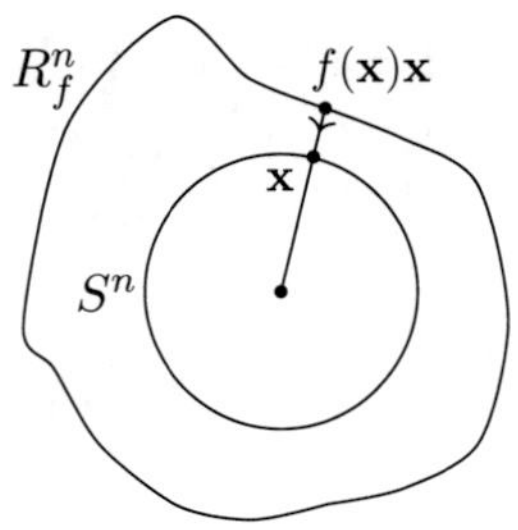

Figure 1.3.11. A radial graph R^n_f projecting onto S^n.

Via the bijection π, we can think of R^n_f as a sphere S^n. This is applicable even if f is not a continuous function. This is thinking of R^n_f intrinsically via the map π. However, if we think of R^n_f extrinsically as a subset of $\mathbb{R}^{n+1}$, then the function f affects the topology and geometry of R^n_f.

1.3.4. Surfaces of revolution.

Surfaces of revolution are surfaces that possess rotational symmetry around an axis.

1.3.4.1. *Tori.* Two-dimensional submanifolds of $\mathbb{R}^3$ are easy to visualize since we see them often in everyday life.

A **torus** T^2 is a doughnut shaped surface in $\mathbb{R}^3$. See Figure 1.3.12. In general, 2-dimensional submanifolds are called **surfaces**.

To understand tori beyond the intuitive level, we consider the class of surfaces in the next subsection.

1.3.4.2. *General surfaces of revolution.* Let C^1 be a **curve** in $\mathbb{R}^2$. We can imagine $\mathbb{R}^2$ as a vertical hyperplane in $\mathbb{R}^3$. Then we can rotate C^1 about the x^3-axis to obtain a **surface of revolution**.

Example 1.6 (Torus). *If we are given a circle C^1_{cir} of radius r centered at a point $(c, 0)$, where $0 < r < c$, then the surface of revolution T^2 obtained by rotating C^1_{cir} about the x^3-axis is called a torus.*

Example 1.7 (Catenoid). *Let*

$$C^1_{\text{cat}} := \{(r, x^3) \colon r = \cosh(x^3)\}. \tag{1.59}$$

*The surface of revolution obtained by rotating C^1_{cat} about the x^3-axis is called a **catenoid**; see Figures* 1.3.13 *and* 1.3.14. *Consequently, the catenoid is*

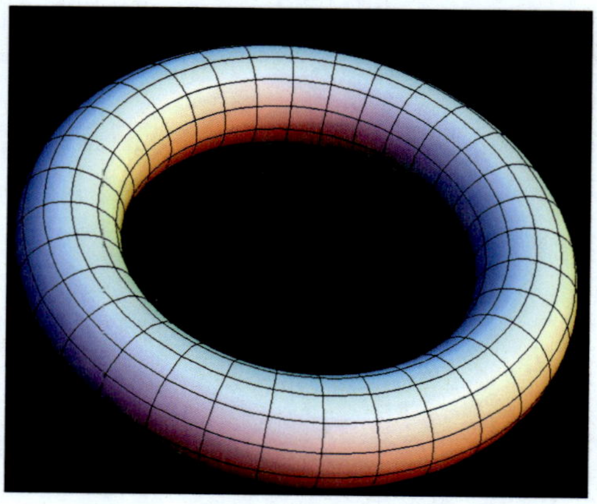

Figure 1.3.12. A torus T^2. Credit: 3DXM Virtual Mathematics Museum by The 3D-XplorMath Consortium. © Copyright 3DXM Consortium. `https://3d-xplormath.org/`

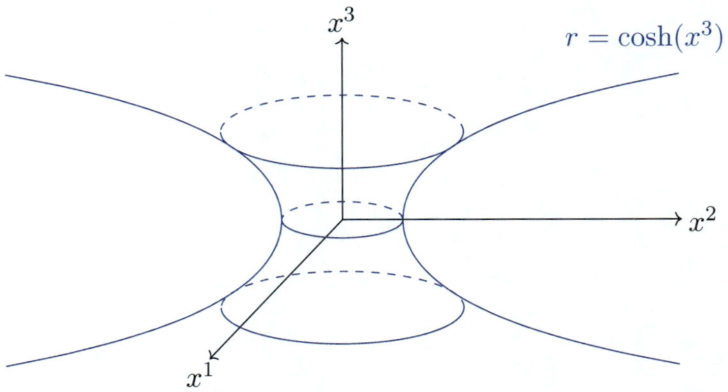

Figure 1.3.13. The catenoid surface is obtained by rotating the curve $r = \cosh(x^3)$ about the x^3-axis.

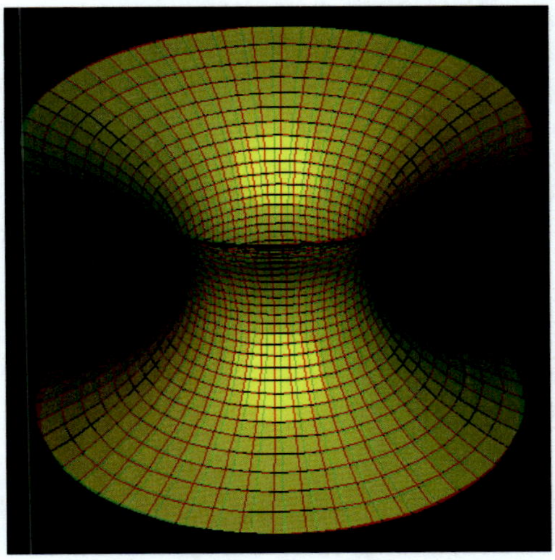

Figure 1.3.14. A catenoid. Credit: 3DXM Virtual Mathematics Museum by The 3D-XplorMath Consortium. © Copyright 3DXM Consortium. `https://3d-xplormath.org/`

given by the equation

$$(x^1)^2 + (x^2)^2 = \cosh^2(x^3). \tag{1.60}$$

For the following, we denote the Euclidean coordinates by x, y, z instead of x^1, x^2, x^3. In general, given a positive function $r = f(z)$ defined for $a \leq z \leq b$, we can rotate its graph around the z-axis to obtain a surface of revolution M^2. Here, we are treating r as the radial polar coordinate, so that the equation for the surface of revolution M^2 is

$$x^2 + y^2 = f(z)^2. \tag{1.61}$$

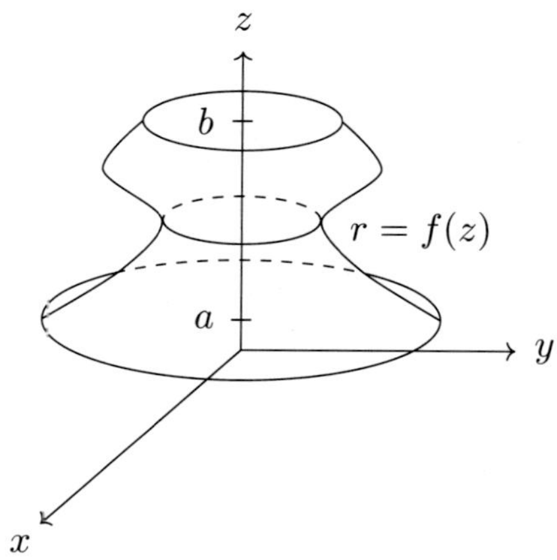

Figure 1.3.15. A surface of revolution.

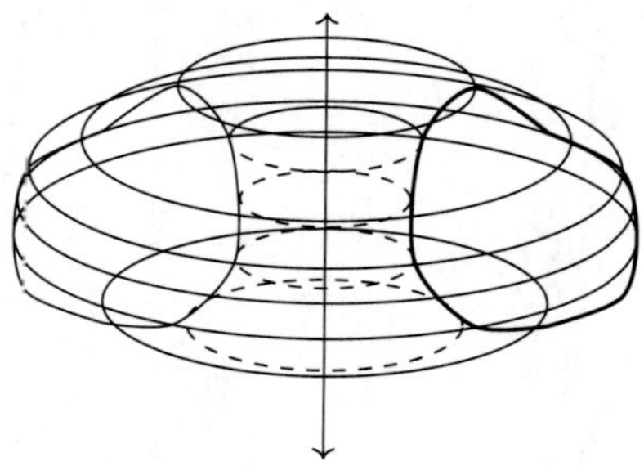

Figure 1.3.16. The surface of revolution obtained by revolving about the z-axis a closed curve (thickened in the diagram) in the rz-plane.

See Figure 1.3.15. We can parametrize the surface of revolution M^2 by the map $F : [a, b] \times [0, 2\pi] \to \mathbb{R}^3$ defined by

$$F(z, \theta) := \big(f(z)\cos(\theta), f(z)\sin(\theta), z\big) \tag{1.62}$$

for $z \in [a, b]$ and $\theta \in [0, 2\pi]$. This is simply because $r = f(z)$ and since $x = r\cos(\theta)$ and $y = r\sin(\theta)$ in polar coordinates.

Now suppose, more generally, that we want to revolve a curve in the plane to obtain a surface. For example, revolve a circle to obtain a torus. How do we parametrize the surface? Let $\gamma : (a, b) \to \mathbb{R}^2$ be an 'embedded' path, where we write $\gamma(t) = (r(t), z(t))$. Assume that the 'radius' $r(t)$ is positive. We may revolve the curve $\gamma\big((a, b)\big)$ to obtain a surface of revolution M^2. See Figure 1.3.16. A parametrization of M^2 minus the curve $\{(r(t), 0, z(t)) : t \in (a, b)\}$ is given by the map $F : (a, b) \times (0, 2\pi) \to \mathbb{R}^3$ defined by

$$F(t, \theta) := \big(r(t)\cos(\theta), r(t)\sin(\theta), z(t)\big). \tag{1.63}$$

In general, surfaces of revolution provide a wonderful class of surfaces on which to study geometry.

Level sets are discussed in §2.2 below. We will discuss parametrized submanifolds in §2.3 below. These provide additional important examples of submanifolds.

1.4. Notes and commentary

Euclidean space and its subsets and functions form the setting of real analysis; for a classic book, see Rudin [**Rud76**]. The are many books on the differential geometry of curves and surfaces; for example, do Carmo [**dC16**] is a classic.

1.5. Exercises

Exercise 1.1. *Write down the definition of an inner product (vector) space. Check with some standard source such as Hoffman and Kunze* [**HK71**] *that you have written down the correct definition.*

Exercise 1.2. *Using the bilinearity of the function A in* (1.7)*, prove* (1.11).

Exercise 1.3. *Write down the definition of a metric space $(\mathfrak{M}, d)$. Check with some standard source such as Rudin* [**Rud76**] *or Munkres* [**Mun00**] *that you have written down the correct definition.*

Exercise 1.4. *Prove the following version of the triangle inequality: For all vectors $\mathbf{V}, \mathbf{W} \in \mathbb{R}^n$,*

$$\|\mathbf{V} + \mathbf{W}\| \leq \|\mathbf{V}\| + \|\mathbf{W}\|. \tag{1.64}$$

Prove that (1.15) *follows from this. By definition, this implies that $(\mathbb{R}^n, d_{\mathbb{E}})$ is a metric space.*

Exercise 1.5. *Prove the Cauchy–Schwarz inequality* (1.16).
Hint: Show that for all $\lambda \in \mathbb{R}$,

$$\|\mathbf{V} - \lambda\mathbf{W}\|^2 = \|\mathbf{V}\|^2 + \lambda^2\|\mathbf{W}\|^2 - 2\lambda\,\mathbf{V} \cdot \mathbf{W}. \tag{1.65}$$

Then take $\lambda = \frac{\mathbf{V}\cdot\mathbf{W}}{\|\mathbf{W}\|^2}$.

Exercise 1.6. *Let $\mathbf{X}, \mathbf{Y} \in \mathbb{R}^n$. Verify that the vectors $\mathbf{Y}$ and $\mathbf{X} - \frac{\mathbf{X}\cdot\mathbf{Y}}{\|\mathbf{Y}\|^2}\mathbf{Y}$ are perpendicular. Then, using the right triangle in Figure* 1.2.1*, prove* (1.17).

Exercise 1.7. *Prove that the map $\pi_i : \mathbb{A} \to \mathbb{R}^n_i$ defined by* (1.21) *is a bijection, where $\mathbb{A}$ is the affine hyperplane defined by* (1.20) *and $N^i \neq 0$. Give the formula for its inverse $\pi_i^{-1} : \mathbb{R}^n_i \to \mathbb{A}$.*

Exercise 1.8. *Show that $\phi_{\mathbb{A}}$ defined by* (1.23) *is a bijection.*

Exercise 1.9. *Show that each map* $\phi_{i,\pm} : S^n_{i,\pm} \to B_i$, *as defined by* (1.30), *is a bijection. Give the formula for its inverse* $\phi_{i,\pm}^{-1} : B_i \to S^n_{i,\pm}$.

Exercise 1.10. *Prove that the stereographic projection map is given by* (1.33).

Exercise 1.11. *Let* G *be the stereographic projection map.*

(i) *Prove that* $G(\mathbf{x}) \cdot \mathbf{N} = -1$ *for all* $\mathbf{x} \in S^n \setminus \{\mathbf{N}\}$. *This proves that* G *is well-defined.*

(ii) *Prove that the map* $F : \mathbb{L} \to S^n \setminus \{\mathbf{N}\}$ *defined by*

$$F(\mathbf{y}) = \frac{4\mathbf{y} + (\|\mathbf{y}\|^2 - 1)\,\mathbf{N}}{\|\mathbf{y}\|^2 + 3} \tag{1.66}$$

is the inverse of the map G. *Thus, the map* F *parametrizes the sphere minus a point.*

Exercise 1.12. *Explain why the domain of the function* f *defining Scherk's surface in Example* 1.4 *is like the dark squares of an infinite chessboard.*

Exercise 1.13. *Show that the map* $\pi_{\mathbb{A}}|_{G^n}$ *defined by* (1.44) *is a bijection. Give the formula for its inverse* $\pi_{\mathbb{A}}|_{G^n}^{-1} : \mathbb{A} \to G^n$.

Exercise 1.14. *Let* $\mathbb{A}$ *be the affine hyperplane given by* (1.39). *Let* $\mathbf{X} \in \mathbb{R}^n$. *Find the formulas for the vectors* $\mathbf{Y}$ *and* $\mathbf{Z}$ *such that* $\mathbf{Y} \in \mathbb{A}$ *is the orthogonal projection of* $\mathbf{X}$ *onto* $\mathbb{A}$ *and* $\mathbf{X} = \mathbf{Y} + \mathbf{Z}$.

Exercise 1.15. *Let* $\phi_{i,\pm} : S^n_{i,\pm} \to B_i$ *be as in Exercise* 1.9. *What is the formula for the function, denoted by* $f_{i,\pm} : B_i \to \mathbb{R}$, *such that the map* $\phi_{i,\pm}^{-1}$ *parametrizes its graph?*

Exercise 1.16. *Show that the map from* $\mathbb{R}/2\pi\mathbb{Z}$ *to* S^1 *defined by* (1.55) *is well-defined and a bijection.*

Exercise 1.17. *Show that the map* π *given by* (1.58) *is a well-defined bijection. Give the formula for its inverse* $\pi^{-1} : S^n \to R^n_f$.

Chapter 2

Differential Calculus of Submanifolds

In the previous chapter, we provided an intuitive definition of a submanifold and considered various examples to illustrate its locally Euclidean nature. In this chapter, we will initiate a rigorous study of submanifolds, applying methods from differential vector calculus. The readers may skip any preliminary material with which they are familiar.

2.1. Point-set topology and vector calculus

To give a rigorous definition of submanifold, we need to discuss some basic notions from point-set topology and vector calculus.

2.1.1. Topology. Given $\mathbf{x} \in \mathbb{R}^n$ and $r > 0$, the **open ball** of radius r centered at $\mathbf{x}$ is

$$B_r(\mathbf{x}) := \{\mathbf{y} \in \mathbb{R}^n : \|\mathbf{y} - \mathbf{x}\| < r\}. \tag{2.1}$$

Definition 2.1. We say that a subset $\mathcal{U}$ of $\mathbb{R}^n$ is **open** if for every $\mathbf{x} \in \mathcal{U}$ there exists $\epsilon > 0$ such that $B_\epsilon(\mathbf{x}) \subset \mathcal{U}$.

We remark that unless $\mathcal{U} = \mathbb{R}^n$ or the empty set, the positive real number ϵ above depends on the point $\mathbf{x}$ in $\mathcal{U}$. In particular, if $\mathbf{z} \notin \mathcal{U}$, then $\epsilon \leq \|\mathbf{x}-\mathbf{z}\|$. We have that $B_r(\mathbf{x})$ is an open set; see Exercise 2.1 and Figure 2.1.1.

Example 2.2. *Let* $\mathbf{N}$ *be a unit vector in* $\mathbb{R}^n$ *and let* $c \in \mathbb{R}$. *Then we have the open* ***half-space***

$$\mathcal{H}^n := \{\mathbf{x} \in \mathbb{R}^n : \mathbf{x} \cdot \mathbf{N} > c\}. \tag{2.2}$$

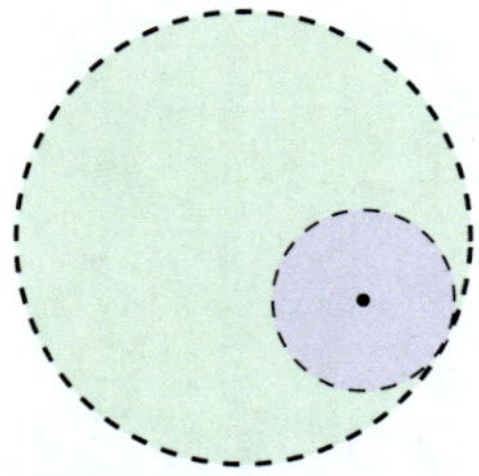

Figure 2.1.1. A blue ball centered at a point $\mathbf{y} \in B_r(\mathbf{x})$ which just fits inside the green ball $B_r(\mathbf{x})$.

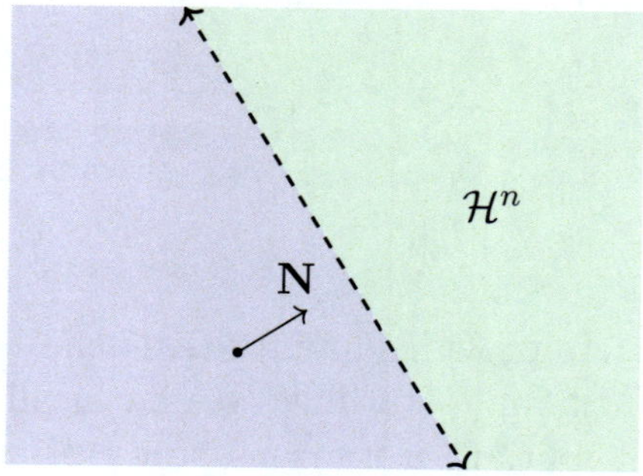

Figure 2.1.2. The open half-space $\mathcal{H}^n$ is the green region.

The half-space $\mathcal{H}^n$ is an open set; see Exercise 2.2 *and Figure* 2.1.2.

Open sets in $\mathbb{R}^n$ satisfy the following properties (see Exercise 2.3).

Properties 2.3 (Topology of $\mathbb{R}^n$)**.**

(1) $\emptyset$ *and* $\mathbb{R}^n$ *are each open sets, where* $\emptyset$ *denotes the empty set.*

(2) *Arbitrary unions of open sets are open sets.*

(3) *Finite intersections of open sets are open sets.*

Let $\mathcal{T}_{\mathbb{R}^n}$ denote the collection of open sets in $\mathbb{R}^n$. This collection is called the standard topology of $\mathbb{R}^n$.

We say that a subset $\mathcal{C}$ of $\mathbb{R}^n$ is **closed** if it is equal to the complement of an open subset of $\mathbb{R}^n$.

The **Sierpiński Carpet** S is a closed and bounded subset of $\mathbb{R}^2$; its complement $\mathbb{R}^2 \setminus S$ is an open set (see Heinonen [**Hei01**, p. 125]). See Figure 2.1.3.

The notion of open sets is easily generalized to metric spaces. Let $(\mathfrak{M}, d)$ be a metric space. The definition of an open ball is exactly the same: For $x \in \mathfrak{M}$ and $r > 0$,

$$B_r(x) := \{y \in \mathfrak{M}\colon d(y, x) < r\}. \tag{2.3}$$

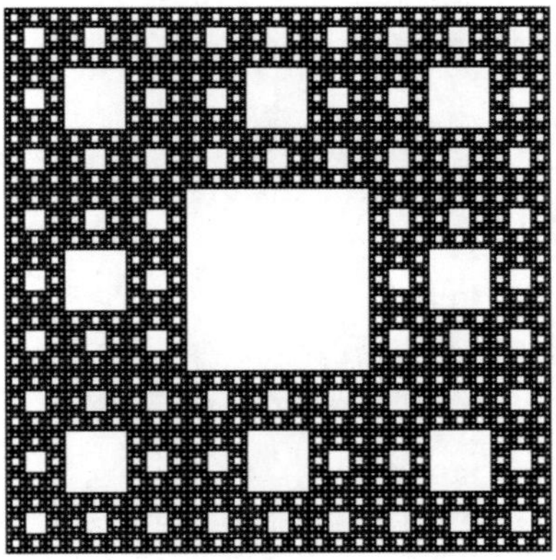

Figure 2.1.3. An approximation of the Sierpiński Carpet.

Definition 2.4. We say that $\mathcal{U} \subset \mathfrak{M}$ is an **open set** if for every $x \in \mathcal{U}$ there exists $\epsilon > 0$ such that $B_\epsilon(x) \subset \mathcal{U}$.

Let $\mathcal{T}_{\mathfrak{M}}$ denote the collection of open sets in $(\mathfrak{M}, d)$.

In general, a **topological space** $(S, \mathcal{T})$ is defined to be a set S with a collection $\mathcal{T}$ of subsets of S satisfying the three properties above. An element of $\mathcal{T}$ is called an open set. See Exercise 2.4 for a proof that a metric space, with the collection of its open sets, is a topological space.

Let S be a subset of $\mathbb{R}^n$. The **subspace topology** (a.k.a. relative topology) on S is defined by

$$\mathcal{T}_S := \{\mathcal{U} \cap S \colon \mathcal{U} \in \mathcal{T}_{\mathbb{R}^n}\}. \tag{2.4}$$

That is, the relatively open subsets of S are the intersections of open subsets of $\mathbb{R}^n$ with S. See Exercise 2.5 for a characterization of relatively open subsets.

Definition 2.5. Let $(S, \mathcal{T})$ be a topological space and suppose that $\sim$ is an equivalence relation on the set S. Let $S/\sim$ denote the quotient set of equivalence classes of $\sim$ and let $\pi : S \to S/\sim$ denote the quotient map. The **quotient space** is the set $S/\sim$ with the topology defined by a subset U of $S/\sim$ is open if and only if $\pi^{-1}(U)$ is open in S.

For further background on point-set topology, the reader may consult Munkres [**Mun00**] or Rudin [**Rud76**], where the focus of the latter book is real analysis.

2.1.2. Differentiability. Since we are interested in smooth submanifolds, we need to first consider the basic notions related to differentiability. Let $f : \mathbb{R}^n \to \mathbb{R}^k$ be a map. We say that f is **differentiable at** $\mathbf{x}_0 \in \mathbb{R}^n$ if there

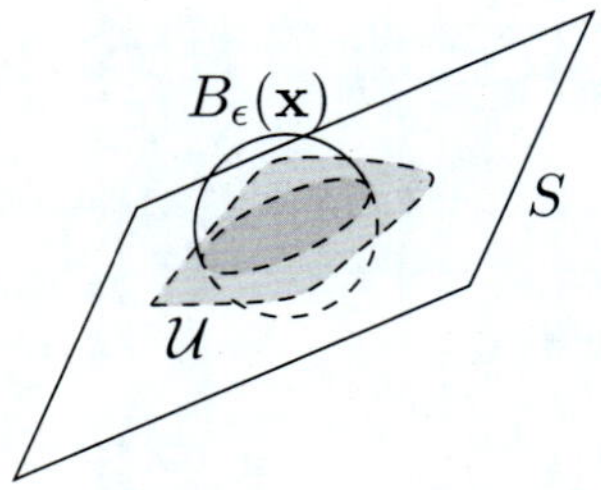

Figure 2.1.4. The darker shaded region is $B_\epsilon(\mathbf{x}) \cap S$, which is contained in the lighter shaded region $\mathcal{U}$.

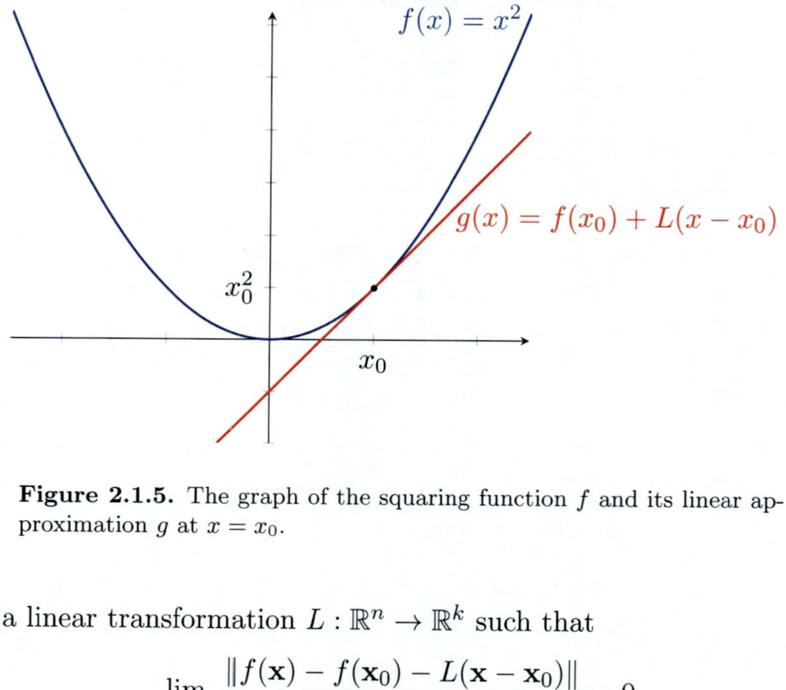

Figure 2.1.5. The graph of the squaring function f and its linear approximation g at $x = x_0$.

exists a linear transformation $L : \mathbb{R}^n \to \mathbb{R}^k$ such that

$$\lim_{\mathbf{x} \to \mathbf{x}_0} \frac{\|f(\mathbf{x}) - f(\mathbf{x}_0) - L(\mathbf{x} - \mathbf{x}_0)\|}{\|\mathbf{x} - \mathbf{x}_0\|} = 0. \tag{2.5}$$

We call $L = df_{\mathbf{x}_0}$ the (**total**) **derivative** of f at $\mathbf{x}_0$. One can write this as

$$f(\mathbf{x}) = f(\mathbf{x}_0) + L(\mathbf{x} - \mathbf{x}_0) + \mathrm{o}(\|\mathbf{x} - \mathbf{x}_0\|). \tag{2.6}$$

Here, we denote $g(\mathbf{x}) = \mathrm{o}(\|\mathbf{x} - \mathbf{x}_0\|)$ if $\lim_{\mathbf{x} \to \mathbf{x}_0} \frac{g(\mathbf{x})}{\|\mathbf{x} - \mathbf{x}_0\|} = 0$.

So f being differentiable at $\mathbf{x}_0$ means that the generally nonlinear function f is well-approximated by some linear map L, which is called the derivative $df_{\mathbf{x}_0}$ of f at $\mathbf{x}_0$. See Figure 2.1.5.

In the definition above, we just need that f is defined on an open set $\mathcal{U}$ containing $\mathbf{x}_0$. We say that f is **differentiable** if f is differentiable at each point $\mathbf{x}_0$ in its domain $\mathcal{U}$.

Denote the space of linear transformations from $\mathbb{R}^n$ to $\mathbb{R}^k$ by $\mathrm{Lin}(\mathbb{R}^n, \mathbb{R}^k)$. If $f : \mathcal{U} \subset \mathbb{R}^n \to \mathbb{R}^k$ is differentiable, then

$$df : \mathcal{U} \to \mathrm{Lin}(\mathbb{R}^n, \mathbb{R}^k), \tag{2.7}$$

where $df(\mathbf{x}_0) = df_{\mathbf{x}_0}$.

Let $\{\mathbf{e}_i\}_{i=1}^m$ be the standard Euclidean basis for $\mathbb{R}^m$ defined by

$$\mathbf{e}_i := (0, \dots, 0, 1, 0, \dots, 0), \tag{2.8}$$

where the 1 is the i-th slot. Using the standard bases for $\mathbb{R}^n$ and $\mathbb{R}^k$, we can identify a linear transformation $L : \mathbb{R}^n \to \mathbb{R}^k$ with a $k \times n$ matrix:

$$L_{ij} := L(\mathbf{e}_j) \cdot \mathbf{e}_i. \tag{2.9}$$

Thus, we can identify $\mathrm{Lin}(\mathbb{R}^n, \mathbb{R}^k)$ with the vector space of $k \times n$ matrices, which we in turn can identify with the Euclidean space $\mathbb{R}^{kn}$.

Hence, the derivative of f, as a function, can be considered as a map

$$df : \mathcal{U} \to \mathbb{R}^{kn} \cong \mathrm{Lin}(\mathbb{R}^n, \mathbb{R}^k). \tag{2.10}$$

We identify the linear transformation df with the $k \times n$ matrix $\big((df)_{ij}\big)$ whose entries are given by

$$(df)_{ij} := df(\mathbf{e}_j) \cdot \mathbf{e}_i = \frac{\partial f^i}{\partial x^j}. \tag{2.11}$$

We call this matrix $\left(\frac{\partial f^i}{\partial x^j}\right)$ representing the derivative df the Jacobian matrix of f. It is defined at each point in $\mathcal{U}$ and in general depends on the point.

Example 2.6. *Define the cubing function $f : \mathbb{R} \to \mathbb{R}$ by*

$$f(x) = x^3. \tag{2.12}$$

The function f is a differentiable function. Of course, $df_{x_0} = f'(x_0) = 3x_0^2$ for all $x_0 \in \mathbb{R}$. As a linear transformation, $L = df_{x_0} : \mathbb{R} \to \mathbb{R}$ is given by $L(v) = 3x_0^2 v$. And in this case (2.6) *becomes*

$$x^3 = x_0^3 + 3x_0^2(x - x_0) + \mathrm{o}(|x - x_0|). \tag{2.13}$$

We can independently verify this by the binomial formula:

$$x^3 = (x_0 + (x - x_0))^3 = x_0^3 + 3x_0^2(x - x_0) + 3x_0(x - x_0)^2 + (x - x_0)^3. \tag{2.14}$$

2.1.3. Continuity.

Continuity of functions is naturally a metric space notion, which, as we will see, generalizes to topological spaces.

Definition 2.7 (Continuity in the metric space sense). Let (X, d_X) and (Y, d_Y) be metric spaces. We say that a map $f : X \to Y$ is **continuous at** $x_0 \in X$ if for every $\epsilon > 0$ there exists $\delta > 0$ such that $d(x, x_0) < \delta$ implies $d(f(x), f(x_0)) < \epsilon$.

If f is continuous for all $x_0 \in A$, where A is a subset of X, then we say that f is continuous on A. If $A = X$, then we simply say that f is continuous.

The notion of a continuous map generalizes to maps between topological spaces.

Definition 2.8 (Continuity in the topological space sense)**.** Let $(X, \mathcal{T}_X)$ and $(Y, \mathcal{T}_Y)$ be topological spaces. We say that a map $f : X \to Y$ is **continuous** if for all open subsets $\mathcal{V}$ of Y, its preimage $f^{-1}(\mathcal{V})$ is an open subset of X.

Recall that a metric d_X on a set X induces a topology $\mathcal{T}_X$ on the set X. We have that $f : (X, d_X) \to (Y, d_Y)$ is continuous in the metric space sense if and only if $f : (X, \mathcal{T}_X) \to (Y, \mathcal{T}_Y)$ is continuous in the topological space sense; see Exercise 2.6.

2.1.4. Smoothness. For functions of a single variable, it is easy to define what it means to be k-times differentiable. Namely, let $f : \mathcal{U} \to \mathbb{R}$, where $\mathcal{U}$ is an open subset of $\mathbb{R}$. We say that f is once differentiable if $f'(x)$ exists for all $x \in \mathcal{U}$. We say that f is twice differentiable if $f''(x) = (f')'(x)$ exists for all $x \in \mathcal{U}$. We say that f is 3-times differentiable if $f'''(x) = (f'')'(x)$ exists for all $x \in \mathcal{U}$. We denote $f^{(1)} = f'$, $f^{(2)} = f''$, $f^{(3)} = f'''$, etc.

Given a nonnegative integer ℓ, if $f^{(\ell)}$ exists and is a continuous function, then we say that f is a C^ℓ function. Here, $f^{(0)} = f$. If f is C^ℓ for all $\ell \geq 0$, then we say that f is C^∞.

Now let $f : \mathcal{U} \subset \mathbb{R}^n \to \mathbb{R}^k$ be a (multivariable vector-valued) differentiable function. Write the derivative map in component form as

$$df = \big((df)^1, (df)^2, \ldots, (df)^{kn}\big) : \mathcal{U} \to \mathbb{R}^{kn}. \tag{2.15}$$

Note that, by considering df as a map in this way, df is in general nonlinear. Given $\mathbf{x}_0 \in \mathcal{U}$, $df(\mathbf{x}_0) \in \mathbb{R}^{kn}$ is considered as a point, although it corresponds to a linear transformation from $\mathbb{R}^n$ to $\mathbb{R}^k$, i.e., a $k \times n$ matrix.

We say that the derivative map df itself is differentiable at $\mathbf{x}_0 \in \mathcal{U}$ if $(df)^i$ is differentiable at $\mathbf{x}_0$ for each $1 \leq i \leq kn$. Then

$$(d^2 f)_{\mathbf{x}_0} := d(df)_{\mathbf{x}_0} : \mathbb{R}^n \to \mathbb{R}^{kn} \tag{2.16}$$

is a linear map. That is, $(d^2 f)_{\mathbf{x}_0} \in \mathrm{Lin}(\mathbb{R}^n, \mathbb{R}^{kn})$.[1]

If df is differentiable at each $\mathbf{x}_0$ in $\mathcal{U}$, then we say that f is twice differentiable. We can continue in this way to define what it means to be k-times differentiable, for any integer $k \geq 0$. As before, we say that f is C^k if it is k-times differentiable and its k-th derivative is continuous. If f is k-times

[1] Here, $d^2 f$ is not to be confused with the second 'exterior derivative' of f, which would be equal to zero; see §8.2.3 below. On a Riemannian manifold, we use $\nabla^2 f$ to denote the Hessian of a function f; see §5.4.2 and §8.5 below.

differentiable for all $k \geq 0$, then we say that f is **infinitely differentiable** or **smooth**. In this case, we also say that f is C^∞. We remark that it turns out that smoothness is equivalent to the condition of all partial derivatives of any order existing.

An even stronger condition than smoothness is *real analyticity*. A function $f : \mathcal{U} \subset \mathbb{R}^n \to \mathbb{R}$ is **real analytic** if for each $\mathbf{x} \in \mathcal{U}$, f is equal in a neighborhood of $\mathbf{x}$ to a convergent power series centered at $\mathbf{x}$.

Example 2.9. *Assume that $k = 1$, so that $f : \mathcal{U} \subset \mathbb{R}^n \to \mathbb{R}$. Assume also that f is twice differentiable. The derivative of f is*

$$df : \mathcal{U} \to \operatorname{Lin}(\mathbb{R}^n, \mathbb{R}) = (\mathbb{R}^n)^*, \tag{2.17}$$

where $(\mathbb{R}^n)^*$ is the vector space dual to $\mathbb{R}^n$. In particular, $(\mathbb{R}^n)^*$ is isomorphic to $\mathbb{R}^n$, and an example of an isomorphism is the map induced by the Euclidean inner product:

$$\mathbf{X} \in \mathbb{R}^n \mapsto (\mathbf{V} \mapsto \mathbf{V} \cdot \mathbf{X}) \in (\mathbb{R}^n)^*. \tag{2.18}$$

For each $\mathbf{x}_0 \in \mathcal{U}$,

$$(d^2 f)_{\mathbf{x}_0} = d(df)_{\mathbf{x}_0} : \mathbb{R}^n \to (\mathbb{R}^n)^* \tag{2.19}$$

*is a linear map. We call $(d^2 f)_{\mathbf{x}_0}$ the **Hessian** of f at $\mathbf{x}_0$.*

*Using the standard basis $\{\mathbf{e}_i\}_{i=1}^n$ of $\mathbb{R}^n$, we obtain the **Hessian matrix** of f at $\mathbf{x}_0$, defined by*

$$(f_{ij})_{\mathbf{x}_0} := \big((d^2 f)_{\mathbf{x}_0}(\mathbf{e}_i)\big)(\mathbf{e}_j) \tag{2.20}$$

for $1 \leq i, j \leq n$. The understanding is that since $\mathbf{e}_i \in \mathbb{R}^n$, we have $(d^2 f)_{\mathbf{x}_0}(\mathbf{e}_i) \in (\mathbb{R}^n)^*$, so that $\big((d^2 f)_{\mathbf{x}_0}(\mathbf{e}_i)\big)(\mathbf{e}_j) \in \mathbb{R}$. The Hessian matrix is the matrix of second partial derivatives; see Exercise 2.7.

What are examples of smooth functions? Examples include polynomials, trigonometric functions, exponentials, logarithms, and their compositions. Any function considered in this book is assumed to be smooth unless otherwise stated.

Let $\mathcal{U}$ be an open subset of $\mathbb{R}^n$ and let $\mathbf{x}_0 \in \mathcal{U}$. Let $f : \mathcal{U} \to \mathbb{R}$ be a 2-times continuously differentiable function.[2] Denote $(f_i)_{\mathbf{x}_0} := \frac{\partial f}{\partial x^i}(\mathbf{x}_0)$. By Taylor's theorem, we have

$$\begin{aligned} f(\mathbf{x}) = f(\mathbf{x}_0) &+ \sum_{i=1}^n (f_i)_{\mathbf{x}_0}(\mathbf{x} - \mathbf{x}_0)^i + \frac{1}{2} \sum_{i,j=1}^n (f_{ij})_{\mathbf{x}_0}(\mathbf{x} - \mathbf{x}_0)^i (\mathbf{x} - \mathbf{x}_0)^j \\ &+ o(\|\mathbf{x} - \mathbf{x}_0\|^2). \end{aligned} \tag{2.21}$$

[2]That is, f is twice differentiable and its second derivative is a continuous function.

Here, we say that a function

$$h(\mathbf{x}) = \mathrm{o}(\|\mathbf{x} - \mathbf{x}_0\|^2)$$

if $\lim_{\mathbf{x}\to\mathbf{x}_0} \frac{h(\mathbf{x})}{\|\mathbf{x}-\mathbf{x}_0\|^2} = 0$.

Recall that $\mathbf{x}_0$ is a **critical point** of f if $(df)_{\mathbf{x}_0} = 0$ is the zero function. In this case, we have

$$f(\mathbf{x}) = f(\mathbf{x}_0) + \frac{1}{2}\sum_{i,j=1}^{n} (f_{ij})_{\mathbf{x}_0}(\mathbf{x} - \mathbf{x}_0)^i(\mathbf{x} - \mathbf{x}_0)^j + \mathrm{o}(\|\mathbf{x} - \mathbf{x}_0\|^2). \tag{2.22}$$

In particular, if $\mathbf{x}_0 = \mathbf{0}$, then

$$f(\mathbf{x}) = f(\mathbf{0}) + \frac{1}{2}\sum_{i,j=1}^{n} (f_{ij})_{\mathbf{0}} X^i X^j + \mathrm{o}(\|\mathbf{x}\|^2). \tag{2.23}$$

Thus, $f(\mathbf{x}) - f(\mathbf{0})$ is well-approximated by a quadratic form.

2.2. The implicit function theorem

Given a system of equations, the implicit function theorem provides conditions under which one can 'implicitly' express some of the variables of the system as functions of the remaining variables. Effectively, it is a way to show that certain level sets are locally the graphs of functions.

2.2.1. Level sets. Let $f : \mathbb{R}^{n+1} \to \mathbb{R}$ be a function. Given $c \in \mathbb{R}$, let

$$M_c^n := f^{-1}(c) = \{\mathbf{x} \in \mathbb{R}^{n+1} \colon f(\mathbf{x}) = c\}. \tag{2.24}$$

The set M_c^n, which is the solution set to the equation $f(\mathbf{x}) = c$, is called a **level set**. If we think of $f(\mathbf{x})$ as indicating the level of $\mathbf{x}$, then M_c^n represents the set of points in $\mathbb{R}^{n+1}$ of level c.

Example 2.10. *Let $f : \mathbb{R}^{n+1} \to \mathbb{R}$ be the distance squared to the origin:*

$$f(\mathbf{x}) = \|\mathbf{x}\|^2. \tag{2.25}$$

In this case:

(1) *If $c < 0$, then $M_c^n = \emptyset$.*
(2) *If $c = 0$, then $M_c^n = \{\mathbf{0}\}$.*
(3) *If $c > 0$, then M_c^n is an n-dimensional sphere of radius $\sqrt{c}$.*

*The **derivative** of f at $\mathbf{x}$ is the linear function $df_{\mathbf{x}} : \mathbb{R}^{n+1} \to \mathbb{R}$ defined by*

$$df_{\mathbf{x}}(\mathbf{V}) = \frac{d}{dt}\bigg|_{t=0} \|\mathbf{x} + t\mathbf{V}\|^2 = 2\mathbf{x}\cdot\mathbf{V}. \tag{2.26}$$

See Exercise 2.8.

Let $f : \mathbb{R}^n \times \mathbb{R}^m \cong \mathbb{R}^{n+m} \to \mathbb{R}^m$ be a continuously differentiable map, where n, m are positive integers. We are interested in the level sets $f^{-1}(\mathbf{c}_0)$ for $\mathbf{c}_0 \in \mathbb{R}^m$. Such level sets are given by m equations for $n+m$ unknowns. So we expect that they are likely to be n-dimensional submanifolds. In the example in (2.25) above, we have that M_c^n is an n-dimensional submanifold of $\mathbb{R}^{n+1}$ if $c > 0$.

2.2.2. Statement of the implicit function theorem. We write f in component form as $f = (f^1, \ldots, f^m)$, where $f^i : \mathbb{R}^n \times \mathbb{R}^m \to \mathbb{R}$ for each $i = 1, \ldots, m$. Given $\mathbf{x}_0 \in \mathbb{R}^n$, consider the restricted map (redundantly denoted as) $f : \{\mathbf{x}_0\} \times \mathbb{R}^m \cong \mathbb{R}^m \to \mathbb{R}^m$. Given $\mathbf{y}_0 \in \mathbb{R}^m$, let

$$(d_{\mathbb{R}^m} f)_{\mathbf{y}_0} : \mathbb{R}^m \to \mathbb{R}^m \tag{2.27}$$

denote the derivative of this restricted map f at the point $\mathbf{y}_0$.

Theorem 2.11 (Implicit function theorem). *Let $f : \mathbb{R}^n \times \mathbb{R}^m \to \mathbb{R}^m$ be a continuously differentiable map. If the linear transformation $(d_{\mathbb{R}^m} f)_{\mathbf{y}_0}$ is invertible, then there exist an open neighborhood $\mathcal{U}$ of $\mathbf{x}_0$ and an open neighborhood $\mathcal{V}$ of $\mathbf{y}_0$ and a continuously differentiable function $g : \mathcal{U} \subset \mathbb{R}^n \to \mathcal{V} \subset \mathbb{R}^m$ such that for every point $(\mathbf{x}, \mathbf{y}) \in \mathcal{U} \times \mathcal{V}$ we have*

$$f(\mathbf{x}, \mathbf{y}) = \mathbf{c}_0 := f(\mathbf{x}_0, \mathbf{y}_0) \quad \textit{if and only if} \quad \mathbf{y} = g(\mathbf{x}). \tag{2.28}$$

If f is C^∞, then g is C^∞.

Note that the level set equation $f(\mathbf{x}, \mathbf{y}) = \mathbf{c}_0$ is an implicit equation for $\mathbf{x}$ and $\mathbf{y}$. On the other hand, the equation $\mathbf{y} = g(\mathbf{x})$ is an equation for $\mathbf{y}$ explicitly as a function of $\mathbf{x}$. That is, this level set of f is locally a graph.

The implicit function theorem can be generalized from coordinate subspaces to affine subspaces; see Exercise 2.9.

2.2.3. Examples of applications of the implicit function theorem. The implicit function theorem provides a quick way to show that spheres are submanifolds, i.e., locally graphs.

Example 2.12. *Let $f : \mathbb{R}^2 \to \mathbb{R}$ be defined by* (2.25). *We write this as*

$$f(x, y) = \|(x, y)\|^2 = x^2 + y^2. \tag{2.29}$$

Let (x_0, y_0) be a point on the circle $S^1(r)$ of radius r, where $r > 0$. Then $f(x_0, y_0) = r^2$. Consider the restricted function $f : \{x_0\} \times \mathbb{R} \to \mathbb{R}$. We compute that

$$(d_{\mathbb{R}} f)_{y_0} : \mathbb{R} \to \mathbb{R} \tag{2.30}$$

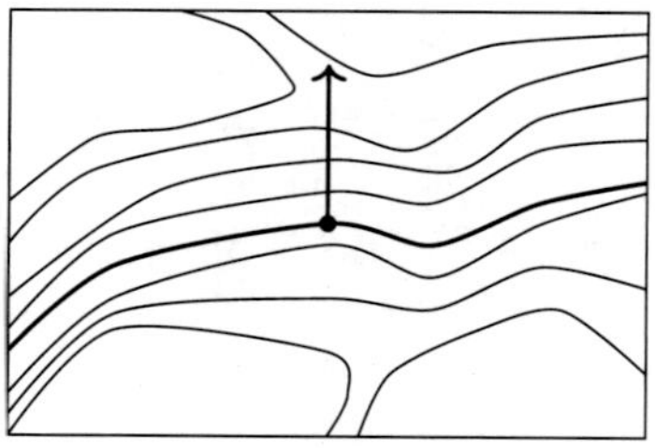

Figure 2.2.1. Level sets $f(x, y) = c$ in $\mathcal{U} \times \mathcal{V}$. The thick curve, passing through the point (x_0, y_0), is the level set $f(x, y) = f(x_0, y_0)$. It is the graph of a function g.

is given by

$$(d_{\mathbb{R}} f)_{y_0}(v) = \left.\frac{d}{dt}\right|_{t=0} \|(x_0, y_0 + tv)\|^2 = 2x_0 v. \tag{2.31}$$

Hence, $(d_{\mathbb{R}} f)_{y_0}$ *is invertible if and only if* $x_0 \neq 0$.

So we assume that $x_0 \neq 0$. *Then, by the implicit function theorem, there exist an open interval* $\mathcal{U}$ *containing* x_0 *and an open interval* $\mathcal{V}$ *containing* y_0 *and a continuously differentiable function* $g : \mathcal{U} \to \mathcal{V}$ *such that for every point* $(x, y) \in \mathcal{U} \times \mathcal{V}$ *we have*

$$x^2 + y^2 = f(x, y) = r^2 \quad \textit{if and only if} \quad y = g(x), \tag{2.32}$$

that is, in a neighborhood $\mathcal{U} \times \mathcal{V}$ *of* (x_0, y_0) *such that in this neighborhood the circle* $S^1(r)$ *is exactly the graph of a function* g.

It is actually easy to explicitly find g, *even though the implicit function theorem only tells us that some such function exists. In particular, if say* $y_0 < 0$, *then we may take* $\mathcal{U} = (-r, r)$ *and* $\mathcal{V} = (-\infty, 0)$ *and* $g : (-r, r) \to (-\infty, 0)$ *defined by*

$$g(x) = -\sqrt{r^2 - x^2}. \tag{2.33}$$

Let us think about what the implicit function theorem says in general when $n = m = 1$. In this case, $f : \mathbb{R} \times \mathbb{R} \cong \mathbb{R}^2 \to \mathbb{R}$. The hypothesis on f becomes: Suppose there is a point (x_0, y_0) such that $\frac{\partial f}{\partial y}(x_0, y_0) \neq 0$. Then there exist an open interval $\mathcal{U}$ containing x_0 and an open interval $\mathcal{V}$ containing y_0 and a continuously differentiable function $g : \mathcal{U} \to \mathcal{V}$ such that for every point $(x, y) \in \mathcal{U} \times \mathcal{V}$ we have

$$f(x, y) = f(x_0, y_0) \quad \text{if and only if} \quad y = g(x). \tag{2.34}$$

See Figure 2.2.1.

Now let us return to the unit n-sphere $S^n \subset \mathbb{R}^{n+1}$. We again consider S^n as a level set of the function f defined by (2.25). In view of the implicit

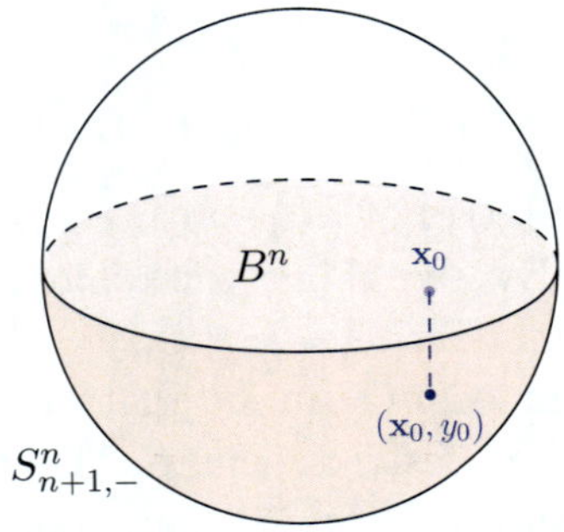

Figure 2.2.2. The southern hemisphere is the graph of the function $g(\mathbf{x}) = -\sqrt{1 - |\mathbf{x}|^2}$.

function theorem, express the function f as $f : \mathbb{R}^n \times \mathbb{R} \to \mathbb{R}$ given by

$$f(\mathbf{x}, y) = |\mathbf{x}|^2 + y^2, \tag{2.35}$$

where $\mathbf{x} \in \mathbb{R}^n$ and $y \in \mathbb{R}$. Given $(\mathbf{x}_0, y_0) \in S^n$, i.e., $|\mathbf{x}|^2 + y^2 = 1$, we have

$$(d_{\mathbb{R}} f)_{y_0} = \frac{\partial f}{\partial y}(\mathbf{x}_0, y_0) = 2y_0. \tag{2.36}$$

By the first equality in the display, we mean that the linear transformation $(d_{\mathbb{R}} f)_{y_0} : \mathbb{R} \to \mathbb{R}$ is given by multiplication by the real number $\frac{\partial f}{\partial y}(\mathbf{x}_0, y_0)$, which by the second equality is equal to $2y_0$. By the implicit function theorem, we have that if $y_0 \neq 0$, then S^n is in a neighborhood of $(\mathbf{x}_0, y_0)$ the graph of a function g. That is, points $(\mathbf{x}, y)$ in S^n near $(\mathbf{x}_0, y_0)$ are characterized as satisfying $y = g(\mathbf{x})$. Indeed, we can see this explicitly as follows. Assume that $y_0 < 0$ (the positive case is similar). Let

$$B^n = \{\mathbf{x} \in \mathbb{R}^n : \|\mathbf{x}\| < 1\} \tag{2.37}$$

and define $g : B^n \to (-\infty, 0)$ by

$$g(\mathbf{x}) = -\sqrt{1 - \|\mathbf{x}\|^2}. \tag{2.38}$$

Then the map $F(\mathbf{x}) := (\mathbf{x}, g(\mathbf{x}))$ parametrizes the graph of g, which is equal to the southern hemisphere $S^n_{n+1,-}$, which contains the point $(\mathbf{x}_0, y_0)$. See Figure 2.2.2.

2.2.4. Level sets of functions whose derivatives have maximal rank. We now return to the general consequences of the implicit function theorem. Suppose that $f : \mathbb{R}^{n+m} \to \mathbb{R}^m$ is a continuously differentiable map, where n, m are positive integers. Suppose that $\mathbf{z}_0 \in \mathbb{R}^{n+m}$ is a point at which the derivative $df_{\mathbf{z}_0} : \mathbb{R}^{n+m} \to \mathbb{R}^m$ has **maximal rank** m. Let (J_{ij}), $1 \leq i \leq m$, $1 \leq j \leq n+m$, be the $m \times (n+m)$ matrix representing the linear transformation $df_{\mathbf{z}_0}$. Since (J_{ij}) has maximal rank, there exist m distinct integers

$$1 \le j_1 \le \cdots \le j_m \le n + m \tag{2.39}$$

such that the $m \times m$ submatrix of (J_{ij}) with columns numbered $j_1, \ldots, j_m$ is an invertible matrix. By switching components, we may assume that $j_1 = n+1, \ldots, j_m = n+m$. Write $\mathbf{z}_0 = (\mathbf{x}_0, \mathbf{y}_0)$, where $\mathbf{x}_0 \in \mathbb{R}^n$ and $\mathbf{y}_0 \in \mathbb{R}^m$. Then the aforementioned $m \times m$ submatrix represents the derivative $(d_{\mathbb{R}^m} f)_{\mathbf{y}_0} : \mathbb{R}^m \to \mathbb{R}^m$ of the restricted map $f : \{\mathbf{x}_0\} \times \mathbb{R}^m \cong \mathbb{R}^m \to \mathbb{R}^m$. Thus, by the implicit function theorem, there exist an open neighborhood $\mathcal{U}$ of $\mathbf{x}_0$ and an open neighborhood $\mathcal{V}$ of $\mathbf{y}_0$ and a continuously differentiable function $g : \mathcal{U} \to \mathcal{V}$ such that for every point $\mathbf{z} = (\mathbf{x}, \mathbf{y}) \in \mathcal{U} \times \mathcal{V}$ we have

$$f(\mathbf{z}) = f(\mathbf{z}_0) \quad \text{if and only if} \quad \mathbf{y} = g(\mathbf{x}). \tag{2.40}$$

For example, if $n = 1$ and $m = 2$, then $f : \mathbb{R}^3 \to \mathbb{R}^2$ and the Jacobian matrix representing *df* and defined by (2.11) is

$$\begin{pmatrix} \dfrac{\partial f^1}{\partial x^1} & \dfrac{\partial f^1}{\partial x^2} & \dfrac{\partial f^1}{\partial x^3} \\[2ex] \dfrac{\partial f^2}{\partial x^1} & \dfrac{\partial f^2}{\partial x^2} & \dfrac{\partial f^2}{\partial x^3} \end{pmatrix}. \tag{2.41}$$

Without loss of generality, we assume that $j_1 = 2$, $j_2 = 3$, and $\mathbf{z}_0 = (x_0, \mathbf{y}_0) \in \mathbb{R}^3$, where $x_0 \in \mathbb{R}$ and $\mathbf{y}_0 \in \mathbb{R}^2$, is a point such that the 2×2 submatrix

$$\begin{pmatrix} \dfrac{\partial f^1}{\partial x^2}(\mathbf{z}_0) & \dfrac{\partial f^1}{\partial x^3}(\mathbf{z}_0) \\[2ex] \dfrac{\partial f^2}{\partial x^2}(\mathbf{z}_0) & \dfrac{\partial f^2}{\partial x^3}(\mathbf{z}_0) \end{pmatrix} \tag{2.42}$$

is invertible. The implicit function theorem says that there exist an open interval $\mathcal{U}$ containing x_0 and an open neighborhood $\mathcal{V}$ of $\mathbf{y}_0$ and a continuously differentiable function $g : \mathcal{U} \to \mathcal{V}$ such that for every point $\mathbf{z} = (x, \mathbf{y}) \in \mathcal{U} \times \mathcal{V}$ we have

$$f(\mathbf{z}) = f(\mathbf{z}_0) \quad \text{if and only if} \quad \mathbf{y} = g(x), \tag{2.43}$$

which describes a curve.

Example 2.13. *Define* $f : \mathbb{R}^4 \to \mathbb{R}^2$ *by* $f(x, y, z, w) = (u, v)$, *where*

$$u = x^2 + y^2 + z^2 + w^2, \qquad v = x^3 + y^3 + z^3 + w^3.$$

Then

$$S := f^{-1}((1, 0))$$

is a 2-dimensional smooth submanifold of $\mathbb{R}^4$.

2.3. Parametrized and higher-codimension submanifolds

We first consider submanifolds that are defined as the *images* of nice maps, called parametrized submanifolds. By a 'nice map', we mean the following.

Definition 2.14. A **smooth embedding** is a smooth map with injective derivatives at each point that is also a topological embedding.

We will discuss this definition, including the notion of topological embedding, in more detail in §2.5.2 below.

So far, the examples we have considered have been codimension-1 submanifolds, a.k.a. hypersurfaces. Here we look at some simple examples of higher-codimension submanifolds. The 'codimension' of a submanifold is defined to be the dimension of the ambient Euclidean space minus the dimension of the submanifold.

2.3.1. Definition of a parametrized submanifold. Assume an open subset $\mathcal{U}$ of $\mathbb{R}^n$ and a smooth embedding $F : \mathcal{U} \to \mathbb{R}^N$. Then $M^n = F(\mathcal{U})$ is a submanifold of $\mathbb{R}^N$. We call M^n a **parametrized submanifold** and we call the map F a **parametrization** of M^n.

2.3.2. Examples of parametrized submanifolds. Many surfaces in $\mathbb{R}^3$ are defined as parametrized submanifolds.

Example 2.15 (Helicoid). *Define the map* $F : \mathbb{R}^2 \to \mathbb{R}^3$ *by*

$$F(u^1, u^2) = \left(u^2 \cos(u^1), u^2 \sin(u^1), u^1\right). \tag{2.44}$$

The ***helicoid*** *is the image* $H^2 := F(\mathbb{R}^2)$ *of the map* F. *We say that* F *parametrizes the helicoid* H^2.

Observe that for each fixed $u^1 \in \mathbb{R}$, *the image curve of the path*

$$u^2 \mapsto \left(u^2 \cos(u^1), u^2 \sin(u^1), u^1\right) \tag{2.45}$$

is a line

$$L_{u_1} := \left\{\left(u^2 \cos(u^1), u^2 \sin(u^1), u^1\right) : u_2 \in \mathbb{R}\right\} \tag{2.46}$$

in the $\{x^3 = u^1\}$*-plane intersecting with the* x^3*-axis, where* L_{u_1} *makes the angle* u^1 *with the positive* x^1*-axis. Note that the helicoid is a union of lines,*

$$H^2 = \bigcup_{u_1 \in \mathbb{R}} L_{u_1}, \tag{2.47}$$

which rotate counterclockwise as u_1 *increases. As such, it is a* ***ruled surface****. See Figure* 2.3.1.

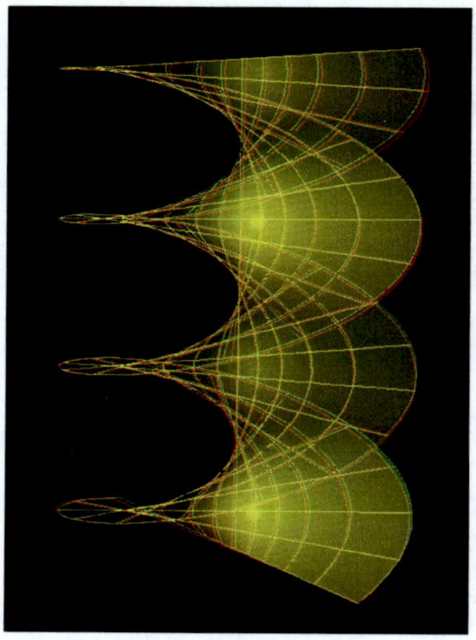

Figure 2.3.1. A helicoid H^2. Credit: 3DXM Virtual Mathematics Museum by The 3D-XplorMath Consortium. © Copyright 3DXM Consortium. `https://3d-xplormath.org/`

For a parametrization of the torus, see Exercise 2.10.

Example 2.16 (Möbius strip). *Define the map* $F : \left[-\frac{1}{2}, \frac{1}{2}\right] \times [0, 2\pi] \to \mathbb{R}^3$, *in component form*

$$F(t,\theta) = \big(x(t,\theta), y(t,\theta), z(t,\theta)\big), \tag{2.48}$$

by

$$\begin{aligned} x(t,\theta) &= \left(1 + t\cos\left(\frac{\theta}{2}\right)\right)\cos(\theta), \\ y(t,\theta) &= \left(1 + t\cos\left(\frac{\theta}{2}\right)\right)\sin(\theta), \\ z(t,\theta) &= t\sin\left(\frac{\theta}{2}\right). \end{aligned}$$

The map F *is an embedding when we restrict the domain to its interior* $\left(-\frac{1}{2}, \frac{1}{2}\right) \times (0, 2\pi)$, *and the image of* F *is a Möbius strip. The surface* M^2 *is obtained by sweeping out a line segment orthogonally along a circle in* $\mathbb{R}^3$ *while rotating the line segment a half turn.*

Example 2.17. *We can also parametrize the* 2*-sphere* $S^2(\rho_0)$ *of radius* ρ_0. *A common way to do this is via spherical coordinates. Let* $\mathbf{x}$ *be a point in* $\mathbb{R}^3$ *that is not equal to the origin* $\mathbf{0}$. *Let* (x^1, x^2, x^3) *be the Cartesian coordinates of* $\mathbf{x}$. *The projection of* $\mathbf{x}$ *to the* x^1x^2*-plane is the map* $\mathbf{x} \to (x^1, x^2, 0) =: \hat{\mathbf{x}}$. *The spherical coordinates of* $\mathbf{x}$ *are typically denoted by* (ρ, θ, ϕ); *see Figure* 2.3.2.

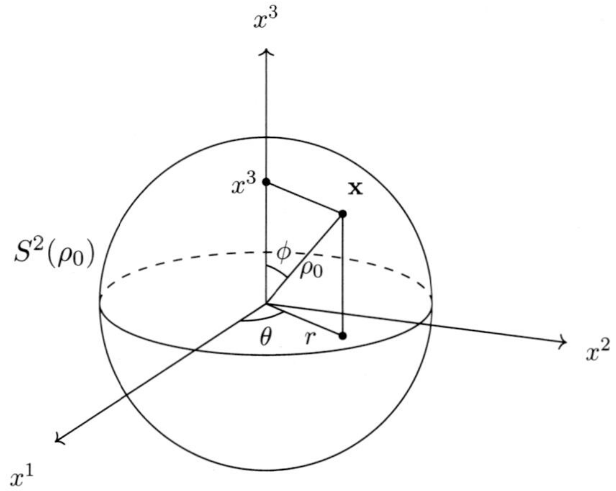

Figure 2.3.2. Picturing spherical coordinates on the 2-sphere of radius ρ_0.

(i) $\rho = \sqrt{(x^1)^2 + (x^2)^2 + (x^3)^2} \in (0, \infty)$ *is the distance from* $\mathbf{x}$ *to the origin* $\mathbf{0}$.

(ii) $\theta \in [0, 2\pi)$ *is the angle between the projection* $\hat{\mathbf{x}}$ *and the positive* x^1*-axis.*

(iii) $\phi \in [0, \pi]$ *is the angle between* $\mathbf{x}$ *and the positive* x^3*-axis.*

It is well known that the Cartesian coordinates (x^1, x^2, x^3) *of a point as a function of the spherical coordinates* (ρ, θ, ϕ) *of that point are given by the spherical-to-Cartesian transformation formula:*

$$x^1 = \rho \sin(\phi) \cos(\theta), \tag{2.49}$$

$$x^2 = \rho \sin(\phi) \sin(\theta), \tag{2.50}$$

$$x^3 = \rho \cos(\phi). \tag{2.51}$$

This coordinate system transformation formula also defines a parametrization of any sphere. Let ρ_0 *be a positive real constant. Let* $S^2(\rho_0)$ *denote the sphere of radius* ρ_0 *centered at the origin* $\mathbf{0}$. *We can define a parametrization of most of the sphere*

$$F : (0, 2\pi) \times (0, \pi) \to \mathbb{R}^3 \tag{2.52}$$

by

$$F(\theta, \phi) := \big(\rho_0 \sin(\phi) \cos(\theta), \rho_0 \sin(\phi) \sin(\theta), \rho_0 \cos(\phi)\big). \tag{2.53}$$

Then the image $F\big((0, 2\pi) \times (0, \pi)\big)$ *is equal to* $S^2(\rho_0)$ *minus a geodesic arc from the north pole to the south pole.*

2.3.3. Examples of higher-codimension submanifolds. So far, the submanifolds we have been considering have been codimension 1. That is, the difference of the ambient dimension and the dimension of the submanifold is equal to 1. We now consider some examples of submanifolds of codimension 2 or greater.

A simple example is a circle in $\mathbb{R}^3$. For example, we may consider

$$S^1 := \{(x^1, x^2, 0) : (x^1)^2 + (x^2)^2 = 1\}. \tag{2.54}$$

This is a 1-dimensional submanifold of $\mathbb{R}^3$, so its codimension is equal to 2. See Figure 2.3.3.

The **trefoil knot** is a circle embedded in $\mathbb{R}^3$. See Figure 2.3.4.

Another knot is the **figure-eight knot**. See Figure 2.3.5.

Example 2.13 is an example of a codimension-2 submanifold of $\mathbb{R}^4$.

A class of higher-codimension submanifolds is given by the following.

Example 2.18. *Let $f : \mathbb{R}^n \to \mathbb{R}$ be a smooth function. Then its derivative may be considered as a map $df : \mathbb{R}^n \to \mathbb{R}^n$ by using the isomorphism of $\mathbb{R}^n$ and its dual vector space $(\mathbb{R}^n)^*$ defined by the Euclidean inner product. The*

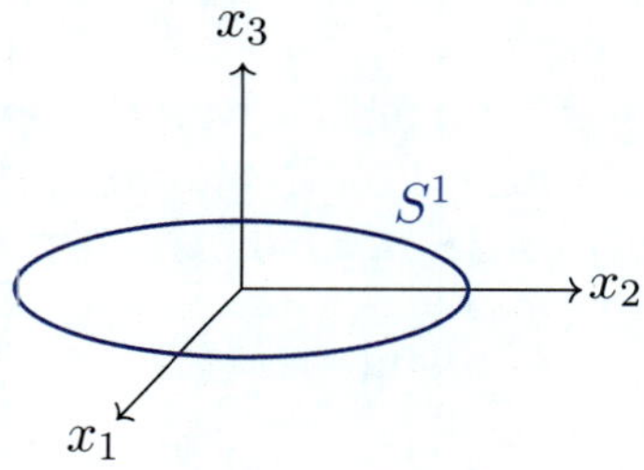

Figure 2.3.3. A circle in $\mathbb{R}^3$. In topology, this is the **unknot**.

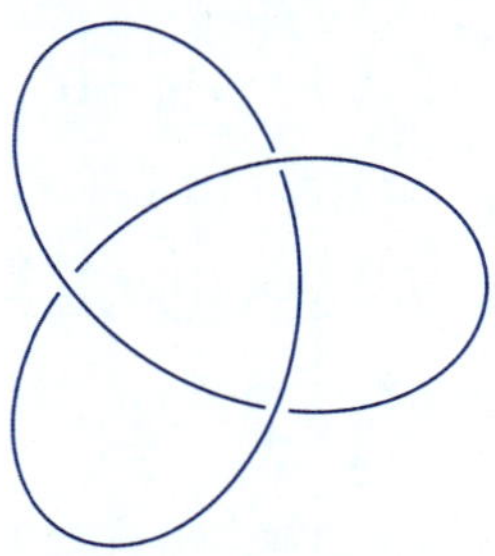

Figure 2.3.4. A trefoil knot.

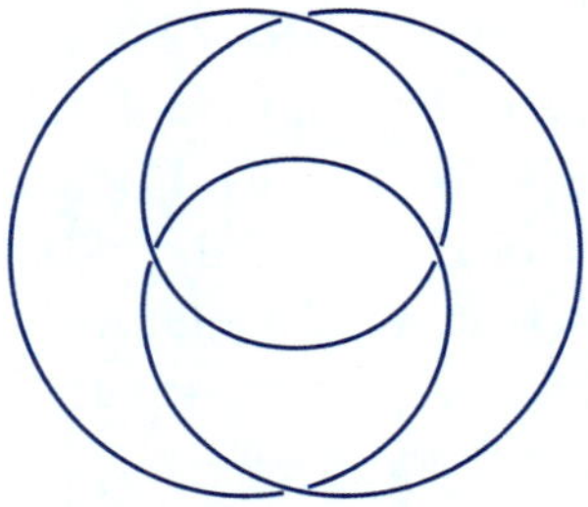

Figure 2.3.5. A figure-eight knot.

graph of df is an n-dimensional submanifold of $\mathbb{R}^{2n}$:

$$M^n := \{(\mathbf{x}, df_{\mathbf{x}}) : \mathbf{x} \in \mathbb{R}^n\} \subset \mathbb{R}^{2n}.$$

The codimension of M^n is also equal to n.

2.4. Examples of coordinate charts for submanifolds

Recall that we have been told that spheres are submanifolds. Now let us look in more detail and rigor at in what sense they are locally Euclidean.

2.4.1. Coordinate charts for spheres. We can express a sphere as the union of two hemispheres: a northern hemisphere and a southern hemisphere. If we do this exactly, then the hemispheres miss exactly the equator (if they are open) or the hemispheres intersect exactly at the equator (if they are closed). Notice that any point on the equator is neither solidly in the red or the blue region. See Figure 2.4.1.

On the other hand, the two subsets $S^n \setminus \{\mathbf{N}\}$ and $S^n \setminus \{\mathbf{S}\}$ overlap (a lot!) and their union is all of S^n. Let

$$G : S^n \setminus \{\mathbf{N}\} \to \mathbb{L}$$

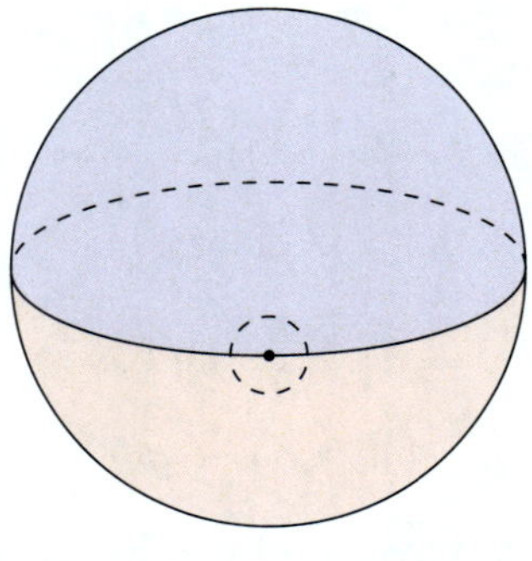

Figure 2.4.1. The southern (red) and northern (blue) hemispheres. The neighborhood of a point on the equator is half red and half blue.

be the stereographic projection map defined by (1.33). The map G is an example of a *coordinate chart.* We can think of G as providing a way of identifying the subset $S^n \setminus \{\mathbf{N}\}$ of S^n with Euclidean space $\mathbb{R}^n$. It is in this sense that any point $\mathbf{x} \neq \mathbf{N}$ in S^n has the 'neighborhood' $S^n \setminus \{\mathbf{N}\}$ which is 'Euclidean' in the sense that it can be identified with Euclidean space via the coordinate chart G. Now, it is pretty clear that even $\mathbf{x} = \mathbf{N}$ is contained in a locally Euclidean neighborhood.

We can define, analogous to $G : S^n \setminus \{\mathbf{N}\} \to \mathbb{L}$, another stereographic projection map $H : S^n \setminus \{\mathbf{S}\} \to \mathbb{K}$. So we can cover S^n by the two coordinate charts G and H, with domains $S^n \setminus \{\mathbf{N}\}$ and $S^n \setminus \{\mathbf{S}\}$, respectively. The **transition map** is defined as

$$H \circ G^{-1} : \mathbb{L} \setminus \{\mathbf{S}\} \to \mathbb{K} \setminus \{\mathbf{N}\}. \tag{2.55}$$

The transition map is a map between open subsets of Euclidean n-spaces. It is a way of describing how the two views G and H of S^n being locally Euclidean differ on the overlap set $(S^n \setminus \{\mathbf{N}\}) \cap (S^n \setminus \{\mathbf{S}\}) = S^n \setminus \{\mathbf{N}, \mathbf{S}\}$.

We remark that, together, G and H form what is called an **atlas** for S^n. By this we mean that their domains $S^n \setminus \{\mathbf{N}\}$ and $S^n \setminus \{\mathbf{S}\}$ cover S^n and that the transition map is a smooth bijective map with injective derivatives and smooth inverse. In Exercise 2.11 you will define H and calculate the transition map $H \circ G^{-1}$.

2.4.2. Alternative coordinate charts for spheres. Instead of using stereographic projection to define charts for S^n, we can use graphical charts for S^n defined in §1.3.2.1, which also form an atlas. The transition maps in this case are easier to calculate, although the final formulas do not look quite as nice.

In particular, recall the maps $\phi_{i,\pm} : S^n_{i,\pm} \to B_i$ defined by (1.31), where B_i is the open unit ball in the i-th coordinate hyperplane $\mathbb{R}^n_i$ defined by (1.29) and where $S^n_{i,\pm}$ is the northern/southern i-th hemisphere defined by (1.26). By (1.28), the union of the coordinate chart domains is all of S^n; i.e., $S^n = \bigcup_{i=1}^{n+1} S^n_{i,\pm}$.

By Exercise 1.9, for each i we have the inverse map

$$\phi^{-1}_{i,\pm} : B_i \to S^n_{i,\pm}. \tag{2.56}$$

In general, given bijections $f : X \to Y$ and $g : Z \to W$, we have the bijection

$$g \circ f : f^{-1}(Y \cap Z) \to g(Y \cap Z). \tag{2.57}$$

Thus, for $1 \leq i, j \leq n+1$, we have the bijective transition map

$$\phi_{j,\pm} \circ \phi^{-1}_{i,\pm} : \phi_{i,\pm}(S^n_{i,\pm} \cap S^n_{j,\pm}) \to \phi_{j,\pm}(S^n_{i,\pm} \cap S^n_{j,\pm}). \tag{2.58}$$

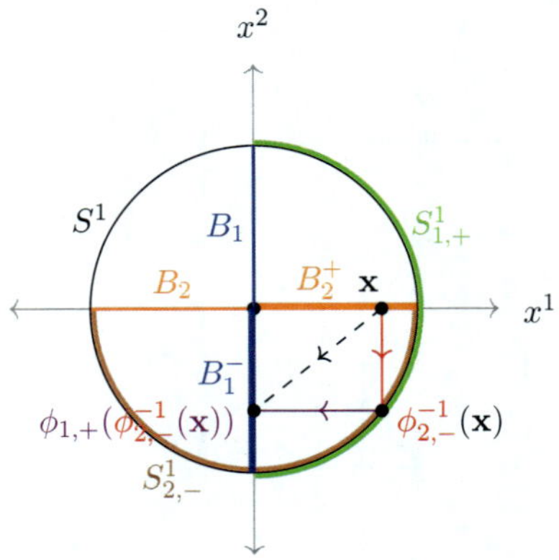

Figure 2.4.2. Visualizing the transition map $\phi_{1,+} \circ \phi_{2,-}^{-1}$ in the case $n = 1$. The domain is the thick orange segment B_2^+ and the codomain is the thick blue segment B_1^-.

For example, for the circle S^1 we have the transition map

$$\phi_{1,+} \circ \phi_{2,-}^{-1} : \phi_{2,-}(S^1_{2,-} \cap S^1_{1,+}) \to \phi_{1,+}(S^1_{2,-} \cap S^1_{1,+}). \tag{2.59}$$

That is,

$$\phi_{1,+} \circ \phi_{2,-}^{-1} : B_2^+ := \{\mathbf{x} \in B_2 \colon x^1 > 0\} \to B_1^- := \{\mathbf{x} \in B_1 \colon x^2 < 0\}. \tag{2.60}$$

See Figure 2.4.2 and see Exercise 2.12 for the formula for this transition map.

2.4.3. Coordinate charts for submanifolds, intuitively. Imagine that we have a subset M^n of $\mathbb{R}^{n+1}$. Suppose that $\{\mathcal{V}_\alpha\}_{\alpha \in A}$ is a collection of subsets of M^n that covers M^n, where A is the index set of the collection. By covering M^n, we mean that

$$M^n = \bigcup_{\alpha \in A} \mathcal{V}_\alpha. \tag{2.61}$$

Now suppose that we have bijections $\mathbf{x}_\alpha : \mathcal{V}_\alpha \to \mathcal{U}_\alpha$, where each $\mathcal{U}_\alpha$ is an open subset of $\mathbb{R}^n$ (we defined 'open subset' in §2.1.1 above). Then, in some sense, we hope that M^n is locally Euclidean and that M^n is a *submanifold*, a concept which we will rigorously define in the next section.

Example 2.19. *Consider an* $(n+1)$*-dimensional solid cube:*

$$C^{n+1} := [-1, 1]^{n+1} \in \mathbb{R}^{n+1}. \tag{2.62}$$

In other words,

$$C^{n+1} = \{\mathbf{x} \in \mathbb{R}^{n+1} \colon -1 \le x^i \le 1 \textit{ for all } 1 \le i \le n+1\}. \tag{2.63}$$

Let $M^n := \partial(C^{n+1})$ *be the boundary of the cube.*

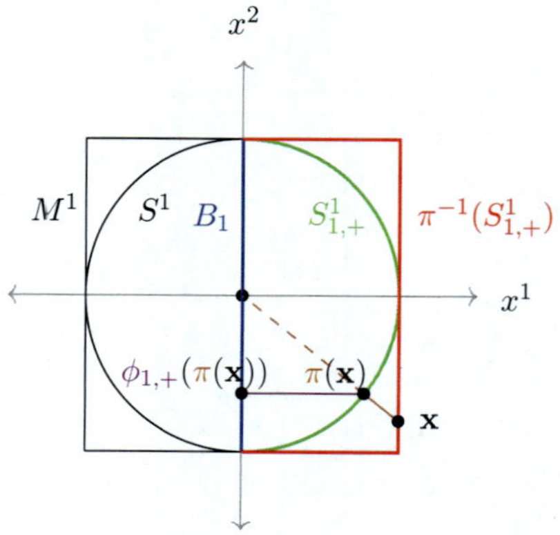

Figure 2.4.3. Visualizing the map $\phi_{1,+} \circ \pi : \pi^{-1}(S^n_{i,\pm}) \to B_i$ in the case $n = 1$.

The subset M^n is an example of a 'topological' submanifold, which is not a 'smooth' submanifold. The intuitive reason is that M^n has corners. In the special case where $n = 1$, $M^1 = \partial([-1,1] \times [-1,1])$ is the boundary of a square and has corners at the four points $(\pm 1, \pm 1)$. For example, the corner point $(1,1)$ in M^1 does not have a neighborhood in M^1 which is locally like $\mathbb{R}^1$ in the smooth sense, but it does have a neighborhood in M^1 which is locally like $\mathbb{R}^1$ in the topological sense since we can continuously bend an L shape into a straight line segment. For the radial projection map, see Exercise 2.13.

Continuing with Example 2.19, the collection $\{\pi^{-1}(S^n_{i,\pm})\}_{i=1}^{n+1}$ covers the boundary of the cube $M^n = \partial(C^{n+1})$. Consider the continuous coordinate charts

$$\phi_{i,\pm} \circ \pi : \pi^{-1}(S^n_{i,\pm}) \to B_i. \tag{2.64}$$

See Figure 2.4.3. The transition maps $\phi_{i,+} \circ \phi_{j,-}^{-1}$ are calculated in Exercise 2.14.

Remark 2.20. Exercise 2.14 shows that the transition maps for the boundary of the square M^1 are the same as those for the unit circle S^1.

Consequently, when we construct these 1-dimensional manifolds M^1 and S^1 using the previously defined atlases of coordinate charts, we establish a bijective map $\pi : M^1 \to S^1$ that, according to these atlases, makes M^1 and S^1 indistinguishable.

This means that, despite M^1 being nonsmooth as embedded in $\mathbb{R}^2$, we can abstractly consider M^1 as a *smooth* 1*-dimensional manifold.*

In Chapter 6, we will delve further into discussing manifolds from this abstract perspective.

2.5. Immersions and embeddings

Up to this point, we have explored parametrized submanifolds without discussing the specific assumptions governing these parametrizations. In this section, we will rigorously examine these assumptions, focusing exclusively on the C^∞ (or smooth) category. In relation to differential geometry, a function is considered C^∞ (pronounced as "C-infinity") if it possesses an infinite number of derivatives.

2.5.1. Immersions.

Definition 2.21. Let $\mathcal{U}$ be an open subset of $\mathbb{R}^n$ and let $F : \mathcal{U} \to \mathbb{R}^N$. We say that F is a **smooth immersion** if the derivative $dF_{\mathbf{x}_0} : \mathbb{R}^n \to \mathbb{R}^N$ is injective for all points $\mathbf{x}_0 \in \mathcal{U}$.

In other words, F is an immersion if the rank of $dF_{\mathbf{x}_0}$ is equal to n for all $\mathbf{x}_0 \in \mathcal{U}$. Note that immersions are allowed to have *self-intersections*; i.e., it is possible that there exist $\mathbf{x}_1 \neq \mathbf{x}_2$ in the domain $\mathcal{U}$ such that $F(\mathbf{x}_1) = F(\mathbf{x}_2)$.

Example 2.22 (Trefoil immersed path in the plane). *Define* $F : \mathbb{R} \to \mathbb{R}^2$ *by*

$$F(t) := (x(t), y(t)) := \big(\cos(t) + 2\cos(2t), \sin(t) - 2\sin(2t)\big). \tag{2.65}$$

See Figure 2.5.1. *We calculate that*

$$F'(t) = (x'(t), y'(t)) = \big(-\sin(t) - 4\sin(2t), \cos(t) - 4\cos(2t)\big). \tag{2.66}$$

The map F *is an immersion; see Exercise* 2.15. *Moreover, it is easy to see that* F *has self-intersections.*

Example 2.23 (A smooth injective map that is not an immersion). *Define the path* $\gamma : \mathbb{R} \to \mathbb{R}^2$ *by*

$$\gamma(t) = (t^3, t^2) =: (x(t), y(t)).$$

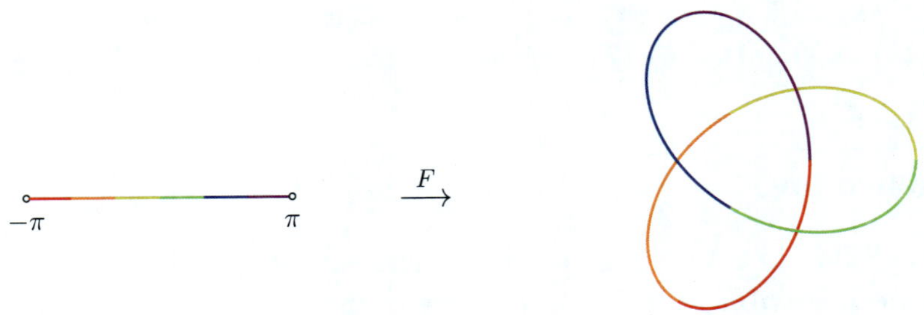

Figure 2.5.1. The immersion $F : \mathbb{R} \to \mathbb{R}^2$. From $-\pi$ to π the color changes from red to violet.

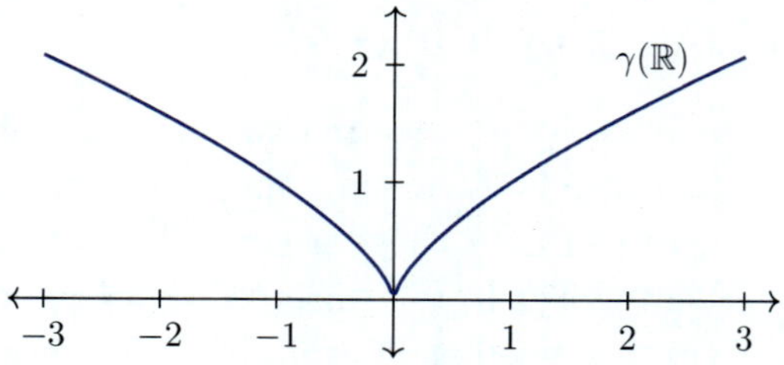

Figure 2.5.2. The curve $\gamma(\mathbb{R})$.

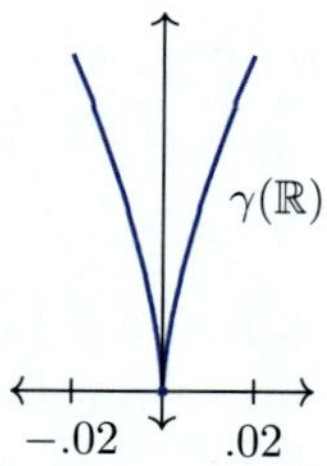

Figure 2.5.3. A close-up picture near the origin $\mathbf{0}$. The curve $\gamma(\mathbb{R})$ forms a cusp at $\mathbf{0}$, and consequently it is not a smooth submanifold.

We compute that $\gamma'(t) = (3t^2, 2t)$. *Thus the derivative is zero,* $\gamma'(t) = \mathbf{0}$, *if and only if* $t = 0$. *In particular,* γ *is not an immersion because of its behavior near* $t = 0$ *in the domain. This corresponds to the cusp at the point* $\gamma(0) = \mathbf{0}$ *in the image curve. Note that* $x(t)^2 = y(t)^3$ *and* $y(t) \geq 0$, *so that* $y(t) = |x(t)|^{2/3}$. *In particular, the image curve* $\gamma(\mathbb{R})$ *lies in the set* S *defined by the equation* $y = |x|^{2/3}$ *(equivalently,* $x^2 = y^3$*). It is not difficult to see that actually* $\gamma(\mathbb{R}) = S$. *That is, the image curve* $\gamma(\mathbb{R})$ *is the graph of a continuous function which is differentiable everywhere except at zero. See Figure* 2.5.2.

We have $\lim_{x\to 0_\pm} \frac{d}{dx}x^{2/3} = \pm\infty$, *where* $\lim_{x\to 0_-}$ *and* $\lim_{x\to 0_+}$ *denote the left and right limits, respectively. Thus the image curve* $\gamma(\mathbb{R})$ *forms a* ***cusp*** *at* $\mathbf{0}$*; see Figure* 2.5.3. *As we will see, curves with cusps (or corners) are not (smooth) submanifolds. This curve is called a* ***semicubical parabola****.*

2.5.2. Embeddings.

Definition 2.24. Let X and Y be topological spaces. A bijection $f : X \to Y$ is a **homeomorphism** if f and f^{-1} are both continuous maps. By the definition of continuity, this is equivalent to the property that for any subset $\mathcal{U}$ of X, its image $f(\mathcal{U})$ is an open subset of Y if and only if $\mathcal{U}$ is an open subset of X.

We say that a continuous map $F : \mathcal{U} \subset \mathbb{R}^n \to \mathbb{R}^N$ is a **topological embedding** if the codomain restricted map $F : \mathcal{U} \to F(\mathcal{U})$ is a homeomorphism, where $F(\mathcal{U})$ is endowed with the subspace topology from $\mathbb{R}^N$. Note that topological embeddings cannot have self-intersections since they are injective.

Example 2.25. *A particularly intriguing example of a topological embedding of the 2-sphere S^2 into 3-dimensional Euclidean space $\mathbb{R}^3$ is known as the* ***Alexander horned sphere****. This construction is remarkable because it highlights the difference between topological and smooth embeddings.*

The construction of the Alexander horned sphere is an infinite multistage process:

(1) ***Start with a torus:*** *Begin with a rotationally symmetric torus T^2 within $\mathbb{R}^3$. This torus is essentially a 2-dimensional surface that resembles a doughnut.*

(2) ***First stage:*** *In the first stage of this infinite construction, we remove an annulus $B^1 \times S^1$ from the torus T^2 and attach two disks $S^0 \times B^2$ to the resulting boundary. This operation is a 1-surgery and leads to the first topological embedding of S^2 into $\mathbb{R}^3$.*

(3) ***Second stage:*** *In the second stage, we remove a smaller disk from each disk $\{\pm 1\} \times B^2$ and attach to the resulting boundary what we colloquially call a 'hook' as pictured in Figure* 2.5.4*. This operation results in a more intricate topological embedding of S^2 into $\mathbb{R}^3$.*

(4) ***Ad infinitum:*** *In the third stage, we add $2^2 = 4$ hooks in a way analogous to the second stage. We continue this process ad infinitum. At each finite stage we have a topological embedding. However, as you move through the stages, the embedded S^2 becomes progressively more tangled within $\mathbb{R}^3$.*

(5) ***The embedding:*** *By taking the limit of this construction, one obtains a topological embedding of S^2 with a Cantor set removed. In particular, the embedding is modified by stage k and after on the interior D_k of 2^k disks, where $D_{k+1} \subset D_k$. Thus, we have a topological embedding of $S^2 \setminus D_\infty$, where $D_\infty := \bigcap_{k=1}^{\infty} D_k$. This embedding extends to all of S^2 as a topological embedding, and this extension is called the Alexander horned sphere.*

(6) ***Properties:*** *The Alexander horned sphere bounds a ball to its interior, but its exterior is not simply connected (see Alexander* [**Ale24**] *and Hatcher* [**Hat02**, Example 2B.2] *for the details of the construction and its properties).*

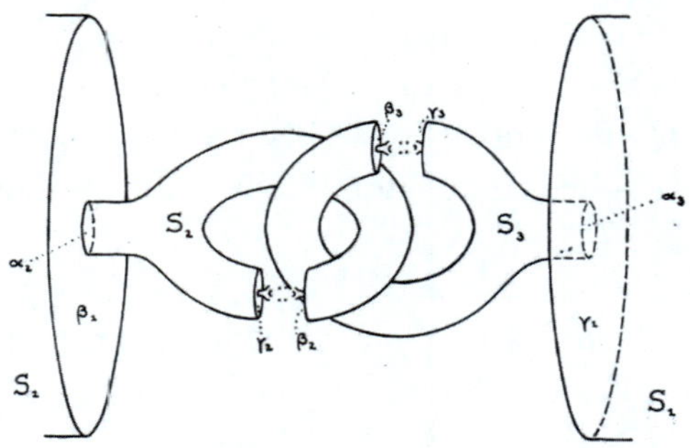

Figure 2.5.4. The Alexander horned sphere. Illustration of the embedding, which is an infinite construction, in Alexander [**Ale24**]. Credit: By Alexander, J. W. – Alexander, J. W. (1924), "An Example of a Simply Connected Surface Bounding a Region which is not Simply Connected", Proceedings of the National Academy of Sciences of the United States of America, National Academy of Sciences, 10 (1): 8–10, Public Domain, `https://commons.wikimedia.org/w/index.php?curid=99685574`

We will avoid pathological examples by staying in the smooth (versus topological) category. We say that F is a **smooth embedding** if F is a topological embedding that is a smooth immersion.

Example 2.26 (The difference between an injective immersion and an embedding). *Define* $H : (-\pi, \pi) \to \mathbb{R}^2$ *by*

$$H(t) := (x(t), y(t)) = \big(\sin(2t), \sin(t)\big). \tag{2.67}$$

H is an injective immersion, but H is not an embedding. See Figure 2.5.5.

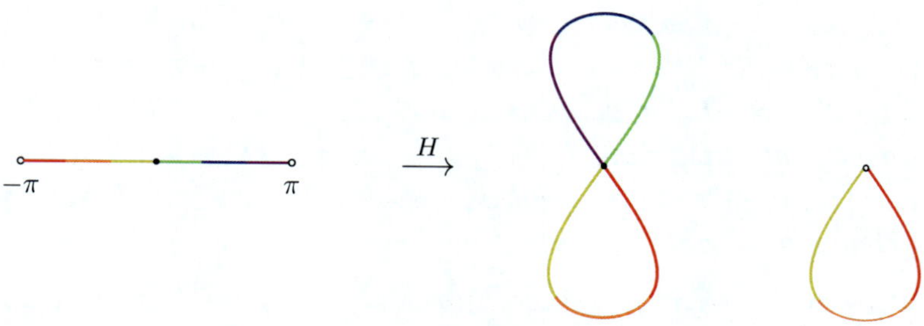

Figure 2.5.5. Left: The domain $(-\pi, \pi)$ of H. Middle: The image of the injective immersion $H : (-\pi, \pi) \to \mathbb{R}^2$. From $-\pi$ to π the color changes from red to violet. Right: The image $H\big((-\pi, 0)\big)$.

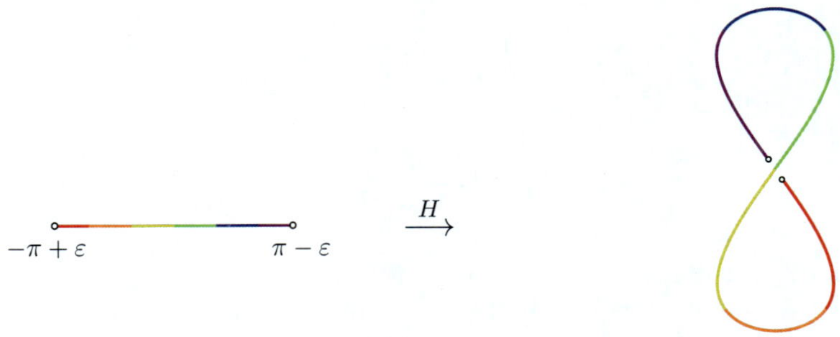

Figure 2.5.6. The restricted map $H : (-\pi+\varepsilon, \pi-\varepsilon) \to \mathbb{R}^2$ is an embedding.

Remark 2.27. If we restrict H to the interval $(-\pi+\varepsilon, \pi-\varepsilon)$, where $\varepsilon \in (0,\pi)$, then the restricted path H is an embedding. See Figure 2.5.6.

Example 2.28. *Let $g : \mathcal{U} \subset \mathbb{R}^n \to \mathbb{R}^\ell$ be a C^∞ map with domain $\mathcal{U}$ an open subset of $\mathbb{R}^n$. Then the* ***graphical map*** *$G : \mathcal{U} \to \mathbb{R}^{n+\ell}$ defined by*

$$G(\mathbf{x}) := (\mathbf{x}, g(\mathbf{x})) \tag{2.68}$$

is a smooth embedding; see Exercise 2.16.

Definition 2.29. We say that a map between metric spaces is **proper** if the preimage of every closed and bounded set is a closed and bounded set.

We state but do not prove the following useful result (see [**Lee13**, Proposition 4.22] for the proof).

Theorem 2.30. *If a smooth injective immersion is also a proper map, then it is a smooth embedding.*

2.5.3. The (rigorous) definition of a submanifold. Roughly speaking, a submanifold is locally the image of a smooth embedding. A precise definition is as follows.

Definition 2.31. A subset M^n of $\mathbb{R}^N$ is a **submanifold** if for each $\mathbf{x}_0 \in M^n$, there exist an open subset $\mathcal{U}$ of $\mathbb{R}^n$, $u_0 \in \mathcal{U}$, and a smooth embedding $F : \mathcal{U} \to \mathbb{R}^N$ such that $F(u_0) = \mathbf{x}_0$, $F(\mathcal{U}) \subset M^n$, and for some $\epsilon > 0$ (locally),

$$F(\mathcal{U}) \cap B_\epsilon(\mathbf{x}_0) = M^n \cap B_\epsilon(\mathbf{x}_0). \tag{2.69}$$

We say that the dimension of the submanifold M^n is n and the codimension of M^n in $\mathbb{R}^N$ is $N-n$. In the literature, M^n is also called an *embedded submanifold.*

See Figure 2.5.7 for a visualization of the definition of a submanifold when $n = 2$ and $N = 3$.

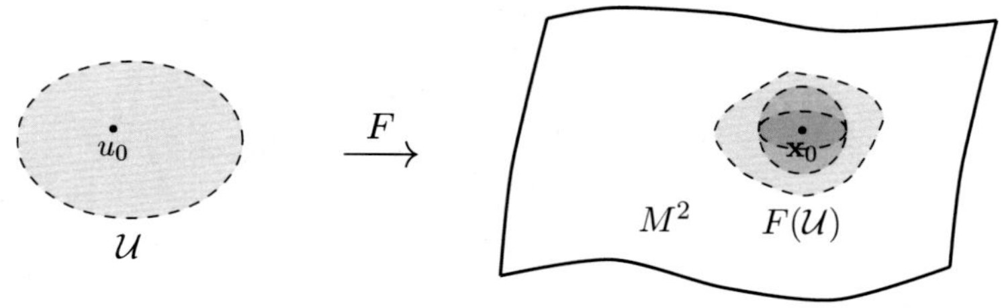

Figure 2.5.7. A 2-dimensional submanifold M^2 of $\mathbb{R}^3$. Left: The point u_0 in the open subset $\mathcal{U}$ of $\mathbb{R}^2$. Right: The light gray 2-dimensional region is the subset $F(\mathcal{U})$ of M^2. The medium gray 3-dimensional region is the open ball $B_\epsilon(\mathbf{x}_0)$. The dark gray 2-dimensional region is the intersection $M^2 \cap B_\epsilon(\mathbf{x}_0)$.

2.5.4. Equivalent definitions of submanifold. A general fact is that any smooth immersion is locally a smooth embedding. That is, if $G : \mathcal{V} \to \mathbb{R}^N$ is a smooth immersion from an open subset $\mathcal{V}$ of $\mathbb{R}^n$ and if $\mathbf{x}_0 \in \mathcal{V}$, then there exists an open subset $\mathcal{U}$ of $\mathbb{R}^n$ with $\mathbf{x}_0 \in \mathcal{U} \subset V$ and such that $G|_{\mathcal{U}} : \mathcal{U} \to \mathbb{R}^N$ is a smooth embedding. Thus, submanifolds are characterized by being locally the image of smooth immersions. Explicitly, a subset M^n of $\mathbb{R}^N$ is a submanifold if and only if for each $\mathbf{x}_0 \in M^n$, there exist an open subset $\mathcal{U}$ of $\mathbb{R}^n$, $u_0 \in \mathcal{U}$, and a smooth immersion $F : \mathcal{U} \to \mathbb{R}^N$ such that $F(u_0) = \mathbf{x}_0$, $F(\mathcal{U}) \subset M^n$, and for some $\epsilon > 0$,

$$F(\mathcal{U}) \cap B_\epsilon(\mathbf{x}_0) = M^n \cap B_\epsilon(\mathbf{x}_0). \tag{2.70}$$

The reason that this definition is equivalent to the original definition of submanifolds is that by passing to a smaller ϵ, we may assume that the immersion F is an embedding (see Theorem 2.35 below).

Submanifolds can also be characterized as being locally a graph. Namely, a subset M^n of $\mathbb{R}^N$ is a submanifold if and only if for each $\mathbf{x}_0 \in M^n$, there exists an open subset $\mathcal{W}$ of $\mathbb{R}^N$ containing $\mathbf{x}_0$ such that $M^n \cap \mathcal{W}$ is an affine graph over some affine hyperplane $\mathbb{A}$.[3] We may in fact assume that each such affine hyperplane is a coordinate hyperplane. See Figure 2.5.8.

A useful characterization of submanifolds is that they are locally the level sets of functions whose derivatives have maximal rank. Namely, a subset M^n of $\mathbb{R}^N$ is a submanifold if and only if for each $\mathbf{x}_0 \in M^n$, there exists an open subset $\mathcal{W}$ of $\mathbb{R}^N$ containing $\mathbf{x}_0$ and a smooth map $f : \mathcal{W} \to \mathbb{R}^{N-n}$ whose derivative has maximal rank at each point in its domain and

$$M^n \cap \mathcal{W} = f^{-1}(\mathbf{c}) \tag{2.71}$$

for some $\mathbf{c} \in \mathbb{R}^{N-n}$. We leave it to the reader to prove the characterization of submanifolds described above.

[3] See §1.3.3.2 for the definition of 'affine graph'.

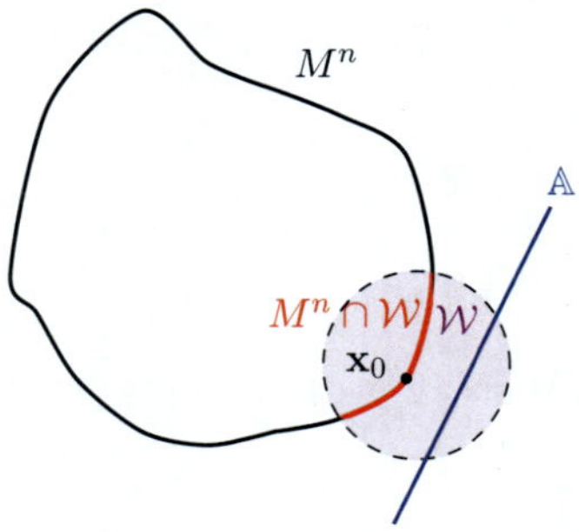

Figure 2.5.8. $M^n \cap \mathcal{W}$ is the affine graph over $\mathbb{A}$.

2.5.5. Examples of submanifolds.

Example 2.32 (Submanifolds we have discussed).

(1) *The affine hyperplane* $\mathbb{A}$ *defined by* (1.20).

(2) *The unit* n*-sphere* S^n *defined by* (1.25).

(3) *The affine graph of a function* $f : \mathbb{A} \to \mathbb{R}$, *where* $\mathbb{A}$ *is an affine hyperplane.*

(4) *A radial graph* R^n *as defined by* (1.57). *This is essentially because the radial projection map* $\pi : R^n \to S^n$ *defined by* (1.58) *and its inverse* π^{-1} *are 'nice maps' and because* S^n *is a submanifold. In particular, we can compose the graphical parametrizations* $\phi_{i,\pm}^{-1} : B_i \to S_{i,\pm}^n$ *of* S^n *with* $\pi^{-1} : S^n \to R^n$ *to obtain smooth embeddings* $\pi^{-1} \circ \phi_{i,\pm}^{-1}$ *into* R^n *whose images cover* R^n.

(5) *Surfaces of revolution are* 2*-dimensional submanifolds of* $\mathbb{R}^3$.

(6) *Open subsets of* $\mathbb{R}^n$ *are* n*-dimensional submanifolds of* $\mathbb{R}^n$.

At this point we leave it as exercises for the reader to check that every set which we asserted to be a submanifold is actually a submanifold; see Exercises 2.17 and 2.18.

2.5.6. Immersions are locally graphs. Let $\mathbb{A}$ be an n-dimensional affine subspace of $\mathbb{R}^N$. Let $\mathbf{a}_0 \in \mathbb{A}$. Then the **parallel vector subspace** is defined by

$$\mathbb{L}_{\mathbb{A}} = \{\mathbf{a} - \mathbf{a}_0 \colon \mathbf{a} \in \mathbb{A}\}. \tag{2.72}$$

Define the orthogonal subspace to $\mathbb{A}$ by

$$\mathbb{A}^{\perp} := \mathbb{L}^{\perp} = \{\mathbf{V} \in \mathbb{R}^N \colon \mathbf{V} \cdot \mathbf{a} = 0 \text{ for all } \mathbf{a} \in \mathbb{A}\}. \tag{2.73}$$

By an **affine graph** over a relatively open subset $\mathcal{W}$ of $\mathbb{A}$ we mean a subset of $\mathbb{R}^N$ of the form

$$\{\mathbf{x} + f(\mathbf{x}) \colon \mathbf{x} \in \mathcal{W}\}, \tag{2.74}$$

where $f : \mathcal{W} \to \mathbb{A}^\perp$.

Example 2.33. *Given a function $f : \mathbb{R}^n \to \mathbb{R}$, recall that earlier we defined its graph as*

$$G^n = \{(\mathbf{x}, f(\mathbf{x})) \colon \mathbf{x} \in \mathbb{R}^n\}. \tag{2.75}$$

We can interpret this as a special case of the general affine graph set-up above as follows. Let $\mathbb{A} = \mathbb{R}^n_{n+1} \subset \mathbb{R}^{n+1}$ be the $(n+1)$-st coordinate hyperplane (defined by (1.18)) and identify $\mathbb{R}^n$ with $\mathbb{R}^n_{n+1}$ in the usual way via $\mathbf{x}$ being identified with $(\mathbf{x}, 0)$ for $\mathbf{x} \in \mathbb{R}^n$. We have $\mathbb{A}^\perp = \mathbb{R}\mathbf{e}_{n+1} = \{t\mathbf{e}_{n+1} \colon t \in \mathbb{R}\}$. Define $\bar{f} : \mathbb{A} \to \mathbb{A}^\perp$ by

$$\bar{f}(\mathbf{x}, 0) = f(\mathbf{x})\mathbf{e}_{n+1}. \tag{2.76}$$

Then the graph of $\bar{f}$ is

$$\{(\mathbf{x}, 0) + \bar{f}(\mathbf{x}, 0) \colon (\mathbf{x}, 0) \in \mathbb{A}\} = \{(\mathbf{x}, f(\mathbf{x})) \colon \mathbf{x} \in \mathbb{R}^n\} = G^n, \tag{2.77}$$

which is the same as the graph of f.

Example 2.34. *Define the path $\gamma : \mathbb{R} \to \mathbb{R}^3$ by*

$$\gamma(t) := (\cos(t), \sin(t), t). \tag{2.78}$$

Then the image curve $\gamma(\mathbb{R})$ is the graph of a function over the X^3-axis $\mathbb{R}\mathbf{e}_3$. Namely, define $f : \mathbb{R}\mathbf{e}_3 \to (\mathbb{R}\mathbf{e}_3)^\perp = \operatorname{span}\{\mathbf{e}_1, \mathbf{e}_2\}$ by

$$f(t\mathbf{e}_3) := \cos(t)\mathbf{e}_1 + \sin(t)\mathbf{e}_2. \tag{2.79}$$

It is easy to see that the curve $\gamma(\mathbb{R})$ is the graph of f.

The following result, which we will take as a fact, says that any immersion is locally the graph of a function over an affine subspace; see Lee [**Lee13**, Theorem 4.25 and Corollary 6.33].

Theorem 2.35 (Smooth immersions are locally smooth embeddings and graphs). *Let $F : \mathcal{U} \subset \mathbb{R}^n \to \mathbb{R}^N$ be an immersion. Then, for any $\mathbf{u}_0 \in \mathcal{U}$, there exists an open subset $\mathcal{V}$ of $\mathcal{U}$ containing $\mathbf{u}_0$ such that $F|_{\mathcal{V}}$ is an embedding and $F(\mathcal{V})$ is equal to the graph of a function over the n-dimensional affine subspace $\mathbb{A} := F(\mathbf{u}_0) + dF_{\mathbf{u}_0}(\mathbb{R}^n)$ of $\mathbb{R}^N$.*

It is easy to see that in the statement of the theorem, we can replace $\mathbb{A}$ by the linear subspace $dF_{\mathbf{u}_0}(\mathbb{R}^n)$.

2.6. Notes and commentary

A classic book on point-set topology is Munkres [**Mun00**]. A classic book on vector calculus, with an emphasis on Stokes's theorem, is Spivak [**Spi65**].

2.7. Exercises

Exercise 2.1. *Prove that for any* $\mathbf{x} \in \mathbb{R}^n$ *and* $r > 0$, $B_r(\mathbf{x})$ *is an open set. Hint: See Figure* 2.1.1.

Exercise 2.2. *Prove that the half-space* $\mathcal{H}^n$ *defined by* (2.2) *is an open set. See Figure* 2.1.2.

Exercise 2.3. *Prove that the three properties in Properties* 2.3 *hold for the open sets in Euclidean space defined by Definition* 2.1.

Exercise 2.4. *Prove that the three properties in Properties* 2.3 *hold, where* $\mathbb{R}^n$ *is replaced by* $\mathfrak{M}$ *in property* (1), *for the Definition* 2.4 *of open sets in any metric space* $(\mathfrak{M}, d)$.

Exercise 2.5. *Prove that a subset* $\mathcal{U}$ *of* S *is open with respect to the subspace topology on* S *(a.k.a. relatively open in* S*) if and only if for every* $\mathbf{x} \in \mathcal{U}$ *there exists* $\epsilon > 0$ *such that*

$$B_\epsilon(\mathbf{x}) \cap S \subset \mathcal{U}.$$

See Figure 2.1.4. *For example, the set* B_i *defined by* (1.29) *is an open subset of the* i*-th coordinate hyperplane* $\mathbb{R}^n_i$ *with respect to the subspace topology.*

Exercise 2.6. *Given a map between metric spaces, prove the equivalence of the continuity notions given by Definitions* 2.7 *and* 2.8.

Exercise 2.7. *Let* $f : \mathcal{U} \subset \mathbb{R}^n \to \mathbb{R}$ *be a twice-differentiable function. Given* $\mathbf{x}_0 \in \mathcal{U}$, *let* $(d^2 f)_{\mathbf{x}_0} \in \mathrm{Lin}(\mathbb{R}^n, \mathbb{R}^n)$ *be its second derivative at* $\mathbf{x}_0$ *as defined by* (2.16). *Denote by* $(f_{ij})_{\mathbf{x}_0}$ *the components of* $(d^2 f)_{\mathbf{x}_0}$ *as a matrix. Show that the*

$$(f_{ij})_{\mathbf{x}_0} = \frac{\partial^2 f}{\partial x^i \partial x^j}(\mathbf{x}_0) \tag{2.80}$$

are the usual second partial derivatives of f.

Exercise 2.8. *Prove* (2.26).

Exercise 2.9. *Prove the following generalization of the implicit function theorem (Theorem* 2.11*). Let* $f : \mathbb{R}^{n+m} \to \mathbb{R}^m$ *be a continuously differentiable map and let* $\mathbb{A}$ *be an* n*-dimensional affine subspace of* $\mathbb{R}^{n+m}$. *Given* $\mathbf{a}_0 \in \mathbb{A}$, *let* $\mathbb{A}^\perp$ *denote the* m*-dimensional affine subspace of* $\mathbb{R}^{n+m}$ *intersecting* $\mathbb{A}$ *orthogonally at* $\mathbf{a}_0$. *Consider the restricted map (redundantly denoted* *as)* $f : \mathbb{A}^\perp \cong \mathbb{R}^m \to \mathbb{R}^m$. *Let*

$$(d_{\mathbb{A}^\perp} f)_{\mathbf{a}_0} : \mathbb{A}^\perp \to \mathbb{R}^m \tag{2.81}$$

denote the derivative of this restricted map f *at the point* $\mathbf{a}_0$.

Prove that if the linear map $(d_{\mathbb{A}^\perp} f)_{\mathbf{a}_0}$ *is invertible, then there exist a relatively open neighborhood* $\mathcal{U}$ *of* $\mathbf{a}_0$ *in* $\mathbb{A}$ *and a relatively open neighborhood* $\mathcal{V}$ *of* $\mathbf{a}_0$ *in* $\mathbb{A}^\perp$ *and a continuously differentiable function* $g : \mathcal{U} \subset \mathbb{A} \to \mathcal{V} \subset \mathbb{A}^\perp$ *such that for every point* $\mathbf{x} + \mathbf{y} \in \mathcal{U} \oplus \mathcal{V} \subset \mathbb{A} \oplus \mathbb{A}^\perp = \mathbb{R}^{n+m}$ *we have*

$$f(\mathbf{x} + \mathbf{y}) = \mathbf{c}_0 := f(\mathbf{a}_0) \quad \text{if and only if} \quad \mathbf{y} = g(\mathbf{x}). \tag{2.82}$$

If f *is* C^∞, *then* g *is* C^∞.

Explain why the implicit function theorem is a special case of this result.

Exercise 2.10. *Parametrize the torus in Example* 1.6.

Exercise 2.11. *Let*

$$\mathbb{K} := \{\mathbf{y} \in \mathbb{R}^{n+1} : y^{n+1} = 1\} \tag{2.83}$$

be the hyperplane tangent to S^n *at* $\mathbf{N}$. *Analogous to the stereographic projection map* $G : S^n \setminus \{\mathbf{N}\} \to \mathbb{L}$ *given by* (1.33), *define another stereographic projection map*

$$H : S^n \setminus \{\mathbf{S}\} \to \mathbb{K}.$$

The map H *is another example of a coordinate chart.*

(i) *Compute a formula for* H.

(ii) *Compute the* ***transition map***

$$H \circ G^{-1} : \mathbb{L} \setminus \{\mathbf{S}\} \to \mathbb{K} \setminus \{\mathbf{N}\}. \tag{2.84}$$

If you like, you can identify both $\mathbb{L} \setminus \{\mathbf{S}\}$ *and* $\mathbb{K} \setminus \{\mathbf{N}\}$ *with* $\mathbb{R}^n \setminus \{\mathbf{0}\}$. *This will make the formula for* $H \circ G^{-1} : \mathbb{R}^n \setminus \{\mathbf{0}\} \to \mathbb{R}^n \setminus \{\mathbf{0}\}$ *look slightly nicer.*

Exercise 2.12. *Work out the formula for* $\phi_{1,+} \circ \phi_{2,-}^{-1}$ *for* $n = 1$ *in* (2.60). *For any* $n \geq 1$, *using your formula, guess the formula for the transition maps*

$$\phi_{i,+} \circ \phi_{j,-}^{-1} : \{\mathbf{x} \in B_j : x^i > 0\} \to \{\mathbf{x} \in B_i : x^j < 0\} \tag{2.85}$$

for $i \neq j$ *in* (2.58). *Finally, prove the formula you guessed. (Alternatively, just derive the formula instead of first guessing it.)*

Exercise 2.13. *Show that the map* $\bar{\pi} : M^n \to S^n$ *defined by*

$$\bar{\pi}(\mathbf{x}) = \frac{\mathbf{x}}{\|\mathbf{x}\|} \tag{2.86}$$

is equal to the radial projection map $\pi : M^n \to S^n$ *defined by* (1.58), *where we think of* M^n *as a radial graph.*

Exercise 2.14. *Let* $M^n = \partial(C^{n+1})$ *be the boundary of the cube in Example* 2.19. *Compute the transition maps*

$$\phi_{i,+} \circ \phi_{j,-}^{-1} : \{\mathbf{x} \in B_j : x^i > 0\} \to \{\mathbf{x} \in B_i : x^j < 0\} \tag{2.87}$$

for $i \neq j$.

Now replace the boundary of the cube M^n *by any radial graph* R^n. *Show that the result for the calculation of* $\phi_{i,-} \circ \phi_{j,+}^{-1}$ *does not change.*

Exercise 2.15. *Prove that the map* F *defined in Example* 2.22 *is an immersion.*

Find $t_1 \neq t_2$ *such that* $F(t_1) = F(t_2)$, *but* $F'(t_1) \neq F'(t_2)$.

Exercise 2.16. *Prove the assertion in Example* 2.28.

Exercise 2.17. *We have considered various ways to show that a sphere is a submanifold. Among these, which is the easiest way to see that a sphere is a submanifold?*

Exercise 2.18. *Prove that the tori given in Example* 1.6 *are* 2*-dimensional submanifolds.*

Chapter 3

Linearizing Submanifolds: Tangent and Tensor Bundles

Submanifolds, like functions, are nonlinear objects in general. A key to understanding them is to consider their linearizations. The linearization of a function is its derivative. Given a point in the domain, the derivative at that point is the best linear approximation to the function for inputs near that point. The linearizations of a submanifold are its *tangent spaces*. Given a point on a submanifold, the tangent space at that point is the linear subspace of the ambient Euclidean space that best approximates the submanifold near that point.

Geometric invariants of submanifolds, such as curvature, are defined in terms of differentiation. Derivatives are linear maps between vector spaces, which include linear functions. Furthermore, given a vector space V, its dual vector space V* comprises all linear functions on V. The dual vector space corresponding to a tangent space is known as a cotangent space.

An inner product on $\mathbb{V}$ is a bilinear function on $\mathbb{V} \times \mathbb{V}$. Multilinear functions on finite products of $\mathbb{V}$ are called tensors on $\mathbb{V}$. Tensors are useful for describing geometric invariants of submanifolds.

A vector field on a submanifold M^n assigns to each point x of M^n a tangent vector at x. In this chapter, besides tangent, normal, and cotangent spaces and bundles, we also consider tensor spaces and bundles on submanifolds; tensor fields assign to each point of the submanifold a tensor on the tangent space of that point.

3.1. Tangent space

Let M^n be an n-dimensional submanifold of $\mathbb{R}^{n+1}$. Recall that the topology of $\mathbb{R}^{n+1}$ determines a subspace topology for M^n where a subset $\mathcal{V}$ of M^n is open in M^n if and only if it is equal to the intersection of some open subset $\mathcal{W}$ of $\mathbb{R}^{n+1}$ with M^n; i.e., $\mathcal{V} = M^n \cap \mathcal{W}$. In this case, we also call $\mathcal{V}$ a **relatively open** subset of M^n.

Remark 3.1. More generally, we may consider ambient Euclidean spaces of dimension N. We consider the codimension-1 case for simplicity of visualization.

3.1.1. Tangent spaces of parametrized submanifolds. Suppose that a mapping $F : \mathcal{U} \subset \mathbb{R}^n \to M^n$ parametrizes a relatively open subset $F(\mathcal{U})$ of a submanifold M^n of $\mathbb{R}^{n+1}$. That is, F is a smooth embedding from $\mathcal{U}$ to $\mathbb{R}^{n+1}$ with its image contained in M^n. See Figure 3.1.1.

Let $\mathbf{u}_0 \in \mathcal{U}$. Then, by hypothesis, the derivative $dF_{\mathbf{u}_0} : \mathbb{R}^n \to \mathbb{R}^{n+1}$ is an injective linear map. As one of a few equivalent definitions, we may define the **tangent space** of M^n at $F(\mathbf{u}_0)$ as the image:

$$T_{F(\mathbf{u}_0)}M := dF_{\mathbf{u}_0}(\mathbb{R}^n), \tag{3.1}$$

which is a linear codimension-1 subspace of $\mathbb{R}^{n+1}$. If we wish to consider the tangent space as based at $F(\mathbf{u}_0)$, we may consider it as the parallel affine hyperplane passing through $F(\mathbf{u}_0)$, called the **tangent plane**:

$$TP_{F(\mathbf{u}_0)}M := F(\mathbf{u}_0) + dF_{\mathbf{u}_0}(\mathbb{R}^n). \tag{3.2}$$

Intuitively, the tangent space at a point $\mathbf{x}_0$ in M^n is the linear subspace of $\mathbb{R}^N$ that best approximates the (nonlinear) submanifold M^n near $\mathbf{x}_0$. See Figure 3.1.2.

Remark 3.2. A way to remember the distinction that we make between tangent *spaces* and tangent *planes* is that the tangent space $T_{\mathbf{x}_0}M$ is a vector subspace and passes through the origin, whereas the tangent plane $TP_{\mathbf{x}_0}M$ is an affine subspace that passes through $\mathbf{x}_0$. See Figure 3.1.2.

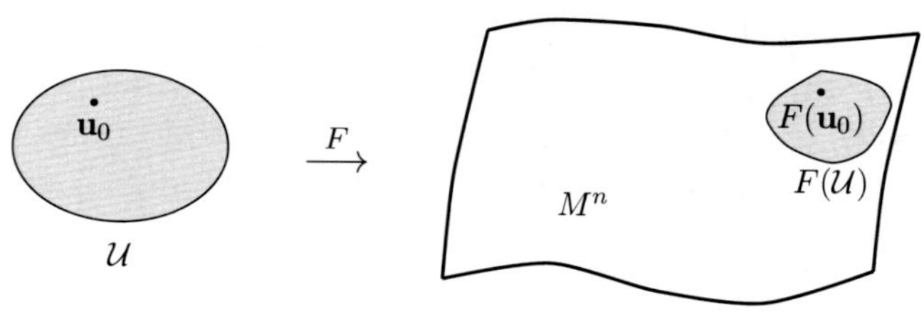

Figure 3.1.1. A local parametrization F of a submanifold M^n.

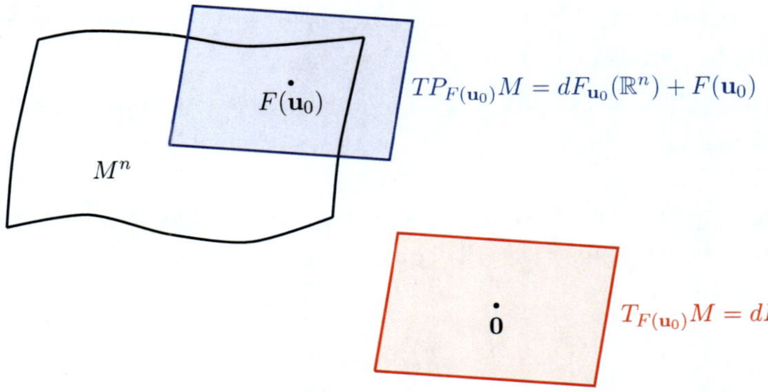

Figure 3.1.2. The tangent space $T_{F(\mathbf{u}_0)}M$ and the tangent plane $TP_{F(\mathbf{u}_0)}M$ of M^n at a point $F(\mathbf{u}_0)$.

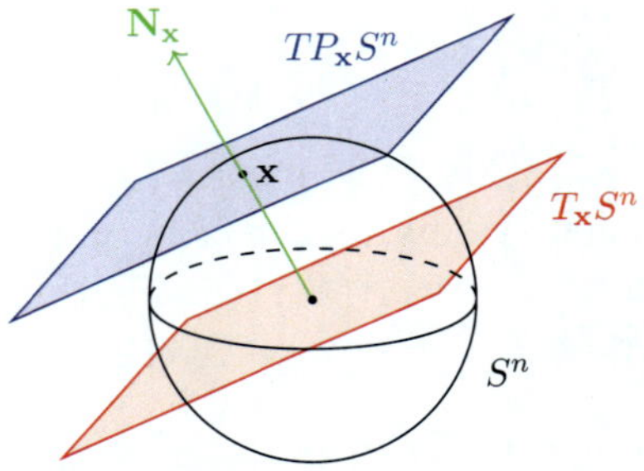

Figure 3.1.3. The tangent space $T_\mathbf{x}S^n$ and tangent plane $TP_\mathbf{x}S^n$ to the unit n-sphere at $\mathbf{x} \in S^n$. The normal is equal to the position vector: $\mathbf{N}_\mathbf{x} = \mathbf{x}$.

Example 3.3 (Tangent spaces of spheres, I). *For the unit sphere S^n in $\mathbb{R}^{n+1}$, we have*

$$T_{\mathbf{x}_0}S^n = \{\mathbf{V} \in \mathbb{R}^{n+1} \colon \mathbf{V} \cdot \mathbf{x}_0 = 0\} =: (\mathbb{R}\mathbf{x}_0)^\perp. \tag{3.3}$$

See Figure 3.1.3. *To see this, we can use the stereographic parametrization defined as follows. Let $\mathbf{N} = \mathbf{e}_{n+1}$ denote the north pole. Define $F : \mathbb{R}^n \to S^n$ by*

$$\begin{aligned} F(\mathbf{u}) &= \frac{4}{\|\mathbf{u}\|^2 + 4}\mathbf{u} + \frac{\|\mathbf{u}\|^2 - 4}{\|\mathbf{u}\|^2 + 4}\mathbf{e}_{n+1} \\ &= \frac{4}{\|\mathbf{u}\|^2 + 4}\left(\mathbf{u} + \frac{\|\mathbf{u}\|^2 - 4}{4}\mathbf{e}_{n+1}\right) \end{aligned} \tag{3.4}$$

for $\mathbf{u} \in \mathbb{R}^n$. Then F parametrizes the relatively open set $S^n \setminus \{\mathbf{N}\}$ in S^n. See Exercise 3.1 *for computing the tangent spaces of S^n as images of dF.*

If you attempted to prove (3.3) using $dF_\mathbf{u}(\mathbb{R}^n) = T_{F(\mathbf{u})}S^n$, you might be hoping that there is a simpler proof. Indeed, there is another way of defining the tangent space which yields a simpler calculation in the case of a sphere. We describe this in general in the next subsection.

See Exercises 3.2 and 3.3 for the tangent spaces of the helicoid and surfaces of revolution, respectively.

3.1.2. Tangent spaces of 2-dimensional graphs. Consider the graph of a function of two variables (we will consider n variables in §3.1.5 below). Let $f : \mathcal{U} \subset \mathbb{R}^2 \to \mathbb{R}$ and let

$$G^2 := \{(x, y, f(x, y)) \colon (x, y) \in \mathcal{U}\}.$$

We consider the graphic parametrization of G^2,

$$F : \mathcal{U} \to F(\mathcal{U}) = G^2 \subset \mathbb{R}^3, \tag{3.5}$$

defined by

$$F(x, y) := (x, y, f(x, y)). \tag{3.6}$$

Let $\mathbf{u}_0 = (x_0, y_0) \in \mathcal{U}$. Consider the path $\beta(t) := (x_0 + t, y_0)$, which is contained in $\mathcal{U}$ for t sufficiently small since $\mathcal{U}$ is open. Note that $\beta(0) = \mathbf{u}_0$ and $\beta'(0) = \mathbf{e}_1$. The path $\gamma(t) := F(\beta(t))$ is contained in G^2. Hence, its derivative $\gamma'(0)$ is tangent to G^2 at $F(\mathbf{u}_0)$. By the chain rule,

$$T_{F(\mathbf{u}_0)}G \ni \gamma'(0) = dF_{\beta(0)}(\beta'(0)) = dF_{\mathbf{u}_0}(\mathbf{e}_1) =: \frac{\partial F}{\partial x}(\mathbf{u}_0). \tag{3.7}$$

Thus,

$$\frac{\partial F}{\partial x}(\mathbf{u}_0) = \mathbf{e}_1 + \frac{\partial f}{\partial x}(\mathbf{u}_0)\mathbf{e}_3 \in T_{F(\mathbf{u}_0)}G. \tag{3.8}$$

Similarly, we derive that

$$\frac{\partial F}{\partial y}(\mathbf{u}_0) = \mathbf{e}_2 + \frac{\partial f}{\partial y}(\mathbf{u}_0)\mathbf{e}_3 \in T_{F(\mathbf{u}_0)}G. \tag{3.9}$$

It is easy to see that $\frac{\partial F}{\partial x}(\mathbf{u}_0)$ and $\frac{\partial F}{\partial y}(\mathbf{u}_0)$ are linearly independent in $\mathbb{R}^3$. Since the vector subspace $T_{F(\mathbf{u}_0)}G$ of $\mathbb{R}^3$ is 2-dimensional, we conclude that $\frac{\partial F}{\partial x}(\mathbf{u}_0)$ and $\frac{\partial F}{\partial y}(\mathbf{u}_0)$ span $T_{F(\mathbf{u}_0)}G$ of $\mathbb{R}^3$. See Figure 3.1.4.

Since the ambient dimension is equal to 3, a normal vector $\mathbf{N}_{F(\mathbf{u}_0)}$ to the graph and to its tangent space at $F(\mathbf{u}_0)$ is given by the cross product of the two tangent vectors above:

$$\begin{aligned} \mathbf{N}_{F(\mathbf{u}_0)} &:= \frac{\partial F}{\partial x}(\mathbf{u}_0) \times \frac{\partial F}{\partial y}(\mathbf{u}_0) \\ &= \left(-\frac{\partial f}{\partial x}(\mathbf{u}_0), -\frac{\partial f}{\partial y}(\mathbf{u}_0), 1\right) \\ &= (-\nabla f(\mathbf{u}_0), 1), \end{aligned} \tag{3.10}$$

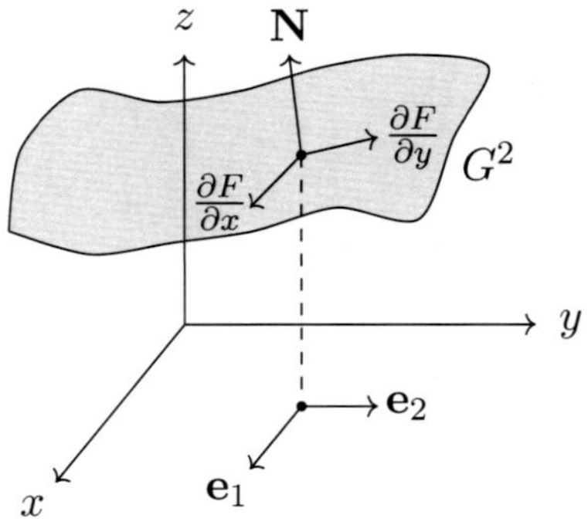

Figure 3.1.4. A normal vector $\mathbf{N}$ to a graph G^2.

where $\nabla f := \left(\frac{\partial f}{\partial x}, \frac{\partial f}{\partial y}\right)$ is the gradient of f. This normal vector $\mathbf{N}$, which is not unit unless $\nabla f = \mathbf{0}$, points upward in the sense that its z component is positive.

Example 3.4 (Two-dimensional spheres). *Let $B^2 := \{(x,y) \in \mathbb{R}^2 \colon x^2+y^2 < 1\}$. Define the function $f : B^2 \to \mathbb{R}$ by*

$$f(x,y) = \sqrt{1-x^2-y^2}. \tag{3.11}$$

Its graph is

$$G^2 = \left\{ \left(x, y, \sqrt{1-x^2-y^2}\right) \colon (x,y) \in B^2 \right\}. \tag{3.12}$$

The graphical parametrization is $F : B^2 \to \mathbb{R}^3$ defined by

$$F(x,y) = \left(x, y, \sqrt{1-x^2-y^2}\right). \tag{3.13}$$

Let $\mathbf{u}_0 = (x_0, y_0)$. Then $\mathbf{x}_0 := F(\mathbf{u}_0) = \left(x_0, y_0, \sqrt{1-x_0^2-y_0^2}\right)$. An upward pointing normal vector is given by (3.10) *as*

$$\begin{aligned} \mathbf{N}_{\mathbf{x}_0} &= (-\nabla f(\mathbf{u}_0), 1) \\ &= \left(\frac{x_0}{\sqrt{1-x_0^2-y_0^2}}, \frac{y_0}{\sqrt{1-x_0^2-y_0^2}}, 1 \right). \end{aligned}$$

Observe that $\mathbf{N}_{\mathbf{x}_0} = \frac{1}{\sqrt{1-x_0^2-y_0^2}} F(\mathbf{u}_0)$, so that $\mathbf{N}_{\mathbf{x}_0}$ and $\mathbf{x}_0 = F(\mathbf{u}_0)$ point in the same direction. In particular, the upward pointing unit *normal at $F(\mathbf{u}_0)$ is equal to $F(\mathbf{u}_0)$ itself. See again Figure* 3.1.3.

3.1.3. Tangent spaces of level sets. Suppose that M^n is defined as the **level set** of a function $f : \mathbb{R}^{n+1} \to \mathbb{R}$. That is,

$$M^n = f^{-1}(c) := \{\mathbf{x} \in \mathbb{R}^{n+1} \colon f(\mathbf{x}) = c\} \tag{3.14}$$

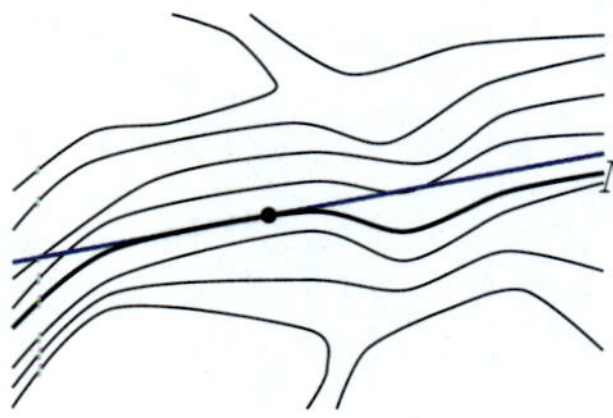

Figure 3.1.5. A tangent space of a level set.

for some $c \in \mathbb{R}$, and for each $\mathbf{x} \in M^n$ we assume that $df_{\mathbf{x}}$ is not the zero function, or equivalently, that $\nabla f(\mathbf{x}) \neq \mathbf{0}$. Then M^n is an n-dimensional submanifold. The tangent space of M^n at a point $\mathbf{x} \in M^n$ is given by

$$T_{\mathbf{x}}M = \ker(df_{\mathbf{x}}) = \{\mathbf{V} \in \mathbb{R}^{n+1} \colon df_{\mathbf{x}}(\mathbf{V}) = 0\} = (\mathbb{R}\nabla f(\mathbf{x}))^{\perp}. \tag{3.15}$$

See Figure 3.1.5.

Example 3.5 (Tangent spaces of spheres, II). *Define $f : \mathbb{R}^{n+1} \to \mathbb{R}$ by*

$$f(\mathbf{x}) := \|\mathbf{x}\|^2 \tag{3.16}$$

as in (2.25)*. The unit n-sphere S^n is equal to the level set $f^{-1}(1)$. Given $\mathbf{x} \in S^n$, we have*

$$df_{\mathbf{x}}(\mathbf{V}) = 2\mathbf{x} \cdot \mathbf{V} \tag{3.17}$$

for $\mathbf{V} \in \mathbb{R}^n$. Hence, by (3.15)*, we have*

$$T_{\mathbf{x}}S^n = \ker(df_{\mathbf{x}}) = \{\mathbf{V} \in \mathbb{R}^{n+1} \colon \mathbf{x} \cdot \mathbf{V} = 0\} = (\mathbb{R}\mathbf{x})^{\perp},$$

which agrees with (3.3)*.*

We can revisit graphs in the context of level sets. For simplicity, consider a 2-dimensional entire graph in $\mathbb{R}^3$. That is, let $f : \mathbb{R}^2 \to \mathbb{R}$. The graph of f is

$$G^2 = \{(x, y, f(x, y)) \colon (x, y) \in \mathbb{R}^2\}. \tag{3.18}$$

Define the function $g : \mathbb{R}^3 \to \mathbb{R}$ by

$$g(x, y, z) = f(x, y) - z. \tag{3.19}$$

We then have that the zero level set of g is the graph G^2; that is,

$$g^{-1}(0) = \{(x, y, z) \in \mathbb{R}^3 \colon z = f(x, y)\} = G^2. \tag{3.20}$$

We compute that the derivative of g is the linear function given by the matrix

$$dg = \begin{bmatrix} \frac{\partial f}{\partial x} & \frac{\partial f}{\partial y} & -1 \end{bmatrix}. \tag{3.21}$$

In other words, the gradient of g is

$$\nabla g = \left(\frac{\partial f}{\partial x}, \frac{\partial f}{\partial y}, -1\right) = (\nabla f, -1). \tag{3.22}$$

Thus, the tangent space to the graph G^2 is given by (we do not bother specifying the point it is at)

$$TG = \ker(dg) = \{\mathbf{V} \in \mathbb{R}^3 \colon \mathbf{V} \cdot \nabla g = 0\}. \tag{3.23}$$

This is the plane perpendicular to $\nabla g = (\nabla f, -1)$. This agrees with our previous calculation.

In Exercise 3.4 you will calculate the tangent spaces of an ellipsoid as a level set.

3.1.4. Agreement of the two definitions of tangent space. Now let us see why the two definitions of tangent space, one definition via parametrizations as an image of a derivative and one definition via level sets as the kernel of a derivative, give the same answer.

Suppose that

$$F(\mathcal{U}) \subset f^{-1}(c);$$

i.e., F parametrizes a relatively open subset of a level set of f. Then $f(F(\mathcal{U})) = \{c\}$; that is,

$$f(F(\mathbf{u})) = c \quad \text{is constant for all } \mathbf{u} \in \mathcal{U}.$$

Thus, for all $\mathbf{v} \in \mathbb{R}^n$,

$$0 = d(f \circ F)_{\mathbf{u}}(\mathbf{v}) = df_{F(\mathbf{u})}\big(dF_{\mathbf{u}}(\mathbf{v})\big), \tag{3.24}$$

where the second equality is true by the chain rule (formula for the derivative of the composition of two maps). This shows that

$$dF_{\mathbf{u}}(\mathbf{v}) \in \ker(df_{F(\mathbf{u})}) \quad \text{for all } \mathbf{v} \in \mathbb{R}^n,$$

and hence the vector subspace containment:

$$dF_{\mathbf{u}}(\mathbb{R}^n) \subset \ker(df_{F(\mathbf{u})}). \tag{3.25}$$

Now, since F is an embedding, we have that

$$\dim dF_{\mathbf{u}}(\mathbb{R}^n) = n. \tag{3.26}$$

Since the derivative $df_{F(\mathbf{u})} : \mathbb{R}^{n+1} \to \mathbb{R}$ is not equal to the zero map, we have $\dim \ker(df_{F(\mathbf{u})}) = n$. From all of this, we conclude that

$$dF_{\mathbf{u}}(\mathbb{R}^n) = \ker(df_{F(\mathbf{u})}). \tag{3.27}$$

Thus, the two definitions of tangent space agree.

Remark 3.6. The linear algebra fact that we used in the argument above is that if $\mathbb{V}$ is a vector space and if $\mathbb{W} \subset \mathbb{X}$ are subspaces of the same dimension, then $\mathbb{W} = \mathbb{X}$.

3.1.5. Tangent spaces of graphs. Now let us consider graphs of scalar-valued functions with domains of arbitrary dimensions. Recall that the graph of a function $f : \mathcal{U} \to \mathbb{R}$, where $\mathcal{U}$ is a nonempty open subset of $\mathbb{R}^n$, is defined by (1.35) as

$$G^n = \{(\mathbf{x}, f(\mathbf{x}))\colon \mathbf{x} \in \mathcal{U}\} \subset \mathbb{R}^{n+1}. \tag{3.28}$$

Recall also that the graphical parametrization of G^n is given by Example 2.28 as the map $F : \mathcal{U} \to F(\mathcal{U}) = G^n$ defined by

$$F(\mathbf{x}) = (\mathbf{x}, f(\mathbf{x})). \tag{3.29}$$

We calculate that

$$dF(\mathbf{e}_i) = \frac{\partial F}{\partial x^i} = \mathbf{e}_i + \frac{\partial f}{\partial x^i}\mathbf{e}_{n+1} \in TG. \tag{3.30}$$

If the derivative is at a point $\mathbf{x} \in \mathcal{U}$, then this vector is in the tangent space at $F(\mathbf{x})$. See Figure 3.1.6. The reason for the containment is that $T_{F(\mathbf{x})}G = dF_{\mathbf{x}}(\mathbb{R}^n)$. We have that the n vectors in $\mathbb{R}^{n+1}$

$$\frac{\partial F}{\partial x^1}(\mathbf{x}), \ldots, \frac{\partial F}{\partial x^n}(\mathbf{x}) \tag{3.31}$$

are linearly independent in $\mathbb{R}^{n+1}$ by (3.30) and since the vectors $\mathbf{e}_1, \ldots, \mathbf{e}_n$ are linearly independent in $\mathbb{R}^n$. These vectors form a basis for the tangent space $T_{F(\mathbf{x})}G$ and

$$T_{F(\mathbf{x})}G = dF_{\mathbf{x}}(\mathbb{R}^n) = \operatorname{span}\left\{\frac{\partial F}{\partial x^1}(\mathbf{x}), \ldots, \frac{\partial F}{\partial x^n}(\mathbf{x})\right\}. \tag{3.32}$$

Given $\mathbf{x} \in \mathcal{U}$, an upward pointing normal vector at $F(\mathbf{x}) = (\mathbf{x}, f(\mathbf{x}))$ is given by

$$\mathbf{N}_{\mathbf{x}} := (-\nabla f(\mathbf{x}), 1) = -\nabla f(\mathbf{x}) + \mathbf{e}_{n+1}, \tag{3.33}$$

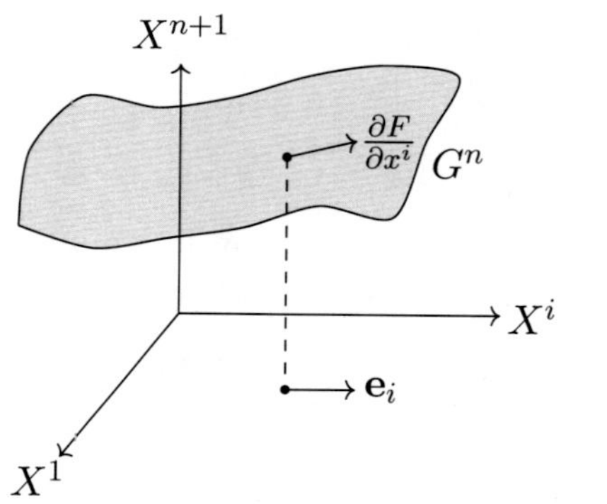

Figure 3.1.6. The coordinate tangent vector $\frac{\partial F}{\partial x^i}$ to the graph G^n.

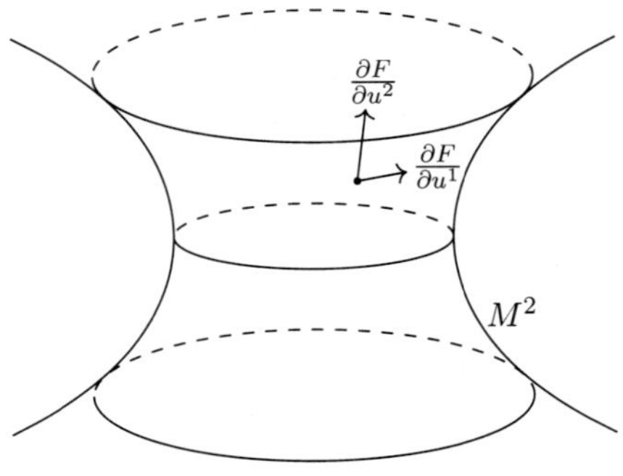

Figure 3.1.7. A surface M^2 and basis vectors $\frac{\partial F}{\partial u^1}$ and $\frac{\partial F}{\partial u^2}$ for a tangent space.

where ∇f is the gradient of f. A few things need to be explained here. Firstly,

$$\nabla f = \left(\frac{\partial f}{\partial x^1}, \dots, \frac{\partial f}{\partial x^n}\right) \in \mathbb{R}^n, \tag{3.34}$$

where $(x^1, \dots, x^n)$ are the Cartesian coordinates on $\mathbb{R}^n$. So the ∇f on the right-hand side of (3.33) is interpreted as being in $\mathbb{R}^{n+1}$ via the inclusion $\mathbb{R}^n \cong \mathbb{R}^n \times \{0\} \subset \mathbb{R}^{n+1}$. By $\mathbf{N}$ being upward pointing, we mean that $\mathbf{N}^{n+1} = \mathbf{N} \cdot \mathbf{e}_{n+1} > 0$. Finally, we make the main observation that

$$\mathbf{N}_x \cdot \frac{\partial F}{\partial x^i}(\mathbf{x}) = -\nabla f(\mathbf{x}) \cdot \mathbf{e}_i + \frac{\partial f}{\partial x^i}(\mathbf{x}) = 0 \tag{3.35}$$

for all $i = 1, \dots, n$. Since $\left\{\frac{\partial F}{\partial x^i}(\mathbf{x})\right\}_{i=1}^n$ is a basis for $T_{F(\mathbf{x})}G$, this implies that $\mathbf{N}_\mathbf{x}$, which is a vector in $\mathbb{R}^{n+1}$ at $F(\mathbf{x})$, is normal to the hyperplane $T_{F(\mathbf{x})}G$. Since $\mathbf{N}_\mathbf{x}$ is a nonzero normal vector to $T_{F(\mathbf{x})}G$, we have

$$T_{F(\mathbf{x})}G = (\mathbb{R}\mathbf{N}_\mathbf{x})^\perp = \{\mathbf{V} \in \mathbb{R}^{n+1} \colon \mathbf{V} \cdot \mathbf{N}_\mathbf{x} = 0\}. \tag{3.36}$$

Compare with Figure 3.1.4. In Exercise 3.5 you will calculate the tangent spaces of a hemisphere as a graph.

If $F \colon \mathcal{U} \subset \mathbb{R}^n \to \mathbb{R}^{n+1}$ is a smooth embedding, then $\left\{\frac{\partial F}{\partial u^1}(\mathbf{u}), \dots, \frac{\partial F}{\partial u^n}(\mathbf{u})\right\}$ is a basis for $T_{F(\mathbf{u})}M$ for each $\mathbf{u} \in \mathcal{U}$; see Exercise 3.6 and Figure 3.1.7.

3.1.6. An intrinsic way to define tangent spaces. Let M^n be an n-dimensional submanifold of $\mathbb{R}^{n+1}$. Let $\gamma : I \to M^n$ be a differentiable path. Assume that 0 lies in the interior of the interval I. Then the vector $\gamma'(0)$ is tangent to M^n at $\gamma(0)$. That is, $\gamma'(0) \in T_{\gamma(0)}M$; see Figure 3.1.8.

We can see this as follows. Let $F : \mathcal{U} \to F(\mathcal{U}) \subset M^n$ be a parametrization of a relatively open subset of M^n which contains the point $\gamma(0)$. Assume that $F^{-1} \circ \gamma : I \to \mathcal{U}$ is also continuously differentiable. Then $(F^{-1} \circ \gamma)'(0)$ is a vector in $\mathbb{R}^n$. By the chain rule, we have

$$\gamma'(0) = \left(F \circ (F^{-1} \circ \gamma)\right)'(0) = dF_{F^{-1}(\gamma(0))}\left((F^{-1} \circ \gamma)'(0)\right) \in T_{\gamma(0)}M. \tag{3.37}$$

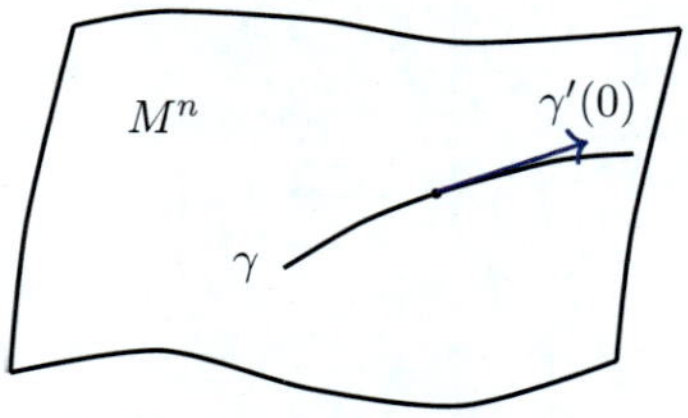

Figure 3.1.8. The tangent space of M^n at a point $F(\mathbf{u}_0)$. The derivative $\gamma'(0)$ of the path γ at 0 is a tangent vector.

Exercise 3.7 describes the tangent spaces of S^n by using paths that are great circles.

Example 3.7 (Graphs). *Let $f : \mathcal{U} \subset \mathbb{R}^n \to \mathbb{R}$ and let $G^n \subset \mathbb{R}^{n+1}$ be the graph of f as defined by* (1.35). *Let $\mathbf{u}_0 \in \mathcal{U}$ and define*

$$\beta_i(t) = \mathbf{u}_0 + t\mathbf{e}_i,$$

for $1 \leq i \leq n$. Note that $\beta_i(0) = \mathbf{u}_0$ and $\beta_i'(t) = \mathbf{e}_i$. Since $\mathcal{U}$ is an open set, there exists $\epsilon > 0$ such that $\beta_i(t) \in \mathcal{U}$ for all $t \in (-\epsilon, \epsilon)$. Let $F : \mathcal{U} \to G^n$ be the graphical parametrization defined by $F(\mathbf{u}) = (\mathbf{u}, f(\mathbf{u}))$; cf. Example 2.28. *Then*

$$\gamma_i(t) := F(\beta_i(t)) = (\beta_i(t), f(\beta_i(t))) \tag{3.38}$$

is a path in G. We compute that

$$\begin{aligned} \gamma_i'(0) &= (F \circ \beta_i)'(0) \\ &= (\beta_i'(0), df(\beta_i'(0))) \\ &= (\mathbf{e}_i, df_{\mathbf{u}_0}(\mathbf{e}_i)) \\ &= \left(\mathbf{e}_i, \frac{\partial f}{\partial u^i}(\mathbf{u}_0)\right). \end{aligned} \tag{3.39}$$

This tangent vector is the same as

$$\frac{\partial F}{\partial u^i}(\mathbf{u}_0) = \left(\mathbf{e}_i, \frac{\partial f}{\partial u^i}(\mathbf{u}_0)\right). \tag{3.40}$$

We recover the result that the collection of tangent vectors

$$\left\{\frac{\partial F}{\partial u^1}(\mathbf{u}_0), \ldots, \frac{\partial F}{\partial u^n}(\mathbf{u}_0)\right\} \tag{3.41}$$

forms a basis for $T_{F(\mathbf{u}_0)}G$.

3.1.7. Parametrizations revisited. Let M^n be a submanifold of $\mathbb{R}^{n+1}$ and let $F : \mathcal{U} \to F(\mathcal{U}) =: \mathcal{V}$ be a parametrization of a relatively open subset of M^n. Let $\mathbf{x}_0 \in \mathcal{V}$. Then there exists a unique point $\mathbf{u}_0 \in \mathcal{U}$ such that $F(\mathbf{u}_0) = \mathbf{x}_0$. The image of the standard Euclidean basis $\{\mathbf{e}_i\}_{i=1}^n$

for $T_{\mathbf{u}_0}\mathbb{R}^n \cong \mathbb{R}^n$ under the derivative $dF_{\mathbf{u}_0}$ is equal to the basis of tangent vectors

$$\left\{ \frac{\partial F}{\partial u^i}_{\mathbf{x}_0} := \frac{\partial F}{\partial u^i}(\mathbf{u}_0) = dF_{\mathbf{u}_0}(\mathbf{e}_i) \right\}_{i=1}^n$$

for $T_{\mathbf{x}_0}M$. Let $\varphi : \mathcal{V} \to \mathcal{U}$ be the inverse map of F; i.e., $\varphi = F^{-1}$. Then $(\mathcal{V}, \varphi)$ are local coordinates on M^n.

Now suppose that $f : M^n \to \mathbb{R}$ is a differentiable function. Define $\phi : \mathcal{U} \to \mathbb{R}$ by

$$\phi := f \circ F. \tag{3.42}$$

We have

$$\frac{\partial F}{\partial u^i}_{\mathbf{x}_0}(f) = \frac{\partial \phi}{\partial u^i}(\mathbf{u}_0). \tag{3.43}$$

Since $\varphi = F^{-1}$ is the set of local coordinates on M^n, we often denote it by

$$\varphi =: (x^1, \ldots, x^n), \tag{3.44}$$

where $x^i : \mathcal{V} \to \mathbb{R}$ for $1 \leq i \leq n$. It is customary to also use the notation

$$\frac{\partial}{\partial x^i}_{\mathbf{x}_0} = \frac{\partial F}{\partial u^i}_{\mathbf{x}_0}. \tag{3.45}$$

So we also write

$$\frac{\partial f}{\partial x^i}(\mathbf{x}_0) := \frac{\partial F}{\partial u^i}_{\mathbf{x}_0}(f). \tag{3.46}$$

We explain this here because there is a slight mismatch between the notation for submanifolds and the notation for abstract manifolds that we will adopt in subsequent chapters.

3.2. Tangent bundle

Can we consider the tangent spaces of a submanifold all together? The answer is yes and we first consider the simplest case of Euclidean space.

Consider Euclidean space $\mathbb{R}^n$ as an n-manifold. Let $\mathbf{x} \in \mathbb{R}^n$. We can naturally identify $T_{\mathbf{x}}\mathbb{R}^n$ with $\mathbb{R}^n$. If we wish not to lose track of which point our tangent space is based at in this identification, we can identify $T_{\mathbf{x}}\mathbb{R}^n$ with $\{\mathbf{x}\} \times \mathbb{R}^n$. We define

$$T\mathbb{R}^n := \bigsqcup_{\mathbf{x} \in \mathbb{R}^n} T_{\mathbf{x}}\mathbb{R}^n, \tag{3.47}$$

where $\bigsqcup$ denotes the disjoint union. By the identification above, we have

$$T\mathbb{R}^n \cong \bigsqcup_{\mathbf{x} \in \mathbb{R}^n} \{\mathbf{x}\} \times \mathbb{R}^n = \mathbb{R}^n \times \mathbb{R}^n. \tag{3.48}$$

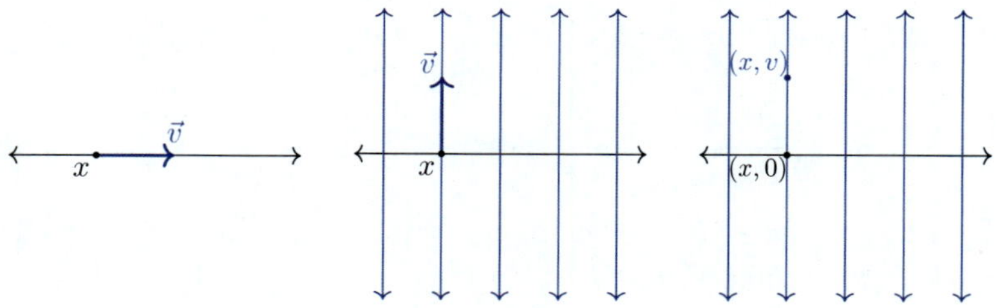

Figure 3.2.1. Visualizing a tangent vector $\vec{v}$ in $T\mathbb{R}$, which is identified with the point (x, v) in $\mathbb{R}^2$.

Example 3.8. *As a special case of our discussion above, the tangent bundle of* $\mathbb{R}$ *is*

$$T\mathbb{R} = \bigsqcup_{x \in \mathbb{R}} T_x\mathbb{R} \cong \mathbb{R} \times \mathbb{R} = \mathbb{R}^2. \tag{3.49}$$

We now discuss how to visualize $T\mathbb{R}$*. Consider a point* (x, v) *in* $\mathbb{R}^2$*. By the identification above, we may consider* v *as a tangent vector to* $\mathbb{R}$ *at the point* x*. Each tangent space* $T_x\mathbb{R}$ *is naturally identified with* $\mathbb{R}$*. So a tangent vector* $\vec{v} \in T_x\mathbb{R}$ *corresponds to a real number, which we call* v*. So we can encode the vector* $\vec{v}$ *by the number* v *and the point that* $\vec{v}$ *is based at. So we write* $\vec{v} = (x, v)$ *via this identification. See Figure* 3.2.1.

Let M^n be an n-dimensional submanifold of $\mathbb{R}^N$. The **tangent bundle** of M^n is defined as the disjoint union of all of the tangent spaces:

$$TM := \bigsqcup_{\mathbf{x} \in M^n} T_{\mathbf{x}}M. \tag{3.50}$$

Similarly to what we did for the tangent bundle of Euclidean space, we may identify each tangent space $T_{\mathbf{x}}M$ with the affine subspace

$$\{\mathbf{x}\} \times T_{\mathbf{x}}M \subset M^n \times \mathbb{R}^N \subset \mathbb{R}^N \times \mathbb{R}^N = \mathbb{R}^{2N}. \tag{3.51}$$

Via this identification, we have

$$TM = \bigcup_{\mathbf{x} \in M^n} \{\mathbf{x}\} \times T_{\mathbf{x}}M \subset \mathbb{R}^{2N}, \tag{3.52}$$

which is a union of disjoint sets. Also via this identification, a tangent vector $\mathbf{V} \in T_{\mathbf{x}}M$ corresponds to $(\mathbf{x}, \mathbf{V}) \in \{\mathbf{x}\} \times T_{\mathbf{x}}M$.

We have the **projection map** $\pi : TM \to M^n$ defined by

$$\pi(\mathbf{V}) := \mathbf{x} \quad \text{if } \mathbf{V} \in T_{\mathbf{x}}M. \tag{3.53}$$

In terms of the identification,

$$\pi : (\mathbf{x}, \mathbf{V}) \mapsto \mathbf{x}. \tag{3.54}$$

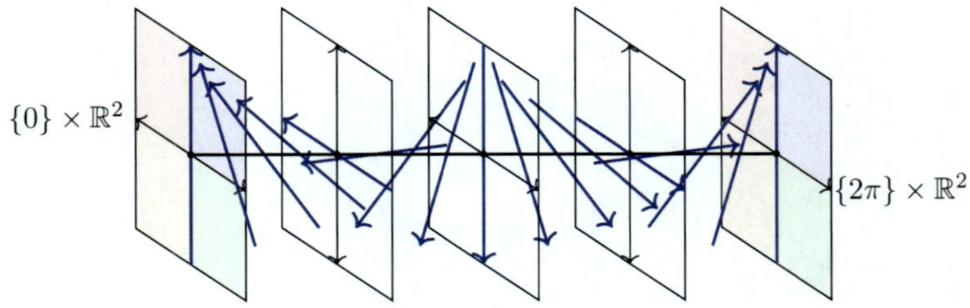

Figure 3.2.2. A model of the tangent bundle TS^1 in $[0, 2\pi] \times \mathbb{R}^2/\sim$, where $(0, \mathbf{Y}) \sim (2\pi, \mathbf{Y})$. The thick blue lines represent tangent spaces T_xS^1. As x increases from 0 to 2π, the tangent spaces rotate counterclockwise, making one full revolution. The tangent bundle to S^1 is topologically a two-sided (non-Möbius) strip, which in turn is topologically the same as $S^1 \times \mathbb{R}$. See Figure 3.2.3. Cf. the Möbius strip in Example 2.16.

Note that

$$\pi^{-1}(\mathbf{x}) = T_{\mathbf{x}}M. \tag{3.55}$$

Note also that

$$TM = \pi^{-1}(M) = \bigsqcup_{\mathbf{x} \in M} \pi^{-1}(\mathbf{x}). \tag{3.56}$$

See Figure 3.2.2 and Exercise 3.8 for the tangent bundle of the unit circle.

The tangent bundle TS^1 is harder to visualize than $T\mathbb{R}$ because TS^1 is a 2-dimensional submanifold of $\mathbb{R}^4$. We can construct a mental model of TS^1 in 3-dimensional space as follows. Recall that we may consider S^1 as $\mathbb{R}/2\pi\mathbb{Z}$, which in turn may be considered as the closed interval $[0, 2\pi]$ with its two endpoints 0 and 2π identified together. So we may consider TS^1 as a subset of the 3-dimensional space $[0, 2\pi] \times \mathbb{R}^2$ where the planes $\{0\} \times \mathbb{R}^2$ and $\{2\pi\} \times \mathbb{R}^2$ are identified with each other by $(0, \mathbf{Y}) \sim (2\pi, \mathbf{Y})$ for $\mathbf{Y} \in \mathbb{R}^2$. See Figure 3.2.2. Topologically, TS^1 is just a cylinder $S^1 \times \mathbb{R}$.[1]

A **vector field** is a map $V : M^n \to TM$ such that $\pi \circ V = \mathrm{id}_{M^n}$. Thus, for each $\mathbf{x} \in M^n$, $V(\mathbf{x}) \in T_{\mathbf{x}}M$. So a vector field on M^n assigns to each point in M^n a tangent vector at that point. See Figure 3.2.4.

For example, a vector field on the unit sphere S^n in $\mathbb{R}^{n+1}$ is equivalent to a map $V : S^n \to \mathbb{R}^{n+1}$ satisfying

$$V(\mathbf{x}) \cdot \mathbf{x} = 0 \quad \text{for all } \mathbf{x} \in S^n.$$

Namely,

$$TS^n = \{(\mathbf{x}, \mathbf{y}) \colon \mathbf{x} \in S^n,\ \mathbf{y} \in \mathbb{R}^{n+1},\ \mathbf{y} \cdot \mathbf{x} = 0\},$$

[1] A belt with a full twist is the same as a cylinder, whereas a belt with a half twist is a Möbius strip.

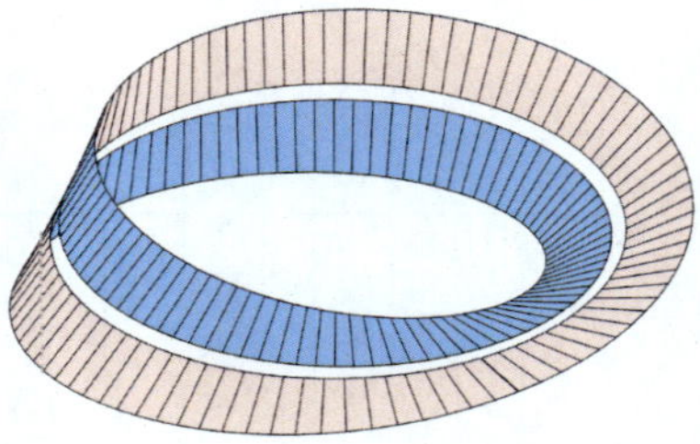

Figure 3.2.3. A two-sided (non-Möbius) strip. Credit: By Ag2gaeh – Own work, CC BY-SA 4.0, https://commons.wikimedia.org/w/index.php?curid=74237150

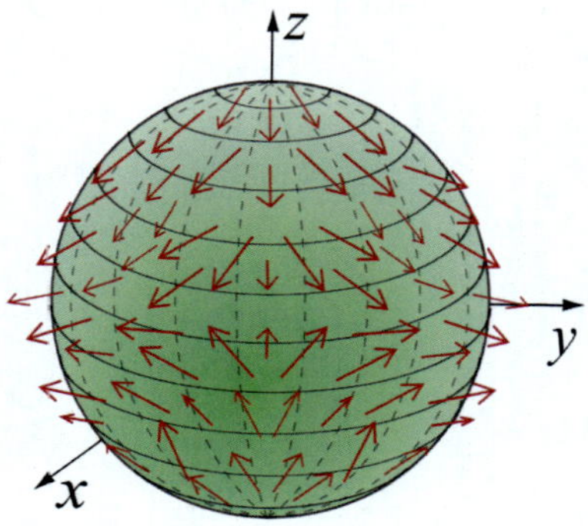

Figure 3.2.4. A vector field on S^2. Credit: By I, Cronholm144, CC BY-SA 3.0, https://commons.wikimedia.org/w/index.php?curid=2304601

where the projection map $\pi : TS^n \to S^n$ is given by $\pi(\mathbf{x}, \mathbf{y}) = \mathbf{x}$. The vector field associated to the map V is given by $\mathbf{x} \mapsto (\mathbf{x}, V(\mathbf{x})) \in T_{\mathbf{x}}S^n$.

In general, if M^n is a hypersurface with a choice of unit normal vector field $\mathbf{N}$, then a vector field on M^n is equivalent to a map $V : M^n \to \mathbb{R}^{n+1}$ satisfying

$$V(\mathbf{x}) \cdot \mathbf{N}_{\mathbf{x}} = 0 \quad \text{for all } \mathbf{x} \in M^n.$$

On odd-dimensional spheres, there are vector fields that are nowhere zero; see Exercise 3.9.

We say that a path $\gamma : I \to M^n$ is an **integral curve** to a vector field V on M^n if for each $t \in I$,

$$\gamma'(t) = V(\gamma(t)) \in T_{\gamma(t)}M. \tag{3.57}$$

Example 3.9. *Let $M^n = \mathbb{R}^n$. Consider the constant vector field defined by $V(\mathbf{x}) = \mathbf{V}_0$ for all $\mathbf{x} \in \mathbb{R}^n$, where $\mathbf{V}_0$ is a fixed vector. Then, for each $\mathbf{x}_0 \in \mathbb{R}^n$, the path $\gamma : \mathbb{R} \to \mathbb{R}^n$ defined by*

$$\gamma(t) := \mathbf{x}_0 + t\mathbf{V}_0$$

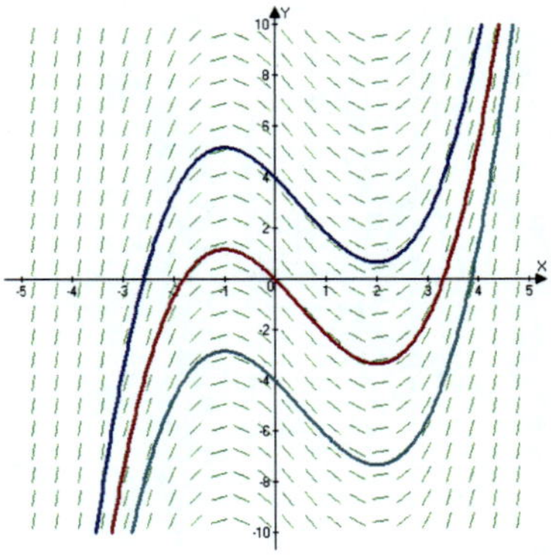

Figure 3.2.5. A vector field on $\mathbb{R}^2$ and three of its integral curves. Credit: By pbroks13talk? – Own work (Original text: self-made), Public Domain, `https://commons.wikimedia.org/w/index.php?curid=4171393`

is an integral curve of the vector field V. *This is simply because* $\gamma'(t) = \mathbf{V}_0 = V(\gamma(t))$ *for all* $t \in \mathbb{R}$.

Consider an ordinary differential equation (ODE) of the form $\frac{dy}{dx} = F(x, y)$, where $F : \mathbb{R}^2 \to \mathbb{R}$ is a smooth function. A function $y = \phi(x)$ is a solution to this equation if and only if

$$\phi'(x) = F\big(x, \phi(x)\big).$$

Define the vector field V on $\mathbb{R}^2$ by

$$V(x, y) = \big(1, F(x, y)\big) = \mathbf{e}_1 + F(x, y)\mathbf{e}_2.$$

The path $\gamma : \mathbb{R} \to \mathbb{R}^2$ defined by

$$\gamma(x) = \big(x, \phi(x)\big)$$

is an integral curve to V. Indeed, we calculate that

$$\gamma'(x) = \big(1, \phi'(x)\big) = \big(1, F(x, \phi(x))\big) = \big(1, F(\gamma(x))\big) = V(\gamma(x)).$$

If $F(x, y) = G(y)$, for some function $G : \mathbb{R} \to \mathbb{R}$, then we say that the ODE is autonomous.

On the other hand, if $F(x, y) = H(x)$, for some function $H : \mathbb{R} \to \mathbb{R}$, then the ODE simply says that a solution $y = \phi(x)$ is the antiderivative of $H(x)$. For example, Figure 3.2.5 shows the vector field V on $\mathbb{R}^2$ and three integral curves for the case where $F(x, y) = x^2 - x - 2$.

Let $F : \mathcal{U} \subset \mathbb{R}^n \to M^n$ parametrize a relatively open subset of a submanifold M^n of $\mathbb{R}^{n+1}$. Then, for each $1 \leq i \leq n$, the map $F(\mathbf{u}) \mapsto \frac{\partial F}{\partial u^i}(\mathbf{u})$ is a vector field on the subset $F(\mathcal{U})$ of M^n. We usually denote this vector field by $\frac{\partial}{\partial x^i}$.

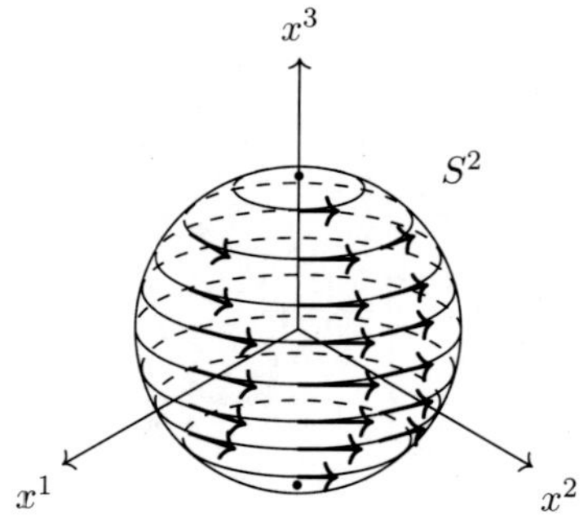

Figure 3.2.6. The vector field V on S^2 defined by (3.58).

For example, if $f : \mathcal{U} \to \mathbb{R}$ is a function, then

$$(\mathbf{u}, f(\mathbf{u})) \mapsto \mathbf{e}_i + \frac{\partial f}{\partial u^i}(\mathbf{u})\mathbf{e}_{n+1}$$

is a vector field on the graph of f.

On S^2 we have the vector field V defined by

$$V(x^1, x^2, x^3) = (-x^2, x^1, 0). \tag{3.58}$$

See Figure 3.2.6.

3.3. The normal and cotangent bundles

Just as we can collect tangent spaces to form the tangent bundle, we can collect normal and cotangents spaces to form the normal and cotangent bundles, respectively.

3.3.1. Normal space and normal bundle.

For a submanifold M^n of $\mathbb{R}^{n+1}$ and $\mathbf{x}_0 \in M^n$, the tangent space $T_{\mathbf{x}_0}M$ is a codimension-1 subspace of $\mathbb{R}^{n+1}$. So there exist exactly two unit vectors, which we will call $\mathbf{N}$ and $-\mathbf{N}$, that are perpendicular to $T_{\mathbf{x}_0}M$. We call each of $\mathbf{N}$ and $-\mathbf{N}$ a **unit normal vector** to M^n at $\mathbf{x}_0$. Note that we have two choices for $\mathbf{N}$.

Example 3.10. *Consider a sphere S of radius r with center $\mathbf{c}_0$ in $\mathbb{R}^{n+1}$; that is,*

$$S = \{\mathbf{x} \in \mathbb{R}^{n+1} \colon \|\mathbf{x} - \mathbf{c}_0\| = r\}. \tag{3.59}$$

For each $\mathbf{x}_0 \in S$, we can choose the unit normal vector

$$\mathbf{N} = \mathbf{N}_{\mathbf{x}_0} = \frac{\mathbf{x}_0 - \mathbf{c}_0}{\|\mathbf{x}_0 - \mathbf{c}_0\|} = \frac{\mathbf{x}_0 - \mathbf{c}_0}{r}. \tag{3.60}$$

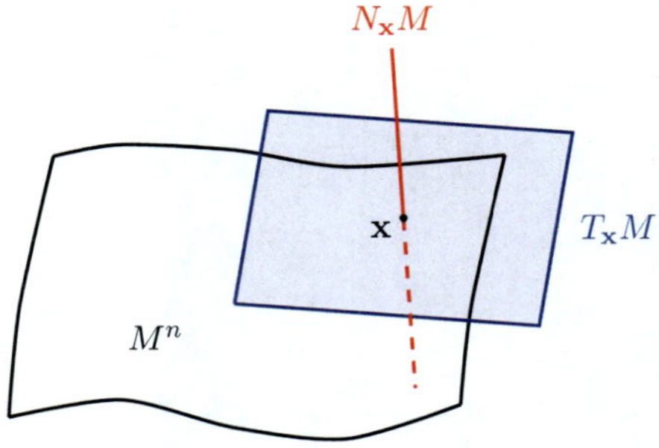

Figure 3.3.1. The normal space $N_{\mathbf{x}}M$ of M^n at a point $\mathbf{x}$ is the orthogonal complement of $T_{\mathbf{x}}M$.

This choice is called the ***outward unit normal vector*** *at* $\mathbf{x}_0$. *The* ***inward unit normal vector*** *at* $\mathbf{x}_0$ *is*

$$-\mathbf{N} = -\mathbf{N}_{\mathbf{x}_0} = -\frac{\mathbf{x}_0 - \mathbf{c}_0}{\|\mathbf{x}_0 - \mathbf{c}_0\|} = -\frac{\mathbf{x}_0 - \mathbf{c}_0}{r}. \tag{3.61}$$

The other choice is to define $\mathbf{N}$ *as the inward unit normal vector. In that case,* $-\mathbf{N}$ *is the outward unit normal vector.*

The normal space $N_{\mathbf{x}}S$ to S at a point $\mathbf{x} \in S$ is defined by

$$N_{\mathbf{x}}S := \mathbb{R}\mathbf{N}_{\mathbf{x}} = \mathbb{R}\frac{\mathbf{x} - \mathbf{c}_0}{r} = \mathbb{R}(\mathbf{x} - \mathbf{c}_0), \tag{3.62}$$

which is a line in the direction $\mathbf{x} - \mathbf{c}_0$. The normal bundle of S is

$$NS := \bigsqcup_{\mathbf{x} \in S} N_{\mathbf{x}}S. \tag{3.63}$$

Now suppose more generally that M^n is an n-dimensional embedded submanifold of $\mathbb{R}^{n+1}$. Let $\mathbf{x} \in M^n$. Then $T_{\mathbf{x}}M$ is a codimension-1 vector subspace of $\mathbb{R}^{n+1}$. The **normal space** of M^n at $\mathbf{x}$ is defined to be

$$N_{\mathbf{x}}M = T_{\mathbf{x}}M^{\perp} = \{\mathbf{V} \in \mathbb{R}^{n+1} \colon \mathbf{V} \cdot \mathbf{W} = 0 \text{ for all } \mathbf{W} \in T_{\mathbf{x}}M\}. \tag{3.64}$$

Since the codimension of M^n in $\mathbb{R}^{n+1}$ is equal to 1, the normal space $N_{\mathbf{x}}M$ is a 1-dimensional vector subspace of $\mathbb{R}^{n+1}$. See Figure 3.3.1.

The **normal bundle** M^n is the disjoint union of the normal spaces:

$$NM := \bigsqcup_{\mathbf{x} \in M} N_{\mathbf{x}}M. \tag{3.65}$$

See Exercises 3.11 and 3.12 for the normal bundles of S^1 and S^2, respectively.

Let M^n be an n-dimensional submanifold of $\mathbb{R}^{n+1}$ and let TM and NM be the tangent and normal bundles of M^n. Let $\mathbf{x} \in M^n$. We have that $T_{\mathbf{x}}M$

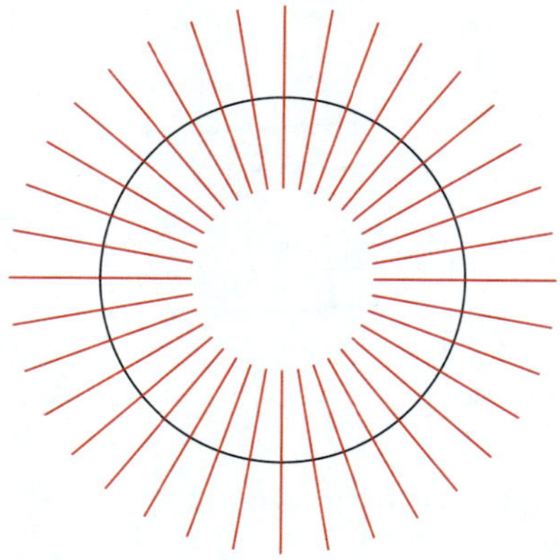

Figure 3.3.2. Picturing the normal bundle of S^1.

and $N_{\mathbf{x}}M$ are orthogonal complements in $\mathbb{R}^{n+1}$. To reflect this, we write this as the direct sum:

$$T_{\mathbf{x}}M \oplus N_{\mathbf{x}}M = \mathbb{R}^{n+1} \quad \text{for each } \mathbf{x} \in M^n. \tag{3.66}$$

We can bundle the formulas above over all $\mathbf{x} \in M^n$ to obtain

$$\begin{aligned} TM \oplus NM &= \bigsqcup_{\mathbf{x}\in M^n} T_{\mathbf{x}}M \oplus \bigsqcup_{\mathbf{x}\in M^n} N_{\mathbf{x}}M \\ &:= \bigsqcup_{\mathbf{x}\in M^n} (T_{\mathbf{x}}M \oplus N_{\mathbf{x}}M) \\ &= \bigsqcup_{\mathbf{x}\in M^n} \mathbb{R}^{n+1} \\ &= M^n \times \mathbb{R}^{n+1}. \end{aligned} \tag{3.67}$$

On the second line the equality is written as $:=$ since this is the (only) line where it is a definition.

We remark that TM and NM are examples of **vector bundles** over M^n. The space $TM \oplus TN$ is called the **Whitney sum** of TM and NM, and it is a vector bundle. Since $TM \oplus NM = M^n \times \mathbb{R}^{n+1}$ by (3.67), we call the result the **trivial bundle**.

3.3.2. Cotangent space and cotangent bundle.

Now let M^n be an n-dimensional submanifold of $\mathbb{R}^N$ and let $\mathbf{x} \in M^n$. Recall that the tangent space $T_{\mathbf{x}}M$ of M^n at $\mathbf{x}$ is an n-dimensional vector space (subspace of $\mathbb{R}^N$).

The **cotangent space** $T^*_{\mathbf{x}}M$ of M^n at $\mathbf{x}$ is simply the dual vector space of the tangent space at $\mathbf{x}$:

$$T^*_{\mathbf{x}}M := (T_{\mathbf{x}}M)^*. \tag{3.68}$$

So $\alpha \in T^*_{\mathbf{x}}M$ means that $\alpha : T_{\mathbf{x}}M \to \mathbb{R}$ is a linear function.

Let $f : M^n \to \mathbb{R}$ be a differentiable function. For each $\mathbf{x} \in M^n$, the derivative of f at $\mathbf{x}$ is denoted by $df_{\mathbf{x}} : T_{\mathbf{x}}M \to \mathbb{R}$. Since $df_{\mathbf{x}}$ is linear, we have that

$$df_{\mathbf{x}} \in T^*_{\mathbf{x}}M$$

is an element of the cotangent space at $\mathbf{x}$. We may think of df as a function $df : TM \to \mathbb{R}$, where

$$df|_{T_{\mathbf{x}}M} := df_{\mathbf{x}}.$$

The **cotangent bundle** of M^n is the disjoint union of the cotangent spaces:

$$T^*M := \bigsqcup_{\mathbf{x}\in M^n} T^*_{\mathbf{x}}M. \tag{3.69}$$

A **1-form** on M^n is a map $\alpha : M^n \to T^*M$ such that $\pi \circ \alpha = \mathrm{id}_{M^n}$. Thus, for each $\mathbf{x} \in M^n$, $\alpha_{\mathbf{x}} \in T_{\mathbf{x}}M$. So a 1-form on M^n assigns to each point in M^n a cotangent vector at that point.

Example 3.11. *If $f : M^n \to \mathbb{R}$ is a differentiable function, then $df : M^n \to T^*M$, defined by $df(\mathbf{x}) := df_{\mathbf{x}}$, is a 1-form on M^n.*

If $\mathcal{V}$ is a subset of M^n, then a 1-form on $\mathcal{V}$ is a map $\alpha : \mathcal{V} \to T^*M$ such that $\pi \circ \alpha = \mathrm{id}_{\mathcal{V}}$. Note that

$$\alpha(\mathcal{V}) \subset T^*M|_{\mathcal{V}} := \bigsqcup_{\mathbf{x}\in \mathcal{V}} T^*_{\mathbf{x}}M. \tag{3.70}$$

Let

$$\varphi_{\mathcal{V}} = (x^1, \dots, x^n) : \mathcal{V} \subset M^n \to \mathbb{R}^n$$

be local coordinates, where $\mathcal{V}$ is a relatively open subset of M^n. By definition, this means that $\varphi_{\mathcal{V}}$ is the inverse function of a local parametrization $F := \varphi_{\mathcal{V}}^{-1} : \varphi_{\mathcal{V}}(\mathcal{V}) \to \mathcal{V}$ of M^n. So $\varphi_{\mathcal{V}} = F^{-1}$. We also call $(\mathcal{V}, \varphi_{\mathcal{V}})$ a *local coordinate chart* or *local coordinate system*.

Let $\mathbf{x} \in \mathcal{V}$ and let $\mathbf{y} := \varphi_{\mathcal{V}}(\mathbf{x})$; then $F(\mathbf{y}) = \mathbf{x}$. We define the tangent vectors

$$\frac{\partial}{\partial x^1}, \dots, \frac{\partial}{\partial x^n} \tag{3.71}$$

at $\mathbf{x}$ by

$$\frac{\partial}{\partial x^i} := d(\varphi_{\mathcal{V}}^{-1})_{\mathbf{y}}(\mathbf{e}_i) = \frac{\partial}{\partial x^i}_{\mathbf{x}} := dF_{\mathbf{y}}(\mathbf{e}_i) = \frac{\partial F}{\partial u^i}(\mathbf{y}) \tag{3.72}$$

for $1 \le i \le n$. Note that each $\frac{\partial}{\partial x^i}$ is a vector field on $\mathcal{V}$.

Given a tangent vector $\mathbf{V}$ at $\mathbf{x}$, we write it in local coordinate component form as

$$\mathbf{V} =: \sum_{i=1}^{n} V^i \frac{\partial}{\partial x^i}. \tag{3.73}$$

That is, given $\mathbf{V}$, this equation defines its components $V^1, \ldots, V^n$ with respect to the local coordinate chart $\varphi_{\mathcal{V}} = (x^1, \ldots, x^n)$.

For each i, $x^i : \mathcal{V} \to \mathbb{R}$ is a differentiable function. Therefore, each derivative dx^i is a 1-form on $\mathcal{V}$.

Lemma 3.12. *At each point $\mathbf{x} \in \mathcal{V}$, the basis of cotangent vectors $\{dx^j\}_{j=1}^n \in T^*_{\mathbf{x}}M$ is equal to the dual basis to the basis of tangent vectors $\left\{\frac{\partial}{\partial x^i}\right\}_{i=1}^n \in T_{\mathbf{x}}M$.*

Proof. We apply the chain rule to the equality of functions $\varphi_{\mathcal{V}} \circ \varphi_{\mathcal{V}}^{-1} = \mathrm{id}_{\varphi_{\mathcal{V}}(\mathcal{V})}$. This yields

$$\mathrm{id}_{\mathbb{R}^n} = d\varphi_{\mathcal{V}} \circ d\varphi_{\mathcal{V}}^{-1}.$$

In particular,

$$\mathbf{e}_i = d\varphi_{\mathcal{V}} \left(d\varphi_{\mathcal{V}}^{-1}(\mathbf{e}_i)\right) = d\varphi_{\mathcal{V}} \left(\frac{\partial}{\partial x^i}\right) = \sum_{j=1}^{n} dx^j \left(\frac{\partial}{\partial x^i}\right) \mathbf{e}_j.$$

Thus,

$$dx^j \left(\frac{\partial}{\partial x^i}\right) = \delta_{ij}$$

for all $1 \leq i, j \leq n$. □

Any cotangent vector α at $\mathbf{x}$ (also called a 1-form at $\mathbf{x}$) may be written in local coordinate component form as

$$\alpha =: \sum_{i=1}^{n} \alpha_i \, dx^i. \tag{3.74}$$

For example, if $f : M^n \to \mathbb{R}$ is a differentiable function, then

$$df = \sum_{i=1}^{n} \frac{\partial f}{\partial x^i} dx^i. \tag{3.75}$$

We can take the tensor product of a cotangent space with itself and consider

$$T^*_{\mathbf{x}}M \otimes T^*_{\mathbf{x}}M.$$

A basis for this vector space is

$$\left\{dx^i \otimes dx^j\right\}_{i,j=1}^n.$$

Thus, any element β of $T^*_{\mathbf{x}}M \otimes T^*_{\mathbf{x}}M$ may be uniquely written as

$$\beta = \sum_{i,j=1}^{n} b_{ij}\, dx^i \otimes dx^j,$$

where $b_{ij} \in \mathbb{R}$ for each i, j and $dx^i = (dx^i)_{\mathbf{x}}$.

For example, an inner product $g_{\mathbf{x}}$ on $T_{\mathbf{x}}M$ may be written as

$$g_{\mathbf{x}} = \sum_{i,j=1}^{n} g_{ij}\, dx^i \otimes dx^j, \tag{3.76}$$

where $g_{ij} = g_{ji} \in \mathbb{R}$ and the matrix (g_{ij}) is positive-definite. For a proof of this, see the derivation of (4.43) below.

3.4. Tensors, tensor products, and tensor bundles

Tensors are, in short, multilinear functions on the products of vector spaces. For a submanifold M^n we will consider multilinear functions on the products of the tangent and cotangent spaces at a point $\mathbf{x}$ for all $\mathbf{x} \in M^n$; we call this a tensor on M^n. The reason why tensors are important is that many geometric invariants of submanifolds are defined as or can be expressed as tensors because they are multilinear functions on the tangent spaces of the submanifolds. In the subsequent chapters we will use tensors extensively.

3.4.1. Linear and bilinear functions. We start with linear and bilinear functions on vector spaces. Let $\mathbb{V}$ be a finite-dimensional real vector space. Recall from (1.5) that its dual space is $\mathbb{V}^* := \mathrm{Lin}(\mathbb{V}, \mathbb{R})$.

What if we consider $\mathbb{V} \times \mathbb{V}$ instead of $\mathbb{V}$? Since $\mathbb{V}$ is a vector space, we can define a vector space structure on $\mathbb{V} \times \mathbb{V}$ and consequently consider linear functions from $\mathbb{V} \times \mathbb{V}$ to $\mathbb{R}$. This is fine, but we are not going to do this. Instead, we consider *bilinear* functions from $\mathbb{V} \times \mathbb{V}$ to $\mathbb{R}$.

Definition 3.13. Let $\mathbb{V}$ and $\mathbb{W}$ be vector spaces. We say that a function $B : \mathbb{V} \times \mathbb{W} \to \mathbb{R}$ is **bilinear** if it has the following two properties:

(1) For each $w_0 \in \mathbb{W}$, the function from $\mathbb{V}$ to $\mathbb{R}$ defined by

$$v \mapsto B(v, w_0) \tag{3.77}$$

is linear.

(2) Similarly, we require that for each $v_0 \in \mathbb{V}$ the function from $\mathbb{W}$ to $\mathbb{R}$ defined by

$$w \mapsto B(v_0, w) \tag{3.78}$$

is linear.

If $\mathbb{W} = \mathbb{V}$, then we call B a **bilinear form** on $\mathbb{V}$.

An **inner product** $\langle\cdot,\cdot\rangle$ on $\mathbb{V}$ is a positive-definite symmetric bilinear form $\langle\cdot,\cdot\rangle : \mathbb{V}\times\mathbb{V}\to\mathbb{R}$ (see Exercise 1.1). For example, the Euclidean inner product on $\mathbb{R}^n$ is defined by

$$\langle\cdot,\cdot\rangle_{\mathbb{E}} = \cdot : (v,w) \mapsto \sum_{i=1}^{n} v^i w^i$$

for $v=(v^1,\dots,v^n), w=(w^1,\dots,w^n)\in\mathbb{R}^n$.

The notion of bilinear function extends to finite sequences of vector spaces; see Exercise 3.13.

Example 3.14. *Given a vector space* $\mathbb{V}$, *we have the* ***evaluation map*** $E:\mathbb{V}^*\times\mathbb{V}\to\mathbb{R}$ *defined by*

$$E(\alpha,v)=\alpha(v) \quad \textit{for } \alpha\in\mathbb{V}^*,\ v\in\mathbb{V}. \tag{3.79}$$

This operation is the foundation for ***tensor contraction****.*

3.4.2. Definition of tensor product. We can define the tensor product of $\mathbb{R}^k$ and $\mathbb{R}^\ell$ in a more or less concrete way as follows. (We forgo the more abstract definitions of tensors because our use of tensors is primarily as a framework for the geometric invariants of submanifolds that we consider and we do not use much more than simply their multilinearity property.) Define their **tensor product** as the vector space of formal sums:

$$\mathbb{R}^k\otimes\mathbb{R}^\ell := \left\{\sum_{i=1}^{k}\sum_{j=1}^{\ell} a^{ij}\,\mathbf{e}_i\otimes\mathbf{e}_j : a^{ij}\in\mathbb{R}\right\}, \tag{3.80}$$

where $\{\mathbf{e}_i\}_{i=1}^k$ and $\{\mathbf{e}_j\}_{j=1}^\ell$ are the standard Euclidean bases for $\mathbb{R}^k$ and $\mathbb{R}^\ell$, respectively, and where the $\mathbf{e}_i\otimes\mathbf{e}_j$ are formal symbols. Note that

$$\dim(\mathbb{R}^k\otimes\mathbb{R}^\ell)=k\ell. \tag{3.81}$$

Let $v=\sum_{i=1}^k v^i\mathbf{e}_i$ and $w=\sum_{j=1}^\ell w^j\mathbf{e}_j$. We define their **tensor product** by

$$\begin{aligned} v\otimes w &= \left(\sum_{i=1}^{k} v^i\mathbf{e}_i\right)\otimes\left(\sum_{j=1}^{\ell} w^j\mathbf{e}_j\right) \\ &:= \sum_{i=1}^{k}\sum_{j=1}^{\ell} v^i w^j\,\mathbf{e}_i\otimes\mathbf{e}_j. \end{aligned} \tag{3.82}$$

This definition is consistent; in particular, the tensor product of $\mathbf{e}_i$ and $\mathbf{e}_j$ is equal to $\mathbf{e}_i\otimes\mathbf{e}_j$.

In general, not every element of $\mathbb{R}^k \otimes \mathbb{R}^\ell$ is of the form $v \otimes w$ for some $v \in \mathbb{R}^k$ and $w \in \mathbb{R}^\ell$. For example, the element

$$\mathbf{e}_1 \otimes \mathbf{e}_1 + \mathbf{e}_2 \otimes \mathbf{e}_2 \in \mathbb{R}^2 \otimes \mathbb{R}^2$$

is not of this form. Another element not of this form is

$$\mathbf{e}_1 \otimes \mathbf{e}_2 - \mathbf{e}_2 \otimes \mathbf{e}_1 \in \mathbb{R}^2 \otimes \mathbb{R}^2.$$

The definition above of tensor product extends to finite-dimensional real vector spaces. Let $\mathbb{V}$ and $\mathbb{W}$ be k- and ℓ-dimensional vector spaces, respectively. Suppose that $\{e_i^{\mathbb{V}}\}_{i=1}^k$ and $\{e_j^{\mathbb{W}}\}_{j=1}^\ell$ are bases for $\mathbb{V}$ and $\mathbb{W}$, respectively. We define the tensor product of these vector spaces by

$$\mathbb{V} \otimes \mathbb{W} := \left\{ \sum_{i=1}^k \sum_{j=1}^\ell a^{ij}\, e_i^{\mathbb{V}} \otimes e_j^{\mathbb{W}} : a^{ij} \in \mathbb{R} \right\}. \tag{3.83}$$

The tensor product of vectors $v \otimes w$, where $v \in \mathbb{V}$ and $w \in \mathbb{W}$, is defined analogously to (3.82).

We may consider linear maps as tensors:

$$\mathbb{V}^* \otimes \mathbb{W} \cong \mathrm{Lin}(\mathbb{V}, \mathbb{W}). \tag{3.84}$$

We can see this isomorphism as follows. Let $\{\omega_{\mathbb{V}}^i\}_{i=1}^k$ be the basis for $\mathbb{V}^*$ dual to the basis $\{e_i^{\mathbb{V}}\}_{i=1}^k$ for $\mathbb{V}$. That is,

$$\omega_{\mathbb{V}}^i(e_m^{\mathbb{V}}) = \delta_{im}. \tag{3.85}$$

Then

$$\mathbb{V}^* \otimes \mathbb{W} := \left\{ \sum_{i=1}^k \sum_{j=1}^\ell a_i^{\,j}\, \omega_{\mathbb{V}}^i \otimes e_j^{\mathbb{W}} : a_i^{\,j} \in \mathbb{R} \right\}. \tag{3.86}$$

We define an isomorphism to be the unique linear map

$$\iota : \mathbb{V}^* \otimes \mathbb{W} \to \mathrm{Lin}(\mathbb{V}, \mathbb{W}) \tag{3.87}$$

satisfying

$$\iota(\omega_{\mathbb{V}}^i \otimes e_j^{\mathbb{W}}) : e_k^{\mathbb{V}} \mapsto \delta_{ik} e_j^{\mathbb{W}}. \tag{3.88}$$

See Exercise 3.14.

We may consider bilinear functions as tensors:

$$\mathbb{V}^* \otimes \mathbb{W}^* \cong \mathrm{Bilin}(\mathbb{V} \times \mathbb{W}, \mathbb{R}), \tag{3.89}$$

where $\mathrm{Bilin}(\mathbb{V} \times \mathbb{W}, \mathbb{R})$ is the vector space of bilinear functions on $\mathbb{V} \times \mathbb{W}$. To see this, first observe that

$$\mathbb{V}^* \otimes \mathbb{W}^* := \left\{ \sum_{i=1}^k \sum_{j=1}^\ell a_{ij}\, \omega_{\mathbb{V}}^i \otimes \omega_{\mathbb{W}}^j : a_{ij} \in \mathbb{R} \right\}. \tag{3.90}$$

Define an isomorphism $\iota : \mathbb{V}^* \otimes \mathbb{W}^* \to \mathrm{Bilin}(\mathbb{V} \times \mathbb{W}, \mathbb{R})$ by

$$\iota : \sum_{i=1}^{k} \sum_{j=1}^{\ell} a_{ij}\, \omega_{\mathbb{V}}^i \otimes \omega_{\mathbb{W}}^j \mapsto \left((v, w) \mapsto \sum_{i=1}^{k} \sum_{j=1}^{\ell} a_{ij} v^i w^j \right). \tag{3.91}$$

See Exercise 3.15.

In particular, we may consider an inner product $\langle \cdot, \cdot \rangle : \mathbb{V} \times \mathbb{V} \to \mathbb{R}$ as an element of $\mathbb{V}^* \otimes \mathbb{V}^*$ that is symmetric and positive-definite.

3.4.3. Multilinear functions.

Definition 3.15. Let $\mathbb{V}$ be a finite-dimensional vector space. Denote $\mathbb{V}^k := \mathbb{V} \times \cdots \times \mathbb{V}$ (k-fold cartesian product). We say that a function $L : \mathbb{V}^k \to \mathbb{R}$ is **multilinear** if for all $1 \leq i \leq k$ and $v_1, \ldots, v_{i-1}, v_{i+1}, \ldots, v_k \in \mathbb{V}$, we have that the function from $\mathbb{V}$ to $\mathbb{R}$ defined by

$$v_i \mapsto L(v_1, \ldots, v_k) \tag{3.92}$$

is linear. We also call L a **covariant k-tensor** on $\mathbb{V}$.

For example, if $k = 2$, then the properties of multilinear and bilinear are the same. If a bilinear function $B : \mathbb{V} \times \mathbb{V} \to \mathbb{R}$ satisfies $B(v_1, v_2) = B(v_2, v_1)$, then we say that B is **symmetric**. Symmetric bilinear functions are especially important in the study of differential geometry.

Using the definition of multilinearity given by Exercise 3.13, we may define a **(k, ℓ)-tensor** on $\mathbb{V}$ as a multilinear function from $\mathbb{V}^k \times (\mathbb{V}^*)^\ell$ to $\mathbb{R}$. As a special case, a $(k, 0)$-tensor on $\mathbb{V}$ is simply a covariant k-tensor on $\mathbb{V}$. A $(0, \ell)$-tensor on $\mathbb{V}$ is called a **contravariant ℓ-tensor** on $\mathbb{V}$. A 0-tensor is simply a real number.

Let $K : \mathbb{V}^k \to \mathbb{R}$ and $L : \mathbb{V}^\ell \to \mathbb{R}$ be multilinear maps. Then their **tensor product** $K \otimes L : \mathbb{V}^{k+\ell} \to \mathbb{R}$ is defined by

$$(K \otimes L)(v_1, \ldots, v_{k+\ell}) := K(v_1, \ldots, v_k) L(v_{k+1}, \ldots, v_{k+\ell}). \tag{3.93}$$

For example, if $K, L : \mathbb{V} \to \mathbb{R}$ are linear maps, then

$$(K \otimes L)(v_1, v_2) = K(v_1) L(v_2). \tag{3.94}$$

Recall from (3.91) that $\mathbb{V}^* \otimes \mathbb{V}^*$ is vector space isomorphic to the space of bilinear functions on $\mathbb{V} \times \mathbb{V}$. Via this isomorphism, the definitions of tensor product given by (3.82) and (3.94) are the same.

The tensor product satisfies the associative law as well as the following distributive laws:

$$c(K \otimes L) = (cK) \otimes L + K \otimes (cL), \quad c \in \mathbb{R}, \tag{3.95}$$

$$(K_1 + K_2) \otimes L = K_1 \otimes L + K_2 \otimes L, \tag{3.96}$$

$$K \otimes (L_1 + L_2) = K \otimes L_1 + K \otimes L_2. \tag{3.97}$$

The tensor product with a 0-tensor is the same as scalar multiplication.

Given a linear map $A : W \to V$ between vector spaces, the **pull-back** of a k-tensor K on V by the map A is defined by

$$A^*(K)(w_1, \ldots, w_k) = K\big(A(w_1), \ldots, A(w_k)\big), \tag{3.98}$$

which is a k-tensor on W. The pull-back satisfies the properties

$$A^*(K \otimes L) = A^*(K) \otimes L + K \otimes A^*(L), \tag{3.99}$$

where L is an ℓ-tensor, and if $B : X \to W$ is a linear map between vector spaces,

$$(A \circ B)^*(K) = B^*\big(A^*(K)\big). \tag{3.100}$$

The definition of tensor product extends from covariant tensors to tensors in general.

3.4.4. Tensors on submanifolds. Let M^n be an n-dimensional submanifold of $\mathbb{R}^N$.

Definition 3.16. A smooth (k,ℓ)**-tensor field** (or (k,ℓ)**-tensor** for short) α on M^n is a smooth assignment to each $\mathbf{x} \in M^n$ of a (k,ℓ)-tensor on the tangent space $T_{\mathbf{x}}M$. That is, for each $\mathbf{x} \in M^n$, $\alpha_{\mathbf{x}} : (T_{\mathbf{x}}M)^k \times (T^*_{\mathbf{x}}M)^\ell \to \mathbb{R}$ is a multilinear map.

We define the tensor product of tensor fields on a submanifold in the same way as the tensor product of tensors on vector spaces. Namely,

$$(\alpha \otimes \beta)_{\mathbf{x}} := \alpha_{\mathbf{x}} \otimes \beta_{\mathbf{x}}. \tag{3.101}$$

Example 3.17. *Define the covariant* 2*-tensor* I *on* M^n *by*

$$\mathrm{I}_{\mathbf{x}} : T_{\mathbf{x}}M \times T_{\mathbf{x}}M \to \mathbb{R},$$

where

$$\mathrm{I}_{\mathbf{x}}(\mathbf{V}, \mathbf{W}) := \mathbf{V} \cdot \mathbf{W} \quad \textit{for } \mathbf{V}, \mathbf{W} \in T_{\mathbf{x}}M \tag{3.102}$$

and where $\cdot$ *denotes the Euclidean inner product on* $\mathbb{R}^N$*. The covariant* 2*-tensor* I *is symmetric and is called the first fundamental form of* M^n*. The first fundamental form is discussed in more detail in* §4.2 *below.*

Many geometric quantities and invariants of submanifolds are tensors.

3.4.5. Tensor spaces and tensor bundles. The **space of (k, ℓ)-tensors at a point $\mathbf{x} \in M^n$** is

$$\left(\bigotimes^k T^*_{\mathbf{x}} M\right) \otimes \left(\bigotimes^\ell T_{\mathbf{x}} M\right).$$

Similarly to the definition of the tangent and cotangent bundles, we define the **bundle of (k, ℓ)-tensors** by

$$\bigotimes^{k,\ell} M := \bigsqcup_{\mathbf{x} \in M^n} \left(\bigotimes^k T^*_{\mathbf{x}} M\right) \otimes \left(\bigotimes^\ell T_{\mathbf{x}} M\right).$$

3.5. Notes and commentary

Tangent and cotangent spaces and bundles as well as tensor calculus, in the setting of smooth manifolds, are discussed for example in Spivak [**Spi79a**] and Lee [**Lee13**].

3.6. Exercises

Exercise 3.1. *For the map $F : \mathbb{R}^n \to S^n$ given in Example 3.3, prove that $dF_{\mathbf{u}}(\mathbb{R}^n) = T_{F(\mathbf{u})} S^n$ gives the same hyperplane as (3.3).*

Hint: Calculate $dF_{\mathbf{u}}(\mathbf{V})$ for $\mathbf{u}, \mathbf{V} \in \mathbb{R}^n$. Consider the equation $dF_{\mathbf{u}}(\mathbf{V}) = \mathbf{0}$. By taking its $(n+1)$-th component, show that $\mathbf{u} \cdot \mathbf{V} = 0$. Then, by considering the first n components, prove that $\mathbf{V} = 0$. This will prove that the kernel of the linear map $dF_{\mathbf{u}}$ is trivial. Next, show that for every $\mathbf{V} \in \mathbb{R}^n$, $dF_{\mathbf{u}}(\mathbf{V}) \cdot F(\mathbf{u}) = 0$. Deduce the result from all of this.

Exercise 3.2. *Find the tangent space at $F(u^1, u^2)$ of the parametrized helicoid H given by Example 2.15. Hint: Compute $\frac{\partial F}{\partial u^1}$, $\frac{\partial F}{\partial u^2}$, and take their cross product to find the normal to $T_{F(u^1,u^2)} H$.*

Exercise 3.3. *Let $\gamma : [a, b] \to \mathbb{R}^2$ be a path, where $\gamma(t) =: (r(t), z(t))$ and $r(t) > 0$. Let M^2 be the surface of revolution in $\mathbb{R}^3$ obtained by revolving the curve $\gamma([a, b])$ about the z-axis. Parametrize M^2 by the map $F : [a, b] \times [0, 2\pi] \to \mathbb{R}^3$ defined by*

$$F(t, \theta) := \big(r(t)\cos(\theta), r(t)\sin(\theta), z(t)\big). \tag{3.103}$$

Find the tangent space of M^2 at $F(t, \theta)$.

Hint: Compute $\frac{\partial F}{\partial t}$, $\frac{\partial F}{\partial \theta}$, and take their cross product to find the normal to $T_{F(t,\theta)} M$.

Exercise 3.4. *Let a, b, c be positive real numbers and let M^2 be the ellipsoid in $\mathbb{R}^3$ given by the equation*

$$\frac{x^2}{a^2} + \frac{y^2}{b^2} + \frac{z^2}{c^2} = 1. \tag{3.104}$$

Let $f(x, y, z)$ be the expression on the left-hand side of the display, so that $M^2 = f^{-1}(1)$. Find the tangent space of M^2 at a point (x, y, z).

Exercise 3.5. *Let $\mathbf{x}_0 = (x_0^1, \dots, x_0^{n+1})$ be a point on the unit n-sphere S^n. Suppose that $1 \leq i \leq n+1$ is such that $x_0^i \neq 0$. Define $\operatorname{sgn}(x_0^i) = +$ if $x_0^i > 0$ and $\operatorname{sgn}(x_0^i) = -$ if $x_0^i < 0$. Then the hemisphere*

$$S^n_{i,\operatorname{sgn}(x_0^i)} = \{\mathbf{x} \in S^n : \operatorname{sgn}(x_0^i) x^i > 0\}$$

is a hemisphere containing $\mathbf{x}_0$.

Show that we can express $S^n_{i,\operatorname{sgn}(x_0^i)}$ as a graph over a subset of the i-th coordinate hyperplane $\mathbb{R}^n_i$. More precisely, show that

$$S^n_{i,\operatorname{sgn}(x_0^i)} = \{(u^1, \dots, u^{i-1}, g(\mathbf{u}), u^{i+1}, \dots, u^{n+1})\}, \tag{3.105}$$

where

$$\mathbf{u} := (u^1, \dots, u^{i-1}, 0, u^{i+1}, \dots, u^{n+1}) \in B_i \subset \mathbb{R}^n_i \tag{3.106}$$

and

$$g(\mathbf{u}) := \operatorname{sgn}(x_0^i)\sqrt{1 - |\mathbf{u}|^2}. \tag{3.107}$$

Use this to compute the tangent space of S^n at an arbitrary point $\mathbf{x}_0 \in S^n$ and verify that this agrees with our previous equation for the tangent space.

Hint: Consider the graphical parametrization $F(\mathbf{u}) = \mathbf{u} + g(\mathbf{u})\mathbf{e}_i$. Show that the normal vector $N_{F(\mathbf{u})} = -\nabla g(\mathbf{u}) + \mathbf{e}_i$ is equal to $\frac{1}{g(\mathbf{u})} F(\mathbf{u})$.

Exercise 3.6. *Let $F : \mathcal{U} \subset \mathbb{R}^n \to \mathbb{R}^{n+1}$ be a smooth embedding. So $M^n = F(\mathcal{U})$ is an n-dimensional submanifold. Prove that*

$$\frac{\partial F}{\partial u^1}(\mathbf{u}), \dots, \frac{\partial F}{\partial u^n}(\mathbf{u}) \tag{3.108}$$

is a basis for $T_{F(\mathbf{u})}M$ for each $\mathbf{u} \in \mathcal{U}$, where $u^1, \dots, u^n$ are the Euclidean coordinates of $\mathbb{R}^n$. See Figure 3.1.7.

Exercise 3.7. *Let $\mathbf{N} \in S^n$, where $\mathbf{N}$ is not necessarily the north pole. Let $\mathbf{V} \in S^n$ satisfy $\mathbf{V} \cdot \mathbf{N} = 0$. Define the path $\gamma_{\mathbf{V}}(t) = \cos(t)\mathbf{N} + \sin(t)\mathbf{V}$. Prove that $\gamma_{\mathbf{V}}(t) \in S^n$ for all $t \in \mathbb{R}$. Compute $\gamma'_{\mathbf{V}}(0)$. Explain why $\gamma'_{\mathbf{V}}(0) \in T_{\mathbf{N}}(S^n)$. Explain why all of the $c\gamma'_{\mathbf{V}}(0)$'s, where $c \geq 0$ and $\mathbf{V} \in S^n$, yield the whole tangent space $T_{\mathbf{N}}(S^n)$.*

Exercise 3.8. *Prove that the tangent bundle of the unit circle S^1 in $\mathbb{R}^2$ is given by*

$$TS^1 = \{(\mathbf{x}, \mathbf{V}) \in \mathbb{R}^2 \times \mathbb{R}^2 : \|\mathbf{x}\| = 1,\ \mathbf{V} \cdot \mathbf{x} = 0\} \subset S^1 \times \mathbb{R}^2 \subset \mathbb{R}^2 \times \mathbb{R}^2. \tag{3.109}$$

Exercise 3.9. *Let S^{2n-1} be the unit sphere in $\mathbb{R}^{2n}$. Prove that the map $V : S^{2n-1} \to \mathbb{R}^{2n}$ given by*

$$V(x^1, \dots, x^{2n}) = (-x^2, x^1, \dots, -x^{2n}, x^{2n-1}) \tag{3.110}$$

defines a vector field on S^{2n-1}.

Exercise 3.10. *Define the (periodic) path $\gamma : \mathbb{R} \to S^{2n-1}$ by*

$$\gamma(t) = \frac{1}{\sqrt{n}}\big(\cos(t+t_1), \sin(t+t_1), \dots, \cos(t+t_n), \sin(t+t_n)\big) \tag{3.111}$$

where $t_1, \dots, t_n$ are constants. Show that γ is an integral curve to the vector field V defined by (3.110).

Let

$$\mathbf{V}_0 = \big(\cos(t_1), \sin(t_1), \dots, \cos(t_n), \sin(t_n)\big),$$
$$\mathbf{W}_0 = \big(-\sin(t_1), \cos(t_1), \dots, -\sin(t_n), \cos(t_n)\big).$$

Show that $\gamma(t) = \cos(t)\,\mathbf{V}_0 + \sin(t)\,\mathbf{W}_0$.

Exercise 3.11. *Explain why the normal bundle NS^1 of the circle is, like the tangent bundle TS^1, a two-sided (non-Möbius) strip. See Figure* 3.3.2.

Exercise 3.12. *Show that the normal bundle NS^2 of the* 2*-sphere is topologically the same as $S^2 \times \mathbb{R}$. In particular, define a homeomorphism between these two spaces.*

Can you generalize this result to a sphere of any dimension?

Exercise 3.13. *Generalize Definition* 3.13 *of a bilinear function to the setting of a finite sequence of vector spaces $\mathbb{V}_1, \dots, \mathbb{V}_k$. Namely, define the notion of a* ***multilinear function*** $B : \mathbb{V}_1 \times \cdots \times \mathbb{V}_k \to \mathbb{R}$.

Exercise 3.14. *Let $\mathbb{V}$ and $\mathbb{W}$ be vector spaces. Prove that $\iota : \mathbb{V}^* \otimes \mathbb{W} \to \operatorname{Lin}(\mathbb{V}, \mathbb{W})$ is a vector space isomorphism.*

Exercise 3.15. *Prove that $\iota : \mathbb{V}^* \otimes \mathbb{W}^* \to \operatorname{Bilin}(\mathbb{V} \times \mathbb{W}, \mathbb{R})$ is a vector space isomorphism.*

Chapter 4

Curvature and the Local Geometry of Submanifolds

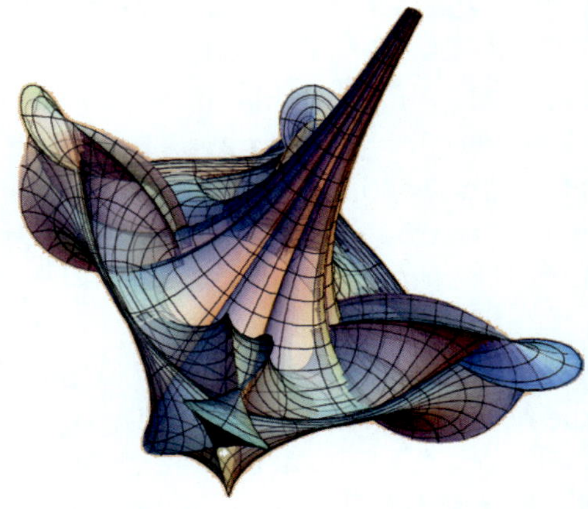

Figure 4.0.1. "Breather Plus Soliton". Credit: 3DXM Virtual Mathematics Museum by The 3D-XplorMath Consortium. © Copyright 3DXM Consortium. `https://3d-xplormath.org/`

We are interested in understanding the *geometry* of n-dimensional embedded submanifolds in an ambient Euclidean space $\mathbb{R}^N$, where $N > n$. The codimension is defined to be the difference $N - n$ between the ambient dimension and the submanifold dimension. When the **codimension** is 1, i.e., $N = n + 1$, submanifolds are called **hypersurfaces**.

In this chapter, we learn about the geometry of embedded submanifolds, and when $n \geq 2$ we will often assume that the codimension is equal to 1; i.e., we are on a hypersurface. In the settings of hypersurfaces and submanifolds, we develop the theory of metric, connection, and curvature. The geometry of a submanifold is governed by the *second fundamental form*, which can be used to define various curvature quantities that are scalar functions. Fundamental results proved in this chapter include the Gauss and Codazzi equations and Gauss's Theorema Egregium. We also discuss the fundamental theorem of hypersurface theory.

4.1. Paths in the plane and in space

Paths in Euclidean space $\mathbb{R}^N$ are simply vector-valued functions of a single variable.

4.1.1. Paths and curves.

The **intervals** we consider may be open, closed, or half-open. So the possibilities are (a, b), $[a, b]$, $[a, b)$, $(a, b]$, $(-\infty, b)$, $(-\infty, b]$, (a, ∞), $[a, \infty)$, and $(-\infty, \infty)$, where $-\infty < a < b < \infty$. In particular, a single point is not considered to be an interval (unless we make an exception for it).

A **path** is a continuously differentiable map $\gamma : I \to \mathbb{R}^N$, where I is an interval. We call its image $c := \gamma(I)$ a **curve**. So we distinguish between the *map* γ, called a path, and the image c of the map, which is a *set* and is called a curve. We remark that in our definition of path, we do not assume that the derivative $\gamma'(t)$ is nonzero, although in practice this is usually the case.

In general, we will consider *immersed paths* that possibly self-intersect; that is, we allow for the possibility that there exist $t_1, t_2 \in I$ with $t_1 \neq t_2$ and $\gamma(t_1) = \gamma(t_2)$. We usually consider the notion of the image curve only when the path is embedded (except possibly self-intersecting at the endpoints of I in the case where I is a closed interval). For simplicity, we always assume that our submanifolds and functions are C^∞. For paths, for example, we should always assume that they are at least twice differentiable.

Example 4.1 (Unit circle in the plane). *Consider the standard parametrization of the unit circle. Define* $\gamma : [0, 2\pi] \to \mathbb{R}^2$ *by*

$$\gamma : z \mapsto \big(\cos(z), \sin(z)\big).$$

We can visualize the graph *of the path* γ *as* $G^1 \subset [0, 2\pi] \times \mathbb{R}^2 \subset \mathbb{R}^3$ *defined by*

$$G^1 = \big\{z, \big(\cos(z), \sin(z)\big) : z \in [0, 2\pi]\big\}.$$

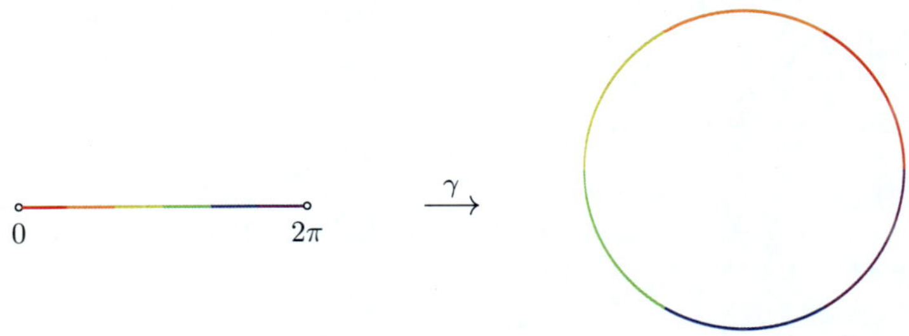

Figure 4.1.1. Visualizing the parametrization γ of the unit circle.

See Figure 4.1.1 *for visualizing the distinction between the path γ, which is a map, and the image curve $\gamma([0,2\pi])$, which is a set, namely the unit circle in the plane.*

4.1.2. Tangent vectors, length of curves, and arc-length parameter. Given a path $\gamma : I \to \mathbb{R}^N$, its **length** is defined as the integral of the speed by

$$\mathrm{L}(\gamma) := \int_I \|\gamma'(t)\|\, dt. \tag{4.1}$$

Example 4.2 (Length of a circle). *The standard parametrization $\gamma : [0, 2\pi] \to \mathbb{R}^2$ of the circle $c = S^1(r)$ of radius r is defined by*

$$\gamma(t) = (r\cos(t), r\sin(t)).$$

We compute that

$$\gamma'(t) = (-r\sin(t), r\cos(t)).$$

Thus, $\|\gamma'(t)\| = r$. We conclude that

$$\mathrm{L}(\gamma) = \int_{[0,2\pi]} r\, dt = 2\pi r.$$

By a **closed path** $\gamma : [a, b] \to \mathbb{R}^N$, we mean that γ is continuously differentiable and $\gamma(a) = \gamma(b)$ and $\gamma'(a) = \gamma'(b)$, where for endpoints we always mean that $\gamma'(a)$ is the right-derivative of γ at a and that $\gamma'(b)$ is the left-derivative of γ at b.

In topology, if we only assume that γ is continuous and $\gamma(a) = \gamma(b)$, then we call γ a **loop**.

One way to obtain a closed path is via a **periodic path**, that is, a path $\beta : \mathbb{R} \to \mathbb{R}^N$ satisfying

$$\beta(t + p) = \beta(t)$$

for all $t \in \mathbb{R}$, where p is some positive number, called a **period** of β. In this case, choosing any $t_0 \in \mathbb{R}$, we obtain that $\gamma : [t_0, t_0 + p] \to \mathbb{R}^n$ is a closed

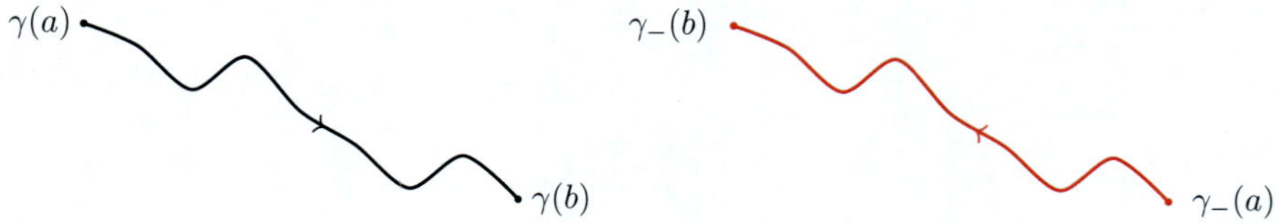

Figure 4.1.2. Left: A parametrized curve γ. Right: Its reversed parametrization γ_-.

path since $\gamma(t_0 + p) = \gamma(t_0)$ and $\gamma'(t_0 + p) = \gamma'(t_0)$. Examples of periodic paths are parametrized circles in $\mathbb{R}^N$; see Exercise 4.1.

Let $\gamma : I \to \mathbb{R}^N$ be a continuously differentiable path. Throughout our discussion, we will assume that $\gamma'(t) \neq \mathbf{0}$ for all $t \in I$. We call $\gamma'(t)$ the **tangent vector** (a.k.a. **velocity vector**) of γ at t. The tangent vector $\gamma'(t)$ is a vector in $\mathbb{R}^N$ at the point $\gamma(t)$, so that $\gamma'(t) \in T_{\gamma(t)}(\mathbb{R}^N)$. The **unit tangent vector field of** γ is defined to be

$$T(t) := \frac{\gamma'(t)}{\|\gamma'(t)\|},$$

which is easily seen to be a unit vector field along γ.

We can reverse the parametrization γ of its image curve as follows. Define the reversed interval

$$I_- := \{t \in \mathbb{R} : -t \in I\}.$$

For example, if $I = [a, b]$, where $a < b$, then $I_- = [-b, -a]$. Define the **reversed parameterized path** $\gamma_- : I_- \to \mathbb{R}^N$ by

$$\gamma_-(t) := \gamma(-t).$$

Note that the image curves of a path and its reverse parametrized path are equal: $\gamma(I) = \gamma_-(-I)$. The unit tangent vector for γ_- at t is equal to the negative of the unit tangent vector for γ at $-t$; see Figure 4.1.2 and Exercise 4.2.

Definition 4.3. A **vector field** $V(t)$ **along** the path $\gamma(t)$ is an assignment to each $t \in I$ of a vector $V(t)$ in $\mathbb{R}^N$ based at $\gamma(t)$; i.e., $V(t) \in T_{\gamma(t)}\mathbb{R}^N$ for all $t \in I$.

Both $\gamma'(t)$ and $T(t)$ are vector fields along $\gamma(t)$.

Define the **differential of the arc-length parameter** by

$$ds := \|\gamma'(t)\|\, dt. \tag{4.2}$$

Up to an additive constant, this equation defines the **arc-length parameter** $s : I \to \mathbb{R}$ by

$$s(t_1) = \int_{t_0}^{t_1} \|\gamma'(t)\|\, dt,$$

where t_0 is some choice of point in the interval I. By hypothesis, we have that the **speed** is positive:

$$s'(t) = \|\gamma'(t)\| > 0.$$

Hence, $s(t)$ is a strictly monotone increasing function of t. This implies that it is invertible, and we denote the inverse function by $s \mapsto t(s)$, which we assume to be a continuously differentiable function.

We may consider γ as a function of arc length by employing the **reparametrization**:

$$\gamma(s) := \gamma(t(s)).$$

In this case, we say that γ is **parameterized by arc length**.

Example 4.4 (Reparametrizing the circle of radius r). *Let $r > 0$. For the parametrization of the circle of radius r defined by $\gamma : [0, 2\pi] \to S^1(r)$, where*

$$\gamma(t) = \big(r\cos(t), r\sin(t)\big),$$

the reparametrization by arc length is the path

$$\gamma(s) = \big(r\cos(s/r), r\sin(s/r)\big),$$

for $r \in [0, 2\pi r]$. Here, we have chosen $s = rt$ (the general arc-length parameter is this plus a constant). This formula reflects that the speed of the original parametrization of γ is the constant r.

Since $\frac{d}{dt}\gamma(t) = \gamma'(t)$, we have

$$\frac{d}{ds}\gamma(s) = \frac{1}{s'(t)}\gamma'(t) = \frac{1}{\|\gamma'(t)\|}\gamma'(t) = T(t), \tag{4.3}$$

which is the unit tangent vector. You may prefer to think of this formula in terms of Leibniz's notation:

$$\frac{d\gamma}{dt} = \frac{d\gamma}{ds}\frac{ds}{dt}.$$

Example 4.5 (Straight lines). *Let $\mathbf{V}$ be a fixed vector in $\mathbb{R}^N$ and let $\mathbf{x}_0$ be a fixed point in $\mathbb{R}^N$. Define the path $\gamma : \mathbb{R} \to \mathbb{R}^N$ by*

$$\gamma(t) = \mathbf{x}_0 + t\mathbf{V}.$$

We have $\gamma'(t) = \mathbf{V}$. We have that the path γ is a constant velocity parametrization of a straight line. Assume that $\mathbf{V} \neq \mathbf{0}$. Then $T(t) \equiv \frac{\mathbf{V}}{\|\mathbf{V}\|}$ is the unit tangent vector. Note that both the tangent vector and the unit tangent vector are constant (independent of t). Note also that we may take the arc-length parameter of γ to be $s = \|\mathbf{V}\|t$.

4.1.3. Product rule for the inner product. Let $\gamma : I \to \mathbb{R}^N$ be a path. Let $V(t)$ and $W(t)$ be vector fields along $\gamma(t)$. Then (see Exercise 4.4)

$$\frac{d}{dt}(V(t) \cdot W(t)) = \frac{dV}{dt}(t) \cdot W(t) + V(t) \cdot \frac{dW}{dt}(t). \tag{4.4}$$

In particular,

$$\frac{d}{dt}\|V(t)\|^2 = 2\frac{dV}{dt}(t) \cdot V(t). \tag{4.5}$$

4.1.4. Curvature of planar curves. What is curvature? To answer this question, we first consider planar curves, that is, paths whose image curves lie in $\mathbb{R}^2$. So, as before, let $\gamma : I \to \mathbb{R}^2$ be a twice-differentiable path with $\gamma'(t) \neq \mathbf{0}$ for all $t \in I$. Let s denote the arc-length parameter, so that the unit tangent vector field is given by $T = \frac{d\gamma}{ds}$.

Now consider the derivative of the unit tangent vector with respect to arc length, which is the second derivative of γ:

$$\frac{dT}{ds} = \frac{d^2\gamma}{ds^2},$$

in Leibniz notation. Denote $\gamma'(s) := \frac{d\gamma}{ds}$, $\gamma''(s) := \frac{d^2\gamma}{ds^2}$, etc. So, equivalently

$$T'(s) = \gamma''(s).$$

We compute that

$$\frac{d}{ds}\|\gamma'(s)\|^2 = \frac{d}{ds}1 = 0.$$

On the other hand, by the product rule

$$\begin{aligned}\frac{d}{ds}\|\gamma'(s)\|^2 &= \frac{d}{ds}(\gamma'(s) \cdot \gamma'(s)) \\ &= \gamma''(s) \cdot \gamma'(s) + \gamma'(s) \cdot \gamma''(s) \\ &= 2\gamma''(s) \cdot \gamma'(s).\end{aligned}$$

Therefore,

$$\frac{dT}{ds} \cdot T = \gamma''(s) \cdot \gamma'(s) = 0. \tag{4.6}$$

That is, the derivative of the unit tangent vector field (with respect to arc length) is perpendicular to the unit tangent vector field at each point along the curve.

So far, everything we have done in this subsection could have been done for paths in $\mathbb{R}^N$ without any change. Now we use the assumption that $N = 2$. Since $\gamma : I \to \mathbb{R}^2$ is continuously differentiable with nonzero speed, the unit vector field $T(t)$ along $\gamma(t)$ is continuous. Choose a continuous unit vector field $N(t)$ along $\gamma(t)$ which is perpendicular to $T(t)$ at each t. We call $N(t)$ a **unit normal vector field** along $\gamma(t)$. Note that there are

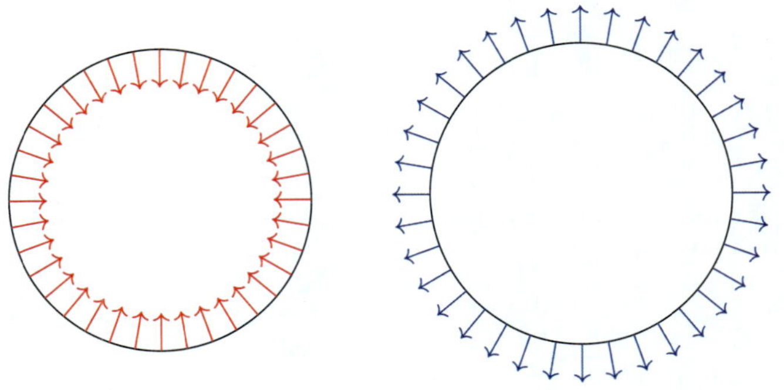

Figure 4.1.3. Two choices of continuous unit normal vector fields to a circle.

exactly two choices of unit normal vector field, a choice and its negative. In particular, given the choice above, $-N(t)$ is the only other choice of unit normal vector field to $\gamma(t)$. See Figure 4.1.3.

Reparametrize γ by arc length s. Since the ambient dimension is equal to 2, we have that

$$\frac{dT}{ds}(s) =: \kappa(s)N(s) \tag{4.7}$$

for some function $\kappa(s)$ of arc length. We call $\kappa(s)$ the **(plane) curvature** of γ at s. Note that the sign of $\kappa(s)$ depends on the choice of unit normal vector field N. By taking the dot product of (4.7) with $N(s)$, we obtain

$$\kappa(s) = \frac{dT}{ds}(s) \cdot N(s). \tag{4.8}$$

A circle of radius r has constant curvature $1/r$ if we use the inward pointing unit normal vector field to define the curvature. See Figure 4.1.4 and Exercise 4.5. Since we like the idea of a circle having positive curvature, one usually uses the inward pointing unit normal vector field to define the curvature of a closed path.

We may ask, What is the formula for the curvature of a path that is not parametrized by arc length? Let $\gamma : I \to \mathbb{R}^2$ be a continuously differentiable path and let $N(t)$ be a choice of a continuous unit normal vector field along $\gamma(t)$. Then by (4.7)

$$\frac{1}{\|\gamma'(t)\|}\frac{d}{dt}\left(\frac{1}{\|\gamma'(t)\|}\frac{d\gamma}{dt}\right)(t) = \kappa(t)N(t).$$

Thus (see Exercise 4.8)

$$\kappa(t) = \frac{1}{\|\gamma'(t)\|^2}\frac{d^2\gamma}{dt^2}(t) \cdot N(t). \tag{4.9}$$

Figure 4.1.4. Left: A small circle with κ large. Middle: A medium circle with κ medium. Right: A large circle with κ small. The smaller the radius of the circle, the faster the direction of the unit outward normal changes.

To arrive at this calculation, we used that

$$\frac{d\gamma}{dt}(t) \cdot N(t) = 0.$$

4.1.5. Curvatures of closed paths. Closed paths in the plane have an 'inside' and an 'outside'. Namely, given a closed smooth path $\gamma : [a, b] \to \mathbb{R}^2$, let c be the image curve $c := \gamma([a, b])$. Then the set $\mathbb{R}^2 \setminus c$ consists of two connected components. (A generalization of this result to *continuous* closed paths is the **Jordan curve theorem**; see [**Mun00**] or [**GP10**, Chapter 2, §5].) Call these two connected components $\mathcal{I}$ and $\mathcal{O}$, where $\mathcal{I}$ is the bounded (inside) component and where $\mathcal{O}$ is the unbounded (outside) component. See Figure 4.1.5. Thus, for closed paths, we can speak of **inward pointing normals** and **outward pointing normals**. For closed embedded paths it is customary to choose the inward pointing unit normal vector field to define the curvature of the path.

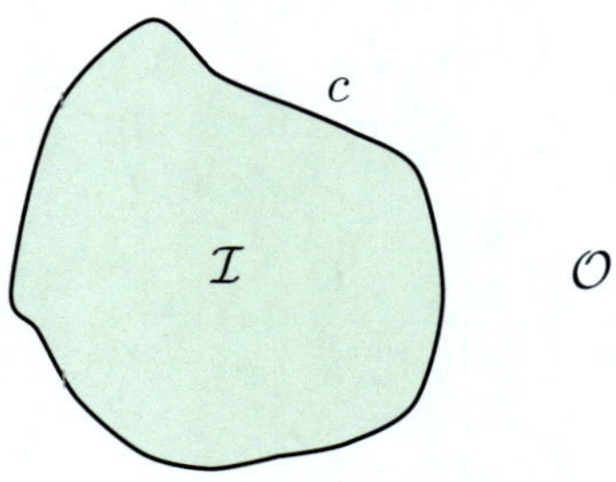

Figure 4.1.5. A closed embedded curve c and its inside component $\mathcal{I}$ (shaded) and its outside component $\mathcal{O}$.

Example 4.6 (Curvature of an ellipse). *Let a, b be positive real numbers. Consider the ellipse given by the set of points $(x, y) \in \mathbb{R}^2$ in the plane satisfying the equation $\frac{x^2}{a^2} + \frac{y^2}{b^2} = 1$. We parametrize this ellipse by the map $F : [0, 2\pi] \to \mathbb{R}^2$ defined by $\gamma(t) = (x(t), y(t))$, where*

$$x(t) = a\cos(t), \quad y(t) = b\sin(t).$$

We calculate that the velocity vector is

$$\gamma'(t) = \big(-a\sin(t), b\cos(t)\big),$$

so that the speed is

$$\|\gamma'(t)\| = \sqrt{a^2\sin^2(t) + b^2\cos^2(t)} =: S.$$

The inward pointing unit normal vector field is given by

$$N(t) = \frac{1}{S}\big(-b\cos(t), -a\sin(t)\big).$$

We compute the curvature with respect to this choice of N as

$$\begin{aligned}\kappa(t) &= \frac{1}{S^2}\big(-a\cos(t), -b\sin(t)\big) \cdot \frac{1}{S}\big(-b\cos(t), -a\sin(t)\big) \\ &= \frac{ab}{S^3} = \frac{ab}{(a^2\sin^2(t) + b^2\cos^2(t))^{3/2}}.\end{aligned}$$

4.1.6. Retrieving a plane curve from its curvature. In this subsection we show that when $N = 2$, a path is determined by its initial position and velocity and its curvature as a function of arc length.

Let $\gamma : [0, L] \to \mathbb{R}^2$ be any unit-speed immersed path in the plane. We write this path in component form as $\gamma(s) =: (x(s), y(s))$. Recall that $T(s) = \gamma'(s)$ denotes the unit tangent vector field and by (4.7) we have $\frac{dT}{ds}(s) = \kappa(s)N(s)$.

Since $T(s)$ is a unit vector field along the path γ, there exists a function $\theta : [0, L] \to \mathbb{R}$ such that

$$T(s) = \big(\cos(\theta(s)), \sin(\theta(s))\big). \tag{4.10}$$

We call $\theta : [0, L] \to \mathbb{R}$ an **angle function**. Without loss of generality, we may assume that

$$\mathbf{N}_0 = \big(-\sin(\theta(0)), \cos(\theta(0))\big).$$

Then

$$N(s) = \big(-\sin(\theta(s)), \cos(\theta(s))\big).$$

Differentiating (4.10) yields

$$\kappa(s)N(s) = T'(s) = \left(-\sin(\theta(s))\frac{d\theta}{ds}(s), \cos(\theta(s))\frac{d\theta}{ds}(s)\right).$$

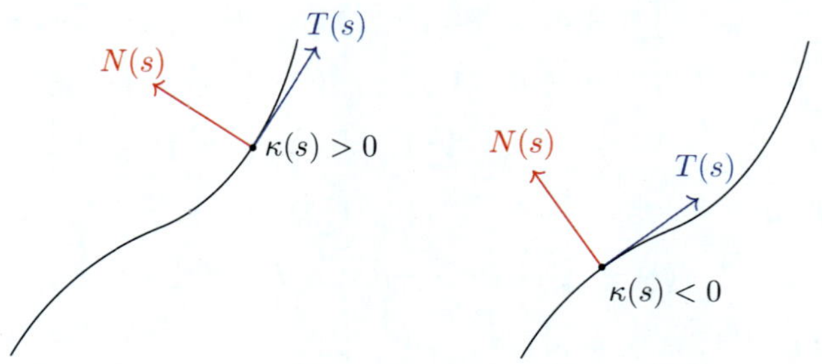

Figure 4.1.6. Left: A point at which the curvature is positive. Right: A point at which the curvature is negative.

We then obtain

$$\kappa(s)N(s) = \frac{d\theta}{ds}(s)N(s),$$

so that

$$\kappa(s) = \frac{d\theta}{ds}(s). \tag{4.11}$$

That is, the curvature is equal to the rate of change of the angle θ per unit arc length. Note that with the choice of N above, the curvature κ is positive at a point on the path if the unit tangent vector field T is curving in the direction of the unit normal field N, whereas the curvature κ is negative at a point on the path if the T is curving in the direction of $-N$. See Figure 4.1.6.

Remark 4.7. Let $\{\mathbf{e}_1, \mathbf{e}_2\}$ be the standard Euclidean orthonormal basis for $\mathbb{R}^2$. We can consider this basis as the constant frame field $\{e_1, e_2\}$ on $\mathbb{R}^2$; that is, $e_i(\mathbf{x}) := \mathbf{e}_i$ for $i = 1, 2$ and $\mathbf{x} \in \mathbb{R}^2$. Then $\theta(s)$ defined above is a smooth function satisfying

$$\theta(s) = \angle\big(T(s), e_1(\gamma(s))\big).$$

The main result of this subsection is the following.

Proposition 4.8 (Fundamental theorem of plane curves)**.** *Suppose that we are given a point* $\mathbf{x}_0 \in \mathbb{R}^2$ *and unit vectors* $\mathbf{T}_0$ *and* $\mathbf{N}_0$ *at the point* $\mathbf{x}_0$ *that are orthogonal to each other. Furthermore, let* $\mathrm{k} : [0, L] \to \mathbb{R}$ *be a function. Then there exists a unique path* $\gamma : [0, L] \to \mathbb{R}^2$ *that is parametrized by arc length and satisfies*

$$\kappa(s) = \mathrm{k}(s) \tag{4.12}$$

for all $s \in [0, L]$, $\gamma(0) = \mathbf{x}_0$, *and* $\gamma'(0) = \mathbf{T}_0$, *where the curvature* $\kappa(s)$ *is defined with respect to the continuous choice of unit normal* $N(s)$ *along* $\gamma(s)$ *with* $N(0) = \mathbf{N}_0$.

Proof. By (4.11), $\frac{d\theta}{ds}(s) = \kappa(s)$. Denote $\theta_0 := \theta(0)$. By the fundamental theorem of calculus, we have

$$\theta(s) = \theta_0 + \int_0^s \kappa(\bar{s})\,d\bar{s}.$$

Therefore

$$\frac{dx}{ds}(s) = \cos(\theta(s)) = \cos\left(\theta_0 + \int_0^s \kappa(\bar{s})\,d\bar{s}\right)$$

and

$$\frac{dy}{ds}(s) = \cos(\theta(s)) = \sin\left(\theta_0 + \int_0^s \kappa(\bar{s})\,d\bar{s}\right).$$

Denote $\mathbf{x}_0 =: (x_0, y_0)$. Integrating the formulas above yields

$$\begin{aligned} x(s) &= x_0 + \int_0^s \frac{dx}{ds}(\bar{s})\,d\bar{s} \\ &= x_0 + \int_0^s \cos\left(\theta_0 + \int_0^{\bar{s}} \kappa(\bar{\bar{s}})\,d\bar{\bar{s}}\right) d\bar{s} \end{aligned} \tag{4.13}$$

and

$$\begin{aligned} y(s) &= y_0 + \int_0^s \frac{dy}{ds}(\bar{s})\,d\bar{s} \\ &= y_0 + \int_0^s \sin\left(\theta_0 + \int_0^{\bar{s}} \kappa(\bar{\bar{s}})\,d\bar{\bar{s}}\right) d\bar{s}. \end{aligned} \tag{4.14}$$

The formulas above show that a path $\gamma : [0, L] \to \mathbb{R}^2$ is uniquely determined by its initial position $\gamma(0) = (x_0, y_0)$, its initial unit tangent vector $\gamma'(0) = (\cos(\theta_0), \sin(\theta_0))$, its curvature function $\kappa(s)$ as a function of arc length, and a choice of unit normal to define the curvature. This proves the uniqueness part of the proposition.

Conversely, if we replace κ by k in (4.13) and (4.14), then we obtain a smooth path whose curvature function satisfies (4.12). This proves the existence part of the proposition. ☐

4.1.7. Total curvature of an embedded plane curve.

The **total curvature** of a path $\gamma : I \to \mathbb{R}^2$ parametrized by arc length is defined as

$$\mathrm{TC}(\gamma) := \int_I \kappa(s)\,ds. \tag{4.15}$$

Theorem 4.9. *The total curvature of an embedded closed planar path is equal to 2π. That is, if $\gamma : I \to \mathbb{R}^2$ is a closed path parametrized by arc* *length, then*

$$\int_I \kappa(s)\,ds = 2\pi, \tag{4.16}$$

where the curvature κ is defined with respect to the inward pointing unit normal vector field.

This theorem is a consequence of the Gauss–Bonnet formula for surfaces with boundary, where we take the surface to be the region enclosed by the closed curve and where the boundary is the closed curve itself; see Theorem 9.20 and Example 9.21 below.

4.1.8. Total curvature of an immersed plane curve. By (4.11), we can calculate the total curvature of an immersed path $\gamma : [0, L] \to \mathbb{R}^2$ in terms of any angle function θ defined by (4.10):

$$\begin{aligned}\mathrm{TC}(\gamma) &= \int_{[0,L]} \kappa(s)\,ds \\ &= \int_{[0,L]} \frac{d\theta}{ds}(s)\,ds \\ &= \theta(L) - \theta(0),\end{aligned}$$

where the last equality is true by the fundamental theorem of calculus.

Now suppose that $\gamma : [0, L] \to \mathbb{R}$ is a closed path. Since $\gamma'(0) = \gamma'(L)$, we have that

$$\theta(L) =: \theta(0) + 2\pi\, w(\gamma) \tag{4.17}$$

for some integer $w(\gamma)$. This integer is called the **winding number** of γ. In this case, the total curvature is given by

$$\mathrm{TC}(\gamma) = \theta(L) - \theta(0) = 2\pi\, w(\gamma).$$

See Figure 4.1.7.

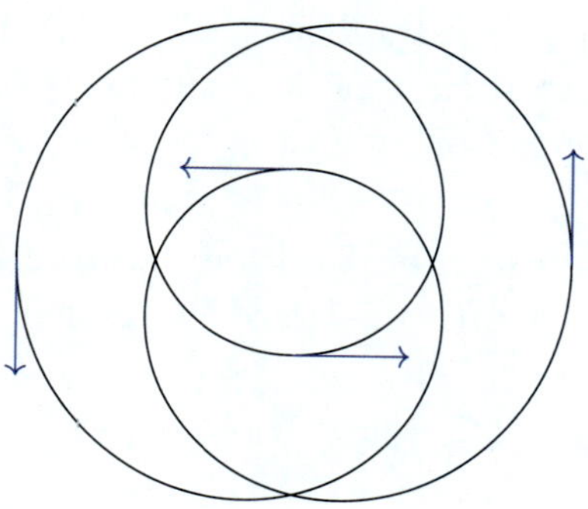

Figure 4.1.7. An immersed path with winding number equal to 3.

Example 4.10. *Let* $r > 0$ *and let* $\gamma(s) = \big(r\cos(s/r), r\sin(s/r)\big)$ *for* $s \in [0, 2\pi r]$. *Then* γ *is a counterclockwise parametrization of the circle of radius* r. *We have*

$$\theta(s) = \frac{s}{r} + \frac{\pi}{2}.$$

Hence the winding number of γ *is equal to*

$$w(\gamma) = \frac{\theta(L) - \theta(0)}{2\pi} = 1.$$

4.1.9. Paths in higher dimensions. We may also consider a path γ in $\mathbb{R}^N$, where $N \geq 3$. Let us consider the case $N = 3$, in which case γ is called a **space path** and its image is called a **space curve**. By reparametrization, we may assume that $\gamma : [0, L] \to \mathbb{R}^3$ is parametrized by arc length. Let $T = \frac{d\gamma}{ds}$ be the unit tangent vector field to γ. Recall from (4.6) that

$$\frac{dT}{ds} \cdot T = 0.$$

Thus, the vector field $\frac{dT}{ds}$ is orthogonal to T along γ.

We define the **curvature (function)** $\kappa : [0, L] \to \mathbb{R}$ of γ by

$$\kappa(s) := \left\| \frac{dT}{ds}(s) \right\|. \tag{4.18}$$

That is, the curvature is equal to the rate of change (in magnitude) of the unit tangent vector field per unit change in arc length. Note that we have defined the curvature so that it is always nonnegative.

Assume that $\frac{dT}{ds}(s) \neq \mathbf{0}$ along $\gamma(s)$; i.e., $\kappa(s) \neq 0$. Then

$$N(s) := \frac{\frac{dT}{ds}(s)}{\left\|\frac{dT}{ds}(s)\right\|} \tag{4.19}$$

is a unit vector field orthogonal to $T(s)$ along $\gamma(s)$. We call N the **(principal) unit normal vector field** along γ. We have

$$\frac{dT}{ds} = \kappa N. \tag{4.20}$$

We can now take advantage of our assumption that the ambient Euclidean dimension is equal to 3. Namely, we define the **(unit) binormal vector field** along γ by

$$B(s) := T(s) \times N(s), \tag{4.21}$$

where $\times$ is the cross product. It is easy to check that the binormal field B is unit and that $\{T, N, B\}$ is an orthonormal frame field along γ.

Now let us consider the rate of change of the principal unit normal vector. Firstly, since N is a unit vector, we have

$$\frac{dN}{ds} \cdot N = 0.$$

Next, using the product rule, we compute that

$$\frac{dN}{ds} \cdot T = \frac{d}{ds}(N \cdot T) - N \cdot \frac{dT}{ds} = -N \cdot \kappa N.$$

Hence, since $N \cdot T \equiv 0$, we obtain

$$\frac{dN}{ds} \cdot T = -\kappa.$$

The **torsion (function)** $\tau : [0, L] \to \mathbb{R}$ of γ is defined by

$$\tau(s) = \frac{dN}{ds}(s) \cdot B(s). \tag{4.22}$$

Since $\{T(s), N(s), B(s)\}$ is an orthonormal basis for $\mathbb{R}^3$ at each point $\gamma(s)$ along the path, we obtain

$$\begin{aligned} \frac{dN}{ds} &= \left(\frac{dN}{ds} \cdot T\right) T + \left(\frac{dN}{ds} \cdot N\right) N + \left(\frac{dN}{ds} \cdot B\right) B \\ &= -\kappa T + \tau B. \end{aligned} \tag{4.23}$$

Since B is a unit vector field along γ, we have

$$\frac{dB}{ds} \cdot B = 0.$$

We compute using the product rule that

$$\frac{dB}{ds} \cdot T = \frac{d}{ds}(B \cdot T) - B \cdot \frac{dT}{ds} = 0.$$

And also

$$\frac{dB}{ds} \cdot N = \frac{d}{ds}(B \cdot N) - B \cdot \frac{dN}{ds} = -\tau.$$

Therefore

$$\begin{aligned} \frac{dB}{ds} &= \left(\frac{dB}{ds} \cdot T\right) T + \left(\frac{dB}{ds} \cdot N\right) N + \left(\frac{dB}{ds} \cdot B\right) B \\ &= -\tau N. \end{aligned} \tag{4.24}$$

Gathering together the calculations above, we obtain that the derivatives with respect to arc length of the unit vector fields $T(s), N(s), B(s)$ are given by the **Frenet–Serret formulas** as

$$
\begin{aligned}
\frac{dT}{ds} &= && \kappa N && , \\
\frac{dN}{ds} &= -\kappa T && && +\tau B \ , \\
\frac{dB}{ds} &= && -\tau N && .
\end{aligned}
\tag{4.25}
$$

Proposition 4.11 (Planarity of vanishing torsion space curves). *Suppose that a space path $\gamma : [0, L] \to \mathbb{R}^3$ satisfies $\tau(s) = 0$ for all $s \in [0, L]$. Then the image curve $c := \gamma([0, L])$ lines in an affine plane.*

Proof. By (4.25) and the vanishing of the torsion, we have $\frac{dB}{ds}(s) \equiv 0$ along $\gamma(s)$. Therefore $\mathbf{B}_0 := B(s)$ is a constant unit vector. From this we obtain

$$0 = \mathbf{B}_0 \cdot T(s) = \mathbf{B}_0 \cdot \frac{d\gamma}{ds}(s) = \frac{d}{ds}\big(\mathbf{B}_0 \cdot \gamma(s)\big).$$

Thus, $\mathbf{B}_0 \cdot \gamma(s) =: C_0$ is a constant. We conclude that

$$\gamma(s) \in \mathbb{A} := \left\{ \mathbf{x} \in \mathbb{R}^3 : \mathbf{x} \cdot \mathbf{B}_0 = C_0 \right\}$$

for all $s \in [0, L]$. □

4.1.10. Retrieving a space curve from its curvature and torsion. When $N = 3$, a path is determined by its initial position, unit tangent, unit normal, and its (positive) curvature and torsion as functions of arc length.

Proposition 4.12 (Fundamental theorem of space curves). *Suppose that we are given a point $\mathbf{x}_0 \in \mathbb{R}^3$ and unit vectors $\mathbf{T}_0$ and $\mathbf{N}_0$ at the point $\mathbf{x}_0$ that are orthogonal to each other. Furthermore, let $\mathrm{k} : [0, L] \to \mathbb{R}$ be a positive function and let $\mathrm{t} : [0, L] \to \mathbb{R}$ be a function. Then there exists a unique path $\gamma : [0, L] \to \mathbb{R}^3$ that is parametrized by arc length and satisfies*

$$
\begin{aligned}
\kappa(s) &= \mathrm{k}(s), \\
\tau(s) &= \mathrm{t}(s)
\end{aligned}
$$

for all $s \in [0, L]$, $\gamma(0) = \mathbf{x}_0$, $\gamma'(0) = \mathbf{T}_0$, and $N(0) := \frac{\gamma''(0)}{\|\gamma''(0)\|} = \mathbf{N}_0$.

For a proof of the proposition, see do Carmo [**dC16**, Appendix to Chapter 4].

For further reading on the geometry of curves, see [**dC16**, Chapter 1], which also includes for example the statement and proof of the **Four-Vertex Theorem** for closed embedded plane curves and the **Crofton formula** in integral geometry.

4.2. First fundamental form of a submanifold

The intrinsic notion of distance on a submanifold of Euclidean space starts with the *first fundamental form*, which is the infinitesimal measure of distance and angle on the submanifold induced from the ambient Euclidean space. This is a special case of the notion of *Riemannian metric*, which will be discussed starting in Part II of this book.

4.2.1. Ambient Euclidean distance. Let M^n be an embedded submanifold of $\mathbb{R}^N$. How do we measure distances in M^n? One answer is: For $\mathbf{x}, \mathbf{y} \in M^n$, consider the ambient Euclidean distance

$$d_{\mathbb{E}}(\mathbf{x}, \mathbf{y}) = \|\mathbf{x} - \mathbf{y}\|.$$

But what if we imagine our whole universe is M^n and disregard the ambient space $\mathbb{R}^N$?[1] Then we could not use $d_{\mathbb{E}}$ and we would have to come up with another notion of distance on M^n. But what notion? The natural answer is the *intrinsic distance.*

4.2.2. Intrinsic distance on a circle. Let us look at the example of the unit circle S^1 in $\mathbb{R}^2$. Let $\mathbf{x}, \mathbf{y} \in S^1$. Using complex numbers, we write $\mathbf{x} = e^{i\theta_{\mathbf{x}}}$ and $\mathbf{y} = e^{i\theta_{\mathbf{y}}}$, where $0 \leq \theta_{\mathbf{x}}, \theta_{\mathbf{y}} < 2\pi$. That is, $\theta_{\mathbf{x}}$ and $\theta_{\mathbf{y}}$ are the smallest positive angles that $\mathbf{x}$ and $\mathbf{y}$ make with the positive x^1-axis, respectively. Without loss of generality, we assume that $\theta_{\mathbf{x}} \leq \theta_{\mathbf{y}}$. Then the intrinsic distance between $\mathbf{x}$ and $\mathbf{y}$ along S^1 is given by

$$d_{S^1}(\mathbf{x}, \mathbf{y}) = \min\{\theta_{\mathbf{y}} - \theta_{\mathbf{x}}, \theta_{\mathbf{x}} + 2\pi - \theta_{\mathbf{y}}\}.$$

This is equal to the angle between $\mathbf{x}$ and $\mathbf{y}$. So, by trigonometry, we also have that

$$d_{S^1}(\mathbf{x}, \mathbf{y}) = 2\sin^{-1}\frac{\|\mathbf{x} - \mathbf{y}\|}{2}. \tag{4.26}$$

See Figure 4.2.1 and Exercise 4.11. Formula (4.26) also gives the formula for the intrinsic distance between points $\mathbf{x}$ and $\mathbf{y}$ along the unit n-sphere S^n in $\mathbb{R}^{n+1}$.

In general, how can we realize the intrinsic distance of a submanifold M^n of $\mathbb{R}^N$? The answer is to consider the **infinitesimal** (in the sense of calculus) notion of distance. Why should we consider things infinitesimally? Take two points $\mathbf{x}$ and $\mathbf{y}$ on M^n. How do we figure out the intrinsic distance between $\mathbf{x}$ and $\mathbf{y}$? A way to do this is to consider all trips along M^n from $\mathbf{x}$ to $\mathbf{y}$. Such a trip is called a **path** and is given by a map $\gamma : [t_1, t_2] \to M^n$ with $\gamma(t_1) = \mathbf{x}$ and $\gamma(t_2) = \mathbf{y}$. We define the distance between $\mathbf{x}$ and $\mathbf{y}$ to

[1] This is what we must do if we consider travel on Earth. That is, we must consider S^2 as our whole universe and disregard the ambient space $\mathbb{R}^3$.

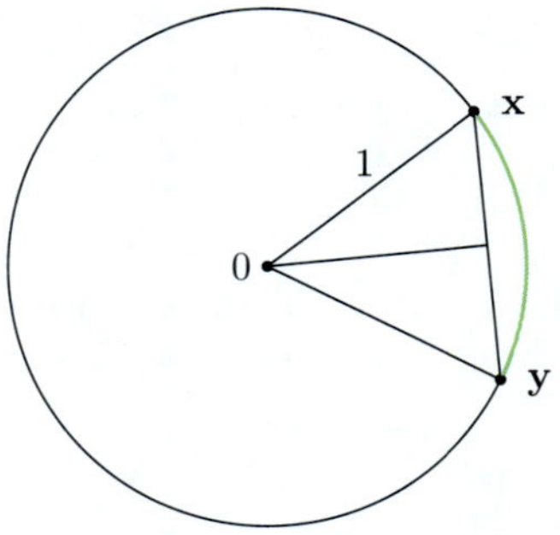

Figure 4.2.1. The distance between $\mathbf{x}$ and $\mathbf{y}$ along S^1 is equal to the angle between $\mathbf{x}$ and $\mathbf{y}$. The shortest path in S^1 from $\mathbf{x}$ to $\mathbf{y}$ is represented by the green arc.

be the length $L(\gamma)$ of the shortest path γ from $\mathbf{x}$ to $\mathbf{y}$. Thus we arrive at the question: What is the length of a path?

4.2.3. Lengths of paths in Euclidean space. In general, suppose that $\gamma : [t_1, t_2] \to \mathbb{R}^N$ is a continuously differentiable, a.k.a. C^1, path. Its length is defined by

$$L(\gamma) = \int_{t_1}^{t_2} \|\gamma'(t)\|\, dt. \tag{4.27}$$

For each $t \in (t_1, t_2)$, $\gamma'(t) \in \mathbb{R}^N$ is called the **tangent vector** of γ at t. And $\|\gamma'(t)\|$ is its length, a.k.a. norm, measured with respect to the Euclidean inner product. All of this holds independently of whether the path γ is in M^n or not. That is, the definition above of length holds for all continuously differentiable paths in the ambient Euclidean space. However, we are interested in paths on a submanifold M^n of $\mathbb{R}^N$. In this case, $\gamma'(t)$ is a tangent vector to M^n; that is, it is in the tangent space $T_{\gamma(t)}M$.

4.2.4. Restricting the Euclidean inner product to a submanifold: The first fundamental form. Since we are ignoring everything outside of M^n, at $\gamma(t) \in M^n$ we consider only the vectors in $T_{\gamma(t)}M$. So we are only interested in the length of vectors in $T_{\mathbf{x}}M$, where $\mathbf{x} \in M^n$. By the polarization identity

$$\mathbf{V} \cdot \mathbf{W} = \frac{1}{2}(\|\mathbf{V} + \mathbf{W}\|^2 - \|\mathbf{V}\|^2 - \|\mathbf{W}\|^2),$$

the norm $\|\cdot\|$ on $T_{\mathbf{x}}M$ determines an inner product on $T_{\mathbf{x}}M$. This is the same as the inner product induced by the inner product on the ambient Euclidean space.

That is, we define the **first fundamental form** by

$$\mathrm{I}(\mathbf{V}, \mathbf{W}) = \mathbf{V} \cdot \mathbf{W} \tag{4.28}$$

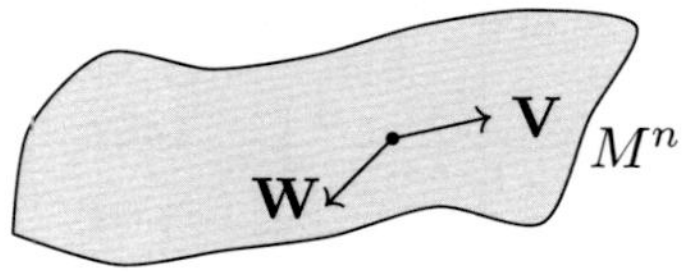

Figure 4.2.2. The inner product $\mathrm{I}(\mathbf{V}, \mathbf{W})$ of tangent vectors $\mathbf{V}$ and $\mathbf{W}$ on M^n is defined to be their Euclidean inner product.

for $\mathbf{V}, \mathbf{W} \in T_{\mathbf{x}}M$ and $\mathbf{x} \in M^n$. As usual, $\mathbf{V} \cdot \mathbf{W}$ denotes the Euclidean inner product of $\mathbf{V}$ and $\mathbf{W}$. The first fundamental form I, which is an inner product on each $T_{\mathbf{x}}M$, $\mathbf{x} \in M^n$, is also called the **induced metric** on M^n. See Figure 4.2.2.

The first fundamental form defines for each $\mathbf{x} \in M^n$ a symmetric bilinear map $\mathrm{I} = \mathrm{I}_{\mathbf{x}}$ from $T_{\mathbf{x}}M \times T_{\mathbf{x}}M$ to $\mathbb{R}$:

$$\mathrm{I}(\mathbf{V}, \mathbf{W}) = \mathrm{I}(\mathbf{W}, \mathbf{V}) \tag{4.29}$$

for all $\mathbf{V}, \mathbf{W} \in T_{\mathbf{x}}M$ and

$$\mathrm{I}(c\mathbf{U} + \mathbf{V}, \mathbf{W}) = c\,\mathrm{I}(\mathbf{U}, \mathbf{W}) + \mathrm{I}(\mathbf{V}, \mathbf{W}), \tag{4.30}$$

where $c \in \mathbb{R}$ and $\mathbf{U} \in T_{\mathbf{x}}M$. Note that it follows that

$$\mathrm{I}(\mathbf{W}, c\mathbf{U} + \mathbf{V}) = c\,\mathrm{I}(\mathbf{W}, \mathbf{U}) + \mathrm{I}(\mathbf{W}, \mathbf{V}). \tag{4.31}$$

See Exercise 4.12. Hence, I is a (covariant) symmetric 2-tensor on M^n, as defined in §3.4.4.

Remark 4.13. Let $\mathbb{V}$ be an n-dimensional vector space and let $\{e_i\}_{i=1}^n$ be a basis for $\mathbb{V}$. If $\cdot$ is an inner product on $\mathbb{V}$, then we may consider $\cdot$ as a quadratic form in n-variables by defining

$$Q(V^1, \ldots, V^n) := V \cdot V = \sum_{i,j=1}^{n} a_{ij} V^i V^j,$$

where $V := \sum_{i=1}^n V^i e_i$ and $a_{ij} := e_i \cdot e_j$. This is the reason for the terminology 'first fundamental *form*'. Note that the basis $\{e_i\}_{i=1}^n$ is **orthonormal** if and only if $a_{ij} = \delta_{ij}$.

Remark 4.14. In general, if $\mathbb{W}$ is a vector subspace of a vector space $\mathbb{V}$ and if $\cdot$ is an inner product on $\mathbb{V}$, then we may consider the restricted inner product I on $\mathbb{W}$ defined by $\mathrm{I}(V, W) = V \cdot W$ for $V, W \in \mathbb{W}$.

By virtue of being an embedded submanifold, M^n is also a *manifold* (which we discuss in subsequent chapters). The first fundamental form I of a submanifold M^n is an example of what is called a **Riemannian metric** (see Riemann [**Rie13**]). The pair (M^n, I) is called a **Riemannian manifold**. Typically, a Riemannian metric is denoted by g:

$$g(\mathbf{V}, \mathbf{W}) := \mathrm{I}(\mathbf{V}, \mathbf{W}) = \mathbf{V} \cdot \mathbf{W} \tag{4.32}$$

for $\mathbf{V}, \mathbf{W} \in T_{\mathbf{x}}M$. The Riemannian metric at a point $\mathbf{x} \in M^n$, as an inner product on $T_{\mathbf{x}}M$, defines distances and angles; see §1.2.4. In particular, the norm of a tangent vector is defined by

$$\|\mathbf{V}\| := \sqrt{g(\mathbf{V}, \mathbf{V})} = \sqrt{\mathrm{I}(\mathbf{V}, \mathbf{V})}. \tag{4.33}$$

The first fundamental form I is a symmetric 2-tensor on M^n. That is, it is a symmetric element of $T^*_{\mathbf{x}}M \otimes T^*_{\mathbf{x}}M$.

4.2.5. Length of a path on a submanifold. It is easy to see that, given a continuously differentiable path $\gamma : [t_1, t_2] \to M^n$, its length is

$$\mathrm{L}(\gamma) = \int_{t_1}^{t_2} \|\gamma'(t)\|\, dt = \int_{t_1}^{t_2} \sqrt{\mathrm{I}(\gamma'(t), \gamma'(t))}\, dt. \tag{4.34}$$

This is the same formula for the length of the path as considered in the ambient Euclidean space. However, we are assuming that the path is in M^n and hence we can express the formula in terms of the first fundamental form I.

4.2.6. Components of the Riemannian metric via local coordinates. Given a submanifold M^n, local coordinates $(\mathcal{V}, \varphi_{\mathcal{V}} := \{x^i\}_{i=1}^n)$ on M^n determine a basis of tangent vector fields $\left\{\frac{\partial}{\partial x^i}\right\}_{i=1}^n$ in $\mathcal{V}$. We can apply the Riemannian metric (a.k.a. the first fundamental form) $g = \mathrm{I}$ to these basis tangent vector fields to obtain a matrix of functions. This makes calculating geometric invariants associated to the Riemannian metric g, such as their curvatures defined later, easier.

Suppose that

$$F : \mathcal{U} \subset \mathbb{R}^n \to F(\mathcal{U}) =: \mathcal{V} \subset M^n \subset \mathbb{R}^N$$

is a parametrization of a submanifold with an inverse map

$$F^{-1} =: \varphi_{\mathcal{V}} =: (x^1, \dots, x^n) : \mathcal{V} \to \varphi_{\mathcal{V}}(\mathcal{V}) = \mathcal{U} \subset \mathbb{R}^n$$

which is a local coordinate system. Recall from §3.1.7 that

$$\left.\frac{\partial}{\partial x^i}\right|_{\mathbf{x}} = \frac{\partial F}{\partial u^i}(\mathbf{u}) = dF_{\mathbf{u}}(\mathbf{e}_i) \in T_{\mathbf{x}}M,$$

where $F(\mathbf{u}) = \mathbf{x}$, is a basis of tangent vectors at each $\mathbf{x} \in \mathcal{V}$. Thus the Riemannian metric g is determined in $\mathcal{V}$ by the square $n \times n$ matrix of its component functions

$$g_{ij}(\mathbf{x}) := g\left(\left.\frac{\partial}{\partial x^i}\right|_{\mathbf{x}}, \left.\frac{\partial}{\partial x^j}\right|_{\mathbf{x}}\right) = g\left(\frac{\partial F}{\partial u^i}(\mathbf{u}), \frac{\partial F}{\partial u^j}(\mathbf{u})\right). \tag{4.35}$$

We call the functions g_{ij} on $\mathcal{V}$ the **components of the Riemannian metric** in the local coordinate system $\{x^i\}$. Since $g = \mathrm{I}$ is symmetric, we have

$$g_{ij} = g_{ji} \quad \text{for all } 1 \leq i, j \leq n.$$

That is, as a matrix, the components $(g_{ij})_{i,j=1}^n$ form a symmetric positive-definite matrix. This matrix, via the local coordinates $\varphi_{\mathcal{V}}$, encodes the information of the Riemannian metric g; see (4.43) below.

Example 4.15. *Let $f : \mathcal{U} \subset \mathbb{R}^n \to \mathbb{R}$ be a smooth function, and let $G^n = \{(\mathbf{u}, f(\mathbf{u}))\colon u \in \mathcal{U}\}$ be its graph. Recall the graphical parametrization $F : \mathcal{U} \to G^n$ defined by $F(\mathbf{u}) = (\mathbf{u}, f(\mathbf{u}))$. Call $\varphi = F^{-1}$ the graphical coordinate system. And recall that $\frac{\partial F}{\partial u^i} = \mathbf{e}_i + \frac{\partial f}{\partial u^i}\mathbf{e}_{n+1}$. Thus,*

$$g_{ij} = g\left(\frac{\partial}{\partial x^i}, \frac{\partial}{\partial x^j}\right) = g\left(\frac{\partial F}{\partial u^i}, \frac{\partial F}{\partial u^j}\right) = \delta_{ij} + \frac{\partial f}{\partial u^i}\frac{\partial f}{\partial u^j}. \tag{4.36}$$

For example, the upper hemisphere, $S^n_{n+1,+}$ defined by (1.26), is the graph of the function $f(\mathbf{u}) := \sqrt{1 - \|\mathbf{u}\|^2}$. So by (4.36) the components of the metric $g = g_{S^n}$, with respect to the graphical coordinate system φ, are given by

$$g_{ij} = g_{ij}(F(\mathbf{u})) = \delta_{ij} + \frac{u^i u^j}{1 - \|\mathbf{u}\|^2}.$$

Does this formula agree with your intuition for the limiting behavior of $g_{ij}(F(\mathbf{u}))$ for $\mathbf{u} = t\mathbf{e}_i$ as $t \to 1$?

Recall that stereographic projection provides local coordinates on S^n. Namely, with $\mathbf{N} := \mathbf{e}_{n+1}$, by (1.33) and identifying the hyperplane $\mathbb{L}$ with $\mathbb{R}^n$ by $(\mathbf{y}, -1) \mapsto \mathbf{y}$, we obtain that $\varphi := G : S^n \setminus \{\mathbf{e}_{n+1}\} \to \mathbb{R}^n$ is given by

$$\varphi(x^1, \ldots, x^{n+1}) = \frac{2(x^1, \ldots, x^n)}{1 - x^{n+1}};$$

that is, $\varphi^i(\mathbf{x}) = \frac{2x^i}{1-x^{n+1}}$ for $1 \leq i \leq n$.

By again using the identification of $\mathbb{L}$ with $\mathbb{R}^n$, we see from (1.66) that the parametrization $F := \varphi^{-1} : \mathbb{R}^n \to S^n \setminus \{\mathbf{e}_{n+1}\}$ is given by

$$F(\mathbf{u}) = \frac{4\mathbf{u} + (\|\mathbf{u}\|^2 - 4)\mathbf{e}_{n+1}}{\|\mathbf{u}\|^2 + 4} = \frac{(4u^1, \ldots, 4u^n, \|\mathbf{u}\|^2 - 4)}{\|\mathbf{u}\|^2 + 4} \tag{4.37}$$

for $\mathbf{u} = (u^1, \ldots, u^n) \in \mathbb{R}^n$.

In these stereographic coordinates, we compute the tangent basis vectors as

$$\begin{aligned}
\frac{\partial F}{\partial u^i}(\mathbf{u}) &= \frac{4\mathbf{e}_i + 2u^i\mathbf{e}_{n+1}}{\|\mathbf{u}\|^2 + 4} - \frac{4\mathbf{u} + (\|\mathbf{u}\|^2 - 4)\mathbf{e}_{n+1}}{(\|\mathbf{u}\|^2 + 4)^2} 2u^i \\
&= \frac{4(\|\mathbf{u}\|^2 + 4)\mathbf{e}_i - 8u^i\mathbf{u} + 16u^i\mathbf{e}_{n+1}}{(\|\mathbf{u}\|^2 + 4)^2}.
\end{aligned}$$

Thus, at $F(\mathbf{u})$, where $\mathbf{u} \in \mathbb{R}^n$, we have (we omit the details of the calculation, which involves some elegant cancellations of terms)

$$g_{ij}(F(\mathbf{u})) = \frac{\partial F}{\partial u^i}(\mathbf{u}) \cdot \frac{\partial F}{\partial u^j}(\mathbf{u}) = \frac{16}{(\|\mathbf{u}\|^2 + 4)^2}\delta_{ij}; \tag{4.38}$$

see Exercise 4.14. For example,

$$g_{ij}(\mathbf{e}_{n+1}) = g_{ij}(F(\mathbf{0})) = \delta_{ij}.$$

We now consider the Riemannian metric components of surfaces of revolution. Let $f : [a, b] \to \mathbb{R}_+$. Define, as in (1.62), a surface of revolution M^2 in $\mathbb{R}^3$ by the map $F : [a, b] \times [0, 2\pi] \to \mathbb{R}^3$ defined by

$$F(u^1, u^2) := \big(f(u^1)\cos(u^2), f(u^1)\sin(u^2), u^1\big) \tag{4.39}$$

for $u^1 \in [a, b]$ and $u^2 \in [0, 2\pi]$. We compute that

$$\frac{\partial F}{\partial u^1} = \big(f'(u^1)\cos(u^2), f'(u^1)\sin(u^2), 1\big) \tag{4.40}$$

and that

$$\frac{\partial F}{\partial u^2} = \big(-f(u^1)\sin(u^2), f(u^1)\cos(u^2), 0\big). \tag{4.41}$$

Thus, the Riemannian metric components $g_{ij} = g_{ij}(F(\mathbf{u})) = \frac{\partial F}{\partial u^i}(\mathbf{u}) \cdot \frac{\partial F}{\partial u^j}(\mathbf{u})$ of (M^2, g) are given by

$$g_{11} = (f'(u^1))^2 + 1, \tag{4.42a}$$

$$g_{22} = (f(u^1))^2, \tag{4.42b}$$

$$g_{12} = g_{21} = 0; \tag{4.42c}$$

see Exercise 4.15.

4.2.7. Riemannian metric in terms of its components. Let (M^n, g) be a smooth embedded submanifold of $\mathbb{R}^N$ with its induced Riemannian metric. Let $(\mathcal{V}, \{x^i\})$ be local coordinates on M^n. For each $\mathbf{x} \in \mathcal{V}$, define the basis of cotangent vectors $\{dx^j\}_{j=1}^n$ at $\mathbf{x}$ as the dual basis to $\left\{\frac{\partial}{\partial x^i}_{\mathbf{x}}\right\}_{i=1}^n$. That is,

$$dx^j\left(\frac{\partial}{\partial x^i}\right) = \delta_{ij}.$$

By definition, $dx^j(\mathbf{x}) \in T^*_{\mathbf{x}}M$ for each $\mathbf{x} \in \mathcal{V}$, where $T^*_{\mathbf{x}}M$ denotes the cotangent space of M^n at $\mathbf{x}$.

Given two cotangent vectors $\alpha, \beta \in T^*_{\mathbf{x}}M$, their **tensor product** is the bilinear form $\alpha \otimes \beta : T_{\mathbf{x}}M \times T_{\mathbf{x}}M \to \mathbb{R}$ defined by

$$(\alpha \otimes \beta)(\mathbf{V}, \mathbf{W}) := \alpha(\mathbf{V})\beta(\mathbf{W}) \quad \text{for } \mathbf{V}, \mathbf{W} \in T_{\mathbf{x}}M.$$

With this, we can express the Riemannian metric g in terms of its components by

$$g = \sum_{i,j=1}^{n} g_{ij}\, dx^i \otimes dx^j. \tag{4.43}$$

Indeed, if $\mathbf{V} = \sum_{i=1}^{n} V^i \frac{\partial}{\partial x^i}$ and $\mathbf{W} = \sum_{j=1}^{n} W^j \frac{\partial}{\partial x^j}$, then

$$\begin{aligned} g(\mathbf{V}, \mathbf{W}) &= g\left(\sum_{i=1}^{n} V^i \frac{\partial}{\partial x^i}, \sum_{j=1}^{n} W^j \frac{\partial}{\partial x^j}\right) \\ &= \sum_{i,j=1}^{n} V^i W^j g\left(\frac{\partial}{\partial x^i}, \frac{\partial}{\partial x^j}\right) \\ &= \sum_{i,j=1}^{n} V^i W^j g_{ij}. \end{aligned}$$

On the other hand, using that

$$dx^i(\mathbf{V}) = dx^i\left(\sum_{k=1}^{n} V^k \frac{\partial}{\partial x^k}\right) = \sum_{k=1}^{n} V^k dx^i\left(\frac{\partial}{\partial x^k}\right) = \sum_{k=1}^{n} V^k \delta_{ik} = V^i,$$

we calculate that

$$\begin{aligned} \left(\sum_{i,j=1}^{n} g_{ij}\, dx^i \otimes dx^j\right)(\mathbf{V}, \mathbf{W}) &= \sum_{i,j=1}^{n} g_{ij}\, dx^i \otimes dx^j(\mathbf{V}, \mathbf{W}) \\ &= \sum_{i,j=1}^{n} g_{ij} dx^i(\mathbf{V})\, dx^j(\mathbf{W}) \\ &= \sum_{i,j=1}^{n} g_{ij} V^i W^j. \end{aligned}$$

Now (4.43) follows.

For example, the Euclidean metric on $\mathbb{R}^n$ may be expressed as

$$g_{\mathbb{E}} = \sum_{i=1}^{n} dx^i \otimes dx^i = \sum_{i,j=1}^{n} \delta_{ij}\, dx^i \otimes dx^j,$$

which holds on all of $\mathbb{R}^n$. Indeed, with this definition, we see that

$$g_{\mathbb{E}}(\mathbf{V}, \mathbf{W}) = \sum_{i=1}^{n} V^i W^i$$

for $\mathbf{V} = (V^1, \ldots, V^n)$ and $\mathbf{W} = (W^1, \ldots, W^n)$, since $dx^i(V) = V^i$.

So, for example, equation (4.42) says that the Riemannian metric of a surface of revolution may be expressed as

$$g(F(\mathbf{u})) = \big((f'(u^1))^2 + 1\big)dx^1 \otimes dx^1 + (f(u^1))^2 dx^2 \otimes dx^2.$$

Another nice example is the spherical metric g_{S^n} in stereographic coordinates. By (4.38), we have

$$g_{S^n}(F(\mathbf{u})) = \frac{16}{(\|\mathbf{u}\|^2+4)^2} \sum_{i=1}^{n} dx^i \otimes dx^i. \tag{4.44}$$

4.3. Euclidean covariant derivative

Before we discuss the differentiation of vector fields on submanifolds, we review the differentiation of vector fields on Euclidean space. This is useful because the notion on submanifolds is obtained by projection onto their tangent spaces.

4.3.1. Definition of Euclidean covariant derivative. Recall that given a differentiable function $f : \mathbb{R}^n \to \mathbb{R}$, we have its *derivative* $df_{\mathbf{x}} : \mathbb{R}^n \to \mathbb{R}$ defined by

$$df_{\mathbf{x}}(\mathbf{V}) = \frac{d}{dt}\bigg|_{t=0} f(\mathbf{x} + t\mathbf{V}) \tag{4.45}$$

for $\mathbf{V} \in T_{\mathbf{x}}\mathbb{R}^n \cong \mathbb{R}^n$, $\mathbf{x} \in \mathbb{R}^n$. Given $\mathbf{V}$, we call this the **directional derivative** of f along $\mathbf{V}$, and we also denote it by

$$\mathbf{V}(f) := df_{\mathbf{x}}(\mathbf{V}). \tag{4.46}$$

If we wish to emphasize that the directional derivative is at the point $\mathbf{x}$, we can denote this by $\mathbf{V}(f)_{\mathbf{x}}$.

Similarly, we can differentiate vector-valued functions and, in particular, vector fields. We write a function $F : \mathbb{R}^n \to \mathbb{R}^N$ in component form as $F = (F^1, \dots, F^N)$, where $F^i : \mathbb{R}^n \to \mathbb{R}$ for each $1 \leq i \leq N$. Then, the derivative of F is

$$dF = (dF^1, \dots, dF^N) : \mathbb{R}^n \to \mathbb{R}^N. \tag{4.47}$$

That is, for $\mathbf{V} \in \mathbb{R}^n$,

$$dF(\mathbf{V}) = \big(dF^1(\mathbf{V}), \dots, dF^N(\mathbf{V})\big). \tag{4.48}$$

In particular, a vector field W on $\mathbb{R}^N$ is simply a function $W : \mathbb{R}^N \to \mathbb{R}^N$. Writing $W = (W^1, \dots, W^N)$ and $\mathbf{V} = (V^1, \dots, V^N)$, we have

$$dW(\mathbf{V}) = \big(dW^1(\mathbf{V}), \dots, dW^N(\mathbf{V})\big), \tag{4.49}$$

where

$$dW^i(\mathbf{V}) = \sum_{j=1}^{N} V^j \frac{\partial W^i}{\partial x^j}. \tag{4.50}$$

We call this the **Euclidean covariant derivative** and we also denote it by

$$D_{\mathbf{V}}W := dW(\mathbf{V}). \tag{4.51}$$

This is just the directional derivative of the vector-valued function W along the direction $\mathbf{V}$.

Example 4.16. *The identity vector field* id : $\mathbb{R}^N \to \mathbb{R}^N$ *is given by*

$$\mathrm{id}_{\mathbf{x}} := \mathrm{id}(\mathbf{x}) := \mathbf{x}. \tag{4.52}$$

We have

$$D_{\mathbf{V}}\,\mathrm{id} = (d\,\mathrm{id})(\mathbf{V}) = \mathbf{V} \tag{4.53}$$

for $\mathbf{V} \in T_{\mathbf{x}}\mathbb{R}^N$, $\mathbf{x} \in \mathbb{R}^N$.

Let W be a vector field on $\mathbb{R}^N$ and let $\mathbf{V} \in T_{\mathbf{x}}\mathbb{R}^N$, where $\mathbf{x} \in \mathbb{R}^N$. Let $\gamma : I \to \mathbb{R}^N$ be a continuously differentiable path with 0 in the interior of I and $\gamma'(0) = \mathbf{V}$. Then

$$D_{\mathbf{V}}W = \left.\frac{d}{dt}\right|_{t=0} (W \circ \gamma)(t) = (W \circ \gamma)'(0). \tag{4.54}$$

This shows that $D_{\mathbf{V}}W$ is determined by W restricted to any path tangent to $\mathbf{V}$.

Let U, V, W be vector fields on $\mathbb{R}^N$. Then (see Exercise 4.16)

$$U(V \cdot W) = D_U V \cdot W + V \cdot D_U W, \tag{4.55}$$

where D is the Euclidean covariant derivative.

4.3.2. Lie bracket and the Jacobi identity. Now, let us introduce the 'Lie bracket,' a basic operation on vector fields. This operation is related to the Euclidean covariant derivative. In the following sections, we will see that the Lie bracket is related to the basic notions of *Levi-Civita connection* and *Riemann curvature tensor.*

4.3.2.1. *Definition of the Lie bracket.* Let V and W be smooth vector fields on $\mathbb{R}^N$. Define a linear map from the space of smooth functions on $\mathbb{R}^N$ to the space of smooth functions on $\mathbb{R}^N$ as follows:

$$L(f) := V(W(f)) - W(V(f)). \tag{4.56}$$

We will now show that L is a **derivation**. By definition, this means that $L : C^\infty(\mathbb{R}^N) \to C^\infty(\mathbb{R}^N)$ is a linear map and if f and g are smooth functions on $\mathbb{R}^n$, then

$$L(fg) = fL(g) + gL(f).$$

By a standard result (see Proposition 6.29 below), this implies that L corresponds to a vector field. We denote this vector field by $[V, W]$ and we call this vector field the **Lie bracket** of V and W. That is, the Lie bracket vector field $[V, W]$, as a derivation, is defined by

$$[V, W](f) := V(W(f)) - W(V(f)). \tag{4.57}$$

To show that $[V, W]$ is a derivation, first recall by definition that

$$L(fg) = V(W(fg)) - W(V(fg)).$$

Regarding the first term on the right-hand side, we compute using the Leibniz rule that

$$W(fg) = fW(g) + gW(f)$$

and hence

$$\begin{aligned} V(W(fg)) &= V(fW(g) + gW(f)) \\ &= fV(W(g)) + gV(W(f)) + V(f)W(g) + V(g)W(f). \end{aligned}$$

Simply by switching V and W, we obtain

$$W(V(fg)) = fW(V(g)) + g(W(V(f)) + W(f)V(g) + W(g)V(f).$$

Therefore,

$$\begin{aligned} L(fg) &= V(W(fg)) - W(V(fg)) \\ &= fV(W(g)) + gV(W(f)) - fW(V(g)) - gW(V(f)) \\ &= fL(g) + gL(f). \end{aligned}$$

This implies that L is a derivation.

A key fact is that if V and W are smooth tangent vector fields on a submanifold M^n of $\mathbb{R}^N$, then we can define their Lie bracket $[V, W]$ exactly as above and $[V, W]$ will be a derivation acting on smooth functions on M^n. Indeed, $W(f)$ is a smooth function on M^n, as is $V(W(f))$. And the same is true for V and W switched. Thus, all of the calculations above go through for $[V, W]$ acting on functions f on M^n. By Proposition 6.29 below, this proves that $[V, W]$ is a tangent vector field on M^n. In summary, the Lie bracket operation restricts to vector fields on a submanifold.

4.3.2.2. *Identities satisfied by the Lie bracket.* We also observe the easy antisymmetry identity:

$$[V, W] = -[W, V]. \tag{4.58}$$

The **Jacobi identity** for the Lie bracket of vector fields says the following. For any vector fields U, V, W on a submanifold M^n of $\mathbb{R}^N$, we have

$$\big[[U, V], W\big] + \big[[V, W], U\big] + \big[[W, U], V\big] = 0; \tag{4.59}$$

see Exercise 4.17. Notice how U, V, W are cyclically permuted in the sum on the left-hand side. The Jacobi identity is a symmetry property that reflects the diffeomorphism invariance of the Lie bracket operation (see, e.g., [**CLN06**, Exercise 1.26(1)]).

4.3.2.3. *Lie bracket in local coordinates.* In the following, we give a formula for the Lie bracket of two vector fields on a submanifold expressed in terms of local coordinates.

Given a parametrization $F : \mathcal{U} \to \mathcal{V}$ of a relatively open subset $\mathcal{V}$ of M^n, we have its inverse coordinate system $\varphi := F^{-1} : \mathcal{V} \to \mathcal{U}$. Recall that $\{\frac{\partial F}{\partial u^i}(\mathbf{u})\}$ is a basis of tangent vectors at $\mathbf{x} := F(\mathbf{u})$. Given a vector field V on M^n, we may express it in the subset $\mathcal{V}$ as

$$V(\mathbf{x}) =: \sum_{i=1}^{n} V^i(\mathbf{x}) \frac{\partial F}{\partial u^i}(\mathbf{u}), \tag{4.60}$$

where $\mathbf{u} = F^{-1}(\mathbf{x}) = \varphi(\mathbf{x})$. Similarly, given a vector field W on M^n, we express it as

$$W(\mathbf{x}) =: \sum_{i=1}^{n} W^i(\mathbf{x}) \frac{\partial F}{\partial u^i}(\mathbf{u}). \tag{4.61}$$

Using that $\frac{\partial F}{\partial u^i}(f) = \frac{\partial}{\partial u^i}(f \circ F)$, we compute that

$$\begin{aligned} W(f) &= \sum_{i=1}^{n} W^i \frac{\partial F}{\partial u^i}(f) \\ &= \sum_{i=1}^{n} W^i \frac{\partial}{\partial u^i}(f \circ F). \end{aligned}$$

We then compute that

$$V(W(f)) = \sum_{j=1}^{n} V^j \frac{\partial F}{\partial u^j} \left(\sum_{i=1}^{n} W^i \frac{\partial}{\partial u^i}(f \circ F) \right)$$
$$= \sum_{i,j=1}^{n} V^j \frac{\partial}{\partial u^j}(W^i \circ F) \frac{\partial}{\partial u^i}(f \circ F)$$
$$+ \sum_{i,j=1}^{n} V^j W^i \frac{\partial^2}{\partial u^j \partial u^i}(f \circ F).$$

Simply by switching V and W, we obtain

$$W(V(f)) = \sum_{i,j=1}^{n} W^j \frac{\partial}{\partial u^j}(V^i \circ F) \frac{\partial}{\partial u^i}(f \circ F)$$
$$+ \sum_{i,j=1}^{n} W^j V^i \frac{\partial^2}{\partial u^j \partial u^i}(f \circ F).$$

Thus, as derivations, we have that

$$[V, W](f) = V(W(f)) - W(V(f))$$
$$= \sum_{i,j=1}^{n} V^j \frac{\partial}{\partial u^j}(W^i \circ F) \frac{\partial}{\partial u^i}(f \circ F)$$
$$- \sum_{i,j=1}^{n} W^j \frac{\partial}{\partial u^j}(V^i \circ F) \frac{\partial}{\partial u^i}(f \circ F)$$
$$= \sum_{i,j=1}^{n} \left(V^j \frac{\partial F}{\partial u^j}(W^i) - W^j \frac{\partial F}{\partial u^j}(V^i) \right) \frac{\partial F}{\partial u^i}(f).$$

In other words, as functions, we have that

$$[V, W] = \sum_{i,j=1}^{n} \left(V^j \frac{\partial F}{\partial u^j}(W^i) - W^j \frac{\partial F}{\partial u^j}(V^i) \right) \frac{\partial F}{\partial u^i}. \tag{4.62}$$

Since the right-hand side is a linear combination of the tangent vectors $\frac{\partial F}{\partial u^i}$, this reproves the fact that if V and W are smooth tangent vector fields on M^n, then so is their bracket $[V, W]$.

4.3.3. Euclidean covariant derivative is torsion-free. The Euclidean covariant derivative D on $\mathbb{R}^N$ satisfies the **torsion-free** condition

$$D_V W - D_W V = [V, W]; \tag{4.63}$$

see Exercise 4.18.

Observe that if $V = (V^1, \dots, V^N)$ and $W = (W^1, \dots, W^N)$ are vector fields on $\mathbb{R}^N$, then

$$[V, W]^i = (D_V W - D_W V)^i = V(W^i) - W(V^i) = \sum_{j=1}^{N} \left(V^j \frac{\partial W^i}{\partial u^j} - W^j \frac{\partial V^i}{\partial u^j} \right). \tag{4.64}$$

Example 4.17. *On $\mathbb{R}^2$ consider the vector fields $V = \frac{\partial}{\partial x}$ and $W = -y\frac{\partial}{\partial x} + x\frac{\partial}{\partial y}$. If we calculate using the definition of Lie bracket, then we get*

$$\begin{aligned}
[V, W](f) &= V(W(f)) - W(V(f)) \\
&= V\left(-y\frac{\partial f}{\partial x} + x\frac{\partial f}{\partial y}\right) - W\left(\frac{\partial f}{\partial x}\right) \\
&= \left(-y\frac{\partial^2 f}{\partial x^2} + \frac{\partial f}{\partial y} + x\frac{\partial^2 f}{\partial x \partial y}\right) - \left(-y\frac{\partial^2 f}{\partial x^2} + x\frac{\partial^2 f}{\partial y \partial x}\right) \\
&= \frac{\partial}{\partial y}(f).
\end{aligned}$$

Therefore,

$$[V, W] = \frac{\partial}{\partial y}. \tag{4.65}$$

On the other hand,

$$\begin{aligned}
(D_V W - D_W V)^1 &= V(W^1) - W(V^1) \\
&= e_1(-y) - (-ye_1 + xe_2)(1) = 0, \\
(D_V W - D_W V)^2 &= V(W^2) - W(V^2) \\
&= e_1(x) - (-ye_1 + xe_2)(0) = 1.
\end{aligned}$$

Thus,

$$D_V W - D_W V = e_2 = \frac{\partial}{\partial y}, \tag{4.66}$$

in agreement with (4.65).

4.4. Second fundamental form

In this section we primarily restrict ourselves to the consideration of the codimension-1 case of embedded submanifolds, i.e., hypersurfaces. The (extrinsic) curvature of a hypersurface M^n in $\mathbb{R}^{n+1}$ is governed by the (infinitesimal) rate of change of the unit normal vector field N. Since N is a (Euclidean) vector field along M^n, we consider its directional derivative along tangent vectors to M^n. Toward understanding this, we first recall some facts about taking such directional derivatives along a hypersurface.

4.4.1. Directional derivatives of functions on submanifolds. Let Y be a Euclidean vector field along M^n; that is, $Y : M^n \to \mathbb{R}^N$. Let $\mathbf{x} \in M^n$ and $\mathbf{V} \in T_{\mathbf{x}}M$. Recall from (4.51) that the directional derivative (a.k.a. Euclidean covariant derivative) of Y along $\mathbf{V}$ is given by

$$D_{\mathbf{V}}Y := \mathbf{V}(Y) := \big(\mathbf{V}(Y^1), \dots, \mathbf{V}(Y^N)\big), \tag{4.67}$$

where $\mathbf{V}(Y^i)$ denotes the directional derivative of the function $Y^i : M^n \to \mathbb{R}$ along $\mathbf{V}$.

Regarding the directional derivatives of functions, we make the following observations. Let $f : \mathbb{R}^N \to \mathbb{R}$ be a differentiable function, let $\mathbf{x} \in \mathbb{R}^N$, and let $\mathbf{V} \in T_{\mathbf{x}}\mathbb{R}^N \cong \mathbb{R}^N$. Let $\gamma : I \to \mathbb{R}^N$ be a differentiable path with $\gamma(0) = \mathbf{x}$ and $\gamma'(0) = \mathbf{V}$. Then, by the chain rule, we have

$$\left.\frac{d}{dt}\right|_{t=0} (f \circ \gamma)(t) = df_{\mathbf{x}}\big(\gamma'(0)\big) = df_{\mathbf{x}}(\mathbf{V}). \tag{4.68}$$

Note that this formula for the directional derivative of f along $\mathbf{V}$ is true for all γ with the property that $\gamma'(0) = \mathbf{V} \in T_{\mathbf{x}}\mathbb{R}^N$.

As a consequence of this, we can define the directional derivative of a function on a submanifold $g : M^n \to \mathbb{R}$ along a tangent vector $\mathbf{V} \in T_{\mathbf{x}}M$. Namely, we define

$$dg(\mathbf{V}) := \mathbf{V}(g) := \left.\frac{d}{dt}\right|_{t=0} (g \circ \gamma)(t), \tag{4.69}$$

where $\gamma : I \to M^n$ is any path with $\gamma'(0) = \mathbf{V}$. Note that if $\bar{g} : \mathbb{R}^N \to \mathbb{R}$ is any differentiable extension of g, that is, the restriction $\bar{g}|_{M^n} = g$, then

$$dg(\mathbf{V}) = d\bar{g}(\mathbf{V}) = \left.\frac{d}{dt}\right|_{t=0} (\bar{g} \circ \bar{\gamma})(t) \tag{4.70}$$

for any $\mathbf{V} \in T_X M$ and $\bar{\gamma} : I \to \mathbb{R}^N$ with $\bar{\gamma}'(0) = \mathbf{V}$. Namely, we can compute the derivative of g by considering any differentiable extension $\bar{g}$ of g to $\mathbb{R}^N$ and computing the directional derivative of $\bar{g}$ along tangent vectors to M^n. Note that the image of $\bar{\gamma}$ need not necessarily be contained in M^n.

The discussion in the preceding two paragraphs applies to vector-valued functions. Namely, if $Y : M^n \to \mathbb{R}^N$ is a vector-valued function on M^n, i.e., in this case a Euclidean vector field on M^n, then the same applies to

$$D_{\mathbf{V}}Y = dY(\mathbf{V}) = \mathbf{V}(Y) = \left.\frac{d}{dt}\right|_{t=0} (\bar{Y} \circ \bar{\gamma})(t) \tag{4.71}$$

for any $\mathbf{V} \in T_{\mathbf{x}}M$, where $\bar{\gamma} : I \to \mathbb{R}^N$ satisfies $\bar{\gamma}'(0) = \mathbf{V}$ and where each of the three expressions on the left of the display means the directional derivative of Y along $\mathbf{V}$, which is the same as the Euclidean covariant derivative of Y in the direction $\mathbf{V}$.

4.4.2. Definition of the Weingarten map. Now we are ready to see how the unit normal of a hypersurface changes. The rate of change of the unit normal reflects the bending and curving of the hypersurface.

Let M^n be an n-dimensional submanifold of $\mathbb{R}^{n+1}$ and assume that there exists a continuous choice of unit normal vector field N on all of M^n. Let $\mathbf{x} \in M^n$ and let $\mathbf{V} \in T_{\mathbf{x}}M$. Note that $\mathbf{V} \cdot N_{\mathbf{x}} = 0$, where $N_{\mathbf{x}}$ denotes N at $\mathbf{x}$. Since there exists a path $\gamma : I \to M^n$ with 0 in the interior of I, $\gamma(0) = \mathbf{x}$, and $\gamma'(0) = \mathbf{V}$, the Euclidean covariant derivative

$$D_{\mathbf{V}}N = \mathbf{V}(N) \tag{4.72}$$

makes sense. Indeed, we may define it by

$$D_{\mathbf{V}}N := \left.\frac{d}{dt}\right|_{t=0} N\big(\gamma(t)\big), \tag{4.73}$$

where the right-hand side is independent of such γ.

Since $\|N\|^2 \equiv 1$ on M^n, by taking the directional derivative, we obtain

$$0 = \mathbf{V}(1) = \mathbf{V}(\|N\|^2) = 2D_{\mathbf{V}}N \cdot N. \tag{4.74}$$

Thus the derivative of the normal $D_{\mathbf{V}}N \in T_{\mathbf{x}}M$ is tangential. In view of the calculation above, for each point $\mathbf{x} \in M^n$, we define

$$\mathrm{L} : T_{\mathbf{x}}M \to T_{\mathbf{x}}M \tag{4.75}$$

by

$$\mathrm{L}(\mathbf{V}) := D_{\mathbf{V}}N. \tag{4.76}$$

This is called the **Weingarten map**. See Figure 4.4.1.

The Weingarten map is a linear map:

$$\mathrm{L}(a\mathbf{V}_1 + \mathbf{V}_2) = a\,\mathrm{L}(\mathbf{V}_1) + \mathrm{L}(\mathbf{V}_2). \tag{4.77}$$

(Exercise: Prove this.) Observe that the function $\mathrm{L} : T_{\mathbf{x}}M \times T_{\mathbf{x}}^*M \to \mathbb{R}$, defined by $\mathrm{L}(\mathbf{V}, \mathbf{W}^*) := \mathbf{W}^*(\mathrm{L}(\mathbf{V}))$, is bilinear. Hence, the Weingarten map L is (equivalent to) a $(1,1)$-tensor on M^n in the sense of §3.4.4.

Example 4.18 (Hyperplanes). *Let $\mathbb{A}^n$ be an affine hyperplane in $\mathbb{R}^{n+1}$. Then any continuous choice of unit normal vector field N to $\mathbb{A}^n$ must be a constant vector field. Hence, for any $\mathbf{x} \in \mathbb{A}^n$, $\mathbf{V} \in T_{\mathbf{x}}(\mathbb{A}^n)$, we have $D_{\mathbf{V}}N = \mathbf{0}$. Thus, the Weingarten map of $\mathbb{A}^n$ is identically zero. See Figure* 4.4.2.

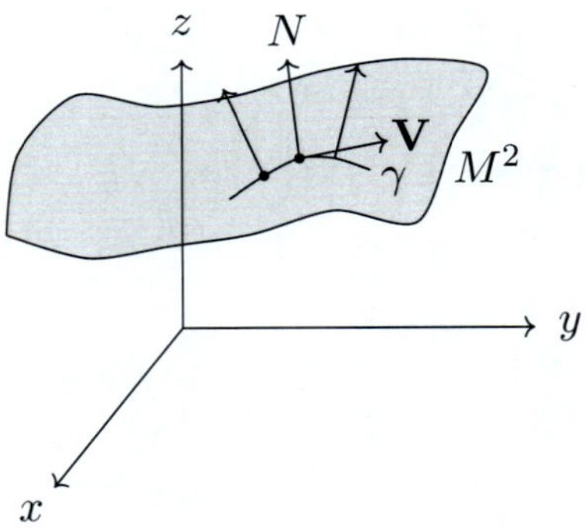

Figure 4.4.1. A unit normal vector N changing along a curve γ in a surface M^2 tangent to $\mathbf{V}$. The rate of change of N in the direction $\mathbf{V}$ is equal to the Weingarten map value $\mathrm{L}(\mathbf{V})$.

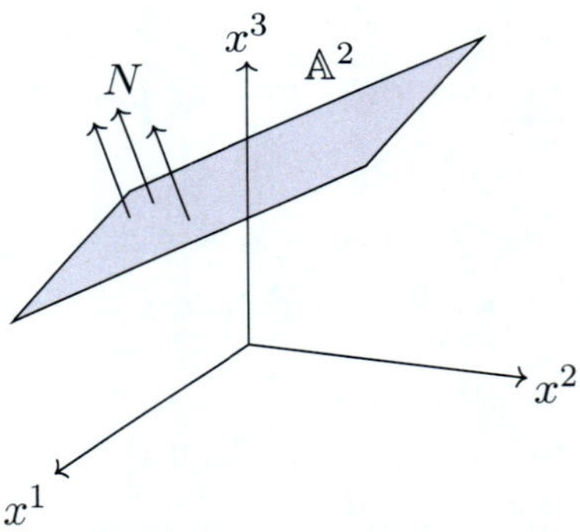

Figure 4.4.2. An affine plane $\mathbb{A}^2$ in $\mathbb{R}^3$ has constant unit normal N and hence trivial Weingarten map.

4.4.3. Definition of the second fundamental form. We consider an equivalent way to view the Weingarten map as a symmetric 2-tensor (i.e., symmetric bilinear form on each tangent space).

We define the **second fundamental form**

$$\mathrm{II} := \mathrm{II}_{\mathbf{x}} : T_{\mathbf{x}}M \times T_{\mathbf{x}}M \to \mathbb{R}$$

of M^n by

$$\mathrm{II}(\mathbf{V}, \mathbf{W}) := \mathrm{L}(\mathbf{V}) \cdot \mathbf{W} = D_{\mathbf{V}}N \cdot \mathbf{W} \tag{4.78}$$

for tangent vectors $\mathbf{V}, \mathbf{W} \in T_{\mathbf{x}}M$.

The second fundamental form II of M^n is a bilinear form; that is,

$$\text{II}(a\mathbf{V}_1 + \mathbf{V}_2, \mathbf{W}) = a\,\text{II}(\mathbf{V}_1, \mathbf{W}) + \text{II}(\mathbf{V}_2, \mathbf{W}), \tag{4.79a}$$

$$\text{II}(\mathbf{V}, a\mathbf{W}_1 + \mathbf{W}_2) = a\,\text{II}(\mathbf{V}, \mathbf{W}_1) + \text{II}(\mathbf{V}, \mathbf{W}_2); \tag{4.79b}$$

see Exercise 4.19. Hence, the second fundamental form II is a covariant 2-tensor on M^n.

In general, given a vector space $\mathbb{V}$ with an inner product $\langle \cdot, \cdot \rangle$ and given a linear transformation $L : \mathbb{V} \to \mathbb{V}$, we may define the associated bilinear function $B : \mathbb{V} \times \mathbb{V} \to \mathbb{R}$ by

$$B(V, W) := \langle L(V), W \rangle. \tag{4.80}$$

In this way, the second fundamental form II is the associated bilinear function to the Weingarten linear transformation L.

Example 4.19. *The Weingarten map of S^n is given by*

$$\text{L}(\mathbf{V}) = \mathbf{V} \tag{4.81}$$

for $\mathbf{V} \in T_{\mathbf{x}}(S^n)$, $\mathbf{x} \in S^n$. To see this, we calculate that

$$\text{L}(\mathbf{V}) = \mathbf{V}(\text{id}) = \mathbf{V},$$

where $\text{id} : S^n \to S^n$ *is the identity map; i.e.,* $\text{id}(\mathbf{x}) = \mathbf{x}$. *Note that if* $\gamma : I \to S^n$ *satisfies* $\gamma'(0) = \mathbf{V}$, *then a (somewhat trivial) calculation gives*

$$\mathbf{V}(\text{id}) = \left.\frac{d}{dt}\right|_{t=0} \text{id}(\gamma(t)) = \left.\frac{d}{dt}\right|_{t=0} \gamma(t) = \gamma'(0) = \mathbf{V}.$$

The product rule for differentiating inner products yields symmetries for the second fundamental form II and the Weingarten map L.

For the Weingarten map and second fundamental form of round spheres, see Exercise 4.20.

4.4.4. Symmetry of the second fundamental form.

We may also express the second fundamental form as follows:

$$\text{II}(\mathbf{V}, \mathbf{W}) = -D_{\mathbf{V}} W \cdot N, \tag{4.82}$$

where W is any vector field on M^n with $W_{\mathbf{x}} = \mathbf{W}$. This is because, by applying the product rule to (4.78), we obtain

$$\begin{aligned} \text{II}(\mathbf{V}, \mathbf{W}) &= D_{\mathbf{V}} N \cdot W_{\mathbf{x}} \\ &= \mathbf{V}(N \cdot W) - N_{\mathbf{x}} \cdot D_{\mathbf{V}} W \end{aligned}$$

and the first term on the last line vanishes since $N \cdot W \equiv 0$ on M^n. Thus, we can think of the second fundamental form as the normal component of the Euclidean covariant derivative of tangent vector fields on M^n. As we will see in §4.4.9 below, this interpretation applies to higher-codimensional submanifolds.

Lemma 4.20. *The second fundamental form is symmetric: For any* $\mathbf{V}, \mathbf{W} \in T_{\mathbf{x}}M$,

$$\mathrm{II}(\mathbf{V}, \mathbf{W}) = \mathrm{II}(\mathbf{W}, \mathbf{V}). \tag{4.83}$$

Hence, the Weingarten map $\mathrm{L} : T_{\mathbf{x}}M \to T_{\mathbf{x}}M$ *is self-adjoint:*

$$\mathrm{L}(\mathbf{V}) \cdot \mathbf{W} = \mathbf{V} \cdot \mathrm{L}(\mathbf{W}). \tag{4.84}$$

4.4.4.1. *Proof of the symmetry of the second fundamental form.*

Proof of Lemma 4.20. Let $V_{\mathbf{x}}, W_{\mathbf{x}} \in T_{\mathbf{x}}M$. Extend $V_{\mathbf{x}}, W_{\mathbf{x}}$ to vector fields V, W, respectively, where they are defined in a neighborhood $\mathcal{V}$ of $\mathbf{x}$ in M^n. We compute that

$$\begin{aligned} \mathrm{II}(V_{\mathbf{x}}, W_{\mathbf{x}}) &= D_{V_{\mathbf{x}}} N \cdot W_{\mathbf{x}} \\ &= V_{\mathbf{x}}(N \cdot W) - N_{\mathbf{x}} \cdot D_{V_{\mathbf{x}}} W \\ &= -N_{\mathbf{x}} \cdot D_{V_{\mathbf{x}}} W \end{aligned}$$

since $V_{\mathbf{x}}(N \cdot W) = V_{\mathbf{x}}(0) = 0$. By switching V and W, we immediately obtain

$$\mathrm{II}(W_{\mathbf{x}}, V_{\mathbf{x}}) = -N_{\mathbf{x}} \cdot D_{W_{\mathbf{x}}} V.$$

Therefore,

$$\begin{aligned} \mathrm{II}(V_{\mathbf{x}}, W_{\mathbf{x}}) - \mathrm{II}(W_{\mathbf{x}}, V_{\mathbf{x}}) &= N_{\mathbf{x}} \cdot (-D_{V_{\mathbf{x}}} W + D_{W_{\mathbf{x}}} V) \\ &= N_{\mathbf{x}} \cdot [W, V]_{\mathbf{x}} \\ &= 0, \end{aligned}$$

where for the last equality we used that the Euclidean vector field $[W, V]$ along M^n is tangent to $\mathcal{V} \subset M^n$ (at $\mathbf{x}$). The lemma follows. □

4.4.5. Principal curvatures of a hypersurface. Because the Weingarten map $\mathrm{L} : T_{\mathbf{x}}M \to T_{\mathbf{x}}M$ is self-adjoint, it has real eigenvalues and there exists an orthonormal basis $\{e_i\}_{i=1}^n$ of $T_{\mathbf{x}}M$ consisting of eigenvectors of L. Let $\{\kappa_i\}_{i=1}^n$ denote the corresponding eigenvalues. Then

$$\mathrm{L}(e_i) = \kappa_i e_i. \tag{4.85}$$

We call each κ_i a **principal curvature** of M^n. We call each e_i a **principal direction** of M^n.

Given $\kappa \in \mathbb{R}$, let

$$E_\kappa := \{\mathbf{V} \in T_{\mathbf{x}}M \colon \mathrm{L}(\mathbf{V}) = \kappa \mathbf{V}\} \tag{4.86}$$

denote the corresponding eigenspace of L. We have $E_\kappa \neq \{\mathbf{0}\}$ if and only if $\kappa \in \{\kappa_1, \ldots, \kappa_n\}$. Note that if $\mathbf{V} \in E_\kappa$ and $\mathbf{W} \in E_\lambda$, then

$$\kappa \mathbf{V} \cdot \mathbf{W} = \mathrm{L}(\mathbf{V}) \cdot \mathbf{W} = \mathbf{V} \cdot \mathrm{L}(\mathbf{W}) = \lambda \mathbf{V} \cdot \mathbf{W}.$$

Thus, if $\kappa \neq \lambda$, then $\mathbf{V} \cdot \mathbf{W} = 0$.

4.4.6. Geometric interpretation of principal curvatures. Let $\mathbf{V}$ be a unit vector in $T_{\mathbf{x}}M$. The unit normal $N_{\mathbf{x}}$ and $\mathbf{V}$ span a plane P in $\mathbb{R}^{n+1}$. The plane P intersects M^n in a curve $C := M^n \cap P$. Then

$$\mathrm{II}(\mathbf{V}, \mathbf{V}) = \mathrm{L}(\mathbf{V}) \cdot \mathbf{V} = \pm(\text{plane}) \text{ curvature } \kappa \text{ of } C, \tag{4.87}$$

where κ is defined by (4.7). In particular, this applies to $\mathbf{V} = e_i$ a principal direction, in which case $\mathrm{II}(e_i, e_i) = \kappa_i$. In this case we call P a **plane of a principal curvature**; see Figure 4.4.3.

The proof of (4.87) is as follows. Let $\mathbf{V}_0$ be a unit vector in $T_{\mathbf{x}}M$. Let $\gamma(s)$ be a unit-speed parametrization of the curve C with $\gamma'(0) = \mathbf{V}_0$. Let $V(s) = \gamma'(s)$. We have that N is normal to $\gamma(s)$ in P at $s = 0$. Thus, the plane curvature κ of $\gamma(s)$ in P at $s = 0$ is given by

$$\kappa = -D_{\mathbf{V}_0} V \cdot N = -\mathbf{V}_0(V \cdot N) + D_{\mathbf{V}_0} N \cdot V.$$

Since $V(s) \cdot N_{\gamma(s)} \equiv 0$, we have $\mathbf{V}_0(V \cdot N) = 0$. Thus,

$$\kappa = D_{\mathbf{V}_0} N \cdot \mathbf{V}_0 = \mathrm{II}(\mathbf{V}_0, \mathbf{V}_0).$$

Remark 4.21. Let $\kappa_1 \leq \kappa_2 \leq \cdots \leq \kappa_n$ denote the principal curvatures of M^n at a point $\mathbf{x}$, in nondecreasing order. We claim that

$$\kappa_1 = \inf_{\mathbf{V}} \mathrm{II}(\mathbf{V}, \mathbf{V}), \tag{4.88}$$

where the infimum is taken over all unit tangent vectors $\mathbf{V}$ in $T_{\mathbf{x}}M$; i.e., $\|\mathbf{V}\| = 1$. To see this, we first note that since the unit sphere $S^{n-1}_{\mathbf{x}}$ in $T_{\mathbf{x}}M$ is compact, a minimizer $\mathbf{V}_1$ of the continuous functional $\mathbf{V} \mapsto \mathrm{II}(\mathbf{V}, \mathbf{V})$ exists. That is, $\|\mathbf{V}_1\| = 1$ and

$$\mathrm{II}(\mathbf{V}_1, \mathbf{V}_1) = \inf_{\mathbf{V}} \mathrm{II}(\mathbf{V}, \mathbf{V}) =: \lambda_1.$$

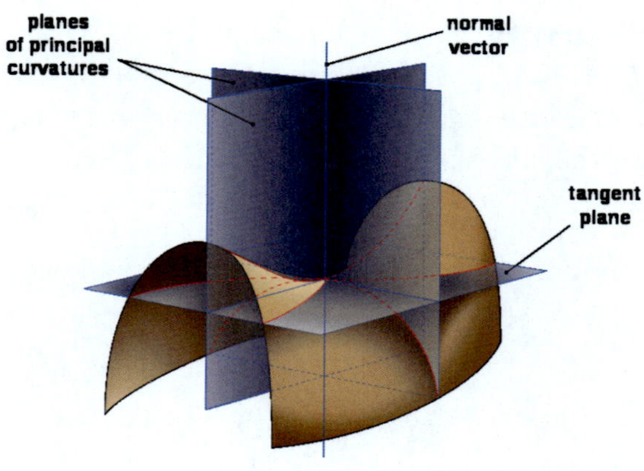

Figure 4.4.3. Illustrating the principal curvatures of a surface in $\mathbb{R}^3$. Credit: By Eric Gaba (Sting) – Based upon a drawing in a book, CC BY-SA 3.0, `https://commons.wikimedia.org/w/index.php?curid=1325452`

Now we will show that $\mathbf{V}_1$ is an eigenvector of the Weingarten map L with eigenvalue equal to λ_1, which in turn implies that $\kappa_1 = \lambda_1$. Suppose that $\mathbf{U} \in T_{\mathbf{x}}M$ satisfies $\mathbf{U} \cdot \mathbf{V}_1 = 0$. Since $\mathbf{V}_1$ is a critical point of the functional, we have

$$\begin{aligned} 0 &= \frac{d}{dt}\bigg|_{t=0} \mathrm{II}(\mathbf{V}_1 + t\mathbf{U}, \mathbf{V}_1 + t\mathbf{U}) \\ &= 2\,\mathrm{II}(\mathbf{V}_1, \mathbf{U}) \\ &= 2\,\mathrm{L}(\mathbf{V}_1) \cdot \mathbf{U}, \end{aligned}$$

where we used that the second fundamental form II is symmetric. Thus,

$$\mathrm{L}(\mathbf{V}_1) = \lambda \mathbf{V}_1$$

for some $\lambda \in \mathbb{R}$. Finally,

$$\lambda = \lambda \|\mathbf{V}_1\|^2 = \mathrm{L}(\mathbf{V}_1) \cdot \mathbf{V}_1 = \mathrm{II}(\mathbf{V}_1, \mathbf{V}_1) = \lambda_1,$$

so that λ_1 is the eigenvalue associated to the eigenvector $\mathbf{V}_1$. It is easy to see that λ_1 is the smallest eigenvalue; hence $\lambda_1 = \kappa_1$.

Similarly, one can show that

$$\kappa_2 = \inf_{\mathbf{V} \perp \mathbf{V}_1} \mathrm{II}(\mathbf{V}, \mathbf{V}), \tag{4.89}$$

where the infimum is taken over all unit tangent vectors $\mathbf{V}$ in $T_{\mathbf{x}}M$ perpendicular to $\mathbf{V}_1$. One can show that κ_2 is an eigenvalue of L.

Suppose that $\mathbf{V}_2$ is an eigenvector associated to the eigenvalue κ_2. Then,

$$\kappa_1 \mathbf{V}_1 \cdot \mathbf{V}_2 = \mathrm{L}(\mathbf{V}_1) \cdot \mathbf{V}_2 = \mathbf{V}_1 \cdot \mathrm{L}(\mathbf{V}_2) = \kappa_2 \mathbf{V}_1 \cdot \mathbf{V}_2.$$

Thus, if $\kappa_1 \neq \kappa_2$ are distinct, then $\mathbf{V}_1 \cdot \mathbf{V}_2 = 0$. On the other hand, if $\kappa_1 = \kappa_2$, then we can still choose a unit eigenvector $\mathbf{V}_2$ associated to κ_2 satisfying $\mathbf{V}_1 \cdot \mathbf{V}_2 = 0$.

By continuing in this way, one can show that for each $2 \leq j \leq n$, if $\mathbf{V}_1, \ldots, \mathbf{V}_{j-1}$ are orthonormal eigenvectors associated to $\kappa_1, \ldots, \kappa_{j-1}$, respectively, then

$$\kappa_j = \inf_{\mathbf{V} \perp \{\mathbf{V}_1, \ldots, \mathbf{V}_{j-1}\}} \mathrm{II}(\mathbf{V}, \mathbf{V}) \tag{4.90}$$

and there exists an associated unit eigenvector $\mathbf{V}_j$ perpendicular to the span of $\{\mathbf{V}_1, \ldots, \mathbf{V}_{j-1}\}$.

4.4.7. Mean curvature and Gauss curvature.

The sum of the principal curvatures is called the **mean curvature** H:

$$H := \kappa_1 + \cdots + \kappa_n. \tag{4.91}$$

This is the same as the trace of the Weingarten map:

$$H = \mathrm{trace}(\mathrm{L}). \tag{4.92}$$

Figure 4.4.4. From left to right: Riemannian surfaces with negative, zero, and positive Gauss curvatures. Credit: By Nicoguaro – Own work, CC0, `https://commons.wikimedia.org/w/index.php?curid=46262902`

Definition 4.22. If a hypersurface M^n has zero mean curvature, i.e., $H \equiv 0$, then we say that M^n is a **minimal hypersurface**.

The reason for the term 'minimal' is that mean curvature zero hypersurfaces locally minimize area; see Chapter 12 below.

The product of the principal curvatures is called the **Gauss curvature** K:

$$K := \kappa_1 \cdots \kappa_n. \tag{4.93}$$

This is the same as the determinant of the Weingarten map:

$$K = \det(\mathrm{L}). \tag{4.94}$$

In Figure 4.4.4, the surfaces at the left, middle, and right have principal curvatures $\kappa_1 \leq \kappa_2$ satisfying $\kappa_1 < 0 < \kappa_2$, $\kappa_1 = 0 < \kappa_2$, and $0 < \kappa_1 = \kappa_2$ and hence Gauss curvatures satisfying $K < 0$, $K = 0$, and $K > 0$, respectively.

If M^2 is a minimal surface in $\mathbb{R}^3$, then $\kappa_2 = -\kappa_1$ at each point in M^2. Hence the Gauss curvature of M^2 is nonpositive

$$K = \kappa_1 \kappa_2 = -\kappa_1^2 \leq 0.$$

4.4.8. Second fundamental form in local coordinates. Local coordinates provide a convenient way to calculate geometric invariants of hypersurfaces such as the second fundamental form, mean curvature, and Gauss curvature.

Let $F : \mathcal{U} \to M^n$ be a local parametrization of M^n. The components of the second fundamental form are defined by

$$\mathrm{II}_{ij} := \mathrm{II}\left(\frac{\partial F}{\partial u^i}, \frac{\partial F}{\partial u^j}\right). \tag{4.95}$$

By this and (4.82), we have

$$\mathrm{II}_{ij} = -D_{\frac{\partial F}{\partial u^i}} \frac{\partial F}{\partial u^j} \cdot N = -\frac{\partial^2 F}{\partial u^i \partial u^j} \cdot N. \tag{4.96}$$

Note that since $N \cdot \frac{\partial F}{\partial u^j} = 0$ we also have

$$\mathrm{II}_{ij} = D_{\frac{\partial F}{\partial u^i}} N \cdot \frac{\partial F}{\partial u^j} = \frac{\partial N}{\partial u^i} \cdot \frac{\partial F}{\partial u^j}. \tag{4.97}$$

The last equality is because for $u \in \mathcal{U} \subset \mathbb{R}^n$,

$$(D_{\frac{\partial F}{\partial u^i}} N)(F(u)) = \frac{\partial}{\partial u^i}(N \circ F)(u) =: \frac{\partial N}{\partial u^i}(u).$$

We remark that similarly to (4.43) for g in terms of g_{ij} and for the same reason, we have

$$\mathrm{II} = \sum_{i,j=1}^{n} \mathrm{II}_{ij}\, dx^i \otimes dx^j. \tag{4.98}$$

Now suppose that the matrices (g_{ij}) and (II_{ij}) are diagonal (with respect to the same coordinate system). That is, $g_{ij} = \mathrm{II}_{ij} = 0$ if $i \neq j$. Then the tangent vectors $e_i := \frac{\frac{\partial F}{\partial u^i}}{\left\|\frac{\partial F}{\partial u^i}\right\|}$ are principal directions with associated principal curvatures

$$\kappa_i = \frac{\mathrm{II}_{ii}}{g_{ii}}. \tag{4.99}$$

Consider the surface of revolution M^2 in $\mathbb{R}^3$ parametrized by the map $F : [a, b] \times [0, 2\pi] \to \mathbb{R}^3$ defined by (4.39); that is,

$$F(u^1, u^2) = \big(f(u^1)\cos(u^2), f(u^1)\sin(u^2), u^1\big) \tag{4.100}$$

for $u^1 \in [a, b]$ and $u^2 \in [0, 2\pi]$. Using (4.40) and (4.41), namely,

$$\frac{\partial F}{\partial u^1} = \big(f'(u^1)\cos(u^2), f'(u^1)\sin(u^2), 1\big)$$

and

$$\frac{\partial F}{\partial u^2} = \big(-f(u^1)\sin(u^2), f(u^1)\cos(u^2), 0\big),$$

we compute that

$$\frac{\partial^2 F}{\partial (u^1)^2} = \big(f''(u^1)\cos(u^2), f''(u^1)\sin(u^2), 0\big), \tag{4.101}$$

$$\text{(4.102)} \qquad \frac{\partial^2 F}{\partial (u^2)^2} = \big(-f(u^1)\cos(u^2), -f(u^1)\sin(u^2), 0\big),$$

$$\text{(4.103)} \qquad \frac{\partial^2 F}{\partial u^1 \partial u^2} = \frac{\partial^2 F}{\partial u^2 \partial u^1} = \big(-f'(u^1)\sin(u^2), f'(u^1)\cos(u^2), 0\big).$$

A unit normal vector field to M^2 is given by

$$\text{(4.104)} \qquad N := N_{F(\mathbf{u})} := \frac{\big(\cos(u^2), \sin(u^2), -f'(u^1)\big)}{\sqrt{1+(f'(u^1))^2}}.$$

We compute the Weingarten map as

$$\text{(4.105)} \qquad \begin{aligned} \mathrm{L}\left(\frac{\partial F}{\partial u^2}\right) &= D_{\frac{\partial F}{\partial u^2}} N = \frac{\partial N}{\partial u^2} \\ &= \frac{\big(-\sin(u^2), \cos(u^2), 0\big)}{\sqrt{1+(f'(u^1))^2}} \\ &= \frac{1}{f(u^1)\sqrt{1+(f'(u^1))^2}} \frac{\partial F}{\partial u^2}, \end{aligned}$$

where the last equality is true by (4.41), and

$$\text{(4.106)} \qquad \mathrm{L}\left(\frac{\partial F}{\partial u^1}\right) = D_{\frac{\partial F}{\partial u^1}} N = \frac{\partial N}{\partial u^1} = \frac{-f''(u^1)}{\big(1+(f'(u^1))^2\big)^{3/2}} \frac{\partial F}{\partial u^1}.$$

Thus, the principal curvatures are

$$\text{(4.107a)} \qquad \kappa_1 = \frac{-f''(u^1)}{\big(1+(f'(u^1))^2\big)^{3/2}},$$

$$\text{(4.107b)} \qquad \kappa_2 = \frac{1}{f(u^1)\sqrt{1+(f'(u^1))^2}},$$

with associated principal directions given by the vectors $e_1 := \frac{\partial F}{\partial u^1} / \left\| \frac{\partial F}{\partial u^1} \right\|$ and $e_2 := \frac{\partial F}{\partial u^2} / \left\| \frac{\partial F}{\partial u^2} \right\|$, respectively.

Remark 4.23. You may be wondering why $D_{\frac{\partial F}{\partial u^i}} N = \frac{\partial N}{\partial u^i}$. To see this in more detail, let $\mathbf{u} \in \mathcal{U}$ and define $\gamma_i(t) = F(\mathbf{u} + t\mathbf{e}_i)$. Then $\gamma_i'(0) = \frac{\partial F}{\partial u^i}(\mathbf{u})$. Thus,

$$\left(D_{\frac{\partial F}{\partial u^i}(\mathbf{u})} N\right)_{F(\mathbf{u})} = \left.\frac{d}{dt}\right|_{t=0} N\big(F(\mathbf{u}+t\mathbf{e}_i)\big) = \frac{\partial}{\partial u^i}(N \circ F)(\mathbf{u}) =: \frac{\partial N}{\partial u^i}(\mathbf{u}).$$

Alternatively, by (4.96) we find that the components of the second fundamental form of a surface of revolution are given by

$$\mathrm{II}_{11} = \frac{-f''(u^1)}{\sqrt{1+(f'(u^1))^2}}, \tag{4.108a}$$

$$\mathrm{II}_{22} = \frac{f(u^1)}{\sqrt{1+(f'(u^1))^2}}, \tag{4.108b}$$

$$\mathrm{II}_{12} = \mathrm{II}_{21} = 0. \tag{4.108c}$$

Thus, by (4.42) and (4.99), we again have

$$\kappa_1 = \frac{\mathrm{II}_{11}}{g_{11}} = \frac{-f''(u^1)}{\left(1+(f'(u^1))^2\right)^{3/2}}, \tag{4.109a}$$

$$\kappa_2 = \frac{\mathrm{II}_{22}}{g_{22}} = \frac{1}{f(u^1)\sqrt{1+(f'(u^1))^2}}. \tag{4.109b}$$

As an example, if $f(u^1) = \sqrt{1-(u^1)^2}$, which yields the unit 2-sphere, then one calculates that $\kappa_1 = \kappa_2 = 1$. Indeed, in this case we have $1+(f'(u^1))^2 = \frac{1}{1-(u^1)^2}$, $f''(u^1) = -\frac{1}{(1-(u^1)^2)^{3/2}}$, and the calculation follows easily.

Example 4.24. *The catenoid in Example* 1.7 *is the case where* $f(u^1) = \cosh(u^1)$. *In this case,*

$$f(u^1) = f''(u^1) = \sqrt{1+(f'(u^1))^2}, \tag{4.110}$$

and hence (4.107) *yields*

$$\kappa_2 = -\kappa_1 = \cosh^{-2}(u^1). \tag{4.111}$$

See Exercise 4.21. *Thus, the catenoid is a minimal surface.*

Another nice class of examples is that of graphs. Recall from Example 4.15, for the graphic parametrization F of the graph G of a function f, we have

$$\frac{\partial F}{\partial u^i} = \mathbf{e}_i + \frac{\partial f}{\partial u^i}\mathbf{e}_{n+1}.$$

Taking a second partial derivative, we have

$$\frac{\partial^2 F}{\partial u^j \partial u^i} = \frac{\partial^2 f}{\partial u^j \partial u^i}\mathbf{e}_{n+1}.$$

Choosing the upward pointing unit normal vector field to G, we have

$$N = \frac{-\nabla f + \mathbf{e}_{n+1}}{\sqrt{\|\nabla f\|^2+1}}. \tag{4.112}$$

Thus, by (4.96), we have that the components of the second fundamental form are given by

$$\mathrm{II}_{ij} = -\frac{\partial^2 F}{\partial u^i \partial u^j} \cdot N = -\frac{1}{\sqrt{\|\nabla f\|^2 + 1}} \frac{\partial^2 f}{\partial u^i \partial u^j}. \tag{4.113}$$

Example 4.25. *Recall from Example 1.4, Scherk's first surface is the graph of the two-variable function*

$$f(u^1, u^2) = \ln \frac{\cos(u^2)}{\cos(u^1)}.$$

We compute that

$$\frac{\partial f}{\partial u^i} = -(-1)^i \tan(u^i) \quad \textit{for } i = 1, 2.$$

Hence, by (4.36), we have

$$\begin{aligned} \mathrm{I}_{ij} &= \delta_{ij} + \frac{\partial f}{\partial u^i} \frac{\partial f}{\partial u^j} \\ &= \delta_{ij} + (-1)^{i+j} \tan(u^i) \tan(u^j). \end{aligned}$$

By Exercise 4.22, the components of the inverse matrix $(\mathrm{I}^{ij}) := ((\mathrm{I}^{-1})_{ij})$ *of the matrix* (I_{ij}) *are given by*

$$\mathrm{I}^{ij} = \delta_{ij} - \frac{(-1)^{i+j} \tan(u^i) \tan(u^j)}{1 + \tan^2(u^1) + \tan^2(u^2)},$$

where $\|\nabla f\|^2 = \tan^2(u^1) + \tan^2(u^2)$.

Taking a second derivative of f*, we have*

$$\frac{\partial^2 f}{\partial u^j \partial u^i} = -(-1)^i \sec^2(u^i)\, \delta_{ij} \quad \textit{for } i = 1, 2.$$

In particular, the Hessian of f *is diagonal, and hence* (II_{ij}) *is also diagonal. The mean curvature of Scherk's first surface is equal to (see Exercise 4.23 for the first equality)*

$$\begin{aligned} H &= \sum_{i,j=1}^{2} \mathrm{I}^{ij}\, \mathrm{II}_{ij} = \sum_{i=1}^{2} \mathrm{I}^{ii}\, \mathrm{II}_{ii} \\ &= \sum_{i=1}^{2} \left(1 - \frac{\tan^2(u^i)}{1 + \tan^2(u^1) + \tan^2(u^2)}\right) \frac{(-1)^i \sec^2(u^i)}{\sqrt{1 + \tan^2(u^1) + \tan^2(u^2)}} \\ &= \left(1 + \tan^2(u^1) + \tan^2(u^2)\right)^{-3/2} \sum_{i=1}^{2} \sec^2(u^{3-i})(-1)^i \sec^2(u^i) \\ &= 0, \end{aligned}$$

where we used the identity:

$$1+\tan^2(u^1)+\tan^2(u^2)=1+\tan^2(u^i)+\tan^2(u^{3-i})=\tan^2(u^i)+\sec^2(u^{3-i})$$

for $i=1,2$. Hence, Scherk's first surface is a minimal surface.

4.4.9. Second fundamental form of a submanifold. Now suppose that M^n is a smooth submanifold of $\mathbb{R}^N$. In this more general setting, the **second fundamental form** at a point $\mathbf{x}\in M^n$ is defined by

$$B=B_{\mathbf{x}}:T_{\mathbf{x}}M\times T_{\mathbf{x}}M\to N_{\mathbf{x}}M, \tag{4.114}$$

where

$$B(\mathbf{V},\mathbf{W}):=(D_{\mathbf{V}}W)^{\perp} \tag{4.115}$$

and $N_{\mathbf{x}}M=(T_{\mathbf{x}}M)^{\perp}$ is the normal space at $\mathbf{x}$. For submanifolds of general codimension, we use B to denote the second fundamental form. For hypersurfaces, i.e., when the codimension equals 1, we use II to denote the second fundamental form, where $B=-\mathrm{II}\,N$ by (4.82). That is, when M^n is a hypersurface, we may use a choice of unit normal vector field N to identify each normal space $N_{\mathbf{x}}M$ with $\mathbb{R}$. Thus, in this case, B is equivalent to II.

Note that, to define B by (4.115), we need to extend $\mathbf{W}$ to a vector field W in a neighborhood of $\mathbf{x}$, but the right-hand side is independent of this extension. This fact is a consequence of the property that B is symmetric:

$$B(\mathbf{V},\mathbf{W})=B(\mathbf{W},\mathbf{V}). \tag{4.116}$$

To see this property, we compute that for any vector fields V and W,

$$B(V,W)-B(W,V)=(D_VW-D_WV)^{\perp}=[V,W]^{\perp}=0,$$

where the last equality is because the Lie bracket of tangential vector fields is tangential. We remark that we can think of the second fundamental form B as a covariant symmetric 2-tensor on M^n with values in the normal bundle NM.

Given $\mathbf{x}\in M^n$, the **mean curvature vector** at x is

$$\vec{H}(\mathbf{x}):=\operatorname{trace}(B_{\mathbf{x}})=\sum_{i=1}^{n}B_{\mathbf{x}}(e_i,e_i), \tag{4.117}$$

where $\{e_i\}_{i=1}^n$ is an orthonormal tangent frame field at $\mathbf{x}$.

If $N=n+1$, i.e., M^n is a hypersurface, and if N is a choice of a smooth unit normal vector field to M^n, then

$$-\mathrm{II}(V,W)N=B(V,W)\qquad\text{and}\qquad -HN=\vec{H}. \tag{4.118}$$

4.5. Induced covariant derivative on a hypersurface

A natural notion of the differentiation of vector fields on a submanifold may be defined by projecting the Euclidean covariant derivative onto the tangent spaces of the submanifold.

4.5.1. Example of great circles on spheres. Let us look at the non-trivial hypersurfaces that feel most familiar to us: spheres. In particular, consider the unit 2-sphere S^2 in $\mathbb{R}^3$. Let $\mathbf{N} = (0,0,1)$ and $\mathbf{S} = (0,0,-1)$ be the north and south poles, respectively. Define a unit vector field U on $S^2 \setminus \{\mathbf{N} \cup \mathbf{S}\}$ by

$$U_{\mathbf{x}} := \left(-\frac{x^1x^3}{\sqrt{1-(x^3)^2}}, -\frac{x^2x^3}{\sqrt{1-(x^3)^2}}, \sqrt{1-(x^3)^2} \right), \tag{4.119}$$

where $\mathbf{x} = (x^1, x^2, x^3)$.

We claim that the integral curves to the vector field U are the great half-circles from $\mathbf{S}$ to $\mathbf{N}$. To see this, let $\mathbf{Y} = (y^1, y^2, 0)$ be a unit vector tangent to S^2 at $\mathbf{S}$, so that $\|\mathbf{Y}\|^2 = (y^1)^2 + (Y^2)^2 = 1$. Define the path $\gamma : [0,\pi] \to S^2$ by

$$\gamma(s) := \cos(s)\mathbf{S} + \sin(s)\mathbf{Y} = \big(\sin(s)Y^1, \sin(s)Y^2, -\cos(s)\big). \tag{4.120}$$

Note that

$$\gamma'(s) := -\sin(s)\mathbf{S} + \cos(s)\mathbf{Y} = \big(\cos(s)Y^1, \cos(s)Y^2, \sin(s)\big). \tag{4.121}$$

Since $\mathbf{Y} \cdot \mathbf{S} = 0$, we have that $\gamma(s)$ and $\gamma'(s)$ are both unit vectors. We also see that γ is a great half-circle joining $\mathbf{S}$ to $\mathbf{N}$. Now let

$$(x^1, x^2, x^3) := \big(\sin(s)Y^1, \sin(s)Y^2, -\cos(s)\big).$$

Note that $\sin(s) \geq 0$ since $s \in [0,\pi]$. Thus,

$$\sin(s) = \sqrt{1-\cos^2(s)} = \sqrt{1-(x^3)^2}.$$

This implies that

$$Y^i = \frac{1}{\sin(s)}x^i = \frac{x^i}{\sqrt{1-(x^3)^2}} \quad \text{for } i = 1,2.$$

Since $\cos(s) = -x^3$, we conclude that

$$\begin{aligned}\gamma'(s) &= \big(\cos(s)Y^1, \cos(s)Y^2, \sin(s)\big) \\ &= \left(-\frac{x^1x^3}{\sqrt{1-(x^3)^2}}, -\frac{x^2x^3}{\sqrt{1-(x^3)^2}}, \sqrt{1-(x^3)^2} \right) \\ &= U_{\mathbf{x}}.\end{aligned}$$

Now let us compute the Euclidean covariant derivative

$$\begin{aligned} D_{\gamma'(s)}\gamma'(s) &= \gamma''(s) \\ &= \frac{d}{ds}\big(\cos(s)Y^1, \cos(s)Y^2, \sin(s)\big) \\ &= \big(-\sin(s)Y^1, -\sin(s)Y^2, \cos(s)\big) \\ &= -N_{\gamma(s)}. \end{aligned} \tag{4.122}$$

Thus, great half-circles on S^2, which are length minimizing geodesics, have that their Euclidean covariant derivative satisfies the property that the tangential component of $D_{\gamma'(s)}\gamma'(s)$ vanishes.

Remark 4.26. Alternatively, one can compute $D_{U_{\mathbf{x}}}U$; but this is more complicated.

4.5.2. Induced covariant derivative (a.k.a. Levi-Civita connection) of a submanifold. Given a submanifold, we will now consider the tangential component of the Euclidean covariant derivative to define covariant differentiation of vector fields.

Let M^n be an embedded smooth submanifold in $\mathbb{R}^N$. Given a tangent vector $\mathbf{V} \in T_{\mathbf{x}}M$ and a vector field W, the **induced covariant derivative** is defined by

$$\nabla_{\mathbf{V}}W := (D_{\mathbf{V}}W)^T \in T_{\mathbf{x}}M, \tag{4.123}$$

where $\cdot^T$ denotes the tangential component of a Euclidean vector. We also call ∇ the **Levi-Civita connection** of the metric $g = \mathrm{I}$. Covariant differentiation is the natural way to differentiate a vector field (as opposed to a function) on a submanifold.

By (4.115), we have

$$D_V W = \nabla_V W + B(V, W), \tag{4.124}$$

where B denotes the second fundamental form of the submanifold M^n.

If M^n is a hypersurface, i.e., $N = n + 1$, then by (4.82) we have that

$$\begin{aligned} \nabla_{\mathbf{V}}W &= D_{\mathbf{V}}W - (N \cdot D_{\mathbf{V}}W)N_{\mathbf{x}} \\ &= D_{\mathbf{V}}W + \mathrm{II}(\mathbf{V}, W_{\mathbf{x}})N_{\mathbf{x}}, \end{aligned} \tag{4.125}$$

where N is a continuous choice of a unit normal vector field (which determines the sign of II).

Example 4.27. *For the unit n-sphere S^n in $\mathbb{R}^{n+1}$, we have by* (4.125) *that*

$$\nabla_V W = D_V W + (V \cdot W)N \tag{4.126}$$

for vector fields V, W on S^n, where $N_{\mathbf{x}} = \mathbf{x}$ is the unit outward normal vector field to S^n.

4.5.3. Properties of the Levi-Civita connection.

Lemma 4.28. *Let M^n be a submanifold in $\mathbb{R}^N$. The Levi-Civita connection has the following* **linear connection** *properties:*

(1) $\nabla_{cV_1+V_2}W = c\nabla_{V_1}W + \nabla_{V_2}W$.

(2) $\nabla_V(cW_1 + W_2) = c\nabla_V W_1 + \nabla_V W_2$.

(3) $\nabla_V(fW) = V(f)W + f\nabla_V W$.

The first two properties are linearity properties, whereas the third property is a product (Leibniz) rule.

If U, V, W are vector fields on a smooth embedded submanifold M^n, then

$$U(V \cdot W) = \nabla_U V \cdot W + V \cdot \nabla_U W, \tag{4.127}$$

where ∇ is the Levi-Civita connection. That is,

$$U\big(\mathrm{I}(V, W)\big) = \mathrm{I}(\nabla_U V, W) + \mathrm{I}(V, \nabla_U W). \tag{4.128}$$

This is known as the **metric compatibility** property or **product rule** for the Levi-Civita connection. As we will see in (7.25) below, this says that the metric is parallel, i.e., has zero covariant derivative.

For any vector fields V, W on a smooth embedded submanifold M^n,

$$\nabla_V W - \nabla_W V = [V, W], \tag{4.129}$$

where $[V, W]$ is the Lie bracket of V and W; see Exercise 4.25. Because of this equation, we say that the Levi-Civita connection is **torsion-free**.

For any smooth manifold, there is the notion of 'Lie derivative', which will be discussed in §6.5.2 below.

4.5.4. Parallel vector fields and geodesics. Let $\gamma : [a, b] \to M^n$ be a smooth path. Let V be a tangent vector field along γ, so that $V(t) \in T_{\gamma(t)}M$ for $t \in [a, b]$. Then the covariant derivative

$$\nabla_{\gamma'(t)}V := (D_{\gamma'(t)}V)^T$$

is well-defined.

Definition 4.29. We say that V is **parallel** if $\nabla_{\gamma'(t)}V = 0$ for all $t \in [a, b]$.

By the existence and uniqueness theorem for systems of linear first-order ordinary differential equations, given $V_a \in T_{\gamma(a)}M$, there exists a unique parallel vector field V along γ such that $V(a) = V_a$.

Suppose that V and W are vector fields along a path γ. Then we have

$$\gamma'(t)\langle V(t), W(t)\rangle = \langle \nabla_{\gamma'(t)}V(t), W(t)\rangle + \langle V(t), \nabla_{\gamma'(t)}W(t)\rangle. \tag{4.130}$$

In particular, if V and W are both parallel, then $\langle V(t), W(t)\rangle$ is constant.

Example 4.30. *If γ is a unit-speed parametrization of a great circle on S^2, then*

$$\nabla_{\gamma'(s)}\gamma'(s) = \mathbf{0}$$

by (4.122).

Definition 4.31. Let $\gamma : I \to M^n$ be a path with $\gamma'(t) \neq \mathbf{0}$ for all $t \in I$. We say that γ is a **geodesic** if

$$\nabla_{\gamma'(t)}\left(\frac{\gamma'(t)}{\|\gamma'(t)\|}\right) = \mathbf{0} \tag{4.131}$$

for all $t \in I$.

Equivalently, if T is the unit tangent vector field of γ, γ is a geodesic if

$$\mathbf{0} = \nabla_T T = (D_T T)^{\perp}.$$

Suppose γ is parametrized by arc length s (i.e., is unit speed). Then

$$D_T T = D_{\gamma'}\gamma' = \gamma''.$$

To see the last equality, suppose that $V(s)$ is a vector field along $\gamma(s)$. That is, for each s, $V(s)$ is in $T_{\gamma(s)}\mathbb{R}^{n+1}$. We have

$$D_{\gamma'(s)}V(s) = \frac{d}{ds}V_{\gamma(s)} = \frac{d}{ds}V(s) = V'(s).$$

In particular, if $V(s) = \gamma'(s)$, then

$$D_{\gamma'(s)}\gamma'(s) = \gamma''(s).$$

By Example 4.30, (unit-speed) parametrizations of great circles on S^2 are geodesics. This easily generalizes to S^n; see Exercise 4.26.

Remark 4.32. Let $C^\infty(TM)$ denote the space of smooth vector fields on M^n.[2] Remarkably, the Levi-Civita connection $(V, W) \mapsto \nabla_V W$ is the unique function from $C^\infty(TM) \times C^\infty(TM)$ to $C^\infty(TM)$ satisfying the linear connection properties of Lemma 4.28, the torsion-free property (4.129), and the metric compatibility property (4.128).

4.5.5. Uniqueness of the Levi-Civita connection.

Denote $g = \mathrm{I}$. Let ∇ denote the Levi-Civita connection of g defined by (4.123). By the metric compatibility condition (4.128) and its cyclic permutations, we have

$$\begin{aligned}
U\big(g(V,W)\big) &= g(\nabla_U V, W) + g(V, \nabla_U W),\\
V\big(g(U,W)\big) &= g(\nabla_V U, W) + g(U, \nabla_V W)\\
&= g(\nabla_U V, W) - g([U,V], W) + g(U, \nabla_V W),\\
-W\big(g(U,V)\big) &= -g(\nabla_W U, V) - g(U, \nabla_W V).
\end{aligned}$$

[2]This notation is a special case of the notion for smooth sections of vector bundles (vector fields are sections of TM); see Definition 10.14 below.

By summing these three formulas, we obtain the following **Koszul formula**:

$$(4.132)\quad \begin{aligned} 2g\left(\nabla_U V, W\right) = &\; U\left(g\left(V, W\right)\right) + V\left(g\left(U, W\right)\right) - W\left(g\left(U, V\right)\right) \\ &+ g\left(\left[U, V\right], W\right) - g\left(\left[U, W\right], V\right) - g\left(\left[V, W\right], U\right), \end{aligned}$$

where we used the torsion-free condition (e.g., $\nabla_U W - \nabla_W U = [U, W]$). This formula uniquely determines $\nabla_U V$ for any vector fields U and V. In particular, if $\{e_i\}_{i=1}^n$ is an orthonormal frame, then we may substitute (4.132) into

$$\nabla_U V = \sum_{i=1}^n g\left(\nabla_U V, e_i\right) e_i$$

to obtain a formula for $\nabla_U V$.

Summarizing, we have shown the following.

Theorem 4.33. *The Levi-Civita connection is the unique linear connection that is both torsion-free and compatible with the Riemannian metric g.*

4.5.6. Levi-Civita connection in local coordinates: Christoffel symbols. In Euclidean space $\mathbb{R}^n$ we have that the Lie brackets of the standard basis vector fields vanish:

$$(4.133)\qquad [\mathbf{e}_i, \mathbf{e}_j] = 0 \quad \text{for all } 1 \leq i, j \leq n.$$

Let $F : \mathcal{U} \to M^n$ parametrize a relatively open subset of a submanifold of $\mathbb{R}^{n+1}$. We have the basis of tangent vector fields $\frac{\partial F}{\partial u^i} = dF(\mathbf{e}_i)$, $1 \leq i \leq n$. The Lie bracket has the following naturality property under the push-forward by derivatives:

$$(4.134)\qquad [dF(V), dF(W)] = dF([V, W])$$

for any vector fields V and W on $\mathcal{U}$. In particular, we have

$$(4.135)\qquad \left[\frac{\partial F}{\partial u^i}, \frac{\partial F}{\partial u^j}\right] = [dF(\mathbf{e}_i), dF(\mathbf{e}_j)] = dF([\mathbf{e}_i, \mathbf{e}_j]) = dF(\mathbf{0}) = \mathbf{0}.$$

Now we choose in (4.132) the vector fields to be coordinate basis tangent vector fields. Namely, let $U = \frac{\partial F}{\partial u^i}$, $V = \frac{\partial F}{\partial u^j}$, and $W = \frac{\partial F}{\partial u^\ell}$. Denote $X \cdot Y := g(X, Y)$. Then (4.132) yields

$$(4.136)\qquad \begin{aligned} &2g\left(\nabla_{\frac{\partial F}{\partial u^i}} \frac{\partial F}{\partial u^j}, \frac{\partial F}{\partial u^\ell}\right) \\ &= \frac{\partial F}{\partial u^i}\left(\frac{\partial F}{\partial u^j} \cdot \frac{\partial F}{\partial u^\ell}\right) + \frac{\partial F}{\partial u^j}\left(\frac{\partial F}{\partial u^i} \cdot \frac{\partial F}{\partial u^\ell}\right) - \frac{\partial F}{\partial u^\ell}\left(\frac{\partial F}{\partial u^i} \cdot \frac{\partial F}{\partial u^j}\right) \end{aligned}$$

since all of the Lie brackets vanish by (4.135). Using the notations $\frac{\partial}{\partial x^i} := \frac{\partial F}{\partial u^i}$ and $g_{ij} = g\left(\frac{\partial}{\partial x^i}, \frac{\partial}{\partial x^j}\right) = \frac{\partial}{\partial x^i} \cdot \frac{\partial}{\partial x^j}$, this says that

$$(4.137)\qquad 2g\left(\nabla_{\frac{\partial}{\partial x^i}} \frac{\partial}{\partial x^j}, \frac{\partial}{\partial x^\ell}\right) = \frac{\partial}{\partial x^i} g_{j\ell} + \frac{\partial}{\partial x^j} g_{i\ell} - \frac{\partial}{\partial x^\ell} g_{ij}.$$

Now we define the **Christoffel symbols** Γ_{ij}^k by

$$\nabla_{\frac{\partial}{\partial x^i}} \frac{\partial}{\partial x^j} =: \sum_{k=1}^{n} \Gamma_{ij}^k \frac{\partial}{\partial x^k}. \tag{4.138}$$

By (4.137), we have

$$2\sum_{k=1}^{n} \Gamma_{ij}^k \, g_{k\ell} = \frac{\partial}{\partial x^i} g_{j\ell} + \frac{\partial}{\partial x^j} g_{i\ell} - \frac{\partial}{\partial x^\ell} g_{ij}.$$

Therefore, we obtain the following well-known formula for the Christoffel symbols:

$$\Gamma_{ij}^k = \frac{1}{2}\sum_{\ell=1}^{n} g^{k\ell}\left(\frac{\partial}{\partial x^i} g_{j\ell} + \frac{\partial}{\partial x^j} g_{i\ell} - \frac{\partial}{\partial x^\ell} g_{ij}\right). \tag{4.139}$$

We further observe that since ∇ and D differ by a normal vector,

$$\begin{aligned} \sum_{k=1}^{n} \Gamma_{ij}^k \, g_{k\ell} &= \left\langle \nabla_{\frac{\partial F}{\partial u^i}} \frac{\partial F}{\partial u^j}, \frac{\partial F}{\partial u^\ell} \right\rangle \\ &= \left\langle D_{\frac{\partial F}{\partial u^i}} \frac{\partial F}{\partial u^j}, \frac{\partial F}{\partial u^\ell} \right\rangle \\ &= \left\langle \frac{\partial^2 F}{\partial u^i \partial u^j}, \frac{\partial F}{\partial u^\ell} \right\rangle. \end{aligned} \tag{4.140}$$

Given vector fields V and W on M^n, we can compute the covariant derivative $\nabla_V W$ using local coordinates in terms of the Christoffel symbols: For tangent vector fields V and W on a parametrized submanifold M^n in $\mathbb{R}^N$

$$\nabla_V W = \sum_{i,j=1}^{n} V^i \frac{\partial W^j}{\partial x^i} \frac{\partial}{\partial x^j} + \sum_{i,j,k=1}^{n} V^i W^j \Gamma_{ij}^k \frac{\partial}{\partial x^k}, \tag{4.141}$$

where the Γ_{ij}^k are the Christoffel symbols; see Exercise 4.27.

In particular, recall that if $\gamma : I \to M^n$ is a path, then $\gamma'(t)$ is a vector field along $\gamma(t)$. We have

$$\gamma'(t) = \sum_{j=1}^{n} \frac{d\gamma^j}{dt}(t) \frac{\partial}{\partial x^j}, \tag{4.142}$$

where $\gamma^j := x^j \circ \gamma$ and where $\frac{\partial}{\partial x^j}$ is evaluated at $\gamma(t)$. We compute that

$$\nabla_{\gamma'(t)}\gamma'(t) = \nabla_{\gamma'(t)}\left(\sum_{j=1}^n \frac{d\gamma^j}{dt}(t)\frac{\partial}{\partial x^j}\right) \tag{4.143}$$
$$= \sum_{j=1}^n \gamma'(t)\left(\frac{d\gamma^j}{dt}\right)\frac{\partial}{\partial x^j} + \sum_{j=1}^n \frac{d\gamma^j}{dt}(t)\nabla_{\gamma'(t)}\frac{\partial}{\partial x^j}$$
$$= \sum_{j=1}^n \frac{d^2\gamma^j}{dt^2}\frac{\partial}{\partial x^j} + \sum_{i,j,k=1}^n \frac{d\gamma^j}{dt}\frac{d\gamma^i}{dt}\Gamma_{ij}^k\frac{\partial}{\partial x^k},$$

where we used the product rule for the Levi-Civita connection. We may rewrite this as

$$\nabla_{\gamma'(t)}\gamma'(t) = \sum_{k=1}^n\left(\frac{d^2\gamma^k}{dt^2} + \sum_{i,j=1}^n \frac{d\gamma^i}{dt}\frac{d\gamma^j}{dt}\Gamma_{ij}^k\right)\frac{\partial}{\partial x^k}. \tag{4.144}$$

Thus, the geodesic equation $\nabla_{\gamma'(t)}\gamma'(t) = 0$ is, in local coordinates,

$$\frac{d^2\gamma^k}{dt^2}(t) + \sum_{i,j=1}^n \frac{d\gamma^i}{dt}(t)\frac{d\gamma^j}{dt}(t)\Gamma_{ij}^k(\gamma(t)) = 0 \quad \text{for all } 1 \leq k \leq n, \tag{4.145}$$

where we have written the formula more explicitly.

Example 4.34. *Let G be the graph of a function $f : \mathbb{R}^n \to \mathbb{R}$. Recall that the graphical parametrization is $F(\mathbf{x}) = (\mathbf{x}, f(\mathbf{x}))$ for $\mathbf{x} \in \mathbb{R}^n$ and that the coordinate tangent vectors are*

$$\frac{\partial F}{\partial x^i} = \mathbf{e}_i + \frac{\partial f}{\partial x^i}\mathbf{e}_{n+1}.$$

The components of the metric are

$$g_{ij} = \frac{\partial F}{\partial x^i}\cdot\frac{\partial F}{\partial x^j} = \delta_{ij} + \frac{\partial f}{\partial x^i}\frac{\partial f}{\partial x^j}. \tag{4.146}$$

By (4.228) *in Exercise* 4.22 *below,*

$$g^{ij} := (g^{-1})_{ij} = \delta_{ij} - \frac{1}{\|\nabla f\|^2 + 1}\frac{\partial f}{\partial x^i}\frac{\partial f}{\partial x^j}. \tag{4.147}$$

From all of this we can compute the Christoffel symbols as

(4.148)

$$\Gamma_{ij}^k = \frac{1}{2}\sum_{\ell=1}^n g^{k\ell}\left(\frac{\partial}{\partial x^i}\left(\frac{\partial f}{\partial x^j}\frac{\partial f}{\partial x^\ell}\right) + \frac{\partial}{\partial x^j}\left(\frac{\partial f}{\partial x^i}\frac{\partial f}{\partial x^\ell}\right) - \frac{\partial}{\partial x^\ell}\left(\frac{\partial f}{\partial x^i}\frac{\partial f}{\partial x^j}\right)\right)$$
$$= \sum_{\ell=1}^n\left(\delta_{k\ell} - \frac{1}{\|\nabla f\|^2 + 1}\frac{\partial f}{\partial x^k}\frac{\partial f}{\partial x^\ell}\right)\frac{\partial^2 f}{\partial x^i \partial x^j}\frac{\partial f}{\partial x^\ell}.$$

Example 4.35. *As in* (4.39), *let* $F : [a,b] \times [0,2\pi] \to \mathbb{R}^3$, *defined by*

$$F(u^1,u^2) := \big(f(u^1)\cos(u^2), f(u^1)\sin(u^2), u^1\big), \tag{4.149}$$

parametrize a surface of revolution M^2 *in* $\mathbb{R}^3$. *Recall from* (4.42) *that the Riemannian metric components* $g_{ij} = \frac{\partial F}{\partial u^i} \cdot \frac{\partial F}{\partial u^j}$ *of* (M^2, g) *are given by*

$$\begin{aligned} g_{11} &= (f'(u^1))^2 + 1, \\ g_{22} &= (f(u^1))^2, \\ g_{12} &= g_{21} = 0. \end{aligned}$$

Thus,

$$\begin{aligned} g^{11} &= \left((f'(u^1))^2 + 1\right)^{-1}, \\ g^{22} &= (f(u^1))^{-2}, \\ g^{12} &= g^{21} = 0. \end{aligned}$$

Just using that $(g^{k\ell})$ *is diagonal and a function of* u^1 *only, we have for all* $1 \le i,j,k \le 2$,

$$\begin{aligned} \Gamma_{ij}^k &= \frac{1}{2} g^{kk}(\partial_i g_{jk} + \partial_j g_{ik} - \partial_k g_{ij}) \\ &= \frac{1}{2}\left(\delta_{jk} g^{kk}\partial_i g_{kk} + \delta_{ik} g^{kk}\partial_j g_{kk} - \delta_{k1} g^{11}\partial_1 g_{ij}\right) \\ &= \frac{1}{2}\left(\delta_{jk}\delta_{i1}\partial_1 \ln g_{kk} + \delta_{ik}\delta_{j1}\partial_1 \ln g_{kk} - \delta_{k1} g^{11}\partial_1 g_{ij}\right), \end{aligned}$$

where $\partial_i := \frac{\partial}{\partial x^i}$ *is the directional derivative along the direction* $\frac{\partial F}{\partial u^i}$ *and where we used that* $\partial_2 g_{ij} = 0$. *From this we obtain for all* $1 \le i,j \le 2$,

$$\begin{aligned} \Gamma_{ij}^1 &= \delta_{i1}\delta_{j1}\partial_1 \ln g_{11} - \frac{1}{2} g^{11}\partial_1 g_{ij}, \\ \Gamma_{ij}^2 &= \frac{1}{2}\left(\delta_{j2}\delta_{i1} + \delta_{i2}\delta_{j1}\right)\partial_1 \ln g_{22}. \end{aligned}$$

Thus,

$$\Gamma_{11}^1 = \frac{1}{2}\partial_1 \ln g_{11}, \tag{4.150a}$$

$$\Gamma_{12}^1 = 0, \tag{4.150b}$$

$$\Gamma_{22}^1 = -\frac{1}{2} g^{11}\partial_1 g_{22} \tag{4.150c}$$

$$\Gamma_{11}^2 = 0, \tag{4.150d}$$

$$\Gamma_{12}^2 = \frac{1}{2}\partial_1 \ln g_{22}, \tag{4.150e}$$

$$\Gamma_{22}^2 = 0. \tag{4.150f}$$

Now, consider a path of the form

$$\gamma(t) := F(h(t), u^2) = \big(f(h(t))\cos(u^2), f(h(t))\sin(u^2), h(t)\big), \tag{4.151}$$

where u^2 *is a constant and* $h(t)$ *is a function. By analogy of the surface of revolution with a 2-sphere, we will call the image curve of the path* γ *a* ***meridian****. Then*

$$\begin{aligned}\frac{d\gamma}{dt}(t) &= \big(f'(h(t))h'(t)\cos(u^2), f'(h(t))h'(t)\sin(u^2), h'(t)\big) \\ &= h'(t)\frac{\partial F}{\partial u^1}(h(t), u^2)\end{aligned}$$

by (4.40)*. By* (4.142)*, in terms of components, this is*

$$\frac{d\gamma^1}{dt}(t) = h'(t), \qquad \frac{d\gamma^2}{dt}(t) = 0.$$

Thus, for paths of the form (4.151) *in a surface of revolution, the geodesic equation* (4.145) *is equivalent to*

$$0 = \frac{d^2\gamma^1}{dt^2}(t) + \left(\frac{d\gamma^1}{dt}\right)^2(t)\,\Gamma^1_{11}\big(\gamma(t)\big) \tag{4.152}$$

since the $k = 2$ *component of* (4.145) *is a vacuous equation* $(0 = 0)$ *and since* $\Gamma^1_{12} = \Gamma^1_{21} = 0$ *and* $\frac{d\gamma^2}{dt} = 0$*. Use the abbreviation* $g_{11}\big(F(u^1, u^2)\big) =: g_{11}(u^1)$*. Since* $\Gamma^1_{11}\big(\gamma(t)\big) = \frac{1}{2}\partial_1 \ln g_{11} = \frac{1}{2}(\ln g_{11})'(h(t))$*, we obtain*

$$0 = h''(t) + \frac{1}{2}h'(t)^2(\ln g_{11})'(h(t)); \tag{4.153}$$

that is,

$$\frac{h''(t)}{h'(t)} = -\frac{1}{2}(\ln g_{11})'(h(t))\,h'(t) - \frac{1}{2}\big(\ln g_{11}(h(t))\big)'$$

by the chain rule. The general solution to this ordinary differential equation satisfies

$$h'(t) = C g_{11}(h(t))^{-\frac{1}{2}},$$

where C *is a constant. Note that this equation is equivalent to the speed* $\|\gamma'(t)\|$ *of the meridian* $\gamma(t)$ *being constant. This leads us to ask whether we could have derived the equation for* $\gamma(t)$ *to be a geodesic in a simpler fashion?*

4.6. Curvature tensor of a submanifold

The most direct notions of the curvatures of hypersurfaces arise from the second fundamental form II and its eigenvalues, which are the principal curvatures $\kappa_1, \ldots, \kappa_n$. Remarkably, there is another, fundamental notion of curvature which arises from the Levi-Civita connection. The idea of measuring the degree of noncommutativity of operations is fundamental and ubiquitous in mathematics. We apply this idea to covariant differentiation to obtain the notion of the Riemann curvature tensor.

4.6.1. Definition of the Riemann curvature tensor. Let U, V, W be vector fields on a submanifold M^n of $\mathbb{R}^N$. Recall that ∇ denotes the Levi-Civita connection. Firstly, we can take a second covariant derivative of a vector field:

$$\nabla_U(\nabla_V W),$$

which is a vector field on M^n. We can now consider the commutator of the covariant derivative operator. It seems natural to just consider

$$\nabla_U(\nabla_V W) - \nabla_V(\nabla_U W).$$

However, as we will presently explain, this is not quite the right quantity to consider.

We modify this slightly and define the **Riemann curvature tensor** by

$$\operatorname{Rm}(U, V)W := \nabla_U(\nabla_V W) - \nabla_V(\nabla_U W) - \nabla_{[U,V]}W \tag{4.154}$$

for vector fields U, V, W on M^n. Since the Levi-Civita connection is uniquely determined by the Riemannian metric I, we call Rm the Riemann curvature tensor of the Riemannian manifold (M^n, I).

Recall from (4.129) that $[U, V] = \nabla_U V - \nabla_V U$. Therefore, we can rewrite the Riemann curvature tensor as

$$\operatorname{Rm}(U, V)W = \nabla_U(\nabla_V W) - \nabla_{\nabla_U V}W - \big(\nabla_V(\nabla_U W) - \nabla_{\nabla_V U}W\big). \tag{4.155}$$

It turns out that the real second covariant derivative is not $\nabla_U(\nabla_V W)$, but rather it is

$$\nabla^2_{U,V}W := \nabla_U(\nabla_V W) - \nabla_{\nabla_U V}W. \tag{4.156}$$

This definition will make more sense later; for now we just think of it as a way to construct a second covariant derivative of W in the directions U and V where we do not include the effect of differentiating V in the direction of U, so we subtract the term $\nabla_{\nabla_U V}W$ from $\nabla_U(\nabla_V W)$. Thus, the Riemann curvature tensor is a true commutator:

$$\operatorname{Rm}(U, V)W = \nabla^2_{U,V}W - \nabla^2_{V,U}W. \tag{4.157}$$

4.6.2. Riemann curvature tensor of Euclidean space. Now let us see how this works for Euclidean space with the Euclidean metric and the associated Euclidean covariant derivative. Namely, consider $\mathbb{R}^N$ with the Euclidean metric given by $g_{\mathbb{E}}(V, W) := V \cdot W$ and the Euclidean covariant derivative defined by

$$D_V W = \big(V(W^1), \ldots, V(W^N)\big).$$

We have that

$$D_U(D_V W) = \big(U(V(W^1)), \dots, U(V(W^N))\big).$$

Thus,

$$\begin{aligned} &D_U(D_V W) - D_V(D_U W) \\ &\quad = \big(U(V(W^1)) - V(U(W^1)), \dots, U(V(W^N)) - V(U(W^N))\big) \\ &\quad = \big([U,V](W^1), \dots, [U,V](W^N)\big) \\ &\quad = D_{[U,V]} W, \end{aligned}$$

where the equality on the third line follows from the definition of the Lie bracket. We conclude that

(4.158)
$$\begin{aligned} \mathrm{Rm}_{\mathbb{E}}(U,V)W &= D_U(D_V W) - D_{D_U V} W - \big(D_V(D_U W) - D_{D_V U} W\big) \\ &= 0. \end{aligned}$$

In general, when the Riemann curvature tensor is identically zero, we say that the Riemannian manifold is **flat**. So Euclidean space, i.e., $\mathbb{R}^N$ with the Euclidean Riemannian metric $g_{\mathbb{E}}$, is flat.

4.6.3. Properties of the Riemann curvature tensor.

The Riemann curvature tensor of a submanifold has some amazing properties.

4.6.3.1. *Multilinearity of* Rm. Firstly, we discuss the routine properties.

Lemma 4.36. *Let M^n be a submanifold of $\mathbb{R}^N$. Let U, V, W, U_i, V_i, W_i, $i = 1, 2$, be vector fields on M^n, and let c be a real number. The Riemann curvature tensor is multilinear:*

(4.159a) $$\mathrm{Rm}(cU_1 + U_2, V)W = c\,\mathrm{Rm}(U_1, V)W + \mathrm{Rm}(U_2, V)W,$$

(4.159b) $$\mathrm{Rm}(U, cV_1 + V_2)W = c\,\mathrm{Rm}(U, V_1)W + \mathrm{Rm}(U, V_2)W,$$

(4.159c) $$\mathrm{Rm}(U, V)(cW_1 + W_2) = c\,\mathrm{Rm}(U, V)W_1 + \mathrm{Rm}(U, V)W_2.$$

The Riemann curvature tensor is antisymmetric in its first two components:

(4.160) $$\mathrm{Rm}(U, V)W = -\,\mathrm{Rm}(V, U)W.$$

The proof of Lemma 4.36 is left as Exercise 4.28.

4.6.3.2. *The first Bianchi identity.* Now we have a cool and fundamental identity:

Theorem 4.37. *The Riemann curvature tensor satisfies the **first Bianchi identity**:*

(4.161) $$\mathrm{Rm}(U, V)W + \mathrm{Rm}(V, W)U + \mathrm{Rm}(W, U)V = 0.$$

Note that for the expression on the left-hand side, we cyclically permuted U, V, W.

Proof. The first Bianchi identity follows from a direct calculation using the definition of the Riemann curvature tensor, the torsion-free property of the Levi-Civita connection, and the Jacobi identity for the Lie bracket of vector fields. We compute that

$$\begin{aligned}\mathrm{Rm}(U,V)W &= \nabla_U(\nabla_V W) - \nabla_V(\nabla_U W) - \nabla_{[U,V]}W,\\ \mathrm{Rm}(V,W)U &= \nabla_V(\nabla_W U) - \nabla_W(\nabla_V U) - \nabla_{[V,W]}U,\\ \mathrm{Rm}(W,U)V &= \nabla_W(\nabla_U V) - \nabla_U(\nabla_W V) - \nabla_{[W,U]}V.\end{aligned}$$

Thus, summing these three equations while rearranging the terms on the right-hand side, we obtain that

$$\begin{aligned}&\mathrm{Rm}(U,V)W + \mathrm{Rm}(V,W)U + \mathrm{Rm}(W,U)V\\ &\quad= \nabla_U(\nabla_V W) - \nabla_U(\nabla_W V) + \nabla_V(\nabla_W U)\\ &\qquad - \nabla_V(\nabla_U W) + \nabla_W(\nabla_U V) - \nabla_W(\nabla_V U)\\ &\qquad - \nabla_{[U,V]}W - \nabla_{[V,W]}U - \nabla_{[W,U]}V\\ &\quad= \nabla_U[V,W] + \nabla_V[W,U] + \nabla_W[U,V]\\ &\qquad - \nabla_{[U,V]}W - \nabla_{[V,W]}U - \nabla_{[W,U]}V\\ &\quad= [U,[V,W]] + [V,[W,U]] + [W,[U,V]]\\ &\quad= 0,\end{aligned}$$

where we used that ∇ is torsion-free (e.g., $\nabla_V W - \nabla_W V = [V,W]$) and where the last equality is by the Jacobi identity (4.59). ☐

For further symmetries satisfied by the Riemann curvature tensor, see Lemma 7.36 below.

4.6.3.3. *Tensorality of* Rm. A remarkable property of the Riemann curvature tensor, denoted as Rm, is its tensorality. Specifically, the vector $\mathrm{Rm}(U,V)W$ at a point $\mathbf{x} \in M^n$, which is constructed using the vector fields U, V, W, depends solely on the vectors U, V, W at the point $\mathbf{x}$ and not their behavior elsewhere.

The first step in proving this is the following property which further justifies the naturality of our definition of Rm.

Theorem 4.38. *Let U, V, W be vector fields on M^n and let f be a function on M^n. Then*

$$f\,\mathrm{Rm}(U,V)W = \mathrm{Rm}(fU,V)W = \mathrm{Rm}(U,fV)W = \mathrm{Rm}(U,V)fW. \tag{4.162}$$

Proof. The proof of this is in fact a straightforward calculation. We just prove the first equality and then leave the other two equalities as exercises. Firstly, we compute that

$$\begin{aligned}\nabla^2_{fU,V}W &= \nabla_{fU}(\nabla_V W) - \nabla_{\nabla_{fU}V}W \\ &= f\nabla_U(\nabla_V W) - \nabla_{f\nabla_U V}W \\ &= f\nabla_U(\nabla_V W) - f\nabla_{\nabla_U V}W \\ &= f\nabla^2_{U,V}W.\end{aligned}$$

Secondly, we compute that

$$\begin{aligned}\nabla^2_{U,fV}W &= \nabla_U(\nabla_{fV}W) - \nabla_{\nabla_U(fV)}W \\ &= \nabla_U(f\nabla_V W) - \nabla_{f\nabla_U(V)+U(f)V}W \\ &= U(f)\nabla_V W + f\nabla_U(\nabla_V W) - \nabla_{f\nabla_U(V)}W - \nabla_{U(f)V}W \\ &= f\nabla_U(\nabla_V W) - f\nabla_{\nabla_U(V)}W \\ &= f\nabla^2_{U,V}W.\end{aligned}$$

We conclude that

$$\begin{aligned}\mathrm{Rm}(fU,V)W &= \nabla^2_{fU,V}W - \nabla^2_{V,fU}W \\ &= f\nabla^2_{U,V}W - f\nabla^2_{V,U}W \\ &= f\,\mathrm{Rm}(U,V)W.\end{aligned}$$

The other two equalities are proved similarly; see Exercise 4.29. □

As a consequence of Theorem 4.38, we have the following, which says that the Riemann curvature tensor is indeed a tensor.

Corollary 4.39 (Tensor property of Rm). *The tangent vector* $\mathrm{Rm}(U,V)W$ *at* $\mathbf{x} \in M^n$ *depends only on* U, V, W *at* $\mathbf{x}$.

Proof. Step 1. We first show that if $U = U'$ in some open neighborhood $\mathcal{V}$ of $\mathbf{x}$, then

$$(\mathrm{Rm}(U,V)W)_{\mathbf{x}} = (\mathrm{Rm}(U',V)W)_{\mathbf{x}}. \tag{4.163}$$

Indeed, to see this, there exists a smooth function $\varphi : M^n \to \mathbb{R}$ with the property that $\varphi \equiv 1$ on $M^n \setminus \mathcal{V}$ and $\varphi(\mathbf{x}) = 0$. (This is because there exists a smooth partition of unity subordinate to the open cover $\{\mathcal{V}, M^n \setminus \{\mathbf{x}\}\}$ of M^n; see Lee [**Lee13**, Theorem 2.23].) Observe that

$$U - U' = \varphi(U - U').$$

So, by Theorem 4.38, we have that

$$\begin{aligned}(\mathrm{Rm}(U,V)W)_{\mathbf{x}} - (\mathrm{Rm}(U',V)W)_{\mathbf{x}} &= (\mathrm{Rm}(U-U',V)W)_{\mathbf{x}} \\ &= (\mathrm{Rm}(\varphi(U-U'),V)W)_{\mathbf{x}} \\ &= \varphi(\mathbf{x})(\mathrm{Rm}(U-U',V)W)_{\mathbf{x}} \\ &= \mathbf{0}\end{aligned}$$

since $\varphi(\mathbf{x}) = 0$.

Step 2. By Step 1, we only need to assume that U is just defined in some neighborhood $\mathcal{W}$ of $\mathbf{x}$. Suppose now that U, U' are vector fields on M^n such that $U_{\mathbf{x}} = U'_{\mathbf{x}}$. Choose local coordinates $(\mathcal{W}, \varphi_{\mathcal{W}})$, where $\varphi_{\mathcal{W}} = (x^1, \ldots, x^n)$. Let $\partial_i := \frac{\partial}{\partial x^i}$. We can express U and U' in $\mathcal{W}$ as

$$U = \sum_{i=1}^{n} U^i \partial_i$$

and

$$U' = \sum_{i=1}^{n} (U')^i \partial_i,$$

where $U^i(\mathbf{x}) = (U')^i(\mathbf{x})$ for $1 \leq i \leq n$. We now compute that

$$\begin{aligned}&(\mathrm{Rm}(U,V)W)_{\mathbf{x}} - (\mathrm{Rm}(U',V)W)_{\mathbf{x}} \\ &\quad = (\mathrm{Rm}(U-U',V)W)_{\mathbf{x}} \\ &\quad = \left(\mathrm{Rm}\left(\textstyle\sum_i U^i\partial_i - \sum_i (U')^i\partial_i, V\right) W\right)_{\mathbf{x}} \\ &\quad = \sum_{i=1}^{n} \left(\mathrm{Rm}\left((U^i - (U')^i)\partial_i, V\right) W\right)_{\mathbf{x}} \\ &\quad = \sum_{i=1}^{n} (U^i(\mathbf{x}) - (U')^i(\mathbf{x}))\left(\mathrm{Rm}\left(\partial_i, V\right) W\right)_{\mathbf{x}}.\end{aligned}$$

Since $U_{\mathbf{x}} = U'_{\mathbf{x}}$, we have $U^i(\mathbf{x}) = (U')^i(\mathbf{x})$ for all $1 \leq i \leq n$. We conclude that $(\mathrm{Rm}(U,V)W)_{\mathbf{x}} = (\mathrm{Rm}(U',V)W)_{\mathbf{x}}$. This proves that the dependence of $(\mathrm{Rm}(U,V)W)_{\mathbf{x}}$ on the vector field U is only in terms of the vector $U(\mathbf{x})$. That the dependence of $(\mathrm{Rm}(U,V)W)_{\mathbf{x}}$ on V, W is only in terms of $V(\mathbf{x}), W(\mathbf{x})$ follows for the same reasons. □

The definition (4.154) of the Riemann curvature tensor is in terms of globally defined quantities, namely, vector fields on M^n. Corollary 4.39 says that the Riemann curvature tensor is actually equivalent to an infinitesimal notion. That is, pointwise, Rm depends only on the values of the vector fields at a point. This and its multilinearity properties imply that the Riemann curvature tensor is a $(3,1)$-tensor in the sense of §3.4.4.

4.7. Gauss and Codazzi equations

Both the second fundamental form and the Riemann curvature tensor involve taking Euclidean or induced covariant derivatives, utilizing either the normal vector field or tangent vector fields. Because of this commonality, these two notions of curvature are related. The precise relations are given by the Gauss and Codazzi equations (see [**Gau65**], [**Cod69**], and [**Mai56**]).[3]

4.7.1. Statements and proofs of the Gauss and Codazzi equations. Let M^n be a smooth embedded submanifold of $\mathbb{R}^N$. By (4.82), the induced covariant derivative satisfies the formula

$$\begin{aligned} D_V W &= \nabla_V W + (D_V W \cdot N) N \\ &= \nabla_V W - \mathrm{II}(V, W) N \end{aligned} \tag{4.164}$$

for tangent vector fields V and W on M^n. Using this, we can decompose the Euclidean Riemann curvature tensor vanishing equation (4.158) into its tangential and normal components. The calculation starts as follows: By (4.164),

$$\begin{aligned} D_U(D_V W) &= D_U(\nabla_V W) - D_U\big(\mathrm{II}(V, W) N\big) \\ &= \nabla_U(\nabla_V W) - \mathrm{II}(U, \nabla_V W) N \\ &\quad - U\big(\mathrm{II}(V, W)\big) N - \mathrm{II}(V, W) D_U N \\ &= \nabla_U(\nabla_V W) - \mathrm{II}(V, W)\, \mathrm{L}(U) \\ &\quad - \big(\mathrm{II}(U, \nabla_V W) + U\big(\mathrm{II}(V, W)\big)\big) N. \end{aligned}$$

By simply switching U and V, we obtain

$$\begin{aligned} D_V(D_U W) &= \nabla_V(\nabla_U W) - \mathrm{II}(U, W)\, \mathrm{L}(V) \\ &\quad - \big(\mathrm{II}(V, \nabla_U W) + V\big(\mathrm{II}(U, W)\big)\big) N. \end{aligned}$$

Next, we observe that

$$D_{[U,V]} W = \nabla_{[U,V]} W - \mathrm{II}([U, V], W) N.$$

[3] The Codazzi equations are also called the Codazzi–Mainardi equations.

Thus, by combining the three formulas above, we may express the vanishing of the Euclidean Riemann curvature tensor as

$$\begin{aligned}\mathbf{0} &= R_{\mathbb{E}}(U,V)W \\ &= D_U(D_V W) - D_V(D_U W) - D_{[U,V]}W \\ &= \nabla_U(\nabla_V W) - \mathrm{II}(V,W)\,\mathrm{L}(U) \\ &\quad - \big(\mathrm{II}(U,\nabla_V W) + U\big(\mathrm{II}(V,W)\big)\big)N \\ &\quad -\nabla_V(\nabla_U W) + \mathrm{II}(U,W)\,\mathrm{L}(V) \\ &\quad + \big(\mathrm{II}(V,\nabla_U W) + V\big(\mathrm{II}(U,W)\big)\big)N \\ &\quad -\nabla_{[U,V]}W + \mathrm{II}([U,V],W)\,N.\end{aligned}$$

Recall from (4.154) that the Riemann curvature tensor of I is given by

$$\mathrm{Rm}(U,V)W = \nabla_U(\nabla_V W) - \nabla_V(\nabla_U W) - \nabla_{[U,V]}W.$$

Thus, we have

$$\begin{aligned}\mathbf{0} = \mathrm{Rm}(U,V)W &- \mathrm{II}(V,W)\,\mathrm{L}(U) + \mathrm{II}(U,W)\,\mathrm{L}(V) \\ &- \big(\mathrm{II}(U,\nabla_V W) + U\big(\mathrm{II}(V,W)\big)\big)N \\ &+ \big(\mathrm{II}(V,\nabla_U W) + V\big(\mathrm{II}(U,W)\big)\big)N \\ &+ \mathrm{II}(\nabla_U V - \nabla_V U, W)\,N,\end{aligned} \tag{4.165}$$

where in the last line we substituted $\nabla_U V - \nabla_V U$ for $[U,V]$. We further rearrange the terms as

$$\begin{aligned}\mathbf{0} = \mathrm{Rm}(U,V)W &- \mathrm{II}(V,W)\,\mathrm{L}(U) + \mathrm{II}(U,W)\,\mathrm{L}(V) \\ &- \big(U\big(\mathrm{II}(V,W)\big) - \mathrm{II}(V,\nabla_U W) - \mathrm{II}(\nabla_U V, W)\big)N \\ &+ \big(V\big(\mathrm{II}(U,W)\big) - \mathrm{II}(U,\nabla_V W) - \mathrm{II}(\nabla_V U, W)\big)N.\end{aligned} \tag{4.166}$$

Notice that the first line of the display is all tangent vectors, whereas the second and third lines of the display are all normal vectors. To extract interesting formulas from this equation, we observe that if $\mathbf{0} = cN_{\mathbf{x}} + \mathbf{Y}$ for some $c \in \mathbb{R}$ and $\mathbf{Y} \in T_{\mathbf{x}}M$, then $c = 0$ and $\mathbf{Y} = \mathbf{0}$. This is simply because $\mathbf{Y} \cdot N_{\mathbf{x}} = 0$. In other words, if a tangential vector and a normal vector sum to zero, then both the tangential vector and the normal vector are zero. Therefore, we conclude the following.

Theorem 4.40. *If M^n is a hypersurface in $\mathbb{R}^{n+1}$, then we have the **Gauss equation***

$$\mathrm{Rm}(U,V)W = \mathrm{II}(V,W)\,\mathrm{L}(U) - \mathrm{II}(U,W)\,\mathrm{L}(V) \tag{4.167}$$

*and the **Codazzi equation***

$$\begin{aligned}&U(\mathrm{II}(V,W)) - \mathrm{II}(V,\nabla_U W) - \mathrm{II}(\nabla_U V, W) \\ &\quad = V(\mathrm{II}(U,W)) - \mathrm{II}(U,\nabla_V W) - \mathrm{II}(\nabla_V U, W).\end{aligned} \tag{4.168}$$

The **Riemann curvature 4-tensor** is defined by

$$\operatorname{Rm}(U,V,W,X) := \operatorname{Rm}(U,V)W \cdot X. \tag{4.169}$$

The Gauss equation (4.167) is equivalent to

$$\operatorname{Rm}(U,V,W,X) = \mathrm{II}(V,W)\,\mathrm{II}(U,X) - \mathrm{II}(U,W)\,\mathrm{II}(V,X). \tag{4.170}$$

Example 4.41. *For an n-dimensional sphere $S^n(r)$ of radius $r > 0$, we have $\mathrm{II} = \frac{1}{r}g$. Thus, by the Gauss equation* (4.170), *we have*

$$\operatorname{Rm}(U,V,W,X) = \frac{1}{r^2}\big(g(V,W)g(U,X) - g(U,W)g(V,X)\big). \tag{4.171}$$

Note that for any orthonormal vectors e_1, e_2, that is, $g(e_i, e_j) = \delta_{ij}$ for $1 \leq i, j \leq 2$, we have

$$\operatorname{Rm}(e_2, e_1, e_1, e_2) = \frac{1}{r^2}. \tag{4.172}$$

We call $\operatorname{Rm}(e_2, e_1, e_1, e_2)$ *the* sectional curvature *of the plane spanned by e_1, e_2 in $T_{\mathbf{x}}(S^n(r))$. The sphere of radius r has constant sectional curvature $\frac{1}{r^2}$.*

We now rewrite the Codazzi equation in a nicer form. To accomplish this, we note that covariant differentiation acts not just on vector fields, but also on tensors, such as the second fundamental form, which is a 2-tensor. The basic principle governing their differentiation is the product rule. Namely, we define the covariant derivative $\nabla_U \mathrm{II}$ of II in the direction U by the **covariant derivative product rule**:

$$U(\mathrm{II}(V,W)) =: (\nabla_U \mathrm{II})(V,W) + \mathrm{II}(V, \nabla_U W) + \mathrm{II}(\nabla_U V, W); \tag{4.173}$$

that is,

$$(\nabla_U \mathrm{II})(V,W) := U(\mathrm{II}(V,W)) - \mathrm{II}(V, \nabla_U W) - \mathrm{II}(\nabla_U V, W). \tag{4.174}$$

We then see that the Codazzi equation (4.168) greatly simplifies to

$$(\nabla_U \mathrm{II})(V,W) = (\nabla_V \mathrm{II})(U,W). \tag{4.175}$$

Defining the covariant derivative of II as a 3-tensor by

$$(\nabla \mathrm{II})(U,V,W) := (\nabla_U \mathrm{II})(V,W), \tag{4.176}$$

we obtain

$$(\nabla \mathrm{II})(U,V,W) = (\nabla \mathrm{II})(V,U,W). \tag{4.177}$$

That is, the covariant 3-tensor $\nabla\,\mathrm{II}$ is symmetric in its first two components. Since $\nabla\,\mathrm{II}$ is also symmetric in its last two components, we see that $\nabla\,\mathrm{II}$ is fully symmetric.

We take this opportunity to observe that we can define the covariant derivative of the first fundamental form $g = \mathrm{I}$ in the same way:

$$(\nabla_U g)(V, W) := U(g(V, W)) - g(V, \nabla_U W) - g(\nabla_U V, W). \tag{4.178}$$

By the metric compatibility condition (4.128), we have that the first fundamental form g is parallel:

$$(\nabla g)(U, V, W) := (\nabla_U g)(V, W) = 0 \tag{4.179}$$

for all U, V, W. We write this simply as $\nabla g = 0$.

The notion of covariant derivative, which was described above for II and g, naturally extends to 2-tensors. When a 2-tensor has zero covariant derivative, we say that it is **parallel**. Thus, the Riemannian metric g is parallel (with respect to the Levi-Civita connection).

Example 4.42. *For an n-dimensional sphere $S^n(r)$ in $\mathbb{R}^{n+1}$, we have $\mathrm{II} = \frac{1}{r}g$. Since $\nabla g = 0$, we thus have $\nabla\,\mathrm{II} = \frac{1}{r}\nabla g = 0$.*

In general, if α is a covariant k-tensor and if $U, V_1, \ldots, V_k$ are vectors at a point, which we extend arbitrarily to vector fields defined in a neighborhood of that point, then the covariant derivative of α is defined by the 'product rule':

$$(\nabla_U \alpha)(V_1, \ldots, V_k) := U\big(\alpha(V_1, \ldots, V_k)\big) - \sum_{i=1}^{k} \alpha(V_1, \ldots, \nabla_U V_i, \ldots V_k). \tag{4.180}$$

This definition is independent of the choices of extensions of the vector fields.

4.7.2. The Gauss and Codazzi equations in local coordinates.

Let $F : \mathcal{U} \to M^n$ be a local parametrization of an embedded hypersurface M^n. Recall from §4.4.8 that the components of the second fundamental form are defined by

$$\mathrm{II}_{ij} = \mathrm{II}(\partial_i F, \partial_j F),$$

where $\partial_i F = \frac{\partial F}{\partial u^i}$.

The covariant derivative $\nabla\,\mathrm{II}$ of the second fundamental form is a 3-tensor. We denote its components by

$$\nabla_i \mathrm{II}_{jk} := (\nabla\,\mathrm{II})(\partial_i F, \partial_j F, \partial_k F).$$

By the Codazzi equation (4.177), we have

$$\nabla_i \mathrm{II}_{jk} = \nabla_j \mathrm{II}_{ik}\,. \tag{4.181}$$

We remark that by using the covariant derivative product rule (4.174), we obtain

$$\nabla_i \operatorname{II}_{jk} = \partial_i \operatorname{II}_{jk} - \sum_{\ell=1}^{n} \Gamma_{ij}^{\ell} \operatorname{II}_{\ell k} - \sum_{\ell=1}^{n} \Gamma_{ik}^{\ell} \operatorname{II}_{j\ell}, \tag{4.182}$$

where $\partial_i = \frac{\partial}{\partial u^i}$.

Let

$$R_{ijk\ell} := \operatorname{Rm}(\partial_i F, \partial_j F, \partial_k F, \partial_\ell F)$$

denote the components of the Riemann curvature 4-tensor defined by (4.169). In local coordinates, the Gauss equation (4.170) is

$$R_{ijk\ell} = \operatorname{II}_{i\ell} \operatorname{II}_{jk} - \operatorname{II}_{ik} \operatorname{II}_{j\ell}\,. \tag{4.183}$$

As a $(3,1)$-tensor, the components of Rm are defined by

$$\operatorname{Rm}(\partial_i, \partial_j)\partial_k =: \sum_{\ell=1}^{n} R_{ijk}^{\ell} \partial_\ell. \tag{4.184}$$

Using the definition of Rm as an expression in the second covariant derivative, one can calculate that (see Exercise 4.31)

$$R_{ijk}^{\ell} = \partial_i \Gamma_{jk}^{\ell} - \partial_j \Gamma_{ik}^{\ell} + \sum_{p=1}^{n} \Gamma_{jk}^{p} \Gamma_{ip}^{\ell} - \sum_{p=1}^{n} \Gamma_{ik}^{p} \Gamma_{jp}^{\ell}. \tag{4.185}$$

4.8. Theorema Egregium

In this section we discuss a fundamental result of Gauss [**Gau65**].

4.8.1. Statement and proof of Gauss's Theorema Egregium. Recall from (4.94) that the Gauss curvature of a hypersurface M^n in $\mathbb{R}^{n+1}$ is defined by

$$K = \det(\mathrm{L}) = \frac{\det \mathrm{II}}{\det \mathrm{I}}. \tag{4.186}$$

When $n = 2$, we can express the Gauss curvature K as follows. Let $\mathbf{x} \in M^2$ and let $\{e_1, e_2\}$ be an orthonormal frame at $\mathbf{x}$. Then

$$K(\mathbf{x}) = \operatorname{II}(e_1, e_1) \operatorname{II}(e_2, e_2) - \operatorname{II}(e_1, e_2)^2. \tag{4.187}$$

Now we can bring the Gauss equation (4.170) into the mix. Namely, it implies that

$$\operatorname{Rm}(e_2, e_1, e_1, e_2) = \operatorname{II}(e_1, e_1) \operatorname{II}(e_2, e_2) - \operatorname{II}(e_1, e_2)^2.$$

We conclude that

$$K(\mathbf{x}) = \mathrm{Rm}(e_2, e_1, e_1, e_2), \tag{4.188}$$

where the right-hand side is independent of the choice of orthonormal frame $\{e_1, e_2\}$ at $\mathbf{x}$.

We say that a geometric quantity or operator on a hypersurface M^n is **intrinsic** if it depends solely on the first fundamental form I (more about this in the next subsection). In particular, intrinsic quantities do not depend on the second fundamental form II. By Theorem 4.33, the Levi-Civita connection ∇ is intrinsic. Since the Riemann curvature tensor Rm can be defined purely in terms of the Levi-Civita connection (see (4.155)), the Riemann curvature tensor is also intrinsic. Thus, by (4.188), we conclude the following.

Theorem 4.43 (Gauss's Theorema Egregium). *The Gauss curvature K is an intrinsic invariant of a surface M^2 embedded in $\mathbb{R}^3$.*

4.8.2. Intrinsic versus extrinsic. Suppose M^n and $\tilde{M}^n$ are hypersurfaces in $\mathbb{R}^{n+1}$. Let I, II and $\tilde{\mathrm{I}}, \tilde{\mathrm{II}}$ denote the first and second fundamental forms of M^n and $\tilde{M}^n$, respectively. A map $\Phi : M^n \to \tilde{M}^n$ is a **diffeomorphism** if it is bijective, $\iota \circ \Phi : M^n \to \mathbb{R}^{n+1}$ is an embedding, and $\tilde{\iota} \circ \Phi^{-1} : \tilde{M}^n \to \mathbb{R}^{n+1}$ is an embedding, where ι and $\tilde{\iota}$ are the inclusion maps of M^n and $\tilde{M}^n$ into $\mathbb{R}^{n+1}$, respectively.

We say a diffeomorphism $\Phi : M^n \to \tilde{M}^n$ is an **isometry** if

$$\tilde{\mathrm{I}}\big(d\Phi(\mathbf{V}), d\Phi(\mathbf{W})\big) = \mathrm{I}(\mathbf{V}, \mathbf{W}) \tag{4.189}$$

for all $\mathbf{V}, \mathbf{W} \in T_{\mathbf{x}}M$, $\mathbf{x} \in M^n$.[4] That is,

$$d\Phi(\mathbf{V}) \cdot d\Phi(\mathbf{W}) = \mathbf{V} \cdot \mathbf{W}. \tag{4.190}$$

In other words, via (the derivative of) the map Φ, the dot products on M^n are the same as the dot products on $\tilde{M}^n$. As we will presently see, this means that Φ preserves infinitesimal distances and angles.

In particular, taking $\mathbf{V} = \mathbf{W}$, we obtain

$$\|\Phi(\mathbf{V})\| = \|\mathbf{V}\|. \tag{4.191}$$

Hence, if $\mathbf{V}$ and $\mathbf{W}$ are nonzero tangent vectors, then $d\Phi(\mathbf{V})$ and $d\Phi(\mathbf{W})$ are nonzero and

$$\frac{d\Phi(\mathbf{V}) \cdot d\Phi(\mathbf{W})}{\|d\Phi(\mathbf{V})\| \, \|d\Phi(\mathbf{W})\|} = \frac{\mathbf{V} \cdot \mathbf{W}}{\|\mathbf{V}\| \, \|\mathbf{W}\|}. \tag{4.192}$$

We conclude that

$$\angle_{\tilde{\mathrm{I}}}(\Phi(\mathbf{V}), \Phi(\mathbf{W})) = \angle_{\mathrm{I}}(\mathbf{V}, \mathbf{W}). \tag{4.193}$$

[4]Another way to think of this formula is that the pull-back of $\tilde{\mathrm{I}}$ by Φ is equal to I; see §6.5.1 for the definition of 'pull-back'.

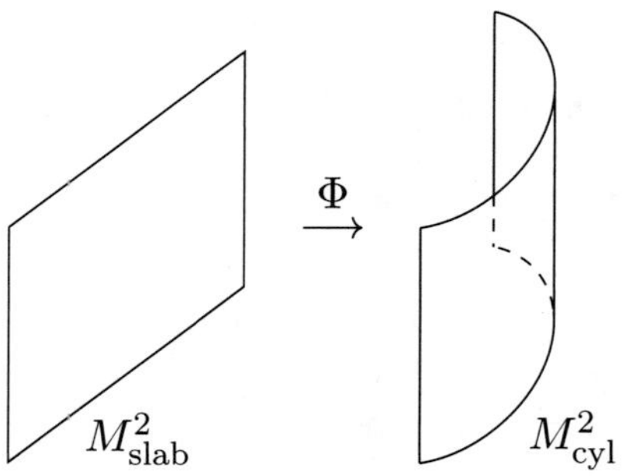

Figure 4.8.1. The isometry $\Phi : M^2_{\text{slab}} \to M^2_{\text{cyl}}$.

Example 4.44. *Consider a half-cylinder in* $\mathbb{R}^3$ *defined by*

$$M^2_{\text{cyl}} = \big\{(x,y,z) \in \mathbb{R}^3 \colon x^2+y^2=1,\ y>0,\ z \in \mathbb{R}\big\}. \tag{4.194}$$

On the other hand, consider a flat 2*-dimensional slab*

$$M^2_{\text{slab}} = \big\{(u,v,w) \in \mathbb{R}^3 \colon 0<u<\pi,\ v=0,\ w \in \mathbb{R}\big\}. \tag{4.195}$$

Define $\Phi : M^2_{\text{slab}} \to M^2_{\text{cyl}}$ *by*

$$\Phi(u,w) := \big(\cos(u), \sin(u), w\big). \tag{4.196}$$

Then Φ *is an isometry. See Figure* 4.8.1 *and Exercise* 4.30.

A consequence of the fact that the Riemann curvature tensor is intrinsic is that if Φ is an isometry, then

$$\widetilde{\text{Rm}}\big(d\Phi(\mathbf{U}), d\Phi(\mathbf{V})\big)d\Phi(\mathbf{W}) = d\Phi\big(\text{Rm}(\mathbf{U},\mathbf{V})\mathbf{W}\big), \tag{4.197}$$

or equivalently,[5]

$$\widetilde{\text{Rm}}\big(d\Phi(\mathbf{U}), d\Phi(\mathbf{V})\big)d\Phi(\mathbf{W}) \cdot d\Phi(\mathbf{Z}) = \text{Rm}(\mathbf{U},\mathbf{V})\mathbf{W}\cdot\mathbf{Z}. \tag{4.198}$$

This fact will be discussed in Chapter 7.

On the other hand, in general we have that $\tilde{\text{II}}\big(d\Phi(\mathbf{V}), d\Phi(\mathbf{W})\big)$ is not equal to $\text{II}(\mathbf{V},\mathbf{W})$.[6] Therefore II is not intrinsic. For example, consider Example 4.44. Here, we have that $\text{II}_{\text{slab}} \equiv 0$, whereas $\text{II}_{\text{cyl}} \not\equiv 0$. In particular,

$$\text{II}_{\text{slab}}(\mathbf{V},\mathbf{W}) \not\equiv \text{II}_{\text{cyl}}\left(d\Phi(\mathbf{V}), d\Phi(\mathbf{W})\right).$$

In fact, at each point of M^2_{cyl} the principal curvatures are 0 and 1. Thus, both M^2_{slab} and M^2_{cyl} have zero Gauss curvature everywhere, whereas M^2_{slab} has zero mean curvature and M^2_{cyl} has mean curvature equal to 1 everywhere. So we also see that the mean curvature is not an intrinsic invariant.

[5]In other words, the pull-back of $\widetilde{\text{Rm}}$ by Φ is equal to Rm.

[6]In other words, in general, the pull-back of $\tilde{\text{II}}$ by Φ is not equal to II.

4.9. Fundamental theorem of hypersurface theory

Recall, for curves in the plane, that there is the fundamental theorem of plane curves (Proposition 4.8), which says that a plane curve is determined, up to a Euclidean isometry, by its curvature κ. In this section, we discuss the fundamental theorem of the differential geometry of hypersurfaces, which is a local result.

4.9.1. Frobenius theorem.

Let M^n be a smooth submanifold of $\mathbb{R}^N$.

Definition 4.45. A k-dimensional **distribution** D on M^n is an assignment to each $\mathbf{x} \in M^n$ of a k-dimensional vector subspace $D_{\mathbf{x}}$ of $T_{\mathbf{x}}M$. We say that D is smooth if the set $D := \bigsqcup_{\mathbf{x}\in M^n} D_{\mathbf{x}}$ is a smooth (k-dimensional) subbundle of TM.

For example, if V is a smooth nonzero vector field on M^n, then $D_x := \mathbb{R}V_{\mathbf{x}}$, $\mathbf{x} \in M^n$, defines a smooth 1-dimensional distribution on M^n. If g is the induced Riemannian metric (a.k.a. first fundamental form) on M^n, then

$$D_{\mathbf{x}}^{\perp} := (\mathbb{R}V_x)^{\perp} = \{W \in T_{\mathbf{x}}M \colon g(W,V) = 0\}, \quad \mathbf{x} \in M^n,$$

defines a smooth $(n-1)$-dimensional distribution on M^n.

We say that a vector field V is a **local section** of a distribution D if V is defined on an open set $\mathcal{U}$ and $V_{\mathbf{x}} \in D_{\mathbf{x}}$ for all $\mathbf{x} \in \mathcal{U}$.

We say that a distribution D on a manifold M^n is **involutive** if for any local sections V and W of D over an open set $\mathcal{U}$ we have that their Lie bracket $[V,W]$ is a local section of D over $\mathcal{U}$.

We say that a k-dimensional distribution D is **integrable** if for every $\mathbf{x}$, there exists an embedded k-dimensional submanifold N^k containing $\mathbf{x}$ such that $T_{\mathbf{y}}N = D_{\mathbf{y}}$ for all $\mathbf{y} \in N^k$. We call each such N^k an **integral submanifold** of D.

Example 4.46. *Let D be a 1-dimensional distribution on a submanifold M^n. Then D is both involutive and integrable. This is because of the following. For any $\mathbf{x} \in M^n$, there exists a nonzero vector field V defined in a neighborhood $\mathcal{U}$ of $\mathbf{x}$ such that $\mathbb{R}V_{\mathbf{y}} = D_{\mathbf{y}}$ for all $\mathbf{y} \in \mathcal{U}$. We have $[V,V] = 0$, so that D is involutive. Furthermore, by standard ODE theory, there exists an integral curve to V passing through $\mathbf{x}$, so that D is integrable.*

In general, suppose that M^n is partitioned into a family of disjoint, connected, smooth k-dimensional embedded submanifolds $\{N_\alpha^k\}_{\alpha\in A}$. We further assume that for each $\mathbf{x} \in M^n$, there exists a local coordinate chart $(\mathcal{U}, \{x^i\}_{i=1}^n)$ such that each $N_\alpha^k \cap \mathcal{U}$ is given by $(x^{k+1}, \dots, x^n)^{-1}(\mathbf{c})$ for some

$\mathbf{c} \in \mathbb{R}^{n-k}$ (depending on α). In this case, we say that $\{N_\alpha^k\}_{\alpha\in A}$ forms a **foliation** of M^n. Then, the distribution D defined by $D_{\mathbf{x}} = T_{\mathbf{x}}N_\alpha$, where $\alpha \in A$ is the unique element such that $\mathbf{x} \in N_\alpha^k$, is a smooth, integral distribution on M^n.

Lemma 4.47. *Every integrable distribution is an involutive distribution.*

See, e.g., [**Lee13**, Proposition 19.3] for a proof.

Proposition 4.48 (Frobenius's theorem). *Every involutive distribution is an integrable distribution.*

See, e.g., [**Lee13**, Theorem 19.12] for a proof.

4.9.2. The Frobenius theorem and local solvability. We have the following consequence of the Frobenius theorem.

Theorem 4.49 (Local solvability of a first-order partial differential equation). *Let $\mathcal{U} \subset \mathbb{R}^n \times \mathbb{R}^m = \mathbb{R}^{n+m}$ be an open set. Let $A_i^j : \mathcal{U} \to \mathbb{R}$, $1 \leq i \leq n$, $1 \leq j \leq m$, be C^∞ functions. Let $\mathbf{x}_0 \in \mathbb{R}^n$ and $\mathbf{y}_0 \in \mathbb{R}^m$ be such that $(\mathbf{x}_0, \mathbf{y}_0) \in \mathcal{U}$. Denote $\mathbf{x} = (x^1, \ldots, x^n)$. Consider the partial differential equation*

$$\frac{\partial f^j}{\partial x^i}(\mathbf{x}) = A_i^j(\mathbf{x}, \mathbf{f}(\mathbf{x})), \quad 1 \leq i \leq n,\ 1 \leq j \leq m, \tag{4.199}$$

for $\mathbf{f} = (f^1, \ldots, f^m) :$ (neighborhood of $\mathbf{x}_0$) $\to \mathbb{R}^m$; that is,

$$\frac{\partial \mathbf{f}}{\partial x^i}(\mathbf{x}) = \mathbf{A}_i(\mathbf{x}, \mathbf{f}(\mathbf{x})), \quad 1 \leq i \leq n, \tag{4.200}$$

where $\mathbf{A}_i = (A_i^1, \ldots, A_i^m) : \mathcal{U} \times \mathbb{R}^m \to \mathbb{R}^m$. Then:

(1) For every $(\mathbf{x}_0, \mathbf{y}_0) \in \mathcal{U}$ there exists a neighborhood $\mathcal{V}$ of $\mathbf{x}_0$ in $\mathbb{R}^n$ and a C^∞ function $\mathbf{f} : \mathcal{V} \subset \mathbb{R}^n \to \mathbb{R}^m$ solving the PDE (4.199) with the initial condition $\mathbf{f}(\mathbf{x}_0) = \mathbf{y}_0$

if and only if

(2) at every point $(\mathbf{x}, \mathbf{y})$ in $\mathcal{U}$,

$$\frac{\partial A_i^j}{\partial x^k} + \sum_{l=1}^m A_k^l \frac{\partial A_i^j}{\partial y^l} = \frac{\partial A_k^j}{\partial x^i} + \sum_{l=1}^m A_i^l \frac{\partial A_k^j}{\partial y^l}, \quad 1 \leq i, k \leq n,\ 1 \leq j \leq m. \tag{4.201}$$

Proof. The forward implication follows from the equality of mixed partial derivatives: The equality $\frac{\partial^2 \mathbf{f}}{\partial x^i \partial x^j} = \frac{\partial^2 \mathbf{f}}{\partial x^j \partial x^i}$ implies (4.201). So we only need to prove the backward implication.

Define pointwise linearly independent C^∞ vector fields on $\mathcal{U}$ by

$$V_i := e_i + \sum_{j=1}^m A_i^j e_{n+j}, \quad 1 \leq i \leq n, \tag{4.202}$$

where $\{e_p\}_{p=1}^{n+m}$ is the standard basis for $\mathbb{R}^{n+m}$. Let $\mathcal{D}$ denote the n-dimensional distribution with basis $\{V_i\}_{i=1}^n$.

We compute the Lie brackets of the basis as

$$\begin{aligned}[V_i, V_j] &= \left[e_i + A_i^j\, e_{n+j}, e_k + A_k^l\, e_{n+l}\right] \\ &= \left(e_i(A_k^l) + A_i^j\, e_{n+j}(A_k^l)\right)e_{n+l} - \left(e_k(A_i^j) + A_k^l\, e_{n+l}(A_i^j)\right)e_{n+j} \\ &= 0,\end{aligned}$$

where the last equality is by (4.201). Thus the distribution is involutive, and hence by Frobenius's theorem (Proposition 4.48) there exists an integral submanifold M^n of $\mathcal{D}$ passing through $(\mathbf{x}_0, \mathbf{y}_0)$ in $\mathcal{U}$. In particular, at each $(\mathbf{x}, \mathbf{y}) \in M^n$, $\{V_i\}_{i=1}^n$ is a basis for $T_{(\mathbf{x},\mathbf{y})}M$.

Define the projection $\pi : M^n \to \mathbb{R}^n$ by $\pi(\mathbf{x}, \mathbf{y}) = \mathbf{x}$. We have $d\pi(V_i) = e_i$. Thus, $\pi^{-1} : \pi(M^n) \to M^n$ is a diffeomorphism of the form

$$\pi^{-1}(\mathbf{x}) = (\mathbf{x}, \mathbf{f}(\mathbf{x})),$$

for some function $\mathbf{f} : \pi(M^n) \to \mathbb{R}^m$. Since $(\mathbf{x}_0, \mathbf{y}_0) \in M^n$, we have $\mathbf{f}(\mathbf{x}_0) = \mathbf{y}_0$. Moreover, since $d\pi(V_i) = e_i$, we have

$$\left(e_i, \frac{\partial \mathbf{f}}{\partial x^i}\right) = d\pi^{-1}(e_i) = V_i = e_i + A_i^j\, e_{n+j},$$

so that $\frac{\partial f^j}{\partial x^i} = A_i^j$. Thus $\mathbf{f}$ is a solution to (4.199) in the neighborhood $\pi(M^n)$ of $\mathbf{x}_0$. □

4.9.3. Statement of the fundamental theorem. For hypersurfaces in Euclidean space, we have the following result, called the **fundamental theorem of the differential geometry of hypersurfaces**.

Theorem 4.50. *Let $\mathcal{U}$ be an open subset of $\mathbb{R}^n$ and let g and h be symmetric 2-tensors on $\mathcal{U}$, where g is positive-definite. Then, for any $\mathbf{u}_0 \in \mathcal{U}$, there exist a neighborhood $\mathcal{V} \subset \mathcal{U}$ of $\mathbf{u}_0$ and an embedding $F : \mathcal{V} \to \mathbb{R}^{n+1}$ which induces g and h as its first and second fundamental forms, respectively; i.e., $F^*\,\mathrm{I} = g$ and $F^*\,\mathrm{II} = h$ on $\mathcal{V}$ if and only if g and h satisfy the Gauss–Codazzi equations on $\mathcal{V}$; that is,*

$$\mathrm{Rm}^g(X, Y, Z, W) = h(X, W)h(Y, Z) - h(X, Z)h(Y, W), \tag{4.203a}$$

$$(\nabla_X^g h)(Y, Z) = (\nabla_Y^g h)(X, Z). \tag{4.203b}$$

This theorem establishes that given a local region and two compatible symmetric 2-tensors g and h satisfying the Gauss–Codazzi equations, there exists an embedding F that realizes g and h as its first and second fundamental forms, respectively. We will prove this theorem using the Frobenius theorem.

4.9.4. Proof of the fundamental theorem of hypersurface theory.

Proof of Theorem 4.50. Let $\{u^i\}_{i=1}^n$ be the Euclidean coordinates on $\mathbb{R}^n$. We have on the open subset $\mathcal{U}$ of $\mathbb{R}^n$ that the symmetric 2-tensors g and h may be expressed as

$$g = g_{ij}du^i \otimes du^j\,, \qquad h = h_{ij}du^i \otimes du^j\,.$$

Define the Christoffel symbols Γ^k_{ij}, $1 \le i,j,k \le n$, of g by the formula (4.139).

Consider the first-order system of linear partial differential equations for maps

$$e_j : \mathcal{U} \to \mathbb{R}^{n+1}, \quad 1 \le j \le n+1, \tag{4.204}$$

given by

$$\frac{\partial e_j}{\partial u^i}(\mathbf{u}) = \sum_{k=1}^{n+1} \boldsymbol{\Gamma}^k_{ij}(\mathbf{u})\, e_k(\mathbf{u})\,, \tag{4.205}$$

where the functions $\boldsymbol{\Gamma}^k_{ij} : \mathcal{U} \to \mathbb{R}$, $1 \le i,j,k \le n+1$, are defined by, for $1 \le a,b,c \le n$,
(4.206)

$$\boldsymbol{\Gamma}^c_{ab} := \Gamma^c_{ab}\,, \quad \boldsymbol{\Gamma}^{n+1}_{ab} := -h_{ab}\,, \quad \boldsymbol{\Gamma}^c_{a(n+1)} := \sum_{d=1}^n g^{cd} h_{da}\,, \quad \boldsymbol{\Gamma}^{n+1}_{a(n+1)} := 0\,,$$

and (g^{cd}) is the inverse matrix of (g_{ab}). Observe that since $\Gamma^k_{ij} = \Gamma^k_{ji}$ and since the 2-tensor h is symmetric, we have

$$\frac{\partial e_j}{\partial u^i} = \frac{\partial e_i}{\partial u^j} \tag{4.207}$$

for $1 \le i,j \le n$. The e_j, for $1 \le j \le n$, will be the coordinate tangent vector fields to the local embedding that we will construct below. The vector field e_{n+1} will be the unit normal to the embedded hypersurface.

Let

$$E := (e_1, \ldots, e_{n+1}) : \mathcal{U} \to \mathbb{R}^{(n+1)^2}.$$

Then the system of equations (4.205) is equivalent to the system

$$\frac{\partial E}{\partial u^i}(\mathbf{u}) = \mathbf{A}_i(\mathbf{u}, E(\mathbf{u}))\,, \tag{4.208}$$

where $\mathbf{A}_i : \mathcal{U} \times \mathbb{R}^{(n+1)^2} \to \mathbb{R}^{(n+1)^2}$ is defined by

$$\mathbf{A}_i(\mathbf{u}, E(\mathbf{u})) := \left(\sum_{k=1}^{n+1} \boldsymbol{\Gamma}^k_{i1}(\mathbf{u})\, e_k(\mathbf{u}), \ldots, \sum_{k=1}^{n+1} \boldsymbol{\Gamma}^k_{i(n+1)}(\mathbf{u})\, e_k(\mathbf{u}) \right). \tag{4.209}$$

Choose a basis $\{e_i^0\}_{i=1}^{n+1}$ for $\mathbb{R}^{n+1}$ so that for $1 \leq i, j \leq n$,

$$\langle e_i^0, e_j^0 \rangle = g_{ij}(\mathbf{u}_0), \quad \langle e_i^0, e_{n+1}^0 \rangle = 0, \quad \langle e_{n+1}^0, e_{n+1}^0 \rangle = 1, \tag{4.210}$$

which is clearly possible since (g_{ij}) is symmetric and positive-definite.

We claim that by the Frobenius theorem (Proposition 4.48), there exists a neighborhood $\mathcal{V}$ of $\mathbf{u}_0$ and a solution

$$E : \mathcal{V} \to \mathbb{R}^{(n+1)^2}$$

to the system (4.208) with $E(\mathbf{u}_0) = (e_1^0, \ldots, e_{n+1}^0)$. By Theorem 4.49, we need to show that the functions

$$\mathbf{A}_i^j(\mathbf{u}, E(\mathbf{u})) := \sum_{k=1}^{n+1} \mathbf{\Gamma}_{ij}^k(\mathbf{u}) e_k(\mathbf{u})$$

satisfy (4.201). In this case, this says that for $1 \leq i, k \leq n$, $1 \leq j \leq n+1$,

$$\frac{\partial \mathbf{A}_i^j}{\partial u^k} + \sum_{l=1}^{n+1} \mathbf{\Gamma}_{ij}^l \mathbf{A}_k^l = \frac{\partial \mathbf{A}_k^j}{\partial u^i} + \sum_{l=1}^{n+1} \mathbf{\Gamma}_{kj}^l \mathbf{A}_i^l. \tag{4.211}$$

That is, for $1 \leq i, k \leq n$, $1 \leq j, m \leq n+1$, we need to show that

$$\frac{\partial \mathbf{\Gamma}_{ij}^m}{\partial u^k} + \sum_{l=1}^{n+1} \mathbf{\Gamma}_{ij}^l \mathbf{\Gamma}_{kl}^m = \frac{\partial \mathbf{\Gamma}_{kj}^m}{\partial u^i} + \sum_{l=1}^{n+1} \mathbf{\Gamma}_{kj}^l \mathbf{\Gamma}_{il}^m. \tag{4.212}$$

This system of equations holds because it is equivalent to the Gauss–Codazzi equations of (4.203) assumed in the hypothesis of Theorem 4.50. Indeed, by (4.206), we have:

(1) The $m = j = n+1$ case of (4.212) is true because $\mathbf{\Gamma}_{a(n+1)}^{n+1} = 0$ and

$$\begin{aligned} \sum_{l=1}^{n} \mathbf{\Gamma}_{i(n+1)}^l \mathbf{\Gamma}_{kl}^{(n+1)} &= \sum_{l,p=1}^{n} g^{lp} h_{pi}(-h_{kl}) \\ &= \sum_{l,p=1}^{n} g^{lp} h_{pk}(-h_{il}) \\ &= \sum_{l=1}^{n} \mathbf{\Gamma}_{k(n+1)}^l \mathbf{\Gamma}_{il}^{(n+1)}. \end{aligned}$$

(2) The $m = n+1$, $j \leq n$ case of (4.212) is equivalent to

$$-\partial_k h_{ij} - \sum_{l=1}^{n} \mathbf{\Gamma}_{ij}^l h_{kl} = -\partial_i h_{kj} - \sum_{l=1}^{n} \mathbf{\Gamma}_{kj}^l h_{il},$$

which is also equivalent to

$$-\partial_k h_{ij} + \sum_{l=1}^{n} \Gamma_{kj}^{l} h_{il} + \sum_{p=1}^{n} \Gamma_{ki}^{p} h_{pj} = -\partial_i h_{kj} + \sum_{l=1}^{n} \Gamma_{ij}^{l} h_{kl} + \sum_{p=1}^{n} \Gamma_{ik}^{p} h_{pj},$$

which in turn by (4.182) is the Codazzi equation $\nabla_k^g h_{ij} = \nabla_i^g h_{kj}$.

(3) The $j = n+1$, $m \leq n$ case of (4.212) is true because the equation (let $h_i^m := g^{mk} h_{ki}$)

$$\partial_k h_i^m + \sum_{l=1}^{n} h_i^l \mathbf{\Gamma}_{kl}^{m} = \partial_i h_k^m + \sum_{l=1}^{n} h_k^l \mathbf{\Gamma}_{il}^{m}$$

is equivalent to the Codazzi equation $\nabla_k h_i^m = \nabla_i h_k^m$.

(4) Finally, the $j, m \leq n$ case of (4.212) says that

$$\frac{\partial \mathbf{\Gamma}_{ij}^{m}}{\partial u^k} + \sum_{l=1}^{n} \mathbf{\Gamma}_{ij}^{l} \mathbf{\Gamma}_{kl}^{m} - h_{ij} h_k^m = \frac{\partial \mathbf{\Gamma}_{kj}^{m}}{\partial u^i} + \sum_{l=1}^{n+1} \mathbf{\Gamma}_{kj}^{l} \mathbf{\Gamma}_{il}^{m} - h_{kj} h_i^m,$$

which by (4.182) is equivalent to

$$R_{kij}^{m} = h_{ij} h_k^m - h_{kj} h_i^m,$$

which in turn is equivalent to the Gauss equation.

Hence, we have verified (4.212). Therefore, by the Frobenius theorem, there exists a solution to (4.205) with $E(\mathbf{u}_0) = (e_1^0, \ldots, e_{n+1}^0)$.

We now claim that the Frobenius theorem implies that the maps $\{e_j\}_{j=1}^{n+1}$ satisfy

$$\langle e_i(\mathbf{u}), e_j(\mathbf{u}) \rangle = g_{ij}(\mathbf{u}) \quad \text{for } 1 \leq i, j \leq n \tag{4.213}$$

and

$$\langle e_i(\mathbf{u}), e_{n+1}(\mathbf{u}) \rangle = 0 \quad \text{and} \quad |e_{n+1}(\mathbf{u})| = 1. \tag{4.214}$$

To see this, let

$$\bar{g}_{ij} := \langle e_i(\mathbf{u}), e_j(\mathbf{u}) \rangle, \quad 1 \leq i, j \leq n+1, \tag{4.215}$$

and let $\partial_l = \frac{\partial}{\partial u^l}$. We compute using (4.205) that for $1 \leq i, j \leq n+1$ and $1 \leq l \leq n$,

$$\begin{aligned} \partial_l \bar{g}_{ij}(\mathbf{u}) &= \langle \partial_l e_i(\mathbf{u}), e_j(\mathbf{u}) \rangle + \langle e_i(\mathbf{u}), \partial_l e_j(\mathbf{u}) \rangle \\ &= \sum_{k=1}^{n+1} \langle \mathbf{\Gamma}_{li}^{k}(\mathbf{u}) e_k(\mathbf{u}), e_j(\mathbf{u}) \rangle + \sum_{k=1}^{n+1} \langle e_i(\mathbf{u}), \mathbf{\Gamma}_{lj}^{k}(\mathbf{u}) e_k(\mathbf{u}) \rangle \\ &= \sum_{k=1}^{n+1} \mathbf{\Gamma}_{li}^{k}(\mathbf{u}) \bar{g}_{kj}(\mathbf{u}) + \sum_{k=1}^{n+1} \mathbf{\Gamma}_{lj}^{k}(\mathbf{u}) \bar{g}_{ik}(\mathbf{u}). \end{aligned} \tag{4.216}$$

On the other hand, since the Γ^k_{ij} are the Christoffel symbols of g_{ij}, by defining $g_{i(n+1)} = g_{(n+1)i} = 0$ and $g_{(n+1)(n+1)} = 1$, we have by $\nabla^g g = 0$ and (4.206) that

$$\partial_l g_{ij} = \sum_{k=1}^{n+1} \mathbf{\Gamma}^k_{li} g_{kj} + \sum_{k=1}^{n+1} \mathbf{\Gamma}^k_{lj} g_{ik} \tag{4.217}$$

for $1 \leq i, j \leq n+1$ and $1 \leq l \leq n$.

One verifies that the first-order system of partial differential equations given by (4.216), and equivalently by (4.217), satisfies the Frobenius integrability condition, and hence we obtain from $\bar{g}_{ij}(\mathbf{u}_0) = g_{ij}(\mathbf{u}_0)$ and from uniqueness that $\bar{g}_{ij}(\mathbf{u}) = g_{ij}(\mathbf{u})$ for $\mathbf{u}$ in a neighborhood of $\mathbf{u}_0$ and for $1 \leq i, j \leq n+1$. This completes the proof of (4.213) and (4.214).

Write

$$e_i(\mathbf{u}) =: \sum_{k=1}^{n+1} B^k_i(\mathbf{u}) \mathbf{e}_k, \quad 1 \leq i \leq n, \tag{4.218}$$

where $\{\mathbf{e}_k\}_{k=1}^{n+1}$ is the standard basis for $\mathbb{R}^{n+1}$. From (4.207) it follows that

$$\frac{\partial B^k_j}{\partial u^i}(\mathbf{u}) = \frac{\partial B^k_i}{\partial u^j}(\mathbf{u}), \quad 1 \leq i, j \leq n, \ 1 \leq k \leq n+1. \tag{4.219}$$

This implies that (4.201) holds for $A^j_i(\mathbf{u}, \mathbf{f}(\mathbf{u})) = B^j_i(\mathbf{u})$. Therefore, by the Frobenius theorem, there exist a neighborhood $\mathcal{W}$ of $\mathbf{u}_0$ and functions $F^j : \mathcal{W} \to \mathbb{R}$, $1 \leq j \leq n+1$, that satisfy

$$\frac{\partial F^j}{\partial u^i}(\mathbf{u}) = B^j_i(\mathbf{u}) \quad \text{for } 1 \leq i \leq n, \tag{4.220}$$

with $F^j(\mathbf{u}_0) = 0$.

Define

$$F := (F^1, \ldots, F^{n+1}) : \mathcal{W} \to \mathbb{R}^{n+1}.$$

By (4.220) and (4.218), we have that

$$\partial_i F(\mathbf{u}) := \frac{\partial F}{\partial u^i}(\mathbf{u}) = \sum_{j=1}^{n+1} \frac{\partial F^j}{\partial u^i}(\mathbf{u}) \mathbf{e}_j = \sum_{j=1}^{n+1} B^j_i(\mathbf{u}) \mathbf{e}_j = e_i(\mathbf{u}) \tag{4.221}$$

for $1 \leq i \leq n$. Thus F is an embedding when restricted to some neighborhood $\mathcal{V}$ of $\mathbf{u}_0$ and satisfies

$$\langle \partial_i F(\mathbf{u}), \partial_j F(\mathbf{u}) \rangle = \langle e_i(\mathbf{u}), e_j(\mathbf{u}) \rangle = g_{ij}(\mathbf{u})$$

by (4.213), so that $g = F^* \mathrm{I}$ is equal to the first fundamental form of the embedded hypersurface $F(\mathcal{V})$.

Finally, we compute using (4.205), (4.206), (4.214), and (4.221) that at any $\mathbf{u} \in \mathcal{V}$ the components of the second fundamental form are given by

$$\begin{aligned}\mathrm{II}\left(\partial_i F, \partial_j F\right) &= -\langle \partial_i \partial_j F, e_{n+1}\rangle \\ &= -\langle \partial_i e_j, e_{n+1}\rangle \\ &= -\sum_{k=1}^{n+1} \mathbf{\Gamma}_{ij}^k \langle e_k, e_{n+1}\rangle \\ &= h_{ij}.\end{aligned}$$

Thus, $h = F^* \mathrm{II}$ is equal to the second fundamental form of $F(\mathcal{V})$. This completes the proof of the theorem. □

4.10. Notes and commentary

An excellent book that discusses both hypersurface theory and Riemannian geometry is Hicks [**Hic65**].

As an alternative to the parametrization (4.39) of a surface M^2 of revolution in $\mathbb{R}^3$, consider the parametrization given by (3.103), i.e., a map $G : I \times [0, 2\pi] \to \mathbb{R}^3$ of the form

$$G(t, \theta) = \big(r(t)\cos(\theta), r(t)\sin(\theta), z(t)\big),$$

where $r, z : I \to (0, \infty)$ satisfy the normalization

$$(r')^2 + (z')^2 = 1.$$

Using

$$\frac{\partial G}{\partial t} = \big(r'(t)\cos(\theta), r'(t)\sin(\theta), z'(t)\big), \quad \frac{\partial G}{\partial \theta} = \big(-r(t)\sin(\theta), r(t)\cos(\theta), 0\big),$$

we have that the components of the metric $g_{ij} := g\big(\frac{\partial}{\partial u^i}, \frac{\partial}{\partial u^j}\big)$, $u^1 := t$, $u^2 := \theta$, are given by

$$g_{11} = 1, \quad g_{22} = r(t)^2, \quad g_{12} = g_{21} = 0.$$

A choice of unit normal N to M^2 is given by

$$N(t, \theta) = \big(z'(t)\cos(\theta), z'(t)\sin(\theta), -r'(t)\big).$$

The components of the second fundamental form $\mathrm{II}_{ij} := \frac{\partial N}{\partial u^i} \cdot \frac{\partial G}{\partial u^j}$ are given by

$$\begin{aligned}\mathrm{II}_{11} &= z''(t)r'(t) - r''(t)z'(t), \\ \mathrm{II}_{22} &= z'(t)r(t), \\ \mathrm{II}_{12} &= \mathrm{II}_{21} = 0.\end{aligned}$$

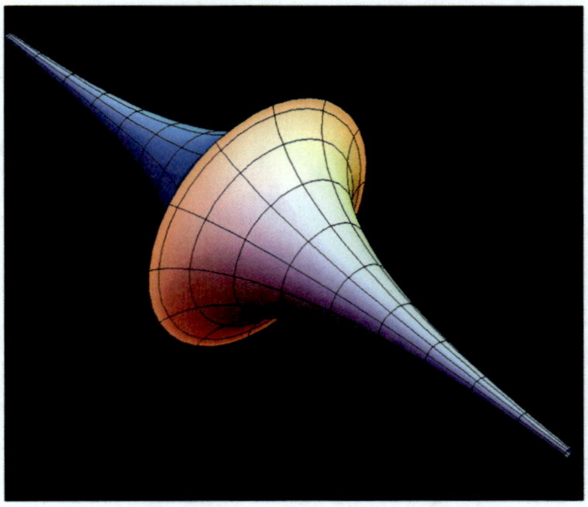

Figure 4.10.1. A pseudosphere. Credit: 3DXM Virtual Mathematics Museum by The 3D-XplorMath Consortium. © Copyright 3DXM Consortium. `https://3d-xplormath.org/`

We calculate that

$$\begin{aligned} \mathrm{II}_{11}\,\mathrm{II}_{22} &= z''(t)z'(t)r'(t)r(t) - r''(t)(z'(t))^2 r(t) \\ &= -r''(t)r'(t)r'(t)r(t) - r''(t)\big(1-(r'(t))^2\big)r(t) \\ &= -r(t)r''(t). \end{aligned}$$

Therefore the Gauss curvature K of M^2 is given by

$$K = \frac{\mathrm{II}_{11}\,\mathrm{II}_{22} - \mathrm{II}_{12}\,\mathrm{II}_{21}}{g_{11}g_{22} - g_{12}g_{21}} = -\frac{r''(t)}{r(t)}. \tag{4.222}$$

For example, if $r''(t) = r(t)$ for $t \in I$, then $K = -1$. In particular, define $r : (0,\infty) \to (0,\infty)$ by

$$r(t) = \mathrm{e}^{-t}.$$

If we choose $z(t)$ so that $z'(t) = \sqrt{1-\mathrm{e}^{-2t}}$, then G parametrizes a surface of revolution with constant Gauss curvature -1. In particular, we may take $z(t) = \int_0^t \sqrt{1-\mathrm{e}^{-2u}}du$. Such a surface is called a **pseudosphere**.[7]

Similarly, we may define $r : (-\infty, 0) \to (0,\infty)$ by $r(t) = \mathrm{e}^t$. Putting these two surfaces together, we obtain the surface in Figure 4.10.1. Observe that taking the closure of this surface adds the circle $S^1 \times \{0\}$.

4.11. Exercises

Exercise 4.1 (Circles in $\mathbb{R}^N$)**.** *Let* $\mathbf{V}$ *and* $\mathbf{W}$ *be unit vectors in* $\mathbb{R}^N$ *satisfying* $\mathbf{V}\cdot\mathbf{W} = 0$. *Let r be a positive real number. Define the path* $\gamma : [0, 2\pi] \to \mathbb{R}^N$

[7]The graph of the curve that is being rotated to obtain the pseudosphere is called a **tractrix**; see Spivak [**Spi79c**].

by

$$\gamma(t) = r\cos(t)\mathbf{V} + r\sin(t)\mathbf{W}.$$

Prove that the length of the path γ *satisfies*

$$\mathrm{L}(\gamma) = 2\pi r.$$

Prove also that γ *is a closed path.*

Exercise 4.2. *Show that the unit tangent vector for* γ_- *at* t *is equal to the negative of the unit tangent vector for* γ *at* $-t$.

Exercise 4.3. *Let* γ *and* γ_- *be as in Exercise* 4.2. *Given* $t_0 \in I_-$, *define* $s_-(t_1) := \int_{t_0}^{t_1} \|\gamma_-'(t)\|\,dt$, *which is an arc-length parameter for* γ_-. *Prove that*

$$s_-(t_1) = -s(-t_1),$$

where

$$s(\tau) := \int_{-t_0}^{\tau} \|\gamma'(t)\|\,dt,$$

which is an arc-length parameter for γ.

Exercise 4.4 (Product rule for the inner product of vector fields along a path). *Prove* (4.4). *What does* (4.5) *say when* $V(t)$ *is a unit vector field?*

Exercise 4.5. *Prove that a circle of radius* r *has constant curvature* $1/r$ *if we use the inward pointing unit normal vector field to define the curvature.*

Exercise 4.6. *Let* $\kappa(s)$ *denote the curvature of a planar path* $\gamma(s)$ *with respect to a choice* $N(s)$ *of unit normal vector field. Prove that the curvature of* $\gamma(s)$ *with respect to the opposite choice* $-N(s)$ *of unit normal vector field is equal to* $-\kappa(s)$.

Exercise 4.7. *Let* γ *be a continuously differentiable path parametrized by arc length. Choose unit normal vector fields to* γ *and* γ_- *so that* $N_{\gamma_-}(s) = N_\gamma(-s)$. *Prove that the curvatures of* γ *and* γ_- *are equal; namely,*

$$\kappa_{\gamma_-}(s) = \kappa_\gamma(-s).$$

Exercise 4.8. *Write in detail the derivation of* (4.9).

Exercise 4.9. *Even though the trefoil immersed path* F *in the plane given by Example* 2.22 *is immersed, it is pretty obvious geometrically that there is a natural choice for the 'inward pointing' unit normal vector field* N *to* F. *Calculate* N. *Using this normal, calculate the curvature* κ *of the path* F.

Exercise 4.10. *Let* $r > w$ *be positive real numbers and let* k *be a positive integer. Compute the curvature, with respect to the inward pointing normal, of the closed path* γ *defined in polar coordinates by*

$$\gamma(\theta) = \big(r + w\cos(k\theta), \theta\big), \quad \theta \in [0, 2\pi].$$

See Figure 4.11.1.

Figure 4.11.1. The closed path γ in the case where $k = 12$.

Exercise 4.11. *Verify* (4.26), *and show that*

$$\|\mathbf{x} - \mathbf{y}\| \leq d_{S^1}(\mathbf{x}, \mathbf{y}),$$

with equality if and only if $\mathbf{x} = \mathbf{y}$, *in which case both the extrinsic and intrinsic distances are zero.*

Exercise 4.12. *Verify* (4.29), (4.30), *and* (4.31). *You may assume that the Euclidean inner product satisfies these properties. So this exercise is easy.*

Also explain why (4.31) *follows directly from* (4.29) *and* (4.30).

Exercise 4.13. *Define the* ***energy*** E *of a path* $\gamma : [t_1, t_2] \to M^n$ *by*

$$\mathrm{E}(\gamma) := \int_{t_1}^{t_2} \|\gamma'(t)\|^2 \, dt = \int_{t_1}^{t_2} \mathrm{I}(\gamma'(t), \gamma'(t)) \, dt. \tag{4.223}$$

The Cauchy–Schwarz inequality (as a special case of ***Hölder's inequality****) says that for any two continuous functions* f *and* g *on the interval* $[t_1, t_2]$, *we have*

$$\left(\int_{t_1}^{t_2} |f(t)g(t)| \, dt \right)^2 \leq \int_{t_1}^{t_2} f(t)^2 dt \int_{t_1}^{t_2} g(t)^2 dt, \tag{4.224}$$

where equality holds if and only if $c_1|f(t)| = c_2|g(t)|$ *on* $[t_1, t_2]$ *for some constants* c_1, c_2, *not both of them zero. Use this inequality to prove that*

$$\mathrm{L}(\gamma)^2 \leq (t_2 - t_1) \, \mathrm{E}(\gamma), \tag{4.225}$$

where equality holds if and only if $\|\gamma'(t)\| \equiv \frac{\mathrm{L}(\gamma)}{t_2 - t_1}$ *on* $[t_1, t_2]$.

Exercise 4.14. *Verify* (4.38).

Exercise 4.15. *Verify* (4.42).

Exercise 4.16. *Prove the product rule for the Euclidean covariant derivative* (4.55).

Exercise 4.17. *Prove the Jacobi identity* (4.59).
Hint: For this exercise we only need to use (4.57)*. Continue the calculation*

$$\big[[U,V],W\big](f) = [U,V]\big(W(f)\big) - W\big([U,V](f)\big),$$

and then write out the analogous results for the other two cyclic permutations of U, V, W.

Exercise 4.18. *Prove the torsion-free condition for the Euclidean covariant derivative* (4.63).
Hint: On one hand, regarding the left-hand side, we have

$$(D_V W)(f) = \sum_{i=1}^{N} (D_V W)^i \frac{\partial f}{\partial u^i} = \sum_{i,j=1}^{N} V^j \frac{\partial W^i}{\partial u^j}\frac{\partial f}{\partial u^i},$$

and also the similar formula with V *and* W *switched. On the other hand, regarding the right-hand side, we have*

$$\begin{aligned} V\big(W(f)\big) &= V\left(\sum_{i=1}^{N} W^i \frac{\partial f}{\partial u^i}\right) \\ &= \sum_{i,j=1}^{N} V^j \frac{\partial W^i}{\partial u^j}\frac{\partial f}{\partial u^i} + \sum_{i,j=1}^{N} V^j W^i \frac{\partial^2 f}{\partial u^j \partial u^i}, \end{aligned}$$

and also the similar formula with V *and* W *switched.*

Exercise 4.19 (II is bilinear)**.** *Prove* (4.79).

Exercise 4.20. *Let* M^n *be the* n*-dimensional sphere of radius* r *centered at the origin in* $\mathbb{R}^{n+1}$.

(i) *Prove that the Weingarten map of* M^n *is given by*

$$\mathrm{L}(\mathbf{V}) = \frac{1}{r}\mathbf{V} \tag{4.226}$$

for $\mathbf{V} \in T_{\mathbf{x}}M$, $\mathbf{x} \in M^n$.

(ii) *Prove that the second fundamental form of* M^n *is given by*

$$\mathrm{II}(\mathbf{V},\mathbf{W}) = \frac{1}{r}\,\mathrm{I}(\mathbf{V},\mathbf{W}), \tag{4.227}$$

where I *is the first fundamental form of* M^n.

Exercise 4.21. *Verify the details of the proof of the assertion in Example* 4.24 *that the catenoid is a minimal surface.*

Exercise 4.22. *Let* $V = (V_1, \dots, V_n) \in \mathbb{R}^n$*. Show that the inverse* A^{-1} *of the* $n \times n$ *matrix* $A = (A_{ij})$*, where* $A_{ij} := \delta_{ij} + V_i V_j$*, has components*

$$(A^{-1})_{ij} = \delta_{ij} - \frac{1}{\|V\|^2 + 1} V_i V_j. \tag{4.228}$$

Exercise 4.23. *Prove that, in general, for a smooth embedded hypersurface M^n in $\mathbb{R}^{n+1}$, the formula for the mean curvature is given by*

$$H = \sum_{i,j=1}^{n} \mathrm{I}^{ij}\, \mathrm{II}_{ij}\,. \tag{4.229}$$

Exercise 4.24. *Analogously to Exercise* 4.16*, prove the metric compatibility property* (4.127) *of the Levi-Civita connection.*

Exercise 4.25. *Prove the torsion-free property* (4.129) *for the Levi-Civita connection.*

Exercise 4.26. *Let* $\mathbf{V}, \mathbf{W}$ *be unit vectors in* $\mathbb{R}^{n+1}$ *that are perpendicular to each other. Prove that the path* $\gamma : \mathbb{R} \to S^n$ *defined by*

$$\gamma(t) = \cos(t)\,\mathbf{V} + \sin(t)\,\mathbf{W}$$

is a geodesic in (S^n, I)*, where* I *is the first fundamental form of* S^n*.*

Exercise 4.27. *Prove the formula* (4.141) *for the Levi-Civita connection in local coordinates.*

Exercise 4.28. *Prove Lemma* 4.36 *that the Riemann curvature tensor is multilinear and antisymmetric in its first two components.*

Exercise 4.29. *Complete the proof of Theorem* 4.38*.*

Exercise 4.30. *Prove that* $\Phi : M^2_{\mathrm{slab}} \to M^2_{\mathrm{cyl}}$ *defined by* (4.196) *is an isometry.*

Exercise 4.31. *Prove equation* (4.185) *for the Riemann curvature* $(3,1)$*-tensor in terms of local coordinates.*

Chapter 5

Global Theorems in the Theory of Submanifolds

Figure 5.0.1. Boy's Surface (Bryant and Kusner), which is an immersion of the real projective plane in $\mathbb{R}^3$. Credit: 3DXM Virtual Mathematics Museum by The 3D-XplorMath Consortium. © Copyright 3DXM Consortium. `https://3d-xplormath.org/`

In this chapter, we discuss a small selection of global results on hypersurfaces in Euclidean space. In §5.1 we consider totally umbilical hypersurfaces and we classify them as round spheres or hyperplanes. In §5.2 we explore closed hypersurfaces and show that there exist points about which they are locally convex; we also discuss the strict convexity in the case of nonvanishing Gauss curvature. In §5.3 we discuss some global results on convex hypersurfaces. In §5.4 we discuss the Hessian, Laplacian, and divergence operators. Integral identities for hypersurfaces are proved in §5.5. In §5.6 we discuss the rigidity of closed convex surfaces in $\mathbb{R}^3$. §5.7 is a short appendix on calculating the Hessian and Laplacian of the position map in local coordinates.

5.1. Totally umbilical hypersurfaces

Spheres and hyperplanes are the most symmetric Euclidean hypersurfaces. In this section we show that we can characterize these hypersurfaces by a simple property that their second fundamental forms satisfy.

5.1.1. The totally umbilical hypersurface equation. Suppose that $n \geq 2$ and that an embedded hypersurface M^n in $\mathbb{R}^{n+1}$ has the property that its second fundamental form is equal to a function times its first fundamental form; that is,

$$\mathrm{II} = f\,\mathrm{I} \tag{5.1}$$

for some function f on M^n. In this case we say that M^n is **totally umbilical**.

By tracing equation (5.1), we have that the mean curvature of M^n satisfies $H = nf$, so that equation (5.1) becomes

$$\mathrm{II} = \frac{1}{n} H\,\mathrm{I}. \tag{5.2}$$

Note that this equation is true for any curve in the plane and this is why we restrict our attention to the case where $n \geq 2$.

The **divergence** of a 2-tensor α on M^n is defined by

$$\operatorname{div}(\alpha)_k := I^{ij}\nabla_i \alpha_{jk}, \tag{5.3}$$

where ∇ denotes the Levi-Civita connection of (M^n, I). Recall that (I^{ij}) is the inverse matrix of (I_{ij}) and that the components of the covariant derivative of the tensor α are denoted in local coordinates by $\nabla_i \alpha_{jk} := (\nabla_{\partial_i}\alpha)(\partial_j, \partial_k)$. More generally, the divergence of a k-tensor is a trace of the covariant derivative of the tensor, which produces a $(k-1)$-tensor. In particular, the divergence of the second fundamental form is the 1-form given by

$$\operatorname{div}(\mathrm{II})_k := I^{ij}\nabla_i \, \mathrm{II}_{jk}\,. \tag{5.4}$$

(See §5.4.2 below for further discussion of the divergence operator.)

5.1.2. Classification of complete totally umbilical hypersurfaces. The main result of this section is the following.

Proposition 5.1 (Complete totally umbilical hypersurfaces are spheres or hyperplanes). *Any complete (as a metric space), connected, totally umbilical embedded hypersurface M^n in $\mathbb{R}^{n+1}$ is either a round sphere (if $H \neq 0$) or a hyperplane (if $H = 0$).*

Proof. We may take the divergence of equation (5.2) and use the Codazzi equation (4.175) to obtain

$$I^{ij}\frac{1}{n}\nabla_i H \,\mathrm{I}_{jk} = I^{ij}\nabla_i \,\mathrm{II}_{jk} = I^{ij}\nabla_k \,\mathrm{II}_{ij};$$

that is,

$$\frac{1}{n}\nabla_k H = \nabla_k H. \tag{5.5}$$

Since $n \neq 1$, this implies that $\nabla H = 0$ on M^n and since we assumed that M^n is connected, this implies that the mean curvature H is constant on M^n. Thus, II is equal to a constant times I on all of M^n.

Denote $\kappa := \frac{1}{n}H$, so that our connected hypersurface M^n in $\mathbb{R}^{n+1}$ has the property that $\mathrm{II} = \kappa\,\mathrm{I}$ on M^n, where κ is a constant. Let N be the unit normal vector field to M^n used to define the second fundamental form. Define the map $F : M^n \to \mathbb{R}^{n+1}$ by

$$F(\mathbf{x}) := \kappa\mathbf{x} - N_{\mathbf{x}}. \tag{5.6}$$

We may think of F as a Euclidean vector field on M^n; that is,

$$F = \kappa\,\mathrm{id} - N, \tag{5.7}$$

where $\mathrm{id}(\mathbf{x}) = \mathbf{x}$ is the identity vector field. For any tangent vectors $\mathbf{V}$ and $\mathbf{W}$ on M^n, using that κ is a constant we compute that

$$\begin{aligned}\langle D_{\mathbf{V}}F, \mathbf{W}\rangle &= \langle D_{\mathbf{V}}(\kappa\,\mathrm{id} - N), \mathbf{W}\rangle \\ &= \langle \kappa\mathbf{V} - D_{\mathbf{V}}N, \mathbf{W}\rangle \\ &= \kappa\,\mathrm{I}(\mathbf{V},\mathbf{W}) - \mathrm{II}(\mathbf{V},\mathbf{W}) \\ &= 0,\end{aligned}$$

where D denotes the Euclidean covariant derivative. We also have

$$\langle D_{\mathbf{V}}F, N\rangle = \langle \kappa\mathbf{V} - D_{\mathbf{V}}N, N\rangle = 0.$$

Thus, $D_{\mathbf{V}}F = \mathbf{0}$ for all tangent vectors $\mathbf{V}$. Since M^n is connected, we obtain that the vector field F on M^n is equal to a constant vector, which we call $\mathbf{X}_0$.

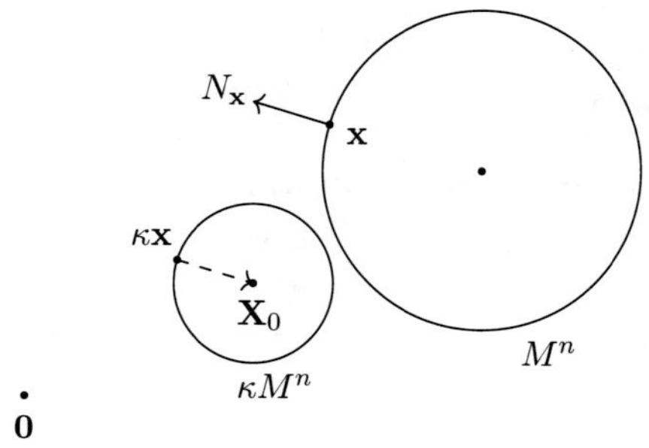

Figure 5.1.1. Visualizing the equation $\kappa\mathbf{x} - N_{\mathbf{x}} = \mathbf{X}_0$. This is an example of rescaling by curvature, which in this case is κ.

(1) Suppose that $\kappa \neq 0$. We then conclude that for all $\mathbf{x} \in M^n$, we have that $\kappa\mathbf{x} - N_{\mathbf{x}} = F(\mathbf{x}) = \mathbf{X}_0$ (see Figure 5.1.1); that is,

$$\mathbf{x} = \frac{1}{\kappa}(\mathbf{X}_0 + N_{\mathbf{x}}).$$

Since each $N_{\mathbf{x}}$ is a unit vector, we obtain from this that M^n is a relatively open subset of the sphere of radius $\frac{1}{|\kappa|}$ centered at the point $\frac{1}{\kappa}\mathbf{X}_0$. Since M^n is assumed to be complete, we conclude that M^n must be equal to the entire sphere.

(2) Suppose that $\kappa = 0$; that is, M^n is totally geodesic. Since M^n is connected, in this case we have that N is a constant vector field on M^n, which we call $\mathbf{N}_0$. For any tangent vector $\mathbf{V}$ of M^n we have that

$$\mathbf{V}(\mathrm{id} \cdot N) = \mathbf{V} \cdot N = \mathbf{V} \cdot \mathbf{N}_0 = 0.$$

Since M^n is assumed to be connected, this implies that the function $\mathrm{id} \cdot N$ is equal to a constant on M^n; see Exercise 5.10. That is, there exists $C \in \mathbb{R}$ such that $\mathbf{x} \cdot \mathbf{N}_0 = C$ for all $\mathbf{x} \in M^n$. Thus, M^n is a relatively open subset of a hyperplane. Since M^n is complete, we have that M^n is equal to the entire hyperplane. This completes the proof of the proposition. □

5.2. Closed hypersurfaces

Hypersurfaces can be categorized by whether they are compact or noncompact. In this section we demonstrate some nice properties of closed (i.e., compact without boundary) hypersurfaces.

5.2.1. Existence of a locally convex neighborhood. Let M^n be a connected closed hypersurface embedded in $\mathbb{R}^{n+1}$. By the Jordan–Brouwer Separation Theorem (see [**GP10**, Chapter 2, §5]), we have that $\mathbb{R}^{n+1} \setminus M^n$ consists of two components, exactly one of which has compact closure; we

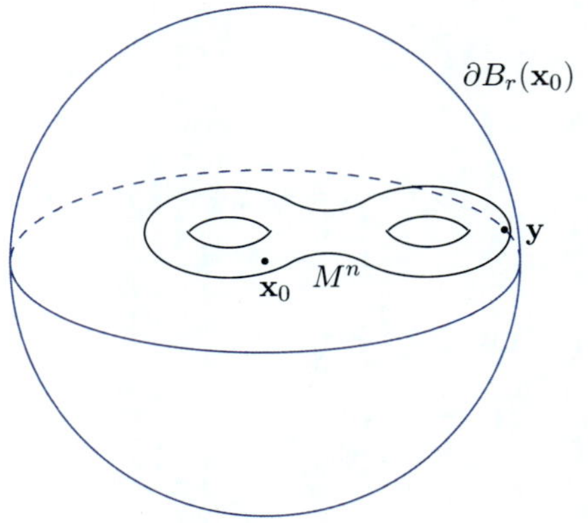

Figure 5.2.1. A closed embedded hypersurface M^n enclosed in the smallest ball $\overline{B_r(\mathbf{x}_0)}$. The point $\mathbf{y}$ is a point on M^n farthest from $\mathbf{x}_0$.

call this component the **interior region**. Let N be the unit **outward normal** to M^n; that is, N points in the direction opposite of the interior region. Note that the unit normal field N is a smooth vector field. In this case, we say that M^n is **oriented** and that N defines an orientation on M^n (see §8.1 below for a more general definition).

Proposition 5.2 (Existence of a locally convex neighborhood in any closed hypersurface). *If M^n is a closed hypersurface embedded in $\mathbb{R}^{n+1}$, then there exists a point on M^n for which the second fundamental form* II *is positive-definite in a neighborhood of this point.*

Proof. Fix a point $\mathbf{x}_0 \in \mathbb{R}^{n+1}$; e.g., we can take $\mathbf{x}_0$ to be the origin. Since the distance function to $\mathbf{x}_0$ is continuous on M^n and since M^n is compact, there exists a point $\mathbf{y} \in M^n$ that maximizes the distance to $\mathbf{x}_0$. We have

$$M^n \subset \overline{B_r(\mathbf{x}_0)},$$

where $r := \|\mathbf{y} - \mathbf{x}_0\| > 0$ and $\mathbf{y} \in \partial B_r(\mathbf{x}_0)$. See Figure 5.2.1.

Observe that the unit outward normal to both M^n and $\partial B_r(\mathbf{x}_0)$ at $\mathbf{y}$ is given by the single vector $N_{\mathbf{y}} = \frac{\mathbf{y}-\mathbf{x}_0}{\|\mathbf{y}-\mathbf{x}_0\|}$. Let $\kappa_i(\mathbf{y})$ be any principal curvature of M^n at $\mathbf{y}$ and let e_i be an associated principal vector, so that

$$\mathrm{L}(e_i) = \kappa_i(\mathbf{y}) e_i,$$

where $\mathrm{L} : T_{\mathbf{y}}M \to T_{\mathbf{y}}M$ is the Weingarten map.

Let P be the 2-plane containing the points $\mathbf{x}_0$ and $\mathbf{y}$ and spanned by the vectors $N_{\mathbf{y}}$ and e_i. Consider the curve C in M^n defined by $M^n \cap P$. Since $\mathbf{y}$ is a farthest point from $\mathbf{x}_0$, we have that C is contained in the 2-disk $\bar{D}_r(\mathbf{x}_0) = \overline{B_r(\mathbf{x}_0)} \cap P$. Near $\mathbf{y}$, parametrize C by the graphical path $\beta : I \to P$ with $\beta'(0) = e_i$ and $\beta(t) = te_i + f(t)N_{\mathbf{y}}$, where $f : I \to \mathbb{R}$,

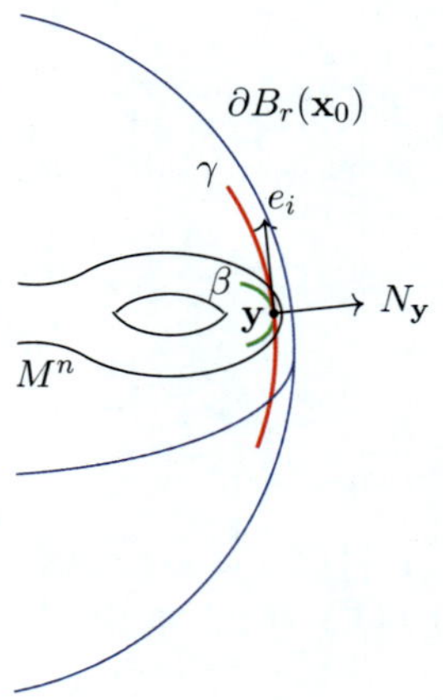

Figure 5.2.2. The path β in M^n and the comparison path γ in $\partial B_r(\mathbf{x}_0)$.

$f(0) = r$, and $f'(0) = 0$. We have

$$\kappa_i(\mathbf{y}) = -\left.\langle D_{e_i}\beta'(t), N_{\mathbf{y}}\rangle\right|_{t=0} = -f''(0). \tag{5.8}$$

On the other hand, in comparison, near $\mathbf{y}$ the round circle $\partial B_r(\mathbf{x}_0) \cap P$ is parametrized by the graphical path $\gamma : I \to P$ defined by $\gamma(t) = te_i + g(t)N_{\mathbf{y}}$, where $g(t) = \sqrt{r^2 - t^2}$. See Figure 5.2.2.

Since $C \subset \bar{D}_r(\mathbf{x}_0)$, we have $f(t) \leq g(t)$ and $f(0) = g(0)$. Thus, by the second derivative test in the sense of calculus, we have

$$\kappa_i(\mathbf{y}) = -f''(0) \geq -g''(0) = r^{-1} > 0.$$

In conclusion, we have shown that $\mathrm{II} \geq r^{-1}\,\mathrm{I}$ at the point $\mathbf{y}$. In particular, $H(\mathbf{y}) \geq nr^{-1}$ and $K(\mathbf{y}) \geq r^{-n}$. Moreover, since the second fundamental form is continuous, this implies that II is positive-definite in a neighborhood of $\mathbf{y}$. □

For the proof of the following, see Exercise 5.11.

Corollary 5.3. *There do not exist closed immersed minimal hypersurfaces in* $\mathbb{R}^{n+1}$.

5.2.2. The Gauss map. Let $F : M^n \to \mathbb{R}^{n+1}$ be an immersed hypersurface. Assume that there exists a smooth unit normal vector field $\mathbf{x} \mapsto N_{F(\mathbf{x})}$, so that $N_{F(\mathbf{x})}$ is perpendicular to $T_{F(\mathbf{x})}M$ for $\mathbf{x} \in M^n$. In particular, M^n is oriented.

Definition 5.4. The **Gauss map** of an oriented immersed hypersurface is the map of a point to its normal; that is, $G : M^n \to S^n \subset \mathbb{R}^{n+1}$ is defined by

$$G(\mathbf{x}) := N_{F(\mathbf{x})}. \tag{5.9}$$

Locally, we may consider M^n as an embedded hypersurface, so that $M^n \subset \mathbb{R}^{n+1}$ and the parametrization $F : M^n \to \mathbb{R}^{n+1}$ is the inclusion map. Then we have that $T_{\mathbf{x}}M$ and $T_{G(\mathbf{x})}S^n$ are both perpendicular to $N_{\mathbf{x}}$. The derivative of the Gauss map is the Weingarten map:

$$dG_{\mathbf{x}} = \mathrm{L} : T_{\mathbf{x}}M \to T_{G(\mathbf{x})}S^n. \tag{5.10}$$

Hence, the Gauss curvature may be expressed as

$$K(\mathbf{x}) = \det(\mathrm{L}) = \det(dG_{\mathbf{x}}). \tag{5.11}$$

Let $d\mu$ denote the volume form on M^n with respect to the first fundamental form $g := \mathrm{I}$ and the orientation on M^n induced by N. Let $d\bar{\mu}$ denote the volume form of the unit-sphere metric $\bar{g}$ on S^n. See §8.2.6 below for a discussion of the volume form of an oriented Riemannian manifold.

Lemma 5.5. *We have*

$$K\, d\mu = G^*(d\bar{\mu}), \tag{5.12}$$

where $G : M^n \to S^n$ is the Gauss map.

Proof. Let $\{e_i\}_{i=1}^n$ be an orthonormal frame at a point $\mathbf{x} \in M^n$ such that $e_i \in T_{F(\mathbf{x})}M$ and $\{N_{F(\mathbf{x})}, e_1, \dots, e_n\}$ is positively oriented with respect to the standard orientation on $\mathbb{R}^{n+1}$.[1] (Denoting $e_0 = N_{F(\mathbf{x})}$, by definition we have that $\langle e_i, e_j\rangle = \delta_{ij}$ for $0 \leq i, j \leq n$.) Since we may identify $T_{F(\mathbf{x})}M$ with $T_{G(\mathbf{x})}S^n$ (as both of these hyperplanes are normal to $N_{F(\mathbf{x})}$), we may consider $\{e_i\}_{i=1}^n$ as an orthonormal tangent frame at $G(\mathbf{x}) \in S^n$. We also have that $\{e_i\}_{i=1}^n$ is positively oriented with respect to the standard orientation on S^n since $N_{F(\mathbf{x})}$ is equal to the unit outward normal to S^n at $G(\mathbf{x})$. Let $\{\omega^j\}_{j=1}^n$ be the dual orthonormal coframe, which is the basis for the cotangent space $T^*_{F(\mathbf{x})}M = T^*_{G(\mathbf{x})}S^n$ satisfying $\omega^j(e_i) = \delta_{ij}$ for $1 \leq i, j \leq n$. Then

$$d\mu_{\mathbf{x}} = \omega^1 \wedge \cdots \wedge \omega^n \qquad \text{on } M^n \tag{5.13}$$

and

$$d\bar{\mu}_{G(\mathbf{x})} = \omega^1 \wedge \cdots \wedge \omega^n \qquad \text{on } S^n. \tag{5.14}$$

[1]That is, the $(n+1) \times (n+1)$ matrix $\begin{pmatrix} N_{F(\mathbf{x})} & e_1 & \cdots & e_n \end{pmatrix}$ has positive determinant.

We compute using (5.14) that

$$\begin{aligned} G^*(d\bar{\mu})(e_1,\dots,e_n) &= d\bar{\mu}\big(dG(e_1),\dots dG(e_n)\big) \\ &= d\bar{\mu}\big(\mathrm{L}(e_1),\dots,\mathrm{L}(e_n)\big) \\ &= \det(\mathrm{L})\, d\bar{\mu}(e_1,\dots,e_n) \\ &= K, \end{aligned}$$

where $\mathrm{L} : T_{F(\mathbf{x})}M \to T_{F(\mathbf{x})}M$ is the Weingarten map. By (5.13), this implies that $K\, d\mu = G^*(d\bar{\mu})$. □

By integrating (5.12) and using the naturality of the integration of differential forms (see (8.22) below), we have:

Corollary 5.6. *If $G : M^n \to S^n$ is a diffeomorphism (which is necessarily orientation preserving), then*

$$\int_{M^n} K\, d\mu = n\omega_{n+1}, \tag{5.15}$$

where $n\omega_{n+1}$ is the volume of the unit n-sphere.

5.3. Convex closed hypersurfaces

Let M^n be an oriented immersed complete hypersurface in $\mathbb{R}^{n+1}$. We say that M^n is **locally uniformly convex** if its second fundamental form is positive-definite; i.e., $\mathrm{II} > 0$ on M^n.

Let M^n be an embedded closed hypersurface and let Ω be its interior region, which has compact closure $\overline{\Omega}$. We say that M^n is **weakly convex** if for every $x, y \in M^n$, the line segment $\overline{xy}$ is contained in $\overline{\Omega}$. We say that M^n is **strictly convex** if for every pair of distinct points $x, y \in M^n$, the interior of the line segment $\overline{xy}$ is contained in Ω. For example, the boundary of a cube, with its edges and vertices rounded off so that the surface is smooth, is weakly convex. On the other hand, the boundary of an ellipsoid is strictly convex.

5.3.1. Convexity of closed hypersurfaces with nonvanishing Gauss curvature.

In this subsection we discuss a result of Hadamard [**Had97**] and Stoker [**Sto36**], which gives a nice characterization of embedded closed convex hypersurfaces.

Lemma 5.7. *If $F : M^n \to \mathbb{R}^{n+1}$ is an orientable immersed closed hypersurface with Gauss curvature never vanishing and $n \geq 1$, then there exists a smooth choice of unit normal field such that $\mathrm{II} > 0$ on all of M^n; that is, M^n is locally uniformly convex.*

Proof. By Exercise 5.11, there exists an open subset $\mathcal{U}$ of M^n, with a choice of a smooth unit normal vector field N on $\mathcal{U}$, on which $\mathrm{II} > 0$. Since M^n is orientable, we can extend N to all of M^n as a smooth unit normal vector field. Since $K \neq 0$ everywhere, by continuity we have that

$$\mathrm{II} > 0 \quad \text{on all of } M^n. \tag{5.16}$$

Indeed, suppose for a contradiction that II were not positive-definite on all of M^n. Let $\kappa_1 : M^n \to \mathbb{R}$ be the smallest principal curvature function. Then

$$\kappa_1(\mathbf{x}) = \inf_{\mathbf{V} \in T_{F(\mathbf{x})}M,\, \|\mathbf{V}\|=1} \mathrm{II}(\mathbf{V}, \mathbf{V}),$$

so that κ_1 would be continuous and change sign. Therefore, by the intermediate value theorem, there would exist a point $\mathbf{x}_0 \in M^n$ such that $\kappa_1(\mathbf{x}_0) = 0$, which would imply that $K(\mathbf{x}_0) = 0$, which is a contradiction. □

Lemma 5.8. *If M^n is an orientable closed hypersurface immersed in $\mathbb{R}^{n+1}$ with $\mathrm{II} > 0$ and $n \geq 2$, then there exists a smooth choice of a unit normal field such that the Gauss map $G : M^n \to S^n$ is a diffeomorphism.*

Proof. Since $\det(dG) = \det(\mathrm{L}) = K > 0$, we have that the Gauss map G is an immersion. This implies that G is locally a diffeomorphism. Hence, for each $\mathbf{y} \in S^n$, its preimage $G^{-1}(\mathbf{y})$ is a finite subset of M^n. So we may write $G^{-1}(\mathbf{y}) =: \{\mathbf{x}_1, \ldots, \mathbf{x}_k\}$, where $k := |G^{-1}(\mathbf{y})|$. Let $r := \frac{1}{2} \min_{i \neq j} \|\mathbf{x}_i - \mathbf{x}_j\|$ and define $\mathcal{U}_i := M^n \cap B_r(\mathbf{x}_i)$. Then $\{\mathcal{U}_i\}_{i=1}^k$ is a collection of disjoint open subsets of M^n and $\mathbf{x}_i \in \mathcal{U}_i$ for each i. We claim that there exists $\epsilon > 0$ such that

$$G^{-1}(B_\epsilon(\mathbf{y}) \cap S^n) \subset \bigcup_{i=1}^k \mathcal{U}_i =: \mathcal{U}. \tag{5.17}$$

If not, then for each positive integer j there would exist $\mathbf{w}_j \in M^n \setminus \mathcal{U}$ such that $G(\mathbf{w}_j) \in B_{1/j}(\mathbf{y})$. Since M^n is compact, by passing to a subsequence, we would have that $\mathbf{w}_j$ converges to some point $\mathbf{w}_\infty \in M^n$. Since $M^n \setminus \mathcal{U}$ is a closed set, we would have that $\mathbf{w}_\infty \in M^n \setminus \mathcal{U}$. By the continuity of G, we would also have that $G(\mathbf{w}_\infty) = \lim_{j\to\infty} G(\mathbf{w}_j) = \mathbf{y}$, which is a contradiction since $\mathbf{w}_\infty \notin \mathcal{U} \supset G^{-1}(\mathbf{y})$. Thus we have proved the existence of $\epsilon > 0$ satisfying (5.17).

Now, since G is a local diffeomorphism, there exists $\epsilon' \in (0, \epsilon)$ such that $G^{-1}(\mathcal{V}')$, where $\mathcal{V}' := B_{\epsilon'}(\mathbf{y}) \cap S^n$, is equal to the disjoint union of sets $\mathcal{U}_i'$, $1 \leq i \leq k$, such that G maps each $\mathcal{U}_i'$ diffeomorphically to $\mathcal{V}'$ and $\mathbf{x}_i \in \mathcal{U}_i'$ for each i. This proves that G is a smooth covering map (see Definition 6.33 below for the definition). Furthermore, because the image of G is S^n, which is simply connected since $n \geq 2$, we conclude that the Gauss map $G : M^n \to S^n$ is a diffeomorphism (see [**Mas77**] for covering space theory). □

Theorem 5.9 (Hadamard and Stoker). *If M^n is an orientable closed hypersurface immersed in $\mathbb{R}^{n+1}$ with Gauss curvature never vanishing and $n \geq 2$, then M^n is strictly convex, embedded, and* $\mathrm{II} > 0$ *on all of M^n.*

Proof. We first show that for each $\mathbf{x} \in M^n$, the immersed hypersurface M^n lies on one side of the tangent hyperplane $T_{F(\mathbf{x})}M^n$. To this end and for a contradiction, suppose that there exists $\mathbf{x}_0 \in M^n$ such that M^n lies on both sides of $T_{F(\mathbf{x}_0)}M^n$. Let $\mathbf{N}_0$ be a choice of unit normal vector at $\mathbf{x}_0$. Define the 'height' function $f : M^n \to \mathbb{R}$ by

$$f(\mathbf{x}) = \langle F(\mathbf{x}), \mathbf{N}_0 \rangle.$$

Observe that for all $\mathbf{V} \in T_{F(\mathbf{x}_0)}M$, we have

$$\mathbf{V}(f) = \langle \mathbf{V}, \mathbf{N}_0 \rangle = 0.$$

Thus the function f has at least three critical points on M^n: one at a maximum point $\mathbf{x}_+$ of f, one at a minimum point $\mathbf{x}_-$ of f, and one at $\mathbf{x}_0$, where the value of f is strictly between its minimum and its maximum since M^n lies on both sides of $T_{F(\mathbf{x}_0)}M^n$. We then have that the unit normal vectors $G(\mathbf{x}_+)$, $G(\mathbf{x}_-)$, and $G(\mathbf{x}_0)$ are each plus or minus one another and hence two of these vectors must be equal, which is a contradiction to the injectivity of the Gauss map from Lemma 5.8. Let us take stock of what we have proven so far. We have shown that for each $\mathbf{x} \in M^n$, the tangent hyperplane $T_{F(\mathbf{x})}M$ divides $\mathbb{R}^{n+1}$ into two open half-spaces $H_{\mathbf{x}}^+$ and $H_{\mathbf{x}}^-$ so that M^n is contained in the closed half-space $\bar{H}_{\mathbf{x}}^- = H_{\mathbf{x}}^- \cup T_x M$.

Since M^n lies on one side of each tangent plane, we have that M^n must be embedded. For simplicity, now assume that F is the inclusion map, so that $F(\mathbf{x}) = \mathbf{x}$. Let Ω denote the compact region that M^n bounds. From the above, one can show that for each pair of distinct points $\mathbf{x}$ and $\mathbf{y}$ in M^n, the line segment $\overline{\mathbf{xy}}$ is contained in Ω. Furthermore, by using that $\mathrm{II} > 0$, one can prove that the open line segment from $\mathbf{x}$ to $\mathbf{y}$ is contained in the interior of Ω. This completes the proof of the theorem (see Exercise 5.12 for a hint of how to fill in the details above). □

5.3.2. Convex hypersurfaces and the support function. Let M^n be an embedded closed hypersurface in $\mathbb{R}^{n+1}$ and let N be the unit outward normal vector field. Its **support function** $\sigma : M^n \to \mathbb{R}$ is defined by

$$\sigma(\mathbf{x}) = \langle \mathbf{x}, N_{\mathbf{x}} \rangle, \tag{5.18}$$

where $N_{\mathbf{x}}$ denotes the unit normal at $\mathbf{x}$. Since the tangent hyperplane $TP_{\mathbf{x}}M$ passes through $\mathbf{x}$ and is perpendicular to $N_{\mathbf{x}}$, this is a signed distance from the origin to the tangent hyperplane $TP_{\mathbf{x}}M$.

The normal component of the position vector is

$$\mathbf{x}^{\perp} = \langle \mathbf{x}, N_{\mathbf{x}} \rangle N_{\mathbf{x}} = \sigma(\mathbf{x}) N_{\mathbf{x}}, \tag{5.19}$$

whereas the tangential component of the position vector is

$$\mathbf{x}^\top = \mathbf{x} - \langle \mathbf{x}, N_\mathbf{x} \rangle N_\mathbf{x} = \mathbf{x} - \sigma(\mathbf{x}) N_\mathbf{x}. \tag{5.20}$$

5.3.3. The support function as a function on the unit n-sphere. Now let M^n be a strictly convex embedded closed hypersurface in $\mathbb{R}^{n+1}$. Since the Gauss map $G : M^n \to S^n$, defined by $G(\mathbf{x}) = N_\mathbf{x}$, is a diffeomorphism, we may parametrize M^n by its inverse $G^{-1} : S^n \to M^n$. So the support function may be considered as a function on S^n:

$$\sigma := \sigma \circ G^{-1} : S^n \to \mathbb{R}. \tag{5.21}$$

That is,

$$\sigma(\mathbf{n}) := \langle G^{-1}(\mathbf{n}), \mathbf{n} \rangle \quad \text{for } \mathbf{n} \in S^n. \tag{5.22}$$

We can recover a convex closed embedded hypersurface M^n from its support function $\sigma : S^n \to \mathbb{R}$ because the compact region Ω that M^n bounds is equal to the intersection of the support half-spaces; that is,

$$\Omega = \bigcap_{\mathbf{y} \in S^n} \bar{H}^-_\mathbf{y} \tag{5.23}$$

(see Exercise 5.13), where

$$\bar{H}^-_\mathbf{y} := \{\mathbf{x} \in \mathbb{R}^{n+1} \colon \langle \mathbf{x}, N_\mathbf{y} \rangle \leq \sigma(\mathbf{y})\}. \tag{5.24}$$

We can show the set equality (5.23) by using the fact that M^n lies on one side of each of its tangent hyperplanes. Consequently, we can recover the hypersurface by $M^n := \partial\Omega$.

Let $\overline{g}$ denote the standard metric on S^n and let $\overline{\nabla}$ denote its associated covariant derivative. Recall that the Weingarten map $\mathrm{L} : T_\mathbf{x}M \to T_\mathbf{x}M$ is a self-adjoint vector space endomorphism for each $\mathbf{x} \in M^n$. We also may naturally identity $T_\mathbf{x}M$ with $T_{G(\mathbf{x})}S^n$. Thus, we may consider the Weingarten map as an endomorphism $\mathrm{L} : T_\mathbf{n}S^n \to T_\mathbf{n}S^n$ for each $\mathbf{n} \in S^n$. Since M^n is strictly convex, L is invertible.

Proposition 5.10. *For any $\mathbf{n} \in S^n$ and $\mathbf{V}, \mathbf{W} \in T_\mathbf{n}S^n$, we have that*

$$A(\mathbf{V}, \mathbf{W}) := (\overline{\nabla}^2 \sigma + \sigma \overline{g})(\mathbf{V}, \mathbf{W}) = \overline{g}(\mathrm{L}^{-1}(\mathbf{V}), \mathbf{W}). \tag{5.25}$$

Proof. The first derivative of σ is given by

$$\mathbf{W}(\sigma) = \mathbf{W}(\langle G^{-1}(\mathbf{n}), \mathbf{n} \rangle) = \langle G^{-1}(\mathbf{n}), \mathbf{W} \rangle,$$

where the second equality is because $\langle D_\mathbf{W}(G^{-1}(\mathbf{n})), \mathbf{n} \rangle = 0$, which in turn is due to $D_\mathbf{W}(G^{-1}(\mathbf{n})) \in T_{G^{-1}(\mathbf{n})}M \cong T_\mathbf{n}S^n$. We compute that the second directional derivative of σ is given by

$$\begin{aligned}\mathbf{V}(\mathbf{W}(\sigma)) &= \mathbf{V}(\langle G^{-1}(\mathbf{n}), \mathbf{W}\rangle)\\ &= \langle d(G^{-1})_{\mathbf{n}}(\mathbf{V}), \mathbf{W}\rangle + \langle G^{-1}(\mathbf{n}), D_{\mathbf{V}}\mathbf{W}\rangle.\end{aligned}$$

We also have that

$$(\overline{\nabla}_{\mathbf{V}}\mathbf{W})(\sigma) = \langle G^{-1}(\mathbf{n}), \overline{\nabla}_{\mathbf{V}}\mathbf{W}\rangle.$$

Thus,

$$\begin{aligned}\overline{\nabla}^2_{\mathbf{V},\mathbf{W}}\sigma &= \mathbf{V}(\mathbf{W}(\sigma)) - (\overline{\nabla}_{\mathbf{V}}\mathbf{W})(\sigma)\\ &= \langle (dG)^{-1}_{\mathbf{n}}(\mathbf{V}), \mathbf{W}\rangle + \langle G^{-1}(\mathbf{n}), D_{\mathbf{V}}\mathbf{W} - \overline{\nabla}_{\mathbf{V}}\mathbf{W}\rangle\\ &= \langle \mathrm{L}^{-1}(\mathbf{V}), \mathbf{W}\rangle + \langle G^{-1}(\mathbf{n}), -\overline{\mathrm{II}}(\mathbf{V},\mathbf{W})\mathbf{n}\rangle\end{aligned}$$

since $dG = \mathrm{L}$ is the Weingarten map and where $\overline{\mathrm{II}}$ is the second fundamental form of $(S^n, \overline{g})$. Since S^n is totally umbilical with principal curvature equal to 1, we have $\overline{\mathrm{II}} = \overline{\mathrm{I}} = \overline{g}$. Therefore

$$(\overline{\nabla}^2_{\mathbf{V},\mathbf{W}}\sigma)(\mathbf{n}) = \overline{g}(\mathrm{L}^{-1}(\mathbf{V}), \mathbf{W}) - \overline{g}(\mathbf{V},\mathbf{W})\sigma(\mathbf{n}). \tag{5.26}$$

The proposition follows. □

Let $\{e_1, \ldots, e_n\}$ be an orthonormal basis of principal directions with associated principal curvatures $\{\kappa_1, \ldots, \kappa_n\}$. By Proposition 5.10, we have that

$$(\overline{\nabla}^2\sigma + \sigma\overline{g})(e_i, e_i) = A(e_i, e_i) = \overline{g}(\mathrm{L}^{-1}(e_i), e_i) = \kappa_i^{-1}.$$

Therefore, the Gauss curvature may be expressed via

$$\begin{aligned}K^{-1} &= \kappa_1^{-1}\cdots\kappa_n^{-1}\\ &= \prod_{i=1}^{n}(\overline{\nabla}^2\sigma + \sigma\overline{g})(e_i, e_i)\\ &= \frac{\det\left(\overline{\nabla}^2\sigma + \sigma\overline{g}\right)}{\det(\overline{g})}.\end{aligned} \tag{5.27}$$

For the expression in the last line, by $\frac{\det(\alpha)}{\det(\beta)}$ for symmetric 2-tensors α and β, we mean $\frac{\det(\alpha_{ij})}{\det(\beta_{ij})}$, where α_{ij} and β_{ij} are the components of α and β with respect to any basis, respectively. It is easy to see that this expression is independent of the choice of basis.

Lemma 5.11.

$$G^*\left(\overline{\nabla}^2\sigma + \sigma\overline{g}\right) = \mathrm{II}. \tag{5.28}$$

Proof. We compute for $X, Y \in TM$ that

$$\begin{aligned} G^*\big(\overline{\nabla}^2\sigma + \sigma\overline{g}\big)(X,Y) &= \big(\overline{\nabla}^2\sigma + \sigma\overline{g}\big)\big(dG(X), dG(Y)\big) \\ &= \overline{g}\big(\mathrm{L}^{-1}(\mathrm{L}(X)), \mathrm{L}(Y)\big) \\ &= g(X, \mathrm{L}(Y)) \\ &= \mathrm{II}(X,Y), \end{aligned}$$

where we used that $\overline{g} = g$ via the natural isomorphism between $T_{G(\mathbf{x})}S^n$ and $T_{\mathbf{x}}M$.[2] □

We also observe that

$$G^*(\overline{g})(X,Y) = g(\mathrm{L}(X), \mathrm{L}(Y)) =: \mathrm{II}^2(X,Y), \tag{5.29}$$

so that

$$G^*(\overline{g}) = \mathrm{II}^2\,. \tag{5.30}$$

We call II^2 the square of the second fundamental form. With respect to an orthonormal frame $\{e_i\}_{i=1}^n$,

$$\mathrm{II}^2(e_i, e_j) = \delta_{ij}(\mathrm{II}(e_i, e_i))^2.$$

With respect to local coordinates $\{x^i\}_{i=1}^n$,

$$(\mathrm{II}^2)_{ij} = \sum_{k,\ell=1}^{n} \mathrm{II}_{ik}\, g^{k\ell}\, \mathrm{II}_{\ell j}\,.$$

We remark that the pointwise norm of the 2-tensor II is defined by

$$|\,\mathrm{II}\,|^2 := \sum_{i,j=1}^{n} g^{ij}(\mathrm{II}^2)_{ij} = \sum_{i,j,k,\ell=1}^{n} g^{ij} g^{k\ell}\, \mathrm{II}_{ik}\, \mathrm{II}_{\ell j}\,. \tag{5.31}$$

Equivalently, with respect to an orthonormal frame $\{e_i\}_{i=1}^n$,

$$|\,\mathrm{II}\,|^2 = \sum_{i,j=1}^{n} \big(\mathrm{II}(e_i, e_j)\big)^2. \tag{5.32}$$

See §5.7 below for the generalization of the notion of pointwise norm to arbitrary covariant tensors.

5.4. The Hessian, Laplacian, and divergence theorem

In this section we introduce two important differential operators: the Hessian and the Laplacian. We consider them both on Euclidean space and for submanifolds.

[2]More formally, if $\iota : T_{G(\mathbf{x})}S^n \to T_{\mathbf{x}}M$ is the natural isomorphism, then $\overline{g} = \iota^* g$.

5.4.1. The Laplacian and divergence theorem on Euclidean space. Given a smooth function $f : \mathcal{U} \to \mathbb{R}$, where $\mathcal{U}$ is an open subset of $\mathbb{R}^n$, its **(Euclidean) Laplacian** is defined by

$$\Delta_{\mathbb{E}} f = \sum_{i=1}^{n} \frac{\partial^2 f}{\partial (u^i)^2}. \tag{5.33}$$

The Laplacian is the trace of the **Hessian**, which is the $n \times n$ matrix of second derivatives:

$$\mathrm{Hess}(f) := \left(\frac{\partial^2 f}{\partial u^i \partial u^j} \right). \tag{5.34}$$

We say that f is a **harmonic function** if f satisfies the **Laplace equation**:

$$\Delta_{\mathbb{E}} f = 0. \tag{5.35}$$

Rotationally symmetric examples of harmonic functions defined on $\mathbb{R}^n \setminus \{\mathbf{0}\}$ are the **Newtonian potentials** (a.k.a. global **Green's functions**) defined as follows. For $n \geq 3$,

$$G(\mathbf{x}) := c_n \|\mathbf{x}\|^{2-n}, \tag{5.36}$$

where $c_n := \frac{1}{n(n-1)\omega_n}$, ω_n is the volume of the Euclidean unit n-ball, and for $n = 2$,

$$G(\mathbf{x}) := -\frac{1}{2\pi} \ln \|\mathbf{x}\|. \tag{5.37}$$

One calculates that

$$\Delta_{\mathbb{E}} G = 0 \quad \text{on } \mathbb{R}^n \setminus \{\mathbf{0}\}. \tag{5.38}$$

On all of $\mathbb{R}^n$, **in the sense of distributions**, the Laplacian of the Newtonian potential function is equal to the Dirac delta function. So the Newtonian potential function is called a **fundamental solution** to the Laplace equation.

Let $V : \mathcal{U} \to \mathbb{R}^{n+1}$ be a smooth vector field on an open subset $\mathcal{U}$ of $\mathbb{R}^{n+1}$. The **(Euclidean) divergence** of $V =: (V^1, \ldots, V^{n+1})$ is defined by

$$\mathrm{div}_{\mathbb{E}}(V)(\mathbf{x}) := \sum_{i=1}^{n+1} \frac{\partial V^i}{\partial u^i}(\mathbf{x}) \tag{5.39}$$

for $\mathbf{x} \in \mathcal{U}$. Observe that the divergence of the gradient of a function is equal to the Laplacian of the function. Namely, recall that if $f : \mathcal{U} \to \mathbb{R}$ is a function, then its Euclidean gradient is given by

$$\nabla_{\mathbb{E}} f = \left(\frac{\partial f}{\partial u^1}, \ldots, \frac{\partial f}{\partial u^{n+1}} \right). \tag{5.40}$$

Thus,

$$\Delta_{\mathbb{E}} f = \operatorname{div}_{\mathbb{E}}(\nabla_{\mathbb{E}} f). \tag{5.41}$$

The **divergence theorem** says the following.

Proposition 5.12 (Euclidean divergence theorem). *Let Ω be a compact domain in $\mathbb{R}^{n+1}$ with smooth boundary $\partial\Omega =: M^n$. If V is a C^1 vector field defined in a neighborhood of Ω, then*

$$\int_\Omega \operatorname{div}_{\mathbb{E}}(V) = \int_{M^n} V \cdot N, \tag{5.42}$$

where N is the unit outward normal vector field to M^n.

For the proof, see Corollary 8.9 below, which is stated in more generality.

By taking V to be equal to the gradient of a function f, we obtain that

$$\int_\Omega \Delta_{\mathbb{E}} f = \int_{M^n} N(f). \tag{5.43}$$

By taking V to be equal to $f\nabla_{\mathbb{E}}h$ for some functions f, h, we obtain that

$$\int_\Omega (f\Delta_{\mathbb{E}}h + \langle \nabla_{\mathbb{E}} f, \nabla_{\mathbb{E}} h\rangle) = \int_{M^n} f N(h). \tag{5.44}$$

A consequence of this is the formula

$$\int_\Omega (f\Delta_{\mathbb{E}}h - h\Delta_{\mathbb{E}} f) = \int_{M^n} (f N(h) - h N(f)). \tag{5.45}$$

5.4.2. The Hessian and Laplacian on a submanifold. Let M^n be a submanifold of $\mathbb{R}^N$. Recall from (4.123) that the Levi-Civita connection ∇ defines for a tangent vector $\mathbf{V}$ on a vector field W on M^n a 'directional derivative' of W along the direction $\mathbf{V}$ by

$$\nabla_{\mathbf{V}} W = (D_{\mathbf{V}} W)^T, \tag{5.46}$$

where $\cdot^T$ denotes the tangential component of a vector in $\mathbb{R}^N$.

Let $f : M^n \to \mathbb{R}$ be a smooth function. Recall that its derivative df is a 1-form on M^n; that is, for each $\mathbf{x} \in M^n$, $df_{\mathbf{x}} : T_{\mathbf{x}}M \to \mathbb{R}$ is a linear function. We can define the covariant derivative of the 1-form df by imposing the product rule:

$$\mathbf{V}(df(W)) =: (\nabla_{\mathbf{V}} df)(W) + df(\nabla_{\mathbf{V}} W); \tag{5.47}$$

that is,

$$(\nabla_{\mathbf{V}} df)(W) := \mathbf{V}(W(f)) - (\nabla_{\mathbf{V}} W)(f). \tag{5.48}$$

For this definition of $\nabla_{\mathbf{V}} df$ to make sense, we need to show that the right-hand side of (5.48) at a point $\mathbf{x} \in M^n$ depends only on the vector field

W at $\mathbf{x}$. To see this, we first note that the right-hand side is linear in W. Secondly, let h be a function on M^n. We compute that

(5.49)
$$\begin{aligned}\mathbf{V}((hW)(f)) - (\nabla_{\mathbf{V}}(hW))(f) &= \mathbf{V}(hW(f)) - \mathbf{V}(h)W(f) - h(\nabla_{\mathbf{V}}W)(f) \\ &= h\mathbf{V}(W(f)) - h(\nabla_{\mathbf{V}}W)(f).\end{aligned}$$

By Corollary 4.39 (the proof applies equally well to the present situation), we conclude that the right-hand side depends only on W at $\mathbf{x}$.

The (covariant) **Hessian** $\nabla^2 f$ of f on M^n is defined by

$$(\nabla^2 f)(\mathbf{V}, \mathbf{W}) := (\nabla_{\mathbf{V}} df)(\mathbf{W}). \tag{5.50}$$

The Hessian $\nabla^2 f$ is a symmetric covariant 2-tensor on M^n. Since the Levi-Civita connection ∇ is torsion-free, we have that $\nabla^2 f$ is symmetric. Namely, we compute that for any tangent vectors $\mathbf{V}, \mathbf{W}$ at a point $\mathbf{x}$ and extensions to vector fields V, W in a neighborhood of $\mathbf{x}$,

$$\begin{aligned}(\nabla^2 f)(\mathbf{V}, \mathbf{W}) - (\nabla^2 f)(\mathbf{W}, \mathbf{V}) &= \mathbf{V}(W(f)) - (\nabla_{\mathbf{V}}W)(f) \\ &\quad - \mathbf{W}(V(f)) + (\nabla_{\mathbf{W}}V)(f) \\ &= [V, W](f) - [V, W](f) \\ &= 0,\end{aligned} \tag{5.51}$$

where we used (4.25) for the penultimate equality.

The **Laplacian** is the trace of the Hessian; that is,

$$\Delta_g f(\mathbf{x}) := \sum_{i=1}^n (\nabla^2 f)_{\mathbf{x}}(e_i, e_i), \tag{5.52}$$

where $\{e_i\}$ is any orthonormal basis of tangent vectors at $\mathbf{x}$.

The **divergence** of a tangent vector field V on M^n is defined by

$$\operatorname{div}_g(V) := \sum_{i=1}^n \langle \nabla_{e_i} V, e_i \rangle. \tag{5.53}$$

The divergence of a 1-form α on M^n is defined by

$$\operatorname{div}_g(\alpha) := \sum_{i=1}^n (\nabla_{e_i} \alpha)(e_i). \tag{5.54}$$

The Laplacian is equal to the divergence of the gradient or derivative:

$$\Delta_g f = \operatorname{div}_g(\nabla f) = \operatorname{div}_g(df). \tag{5.55}$$

With respect to local coordinates $(x^1, \ldots, x^n) = F^{-1}$ induced by a parametrization F, we have that

$$\nabla_i f := df(\partial_i F) = \frac{\partial f}{\partial x^i} = \frac{\partial (f \circ F)}{\partial u^i}. \tag{5.56}$$

The components of the Hessian are defined by

$$\nabla_i \nabla_j f := (\nabla^2 f)(\partial_i F, \partial_j F). \tag{5.57}$$

By definition, we may write the components of the Hessian, in terms of the second partial derivatives and the Christoffel symbols, as

$$\begin{aligned} \nabla_i \nabla_j f &= \partial_i F(\partial_j F(f)) - (\nabla_{\partial_i F} \partial_j F)(f) \\ &= \frac{\partial^2 f}{\partial x^i \partial x^j} - \sum_{k=1}^{n} \Gamma_{ij}^k \frac{\partial f}{\partial x^k}. \end{aligned} \tag{5.58}$$

With respect to local coordinates, the Laplacian is given by

$$\Delta_g f = g^{ij} \nabla_i \nabla_j f = \sum_{i,j=1}^{n} g^{ij} \left(\frac{\partial^2 f}{\partial x^i \partial x^j} - \sum_{k=1}^{n} \Gamma_{ij}^k \frac{\partial f}{\partial x^k} \right). \tag{5.59}$$

For closed submanifolds, we have the following, which is a special case of Corollary 8.9 below.

Proposition 5.13 (Divergence theorem for closed submanifolds). *Let M^n be a compact hypersurface without boundary in $\mathbb{R}^{n+1}$. Let V be a C^1 vector field on M^n. Then*

$$\int_{M^n} \operatorname{div}_g(V) = 0. \tag{5.60}$$

In particular, by taking $V = \nabla f$ to be a gradient vector field, we obtain

$$\int_{M^n} \Delta_g f = 0. \tag{5.61}$$

5.5. Hypersurface integrals

Using that the Laplacian of the position map of a hypersurface is equal to the mean curvature vector, we derive integral formulas relating the area of a hypersurface, the mean and scalar curvatures, and the support function.

5.5.1. The Hessian and Laplacian of the position map. Let M^n be an embedded submanifold in $\mathbb{R}^N$. Define the distance-squared function $f : M^n \to \mathbb{R}$ by

$$f : \mathbf{x} \mapsto \|\mathbf{x}\|^2.$$

Let V and W be tangent vector fields on M^n. The first derivative of f is given by

$$\frac{1}{2} V(f) = \frac{1}{2} V(\|\mathbf{x}\|^2) = \langle V, \mathbf{x} \rangle = \langle V, \mathbf{x}^\top \rangle, \tag{5.62}$$

where the last equality is because the vector field V is tangential to M^n. In other words, the gradient on M^n of f is given by

$$(\nabla f)_{\mathbf{x}} = 2\mathbf{x}^\top. \tag{5.63}$$

Now let M^n be an embedded hypersurface in $\mathbb{R}^{n+1}$. The second derivative of f is

$$\frac{1}{2}W(V(f)) = W(\langle V, \mathbf{x}\rangle) = \langle D_W V, \mathbf{x}\rangle + \langle V, W\rangle.$$

We also have

$$\frac{1}{2}(\nabla_W V)(f) = \langle \nabla_W V, \mathbf{x}\rangle.$$

Therefore, one-half times the Hessian of f is given by

$$\begin{aligned}\frac{1}{2}(\nabla^2 f)(W, V) &= \frac{1}{2}W(V(f)) - \frac{1}{2}(\nabla_W V)(f)\\ &= \langle D_W V - \nabla_W V, \mathbf{x}\rangle + \langle V, W\rangle\\ &= -\,\mathrm{II}(W, V)\langle N, \mathbf{x}\rangle + \langle V, W\rangle.\end{aligned} \tag{5.64}$$

That is,

$$\frac{1}{2}\nabla^2 f = -\sigma\,\mathrm{II} + \mathrm{I}. \tag{5.65}$$

Tracing this, we obtain

$$\frac{1}{2}\Delta f = -\sigma H + n, \tag{5.66}$$

where Δ is the Laplacian of (M^n, I).

We can also perform a similar calculation for the inclusion map $\mathbf{x} : M^n \to \mathbb{R}^{n+1}$. We compute that

$$V(\mathbf{x}) = V$$

and

$$W(V(\mathbf{x})) = D_W V.$$

Furthermore,

$$\nabla_W V(\mathbf{x}) = \nabla_W V.$$

The Hessian of the vector-valued function $\mathbf{x} =: (x^1, \ldots, x^{n+1})$ is defined componentwise:

$$\nabla^2 \mathbf{x} := (\nabla^2 x^1, \ldots, \nabla^2 x^{n+1}). \tag{5.67}$$

Therefore,

$$\begin{aligned}\nabla^2_{W,V}\mathbf{x} := (\nabla^2 \mathbf{x})(W, V) &= W(V(\mathbf{x})) - (\nabla_W V)(\mathbf{x})\\ &= D_W V - \nabla_W V\\ &= -\,\mathrm{II}(W, V)N.\end{aligned} \tag{5.68}$$

We leave it as an exercise that one can use this formula to give an alternate derivation of (5.65). Note also that by tracing, we have that

$$\Delta \mathbf{x} = -HN \tag{5.69}$$

is the mean curvature vector.

5.5.2. Integral formulas on closed hypersurfaces. Assume now that M^n is a closed embedded hypersurface. Integrating (5.119) while using the divergence theorem (Proposition 5.13) on M^n yields the **Minkowski formula**:

$$n\,\mathrm{Area}(M^n) = \int_{M^n} \sigma H \, d\mu_g, \tag{5.70}$$

where $d\mu_g$ denotes the (n-dimensional) volume form of (M^n, g).[3] For example, if $n = 1$, then we obtain the length formula:

$$\mathrm{L} = \int_{M^1} \sigma \kappa \, ds. \tag{5.71}$$

We have

$$\frac{1}{2}\Delta_{\mathbb{R}^{n+1}} \|\mathbf{x}\|^2 = n + 1. \tag{5.72}$$

Let Ω be the compact region bounded by M^n. By (5.72) and the divergence theorem on $\mathbb{R}^{n+1}$, we have that

$$\begin{aligned}(n+1)\,\mathrm{Vol}(\Omega) &= \int_\Omega \frac{1}{2}\Delta_{\mathbb{R}^{n+1}} \|\mathbf{x}\|^2 \, d\mu_{\mathbb{R}^{n+1}}(\mathbf{x}) \\ &= \int_{M^n} \frac{1}{2} N(\|\mathbf{x}\|^2)\, d\mu_g(\mathbf{x}) \\ &= \int_{M^n} \langle \mathbf{x}, N_{\mathbf{x}} \rangle \, d\mu_g(\mathbf{x}),\end{aligned}$$

where $d\mu_g = d\mu$ is the volume form of (M^n, I). That is,

$$\mathrm{Vol}(\Omega) = \frac{1}{n+1} \int_{M^n} \sigma(\mathbf{x})\, d\mu_g(\mathbf{x}). \tag{5.73}$$

We now compute the Hessian and Laplacian of the support function σ. Differentiating σ yields

$$V(\sigma) = \langle \mathbf{x}, D_V N \rangle = \langle \mathbf{x}, \mathrm{L}(V) \rangle = \langle \mathbf{x}^\top, \mathrm{L}(V) \rangle = \mathrm{II}(V, \mathbf{x}^\top),$$

where we used $\langle V, N \rangle = 0$ for the first equality. Since the second derivative of the support function σ is

$$W(V(\sigma)) = (\nabla_W \mathrm{II})(V, \mathbf{x}^\top) + \mathrm{II}(\nabla_W V, \mathbf{x}^\top) + \mathrm{II}(V, \nabla_W(\mathbf{x}^\top))$$

by the product rule, we obtain that the covariant Hessian of σ on M^n is given by

$$\begin{aligned}\nabla^2_{W,V}\sigma &= W(V(\sigma)) - (\nabla_W V)(\sigma) \\ &= (\nabla_W \mathrm{II})(V, \mathbf{x}^\top) + \mathrm{II}\left(V, \nabla_W(\mathbf{x}^\top)\right).\end{aligned} \tag{5.74}$$

[3]We use Area to denote the n-dimensional volume of M^n to distinguish it from the $(n+1)$-dimensional volume of the region interior to M^n.

On the other hand, using (4.125) and the product rule, we compute that

$$\begin{aligned}\nabla_W(\mathbf{x}^\top) - \mathrm{II}(W, \mathbf{x}^\top)N &= D_W(\mathbf{x}^\top) \\ &= D_W(\mathbf{x} - \langle \mathbf{x}, N\rangle N) \\ &= W - \langle \mathbf{x}, \mathrm{L}(W)\rangle N - \langle \mathbf{x}, N\rangle\, \mathrm{L}(W),\end{aligned}$$

so that by taking the tangent components we arrive at the formula

$$\nabla_W(\mathbf{x}^\top) = W - \langle \mathbf{x}, N\rangle\, \mathrm{L}(W). \tag{5.75}$$

Therefore, (5.74) yields

$$\nabla^2_{W,V}\sigma = (\nabla_{\mathbf{x}^\top}\mathrm{II})(V, W) + \mathrm{II}(V, W) - \langle \mathbf{x}, N\rangle\, \mathrm{II}^2(V, W), \tag{5.76}$$

where we also used the Codazzi equation (4.175).

Tracing this equation (i.e., taking $W = V = e_i$, where $\{e_i\}_{i=1}^n$ is an orthonormal frame tangent to M^n, and summing over i) yields

$$\begin{aligned}(\Delta_g\sigma)(\mathbf{x}) &= \langle \nabla H(\mathbf{x}), \mathbf{x}^\top\rangle + H(\mathbf{x}) - \langle \mathbf{x}, N_{\mathbf{x}}\rangle |\,\mathrm{II}\,|^2(\mathbf{x}) \\ &= \mathrm{div}(H\mathbf{x}^\top) - H(\mathbf{x})\,\mathrm{div}(\mathbf{x}^\top) + H(\mathbf{x}) - \langle \mathbf{x}, N_{\mathbf{x}}\rangle |\,\mathrm{II}\,|^2(\mathbf{x}),\end{aligned} \tag{5.77}$$

where $|\,\mathrm{II}\,|^2$ is defined by (5.31). On the other hand, by (5.75),

$$\mathrm{div}(X^\top)(\mathbf{x}) = \sum_{i=1}^n (\nabla_{e_i}(\mathbf{x}^\top)\cdot e_i) = n - \langle \mathbf{x}, N_{\mathbf{x}}\rangle H(\mathbf{x}). \tag{5.78}$$

Therefore, by integrating (5.77), applying the divergence theorem, and (5.78), we obtain

$$\begin{aligned}0 &= \int_{M^n} (\Delta_g\sigma)(\mathbf{x})\,d\mu(\mathbf{x}) \\ &= \int_{M^n} \left((1-n)H(\mathbf{x}) + \langle \mathbf{x}, N_{\mathbf{x}}\rangle (H^2 - |\,\mathrm{II}\,|^2)(\mathbf{x})\right) d\mu(\mathbf{x});\end{aligned}$$

that is, we have the Minkowski formula

$$\int_{M^n} R(\mathbf{x})\sigma(\mathbf{x})\,d\mu(\mathbf{x}) = (n-1)\int_{M^n} H(\mathbf{x})\,d\mu(\mathbf{x}), \tag{5.79}$$

where $R = H^2 - |\,\mathrm{II}\,|^2$ is the scalar curvature of M^n. In particular, if $n = 2$, then we have

$$2\int_{M^2} K(\mathbf{x})\sigma(\mathbf{x})\,d\mu(\mathbf{x}) = \int_{M^2} H(\mathbf{x})\,d\mu(\mathbf{x}), \tag{5.80}$$

where K is the Gauss curvature of M^2.

5.5.3. Integral formulas for convex embedded hypersurfaces. Let $A_0 \in \mathbb{R}^{n+1}$ be a constant vector field on Ω. By the Euclidean divergence theorem, we have

$$\int_{M^n} \langle N, A_0 \rangle \, d\mu_g = \int_\Omega \operatorname{div}_{\mathbb{R}^{n+1}}(A_0) = 0. \tag{5.81}$$

Now assume that M^n is a strictly convex closed hypersurface. Then the Gauss map G is a diffeomorphism and, by the change of variables formula for the Gauss map transformation, we obtain

$$0 = \int_{M^n} \langle N, A_0 \rangle \, d\mu_g = \int_{S^n} \langle N, A_0 \rangle \, K^{-1} \, d\mu_{S^n}(N), \tag{5.82}$$

where K is the Gauss curvature at $X = G^{-1}(N)$. Since this is true for any $A_0 \in \mathbb{R}^{n+1}$, as a vector equation, this says that

$$\int_{S^n} \frac{1}{K(\mathbf{n})} \mathbf{n} \, d\mu_{S^n}(\mathbf{n}) = \mathbf{0}. \tag{5.83}$$

By Lemma 5.5, the Jacobian determinant of the inverse of the Gauss map is equal to K^{-1}. Therefore

$$\operatorname{Area}(M^n) = \int_{S^n} (G^{-1})^*(d\mu_g) = \int_{S^n} \frac{1}{K(\mathbf{n})} d\mu_{S^n}(\mathbf{n}). \tag{5.84}$$

Moreover, by (5.73), we have that

$$\operatorname{Vol}(\Omega) = \frac{1}{n+1} \int_{S^n} \frac{\langle G^{-1}(\mathbf{n}), \mathbf{n} \rangle}{K(\mathbf{n})} d\mu_{S^n}(\mathbf{n}). \tag{5.85}$$

5.6. Highlights of submanifold theory

This section discusses some well-known classical results on submanifolds of Euclidean spaces. It requires some knowledge of abstract smooth manifolds, Riemannian manifolds, and moving frames. As such, the readers who are not familiar with those topics may wish to skip this section until after they have learned those prerequisites, which are in Part II of this book.

5.6.1. Rigidity of convex surfaces. In this subsection we discuss the rigidity theorem for convex surfaces of Cohn-Vossen [**CV27**, **CV36**]. As with most expositions, such as Chern [**Che51**], Spivak [**Spi79e**], and Han and Hong [**HH06**], the proof of Cohn-Vossen's theorem that we give is due to Herglotz [**Her43**].

First, we prove a rigidity theorem for isometries of hypersurfaces (not necessarily convex) that preserve the second fundamental form.

Lemma 5.14. *Let M^n and M'^n be two connected hypersurfaces without boundary in $\mathbb{R}^{n+1}$. Let* I, II *and* I′, II′ *denote their first, second fundamental forms, respectively. If $f : M^n \to M'^n$ is a diffeomorphism such that*

$$f^* \mathrm{I}' = \mathrm{I} \quad \textit{and} \quad f^* \mathrm{II}' = \mathrm{II}, \tag{5.86}$$

then f extends to an ambient isometry from $\mathbb{R}^{n+1}$ to $\mathbb{R}^{n+1}$.

Proof. Let N and N' denote the unit normal vector fields to M^n and M'^n inducing the second fundamental forms II and II′, respectively. In general, geometric quantities on M'^n and M^n will be denoted with and without primes, respectively. Let $\mathbf{x} \in M^n$. We have that $df := df_{\mathbf{x}} : T_{\mathbf{x}}M \to T_{f(\mathbf{x})}M'$ is a linear isometry.

By (5.86), we have

$$\big\langle \mathrm{L}'(df(V)), df(W)\big\rangle = \langle \mathrm{L}(V), W\rangle = \big\langle df(\mathrm{L}(V)), df(W)\big\rangle$$

for all $V, W \in T_{\mathbf{x}}M$, so that the Weingarten maps L and L′ commute with the map df:

$$\mathrm{L}' \circ df = df \circ \mathrm{L} : T_{\mathbf{x}}M \to T_{f(\mathbf{x})}M'. \tag{5.87}$$

Define $\ell : M^n \to \mathrm{O}(n+1)$ such that

$$\mathrm{O}(n+1) \ni \ell_{\mathbf{x}} := \ell(\mathbf{x}) : \mathbb{R}^{n+1} \to \mathbb{R}^{n+1}$$

is the unique linear isometry satisfying

$$\ell_{\mathbf{x}}|_{T_{\mathbf{x}}M} = df_{\mathbf{x}}, \quad \ell_{\mathbf{x}}(N_{\mathbf{x}}) = N'_{f(\mathbf{x})}. \tag{5.88}$$

We will show that $V(\ell) = D_V \ell = 0$ for $V \in TM$, so that the map $\mathbf{x} \mapsto \ell_{\mathbf{x}}$ is constant.

Let $Y : M^n \to \mathbb{R}^{n+1}$ be a smooth vector-valued function. Corresponding to the decomposition $M^n \times \mathbb{R}^{n+1} = TM \oplus NM$ given by (3.67), we may write

$$Y = W + aN,$$

where W is a tangential vector field on M^n and $a : M^n \to \mathbb{R}$ is a smooth function. We compute for $V \in T_{\mathbf{x}}M$ that

$$\begin{aligned} D_V Y &= D_V W + V(a)N + a(\mathbf{x}) D_V N \\ &= \nabla_V W - \langle \mathrm{L}(V), W\rangle N + V(a)N + a(\mathbf{x})\,\mathrm{L}(V). \end{aligned}$$

Using that f preserves the first and second fundamental forms, and hence preserves the Levi-Civita connection and Weingarten map, and using (5.87),

we compute that

$$\begin{aligned}\ell(D_V Y) &= df(\nabla_V W) - \langle \mathrm{L}(V), W\rangle N' + V(a)N' + a\,df(\mathrm{L}(V))\\ &= \nabla'_{df(V)} df(W) - \langle \mathrm{L}'(df(V)), df(W)\rangle N'\\ &\quad + df(V)(a\circ f^{-1})N' + (a\circ f^{-1})\,\mathrm{L}'(df(V))\\ &= D_{df(V)} df(W) + D_{df(V)}((a\circ f^{-1})N')\\ &= D_{df(V)}(\ell(Y)).\end{aligned}$$

Therefore

$$(D_V\ell)(Y) = D_{df(V)}(\ell(Y)) - \ell(D_V Y) = 0.$$

We have proved that $\ell := \ell_{\mathbf{x}} \in \mathrm{O}(n+1)$ is independent of $\mathbf{x}\in M^n$.

Now choose any point $\mathbf{x}_0 \in M^n$. Consider the Euclidean isometry $G : \mathbb{R}^{n+1} \to \mathbb{R}^{n+1}$ given by translation from $\mathbf{x}_0$ to $f(\mathbf{x}_0)$, followed by the orthogonal transformation ℓ. We have for $\mathbf{x}\in\mathbb{R}^{n+1}$ that $dG_{\mathbf{x}} = \ell$. By this and (5.88), for $\mathbf{x}\in M^n$, we have that

$$dG_{\mathbf{x}}|_{T_{\mathbf{x}}M} = df_{\mathbf{x}} : T_{\mathbf{x}}M \to \mathbb{R}^{n+1}$$

for all $\mathbf{x}\in M^n$. Since $G(\mathbf{x}_0) = f(\mathbf{x}_0)$ and M^n is connected, we conclude from this that G restricted to M^n is equal to f. ☐

Recall from Example 4.44 that a half-cylinder M^2_{cyl} and a flat slab M^2_{slab} in $\mathbb{R}^3$ are isometric via a map $\Phi : M^2_{\text{slab}} \to M^2_{\text{cyl}}$. While these specific spaces are isometric in this way, it is notable that Φ does not extend to an isometry from $\mathbb{R}^3$ to itself.

This brings us to an interesting distinction: The main result of this section presents a rigidity result for closed strictly convex surfaces embedded in $\mathbb{R}^3$.

Theorem 5.15 (Cohn-Vossen: Rigidity of convex surfaces). *Let M^2 and M'^2 be two closed surfaces in $\mathbb{R}^3$ with positive Gauss curvature. Let* I *and* I′ *denote their first fundamental forms, respectively. If there exists an isometry from (M^2, I) to (M'^2, I'), then there exists an isometry from $\mathbb{R}^3$ to $\mathbb{R}^3$ which maps M^2 onto M'^2.*

Proof. Let $J : TM \to TM$ be the almost complex structure of M^2; see (10.103) below.[4] Let $\{e_1, e_2\}$ be a positively oriented local orthonormal tangential frame field defined on an open subset $\mathcal{U}$ of M^2. Then $J(e_1) = e_2$ and $J(e_2) = -e_1$. In particular, $J^2 = -\mathrm{id}_{TM}$. Define the 1-form α on M^2 by

$$\alpha(V) = \langle J(V), \mathbf{x}\rangle \quad \text{for } V \in T_{\mathbf{x}}M. \tag{5.89}$$

[4]For now, we can just think of J as counterclockwise rotation by 90 degrees.

Observe that on $\mathcal{U}$,

$$\alpha(V) = \langle \mathbf{x}, e_2 \rangle \langle V, e_1 \rangle - \langle \mathbf{x}, e_1 \rangle \langle V, e_2 \rangle.$$

Let $\{\omega^1, \omega^2\}$ be the local orthonormal coframe dual to $\{e_1, e_2\}$. We have

$$\alpha = \langle \mathbf{x}, e_2 \rangle \omega^1 - \langle \mathbf{x}, e_1 \rangle \omega^2. \tag{5.90}$$

Let D be the Euclidean covariant derivative. Let ω_i^j denote the connection 1-forms associated to the orthonormal frame field $\{\omega^1, \omega^2\}$; see (8.55) below. For $i = 1, 2$ and $V \in T_{\mathbf{x}}M$, we have

$$\begin{aligned} V\langle \mathbf{x}, e_i \rangle &= \langle D_V \mathbf{x}, e_i \rangle + \langle D_V e_i, \mathbf{x} \rangle \\ &= \langle V, e_i \rangle + \langle \nabla_V e_i, \mathbf{x} \rangle + \langle D_V e_i, N_{\mathbf{x}} \rangle \langle \mathbf{x}, N_{\mathbf{x}} \rangle \\ &= \langle V, e_i \rangle + \omega_i^j(V) \langle \mathbf{x}, e_j \rangle - \mathrm{II}(V, e_i) \sigma(\mathbf{x}), \end{aligned} \tag{5.91}$$

where we used $\sigma(\mathbf{x}) = \langle \mathbf{x}, N_{\mathbf{x}} \rangle$ is the support function; cf. (5.18). Thus

$$d\langle \mathbf{x}, e_i \rangle = \omega^i + \langle \mathbf{x}, e_j \rangle \omega_i^j - \sigma(\mathbf{x})\, \mathrm{L}(e_i)^*,$$

where L is the Weingarten map.

We compute that

$$\begin{aligned} d\alpha &= d\langle \mathbf{x}, e_2 \rangle \wedge \omega^1 - d\langle \mathbf{x}, e_1 \rangle \wedge \omega^2 + \langle \mathbf{x}, e_2 \rangle\, d\omega^1 - \langle \mathbf{x}, e_1 \rangle\, d\omega^2 \\ &= \big(-2 + \sigma(\mathbf{x})(\mathrm{II}(e_1, e_1) + \mathrm{II}(e_2, e_2)) \big)\, \omega^1 \wedge \omega^2 \end{aligned}$$

since by the first Cartan structure equation (8.71) below, $d\omega^1 - \omega_1^2 \wedge \omega^2 = 0$ and $d\omega^2 - \omega_2^1 \wedge \omega^1 = 0$. Therefore we have the global equation

$$d\alpha = (\sigma H - 2)\, d\mu, \tag{5.92}$$

where H is the mean curvature of M^2. Note that by integrating this formula, we obtain the $n = 2$ case of (5.70); namely,

$$\int_{M^2} \sigma H d\mu = 2\, \mathrm{Area}(M^2).$$

Define the 1-form β on M^2 by

$$\beta(W) = \alpha(\mathrm{L}(W)). \tag{5.93}$$

Claim.

$$d\beta = (2\sigma K - H)\, d\mu, \tag{5.94}$$

where K is the Gauss curvature of M^2.

Differentiating (5.90), we have

$$\nabla_V \alpha = V(\langle \mathbf{x}, e_2 \rangle)\omega^1 - V(\langle \mathbf{x}, e_1 \rangle)\omega^2 + \langle \mathbf{x}, e_2 \rangle\, \nabla_V \omega^1 - \langle \mathbf{x}, e_1 \rangle\, \nabla_V \omega^2,$$

so by applying (5.91) and (8.61), i.e., $\nabla_V \omega^i = -\omega^i_j(V)\,\omega^j$, we obtain

$$(5.95)\quad \begin{aligned} \nabla_V \alpha &= \big(\langle V, e_2\rangle + \omega^1_2(V)\langle \mathbf{x}, e_1\rangle - \mathrm{II}(V,e_2)\sigma(\mathbf{x})\big)\,\omega^1 \\ &\quad - \big(\langle V, e_1\rangle + \omega^2_1(V)\langle \mathbf{x}, e_2\rangle - \mathrm{II}(V,e_1)\sigma(\mathbf{x})\big)\,\omega^2 \\ &\quad - \langle \mathbf{x}, e_2\rangle \omega^1_2(V)\omega^2 + \langle \mathbf{x}, e_1\rangle \omega^2_1(V)\omega^1 \\ &= \big(\langle V, e_2\rangle - \mathrm{II}(V,e_2)\sigma(\mathbf{x})\big)\,\omega^1 - \big(\langle V, e_1\rangle - \mathrm{II}(V,e_1)\sigma(\mathbf{x})\big)\,\omega^2. \end{aligned}$$

Note that we may rederive (5.92) from this formula.

Now we compute

$$\begin{aligned} 2d\beta(V,W) &= (\nabla_V\beta)(W) - (\nabla_W\beta)(V) \\ &= (\nabla_V\alpha)(\mathrm{L}(W)) - (\nabla_W\alpha)(\mathrm{L}(V)) \end{aligned}$$

since $\alpha((\nabla_V \mathrm{L})(W)) = \alpha((\nabla_W \mathrm{L})(V))$. For any tangential vector field V, denote $V_i = \langle V, e_i\rangle$ for $i = 1, 2$. By applying (5.95), we obtain

$$\begin{aligned} 2d\beta(V,W) &= \big(V_2 - \mathrm{II}(V,e_2)\sigma(\mathbf{x})\big)\,\mathrm{L}(W)_1 - \big(V_1 - \mathrm{II}(V,e_1)\sigma(\mathbf{x})\big)\,\mathrm{L}(W)_2 \\ &\quad - \big(W_2 - \mathrm{II}(W,e_2)\sigma(\mathbf{x})\big)\,\mathrm{L}(V)_1 + \big(W_1 - \mathrm{II}(W,e_1)\sigma(\mathbf{x})\big)\,\mathrm{L}(V)_2. \end{aligned}$$

In particular,

$$(5.96)\quad \begin{aligned} 2d\beta(e_1,e_2) &= \big(-\mathrm{II}(e_1,e_2)\sigma(\mathbf{x})\big)\,\mathrm{L}(e_2)_1 - \big(1 - \mathrm{II}(e_1,e_1)\sigma(\mathbf{x})\big)\,\mathrm{L}(e_2)_2 \\ &\quad - \big(1 - \mathrm{II}(e_2,e_2)\sigma(\mathbf{x})\big)\,\mathrm{L}(e_1)_1 + \big(-\mathrm{II}(e_2,e_1)\sigma(\mathbf{x})\big)\,\mathrm{L}(e_1)_2 \\ &= -H + 2\sigma(\mathbf{x})K. \end{aligned}$$

This proves (5.94).

Observe that the integral of (5.94) yields (5.80); namely,

$$2\int_{M^2} K\sigma d\mu = \int_{M^2} H d\mu.$$

Let $\varphi : (M^2, \mathrm{I}) \to (M'^2, \mathrm{I}')$ be an isometry. Let L and L′ be the Weingarten maps of M^2 and M'^2, respectively. Define

$$\mathrm{L}'' := d(\varphi^{-1}) \circ \mathrm{L}' \circ d\varphi : T_pM \to T_pM.$$

We have

$$\det \mathrm{L}''(\mathbf{x}) = \det \mathrm{L}'(\varphi(\mathbf{x})) = K'(\varphi(\mathbf{x})) = K(\mathbf{x}) = \det \mathrm{L}(\mathbf{x}).$$

Define the 1-form β' on M^2 by

$$(5.97)\qquad \beta'(W) = \alpha(\mathrm{L}''(W)).$$

One computes that

$$(5.98)\qquad d\beta' = \big(\sigma(2K - \det(\mathrm{L} - \mathrm{L}'')) - H'\big)d\mu,$$

where H' is the mean curvature of M'^2. Indeed, by (5.96), we have that

$$\begin{aligned}2d\beta'(e_1, e_2) &= -\sigma(\mathbf{x})\,\mathrm{L}(e_1)_2\,\mathrm{L}''(e_2)_1 + \sigma(\mathbf{x})\,\mathrm{L}(e_1)_1\,\mathrm{L}''(e_2)_2 \\ &\quad + \sigma(\mathbf{x})\,\mathrm{L}(e_2)_2\,\mathrm{L}''(e_1)_1 - \sigma(\mathbf{x})\,\mathrm{L}(e_2)_1\,\mathrm{L}''(e_1)_2 \\ &\quad - \mathrm{L}''(e_2)_2 - \mathrm{L}''(e_1)_1.\end{aligned}$$

This yields (5.98) since φ being an isometry implies that $K = K'$, the Gauss curvature of M'^2, so that

$$\mathrm{L}(e_1)_1\,\mathrm{L}(e_2)_2 - (\mathrm{L}(e_1)_2)^2 = \mathrm{L}''(e_1)_1\,\mathrm{L}''(e_2)_2 - (\mathrm{L}''(e_1)_2)^2.$$

Now, by Stokes's theorem (see Theorem 8.7 below), we obtain the **Herglotz integral formula**:

$$\begin{aligned}\int_{M^2} H'd\mu &= \int_{M^2} \sigma(2K - \det(\mathrm{L} - \mathrm{L}''))d\mu \\ &= \int_{M^2} H d\mu - \int_{M^2} \sigma \det(\mathrm{L} - \mathrm{L}'')d\mu.\end{aligned} \tag{5.99}$$

Without loss of generality, we may assume that the origin is in the region interior to M^2, so that $\sigma > 0$. By the first statement of Lemma 5.16 below, we have $\det(\mathrm{L} - \mathrm{L}'') \le 0$. Hence

$$\int_{M^2} H'd\mu \ge \int_{M^2} H d\mu. \tag{5.100}$$

By symmetry and since φ is an isometry, we have the reverse inequality. Therefore

$$\int_{M^2} H'd\mu = \int_{M^2} H d\mu, \tag{5.101}$$

which implies that

$$\int_{M^2} \sigma \det(\mathrm{L} - \mathrm{L}'')\,d\mu = 0. \tag{5.102}$$

Since $\det(\mathrm{L} - \mathrm{L}'') \le 0$, we obtain that $\det(\mathrm{L} - \mathrm{L}'') = 0$, from which we conclude that $\mathrm{L} = \mathrm{L}''$ by the second statement in Lemma 5.16 below. Theorem 5.15 now follows from Lemma 5.14. □

Lastly, we prove the algebraic lemma that we used in the proof above.

Lemma 5.16. *If* L *and* L'' *are positive-definite symmetric* 2×2 *matrices satisfying* $\det \mathrm{L} = \det \mathrm{L}''$, *then* $\det(\mathrm{L} - \mathrm{L}'') \le 0$. *If, in addition,* $\det(\mathrm{L} - \mathrm{L}'') = 0$, *then* $\mathrm{L} = \mathrm{L}''$.

Proof. By a change of basis, we may assume that

$$\mathrm{L} = \begin{pmatrix} \kappa_1 & 0 \\ 0 & \kappa_2 \end{pmatrix}, \quad \mathrm{L}'' = \begin{pmatrix} a & b \\ b & d \end{pmatrix},$$

where $\kappa_1, \kappa_2, a, d > 0$ and $b \in \mathbb{R}$. Since $\det \mathrm{L} = \det \mathrm{L}''$, we have

$$\kappa_1 \kappa_2 = ad - b^2 \leq ad. \tag{5.103}$$

We have

$$\det(\mathrm{L} - \mathrm{L}'') = \kappa_1(\kappa_2 - d) + \kappa_2(\kappa_1 - a) \tag{5.104a}$$

$$= (a - \kappa_1)(d - \kappa_2) - b^2. \tag{5.104b}$$

If $a \leq \kappa_1$, then (5.103) implies that $d \geq \kappa_2$. Here, (5.104b) implies that $\det(\mathrm{L} - \mathrm{L}'') \leq 0$. Similarly, we obtain the same result if $d \leq \kappa_2$. Finally, if $a > \kappa_1$ and $d > \kappa_2$, then (5.104a) implies that $\det(\mathrm{L} - \mathrm{L}'') < 0$. We conclude that $\det(\mathrm{L} - \mathrm{L}'') \leq 0$ in all cases.

Now, suppose that $\det(\mathrm{L} - \mathrm{L}'') = 0$. Then by (5.104a) there exists $c \in \mathbb{R}$ such that

$$d - \kappa_2 = c\kappa_2, \quad a - \kappa_1 = -c\kappa_1,$$

which implies that

$$ad = (1 - c^2)\kappa_1\kappa_2 \leq \kappa_1\kappa_2.$$

We conclude that $b = c = 0$ and hence $\mathrm{L} = \mathrm{L}''$. □

5.6.2. Existence of isometric embeddings. In the next chapter we will define abstract smooth manifolds, and in particular this includes smooth surfaces. Generalizing the first fundamental form of a submanifold, we can define the notion of a Riemannian metric on a smooth manifold. The Weyl embedding problem asks whether every closed oriented Riemannian surface with positive Gauss curvature can be isometrically embedded in $\mathbb{R}^3$.

Theorem 5.17 (Weyl embedding problem). *Let (M^2, g) be a closed oriented Riemannian surface with positive Gauss curvature.*

(1) *(Existence) There exists a smooth isometric embedding of (M^2, g) into $\mathbb{R}^3$.*

(2) *(Uniqueness) The isometric embedding is unique up to the composition with an isometry of $\mathbb{R}^3$.*

Major progress on the proof of the Weyl embedding problem was made by Weyl [**Wey16**], including a fundamental C^2 estimate in the form of a geometric Bochner-type estimate. Lewy [**Lew38**] solved the case where g is real analytic. Notable work includes that of Aleksandrov [**Ale48**]. The proof was finally completed by Nirenberg [**Nir53**] and Pogorelov [**Pog52**] independently. Further work is by Heinz [**Hei59**, **Hei62**] and Guan and Li [**GL94**].

The uniqueness part of Theorem 5.17 is Theorem 5.15.

The existence part of Theorem 5.17 is proved by the *continuity method* in partial differential equations. From Theorem 8.27 below, there exists a function u on M^2 such that the pointwise conformally related metric $e^{2u}g$ has constant positive Gauss curvature equal to 1. Moreover, since g and $e^{2u}g$ both have positive Gauss curvature, the metrics

$$g_t := e^{2tu}g \tag{5.105}$$

have positive Gauss curvature for $0 \le t \le 1$. Indeed, by Lemma 8.21 below, we have that

$$\begin{aligned} K(e^{2tu}g) &= e^{-2tu}\left(K(g) - t\Delta_g u\right) \\ &= e^{-2tu}\left(tK(g) - t\Delta_g u\right) + e^{-2tu}(1-t)K(g) \\ &= te^{2(1-t)u}K(e^{2u}g) + e^{-2tu}(1-t)K(g). \end{aligned}$$

Now, since (M^2, g_0) has constant positive curvature and hence admits an isometric embedding in $\mathbb{R}^3$, the existence part of Theorem 5.17 is a consequence of the following two propositions.

Proposition 5.18 (Openness). *The set of $t \in [0,1]$ such that (M^2, g_t) admits an isometric embedding in $\mathbb{R}^3$ is open.*

Proposition 5.19 (Closedness). *The set of $t \in [0,1]$ such that (M^2, g_t) admits an isometric embedding in $\mathbb{R}^3$ is closed.*

The proof of these propositions is beyond the scope of this book. We refer the reader to Han and Hong [**HH06**].

5.6.3. The Minkowski problem.

The Minkowski problem asks under what condition is a positive function on S^n equal to the Gauss curvature of some convex embedded hypersurface. The following supplies a necessary and sufficient condition.

Theorem 5.20 (Minkowski problem). *Let κ be a positive smooth function on the unit n-sphere S^n satisfying the integral condition*

$$\int_{S^n} \frac{1}{\kappa(\mathbf{n})}\mathbf{n}\, d\mu_{S^n}(\mathbf{n}) = \mathbf{0}. \tag{5.106}$$

Then there exists a closed strictly convex smooth hypersurface M^n embedded in $\mathbb{R}^{n+1}$ with Gauss curvature satisfying

$$K(G^{-1}(\mathbf{n})) = \kappa(\mathbf{n}) \quad \textit{for all } \mathbf{n} \in S^n, \tag{5.107}$$

where $G : M^n \to S^n$ is the Gauss map and G^{-1} is its inverse. Moreover, M^n is unique up to translation.

By (5.83), condition (5.106) is a necessary condition to solve (5.107).

5.6.4. The Nash C^∞ isometric embedding theorem. The Nash isometric embedding [**Nas56**] is the following.

Theorem 5.21 (Nash isometric embedding). *For any n there exists N such that if M^n is a closed C^k manifold, where $3 \leq k \leq \infty$, then there exists a C^k isometric embedding of M^n into $\mathbb{R}^N$.*

Nash proved this result for ambient dimension $N = \frac{n(3n+11)}{2}$. Gromov [**Gro70**] later improved the dimension to $N = \frac{(n+2)(n+3)}{2}$ for $k \geq 4$. Günther [**G89a, G89b, G91**] discovered that one can circumvent some of the difficulties which Nash encountered. In particular, the remarkable Nash–Moser iteration method is not required. Moreover, Günther improved the dimension to $N = \max\left\{\frac{n(n+3)}{2} + 5, \frac{n(n+5)}{2}\right\}$.

5.7. Appendix

In this appendix we collect a few supplementary materials related to the discussion in this chapter.

5.7.1. A useful algebraic inequality. We recall an elementary algebraic inequality which is frequently used in geometric analysis. For any symmetric 2-tensor $T = T_{ij} dx^i \otimes dx^j$ on a submanifold M^n of $\mathbb{R}^N$, we have the inequality

$$|T|^2 \geq \frac{1}{n}\big(\operatorname{trace}(T)\big)^2. \tag{5.108}$$

Here, the pointwise norm of a 2-tensor is defined by

$$|T|^2 := \sum_{i,j,k,l=1}^{n} \mathrm{I}^{ik}\,\mathrm{I}^{jl}\,T_{ij}T_{kl}. \tag{5.109}$$

Equivalently, $|T|^2 = \sum_{i,j=1}^n \big(T(e_i,e_j)\big)^2$, where $\{e_i\}_{i=1}^n$ is an orthonormal frame. Note that these definitions generalize (5.31) and (5.32).

We can see inequality (5.108) by diagonalizing the matrix (T_{ij}), i.e., choosing local coordinates $\{x^i\}_{i=1}^n$ so that at a point we have $\mathrm{I}_{ij} = \delta_{ij}$ and $T_{ij} = T_{ii}\delta_{ij}$ (diagonal), where I is the first fundamental form. Then

$$\begin{aligned}
|T|^2 - \frac{1}{n}\big(\operatorname{trace}(T)\big)^2 &= \sum_i (T_{ii})^2 - \frac{1}{n}\Big(\sum_j T_{jj}\Big)^2 \\
&= \sum_i \Big(T_{ii} - \frac{1}{n}\sum_j T_{jj}\Big)^2 \\
&\geq 0.
\end{aligned}$$

Observe by this that we have equality in (5.108) at a point if and only if at that point $T = \frac{1}{n}\operatorname{trace}(T)\,\mathrm{I}$; cf. Exercise 5.14.

5.7.2. Inner products on spaces of tensors at a point. More generally, given covariant p-tensors α and β on M^n, their pointwise inner product is defined by

$$\langle \alpha, \beta \rangle := \mathrm{I}^{i_1 j_1} \cdots \mathrm{I}^{i_p j_p} \alpha_{i_1 \cdots i_p} \beta_{j_1 \cdots j_p}. \tag{5.110}$$

Equivalently, with respect to an orthonormal frame,

$$\langle \alpha, \beta \rangle = \alpha(e_{i_1}, \ldots, e_{i_p}) \beta(e_{i_1}, \ldots, e_{i_p}). \tag{5.111}$$

The pointwise norm of a p-tensor α is defined by

$$|\alpha| := \sqrt{\langle \alpha, \alpha \rangle}. \tag{5.112}$$

Note that definition (5.109) of the norm of a 2-tensor is a special case of this.

5.7.3. Calculating the Hessian and the Laplacian of the position map using local coordinates. In this subsection we give alternate calculations for some of the formulas in §5.5.2. Let X denote the position vector field (a.k.a. position map) on M^n; that is, X is the identity vector field id of $\mathbb{R}^{n+1}$ restricted to M^n. (We can also think of X as the inclusion map of M^n into $\mathbb{R}^{n+1}$.) By definition, $X_{\mathbf{x}} = \mathbf{x}$ for all $\mathbf{x} \in M^n$. We compute, using local coordinates $\{x^i\}_{i=1}^n$, that

$$\frac{1}{2} \nabla_i \|X\|^2 = \frac{1}{2} \frac{\partial}{\partial x^i} \langle X, X \rangle = \left\langle X, \frac{\partial X}{\partial x^i} \right\rangle, \tag{5.113}$$

where $\frac{\partial X}{\partial x^i} = \frac{\partial F}{\partial u^i}$ (F is the local parametrization inducing the local coordinates), so that

$$\frac{1}{2} \nabla \|X\|^2 = \frac{1}{2} \sum_{i=1}^n \nabla_i \|X\|^2 g^{ij} \frac{\partial X}{\partial x^j} = \sum_{i,j=1}^n \left\langle X, \frac{\partial X}{\partial x^i} \right\rangle g^{ij} \frac{\partial X}{\partial x^j} = X^\top;$$

that is,

$$\frac{1}{2} \nabla \|X\|^2 = X^\top. \tag{5.114}$$

Observe that for all $1 \leq i, j, l \leq n$,

$$\frac{\partial^2 X}{\partial x^i \partial x^j} \cdot \frac{\partial X}{\partial x^l} = \sum_{k=1}^n \Gamma_{ij}^k g_{kl} = \sum_{k=1}^n \Gamma_{ij}^k \nabla_k X \cdot \frac{\partial X}{\partial x^l}, \tag{5.115}$$

where the first equality follows from (4.140). Thus, the tangential component of $\frac{\partial^2 X}{\partial x^i \partial x^j}$ is equal to $\sum_{k=1}^n \Gamma_{ij}^k \nabla_k X$. Using this fact, we calculate that

$$\nabla_i \nabla_j X := (\nabla^2 X)(\partial_i, \partial_j) = \frac{\partial^2 X}{\partial x^i \partial x^j} - \sum_{k=1}^{n} \Gamma_{ij}^k \nabla_k X \tag{5.116}$$
$$= \left(\frac{\partial^2 X}{\partial x^i \partial x^j} \right)^{\perp}$$
$$= -\,\mathrm{II}_{ij}\, N_X,$$

where for the last inequality we used (4.96). That is, the Hessian of the position map (with respect to the first fundamental form I) is equal to minus the second fundamental form times the unit normal. Tracing this yields

$$\Delta X = -H N_X. \tag{5.117}$$

That is, the Laplacian of the position map (with respect to I) is equal to the mean curvature vector.

We further compute that

$$\frac{1}{2} \nabla_i \nabla_j \|X\|^2 = \nabla_i X \cdot \nabla_j X + X \cdot \nabla_i \nabla_j X \tag{5.118}$$
$$= g_{ij} - \mathrm{II}_{ij} \langle X, N_X \rangle.$$

Tracing this yields the formula

$$\frac{1}{2} \Delta_g \|X\|^2 = n - H \langle X, N_X \rangle = n - H\sigma \tag{5.119}$$

on M^n.

5.8. Notes and commentary

A well-known area of study in submanifold theory regards the Willmore conjecture [**Wil65**]. Given an oriented closed surface M^2 and an immersion $F : M^2 \to \mathbb{R}^3$, the **Willmore energy** of F is defined as

$$W(F) = \int_{M^2} H^2 \, d\mu. \tag{5.120}$$

The **Willmore conjecture** posits that

$$W(F) \geq 2\pi^2. \tag{5.121}$$

This conjecture was proved by Marques and Neves [**MN14**]. Their proof employs the min-max theory for minimal surfaces of Almgren and Pitts. Notable earlier works on the Willmore conjecture were done by Li and Yau [**LY82**] and Simon [**Sim93**].

Another well-known area of study in submanifold theory regards convex bodies. See the book by Schneider [**Sch14b**].

5.9. Exercises

Exercise 5.1. *Let M^n be an embedded hypersurface in $\mathbb{R}^{n+1}$. Extending the definitions of the first and second fundamental forms, the* ***k-th fundamental form*** *is the bilinear map on the tangent space $T_{\mathbf{x}}M$ defined by*

$$(X,Y) \mapsto \langle \mathrm{L}^k(X), Y\rangle. \tag{5.122}$$

We denote the third fundamental form by $\mathrm{III}(X,Y) = \langle \mathrm{L}^3(X), Y\rangle$. *Show that for a surface M^2 embedded in $\mathbb{R}^3$,*

$$\mathrm{L}^2 - H\,\mathrm{L} + K\,\mathrm{id} = 0, \tag{5.123}$$

and hence

$$\mathrm{III} - H\,\mathrm{II} + K\,\mathrm{I} = 0. \tag{5.124}$$

Exercise 5.2. *Let M^n be an embedded closed hypersurface in $\mathbb{R}^{n+1}$. Suppose that* $\mathrm{III} = f\,\mathrm{II}$ *for some function f on M^n. Show that M^n is a round sphere.*

Exercise 5.3. *Let M^2 be a surface embedded in $\mathbb{R}^3$. Let $\kappa_1 \le \kappa_2$ be the principal curvatures at a point in M^2. Show that*

$$\kappa_1 = \frac{H - \sqrt{H^2 - 4K}}{2}, \qquad \kappa_2 = \frac{H + \sqrt{H^2 - 4K}}{2}, \tag{5.125}$$

where H and K are the mean and Gauss curvatures at that point, respectively.

Exercise 5.4. *Let M^n be a hypersurface embedded in $\mathbb{R}^{n+1}$. Prove that for any vector fields X and Y on M^n,*

$$\nabla_X\big(\mathrm{L}(Y)\big) - \nabla_Y\big(\mathrm{L}(X)\big) = \mathrm{L}([X,Y]), \tag{5.126}$$

where L *denotes the Weingarten map of M^n.*

Exercise 5.5. *Let M^2 be a surface embedded in $\mathbb{R}^3$. Let U denote the set of umbilical points in M^2. Show that the principal curvature functions $\kappa_i : M^2 \to \mathbb{R}$, $i = 1, 2$, are continuous on M^2. Show that they are smooth on the open subset $M^2 \setminus U$.*

Exercise 5.6. *Let M^2 be a hypersurface embedded in $\mathbb{R}^3$.*

(1) *Prove that on $M^2 \setminus U$ there exist functions a and b such that*

$$\nabla_{e_1} e_2 = a e_1, \qquad \nabla_{e_2} e_1 = b e_2. \tag{5.127}$$

Show that

$$\nabla_{e_1} e_1 = -a e_2, \qquad \nabla_{e_2} e_2 = -b e_1. \tag{5.128}$$

(2) *Show that on $M^2 \setminus U$,*

$$\nabla_{e_1}\big(\mathrm{L}(e_2)\big) - \nabla_{e_2}\big(\mathrm{L}(e_1)\big) = a\kappa_1 e_1 - b\kappa_2 e_2. \tag{5.129}$$

(3) *Show that on* $M^2 \setminus U$,

$$a = -\frac{e_2(\kappa_1)}{\kappa_1 - \kappa_2}, \qquad b = \frac{e_1(\kappa_2)}{\kappa_1 - \kappa_2}. \tag{5.130}$$

Exercise 5.7. *Continuing with the previous exercise:*

(1) *Using the definition* (4.154) *of* Rm, *show that*

$$\operatorname{Rm}(e_1, e_2)e_2 = \big(-e_1(b) - e_2(a) - a^2 - b^2\big)e_1. \tag{5.131}$$

From this we conclude that the Gauss curvature at a nonumbilic point is given by

$$K = -e_1(b) - e_2(a) - a^2 - b^2. \tag{5.132}$$

(2) *Let* $\mathbf{x} \in M^2$ *be a nonumbilic point. Suppose* $\mathbf{x}$ *is a critical point of both* κ_1 *and* κ_2*. Show that*

$$K = -e_1(b) - e_2(a) = -\frac{e_1(e_1(\kappa_2))}{\kappa_1 - \kappa_2} + \frac{e_2(e_2(\kappa_1))}{\kappa_1 - \kappa_2}. \tag{5.133}$$

Exercise 5.8. *Prove that if* M^2 *is a connected embedded minimal surface in* $\mathbb{R}^3$ *with constant Gauss curvature, then* M^2 *is subset of a plane.*

Exercise 5.9.

(1) *Let* M^2 *be an embedded surface in* $\mathbb{R}^3$ *with constant Gauss curvature. Show that if* $\mathbf{x} \in M^2$ *has the property that* $\kappa_1(\mathbf{x}) \neq \kappa_2(\mathbf{x})$, $\mathbf{x}$ *is a relative minimum for* κ_1*, and* $\mathbf{x}$ *is a relative maximum for* κ_2*, then* $K(\mathbf{x}) \leq 0$.

(2) *Show that if* M^2 *is a closed surface with constant positive Gauss curvature, then* M^2 *is a round sphere.*

Exercise 5.10. *Prove the fact that if a function* $f : M^n \to \mathbb{R}$ *satisfies* $\nabla f = \mathbf{0}$ *on* M^n*, where* M^n *is connected, then* f *is constant on* M^n.

Exercise 5.11 (Locally convex neighborhoods in immersed hypersurfaces)**.** *Prove that if* M^n *is a closed immersed hypersurface in* $\mathbb{R}^{n+1}$*, then there exists a point* $\mathbf{y}$ *on* M^n *and a choice of a smooth unit normal vector field in a neighborhood of* $\mathbf{y}$ *such that the second fundamental form* II *of* M^n *is positive definite in this neighborhood. In particular, the mean curvature* H *at* $\mathbf{y}$ *is not equal to zero.*

Exercise 5.12. *Fill in the details for the assertions in the last paragraph of the proof of Theorem* 5.9.
Hint: For example, to show that M^n *is embedded, suppose for a contradiction that it were not and let* $\mathbf{X}_0 \in \mathbb{R}^{n+1}$ *be a point for which there would exist distinct points* $\mathbf{x}_1$ *and* $\mathbf{x}_2$ *in* M^n *such that* $F(\mathbf{x}_1) = F(\mathbf{x}_2) = \mathbf{X}_0$*. Obtain a contradiction about whether* $T_{F(\mathbf{x}_1)}M$ *and* $T_{F(\mathbf{x}_2)}M$ *are transverse or not.*

Exercise 5.13. *Prove* (5.23).
Hint: Given $\mathbf{x} \notin \Omega$, *let* $\mathbf{y}$ *be the closest point on* M^n. *Show that* $\mathbf{x} \notin \bar{H}_{\mathbf{y}}^{-}$.

Exercise 5.14. *Show that if* T *is a symmetric* 2*-tensor on a submanifold* M^n *of* $\mathbb{R}^N$, *then*

$$\left|T - \frac{1}{n}\operatorname{trace}(T)\,\mathrm{I}\right|^2 = |T|^2 - \frac{1}{n}\big(\operatorname{trace}(T)\big)^2.$$

Exercise 5.15. *Prove that if* T *is a symmetric* 2*-tensor on a submanifold* M^n *of* $\mathbb{R}^N$, *then*

$$|\nabla T|^2 \geq \frac{1}{n}|\operatorname{div}(T)|^2. \tag{5.134}$$

Exercise 5.16 (Huisken [**Hui84**, Lemma 2.2]). *Let* M^n *be a hypersurface embedded in* $\mathbb{R}^{n+1}$. *Define the covariant* 3*-tensor* E *by*

$$E_{ijk} = \frac{1}{n+2}\big(\nabla_i H\, g_{jk} + \nabla_j H\, g_{ik} + \nabla_k H\, g_{ij}\big). \tag{5.135}$$

Define the covariant 3*-tensor* F *by* $\nabla_i \mathrm{II}_{jk} = E_{ijk} + F_{ijk}$. *Show that*

$$\langle E, F\rangle = 0 \tag{5.136}$$

and

$$|E|^2 = \frac{3}{n+2}|\nabla H|^2. \tag{5.137}$$

Conclude that

$$\left|\nabla \mathrm{II} - \frac{1}{n}\nabla H \otimes g\right|^2 \geq \frac{2(n-1)}{n(n+2)}|\nabla H|^2. \tag{5.138}$$

Give another proof that if M^n *is connected, totally umbilical, and* $n \neq 1$, *then* H *is constant; cf.* (5.5).

Exercise 5.17. *Let* M^n *be a smooth embedded submanifold in* $\mathbb{R}^N$. *Define* $f : M^n \to \mathbb{R}$ *by* $f(\mathbf{x}) = \|\mathbf{x}\|^2$. *Prove that for* $V, W \in T_{\mathbf{x}}M$,

$$\frac{1}{2}(\nabla^2 f)(V, W) = \langle B(V, W), \mathbf{x}^{\perp}\rangle + \langle V, W\rangle, \tag{5.139}$$

where ∇ *is the Levi-Civita connection of* M^n *and where* $B = B_{\mathbf{x}} : T_{\mathbf{x}}M \times T_{\mathbf{x}}M \to N_{\mathbf{x}}M$ *is the second fundamental form of* M^n *defined by* (4.115). *This extends formula* (5.64) *for hypersurfaces.*

Moreover, show that

$$\frac{1}{2}\Delta_{M^n} f = \langle \vec{H}, \mathbf{x}^{\perp} \rangle + n. \tag{5.140}$$

Exercise 5.18. *Let $M^2 \subset \mathbb{R}^3$ be an embedded minimal surface with a smooth unit normal vector field N. Let $G : M^2 \to S^2$ be its Gauss map. Assume that the principal curvatures $\kappa_1 = -\kappa_2$ are nonzero everywhere on M^2. Show that there exists a positive function $f : M^2 \to \mathbb{R}$ such that*

$$\langle dG(X), dG(Y) \rangle = f(\mathbf{x})\langle X, Y \rangle$$

for $X, Y \in T_{\mathbf{x}}M$, $\mathbf{x} \in M^2$.

Show that if $\{e_1, e_2\}$ is an orthonormal tangent frame at $\mathbf{x} \in M^2$, then $\{dG(e_1), dG(e_2), N_{\mathbf{x}}\}$ has the opposite orientation from $\{e_1, e_2, N_{\mathbf{x}}\}$.

Exercise 5.19. *The m-**th mean curvature** H_m, $0 \leq m \leq n$, is defined to be the m-th elementary symmetric polynomial in the principal curvatures:*

$$H_m := \sum_{1 \leq i_1 < i_2 < \cdots < i_m \leq n} \kappa_{i_1} \cdots \kappa_{i_m} \tag{5.141}$$

for $1 \leq m \leq n$, and $H_0 = 1$. Show H_1 is the mean curvature, H_2 is the scalar curvature, and H_n is the Gauss curvature.

Exercise 5.20. *Let II^k be the symmetric 2-tensor corresponding to the k-th power of the Weingarten map L^k; in other words,*

$$\mathrm{II}^k_{ij} := \mathrm{II}_{i\ell_1}\, \mathrm{II}_{\ell_1\ell_2} \cdots \mathrm{II}_{\ell_{k-1}j} \quad \textit{(there are k factors of } \mathrm{II}\textit{)}$$

for $k \geq 1$, and $\mathrm{II}^0 = g$. Define the symmetric 2-tensor A_m by

$$A_m = \sum_{k=1}^{m} (-1)^{k-1} H_{m-k}\, \mathrm{II}^{k-1} \tag{5.142}$$

for $1 \leq m \leq n$, and $A_0 = 0$. Prove, for $1 \leq m \leq n$, that we have the algebraic identities

$$\mathrm{trace}_g(A_m) = (n - m + 1)H_{m-1} \tag{5.143}$$

and

$$\langle A_m, \mathrm{II} \rangle = mH_m. \tag{5.144}$$

Exercise 5.21.

(1) *Prove the differential identity*

$$\mathrm{div}(A_m) = 0. \tag{5.145}$$

(2) *Prove the variation formula*

$$\frac{\partial}{\partial \mathrm{II}} H_m = A_m, \tag{5.146}$$

where the symmetric 2-tensor $\frac{\partial}{\partial \mathrm{II}} H_m$ *is defined by*

$$\left\langle \frac{\partial}{\partial \mathrm{II}} H_m, \alpha \right\rangle = \left.\frac{d}{dt}\right|_{t=0} H_m(\mathrm{II} + t\alpha, g^{-1}) \tag{5.147}$$

for symmetric 2-tensors α.

Exercise 5.22. *Prove the Minkowski formulas*

$$\int_{M^n} H_m \langle X, N\rangle \, d\mu = \frac{n-m+1}{m} \int_{M^n} H_{m-1} \, d\mu \tag{5.148}$$

for $1 \leq m \leq n$.
Hint: Consider

$$\int_{M^n} \mathbf{x}(A_m)_{ij} \nabla_i \nabla_j \mathbf{x} \, d\mu(\mathbf{x})$$

and calculate alternatively by integration by parts and by using the formula $\nabla_i \nabla_j \mathbf{x} = -\mathrm{II}_{ij} N$.

Exercise 5.23. *Let* M^n *be an embedded strictly convex hypersurface. Let* II^{ij} *denote the components of the inverse of the second fundamental form* II^{-1}, *so that* $\sum_{j=1}^n \mathrm{II}^{ij} \mathrm{II}_{jk} = \delta_{ik}$. *Define*

$$\mathcal{L} = K \, \mathrm{II}^{ij} \nabla_i \nabla_j,$$

which is an elliptic operator since $\mathrm{II} > 0$. *Prove that*

$$\nabla_p K = K \, \mathrm{II}^{ij} \nabla_p \mathrm{II}_{ij} . \tag{5.149}$$

Hint: Given any $\mathbf{x} \in M^n$, *choose a local parametrization so that* $\Gamma_{ij}^k(\mathbf{x}) = 0$. *Prove the formula in these coordinates and use that the formula is tensorial.*

Exercise 5.24.

(1) *Prove that*

$$\Delta K = K \, \mathrm{II}^{ij} \Delta \mathrm{II}_{ij} - K \, \mathrm{II}^{ik} \mathrm{II}^{j\ell} \nabla_p \mathrm{II}_{k\ell} \nabla_p \mathrm{II}_{ij} + \frac{1}{K} |\nabla K|^2.$$

(2) *Using Simons's identity (see* (12.71) *below)*

$$\nabla_i \nabla_j H = \Delta \mathrm{II}_{ij} - \mathrm{II}_{ij}^2 H + |\mathrm{II}|^2 \mathrm{II}_{ij},$$

prove that

$$\begin{aligned} \Delta K = {} & K \, \mathrm{II}^{ij} \nabla_i \nabla_j H - K \, \mathrm{II}^{ik} \mathrm{II}^{j\ell} \nabla_p \mathrm{II}_{k\ell} \nabla_p \mathrm{II}_{ij} + \frac{1}{K} |\nabla K|^2 \\ & - nK \left(|\mathrm{II}|^2 - \frac{1}{n} H^2 \right). \end{aligned} \tag{5.150}$$

Exercise 5.25. *Prove that if $M^2 \subset \mathbb{R}^3$ is a strictly convex surface, then*

$$\Delta K = \mathcal{L}H - (\nabla_p \mathrm{II}_{12})^2 + \nabla_p \mathrm{II}_{11} \nabla_p \mathrm{II}_{22} + K(4K - H^2). \tag{5.151}$$

This formula was discovered by Weyl and used in his approach to the Weyl embedding problem; see the next exercise for its application to obtain an a priori estimate.

Exercise 5.26. *Prove that if $M^2 \subset \mathbb{R}^3$ is a strictly convex surface, then*

$$\max_M H^2 \leq \max_M \left(4K - \frac{\Delta K}{K}\right). \tag{5.152}$$

This is Weyl's estimate. Observe that the right-hand side is an intrinsic quantity; i.e., it depends only on the Riemannian metric g. This estimate yields an intrinsic bound for the second fundamental form since the positivity of the Gauss curvature implies that $|\,\mathrm{II}\,|^2 = H^2 - 2K < H^2$.

Exercise 5.27. *Let $A = \overline{\nabla}^2 \sigma + \sigma \overline{g}$ as in Proposition 5.10. Prove, using the Codazzi equation, that*

$$\overline{\nabla}_i A_{jk} = \overline{\nabla}_j A_{ik}. \tag{5.153}$$

Exercise 5.28. *Prove the commutator formula*

$$\begin{aligned} &\overline{\nabla}_i \overline{\nabla}_j A_{k\ell} + \overline{\nabla}_j \overline{\nabla}_i A_{k\ell} \\ &\quad = \overline{\nabla}_k \overline{\nabla}_\ell A_{ij} + \overline{\nabla}_\ell \overline{\nabla}_k A_{ij} + 2\overline{g}_{ij} A_{k\ell} - 2\overline{g}_{k\ell} A_{ij}. \end{aligned} \tag{5.154}$$

Exercise 5.29.

(1) *Prove that*

$$\overline{\nabla}_i \overline{\nabla}_j (H_{-1})^{-1} = \overline{\Delta} A_{ij} + (H_{-1})^{-1} \overline{g}_{ij} - n A_{ij}, \tag{5.155}$$

where

$$(H_{-1})^{-1} := \kappa_1^{-1} + \cdots + \kappa_n^{-1} = \overline{\Delta}\sigma + n\sigma \tag{5.156}$$

is the inverse of the ***harmonic mean curvature****.*

(2) *Prove that*

$$\begin{aligned} &K^{-1} A_{ij}^{-1} \overline{\nabla}_i \overline{\nabla}_j (H_{-1})^{-1} \\ &\quad = K^{-1} A_{ij}^{-1} \overline{\Delta} A_{ij} + K^{-1}((H_{-1})^{-1} H - n^2). \end{aligned} \tag{5.157}$$

Exercise 5.30. *Let β be a positive real number. Prove that*

$$\begin{aligned} \overline{\Delta} K^{-\beta} &= \beta K^{-\beta} A_{ij}^{-1} \overline{\Delta} A_{ij} + \beta \overline{g}^{k\ell} \overline{\nabla}_\ell (K^{-\beta} A_{ij}^{-1}) \overline{\nabla}_k A_{ij} \\ &= \beta K^{-\beta} A_{ij}^{-1} \overline{\Delta} A_{ij} + \beta^2 K^{-\beta} \overline{g}^{k\ell} A_{pq}^{-1} A_{ij}^{-1} \overline{\nabla}_\ell A_{pq} \overline{\nabla}_k A_{ij} \\ &\quad - \beta K^{-\beta} \overline{g}^{k\ell} A_{ip}^{-1} A_{jq}^{-1} \overline{\nabla}_\ell A_{pq} \overline{\nabla}_k A_{ij}. \end{aligned} \tag{5.158}$$

Exercise 5.31. *Let* $\mathcal{L} = K^{-1/n} A_{ij}^{-1} \overline{\nabla}_i \overline{\nabla}_j$.

(1) *Prove that*

$$\mathcal{L}(H_{-1})^{-1} \geq n\overline{\Delta} K^{-1/n} + K^{-1/n}((H_{-1})^{-1} H - n^2). \tag{5.159}$$

(2) *Let* $\mathbf{x}_0 \in S^n$ *be a point at which* $(H_{-1})^{-1}$ *attains its maximum. Prove that at* $\mathbf{x}_0$

$$(H_{-1})^{-1} \leq nK^{-1/n} - \overline{\Delta}(K^{-1/n}). \tag{5.160}$$

Since $\mathbf{x}_0$ *is a maximum point of* $(H_{-1})^{-1}$, *we have obtained an estimate for* $(H_{-1})^{-1}$ *in terms of* K *and its derivatives.*

This is a key estimate for the solution to the Minkowski problem.

Part II

Differential Topology and Riemannian Geometry

Chapter 6

Smooth Manifolds

In calculus, we are acquainted with the fact that Euclidean space serves well as a ground to develop many useful algebraic, geometric, and analytic notions: vector space structures, distance between points, total and partial derivatives, integration, etc. However, n-dimensional Euclidean space $\mathbb{R}^n$ is a very special space. For example, on it we have the special algebraic property of 'linearity'; i.e., the n-tuples associated with 'points' in $\mathbb{R}^n$ can be 'added' componentwise in the arithmetic sense. We can also multiply any n-tuple by a real number in that space.

When out of necessity we have to abandon Euclidean space, it would mean the loss of some of the most convenient structures attached to that space. Under such circumstances, we must decide what are the reasonable things to do so that the generalized 'spaces' that we create will somehow retain the important features of Euclidean space. Such spaces of interest here will be 'smooth manifolds'; some features are retained here through the crucial use of homeomorphisms and the concept of locality.

Homeomorphisms, being topological in nature, loosen the grip of the Euclidean metric properties. They thus provide us the option of introducing new metrics (e.g., Riemannian) on the generalized spaces, *smooth manifolds*. On the other hand, the concept of 'locality' is, in a sense, an imitation of the spirit of 'open sets' in a topology: A 'smooth manifold' is required to be covered by a collection of 'local' patches. Together with homeomorphisms, these *local patches* are called 'charts' by satisfying some required (smoothness) conditions. These 'charts' basically fix the 'structure' (i.e., a 'smooth structure') of the space just as a collection of 'open sets' fixes the (topological) structure of a topological space. Needless to say, the merit of introducing 'smooth manifolds' rests on the ability of the new structure to host the development of some useful algebraic, geometric, and analytic properties.

In this chapter, we discuss the theory of smooth manifolds, including the notions of tangent space and bundle, smooth maps, derivatives, diffeomorphisms, vector fields, Lie brackets, tensors, push-forwards, pull-backs, and Lie derivatives. We also discuss categories of manifolds.

6.1. Manifolds

The version of manifolds we have studied in Part I is called *submanifolds* of Euclidean space, because they are manifolds contained in a higher-dimensional Euclidean space. In this section, we begin the discussion of manifolds in the abstract.

6.1.1. The definition of manifold. The definition of a *smooth manifold* requires, roughly speaking, the coverage of a space by 'Euclidean' open sets that are related to each other, in the overlap regions, in a smooth, a.k.a. C^∞, way. The following definition of 'coordinate chart' formalizes the concept of being *locally Euclidean*.

The given underlying topological space M^n will always be assumed to be Hausdorff and have a countable basis (i.e., second countable) topologically. See the Wikipedia page for 'non-Hausdorff manifold' for a tautological example of a line with two origins, which we do not allow. The condition of having a countable basis is useful for the existence of a *partition of unity*, which in turn is needed to prove the existence of a *Riemannian metric* on a manifold. General references for the point-set topology concepts that we will use are Munkres [**Mun00**] and Engelking [**Eng89**].

Definition 6.1. An n-dimensional **topological coordinate chart** or **local coordinates** $(\mathcal{U}, \varphi_{\mathcal{U}})$ on a topological space M^n consists of an open subset $\mathcal{U}$ of M^n and a homeomorphism $\varphi_{\mathcal{U}}$ from $\mathcal{U}$ onto an open subset $\varphi_{\mathcal{U}}(\mathcal{U})$ of $\mathbb{R}^n$. We often just call $(\mathcal{U}, \varphi_{\mathcal{U}})$ a **chart**. The set $\mathcal{U}$ is called a **(local) coordinate neighborhood** or a 'coordinate patch' and $\varphi_{\mathcal{U}}$ is called a **(local) coordinate map**. See Figure 6.1.1.

Since $\varphi_{\mathcal{U}}$ is a homeomorphism, we think of $\varphi_{\mathcal{U}}(\mathcal{U})$ as a Euclidean model for $\mathcal{U} \subset M^n$. We will often use the notation $(x^1, \ldots, x^n) =: \mathbf{x}$ for $\varphi_{\mathcal{U}}$, even though it is redundant with the notation for a point in $\mathbb{R}^n$. Let $\pi_i : \mathbb{R}^n \to \mathbb{R}$ be the projection onto the i-th coordinate. Then

$$x^i := \pi_i \circ \varphi_{\mathcal{U}} : \mathcal{U} \to \mathbb{R} \tag{6.1}$$

is called the **i-th coordinate function**.

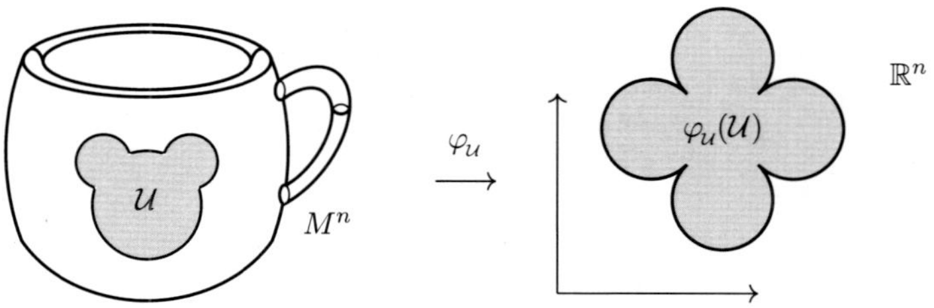

Figure 6.1.1. A local coordinate chart $(\mathcal{U}, \varphi_{\mathcal{U}})$ on a coffee cup. Mickey is mapped to a four-leaf clover.

Let $\mathfrak{U}$ be an open cover of M^n. Let $(\mathcal{U}, \varphi_{\mathcal{U}})$, $\mathcal{U} \in \mathfrak{U}$, be a collection of *charts* on M^n of the same dimension n. Since $\mathfrak{U}$ is a cover of M^n, every point in M^n is contained in a neighborhood that is 'Euclidean' in the topological or smooth sense. But are the notions for Euclidean compatible for different charts where they overlap? By imposing further the 'smooth' compatibility condition on these charts, we arrive at the definition of a *smooth structure* or, in other words, an 'atlas'. A topological space with a 'smooth structure' is then called a 'smooth manifold' or 'manifold' in short, as we now discuss in detail.

Definition 6.2. A collection of n-dimensional charts $\mathbf{A} := \{(\mathcal{U}, \varphi_{\mathcal{U}})\}_{\mathcal{U} \in \mathfrak{U}}$ on a topological space M^n, where $\mathfrak{U}$ is an open cover, forms an n-dimensional C^k **atlas** (or equivalently a C^k **structure**) on M^n if, for any overlapping charts $\mathcal{U}, \mathcal{V} \in \mathfrak{U}$, the **transition function** (see Figure 6.1.2)

$$\varphi_{\mathcal{U}\mathcal{V}} := \varphi_{\mathcal{U}} \circ \varphi_{\mathcal{V}}^{-1} : \varphi_{\mathcal{V}}(\mathcal{U} \cap \mathcal{V}) \to \varphi_{\mathcal{U}}(\mathcal{U} \cap \mathcal{V}) \tag{6.2}$$

is a C^k map. Here, C^k has the usual meaning of k-times continuously differentiable, where k is a positive integer, 0, ∞, or ω. On M^n, a C^0 atlas is called a **topological manifold structure**, a C^∞ atlas is called a **smooth manifold structure**, and a C^ω atlas is called a **real analytic manifold structure**.

Example 6.3. *Suppose that $(\mathcal{U}_1, \varphi_{\mathcal{U}_1})$ and $(\mathcal{U}_2, \varphi_{\mathcal{U}_2})$ are topological charts on a topological space M^n and $\mathcal{U}_1 \cup \mathcal{U}_2 = M^n$. Then the collection $\mathbf{A} := \{(\mathcal{U}_1, \varphi_{\mathcal{U}_1}), (\mathcal{U}_2, \varphi_{\mathcal{U}_2})\}$ is a C^k atlas if and only if $\varphi_{\mathcal{U}_1\mathcal{U}_2}$ and $\varphi_{\mathcal{U}_2\mathcal{U}_1}$ are both C^k maps.*

If an atlas $\mathbf{A}$ is a subset of an atlas $\mathbf{A}'$, then we say that $\mathbf{A}$ is a **subatlas** of $\mathbf{A}'$.

Example 6.4. *Suppose that $\mathbf{A}$ is a C^k atlas on M^n. Let $(\mathcal{V}, \varphi_{\mathcal{V}})$ be a topological coordinate chart on M^n. If $\varphi_{\mathcal{U}\mathcal{V}}$ and $\varphi_{\mathcal{V}\mathcal{U}}$ are both C^k maps for all $(\mathcal{U}, \varphi_{\mathcal{U}}) \in \mathbf{A}$, then the collection $\mathbf{A}' := \mathbf{A} \cup \{(\mathcal{V}, \varphi_{\mathcal{V}})\}$ is a C^k atlas and $\mathbf{A}$ is a subatlas of $\mathbf{A}'$.*

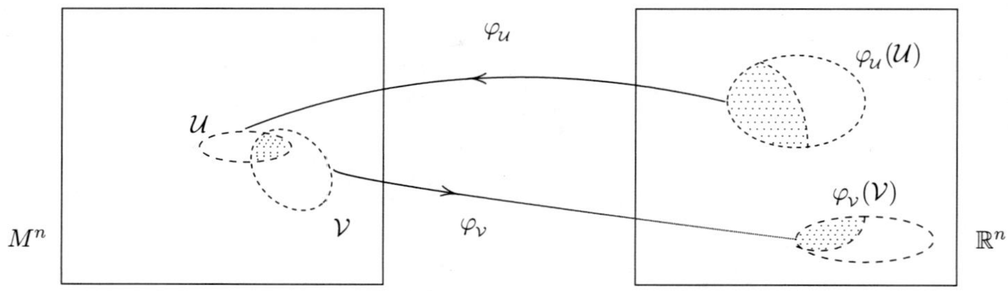

Figure 6.1.2. The transition function $\varphi_{\mathcal{UV}}$.

To avoid repetitive discussions of 'indistinguishable' differentiable structures, one can introduce the notion of an 'equivalence class' of atlases: Two C^k atlases $\mathbf{A}$ and $\mathbf{A}'$ (of the same dimension) are **equivalent** if, for any chart $(\mathcal{U}, \varphi_\mathcal{U})$ in $\mathbf{A}$ and any chart $(\mathcal{V}, \varphi_\mathcal{V})$ in $\mathbf{A}'$, with $\mathcal{U} \cap \mathcal{V} \neq \emptyset$, condition (6.2) is satisfied and also with $\mathcal{U}$ and $\mathcal{V}$ exchanged in (6.2). It is easy to check that this notion of equivalent atlases defines an equivalence relation on the set of all atlases on M^n.

One can also combine all those charts in an equivalence class of atlases to form a single atlas, to be called a **maximal atlas**. By definition, each equivalence class of atlases on M^n contains a unique maximal atlas. This maximal atlas contains each atlas in its equivalence class as a subatlas.

Definition 6.5 (Manifold)**.** An n-dimensional **differentiable manifold** (or simply a **manifold**) is defined as a second countable Hausdorff topological space M^n with an n-dimensional differentiable structure (i.e., an 'atlas' $\mathbf{A}$). A manifold is said to be C^k if its differentiable structure is C^k. A C^0 manifold will be referred to as a **topological manifold**, a C^∞ manifold will be referred to as a **smooth (differentiable) manifold**, and a C^ω manifold will be referred to as a **real analytic (differentiable) manifold**. We will always assume that the atlas $\mathbf{A}$ is maximal.

Using the convention that $0 < k < \infty < \omega$ for any positive integer k, we have that any C^β manifold is a C^α manifold for $\alpha < \beta$. This is simply by applying to the transition functions the fact that any C^β map is a C^α map; in particular, any differentiable manifold M^n is a topological manifold, which we call the **underlying topological manifold**. In more detail, if M^n is a manifold with a C^β atlas $\mathbf{A}_\beta$, then $\mathbf{A}_\beta$ is contained in a unique maximal C^α atlas $\mathbf{A}_\alpha$. This atlas $\mathbf{A}_\alpha$ is equal to the collection of all charts $(\mathcal{U}, \varphi_\mathcal{U})$ on M^n such that the transition functions $\varphi_{\mathcal{UV}}$ and $\varphi_{\mathcal{VU}}$ are C^α for all $(\mathcal{V}, \varphi_\mathcal{V}) \in \mathbf{A}_\beta$. For example, if $(M^n, \mathbf{A}_\beta)$ is a C^β manifold, then $(M^n, \mathbf{A}_0)$ is the underlying topological manifold.

Remark 6.6. If in Definition 6.5 we replace the model space $\mathbb{R}^n$ in defining manifolds by the **complex Euclidean space** $\mathbb{C}^n$ and if we require the transition functions to be **holomorphic maps** between open subsets of $\mathbb{C}^n$, then we obtain the definition of a **complex manifold**.[1]

6.1.2. Examples of manifolds.

Example 6.7 (Vector spaces). *Any n-dimensional real vector space $\mathbb{V}$ can be regarded as a smooth manifold since, for any basis $(v_1, \ldots, v_n)$ of $\mathbb{V}$, we can choose a one-chart atlas $(\mathbb{V}, \varphi_{\mathbb{V}})$ with*

$$\varphi_{\mathbb{V}} : \mathbb{V} \to \mathbb{R}^n \tag{6.3}$$

sending

$$\sum_{i=1}^{n} a^i v_i \longmapsto (a^1, \ldots, a^n). \tag{6.4}$$

This map is smooth, in fact, real analytic. The maximal atlas on $\mathbb{V}$ in which this one-chart atlas is contained is the collection of all charts $(\mathcal{U}, \varphi_{\mathcal{U}})$ with the properties that $\varphi_{\mathcal{U}} \circ \varphi_{\mathbb{V}}^{-1} : \varphi_{\mathbb{V}}(\mathcal{U}) \to \mathbb{R}^n$ and $\varphi_{\mathbb{V}} \circ \varphi_{\mathcal{U}}^{-1} : \varphi_{\mathcal{U}}(\mathcal{U}) \to \mathbb{R}^n$ are C^k.

We remark that our discussion in §1.3.1.2 of affine hyperplanes is similar, but there we are considering manifolds in the context of Euclidean hypersurfaces, whereas here we are considering manifolds in the abstract.

Example 6.8. *Given a field $\mathbb{F}$ equal to $\mathbb{R}$ or $\mathbb{C}$, let $\mathrm{M}(r, s, \mathbb{F})$ denote the set of $r \times s$ matrices over $\mathbb{F}$, that is, the set of all r by s rectangular arrays with entries in $\mathbb{F}$. Since $\mathrm{M}(r, s, \mathbb{F})$ is a vector space, it can be regarded as a smooth manifold. For $\mathbb{F} = \mathbb{R}$, its dimension is rs. For $\mathbb{F} = \mathbb{C}$, which is isomorphic to $\mathbb{R}^2$, its real dimension is $2rs$.*

Example 6.9 (Spheres). *As a differentiable manifold, an n-sphere S^n may be defined via an atlas with only two charts as follows. Consider the disjoint union of two copies of Euclidean space $X := \mathbb{R}^n_1 \sqcup \mathbb{R}^n_2$ (we put subscripts on the Euclidean spaces to distinguish them). Define the inversion $\iota : \mathbb{R}^n_1 \setminus \{\mathbf{0}_1\} \to \mathbb{R}^n_2$, where $\mathbf{0}_1$ denotes the origin of $\mathbb{R}^n_1$, by*

$$\iota(\mathbf{x}) := \frac{\mathbf{x}}{\|\mathbf{x}\|^2}. \tag{6.5}$$

Define the equivalence relation $\sim$ on X by $\mathbf{x} \sim \iota(\mathbf{x})$ for $\mathbf{x} \in \mathbb{R}^n_1 \setminus \{\mathbf{0}_1\}$ together with the stipulation that $\sim$ is reflexive and symmetric (transitivity follows). Let

$$S^n := X/\sim \tag{6.6}$$

[1] By definition, a map $f = (f^1, \ldots, f^n)$ between open subsets of $\mathbb{C}^n$ is a holomorphic map if each component function f^i is a holomorphic function.

be the quotient topological space under this equivalence relation. That is, we are identifying in X *the sets* $\mathbb{R}^n_1 \setminus \{\mathbf{0}_1\}$ *and* $\mathbb{R}^n_2 \setminus \{\mathbf{0}_2\}$ *via the map* ι*. Let*

$$\pi : X \to S^n \tag{6.7}$$

denote the canonical projection map. Let $i_1 : \mathbb{R}^n_1 \to \mathbb{R}^n_1 \sqcup \mathbb{R}^n_2$ *and* $i_2 : \mathbb{R}^n_2 \to \mathbb{R}^n_1 \sqcup \mathbb{R}^n_2$ *denote the inclusion maps. Then we have that the maps*

$$\psi_1 := \pi \circ i_1 : \mathbb{R}^n_1 \to S^n \setminus \{[\mathbf{0}_2]\} =: \mathcal{U}_1 \tag{6.8}$$

and

$$\psi_2 := \pi \circ i_2 : \mathbb{R}^n_2 \to S^n \setminus \{[\mathbf{0}_1]\} =: \mathcal{U}_2 \tag{6.9}$$

are both homeomorphisms. We claim that the collection of two coordinate charts $\{(\mathcal{U}_1, \psi_1^{-1}), (\mathcal{U}_2, \psi_2^{-1})\}$ *is an atlas for* S^n*. Indeed, one calculates that the transition function* $\psi_2^{-1} \circ \psi_1 : \mathbb{R}^n_1 \setminus \{\mathbf{0}_1\} \to \mathbb{R}^n_2 \setminus \{\mathbf{0}_2\}$ *is equal to the inversion* ι*. Similarly,* $\psi_1^{-1} \circ \psi_2 : \mathbb{R}^n_2 \setminus \{\mathbf{0}_2\} \to \mathbb{R}^n_1 \setminus \{\mathbf{0}_1\}$ *is also an inversion. Since inversions are real analytic, by Example* 6.3 *this shows that* S^n *is a real analytic manifold.*

Remark 6.10. In the description above, S^n is constructed more abstractly. A concrete construction of S^n is to take it to be the unit sphere in $\mathbb{R}^{n+1}$. In this case, the stereographic projections from the north and south poles yield two coordinate charts covering S^n. If you solved Exercise 2.11, you would have found that the transition function is equal to 4 times the inversion $\iota : \mathbb{R}^n \setminus \{\mathbf{0}\} \to \mathbb{R}^n \setminus \{\mathbf{0}\}$ defined by (6.5).

Affine graphs, as discussed in §1.3.3.2, can be defined as manifolds using a single coordinate chart (which uniquely extends to a maximal atlas). Radial graphs, as discussed in §1.3.3.3, can be defined as manifolds using just two coordinate charts (as for spheres).

Example 6.11 (Product manifolds)**.** *Suppose that* $M_1^{n_1}$ *and* $M_2^{n_2}$ *are* C^k *manifolds. Then their Cartesian product* $M_1 \times M_2$ *has a natural manifold structure which makes it an* (n_1+n_2)*-dimensional* C^k *manifold. Namely, let* $\mathbf{A}_1$ *and* $\mathbf{A}_2$ *be the atlases of* $M_1^{n_1}$ *and* $M_2^{n_2}$*, respectively. A (not maximal)* C^k *atlas for* $M_1 \times M_2$ *is defined by taking its charts to be* $(\mathcal{U}_1 \times \mathcal{U}_2, \varphi_{\mathcal{U}_1} \times \varphi_{\mathcal{U}_2})$*, where* $(\mathcal{U}_1, \varphi_{\mathcal{U}_1}) \in \mathbf{A}_1$ *and* $(\mathcal{U}_2, \varphi_{\mathcal{U}_2}) \in \mathbf{A}_2$*.*

The definition of the product of 2 manifolds extends to products of ℓ-many manifolds, where $\ell \geq 2$ is an integer; see Exercise 6.2.

Example 6.12. *Let* M^n *be a* C^k *manifold and let* $\mathcal{U}$ *be an open subset of* M^n*. Then* $\mathcal{U}$ *is naturally a* C^k *manifold. Namely, we define the atlas* $\mathbf{A}$ *for* $\mathcal{U}$ *to be the collection* $\mathbf{A}$ *of all charts of the form* $(\mathcal{V} \cap \mathcal{U}, \varphi_{\mathcal{V}}|_{\mathcal{V} \cap \mathcal{U}})$*, where* $(\mathcal{V}, \varphi_{\mathcal{V}})$ *is a chart for* M^n*. We leave it to the reader to check that* $\mathbf{A}$ *is a well-defined atlas for* $\mathcal{U}$*.*

Example 6.13 (Real projective spaces). *Denote* $\mathbb{R}_*^{n+1} := \mathbb{R}^{n+1} \setminus \{\mathbf{0}\}$. *On* $\mathbb{R}_*^{n+1}$ *define the equivalence relation*

$$\mathbf{x} \sim a\mathbf{x}, \quad a \in \mathbb{R}_*, \tag{6.10}$$

where $\mathbb{R}_* := \mathbb{R}_*^1$. *(It is easy to see that* $\sim$ *is an equivalence relation.) As a topological space,* ***real projective space*** $\mathbb{R}P^n$ *is defined as the quotient space associated to this equivalence relation:*

$$\mathbb{R}P^n := \{\,[\mathbf{x}] \mid \mathbf{x} \in \mathbb{R}_*^{n+1}\}, \tag{6.11}$$

where $[\mathbf{x}]$ *denotes the equivalence class of* $\mathbf{x}$.

The standard manifold structure on $\mathbb{R}P^n$ *is defined by the atlas with the* $n+1$ *charts*

$$(\mathcal{U}_i, \varphi_i), \quad i = 1, \dots, n+1,$$

having the patches defined by

$$\mathcal{U}_i := \big\{[\mathbf{x}] \in \mathbb{R}P^n \mid x^i \neq 0\big\}, \quad i = 1, \dots, n+1,$$

and the coordinate maps defined by

$$\varphi_i : [\mathbf{x}] \mapsto \left(\frac{x^1}{x^i}, \dots, \frac{x^{i-1}}{x^i}, \frac{x^{i+1}}{x^i}, \dots, \frac{x^{n+1}}{x^i}\right),$$

which sends $\mathcal{U}_i$ *into* $\mathbb{R}^n$. *We leave it as an exercise to show that this is indeed an atlas defining a (real analytic) manifold structure on* $\mathbb{R}P^n$. *Henceforth, we denote*

$$[\mathbf{x}] =: [x^1 : x^2 : \cdots : x^{n+1}]. \tag{6.12}$$

Remark 6.14. If $\sim$ is an equivalence relation on a set X, we have its quotient set $X/\sim$. Suppose that Y is a subset of X such that for every equivalence class $[x] \in X/\sim$, there exists $y \in Y$ such that $y \in [x]$. Then $\sim$ restricted to Y is an equivalence relation on Y and we have $X/\sim = Y/\sim$.

Regarding Example 6.13, we choose $X = \mathbb{R}_*^{n+1}$ and $Y = S^n$, the unit sphere. Let $[\mathbf{x}] \in \mathbb{R}_*^{n+1}/\sim$. Then $\frac{\mathbf{x}}{\|\mathbf{x}\|} \in S^n$ and $\frac{\mathbf{x}}{\|\mathbf{x}\|} \in [\mathbf{x}]$. So we have that $\mathbb{R}P^n = S^n/\sim$. Now suppose $\mathbf{x} \in S^n$ and $a\mathbf{x} \in S^n$, where $a \in \mathbb{R}_*$. Then we have $\|\mathbf{x}\| = 1$ and $\|a\mathbf{x}\| = 1$, which imply that $|a| = 1$; that is, $a = \pm 1$. Thus, the equivalence relation $\sim$, when restricted to S^n, is given by $\mathbf{x} \sim a\mathbf{x}$, where $a = \pm 1$. Thus, the canonical projection map $\pi : S^n \to \mathbb{R}P^n$ is a two-to-one map.

Example 6.15 (Complex projective spaces). *Denote* $\mathbb{C}_*^{n+1} := \mathbb{C}^{n+1} \setminus \{\mathbf{0}\}$. *On* $\mathbb{C}_*^{n+1}$ *define the equivalence relation*

$$\mathbf{z} \sim \lambda\mathbf{z}, \quad \lambda \in \mathbb{C}_*, \tag{6.13}$$

where $\mathbb{C}_* := \mathbb{C}^1_*$. *As a topological space,* ***complex projective space*** $\mathbb{C}P^n$ *is defined as the quotient space associated to this equivalence relation:*

$$\mathbb{C}P^n := \{\,[\mathbf{z}] \mid \mathbf{z} \in \mathbb{C}^{n+1}_*\}, \tag{6.14}$$

where $[\mathbf{z}]$ *denotes the equivalence class of* $\mathbf{z}$.

The standard manifold structure on $\mathbb{C}P^n$ *is defined by the atlas with the* $n+1$ *charts*

$$(\mathcal{U}_i, \varphi_i), \quad i = 1, \ldots, n+1, \tag{6.15}$$

having the patches defined by

$$\mathcal{U}_i := \{[\mathbf{z}] \in \mathbb{C}P^n \mid z^i \neq 0\}, \quad i = 1, \ldots, n+1, \tag{6.16}$$

and the coordinate maps defined by

$$\varphi_i : [\mathbf{z}] \mapsto \left(z^1(z^i)^{-1}, \ldots, z^{i-1}(z^i)^{-1}, z^{i+1}(z^i)^{-1}, \ldots, z^{n+1}(z^i)^{-1}\right), \tag{6.17}$$

which sends $\mathcal{U}_i$ *into* $\mathbb{C}^n$. *We leave it as an exercise to show that this is indeed an atlas defining a (real analytic) manifold structure on* $\mathbb{C}P^n$*; see Exercise* 6.3. *Henceforth, we denote*

$$[\mathbf{z}] =: [z^1 : z^2 : \cdots : z^{n+1}]. \tag{6.18}$$

We remark that $\mathbb{C}P^n$ *is a complex manifold (see Remark* 6.6*) since one can easily show that the transition functions are holomorphic maps.*

Example 6.16 (Grassmannian). *The* ***Grassmannian*** $\mathrm{Gr}(k, n)$ *of* k*-dimensional vector subspaces of* $\mathbb{R}^n$ *may be given the structure of a smooth manifold of dimension* $k(n-k)$ *(see, e.g., Lee* [**Lee13**, Example 1.36] *or see Exercise* 6.6 *below). The Grassmannian* $\mathrm{Gr}(1, n)$ *is naturally diffeomorphic to* $\mathbb{R}P^{n-1}$.

Example 6.17. *A smooth embedded submanifold* M^n *of Euclidean space* $\mathbb{R}^N$ *is a smooth manifold.*

See §8.8 below for a discussion on the classification of 2-dimensional manifolds, a.k.a. surfaces.

The manifolds that we defined and discussed so far are usually referred to as 'manifolds *without* boundaries' in contrast to 'manifolds *with* boundaries' defined below.

6.1.3. Manifolds with boundary.

Definition 6.18. An n-dimensional **differentiable manifold** M^n **with boundary** is a Hausdorff topological space with a differentiable structure, or atlas, $\{(\mathcal{U}, \varphi_{\mathcal{U}})\}$ with each chart $\mathcal{U}$ being homeomorphic to *either* an open set in $\mathbb{R}^n$ *or* in the space of the upper-half n-plane $\overline{\mathbb{R}}^n_+$ defined by

$$\overline{\mathbb{R}}^n_+ := \{\mathbf{x} \in \mathbb{R}^n \,|\, x^n \geq 0\}.$$

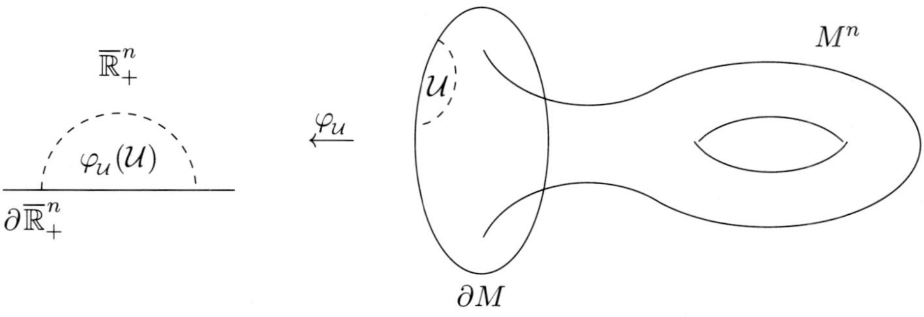

Figure 6.1.3. A boundary coordinate chart $(\mathcal{U}, \varphi_{\mathcal{U}})$.

See Figure 6.1.3. The **boundary** of M^n, denoted by ∂M, consists of those points (in M^n) that are mapped under $\varphi_{\mathcal{U}}$ to $\partial\overline{\mathbb{R}}^n_+$, for some $\mathcal{U}$.

It can be shown that if a point $p \in \mathcal{U} \cap \mathcal{V}$ is such that $\varphi_{\mathcal{U}}(p) \in \partial\overline{\mathbb{R}}^n_+$, then $\varphi_{\mathcal{V}}(p) \in \partial\overline{\mathbb{R}}^n_+$. Hence ∂M is well-defined. It is also a fact that ∂M is an $(n-1)$-dimensional manifold without boundary; i.e., a boundary has no boundary! In other words, the boundary of the boundary is empty; see Exercise 6.7.

Example 6.19. *The **unit** n**-ball***

$$B^n := \{\mathbf{x} \in \mathbb{R}^n \colon \|\mathbf{x}\|^2 = (x^1)^2 + \cdots + (x^n)^2 \leq 1\}$$

is an n-dimensional manifold with boundary $\partial B = S^{n-1}$, which is an $(n-1)$-dimensional manifold without boundary.

Example 6.20. *Let M^n be a C^k manifold with an atlas $\mathbf{A}$. Let $(\mathcal{U}, \varphi_{\mathcal{U}})$ be a chart in $\mathbf{A}$. Choose a point $\mathbf{x} \in \varphi_{\mathcal{U}}(\mathcal{U}) \subset \mathbb{R}^n$. Let $r > 0$ be such that $B_r(\mathbf{x}) \subset \varphi_{\mathcal{U}}(\mathcal{U})$. Then*

$$M^n \setminus \varphi_{\mathcal{U}}^{-1}(B_r(\mathbf{x}))$$

is a manifold with boundary.

Note that the product of two manifolds with boundary is not a manifold with boundary. For example, the closed unit interval $[0, 1]$ is a 1-dimensional C^∞ manifold with boundary $\partial[0, 1] = \{0, 1\}$. However, its product with itself $[0, 1] \times [0, 1]$ is not a 2-dimensional C^∞ manifold with boundary because it has (four) corners.

In general, the product of manifolds with boundary are examples of 'manifolds with corners'; see, e.g., Michor [**Mic80**].

6.2. Tangent space

Recall that we defined tangent vectors and tangent spaces for submanifolds of Euclidean space in Chapter 3. These are the most fundamental algebraic structures that one can first derive from a manifold. In this section, we introduce the definition of tangent spaces for abstract manifolds in two different but equivalent ways, which have varied geometric and algebraic features. Our definitions require no extra geometric structure(s) on M^n.

Let M^n be a smooth manifold. (If you prefer a less-differentiability assumption, you may replace 'smooth' by 'C^1'.) Let $x \in M^n$.

6.2.1. Differentiability of maps between manifolds. Tangent spaces can be defined in terms of paths and functions. Since paths and functions are special cases of maps, we first discuss the differentiability of maps.

Definition 6.21. Let N^m and M^n be C^k manifolds. A continuous map $\phi : N^m \to M^n$ is **C^k-differentiable** (or C^k) if, for all charts $(\mathcal{U}, \varphi_{\mathcal{U}})$ on N^m and $(\mathcal{V}, \varphi_{\mathcal{V}})$ on M^n, the map $\varphi_{\mathcal{V}} \circ \phi \circ \varphi_{\mathcal{U}}^{-1}$ is C^k. Observe that the domain of this map is $\varphi_{\mathcal{U}}(\phi^{-1}(\mathcal{V}) \cap \mathcal{U})$. Since ϕ is continuous, we have that $\phi^{-1}(\mathcal{V})$ is an open set, and hence the domain of $\varphi_{\mathcal{V}} \circ \phi \circ \varphi_{\mathcal{U}}^{-1}$ is an open subset of $\mathbb{R}^m$. We can choose the codomain of $\varphi_{\mathcal{V}} \circ \phi \circ \varphi_{\mathcal{U}}^{-1}$ to be $\mathbb{R}^n$.

We only need to check the condition in Definition 6.21 for $(\mathcal{V}, \varphi_{\mathcal{V}})$ contained in a subatlas of M^n. In particular, by taking $M^n = \mathbb{R}$, we see that a continuous *function* $f : N^m \to \mathbb{R}$ is C^k-differentiable if for all charts $(\mathcal{U}, \varphi_{\mathcal{U}})$ on N^m we have that the function $f \circ \varphi_{\mathcal{U}}^{-1}$ is C^k.

More generally, we just need the differentiability of each of N^m and M^n to be at least C^k to be able to define a C^k map. If $N^m = I$ is an open interval in $\mathbb{R}$, then we call a C^k map $\gamma : I \to M^n$ a C^k *path*.

Example 6.22 (Coordinate maps have the differentiability of the manifold). *Let M^n be a C^k manifold and let $(\mathcal{V}, \varphi_{\mathcal{V}})$ be a chart on M^n. Then $\varphi_{\mathcal{V}} : \mathcal{V} \to \mathbb{R}^n$ is a C^k map. Moreover, its inverse map $\varphi_{\mathcal{V}}^{-1} : \varphi_{\mathcal{V}}(\mathcal{V}) \to M^n$ is also a C^k map.*

6.2.2. The tangent space as a set of equivalence classes of paths. We start with perhaps the most intuitive definition of the tangent space. This is based on the idea that it is useful to consider paths and their actions on functions via differentiation. Let $x \in M^n$ be a point and let $\gamma : (-\epsilon, \epsilon) \to M^n$ be a C^∞ path satisfying $\gamma(0) = x$. Given a function $f : M^n \to \mathbb{R}$ (say at least C^1), we can differentiate f along the 'direction of γ at x' by considering

$$(f \circ \gamma)'(0) = \left.\frac{d}{dt}\right|_{t=0} f(\gamma(t)). \tag{6.19}$$

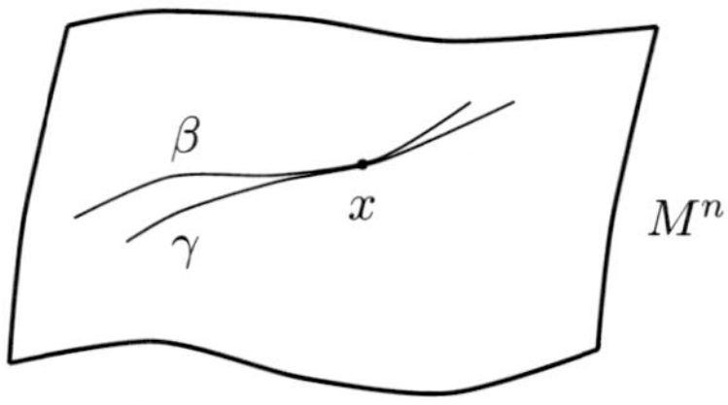

Figure 6.2.1. Two equivalent paths β and γ. We have $[\beta] = [\gamma] \in T_xM$.

We say that two paths $\gamma : (-\epsilon, \epsilon) \to M^n$ and $\beta : (-\delta, \delta) \to M^n$ with $\gamma(0) = \beta(0) = x$ are **equivalent at** x if for all smooth functions f,

$$(f \circ \gamma)'(0) = (f \circ \beta)'(0). \tag{6.20}$$

That is, two paths are equivalent at x if they act on functions in the same way. In this case we write $\gamma \sim \beta$. It is easy to see that $\sim$ is an equivalence relation; we leave it to the reader to check this. Let $[\gamma]$ denote the equivalence class of γ under the equivalence relation $\sim$. See Figure 6.2.1.

Definition 6.23. The **tangent space of** M^n **at** x is defined by

$$T_xM := \{[\gamma] \colon \gamma \text{ is a path with } \gamma(0) = x\}. \tag{6.21}$$

A **tangent vector** at x is an element of the tangent space T_xM.

Because of the definition of $\sim$ and (6.22), we think of $[\gamma]$ as the 'velocity vector' of $\gamma(t)$ at $t = 0$. This is made rigorous with (6.29) below.

Example 6.24 (Tangent vectors of Euclidean space). *Consider the special case where $M^n = \mathbb{R}^n$. Let $\mathbf{x} \in \mathbb{R}^n$ and let $\gamma : (-\epsilon, \epsilon) \to \mathbb{R}^n$ be a C^∞ path with $\gamma(0) = \mathbf{x}$. Then we have $\gamma'(0) = \frac{d\gamma}{dt}(0) \in \mathbb{R}^n$ and*

$$(f \circ \gamma)'(0) = df(\gamma'(0)). \tag{6.22}$$

It is easy to see (we leave it to the reader to check this) that paths $\gamma, \beta : (-\epsilon, \epsilon) \to \mathbb{R}^n$ satisfy $\gamma \sim \beta$ if and only if $\gamma'(0) = \beta'(0)$. Thus, we can equate the formally defined tangent vector $[\gamma]$ with the vector $\gamma'(0)$ at $\mathbf{x} \in \mathbb{R}^n$. The point is that $[\gamma]$ is defined for a differentiable manifold M^n in general, whereas $\gamma'(0)$ is not. In any case, for any $\mathbf{x} \in \mathbb{R}^n$, the tangent space $T_{\mathbf{x}}\mathbb{R}^n$ can be naturally identified with $\mathbb{R}^n$ itself.

In fact, each equivalence class of paths in $\mathbb{R}^n$ can be represented by a linear path. Namely, let $V = (V^1, \ldots, V^n) \in \mathbb{R}^n$, considered as a vector at $\mathbf{x}$. Define the associated path $\gamma_V(t) := \mathbf{x} + tV$. We calculate that

$$\left.\frac{d}{dt} f(\gamma_V(t))\right|_{t=0} = \sum_{i=1}^{n} V^i \frac{\partial f}{\partial x^i}.$$

For example, taking $f(\mathbf{x}) = x^j$, we obtain

$$\frac{d}{dt} f(\gamma_V(t))\bigg|_{t=0} = V^j.$$

So, a vector V is determined just by how γ_V acts on the coordinate functions x^j.

With this first definition of the tangent space, it is not obvious that T_xM is a vector space and what dimension it is. As we will see later, T_xM is a vector space of dimension n.

Let $(\mathcal{U}, \varphi_{\mathcal{U}})$ be local coordinates. Denote

$$\mathbf{x} = (x^1, \ldots, x^n) := \varphi_{\mathcal{U}}. \tag{6.23}$$

Let $x_0 \in \mathcal{U}$ and denote $\mathbf{x}_0 := \mathbf{x}(x_0)$. Given a Euclidean vector $\mathbf{V} = (V^1, \ldots, V^n) \in \mathbb{R}^n$, we have the path $\gamma_{\mathbf{V}} : (-\epsilon, \epsilon) \to M^n$ defined by

$$\gamma_{\mathbf{V}}(t) =: \mathbf{x}^{-1}(\mathbf{x}_0 + t\mathbf{V}). \tag{6.24}$$

Here, $\epsilon > 0$ is chosen small enough so that $\mathbf{x}_0 + tV \in \mathbf{x}(\mathcal{U})$ for $t \in (-\epsilon, \epsilon)$. Note that $\gamma_{\mathbf{V}}(0) = x_0$. We have that $[\gamma_{\mathbf{V}}] \in T_{x_0}M$.

Associated to the local coordinates $(\mathcal{U}, \mathbf{x} = \varphi_{\mathcal{U}})$, the **coordinate tangent vectors** are defined by

$$\frac{\partial}{\partial x^i} := \partial_i := [\gamma_{\mathbf{e}_i}] \in T_{x_0}M, \tag{6.25}$$

where $\{\mathbf{e}_1, \ldots, \mathbf{e}_n\}$ is the standard Euclidean basis for $\mathbb{R}^n$. We also observe that by definition, the set

$$[\gamma_{\mathbb{R}^n}] := \{[\gamma_{\mathbf{V}}] \colon \mathbf{V} \in \mathbb{R}^n\} \subset T_{x_0}M. \tag{6.26}$$

Let $\mathbf{V} \in \mathbb{R}^n$. For each i, we have

$$(x^i \circ \gamma_{\mathbf{V}})'(0) = \frac{d}{dt}\bigg|_{t=0} x^i(\mathbf{x}^{-1}(\mathbf{x}_0 + t\mathbf{V})) = \frac{d}{dt}\bigg|_{t=0} (x_0^i + tV^i) = V^i. \tag{6.27}$$

Thus, given $\mathbf{V}, \mathbf{V}' \in \mathbb{R}^n$, we have that $(x^i \circ \gamma_{\mathbf{V}})'(0) = (x^i \circ \gamma_{\mathbf{V}'})'(0)$ if and only if $\mathbf{V} = \mathbf{V}'$. We conclude that $[\gamma_{\mathbf{V}}] = [\gamma_{\mathbf{V}'}]$ if and only if $\mathbf{V} = \mathbf{V}'$. Thus, given local coordinates $(\mathcal{U}, \{x^i\})$, the map from $\mathbb{R}^n$ to T_xM defined by $\mathbf{V} \mapsto [\gamma_{\mathbf{V}}]$ is an injective map.

Example 6.25. *For a submanifold M^n of Euclidean space $\mathbb{R}^N$, as a manifold its tangent spaces can be naturally identified with the tangent spaces we defined for submanifolds.*

6.2.3. Derivations. A tangent vector $[\gamma] \in T_xM$ acts on functions via defining

$$[\gamma] : C^\infty(M^n) \to \mathbb{R} \tag{6.28}$$

by

$$[\gamma](f) := (f \circ \gamma)'(0). \tag{6.29}$$

We have that for any functions $f, g : M^n \to \mathbb{R}$ and real number $c \in \mathbb{R}$, we have the linearity property

$$[\gamma](cf + g) = c[\gamma](f) + [\gamma](g) \tag{6.30}$$

and the Leibniz (product) rule

$$[\gamma](fg) = g(x)[\gamma](f) + f(x)[\gamma](g). \tag{6.31}$$

These properties simply follow from the corresponding properties for functions of one variable.

By making these properties axioms, we have the following.

Definition 6.26. We say that a map $D : C^\infty(M^n) \to \mathbb{R}$ is a **derivation at** $x \in M^n$ if it satisfies the following properties:

(1) D is a linear function. That is, for any functions $f, g : M^n \to \mathbb{R}$ and real number $c \in \mathbb{R}$, we have

$$D(cf + g) = cD(f) + D(g). \tag{6.32}$$

(2) D satisfies the **Leibniz rule**:

$$D(fg) = g(x)D(f) + f(x)D(g). \tag{6.33}$$

Let

$$\mathcal{D}_x := \{D \colon D \text{ is a derivation at } x\} \tag{6.34}$$

be the **set of derivations at** x.

The set of derivations $\mathcal{D}_x$ has a natural vector space structure:

Lemma 6.27. *Given two derivations $D_1, D_2 : C^\infty(M^n) \to \mathbb{R}$ at $x \in M^n$ and a real number c, we have that the function $cD_1 + D_2 : C^\infty(M^n) \to \mathbb{R}$ defined by*

$$(cD_1 + D_2)(f) := cD_1(f) + D_2(f) \tag{6.35}$$

is a derivation at x. The scalar and vector multiplications defined this way yield a vector space structure on the set $\mathcal{D}_x$ of derivations at x.

We leave it to the reader to give a proof of this lemma (see Exercise 6.8).

For each $[\gamma] \in T_xM$, the associated function $[\gamma] : C^\infty(M^n) \to \mathbb{R}$ as defined by (6.28) is a derivation. In this way, we have a map

$$\iota : T_xM \to \mathcal{D}_x, \tag{6.36}$$

where ι takes a tangent vector to its corresponding function from $C^\infty(M^n)$ to $\mathbb{R}$ defined by (6.29). By the argument culminating at the end of §6.2.2, we have that ι is an injection.

Lemma 6.28. *Let $(\mathcal{U}, \mathbf{x})$ be local coordinates. Two derivations D and D' at $x_0 \in \mathcal{U}$ are equal if and only if $D(x^i) = D'(x^i)$ for all $1 \leq i \leq n$.*

Proof. Clearly if $D = D'$ as derivations, then $D(x^i) = D'(x^i)$ for all $1 \leq i \leq n$. So we prove the converse.

Let $f \in C^\infty(M)$. By Taylor's theorem from vector calculus, in $\mathcal{U}$ we may write f as

$$f(x) = f(x_0) + \sum_{i=1}^{n} a_i(x^i - x_0^i) + g(x), \tag{6.37}$$

where $a_i \in \mathbb{R}$ and $g : \mathcal{U} \to \mathbb{R}$ has the property that

$$g(x) = \sum_{i=1}^{n} b_i(x)(x^i - x_0^i) \tag{6.38}$$

and where each $b_i : \mathcal{U} \to \mathbb{R}$ satisfies $b_i(x_0) = 0$. We compute that

$$D(g) = \sum_{i=1}^{n} \left(D(b_i)(x_0^i - x_0^i) + b_i(x_0)D(x^i)\right) = 0.$$

Using this, we compute that

$$D(f) = \sum_{i=1}^{n} a_i D(x^i - x_0^i) + D(g) = \sum_{i=1}^{n} a_i D(x^i) \tag{6.39}$$

since D of a constant is equal to zero. Hence, if $D(x^i) = D'(x^i)$ for all $1 \leq i \leq n$, then $D = D'$ as derivations. □

The following says that the tangent space T_xM at a point x has a vector space structure defined by its identification with the set of derivations at x. We will always consider the tangent space with this vector space structure.

Proposition 6.29 (The tangent space is the set of derivations)**.** *The map ι given by (6.36) is a bijection from the tangent space T_xM to the vector space $\mathcal{D}_x$ of derivations at x. Thus, the vector space structure on $\mathcal{D}_x$ induces via the map ι a vector space structure on T_xM.*

Proof. It suffices to show that the map ι is a surjection. To this end, let D be a derivation at x. Let $d^i := D(x^i)$. We claim that

$$D = \left[\gamma_{\sum_{i=1}^n d^i \mathbf{e}_i}\right] : C^\infty(M^n) \to \mathbb{R}, \tag{6.40}$$

from which the proposition follows. To see this, using (6.27) we compute that

$$\left[\gamma_{\sum_{j=1}^n d^j \mathbf{e}_j}\right] = (x^i \circ \gamma_{\sum_{j=1}^n d^j \mathbf{e}_j})'(0) = d^i = D(x^i).$$

The proposition now follows from Lemma 6.28. □

Let $\mathbf{V} = (V^1, \ldots, V^n) \in \mathbb{R}^n$. Let $f \in C^\infty(M^n)$ and assume that f locally has the form (6.37). Then, by (6.39) we have that

$$[\gamma_\mathbf{V}](f) = \sum_{i=1}^n a_i [\gamma_\mathbf{V}](x^i) = \sum_{i=1}^n a_i V^i,$$

where the last equality is by (6.27). In particular, taking $\mathbf{V} = \mathbf{e}_j$, we obtain

$$[\gamma_{\mathbf{e}_j}](f) = a_j.$$

Thus, we have that

$$[\gamma_\mathbf{V}](f) = \sum_{i=1}^n V^i [\gamma_{\mathbf{e}_i}](f).$$

Since this formula holds for all $f \in C^\infty(M^n)$, we conclude that

$$[\gamma_\mathbf{V}] = \sum_{i=1}^n V^i [\gamma_{\mathbf{e}_i}] = \sum_{i=1}^n V^i \partial_i. \tag{6.41}$$

This shows that $\{\partial_1, \ldots, \partial_n\}$ is a basis for $T_x M$.

6.3. Maps between manifolds and their derivatives

For the definitions of differentiable C^k, C^∞ (smooth), and real analytic vector-valued functions on Euclidean space, see §2.1.2–2.1.4. In §6.2.1 we discussed the differentiability of maps between manifolds. In this section, we will discuss diffeomorphisms, examples, and derivatives of maps.

6.3.1. Smooth embeddings. A smooth map $F : M^n \to N^m$ is a **smooth embedding** if F is both a smooth immersion and a topological embedding; i.e., F is a homeomorphism onto its image $F(M^n)$, where $F(M^n)$ has the subspace topology induced from the topology of N^m. A smooth embedding that is a proper map (see Definition 2.29) is called a **proper embedding**.

Definition 6.30. A subset Σ of M^n is an **embedded submanifold** if there exists a smooth manifold structure $\mathbf{A}$ on Σ whose associated topology on Σ is the same as the subspace topology induced from M^n and such that the inclusion map $i : \Sigma \to M^n$ is a smooth embedding. We say that Σ is a **properly embedded submanifold** if i is a proper smooth embedding.

6.3.2. Diffeomorphisms. We say that a map $\phi : N^n \to M^n$ is a C^k **diffeomorphism** if ϕ is a bijection and both ϕ and ϕ^{-1} are C^k maps. Given two C^k manifolds N^n and M^n, if there exists a C^k diffeomorphism between them, then we say that N^n and M^n are C^k **diffeomorphic**. Usually, we will be in the smooth category, in which case we just say that N^n and M^n are diffeomorphic.

Observe that if $(N^n, \mathbf{A}_N)$ and $(M^n, \mathbf{A}_M)$ are C^k manifolds and if $\phi : N^n \to M^n$ is a C^k diffeomorphism, then $(\mathcal{U}, \varphi_{\mathcal{U}}) \in \mathbf{A}_M$ if and only if $(\phi^{-1}(\mathcal{U}), \varphi_{\mathcal{U}} \circ \phi) \in \mathbf{A}_N$. We can think of this as saying that the diffeomorphism ϕ preserves the atlases of N^n and M^n.

Example 6.31. *Any set M^n with an atlas consisting of a single chart (M^n, φ) is a real analytic manifold. This is true for a rather tautological reason: The only transition function $\varphi \circ \varphi^{-1} : \varphi(M^n) \to \varphi(M^n)$ is the identity map.*

We have that $\mathbb{R}$ is a 1-dimensional smooth manifold with the maximal atlas $\mathbf{A}$ containing the single chart $(\mathbb{R}, \mathrm{id}_{\mathbb{R}})$. We also can consider $\mathbb{R}$ as a 1-dimensional smooth manifold with the maximal atlas $\mathbf{A}'$ containing the single chart $(\mathbb{R}, f)$, where $f : \mathbb{R} \to \mathbb{R}$ is the cubing function: $f(x) = x^3$. Since f^{-1} is not smooth, we have $(\mathbb{R}, f) \notin \mathbf{A}$; thus $\mathbf{A} \neq \mathbf{A}'$. That is, the differentiable structures $\mathbf{A}$ and $\mathbf{A}'$ on $\mathbb{R}$ are not equal. On the other hand, the map f^{-1} from $(\mathbb{R}, \mathbf{A})$ to $(\mathbb{R}, \mathbf{A}')$ is a diffeomorphism. Indeed, we have that $f \circ f^{-1} \circ \mathrm{id}_{\mathbb{R}}^{-1} = \mathrm{id}_{\mathbb{R}}$ is a smooth map from $\mathbb{R}$ to $\mathbb{R}$ and so is its inverse.

Example 6.32 ($\mathbb{C}P^1$ is diffeomorphic to S^2). *Firstly, recall from Example* 6.15 *that* 1*-dimensional complex projective space is given by*

$$\mathbb{C}P^1 = \{ [z_1 : z_2] \colon (z_1, z_2) \in \mathbb{C}^2_* \}, \tag{6.42}$$

where the equivalence class $[z_1 : z_2]$ *in* (6.42) *is defined by*

$$(z_1, z_2) \sim (\lambda z_1, \lambda z_2) \quad \textit{for all } \lambda \in \mathbb{C}_*. \tag{6.43}$$

By (6.16) *and* (6.17), *we have*

$$\mathcal{U}_1 = \{ [z^1 : z^2] \in \mathbb{C}P^1 \mid z^1 \neq 0 \}, \quad \mathcal{U}_2 = \{ [z^1 : z^2] \in \mathbb{C}P^1 \mid z^2 \neq 0 \} \tag{6.44}$$

and the coordinate maps $\varphi_1 : \mathcal{U}_1 \to \mathbb{C} \cong \mathbb{R}^2$, $\varphi_2 : \mathcal{U}_2 \to \mathbb{C} \cong \mathbb{R}^2$ *are defined by*

$$\varphi_1([z^1 : z^2]) = z^2 (z^1)^{-1}, \quad \varphi_2([z^1 : z^2]) = z^1 (z^2)^{-1}, \tag{6.45}$$

respectively. We have an atlas for $\mathbb{C}P^1$ *consisting of two charts* $\{(\mathcal{U}_1, \varphi_1), (\mathcal{U}_2, \varphi_2)\}$*. The two transition functions are*

(6.46)
$$\varphi_{12} = \varphi_1 \circ \varphi_2^{-1} : \mathbb{C}_* \to \mathbb{C}_* \quad \textit{and its inverse} \quad \varphi_{21} = \varphi_2 \circ \varphi_1^{-1} : \mathbb{C}_* \to \mathbb{C}_*.$$

Note that $\mathbb{C}_* \cong \mathbb{R}^2_*$*. We compute that*

$$\varphi_{12}(z) = \varphi_1([z:1]) = z^{-1}, \quad \varphi_{21}(w) = \varphi_2([1:w]) = w^{-1}. \tag{6.47}$$

Secondly, recall from Example 6.9 *that the* 2*-dimensional sphere is defined as the set* $S^2 := \mathbb{R}^2_1 \sqcup \mathbb{R}^2_2/\sim$ *with the two-chart atlas* $\{(\mathcal{V}_1, \psi_1^{-1}), (\mathcal{V}_2, \psi_2^{-1})\}$*, where*

$$\psi_1 = \pi \circ i_1 : \mathbb{R}^2_1 \to S^2 \setminus \{[\mathbf{0}_2]\} = \mathcal{V}_1,$$
$$\psi_2 = \pi \circ i_2 : \mathbb{R}^2_2 \to S^2 \setminus \{[\mathbf{0}_1]\} = \mathcal{V}_2$$

and where $\pi : \mathbb{R}^2_1 \sqcup \mathbb{R}^2_2 \to S^2$ *is the canonical projection map and* $i_a : \mathbb{R}^2_a \to \mathbb{R}^2_1 \sqcup \mathbb{R}^2_2$*,* $a = 1, 2$*, is the inclusion map. In Example* 6.9*,* $\mathbb{R}^2_1 \setminus \{\mathbf{0}_1\}$ *and* $\mathbb{R}^2_2 \setminus \{\mathbf{0}_2\}$ *are identified by the inversion defined by*

$$\iota(z) = \frac{z}{|z|^2} = \overline{z}^{-1}. \tag{6.48}$$

Thus, we have that $\psi_{12} := \psi_1^{-1} \circ \psi_2 : \mathbb{C}_* \to \mathbb{C}_*$ *and* $\psi_{21} := \psi_2^{-1} \circ \psi_1 : \mathbb{C}_* \to \mathbb{C}_*$ *are given by*

$$\psi_{12}(z) = \overline{z}^{-1} \quad \textit{and} \quad \psi_{21}(w) = \overline{w}^{-1}, \tag{6.49}$$

respectively, where we identified each of $\mathbb{R}^2_1$ *and* $\mathbb{R}^2_2$ *with* $\mathbb{C}$*. In view of this, we replace the chart* $(\mathcal{V}_2, \psi_2)$ *by the chart* $(\mathcal{V}_2, \overline{\psi}_2)$*, where* $\overline{\psi}_2(z) := \psi_2(\overline{z})$*. Then, for the atlas* $\{(\mathcal{V}_1, \psi_1), (\mathcal{V}_2, \overline{\psi}_2)\}$ *for* S^2 *we have that the transition functions are given by*

$$\psi_1^{-1} \circ \overline{\psi}_2(z) = z^{-1}, \quad \overline{\psi}_2^{-1} \circ \psi_1(w) = w^{-1}. \tag{6.50}$$

In effect, we have shown that both $\mathbb{C}P^1$ *and* S^2 *have atlases consisting of two charts each and that their transition functions are the same. Thus, by identifying the copies of* $\mathbb{R}^2$*, we can use the charts to define a diffeomorphism from* $\mathbb{C}P^1$ *to* S^2*. In detail, we define* $F : \mathbb{C}P^1 \to S^2$ *as follows. We define*

$$F|_{\mathcal{U}_1} = \psi_1 \circ \varphi_1 \quad \textit{and} \quad F|_{\mathcal{U}_2} = \overline{\psi}_2 \circ \varphi_2. \tag{6.51}$$

This map is well-defined since $F|_{\mathcal{U}_1}([z^1 : z^2]) = F|_{\mathcal{U}_2}([z^1 : z^2])$ *for* $[z^1 : z^2] \in \mathcal{U}_1 \cap \mathcal{U}_2$*. It is also easy to see that the map* F *is a diffeomorphism.*

6.3.3. Examples of maps.

Definition 6.33. Let $\pi : \widetilde{M}^n \to M^n$ be a smooth map between n-dimensional manifolds. We say that π is a **smooth covering map** if for every $x \in M^n$ there exists an open neighborhood U of x such that $\pi^{-1}(U)$ is equal to some disjoint union $\bigcup_{\alpha \in A} V_\alpha$, where $\pi|_{V_\alpha} : V_\alpha \to U$ is a diffeomorphism for all $\alpha \in A$.

(1) Define the map $\pi : \mathbb{R} \to S^1$ by $\pi(x) = (\cos(x), \sin(x))$. If we use complex numbers, then we can write this map as $\pi(x) = \mathrm{e}^{ix} = \cos(x) + i\sin(x)$. This is a smooth map and is an example of a smooth covering map. For each $\mathbf{x} \in S^1$, its preimage (a.k.a. inverse image) $\pi^{-1}(\mathbf{x})$ is a countably infinite set. Indeed, let $x \in \mathbb{R}$ be such that $\mathrm{e}^{ix} = \mathbf{x}$. Then

$$\pi^{-1}(\mathbf{x}) = \{x + 2\pi k \colon k \in \mathbb{Z}\}.$$

Complex multiplication of elements of S^1 is defined by

$$\mathrm{e}^{ix_1} \cdot \mathrm{e}^{ix_2} = \mathrm{e}^{i(x_1+x_2)}.$$

Moreover, $\cdot : S^1 \times S^1 \to S^1$ is a smooth map. The identity element of S^1 is $1 = \mathrm{e}^{i0}$. The inverse of an element e^{ix} is e^{-ix}. One can easily show that $(S^1, \cdot)$ is a group. Since S^1 is also a manifold and since the multiplication and inverse operations are smooth maps, we say that $(S^1, \cdot)$ is a *Lie group*, which we also denote by $\mathrm{U}(1)$. We will discuss the concept of Lie group in Chapter 10.

(2) Define the map $\pi : S^n \to \mathbb{R}P^n$ by $\pi(\mathbf{x}) = [\mathbf{x}]$. This is also a smooth covering map and is a two-to-one covering. Let $[\mathbf{x}] \in \mathbb{R}P^n$. Without loss of generality, we may assume that $\mathbf{x} \in S^n$. We have

$$\pi^{-1}([\mathbf{x}]) = \{\mathbf{x}, -\mathbf{x}\}.$$

Example 6.34 (Hopf fibration). *To describe this map, it is convenient to describe the 3-sphere S^3 using complex coordinates:*

$$S^3 = \{z := (z_1, z_2) \in \mathbb{C}^2 \colon |z|^2 := z_1\bar{z}_1 + z_2\bar{z}_2 = 1\}. \tag{6.52}$$

We define

$$\pi : S^3 \to \mathbb{C}P^1 \cong S^2 \tag{6.53}$$

by

$$\pi(z^1, z^2) = [z^1 : z^2], \tag{6.54}$$

*where the right-hand side is defined by (6.18). This is the **Hopf fibration**.*[2] *Recall from (6.13) that $[w^1 : w^2] = [z^1 : z^2]$ if and only if $(w^1, w^2) = \lambda(z^1, z^2)$*

[2]The Hopf fibration is an example of a 'fiber bundle', which in turn is a special case of a 'fibration'; see §10.5 below.

for some $\lambda \in \mathbb{C}_*$. *Since* $(w^1, w^2), (z^1, z^2) \in S^3$, *we have* $\lambda \in S^1 \cong \mathrm{U}(1)$. *For any* $[z^1 : z^2] \in \mathbb{C}P^1$, *we have*

$$\pi^{-1}([z^1 : z^2]) = \{(\lambda z^1, \lambda z^2) \colon \lambda \in S^1\}, \tag{6.55}$$

which is diffeomorphic to a circle. In fact, $\mathbb{C}P^1$ *is the quotient space of* S^3 *under the group action of* S^1 *on* S^3 *defined by the map* $\sigma : S^1 \times S^3 \to S^3$, *where*

$$\sigma(\lambda, (z^1, z^2)) := (\lambda z^1, \lambda z^2). \tag{6.56}$$

6.3.4. The derivative of a map. Let $\phi : N^m \to M^n$ be a map. The **derivative** of ϕ at a point $x \in N^m$ is defined by

$$d\phi_x : T_x N \to T_{\phi(x)} M, \tag{6.57}$$

where

$$d\phi_x([\gamma]) := [\phi \circ \gamma] \tag{6.58}$$

for any path $\gamma : (-\epsilon, \epsilon) \to N^m$ with $\gamma(0) = x$. That is, $d\phi_x$ maps the equivalence class of γ to the equivalence class of $\phi \circ \gamma$.

Equivalently, let D be a derivation at $x \in N^m$. We define $d\phi_x(D) : C^\infty(M^n) \to \mathbb{R}$ by

$$d\phi_x(D)(f) := D(f \circ \phi). \tag{6.59}$$

One can show that $d\phi_x(D)$ is a derivation at $\phi(x) \in M^n$; see Exercise 6.9.

We leave it to the reader to check that these two definitions of the derivative $d\phi_x$ are equivalent; see Exercise 6.10.

Let M^n, L^k, and N^m be smooth manifolds and let $f : M^n \to L^k$ and $g : N^m \to M^n$ be smooth maps. We have their composition:

$$f \circ g : N^m \to L^k.$$

We can visualize this via the following commutative diagram:

$$\begin{array}{ccc} & M^n & \\ {\scriptstyle g}\nearrow & & \searrow{\scriptstyle f} \\ N^m & \xrightarrow{\;f\circ g\;} & L^k \end{array}$$

The chain rule says that (see Exercise 6.11)

$$d(f \circ g) = df \circ dg = g^*(df) : T_x N \to T_{(f \circ g)(x)} L. \tag{6.60}$$

As a commutative diagram,

$$\begin{array}{ccc} & T_{g(x)} M & \\ {\scriptstyle dg}\nearrow & & \searrow{\scriptstyle df} \\ T_x N & \xrightarrow{\;d(f\circ g)\;} & T_{f(g(x))} L \end{array}$$

To see the chain rule, let $\gamma : I \to N^m$ be a smooth path with $\gamma(0) = x$. We compute that

$$d(f \circ g)[\gamma] = [(f \circ g) \circ \gamma] = [f \circ (g \circ \gamma)] = df[g \circ \gamma] = df(dg([\gamma])).$$

6.4. Vector fields

We have encountered vector fields for submanifolds in §3.2. In this section, we discuss vector fields for smooth manifolds.

6.4.1. Definition of a vector field.

Let M^n be a smooth manifold. A **vector field** on a subset $\mathcal{S}$ of a manifold M^n is simply the assignment of a tangent vector to each point in $\mathcal{S}$. That is, a vector field X sends

$$p \mapsto X_p := X(p) \in T_pM$$

for $p \in \mathcal{S}$. See Figure 6.4.1.

Example 6.35. *Let $\mathcal{U}$ be an open subset of $\mathbb{R}^n$ and let $f : \mathcal{U} \to \mathbb{R}$ be a differentiable function. Then its **gradient** is the vector field on $\mathcal{U}$ defined by*

$$\nabla f_{\mathbf{x}} := \left(\frac{\partial f}{\partial x^1}(\mathbf{x}), \ldots, \frac{\partial f}{\partial x^n}(\mathbf{x}) \right) \in T_{\mathbf{x}}\mathbb{R}^n \cong \mathbb{R}^n \tag{6.61}$$

for $\mathbf{x} \in \mathcal{U}$. For example, if $f(\mathbf{x}) = |\mathbf{x}|^2$, then $\nabla f_{\mathbf{x}} = 2\mathbf{x}$.

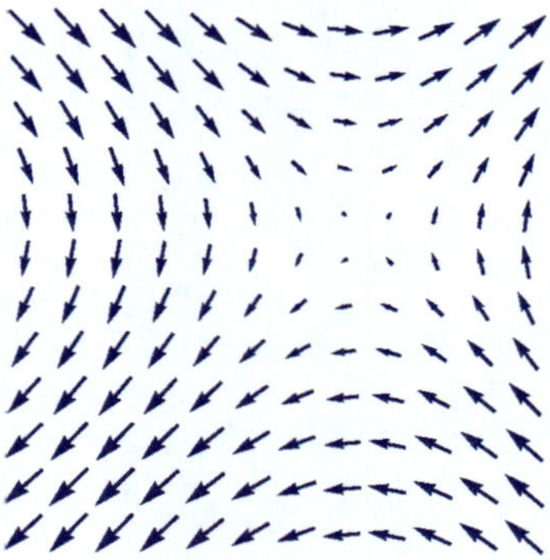

Figure 6.4.1. The vector field $(x, y) \mapsto \sin(y)\mathbf{e}_1 + \sin(x)\mathbf{e}_2$ on $\mathbb{R}^2$. In this figure, all of the tangent spaces $T_{\mathbf{x}}\mathbb{R}^n$ are superimposed on $\mathbb{R}^n$ itself. Credit: By Jim.belk – Own work, Public Domain, `https://commons.wikimedia.org/w/index.php?curid=8008790`

6.4.2. Tangent bundle. It is convenient to consider the **tangent bundle** TM. Recall that in §3.2, we have considered TM for an n-dimensional submanifold M^n of $\mathbb{R}^N$, where TM is defined as a $2n$-dimensional submanifold of $\mathbb{R}^{2N}$. For abstract manifolds, as a set the definition is the same (cf. (3.52)):

$$TM := \bigcup_{x \in M^n} \{x\} \times T_x M. \tag{6.62}$$

As we will see in more detail in Chapter 10, the tangent bundle TM can naturally be given the structure of a smooth $2n$-dimensional manifold. As such, a C^k vector field X on an open subset $\mathcal{U}$ of M^n can be defined as a C^k map

$$X : \mathcal{U} \to TM.$$

We use the notation $\mathcal{X}^k(\mathcal{U})$ to denote the set of all C^k vector fields on $\mathcal{U}$.

6.4.3. Smooth vector fields. For now, we can define what it means for a vector field X on M^n to be C^k as follows. Suppose that $(\mathcal{U}, \varphi_{\mathcal{U}})$ are local coordinates. We may uniquely write on $\mathcal{U}$

$$X =: \sum_{i=1}^{n} X^i \frac{\partial}{\partial x^i}, \tag{6.63}$$

where each X^i is a real-valued function on $\mathcal{U}$. We say that X is C^k if the functions X^i are C^k for any local coordinates. This definition of C^k vector field is equivalent to the one given in the previous subsection.

Using the local expression for X, for any differentiable function f on $\mathcal{U}$ we can write the directional derivative as

$$X_p(f) = \sum_{i=1}^{n} X^i(p) \frac{\partial f}{\partial x^i}(p) \tag{6.64}$$

for $p \in \mathcal{U}$.

The concept of a tangent bundle extends naturally to the broader framework of vector bundles. Similarly, the notion of a vector field generalizes to that of a section of a vector bundle. For more details, refer to Chapter 10.

6.4.4. 1-parameter groups of diffeomorphisms. Given a C^1 vector field, we can construct a 1-parameter group at least *locally*. Conversely, given a global 1-parameter group, where the diffeomorphisms depend smoothly on t, we can construct a *globally defined* smooth vector field. We begin by considering some basic definitions in the global setting.

Let $(\mathbb{R}, +)$ be the additive group of real numbers, that is, the set of real numbers $\mathbb{R}$ together with the binary operation of addition $+ : \mathbb{R} \times \mathbb{R} \to \mathbb{R}$. A **1-parameter group action** on a manifold M^n is defined to be a smooth map

$$\sigma : \mathbb{R} \times M^n \to M^n$$

satisfying

$$\sigma\big(s, \sigma(t,p)\big) = \sigma(s+t, p) \tag{6.65}$$

and

$$\sigma(0,p) = p. \tag{6.66}$$

Observe that the second condition is consistent with the first condition because $\sigma(0, \sigma(0,p)) = \sigma(0,p)$.

Example 6.36. *Let $\mathbf{v} \in \mathbb{R}^n$. The map $\sigma : \mathbb{R} \times \mathbb{R}^n \to \mathbb{R}^n$, given by*

$$\sigma(t, \mathbf{x}) := \mathbf{x} + t\mathbf{v},$$

defines a 1-parameter group action on $\mathbb{R}^n$.

Each map $\phi_t : M^n \to M^n$, $t \in \mathbb{R}$, defined by

$$\phi_t : p \mapsto \sigma(t,p) \tag{6.67}$$

and sending M^n into itself, is thus also smooth. Note that the conditions (6.65) and (6.66) say that

$$\phi_s \circ \phi_t = \phi_{s+t} \quad \text{and} \quad \phi_0 = \mathrm{id}_{M^n}, \tag{6.68}$$

respectively. By taking $s = -t$ in (6.68), we see that

$$\phi_{-t} \circ \phi_t = \mathrm{id}_{M^n}.$$

Similarly, $\phi_t \circ \phi_{-t} = \mathrm{id}_{M^n}$. Therefore, each ϕ_t is a diffeomorphism of M^n and

$$\phi_t^{-1} = \phi_{-t}. \tag{6.69}$$

Thus, $\{\phi_t\}_{t\in\mathbb{R}}$ is said to be a **1-parameter group of diffeomorphisms** of M^n. Via the correspondence $\sigma \longleftrightarrow \{\phi_t\}_{t\in\mathbb{R}}$, 1-parameter group actions and 1-parameter groups of diffeomorphisms are the same.

Given a 1-parameter group action σ on M^n, we can define, for any point $p \in M^n$, a *path* by

$$\gamma_p : \mathbb{R} \to M^n$$

with

$$\gamma_p(t) := \phi_t(p). \tag{6.70}$$

The **vector field associated** to the 1-parameter group of diffeomorphisms $\{\phi_t\}_{t\in\mathbb{R}}$ is defined by

$$V_p := [\gamma_p] \in T_pM \tag{6.71}$$

for all $p \in M^n$, where $[\gamma_p]$ is defined by (6.29). In other words, as a derivation,

$$V_p(f) = (f \circ \gamma_p)'(0). \tag{6.72}$$

For example, the vector field associated to the 1-parameter group action on $\mathbb{R}^n$ in Example 6.36 is the constant vector field $\mathbf{v}$.

Conversely, suppose that V is a vector field on M^n. By the standard existence and uniqueness theorem for systems of first-order ODEs, for each $p \in M^n$ there exist $\epsilon_p > 0$ (in general, depending on p) and a path $\gamma_p : (-\epsilon_p, \epsilon_p) \to M^n$ satisfying

$$\gamma_p'(t) = V_{\gamma_p(t)} \tag{6.73}$$

for all $t \in (-\epsilon_p, \epsilon_p)$ and $\gamma_p(0) = p$. We have the associated *local* 1-parameter group action σ on M^n defined by

$$\sigma : (p, t) \mapsto \gamma_p(t) \tag{6.74}$$

for $p \in M^n$ and $t \in (-\epsilon_p, \epsilon_p)$. For each $p \in M^n$, define (this reverses the definition in (6.70))

$$\phi_t(p) := \gamma_p(t) \tag{6.75}$$

for $t \in (-\epsilon_p, \epsilon_p)$. Given $p \in M^n$, since $\epsilon_p > 0$ and by the continuous dependence of solutions to first-order ODEs (and its implication for the time intervals of existence), there exist $\epsilon > 0$ and a neighborhood $\mathcal{U}$ of p such that $\gamma_x : (-\epsilon, \epsilon) \to M^n$ is defined and satisfies $\gamma_x'(t) = V_{\gamma_x(t)}$ for all $x \in \mathcal{U}$ and $t \in (-\epsilon, \epsilon)$ and $\gamma_x(0) = x$. Thus, we have that $\phi_t : \mathcal{U} \to M^n$ is defined for at least all $t \in (-\epsilon, \epsilon)$. So we call $\{\phi_t\}$ the **local 1-parameter group of local diffeomorphisms** generated by V.

6.4.5. Lie bracket.

6.4.5.1. *Geometric interpretation of the Lie bracket.* The Lie bracket has a geometric interpretation as follows. For simplicity, assume that M^n is a compact manifold without boundary. Let U and V be vector fields on M^n. Let $x \in M^n$ and denote $\gamma_{U,x} : \mathbb{R} \to M^n$ to be the unique integral curve to U satisfying the initial condition $\gamma_{U,x}(0) = x$. The vector field U defines its **flow**, namely, maps $\Phi_{U,t} : M^n \to M^n$ defined by

$$\Phi_{U,t}(X) = \gamma_{U,x}(t). \tag{6.76}$$

That is, the maps $\Phi_{U,t}$ flow for time t points in M^n along integral curves to U. We then have the following fact, which demonstrates that the Lie bracket of two vector fields measures the noncommutativity of the flows induced by the vector fields (see Exercise 6.13):

$$[U,V](x) = \frac{d}{dt}\bigg|_{t=0} \big(\Phi_{-V,\sqrt{t}} \circ \Phi_{-U,\sqrt{t}} \circ \Phi_{V,\sqrt{t}} \circ \Phi_{U,\sqrt{t}}\big)(x). \tag{6.77}$$

Note that $\Phi_{-U,\sqrt{t}} = \Phi_{U,-\sqrt{t}}$ since $\gamma_{-U,x}(\sqrt{t}) = \gamma_{U,x}(-\sqrt{t})$.

6.5. Tensors

Tensors on smooth manifolds generalize vector fields and covector fields. At each point of our manifold, a tensor is a multilinear function on the product of tangent and cotangent spaces. We discussed tensors for submanifolds of Euclidean spaces in §3.4. Tensors for smooth manifolds are defined in exactly the same way. Here, we will continue the discussion by considering push-forwards and pull-backs of tensors, as well as the notion of Lie derivative.

6.5.1. Push-forward and pull-back. Let $\phi : N^m \to M^n$ be a map and let α be a covariant r-tensor on M^n. The **pull-back** by ϕ of α is the covariant r-tensor on N^m defined by

$$\phi^*\alpha(X_1,\ldots,X_k) := \alpha\big(d\phi(X_1),\ldots,d\phi(X_k)\big) \tag{6.78}$$

for $X_1,\ldots,X_k \in T_xN$, $x \in N^m$.

For example, if $f : M^n \to \mathbb{R}$ is a function, then

$$\phi^* df(X) = df(d\phi(X)) = d(f\circ\phi)(X). \tag{6.79}$$

So

$$\phi^*(df) = d(f\circ\phi). \tag{6.80}$$

Now suppose that $\phi : N^n \to M^n$ is a diffeomorphism. The **push-forward** to M^n by ϕ of a vector field V on N^m is defined by

$$(\phi_*(V))_y := d\phi(V_x) \tag{6.81}$$

for $y \in M^n$ and where $x \in N^m$ is the unique point such that $\phi(x) = y$; i.e., $x = \phi^{-1}(y)$. We have that $\phi_* V$ is a vector field on M^n. Observe that if α is a 1-form on M^n, then

$$(\phi^*\alpha)(X) = \alpha(\phi_*(X)). \tag{6.82}$$

If $\phi : N^n \to M^n$ is a diffeomorphism, then we can define the push-forward of a vector field V on N^m, which yields the vector field $\phi_*(V)$ defined by (6.81). On the other hand, if ϕ is not one-to-one, then there exist distinct points $p, q \in N^m$ such that $\phi(p) = \phi(q) =: x$. In this case, $d\phi(V_p)$ and $d\phi(V_a)$ may be nonequal vectors in T_xM. Thus, in this case we do not have a well-defined push-forward vector field. See Figure 6.5.1.

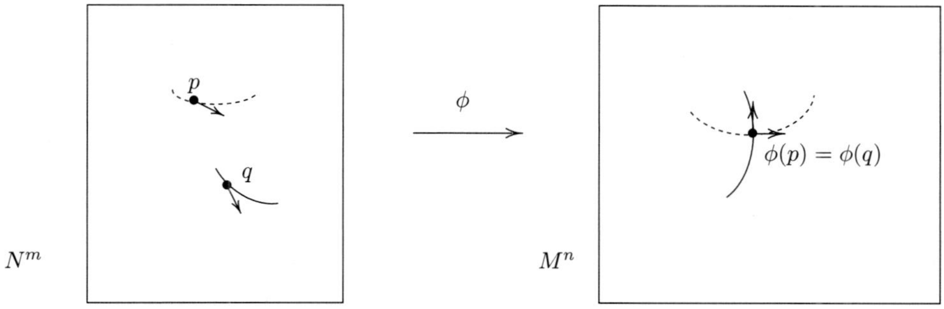

Figure 6.5.1. Visualizing the case when $\phi(p) = \phi(q)$, but $d\phi(V_p) \neq d\phi(V_q) \in T_x M$.

6.5.2. Lie derivative. Let V be a vector field on a differentiable manifold M^n and let $\{\phi_t\}$ be the local 1-parameter group of local diffeomorphisms generated by V. Let α be a covariant k-tensor. The **Lie derivative** of α with respect to V is defined by

$$\mathcal{L}_V \alpha := \partial_t(\phi_t^* \alpha) = \lim_{t\to 0} \frac{1}{t}\big(\phi_t^* \alpha - \alpha\big), \tag{6.83}$$

where $\partial_t := \partial_t|_{t=0}$ is evaluated at $t = 0$. Equivalently,

$$\mathcal{L}_V \alpha = \partial_t((\phi_{-t})_* \alpha) = \lim_{t\to 0} \frac{1}{t}\big(\,(\phi_{-t})_*\, \alpha - \alpha\big). \tag{6.84}$$

The Lie derivative of a tensor represents the infinitesimal action by pull-back of the diffeomorphism group on the tensor. In particular, it captures how the tensor changes infinitesimally under the flow generated by a vector field.

Lemma 6.37.

(1) *If f is a function, then $\mathcal{L}_V f = V(f)$.*

(2) *If W is a vector field, then $\mathcal{L}_V W = [V, W]$.*

(3) *If α and β are tensors, then*

$$\mathcal{L}_V (\alpha \otimes \beta) = (\mathcal{L}_V \alpha) \otimes \beta + \alpha \otimes (\mathcal{L}_V \beta)\,.$$

Proof.

(1) We compute that

$$\mathcal{L}_V f = \partial_t (f \circ \phi_t) = V(f).$$

(2) We compute that

$$\begin{aligned}(\mathcal{L}_V W)_p &= \partial_t\big((d\phi_{-t}(W))_p\big)\\ &= \partial_t\big(d\phi_{-t}(W_{\phi_t(p)})\big),\end{aligned}$$

where we recall that ∂_t is evaluated at $t=0$. Let $f \in C^\infty(M^n)$. Then

$$\begin{aligned}(\mathcal{L}_V W)_p(f) &= \partial_t\big(d\phi_{-t}(W_{\phi_t(p)})\big)(f)\\ &= \partial_t\big(W_{\phi_t(p)}(f\circ\phi_{-t})\big)\\ &= \partial_t\big(W_{\phi_t(p)}(f\circ\phi_0)\big) + \partial_t\big(W_{\phi_0(p)}(f\circ\phi_{-t})\big)\\ &= \partial_t\big(W_{\phi_t(p)}(f)\big) + \partial_t\big(W_p(f\circ\phi_{-t})\big)\\ &= \partial_t\big(W(f)(\phi_t(p))\big) + W_p\big(\partial_t(f\circ\phi_{-t})\big)\\ &= V_p(W(f)) - W_p(V(f))\\ &= [V,W]_p(f).\end{aligned}$$

(3) This follows from simply differentiating the following equation while using the product rule for derivatives:

$$\phi_t^*(\alpha\otimes\beta) = \phi_t^*\alpha\otimes\phi_t^*\beta. \tag{6.85}$$ ☐

Lemma 6.38. *If α is a covariant k-tensor on a smooth manifold M^n, then for any vector fields $V, W_1, \ldots, W_k$,*

(6.86)

$$\begin{aligned}(\mathcal{L}_V\alpha)(W_1,\ldots,W_k) &= V\left(\alpha\left(W_1,\ldots,W_k\right)\right)\\ &\quad - \sum_{i=1}^k \alpha\left(W_1,\ldots,W_{i-1},\left[V,W_i\right],W_{i+1},\ldots,W_k\right).\end{aligned}$$

Proof. This follows from the product rule for the Lie derivative:

(6.87)

$$\begin{aligned}\mathcal{L}_V\left(\alpha\left(W_1,\ldots,W_k\right)\right) &= (\mathcal{L}_V\alpha)(W_1,\ldots,W_k)\\ &\quad + \sum_{i=1}^k \alpha\left(W_1,\ldots,W_{i-1},\mathcal{L}_V W_i,W_{i+1},\ldots,W_k\right).\end{aligned}$$ ☐

Define the **contraction** map

$$C : T_x^*M\otimes T_xM \to \mathbb{R} \tag{6.88}$$

by

$$C(\alpha\otimes X) := \alpha(X). \tag{6.89}$$

So, if α is a 1-form and X is a vector field, then $C(\alpha \otimes X) = \alpha(X)$ is a function. We have that the Lie derivative commutes with contraction; that is,

$$\mathcal{L}_V\big(C(\alpha \otimes X)\big) = C\big(\mathcal{L}_V(\alpha \otimes X)\big). \tag{6.90}$$

Using Lemma 6.37(1) and (3), we calculate that

$$V(\alpha(X)) = \mathcal{L}_V(\alpha(X)) = (\mathcal{L}_V\alpha)(X) + \alpha(\mathcal{L}_V X).$$

Therefore, by Lemma 6.37(2), we obtain

$$(\mathcal{L}_V\alpha)(X) = V(\alpha(X)) - \alpha([V, X]). \tag{6.91}$$

Now let $\phi_t : M^n \to M^n$, $t \in I$, be a 1-parameter family of diffeomorphisms; in particular, the interval I is not necessarily all of $\mathbb{R}$ and $\{\phi_t\}_{t\in I}$ is not necessarily a group. Let

$$V(t) := d\big((\phi_t)^{-1}\big)\left(\frac{d\phi_t}{dt}\right). \tag{6.92}$$

If α is a tensor, then

$$\frac{\partial}{\partial t}(\phi_t^*\alpha) = \mathcal{L}_{V(t)}(\phi_t^*\alpha). \tag{6.93}$$

6.6. Categories of manifolds

In this section, we briefly survey some main directions in the topology of manifolds, paying particular attention to dimension 4. Parts of this section require knowledge of the concepts of homotopy, (co)homology, intersection form, signature, and characteristic classes; we provide references for these concepts. The reader who is not familiar with these concepts may skip this section.

Recall from Definition 6.2 that a C^k manifold M^n is a second countable Hausdorff topological space with a maximal C^k atlas $\mathbf{A}$. Let $(M^n, \mathbf{A}_\infty)$ be a smooth manifold. Its underlying topological manifold $(M^n, \mathbf{A}_0)$ is defined at the end of §6.1.1.

6.6.1. Diffeomorphism and homeomorphism types of manifolds.

Let $\mathfrak{Man}(n,k)$ denote the set of all n-dimensional C^k manifolds. Define the equivalence relation $\sim$ on $\mathfrak{Man}(n,k)$ by $N^n \sim M^n$ if N^n is C^k *diffeomorphic* to M^n. Let $\mathfrak{DM}(n,k)$ denote the quotient space of this equivalence relation; that is, the elements of $\mathfrak{DM}(n,k)$ are the equivalence class of manifolds $[M^n]$. We call each element of $\mathfrak{DM}(n,k)$ a C^k **diffeomorphism type**.

Similarly, we can define the equivalence relation $\approx$ on $\mathfrak{Man}(n,k)$ by $N^n \approx M^n$ if N^n is *homeomorphic* to M^n. Let $\mathfrak{TM}(n,k)$ denote the quotient space of the equivalence relation $\approx$; that is, the elements of $\mathfrak{TM}(n,k)$ are the equivalence classes of manifolds $[\![M^n]\!]$. We call each element of $\mathfrak{TM}(n,k)$

a **homeomorphism type**. So $\mathfrak{TM}(n,k)$ is the space of homeomorphism types of n-dimensional C^k manifolds.

Observe that if manifolds $(M^n, \mathbf{A}_\infty) \sim (\overline{M}^n, \overline{\mathbf{A}}_\infty)$ are diffeomorphic, then $(M^n, \mathbf{A}_0) \approx (\overline{M}^n, \overline{\mathbf{A}}_0)$ are homeomorphic. That is, if two smooth manifolds are diffeomorphic, then their underlying topological manifolds are homeomorphic. This is simply because any diffeomorphism between two smooth manifolds is necessarily a homeomorphism of the underlying topological manifolds, which in turn follows from any smooth map necessarily being continuous.

Conversely, we may ask the question:

> If $(M^n, \mathbf{A}_0) \approx (\overline{M}^n, \overline{\mathbf{A}}_0)$, is it necessarily true that $(M^n, \mathbf{A}_\infty) \sim (\overline{M}^n, \overline{\mathbf{A}}_\infty)$?

Another way to rephrase this question is: If two smooth manifolds are homeomorphic, are they necessarily diffeomorphic? In other words, does two manifolds having the same homeomorphism type imply that they have the same diffeomorphism type?

Theorem 6.39. *Suppose $n = 2$ or $n = 3$ and let $(M^n, \mathbf{A}_\infty)$ and $(\overline{M}^n, \overline{\mathbf{A}}_\infty)$ be smooth closed manifolds. If $(M^n, \mathbf{A}_0) \approx (\overline{M}^n, \overline{\mathbf{A}}_0)$, then $(M^n, \mathbf{A}_\infty) \sim (\overline{M}^n, \overline{\mathbf{A}}_\infty)$. In other words, for closed surfaces and 3-dimensional manifolds, if two smooth manifolds are homeomoprhic, then they are diffeomorphic.*

This theorem is false if $n = 4$! Namely, there exist many examples of pairs of 4-dimensional smooth closed manifolds M_1^4 and M_2^4 that are homeomorphic but not diffeomorphic.

We can frame the discussion above in terms of 'category theory'. Namely, we have a map from $\mathfrak{DM}(n,k)$ to $\mathfrak{TM}(n,k)$ which takes a diffeomorphism type $[M^n]$ of a C^k manifold M^n to its homeomorphism type $[\![M^n]\!]$. This map is called a **forgetful functor** since it 'forgets' the differentiable structure and only remembers the topological structure of the manifold.[3] So Theorem 6.39 says that the forgetful functor from the diffeomorphism types of smooth manifolds $\mathfrak{DM}(n,\infty)$ to the homeomorphism types of smooth manifolds $\mathfrak{TM}(n,\infty)$ is an injection when n is equal to 2 or 3.

6.6.2. Fundamental groups of 4-manifolds.

Based on the fact that there does not exist an algorithm to decide whether a finitely presented group is trivial, Markov proved that there does not exist an algorithm to distinguish pairs of smooth closed 4-manifolds [**Mar58, Mar60**]; see Kirby [**Kir**] for an exposition following the line of reasoning of Scharlemann.

[3] See Mac Lane [**Mac71**] for a discussion of the forgetful functor.

6.6.3. Classification of simply connected 4-manifolds.

Definition 6.40. Two topological spaces X and Y have the **same homotopy type** if there exist continuous functions $f : X \to Y$ and $g : Y \to X$ such that $f \circ g$ is homotopic to id_Y and $g \circ f$ is homotopic to id_X. In this case we say that f and g are **homotopy equivalences**.

J. H. C. Whitehead [**Whi49**] and John Milnor [**Mil58**] proved the following homotopy classification theorem.

Theorem 6.41. *The homotopy type of a closed simply connected* 4*-manifold is uniquely determined by the intersection form.*

A fundamental result in 4-dimensional topology is Michael Freedman's 4-dimensional classification theorem [**Fre82**].

Theorem 6.42. *The homeomorphism type of a closed simply connected* 4*-manifold is uniquely determined by the intersection form and the Kirby–Siebenmann* $\mathbb{Z}_2$*-invariant.*

Freedman's theorem subsumes the 4-dimensional topological Poincaré conjecture:

Corollary 6.43 (4-dimensional Poincaré conjecture [**Fre82**, Theorem 1.6]). *Any closed toplogical* 4*-manifold of the same homotopy type as* S^4 *must be homeomorphic to* S^4.

6.6.4. Smoothability of topological manifolds. A fundamental question is that of smoothability. Suppose that $(M^n, \mathbf{A}_0)$ is a closed topological manifold. Does there exist a smooth atlas $\mathbf{A}'_\infty$ on M^n such that $\mathbf{A}'_0 = \mathbf{A}_0$? In other words, given a closed topological manifold, does it admit a smooth structure that has it as its underlying topological manifold? The answer is yes in dimensions 2 and 3. However, not all closed topological 4-manifolds are smoothable, i.e., admit a smooth structure!

Vladimir A. Rohlin [**Roh52**] proved the following.

Theorem 6.44. *If* M^4 *is a closed orientable smooth* 4*-manifold that admits a spin structure (equivalently, its second Stiefel–Whitney class* $w_2(M^4) = 0$*), then its signature* $\tau(M^4)$ *is divisible by* 16.

Hence, if M^4 *is a spin closed orientable topological* 4*-manifold and if its signature* $\tau(M^4)$ *is not divisible by* 16*, then* M^4 *is not smoothable.*

Many examples of smooth 4-manifolds are given by complex surfaces.

Using gauge theory, Simon Donaldson [**Don83, Don87**] proved the following.

Theorem 6.45. *If M^4 is a smooth closed orientable 4-manifold that has definite intersection form, then the intersection form is diagonalizable over the integers.*

The proof hinges on understanding the Yang–Mills and self-dual connection equations and incorporates works of Taubes [**Tau82**] and Uhlenbeck [**Uhl82a, Uhl82b**]. See also the books by Freed and Uhlenbeck [**FU91**] and Donaldson and Kronheimer [**DK90**]. The Seiberg–Witten equations simplified some results that were previous proved by gauge theory as well as leading to new discoveries; see the books by Morgan [**Mor96**] and Moore [**Moo01**]. A survey of 4-dimensional smooth topology is by Manolescu [**Man**], which also discusses Heegaard Floer theory and Khovanov homology. A book on 4-manifolds is by Gompf and Stipsicz [**GS99**].

6.6.5. The 11/8 conjecture. Donaldson's theorem, stated above, answers the question of which smooth closed oriented 4-manifolds with definite intersection forms are smoothable.

Now assume that our topological 4-manifolds are also simply connected. If the intersection form is indefinite and odd, then it is always smoothable.

If the intersection form is indefinite and even, then we have the following (see Matsumoto [**Mat82**]).

Conjecture 6.46 (11/8 conjecture)**.** *If M^4 is a closed simply connected smooth 4-manifold that has indefinite and even intersection form, then its signature is controlled by its second Betti number in the sense that*[4]

$$|\tau(M^4)| \leq \frac{8}{11} b_2(M^4).$$

It is known that a closed simply connected topological 4-manifold M^4 with indefinite and even intersection form and $|\tau(M^4)| \leq \frac{8}{11} b_2(M^4)$ is necessarily smoothable. Thus, the 11/8 conjecture says that a closed simply connected topological 4-manifold M^4 with indefinite and even intersection form is smoothable if and only if $|\tau(M^4)| \leq \frac{8}{11} b_2(M^4)$. Consequently, an affirmative solution to the 11/8 conjecture would completely resolve the question of which closed simply connected topological 4-manifolds are smoothable.

6.6.6. 4-dimensional smooth Poincaré conjecture. Theorem 6.39 says in particular that a 2- or 3-dimensional smooth closed manifold that is homeomorphic to the standard sphere must be diffeomorphic to the standard sphere. The corresponding result in dimension 4 is unknown. Namely, we have the following.

[4]The second Betti number $b_2(M^4)$ is defined as the rank of the second homology group $H_2(M^4;\mathbb{Z})$. The intersection form is a pairing on $H_2(M^4;\mathbb{Z})$. See Kirby [**Kir89**] for background on these concepts and for the definitions of indefinite and even intersection forms.

Conjecture 6.47 (Smooth Poincaré conjecture). *Any 4-dimensional smooth closed manifold that is homeomorphic to the standard 4-sphere must be diffeomorphic to the standard 4-sphere.*

6.6.7. Higher-dimensional smooth topology. An **exotic sphere** is a smooth manifold homeomorphic to, but not diffeomorphic to, the standard sphere. In dimensions at most 3, there are no exotic spheres. On the other hand, in [**Mil56**], Milnor proved the existence of exotic 7-spheres. If $n \neq 4$, Milnor [**Mil59**] proved that the set of exotic n-spheres, together with the operation of connected sum, forms a finite abelian group minus its identity element. In [**KM63**], Kervaire and Milnor gave a systematic characterization of exotic spheres. By the additional works of Moise [**Moi52**], Browder [**Bro69**], Hill, Hopkins, and Ravenel [**HHR16**], and Wang and Xu [**WX17**], the only odd dimensions n for which there are no exotic spheres (i.e., the smooth structure on S^n is unique) are $n = 1, 3, 5, 61$.

Important books on topological, piecewise linear, and smooth manifolds in dimensions 5 and higher are by Milnor [**Mil65**], Wall [**Wal70**], Browder [**Bro72**], Kirby and Siebenmann [**KS77**], and Ranicki [**Ran81**].

6.7. Notes and commentary

Books on complex and Kähler manifolds include Chern [**Che95**], Morrow and Kodaira [**MK06**], Griffiths and Harris [**GH94**], Zheng [**Zhe00**], and Huybrechts [**Huy05**].

6.8. Exercises

Exercise 6.1. *Show that a surface of revolution, as discussed in* §1.3.4.2, *can be defined as a manifold using two coordinate charts.*

Exercise 6.2. *Extend the definition of product manifold to define the product of ℓ-many manifolds $M_1^{n_1}, \ldots, M_\ell^{n_\ell}$.*

Exercise 6.3. *Prove that the $n+1$ charts in* (6.15) *form an atlas for complex projective space $\mathbb{C}P^n$.*

Exercise 6.4. *Prove the assertions in Example* 6.16 *regarding Grassmannians.*

Exercise 6.5. *Using the geometric interpretation of bracket in* §6.4.5.1, *show that the bracket $[U, V]$ in $\mathbb{R}^n$ of two constant vector fields $U(\mathbf{x}) := U_0$ and $V(\mathbf{x}) := V_0$ is zero.*

Exercise 6.6. *Prove that the Grassmannian* $\mathrm{Gr}(k, n)$ *in Example* 6.16 *may be given the structure of a smooth manifold.*

Exercise 6.7. *Prove that if M^n is a smooth manifold with boundary, then its boundary ∂M may be given the structure of an $(n-1)$-dimensional smooth manifold without boundary, where the underlying topology of ∂M is the same as the induced topology on ∂M as a subspace of M^n.*

Exercise 6.8. *Prove Lemma* 6.27.

Exercise 6.9. *Show that $d\phi_x(D)$ defined by* (6.59) *is a derivation at $\phi(x) \in M^n$.*

Exercise 6.10. *Show that definitions* (6.58) *and* (6.59) *of the derivative $d\phi_x$ are equivalent.*

Exercise 6.11. *Prove the chain rule* (6.60)*: $d(f \circ g) = df \circ dg$.*

Exercise 6.12.

(1) *Compute the* 1*-parameter group of diffeomorphisms of $\mathbb{R}^n$ generated by the vector field $\mathbf{x} \mapsto \mathbf{x}$.*

(2) *Compute the* 1*-parameter group of diffeomorphisms of $\mathbb{R}^{2n}$ generated by the vector field*

$$(x^1, x^2, \ldots, x^{2n-1}, x^{2n}) \mapsto (-x^2, x^1, \ldots, -x^{2n}, x^{2n-1}).$$

Exercise 6.13. *Prove* (6.77) *regarding a geometric interpretation of the Lie bracket.*

Exercise 6.14 (Commutator formula for Lie derivatives)**.** *Prove that for any vector fields X and Y and any tensor α on a smooth manifold M^n,*

$$\mathcal{L}_X(\mathcal{L}_Y\alpha) - \mathcal{L}_Y(\mathcal{L}_X\alpha) = \mathcal{L}_{[X,Y]}\alpha. \tag{6.94}$$

Chapter 7

Riemannian Manifolds

In this chapter, we give an introduction to selected basic topics in Riemannian geometry. Our emphasis is on those aspects of Riemannian geometry necessary for the general study of geometric analysis.

Topics that we cover include metrics, connections, geodesics, variations of arc length, the exponential map, completeness, the curvature tensor and its decomposition, constant curvature metrics, Jacobi fields, Myers's theorem, the Rauch comparison theorem, the cut locus, the injectivity radius, and holonomy. We also state the Cheeger–Gromov compactness theorem.

7.1. Riemannian metrics

The notion of a Riemannian metric on submanifolds of Euclidean space, discussed in Part I of this book, extends to smooth manifolds.

7.1.1. Definition of a Riemannian metric.

Let M^n be a smooth manifold, as defined in Chapter 6. The following is our basic notion of infinitesimal distances and angles on the tangent spaces of M^n.

Definition 7.1. A **Riemannian metric** g on M^n is an assignment, to each point $x \in M^n$, of an inner product g_x on the tangent space T_xM. By g_x being an inner product, we mean that $g_x : T_xM \times T_xM \to \mathbb{R}$ is symmetric, bilinear, and positive-definite; that is,

(Symmetric) $$g_x(W, V) = g_x(V, W),$$

(Bilinear) $$g_x(aV + bW, X) = a g_x(V, X) + b g_x(W, X),$$

(Positive-definite) $$g_x(V, V) > 0 \qquad \text{for } V \in T_xM \setminus \{\vec{0}\},$$

where $\vec{0}$ denotes the origin of T_xM; cf. §1.2.2. We require that g_x depends smoothly on x; that is, for any smooth vector fields V and W, the function $g(V, W)$ is smooth.

A **Riemannian manifold** is defined as a pair (M^n, g) of a smooth manifold and a Riemannian metric on the manifold. Note that a Riemannian metric on M^n is the same as a positive-definite symmetric covariant 2-tensor on M^n.

Example 7.2. *If M^n is a smooth submanifold of $\mathbb{R}^N$, then the first fundamental form defined by* (4.28),

$$g(V, W) := \mathrm{I}(V, W) = V \cdot W,$$

is a Riemannian metric on M^n, where $\cdot$ denotes the Euclidean inner product on $\mathbb{R}^N$; see (4.32). *Thus, all of the examples in Part* I *of submanifolds of Euclidean spaces are naturally endowed with Riemannian metrics induced by their ambient Euclidean spaces.*

Infinitesimal **distances** and **angles** on a Riemannian manifold are defined by

$$|V| := \sqrt{g(V, V)} \quad \text{for } V \in TM, \tag{7.1}$$

called the **norm** of V, and

$$\angle(V, W) := \cos^{-1}\left(\frac{g(V, W)}{|V||W|}\right), \tag{7.2}$$

respectively; cf. §1.2.4.

7.1.2. The metric in local coordinates.

Let $(\mathcal{U}, \{x^i\}_{i=1}^n)$ be local coordinates on M^n. For each $x \in \mathcal{U}$, the vectors $\left\{\frac{\partial}{\partial x^i}(x)\right\}_{i=1}^n$ form a basis for T_xM. The 1-forms $\{dx^i(x)\}_{i=1}^n$ form a dual basis for T_x^*M; that is, $dx^i\left(\frac{\partial}{\partial x^j}\right) = \delta^i_j$. The components of the metric are defined by

$$g_{ij} := g\left(\frac{\partial}{\partial x^i}, \frac{\partial}{\partial x^j}\right); \tag{7.3}$$

cf. (4.35). The metric may be expressed in local coordinates as (cf. (4.43))

$$g = \sum_{i,j=1}^n g_{ij}\, dx^i \otimes dx^j. \tag{7.4}$$

7.1.3. Riemannian length and distance. Given a path $\gamma : [a, b] \to M^n$, its (**Riemannian**) **length** (or **arc length**) is defined as the integral of the speed of γ by

$$\mathrm{L}(\gamma) = \int_a^b |\gamma'(t)|\, dt; \tag{7.5}$$

cf. (4.34).

The (**Riemannian**) **distance** between two points x and y in M^n is defined by

$$d(x, y) = \inf_\gamma \mathrm{L}(\gamma), \tag{7.6}$$

where the infimum is taken over all paths γ joining x and y. We will also use the notations

$$d_x(y) = d_y(x) = d(x, y).$$

We leave it as an exercise to show that (M^n, d) is a metric space (see Definition 1.1).

Example 7.3. *Let S^n be the unit sphere centered at the origin in $\mathbb{R}^{n+1}$ with the induced Riemannian metric g_{S^n}. Then its Riemannian distance is given by*

$$d_{S^n}(\mathbf{x}, \mathbf{y}) = 2 \sin^{-1} \left(\frac{\|\mathbf{x} - \mathbf{y}\|}{2} \right) \in [0, \pi],$$

where $\|\cdot\|$ denotes the Euclidean norm. This follows from trigonometry, the relationship between distance and angle, and the fact that shortest paths in S^n are arcs of great circles with length at most π.

7.1.4. Isometries. Let $F : N^n \to M^n$ be a diffeomorphism between smooth manifolds and let g be a Riemannian metric on M^n. The **pull-back** of g by F is defined by

$$(F^*g)(V, W) := g\big(dF(V), dF(W)\big) \tag{7.7}$$

for all $V, W \in T_yN$, $y \in N^n$. (See §6.5.1 for the general definition of the pull-back of tensors.) It is easy to see that F^*g is a positive-definite symmetric 2-tensor on N^n and hence F^*g is a Riemannian metric on N^n.

As we will see, a Riemannian metric g and its pull-back F^*g are geometrically indistinguishable.

Definition 7.4. Let (M^n, g) and (N^n, h) be Riemannian manifolds. We say that a diffeomorphism $F : N^n \to M^n$ is an **isometry** if $F^*g = h$. In this case we say that (M^n, g) and (N^n, h) are **isometric**.

Example 7.5. *Let $\beta : I \to \mathbb{R}^N$ and $\gamma : J \to \mathbb{R}^P$ be embedded paths with nonzero speeds, where I and J are either both closed intervals or both topological circles. If their lengths are equal, $\mathrm{L}(\beta) = \mathrm{L}(\gamma)$, then the 1-dimensional Riemannian manifolds $\beta(I)$ and $\gamma(J)$ with their induced metrics are isometric.*

Example 7.6 (Linear isometries of Euclidean space)**.** *Let A be an orthogonal transformation of $(\mathbb{R}^n, g_{\mathbb{E}})$, where $g_{\mathbb{E}}$ is the Euclidean metric on $\mathbb{R}^n$. That is,*

$$g_{\mathbb{E}}(A(\mathbf{V}), A(\mathbf{W})) = g_{\mathbb{E}}(\mathbf{V}, \mathbf{W}) \tag{7.8}$$

for all $\mathbf{V}, \mathbf{W} \in \mathbb{R}^n$. Then A is an isometry of $(\mathbb{R}^n, g_{\mathbb{E}})$. The reason is simply because, for any $\mathbf{x} \in \mathbb{R}^n$, we have that

$$dA_{\mathbf{x}} = A : T_{\mathbf{x}}\mathbb{R}^n \cong \mathbb{R}^n \to T_{A(\mathbf{x})}\mathbb{R}^n \cong \mathbb{R}^n$$

and so $dA_{\mathbf{x}}$ is an orthogonal transformation.

Example 7.7 (Pull-back of the Euclidean metric by vector space isomorphisms)**.** *Let $T : \mathbb{R}^n \to \mathbb{R}^n$ be a nonsingular linear transformation, i.e., a vector space self-isomorphism of $\mathbb{R}^n$. Let $T(\mathbf{e}_i) = \sum_{j=1}^n T_i^j \mathbf{e}_j$, where $\{\mathbf{e}_i\}_{i=1}^n$ is the standard Euclidean basis. Then, for $V = \sum_{i=1}^n V^i \mathbf{e}_i$ and $W = \sum_{i=1}^n W^i \mathbf{e}_i$,*

$$\begin{aligned} (T^* g_{\mathbb{E}})(V, W) &= T(V) \cdot T(W) \\ &= T^t T(V) \cdot W \\ &= \sum_{i,j,k=1}^n T_k^i T_j^i V^j W^k. \end{aligned} \tag{7.9}$$

Conversely, suppose that $A = (A_{jk})_{j,k=1}^n$ is a symmetric positive-definite matrix. Then

$$g_A := \sum_{j,k=1}^n A_{jk} dx^j \otimes dx^k$$

is a Riemannian metric on $\mathbb{R}^n$. An elementary linear algebra fact is that there exists a nonsingular linear transformation $T : \mathbb{R}^n \to \mathbb{R}^n$ such that $A_{jk} = \sum_{i=1}^n T_k^i T_j^i$. Thus,

$$g_A = T^* g_{\mathbb{E}}.$$

Thus, g_A is isometric to $g_{\mathbb{E}}$.

7.1.5. Metric duality between the tangent and cotangent bundles. Given a tangent vector $V \in T_xM$, its **metrically dual cotangent vector** $V^\flat \in T_x^*M$ is defined by

$$V^\flat(W) := g(V, W) \tag{7.10}$$

for all $W \in T_xM$.

Conversely, given a cotangent vector $\alpha \in T_x^*M$, its **metrically dual tangent vector** $\alpha^\natural$ is defined via the Riesz representation theorem by

$$\alpha(W) = g(\alpha^\natural, W) \tag{7.11}$$

for all $W \in T_xM$.

Example 7.8. *Suppose that $\{e_i\}_{i=1}^n$ is an orthonormal basis for T_xM. Denote $\omega^i := e_i^\flat$ for each i. We then have for all i, j that*

$$\omega^j(e_i) = \delta_{ij}. \tag{7.12}$$

Indeed, we calculate that

$$\omega^j(e_i) = e_j^\flat(e_i) = g(e_j, e_i) = \delta_{ij}.$$

Thus, the collection of metrically dual covectors $\{\omega^j\}_{j=1}^n$ is the same as the dual basis of $\{e_i\}_{i=1}^n$. This last fact only holds when the basis $\{e_i\}_{i=1}^n$ is orthonormal, as we have assumed above.

The metric duality between the tangent and cotangent bundles extends to contravariant and covariant tensor bundles. This will be discussed in Chapter 10.

7.2. Covariant differentiation

Unlike in Euclidean space, tangent vectors at different points on a smooth manifold cannot be naturally identified. As a consequence, there is no natural definition of differentiating vector fields on a smooth manifold absent a geometric structure such as a metric. However, when a smooth manifold is endowed with a Riemannian metric, there exists a natural and uniquely defined notion of differentiating vector fields, which we now describe.

7.2.1. Definition of the Levi-Civita connection. For a submanifold of Euclidean space, we defined the Levi-Civita connection in (4.123) as the tangential component of the Euclidean covariant derivative. We then considered the properties of the Levi-Civita connection of a submanifold in Lemma 4.28 and Exercises 4.24 and 4.25. It turns out that for a general Riemannian manifold, these properties uniquely characterize a linear connection, again called the **Levi-Civita connection**. That is, we have the following result.

Proposition 7.9 (Levi-Civita connection). *Let (M^n, g) be a Riemannian manifold. Then there exists a unique linear connection*

$$\nabla : C^\infty(TM) \times C^\infty(TM) \to C^\infty(TM)$$

on the tangent bundle TM that is compatible with the metric g and is torsion-free. That is, as a ***linear connection****, $\nabla_V W := \nabla(V, W)$ satisfies the following:*

(1) $\nabla_{fV_1+V_2} W = f\nabla_{V_1} W + \nabla_{V_2} W$.

(2) $\nabla_V (cW_1 + W_2) = c\nabla_V W_1 + \nabla_V W_2$.

(3) $\nabla_V (fW) = V(f)W + f\nabla_V W$

for all vector fields V, W, V_1, V_2, W_1, W_2, scalar functions f, and constants c. By being ***compatible with the metric*** *(a.k.a. the* ***product rule****), we mean that*

$$U\big(g(V, W)\big) = g(\nabla_U V, W) + g(V, \nabla_U W) \tag{7.13}$$

for all vector fields U, V, W. By being ***torsion-free****, we mean that*

$$\nabla_V W - \nabla_W V = [V, W] \tag{7.14}$$

for all vector fields V, W.

Proof. We need to prove existence and uniqueness. The proof of this was implicitly given in §4.5.5. The reason is that when considering the Levi-Civita connection ∇ for a submanifold and deriving a formula for ∇ (see (4.132) for the derivation),

$$\begin{aligned} 2g\left(\nabla_U V, W\right) = {} & U\left(g\left(V, W\right)\right) + V\left(g\left(U, W\right)\right) - W\left(g\left(U, V\right)\right) \\ & + g\left(\left[U, V\right], W\right) - g\left(\left[U, W\right], V\right) - g\left(\left[V, W\right], U\right), \end{aligned} \tag{7.15}$$

we only used the properties of the connection ∇ stated above and not its definition (4.123) as the tangential projection of the Euclidean covariant derivative. By deriving a formula for ∇ (since (7.15) holds for all W, it determines $\nabla_U V$), we proved its uniqueness.

Conversely, by defining the Levi-Civita connection ∇ via the formula (4.132), it is straightforward to check that ∇ satisfies all of the properties stated above in the proposition (we leave this as an exercise). This proves its existence. □

Remark 7.10. By property (1) in Proposition 7.9 of a linear connection, we have that $(\nabla_V W)_x$ depends only on V_x.

Example 7.11 (Euclidean Levi-Civita connection). *As in* (4.51), *the Euclidean covariant derivative D on $(\mathbb{R}^n, g_{\mathbb{E}})$ is given by*

$$D_V W = dW(V) = \big(V(W^1), \ldots, V(W^n)\big). \tag{7.16}$$

So the Euclidean covariant derivative along a certain vector is the same as the directional derivative along the corresponding direction.

Example 7.12 (Levi-Civita connection of S^n). *For the unit n-sphere S^n in $\mathbb{R}^{n+1}$, the Levi-Civita connection ∇ of the induced metric g_{S^n} on S^n is given by*

$$\nabla_V W = D_V W + (V \cdot W)\mathbf{x} \tag{7.17}$$

for any tangent vector V at $\mathbf{x}$ and tangent vector field W on S^n; cf. (4.126).

7.2.2. The Levi-Civita connection in local coordinates. Let $(\mathcal{U}, \{x^i\})$ be local coordinates on M^n. The **Christoffel symbols** Γ_{ij}^k, $1 \leq i, j, k \leq n$, are the functions on $\mathcal{U}$ defined by

$$\nabla_{\frac{\partial}{\partial x^i}} \frac{\partial}{\partial x^j} =: \sum_{k=1}^n \Gamma_{ij}^k \frac{\partial}{\partial x^k}; \tag{7.18}$$

cf. (4.138). We can think of Γ_{ij}^k as the components of the Levi-Civita connection. Just as for (4.139), we can derive the following equation for the Christoffel symbols:

$$\Gamma_{ij}^k = \frac{1}{2} \sum_{\ell=1}^n g^{k\ell} \left(\frac{\partial}{\partial x^i} g_{j\ell} + \frac{\partial}{\partial x^j} g_{i\ell} - \frac{\partial}{\partial x^\ell} g_{ij} \right). \tag{7.19}$$

Indeed, we just apply (7.15) with $U = \frac{\partial}{\partial x^i}$, $V = \frac{\partial}{\partial x^j}$, $W = \frac{\partial}{\partial x^\ell}$ and multiply the result by $g^{k\ell}$ and sum over ℓ.

By Exercise 4.27, if $V = \sum_{i=1}^n V^i \partial_i$ and $W = \sum_{j=1}^n W^j \partial_j$, then

$$\begin{aligned} \nabla_V W &= \sum_{i,j=1}^n V^i \frac{\partial W^j}{\partial x^i} \frac{\partial}{\partial x^j} + \sum_{i,j,k=1}^n V^i W^j \Gamma_{ij}^k \frac{\partial}{\partial x^k} \\ &= \sum_{i,k=1}^n V^i \left(\frac{\partial W^k}{\partial x^i} + \sum_{j=1}^n W^j \Gamma_{ij}^k \right) \frac{\partial}{\partial x^k}. \end{aligned} \tag{7.20}$$

7.2.3. The covariant derivative of covariant tensors. Let α be a covariant k-tensor, i.e., a section of the tensor bundle $\bigotimes^k T^*M$. The product (Leibniz) rule for covariant differentiation says that

$$\begin{aligned} \nabla_V \left(\alpha\left(W_1, \ldots, W_k\right)\right) &= \left(\nabla_V \alpha\right)\left(W_1, \ldots, W_k\right) \\ &\quad + \sum_{i=1}^k \alpha\left(W_1, \ldots, W_{i-1}, \nabla_V W_i, W_{i+1}, \ldots, W_k\right), \end{aligned} \tag{7.21}$$

where $V, W_1, \ldots, W_k$ are vector fields. Although $\alpha(W_1, \ldots, W_k)$ is technically equal to k 'contractions' of the tensor $\alpha \otimes W_1 \otimes \cdots \otimes W_k$, we consider it as a product for the purpose of making an analogy with the Leibniz rule for differentiation. This formula is also characterized as saying that ∇ commutes with contractions. To understand this, the reader will do well to first consider the case $k = 1$.

Since the covariant derivative acting on functions is just the directional derivative, i.e., $\nabla_V f = V(f)$, formula (7.21) is equivalent to

(7.22)

$$(\nabla_V \alpha)(W_1, \ldots, W_k) = V(\alpha(W_1, \ldots, W_k)) - \sum_{i=1}^{k} \alpha(W_1, \ldots, W_{i-1}, \nabla_V W_i, W_{i+1}, \ldots, W_k).$$

For example, if α is a 1-tensor, a.k.a. 1-form, then by (7.22), we have

$$(\nabla_V \alpha)(W) = V(\alpha(W)) - \alpha(\nabla_V W). \tag{7.23}$$

As another example, if α is a 2-tensor, then by (7.22), we have

$$(\nabla_V \alpha)(W_1, W_2) = V(\alpha(W_1, W_2)) - \alpha(\nabla_V W_1, W_2) - \alpha(W_1, \nabla_V W_2). \tag{7.24}$$

7.2.4. Parallel tensors. Generalizing the notion of a parallel vector field in Definition 4.29 and parallel 2-tensor in §4.7, we have the following.

Definition 7.13. A tensor α on (M^n, g) is **parallel** if for every $V \in TM$, $\nabla_V \alpha = 0$.

As a special case of (7.24), by taking $\alpha = g$ to be the Riemannian metric, we obtain

(7.25)

$$(\nabla_V g)(W_1, W_2) = V(g(W_1, W_2)) - g(\nabla_V W_1, W_2) - g(W_1, \nabla_V W_2) = 0$$

by the metric compatibility condition (cf. (4.179)). That is:

Lemma 7.14. *The Riemannian metric g, as a 2-tensor, is parallel.*

7.2.5. The Lie derivative on Riemannian manifolds. Recall that the Lie derivative of tensors is defined for differentiable manifolds, without the need of a Riemannian structure. When we are on a Riemannian manifold, we can relate the Lie derivative to the covariant derivative as follows.

Lemma 7.15 (Riemannian formula for Lie derivative). *Let (M^n, g) be a Riemannian manifold. If α is a covariant k-tensor, then for any vector fields $V, W_1, \ldots, W_k$ on M^n,*

(7.26)

$$(\mathcal{L}_V\alpha)(W_1, \ldots, W_k) = (\nabla_V\alpha)(W_1, \ldots, W_k) + \sum_{i=1}^{k} \alpha(W_1, \ldots, W_{i-1}, \nabla_{W_i}V, W_{i+1}, \ldots, W_k).$$

Proof. The idea of the proof is to compare the product rules for the Lie derivative and the covariant derivative. By (6.86), we have the product rule for the Lie derivative:

$$(\mathcal{L}_V\alpha)(W_1, \ldots, W_k) = V(\alpha(W_1, \ldots, W_k)) - \sum_{i=1}^{k} \alpha(W_1, \ldots, W_{i-1}, [V, W_i], W_{i+1}, \ldots, W_k).$$

By applying the product rule for covariant differentiation (7.22) to the first term on the right-hand side and by using that the Levi-Civita connection is torsion-free, we obtain

$$\begin{aligned}(\mathcal{L}_V\alpha)&(W_1, \ldots, W_k)\\ &= (\nabla_V\alpha)(W_1, \ldots, W_k)\\ &\quad + \sum_{i=1}^{k} \alpha(W_1, \ldots, W_{i-1}, \nabla_V W_i, W_{i+1}, \ldots, W_k)\\ &\quad - \sum_{i=1}^{k} \alpha(W_1, \ldots, W_{i-1}, \nabla_V W_i - \nabla_{W_i}V, W_{i+1}, \ldots, W_k).\end{aligned}$$

The lemma follows from cancelling terms. □

7.2.6. The Lie derivative of the Riemannian metric. For example, if α is a 2-tensor, then by Lemma 7.15, we have

$$(\mathcal{L}_V\alpha)(W_1, W_2) = (\nabla_V\alpha)(W_1, W_2) + \alpha(\nabla_{W_1}V, W_2) + \alpha(W_1, \nabla_{W_2}V). \tag{7.27}$$

As a further special case, by taking $\alpha = g$ to be the Riemannian metric, we obtain

$$(\mathcal{L}_V g)(W_1, W_2) = g(\nabla_{W_1}V, W_2) + g(W_1, \nabla_{W_2}V), \tag{7.28}$$

where we used that the metric is parallel by (7.25); i.e., $\nabla g = 0$. If V is a gradient vector field, i.e., $V = \nabla f$ for some function $f : M^n \to \mathbb{R}$, then we have

$$(\mathcal{L}_{\nabla f}g)(W_1, W_2) = g(\nabla_{W_1}\nabla f, W_2) + g(W_1, \nabla_{W_2}\nabla f);$$

that is:

Lemma 7.16 (Lie derivative of a Riemannian metric)**.** *For any C^2 function f on a Riemannian manifold,*

$$\mathcal{L}_{\nabla f} g = 2\nabla^2 f. \tag{7.29}$$

That is, the Lie derivative of the metric with respect to the gradient of a function is equal to twice the Hessian of the function.

7.2.7. Killing vector fields.

Definition 7.17. A vector field V on a Riemannian manifold (M^n, g) is a **Killing vector field** if $\mathcal{L}_V g = 0$.

Let V be a Killing vector field. By (6.83),

$$\mathcal{L}_V g := \partial_t(\phi_t^* g) = \lim_{t\to 0} \frac{1}{t}(\phi_t^* g - g), \tag{7.30}$$

where $\{\phi_t\}$ is the local 1-parameter group of local diffeomorphisms generated by V. In this sense, a Killing vector field represents an **infinitesimal isometry** of g.

Furthermore, by (7.28), we have for any vector fields X and Y that

$$0 = \mathcal{L}_V g(X, Y) = g(\nabla_X V, Y) + g(X, \nabla_Y V). \tag{7.31}$$

7.3. Geodesics

Geodesics in Riemannian manifolds are generalizations of straight lines in Euclidean spaces. Straight line segments in $\mathbb{R}^n$ minimize the arc length between their endpoints, and in particular, they are critical points of the arc-length functional on the space of paths between two fixed points. To study this arc-length functional on Riemannian manifolds, we consider families of paths. These families of paths form maps and we need to study covariant differentiation along these maps to understand the variations of arc length.

7.3.1. Covariant derivative along a map. We first discuss the notion of the covariant derivative along a map. As a special case, we have the covariant derivative along a path. This will be used in the study of geodesics below.

Let N^m be a differentiable manifold and let (M^n, g) be a Riemannian manifold.

Definition 7.18. By a **vector field V along a map** $F : N^m \to M^n$ we mean a map

$$V : N^m \to TM$$

satisfying $\pi \circ V = F$, where $\pi : TM \to M^n$ is the projection map; that is, $V(x) \in T_{F(x)}M$ for each point $x \in N^m$.

For example, a vector field along the identity map $\mathrm{id}_{M^n} : M^n \to M^n$ is simply a vector field on M^n.

If $\gamma : I \to M^n$, then its tangent vector field $t \mapsto \gamma'(t)$ is a vector field along γ.

Let $\mathfrak{X}(F)$ denote the set of C^∞ vector fields along F (see below for the definition of C^∞ in this context). Given a point $p_0 \in N^m$ and a tangent vector $X \in T_{p_0}N$, we define

$$\nabla_X^F : \mathfrak{X}(F) \to T_{F(p_0)}M,$$

called the **covariant derivative along the map** F, as follows. Choose any local coordinates $\mathbf{x} = (x^1, \ldots, x^n)$ defined in a neighborhood $\mathcal{U}$ of $F(p_0)$. Given a vector field V along the map F, we may express it in local coordinates as

$$V(p) =: \sum_{i=1}^n V^i(p) \left.\frac{\partial}{\partial x^i}\right|_{F(p)}$$

for p in a neighborhood of p_0. We say that V is C^∞ if each function V^i is C^∞. We define

(7.32)

$$\nabla_X^F V := \sum_{k=1}^n \left(X(V^k) + \sum_{i,j=1}^n V^i(p_0)\, dF(X)^j\, \Gamma_{ij}^k(F(p_0)) \right) \left.\frac{\partial}{\partial x^k}\right|_{F(p_0)},$$

where Γ_{ij}^k are the Christoffel symbols of g with respect to the local coordinates $\{x^i\}_{i=1}^n$. The expression on the right-hand side of (7.32) is independent of the local coordinates $\{x^i\}_{i=1}^n$. In particular, given ℓ, by taking $V^i := \delta_{i\ell}$, i.e., $V := \frac{\partial}{\partial x^\ell}$, we have that

$$\nabla_X^F \frac{\partial}{\partial x^\ell} = \sum_{j,k=1}^n dF(X)^j\, \Gamma_{\ell j}^k(F(p_0)) \left.\frac{\partial}{\partial x^k}\right|_{F(p_0)}.$$

In the case that F is an immersion, an equivalent definition of the covariant derivative along the map F is as follows. Let $\mathcal{V}$ be a neighborhood of p_0 on which F is an embedding, so that $F(\mathcal{V})$ is an m-dimensional submanifold of M^n. Given $V \in \mathfrak{X}(F)$, $\overline{V}(x) := V\big((F|_{\mathcal{V}})^{-1}(x)\big)$ is a vector field on $F(\mathcal{V})$. Extend $\overline{V}$ to a vector field in a neighborhood of $F(p_0)$ in M^n, denoted also by $\overline{V}$. Then

$$(\nabla_X^F V)(p_0) = \big(\nabla_{dF(X)}\overline{V}\big)(F(p_0)),$$

where ∇ denotes the Levi-Civita connection of (M^n, g). The right-hand side is independent of the extension $\overline{V}$.

We define

$$\nabla^F : C^\infty(TN) \times \mathfrak{X}(F) \to \mathfrak{X}(F)$$

by $\nabla^F(X,V) := \nabla^F_X V$ for any vector field X on N^m and vector field V along F. Note that $\nabla^F V \in C^\infty(T^*N \otimes F^*TM)$.

For example, the covariant derivative along the identity map $\mathrm{id}_{M^n} : M^n \to M^n$ is the same as the Levi-Civita connection; this is because in this case (7.32) reduces to (7.20).

Analogous to the Levi-Civita connection, the covariant derivative along a map has the following properties, which we leave as an exercise.

Lemma 7.19 (Properties of ∇^F). *Let $F : N^m \to M^n$. The covariant derivative along F is the unique map $\nabla^F : C^\infty(TN) \times \mathfrak{X}(F) \to \mathfrak{X}(F)$ which is bilinear over $\mathbb{R}$ and satisfies the following properties:*
For any $X \in C^\infty(TN)$, $V, W \in \mathfrak{X}(F)$, and C^∞ function $f : N^m \to \mathbb{R}$:

(i) $\nabla^F_{fX} V = f \nabla^F_X V$ *and* (Leibniz rule)

$$\nabla^F_X (fV) = f \nabla^F_X V + X(f) V.$$

(ii) (Ambient vector fields) *For any $\overline{V} \in C^\infty(TM)$ we have*

$$\nabla^F_X(\overline{V} \circ F) = \nabla_{dF(X)} \overline{V},$$

where ∇ denotes the Levi-Civita connection of g.

(iii) (Compatible with g, a.k.a. metric product rule)

$$X\big(g(V,W)\big) = g(\nabla^F_X V, W) + g(V, \nabla^F_X W). \tag{7.33}$$

(iv) (Torsion-free) *For any $X, Y \in C^\infty(TN)$, we have*

$$\nabla^F_X\big(dF(Y)\big) - \nabla^F_Y\big(dF(X)\big) = dF([X,Y]). \tag{7.34}$$

(v) *If $V = W$ in a neighborhood of p_0, then $\nabla^F_X V = \nabla^F_X W$.*

Since we are interested in geodesics, we consider paths in manifolds. If $\gamma : I \subset \mathbb{R} \to M^n$ is a path, then we have covariant differentiation along γ:

$$\nabla^\gamma_{\frac{d}{dt}} : \mathfrak{X}(\gamma) \to \mathfrak{X}(\gamma), \tag{7.35}$$

where $\frac{d}{dt} = \mathbf{e}_1 \in T_t I$. In this case, property (iii) of Lemma 7.19 says that

$$\frac{d}{dt}\langle V(t), W(t)\rangle = \left\langle \nabla^\gamma_{\frac{d}{dt}} V(t), W(t)\right\rangle + \left\langle V(t), \nabla^\gamma_{\frac{d}{dt}} W(t)\right\rangle. \tag{7.36}$$

Let $T(t) = \gamma'(t)$ and let $t_0 \in I$ be such that $T(t_0) \neq \vec{0}$. Choose $\epsilon > 0$ such that $\gamma|_{I\cap(t_0-\epsilon,t_0+\epsilon)}$ is an embedding. If $V(t)$ is a vector field along $\gamma(t)$, then for any extension $\overline{V}$ to M^n of $\gamma(t) \mapsto V(t)$, $t \in I \cap (t_0-\epsilon, t_0+\epsilon)$, we have that

$$\nabla^{\gamma}_{\frac{d}{dt}} V(t)\big|_{t=t_0} = \nabla_{T(t_0)}\overline{V} =: \nabla_T V, \tag{7.37}$$

where the last equality is by definition and to simplify notation.

7.3.2. Parallel vector fields along maps.

We say that a vector field V along a map F is **parallel** if

$$\nabla^F V \equiv 0. \tag{7.38}$$

If V and W are parallel along F, then $\langle V, W\rangle$ is constant on N^m. In particular, if $V(t)$ is parallel along a path $\gamma(t)$, then the function $|V(t)|$ is constant.

Needless to say, a vector field V on M^n is parallel along the identity map id_{M^n} if and only if V is parallel in the usual sense; i.e., $\nabla V \equiv 0$.

Lemma 7.20. *Given a path $\gamma : [a,b] \to M^n$ and a vector $V_a \in T_{\gamma(a)}M$, there exists a unique parallel vector field $V(t)$ along $\gamma(t)$ with $V(a) = V_a$.*

Definition 7.21. We call $V(t)$ the **parallel transport** of V_a along $\gamma(t)$. See Figure 7.3.1.

Proof of Lemma 7.20. In local coordinates, we obtain an initial-value problem for a first-order system of ordinary differential equations:

$$0 = \left(\nabla^{\gamma}_{\frac{d}{dt}} V(t)\right)^k = \frac{dV^k}{dt}(t) + \sum_{i,j=1}^{n} \Gamma^k_{ij}(\gamma(t))\gamma'(t)^i V^j(t), \tag{7.39}$$

where $V(t) = \sum_{i=1}^n V^i(t)\, \frac{\partial}{\partial x^i}\big|_{\gamma(t)}$. The lemma follows from the existence and uniqueness theorem for ordinary differential equations; see, e.g., Arnold's book [**Arn06**, Chapter 2, §8]. □

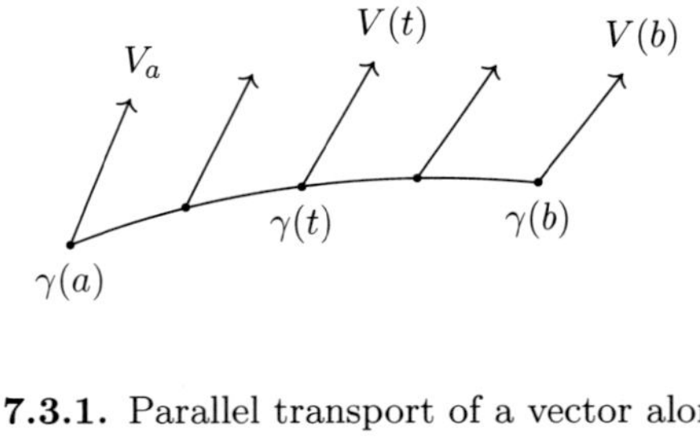

Figure 7.3.1. Parallel transport of a vector along a path.

7.3.3. Definition of geodesic.

Definition 7.22. A path $\gamma : I \to M^n$ with nonzero velocity $\gamma'(t)$ is said to be a **geodesic** if

$$\nabla^\gamma_{\frac{d}{dt}} \left(\frac{\gamma'(t)}{|\gamma'(t)|} \right) = \vec{0} \tag{7.40}$$

for all $t \in I$; cf. Definition 4.31. Writing this in the simplified notation of (7.37), the geodesic equation is

$$\nabla_T \left(\frac{T}{|T|} \right) = \vec{0},$$

where $T(t) := \gamma'(t)$.

In local coordinates, the geodesic equation $\nabla_{\gamma'(t)}\gamma'(t) = \vec{0}$ is

$$\frac{d^2\gamma^k}{dt^2}(t) + \sum_{i,j=1}^{n} \frac{d\gamma^i}{dt}(t)\frac{d\gamma^j}{dt}(t)\Gamma^k_{ij}\big(\gamma(t)\big) = 0 \quad \text{for all } 1 \leq k \leq n, \tag{7.41}$$

where $\gamma^i := x^i \circ \gamma$; cf. (4.145).

7.3.4. Geodesics in Euclidean space. It is easy to see that a path in Euclidean space is a geodesic if and only if it parametrizes a straight line segment. Given $\mathbf{x} \in \mathbb{R}^n$ and $\mathbf{V} \in \mathbb{R}^n$,

$$\gamma(t) := \mathbf{x} + t\mathbf{V}$$

is the unique constant-speed geodesic in $\mathbb{R}^n$ with $\gamma(0) = \mathbf{x}$ and $\gamma'(0) = \mathbf{V}$.

7.3.5. Geodesics in spheres. By Exercise 4.26, if V, W are orthogonal unit vectors in $\mathbb{R}^{n+1}$, then the path $\gamma : \mathbb{R} \to S^n$ defined by

$$\gamma(t) = \cos(t)\,V + \sin(t)\,W$$

is a geodesic in (S^n, g_{S^n}). This is the unique entire unit-speed geodesic on S^n with $\gamma(0) = V$ and $\gamma'(0) = W$; note that $W \in T_V S^n$. Thus we have described all unit-speed geodesics on S^n.

7.3.6. Geodesics in hyperbolic space. We will discuss this in §7.9.4.

7.4. First variation of arc length

Given a space of paths on a Riemannian manifold — for instance, all piecewise smooth paths joining two fixed points — we wish to understand the arc-length functional. More specifically, we begin by identifying the critical points of this functional through considering first variations of paths and their associated derivatives of the arc-length functional.

7.4.1. 1-parameter families of paths.

Let

$$\Gamma : [a,b] \times I \to M^n$$

be a smooth map, where I is an open interval containing 0. Such a map corresponds to the smooth 1-parameter family of paths defined by

$$\gamma_v(t) := \Gamma(t,v)$$

for $(t,v) \in [a,b] \times I$. Let W be a vector field along Γ. Then

$$\nabla^{\Gamma}_{\frac{\partial}{\partial t}} W \quad \text{and} \quad \nabla^{\Gamma}_{\frac{\partial}{\partial v}} W$$

are vector fields along Γ. Given $v_0 \in I$, if we define the restriction of W to $\gamma_{v_0}(t)$ by $W_{v_0}(t) := W(t, v_0)$, then

$$\left(\nabla^{\Gamma}_{\frac{\partial}{\partial t}} W\right)(t_0, v_0) = \left(\nabla^{\gamma_{v_0}}_{\frac{d}{dt}} W_{v_0}\right)(t_0),$$

where $\nabla^{\gamma_{v_0}}_{\frac{d}{dt}}$ is the covariant derivative along the path γ_{v_0}.

Let $\gamma_v : [a,b] \to M^n$ be a 1-parameter family of paths defined for v in an open interval I containing 0. Define the map $\Gamma : [a,b] \times I \to M^n$ by

$$\Gamma(t,v) := \gamma_v(t).$$

Define the coordinate vector fields T and V along Γ by

$$T(t,v) := d\Gamma_{(t,v)}\left(\frac{\partial}{\partial t}\right) \quad \text{and} \quad V(t,v) := d\Gamma_{(t,v)}\left(\frac{\partial}{\partial v}\right). \tag{7.42}$$

We call V the **variation vector field** and T the **tangent vector field**. By (7.5), the length of γ_v is given by

$$\mathrm{L}(\gamma_v) = \int_a^b |T(t,v)|\, dt.$$

7.4.2. First variation of arc-length formula.

Lemma 7.23 (First variation of arc-length formula for paths). *We have*

$$\left.\frac{d}{dv}\right|_{v=0} \mathrm{L}(\gamma_v) = -\int_a^b \left\langle V, \nabla_T\left(\frac{T}{|T|}\right)\right\rangle dt + \left.\left\langle V, \frac{T}{|T|}\right\rangle\right|_a^b, \tag{7.43}$$

where $T = \gamma_0'$ and $V = \left.\frac{\partial}{\partial v}\right|_{v=0}\gamma_v$, provided γ_0 has nonzero speed, and where we used the simplified notation of (7.37). *In particular, if γ_0 has unit-speed, then*

$$\left.\frac{d}{dv}\right|_{v=0} \mathrm{L}(\gamma_v) = -\int_a^b \langle V, \nabla_T T\rangle\, dt + \langle V, T\rangle|_a^b. \tag{7.44}$$

Proof. We compute by differentiating under the integral sign and by using the metric product rule for covariant differentiation along the map $\Gamma(t,v) = \gamma_v(t)$ that

$$\frac{d}{dv}\bigg|_{v=0} \mathrm{L}(\gamma_v) = \int_a^b \frac{\partial}{\partial v}\bigg|_{v=0} |T(t,v)|\, dt \tag{7.45}$$
$$= \frac{1}{2}\int_a^b \frac{1}{|T|}\frac{\partial}{\partial v}\bigg|_{v=0} \langle T,T\rangle\, dt$$
$$= \int_a^b \frac{1}{|T|}\left\langle \nabla^\Gamma_{\frac{\partial}{\partial v}} T, T\right\rangle dt$$
$$= \int_a^b \left\langle \nabla^\Gamma_{\frac{\partial}{\partial t}} V, \frac{T}{|T|}\right\rangle dt,$$

where the last equality is true since $\nabla^\Gamma_{\frac{\partial}{\partial v}} T - \nabla^\Gamma_{\frac{\partial}{\partial t}} V = d\Gamma\left(\left[\frac{\partial}{\partial v},\frac{\partial}{\partial t}\right]\right) = d\Gamma(\mathbf{0}) = \vec{0}$ by (7.34). The result now follows from

$$\frac{\partial}{\partial t}\left\langle V, \frac{T}{|T|}\right\rangle = \left\langle V, \nabla^\Gamma_{\frac{\partial}{\partial t}}\left(\frac{T}{|T|}\right)\right\rangle + \left\langle \nabla^\Gamma_{\frac{\partial}{\partial t}} V, \frac{T}{|T|}\right\rangle$$

and the fundamental theorem of calculus (we have effectively integrated by parts). Note that we denote $\nabla^\Gamma_{\frac{\partial}{\partial t}} W =: \nabla_T W$ for a vector field W along γ. □

We now present some consequences of the first variation formula. By applying it to paths that join a fixed point p to a fixed point q, so that V vanishes at the endpoints, we obtain the following.

Corollary 7.24 (First variation for paths with fixed endpoints). *If $\gamma_v(a)$ is independent of v and if $\gamma_v(b)$ is independent of v, then*

$$\frac{d}{dv}\bigg|_{v=0} \mathrm{L}(\gamma_v) = -\int_a^b \left\langle V, \nabla_T\left(\frac{T}{|T|}\right)\right\rangle dt, \tag{7.46}$$

provided that γ_0 has nonzero speed. In particular, if γ_0 has unit-speed, then

$$\frac{d}{dv}\bigg|_{v=0} \mathrm{L}(\gamma_v) = -\int_a^b \langle V, \nabla_T T\rangle\, dt. \tag{7.47}$$

7.4.3. Geodesics are the critical points of the arc-length functional.

Corollary 7.25 (Geodesics are the critical points of the length functional). *A path γ_0 with nonzero speed is a critical point of the length functional on the space of paths joining a fixed point p to a fixed point q if and only if γ_0 is a geodesic.*

Proof. A path γ_0 with nonzero speed is a critical point of L on the space of paths joining points p and q if and only if $\frac{d}{dv}\big|_{v=0} \mathrm{L}(\gamma_v) = 0$ for all variations $V = \frac{\partial}{\partial v}\big|_{v=0} \gamma_v$ of γ_0 vanishing at the endpoints. By (7.46), this holds if and only if $\int_a^b \langle V, \nabla_T(\frac{T}{|T|})\rangle dt = 0$ for all such V, which in turn holds if and only if $\nabla_T(\frac{T}{|T|}) = 0$. □

By Corollary 7.25, if $\gamma : [a, b] \to M^n$ is a piecewise smooth path satisfying $\mathrm{L}(\gamma) = d(\gamma(a), \gamma(b))$, then γ is a smooth geodesic.

We say that a geodesic segment $\gamma : [a, b] \to M^n$ is a **minimal geodesic** if $\mathrm{L}(\gamma) = d(\gamma(a), \gamma(b))$. More generally, we say that a path $\gamma : I \to M^n$ is a minimal geodesic, where I is an interval, if for every $[a, b] \subset I$, $\gamma|_{[a,b]}$ is a minimal geodesic. If $I = \mathbb{R}$, we call γ a **geodesic line**; so geodesic lines are entire minimal geodesics.

On the other hand, by considering variations of a geodesic that fix the initial point, we obtain the following.

Corollary 7.26 (First variation for a geodesic among paths with fixed initial point). *If $\gamma_v : [0, L] \to M^n$, $v \in I \ni 0$, is a 1-parameter family of paths emanating from a fixed point $p \in M^n$ and if $\gamma = \gamma_0$ is a geodesic parametrized by arc length, then*

$$\frac{d}{dv}\bigg|_{v=0} \mathrm{L}(\gamma_v) = \langle V(L), \gamma'(L)\rangle, \tag{7.48}$$

where $V := \frac{\partial}{\partial v}\big|_{v=0} \gamma_v$.

Proof. This follows immediately from (7.43) since by hypothesis we have that $\nabla_T T = \vec{0}$ and $V(0) = \vec{0}$. □

7.5. The exponential map

Let (M^n, g) be a Riemannian manifold. Note that we do not assume that the associated metric space (M^n, d) is complete, where d is the Riemannian distance defined by (7.6). Recall that S_pM denotes the unit tangent space at $p \in M^n$.

7.5.1. Definition of the exponential map. Let $p \in M^n$. Given a unit vector $U \in S_pM$, let $b(U) \in (0, \infty]$ be the supremum of $\bar{b} \in (0, \infty)$ such that the unique constant-speed geodesic $\gamma_U(t)$ emanating from p with $\gamma_U'(0) = V$ is defined for $t \in [0, \bar{b})$. Let

$$\Omega_pM = \{tU : 0 \leq t < b(U),\, U \in S_pM\} \subset T_pM. \tag{7.49}$$

It is easy to see that the function $b : S_pM \to \mathbb{R}_+$ is lower-semicontinuous and that the set Ω_pM is an open subset of T_pM. We will see in Theorem

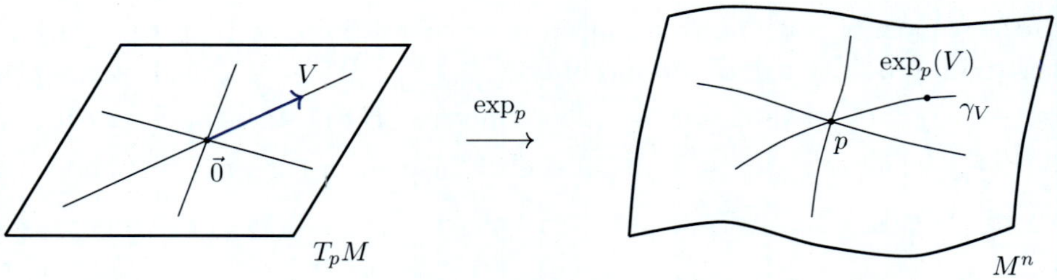

Figure 7.5.1. The exponential map; $\exp_p(V) = \gamma_V(1)$.

7.33 below that Ω_pM is equal to all of T_pM for each $p \in M^n$ if and only if (M^n, d) is a complete metric space.

Example 7.27. *For* $(M^n, g) = (\mathbb{R}^n \setminus \{\mathbf{0}\}, g_{\mathbb{E}})$, $\mathbf{x} \in \mathbb{R}^n \setminus \{\mathbf{0}\}$, *and a unit vector* U *at* $\mathbf{x}$, *we have* $b(U) = \infty$ *if* $U \neq -\frac{\mathbf{x}}{\|\mathbf{x}\|}$ *and* $b(U) = \|\mathbf{x}\|$ *if* $U = -\frac{\mathbf{x}}{\|\mathbf{x}\|}$. *In particular,* γ_U *is defined on* $[0, \infty)$ *if and only if* U *is not pointing toward the origin.*

The **exponential map** $\exp_p : \Omega_pM \to M^n$ is defined by

$$\exp_p(V) := \gamma_V(1), \tag{7.50}$$

where $\gamma_V : [0, \infty) \to M^n$ is the unique constant-speed geodesic emanating from p with $\gamma_V'(0) = V$; see Figure 7.5.1. Observe that Ω_pM is the largest domain on which the map $\exp_p$ given by (7.50) is well-defined. We also note that $\gamma_V(t) = \gamma_{tV}(1)$ for all $V \in T_pM$ and $t \geq 0$ such that $tV \in \Omega_pM$.

By the smooth dependence of solutions to ordinary differential equations on their initial values (see [**Arn06**]), we have the following.

Lemma 7.28. *The exponential map* $\exp_p : \Omega_pM \to M^n$ *is a smooth map.*

Observe that $d(\exp_p)_{\vec{0}} : T_pM \to T_pM$ is the identity map since

$$d(\exp_p)_{\mathbf{0}}(V) = \left.\frac{d}{dt}\right|_{t=0} \exp_p(tV) = \left.\frac{d}{dt}\right|_{t=0} \gamma_{tV}(1) = \left.\frac{d}{dt}\right|_{t=0} \gamma_V(t) = V.$$

Since $d(\exp_p)_{\vec{0}} = \mathrm{id}_{T_pM}$, by the inverse function theorem there exists $\epsilon_p > 0$ such that

$$\exp_p\big|_{B_{\epsilon_p}(\vec{0})} : B_{\epsilon_p}(\vec{0}) \to M^n \tag{7.51}$$

is a diffeomorphism onto an open subset of M^n.

7.5.2. The Gauss lemma. The Gauss lemma, which we state and prove below, says that the derivative of the exponential map preserves distances in the radial direction as well as preserving the hyperplanes perpendicular to the radial direction.

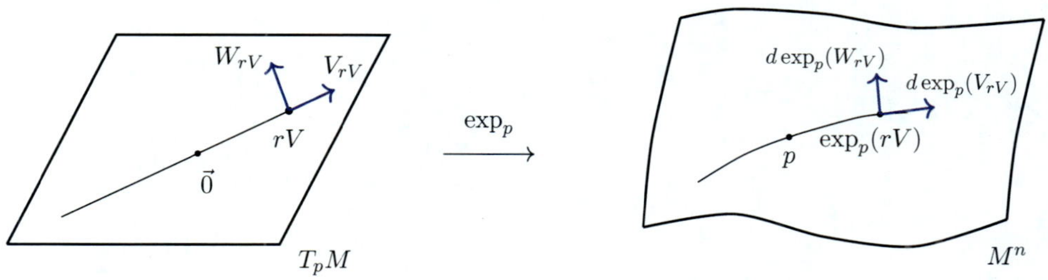

Figure 7.5.2. The statement of the Gauss lemma: $W \perp V$ implies $d\exp_p(W_{rV}) \perp d\exp_p(V_{rV})$.

Regarding notation that we use below, since T_pM is a vector space, for any $Z \in T_pM$ we have that $T_Z(T_pM)$ is naturally isomorphic to T_pM via parallel transport; given any $V \in T_pM$, we denote by V_Z the vector in $T_Z(T_pM)$ which is parallel to V.

Assume that $\Omega_pM = T_pM$, so that $\exp_p$ is defined on all of T_pM. For the following fundamental result, see also [**CE08**, Gauss Lemma 1.8] and [**Gau65**].

Lemma 7.29 (Gauss lemma). *Let $p \in M^n$, $V \in T_pM$, and $r \in [0, \infty)$. If $W \in T_pM \cong T_{rV}(T_pM)$, then*

$$g_{\exp_p(rV)}\big(d\exp_p(W_{rV}), d\exp_p(V_{rV})\big) = g_p(W, V). \tag{7.52}$$

Thus:

(1) $|d\exp_p(V_{rV})| = |V|$.

(2) *If W is perpendicular to V, then the tangent vector $d\exp_p(W_{rV})$ is perpendicular to $d\exp_p(V_{rV}) = \gamma_V'(t)$. See Figure* 7.5.2.

Proof. Given p, V, W, and r in the hypotheses, consider the family of geodesics $\gamma_u : [0,1] \to M^n$ defined by

$$\gamma_u(t) := \exp_p\big(t(rV + uW)\big), \quad 0 \le t \le 1,\ u \in \mathbb{R},$$

emanating from p. Then $\gamma_0(1) = \exp_p(rV) = \gamma_V(r)$,

$$\gamma_0'(1) = r\, d\exp_p(V_{rV}), \quad \text{and} \quad \left.\frac{\partial}{\partial u}\right|_{u=0} \gamma_u(1) = d\exp_p(W_{rV}).$$

We have that the length

$$\mathrm{L}(\gamma_u) = \big|rV + uW\big|_{g_p} = \sqrt{r^2|V|^2 + 2ru\,g(V,W) + u^2|W|^2}.$$

Differentiating this yields

$$\left.\frac{d}{du}\right|_{u=0} \mathrm{L}(\gamma_u) = \frac{1}{|V|}\, g_p(V, W).$$

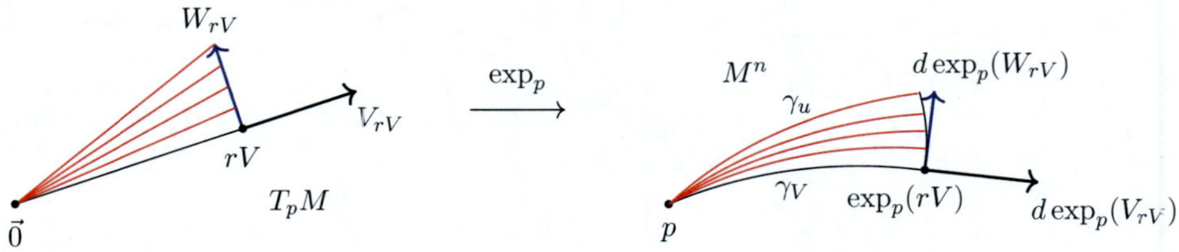

Figure 7.5.3. Proof of the Gauss lemma in the case where V and W are orthogonal: Since the first variation of the length is zero, the endpoint variation is orthogonal to the tangent vector.

On the other hand, by the first variation of arc-length formula (7.48), we have that

$$\begin{aligned}\left.\frac{d}{du}\right|_{u=0} \mathrm{L}(\gamma_u) &= \frac{1}{|\gamma_0'(1)|} g_{\exp_p(rV)}\left(\left.\frac{\partial}{\partial u}\right|_{u=0} \gamma_u(1), \gamma_0'(1)\right) \\ &= \frac{1}{|V|} g_{\exp_p(rV)}\big(d\exp_p(W_{rV}), d\exp_p(V_{rV})\big)\end{aligned}$$

since $\gamma_0'(1) = r\, d\exp_p(V_{rV})$. Hence

$$g_{\exp_p(rV)}\big(d\exp_p(W_{rV}), d\exp_p(V_{rV})\big) = g_p(V, W).$$

The case where V and W are orthogonal is illustrated by Figure 7.5.3. □

7.5.3. Geodesics are locally minimizing.

In addition, the following result says that the exponential map is locally a diffeomorphism near the origin.

Lemma 7.30 (Geodesics are locally minimizing). *For each $p \in M^n$ there exists $\varepsilon_p > 0$ such that:*

(1) *For each $q \in B_{\varepsilon_p}(p)$, there exists a minimal geodesic joining p to q.*
(2) *Every geodesic emanating from p is minimizing up to length ϵ_p.*
(3) *The map $\exp_p : B_{\varepsilon_p}(\vec{0}) \to B_{\varepsilon_p}(p)$ is a diffeomorphism.*

Proof. Let $\epsilon_p > 0$ be as in (7.51). Then we have that

(7.53) $$\left.\exp_p\right|_{B_{\epsilon_p}(\vec{0})} : B_{\epsilon_p}(\vec{0}) \to B := \exp_p(B_{\epsilon_p}(\vec{0}))$$

is a diffeomorphism onto its image. By abuse of notation, let

(7.54) $$\frac{\partial}{\partial r} := d\exp_p\left(\frac{\partial}{\partial r}\right),$$

where $\frac{\partial}{\partial r}$ on the right-hand side is the unit radial vector field on $T_pM \setminus \{\vec{0}\}$ defined by (7.194). For any unit vector $U \in T_pM$ and $0 < t < \epsilon_p$,

$$\frac{\partial}{\partial r}_{\exp_p(tU)} = \gamma_U'(t),$$

so that $\frac{\partial}{\partial r}$ on the left-hand side is a unit vector field on $B \subset M^n$.

Let $q \in B$. We may assume that $q \neq p$. Since $\exp_p|_{B_{\epsilon_p}(\vec{0})} : B_{\epsilon_p}(\vec{0}) \to B$ is a diffeomorphism, there exists a (unique) $V \in T_pM$ with $0 < |V| < \epsilon_p$ such that $\exp_p(V) = q$.

Claim.

$$d(q,p) = |V| < \epsilon_p.$$

Moreover, the path $\gamma : [0, |V|] \to M^n$ *defined by* $\gamma(t) := \exp_p\left(t\frac{V}{|V|}\right)$ *is the unique unit-speed minimal geodesic joining* p *to* q.

Proof of the claim. Let $\beta : [0, L] \to M^n$ be a unit-speed path joining p to q, where $L > 0$. Let $\bar{t} \in (0, L]$ be such that $\beta([0, \bar{t}]) \subset B$ (at the moment we do not yet know that $\beta([0, L]) \subset B$). Then we may write

$$\beta|_{[0,\bar{t}]} =: \exp_p \circ \widetilde{\beta},$$

where $\widetilde{\beta} : [0, \bar{t}] \to B_{\epsilon_p}(\vec{0})$. We compute that

$$\begin{aligned}
\mathrm{L}(\beta|_{[0,\bar{t}]}) = \bar{t} &= \int_0^{\bar{t}} |\beta'(t)|\, dt \\
&\geq \int_0^{\bar{t}} \left\langle \beta'(t), \frac{\partial}{\partial r}_{\beta(t)} \right\rangle dt \\
&= \int_0^{\bar{t}} g_{\beta(t)}\left(d\exp_p(\widetilde{\beta}'(t)), d\exp_p\left(\frac{\partial}{\partial r}\right)\right) dt \\
&= \int_0^{\bar{t}} g_p\left(\widetilde{\beta}'(t), \frac{\partial}{\partial r}\right) dt = \int_0^{\bar{t}} \frac{d}{dt}|\widetilde{\beta}(t)|\, dt \\
&= |\widetilde{\beta}(\bar{t})| = \left|\exp_p^{-1}(\beta(\bar{t}))\right|,
\end{aligned} \tag{7.55}$$

where we used the Gauss lemma. Since the right-hand side depends only on $\beta(\bar{t})$, we have that $\beta|_{[0,\bar{t}]}$ is minimizing if and only if we have equality in (7.55); that is, $\beta'(t) = \frac{\partial}{\partial r}$ for all $t \in [0, \bar{t}]$, which is equivalent to

$$\beta(t) = \gamma_{\beta'(0)}(t) = \exp_p(t\beta'(0)).$$

Inequality (7.55) also implies that if $L < \epsilon_p$, then $\beta([0, L]) \subset B$. So we may now take $\bar{t} = L$ and apply (7.55) again to obtain that

$$\mathrm{L}(\beta) \geq |\exp_p^{-1}(q)| = |V| \in (0, \epsilon_p),$$

with equality if and only if $\beta = \gamma_{V/|V|}\big|_{[0,|V|]}$ and $L = |V|$. This implies that $\gamma(t) = \exp_p(tV/|V|)$, $t \in [0,|V|]$, is the unique unit-speed minimal geodesic joining p to q, and $d(q,p) = |V|$. This proves (1).

Now let $\gamma : [0, L] \to M^n$ be a unit-speed geodesic emanating from p with $L < \epsilon_p$. Since $\gamma(L) \in B$, by the claim we have that γ is a minimal geodesic. This proves (2).

Recall from (7.53) that $\exp_p$ maps $B_{\epsilon_p}(\vec{0})$ diffeomorphically onto its image B. So, to prove (3) we just need to show that $B = B_{\epsilon_p}(p)$. Clearly, $B \subset B_{\epsilon_p}(p)$. Now suppose that $q \notin B$. Let β be any unit-speed path from p to q. Since β must exit B, by (7.55) its length is at least ϵ_p. Thus $q \notin B_{\epsilon_p}(p)$. We have proved that $B_{\epsilon_p}(p) \subset B$, so (3) is proved. □

7.6. The completeness condition

In this subsection we characterize the property of completeness of (M^n, g). This is known as the Hopf–Rinow theorem. As usual, we assume that M^n is connected.

We first prove a pair of lemmas.

Lemma 7.31. *Let $p \in M^n$ and let $\epsilon \in (0, \epsilon_p)$, where ϵ_p is as in (7.51). For every q outside of the ball $B_\epsilon(p)$, there exists a point $q' \in \partial B_\epsilon(p) = S_\epsilon(p)$ such that*

$$d(p,q) = d(p,q') + d(q',q) = \varepsilon + d(q',q). \tag{7.56}$$

Proof. By the definition of distance, for each $k \in \mathbb{N}$, there exists a unit-speed path $\gamma_k : [0, L_k] \to M^n$ from p to q such that $L_k = \mathrm{L}(\gamma_k) \leq d(p,q) + \frac{1}{k}$. Since $q \notin B_\epsilon(p)$, there exists $c_k \in (0, L_k]$ such that $\gamma_k(c_k) \in \partial B_\epsilon(p)$. We may assume that $c_k = \epsilon$ and that $\gamma_k|_{[0,\epsilon]}$ is a minimal geodesic. Since $\epsilon < \epsilon_p$, we have that $\partial B_\epsilon(p)$ is compact. Thus $\gamma_k(\epsilon)$ subconverges to a point $q' \in \partial B_\epsilon(p)$. Since $\beta_k := \gamma_k|_{[\epsilon, L_k]}$ is a path joining $\gamma_k(\epsilon)$ to q of length $L_k - \epsilon \leq d(p,q) + \frac{1}{k} - \epsilon$, we have that $d(q',q) \leq d(p,q) - \epsilon$. We conclude that $d(p,q') + d(q',q) \leq d(p,q)$. By the triangle inequality, we have equality. □

Lemma 7.32. *Let $p \in M^n$. If $\exp_p$ is defined on all of T_pM, then for any $q \in M^n$ there exists a minimal geodesic from p to q.*

Proof. Choose any $\epsilon \in (0, \epsilon_p)$. By Lemma 7.30, we may assume that $q \notin B_\varepsilon(p)$. Then there exists $q' \in \partial B_\varepsilon(p)$ satisfying (7.56), and since $\exp_p$ is defined on all of T_pM we may consider the unit-speed geodesic $\gamma : [0, \infty) \to M^n$ with $\gamma(0) = p$ and $\gamma(\varepsilon) = q'$. Take $t_0 \in [0, d(p,q)]$ to be the largest value of t such that

$$d(p,q) = d(p, \gamma(\bar{t})) + d(\gamma(\bar{t}), q) \tag{7.57}$$

for all $\bar{t} \in [0, t]$. Note that the set of all such t is closed. Since

$$d(p, q) = d(p, \gamma(t)) + d(\gamma(t), q) \quad \text{for } t \in [0, \varepsilon], \tag{7.58}$$

we have $t_0 \geq \varepsilon$.

If $t_0 < d(p, q)$, then there exists $t_1 \in (0, d(\gamma(t_0), q))$ such that $\exp_{\gamma(t_0)}$ is an embedding when restricted to $B_{t_1}(\vec{0}) \subset T_{\gamma(t_0)} M$. This gives us a point $q'' \in \partial B_{t_1}(\gamma(t_0))$ such that

$$d(\gamma(t_0), q) = d(\gamma(t_0), q'') + d(q'', q). \tag{7.59}$$

Now (7.57), (7.59), and the triangle inequality imply that

$$d(p, q'') = d(p, \gamma(t_0)) + d(\gamma(t_0), q'').$$

Let $\sigma : [0, d(\gamma(t_0), q'')] \to M^n$ be the unique unit-speed minimal geodesic from $\gamma(t_0)$ to q''. Therefore, the **concatenated path**

$$\gamma|_{[0,t_0]} \smile \sigma : [0, t_0 + d(\gamma(t_0), q'')] \to M^n, \tag{7.60}$$

obtained by joining σ to $\gamma|_{[0,t_0]}$, from p to q'' is minimal. By the first variation of arc-length formula, $\gamma|_{[0,t_0]} \smile \sigma = \gamma|_{[0,d(p,q'')]}$ is a smooth geodesic. Hence

$$d(p, q) = d(p, \gamma(d(p, q''))) + d(\gamma(d(p, q'')), q),$$

which is a contradiction. Hence $t_0 = d(p, q)$ so that $\gamma|_{[0,d(p,q)]}$ is a minimal geodesic from p to q. $\square$

The **Hopf–Rinow theorem** is the following.

Theorem 7.33 (Hopf–Rinow [**HR31**]). *Let (M^n, g) be a Riemannian manifold. Then the following properties are equivalent:*

(1) *(M^n, d) is a complete metric space.*

(2) *There exists a point $p \in M^n$ such that the exponential map $\exp_p$ is defined on all of T_pM.*

(3) *For all $p \in M^n$, the exponential map $\exp_p$ is defined on all of T_pM.*

Each of these conditions implies the following:

(4) *For any $p, q \in M^n$ there exists a smooth minimal geodesic from p to q.*

Proof. Clearly property (3) implies property (2).

By Lemma 7.32, we have that property (3) implies property (4).

We now show that property (2) implies property (1). Let $\{q_i\}$ be a Cauchy sequence in M^n. By property (2), there exists a sequence of unit-speed minimal geodesics γ_i from p to q_i. Since $\{q_i\}$ is Cauchy, the lengths L_i of γ_i converge to some L. Since the unit $(n-1)$-sphere in T_pM is compact, there exists a subsequence such that $\{\gamma_i'(0)\}$ converges to some unit vector

V at p. Then γ_i subconverges to the geodesic $\gamma_V : [0, L] \to M^n$, where we again used property (2). The continuous dependence of solutions to ordinary differential equations on their initial data implies that q_i converges to $\gamma_V(L)$. This completes the proof that property (2) implies property (1).

Finally we prove that property (1) implies property (3). Given $V \in T_pM$, let $t_0 \in (0, \infty]$ be the largest value of t such that γ is defined on $[0, t)$. Suppose that $t_0 < \infty$. Let $0 \leq t_i \nearrow t_0$. Since (M^n, d) is complete and since the sequence $\{\gamma(t_i)\}$ is Cauchy, it converges to a point, which we define to be $\gamma(t_0)$. Since $\gamma(t)$ is a solution to an ordinary differential equation, the left derivative $\gamma'_-(t_0)$ exists and is a unit vector in $T_{\gamma(t_0)}M$. We may continue $\gamma(t)$ past t_0 by extending it as a geodesic with initial vector $\gamma'_+(t_0) = \gamma'_-(t_0)$. By the theory of ordinary differential equations, $\gamma(t)$ is smooth, including at t_0. This contradicts $t_0 < \infty$. □

Observe that any proper open subset of a connected Riemannian manifold, with the induced Riemannian metric, is incomplete.

7.7. Curvature

In Riemannian geometry, the fundamental tensor from which all notions of curvature are defined is the Riemann curvature tensor. For submanifolds of Euclidean space, we have defined the Riemann curvature tensor and studied some of its properties in §4.6. Many of the proofs of the results in §4.6 are valid without change in the setting of Riemannian manifolds. In such cases, we simply quote these results.

7.7.1. The Riemann curvature tensor.

Let (M^n, g) be a Riemannian manifold.

Definition 7.34. The **Riemann curvature tensor** is defined by (this is the same definition as for submanifolds in Chapter 4; cf. (4.154))

$$\operatorname{Rm}(U, V)W := \nabla_U(\nabla_V W) - \nabla_V(\nabla_U W) - \nabla_{[U,V]}W \tag{7.61}$$

for vector fields U, V, W on M^n, where ∇ is the Levi-Civita connection.

In the following, we quote some results from Chapter 4 regarding the Riemann curvature tensor; the proofs are exactly the same for Riemannian manifolds as for submanifolds. By Lemma 4.36, Rm is multilinear and antisymmetric in its first two components:

$$\operatorname{Rm}(U, V)W = -\operatorname{Rm}(V, U)W. \tag{7.62}$$

By Theorem 4.38, Rm is a $(3, 1)$-tensor; that is, for any vector fields U, V, W on M^n and any function f on M^n, we have

$$f \operatorname{Rm}(U, V)W = \operatorname{Rm}(fU, V)W = \operatorname{Rm}(U, fV)W = \operatorname{Rm}(U, V)fW. \tag{7.63}$$

So by Corollary 4.39, we have that $(\mathrm{Rm}(U,V)W)_x \in T_xM$ for $x \in M^n$ depends only on $U_x, V_x, W_x \in T_xM$.

By Theorem 4.37, Rm satisfies the **first Bianchi identity**:

$$\mathrm{Rm}(U,V)W + \mathrm{Rm}(V,W)U + \mathrm{Rm}(W,U)V = 0. \tag{7.64}$$

We often denote $\mathrm{Rm}(U,V,W) := \mathrm{Rm}(U,V)W$ since we are considering Rm as a tensor of type $(3,1)$.

Proposition 7.35 (Second Bianchi identity). *The covariant derivative of the Riemann curvature tensor satisfies the following symmetry:*

$$(\nabla_U \mathrm{Rm})(V,W,Z) + (\nabla_V \mathrm{Rm})(W,U,Z) + (\nabla_W \mathrm{Rm})(U,V,Z) = 0. \tag{7.65}$$

Proof. We compute that

$$\begin{aligned}(\nabla_U\mathrm{Rm})(V,W)Z &= \nabla_U\big(\nabla_V(\nabla_W Z) - \nabla_W(\nabla_V Z) - \nabla_{[V,W]}Z\big)\\ &\quad - \mathrm{Rm}(\nabla_U V,W)Z - \mathrm{Rm}(V,\nabla_U W)Z - \mathrm{Rm}(V,W)\nabla_U Z\\ &= \nabla_U(\nabla_V(\nabla_W Z)) - \nabla_U(\nabla_W(\nabla_V Z)) - \nabla_U(\nabla_{[V,W]}Z)\\ &\quad - \mathrm{Rm}(\nabla_U V,W)Z - \mathrm{Rm}(V,\nabla_U W)Z - \mathrm{Rm}(V,W)\nabla_U Z.\end{aligned}$$

By symmetry, we also have

$$\begin{aligned}(\nabla_V\mathrm{Rm})(W,U)Z &= \nabla_V(\nabla_W(\nabla_U Z)) - \nabla_V(\nabla_U(\nabla_W Z)) - \nabla_V(\nabla_{[W,U]}Z)\\ &\quad - \mathrm{Rm}(\nabla_V W,U)Z - \mathrm{Rm}(W,\nabla_V U)Z - \mathrm{Rm}(W,U)\nabla_V Z\end{aligned}$$

and

$$\begin{aligned}(\nabla_W\mathrm{Rm})(U,V)Z &= \nabla_W(\nabla_U(\nabla_V Z)) - \nabla_W(\nabla_V(\nabla_U Z)) - \nabla_W(\nabla_{[U,V]}Z)\\ &\quad - \mathrm{Rm}(\nabla_W U,V)Z - \mathrm{Rm}(U,\nabla_W V)Z - \mathrm{Rm}(U,V)\nabla_W Z.\end{aligned}$$

To see that the sum of the three equations above is zero, one uses that

$$\nabla_U(\nabla_V(\nabla_W Z)) - \nabla_V(\nabla_U(\nabla_W Z)) - R(U,V)\nabla_W Z = \nabla_{[U,V]}\nabla_W Z$$

and its two other cyclic permutations, which accounts for 9 of the 18 terms on the right-hand side. We also have that

$$-\mathrm{Rm}(\nabla_U V,W)Z - \mathrm{Rm}(W,\nabla_V U)Z = -\mathrm{Rm}([U,V],W)Z$$

and its cyclic permutations, which accounts for 6 terms. We now leave it to the reader to show that the left-hand side of (7.65) is equal to

$$\nabla_{[[V,W],U]}Z + \nabla_{[[W,U],V]}Z + \nabla_{[[U,V],W]}Z = 0,$$

where the vanishing is by the Jacobi identity. □

We think of the second Bianchi identity as a differential symmetry of the Riemann curvature tensor Rm since it involves the covariant derivatives of Rm. The basic algebraic symmetries of Rm, as a covariant 4-tensor, are given by the following.

Lemma 7.36. *For any tangent vectors U, V, W, Z, we have*

$$\langle \mathrm{Rm}(U,V)W, Z\rangle = -\langle \mathrm{Rm}(V,U)W, Z\rangle, \tag{7.66a}$$

$$\langle \mathrm{Rm}(U,V)W, Z\rangle = -\langle \mathrm{Rm}(U,V)Z, W\rangle, \tag{7.66b}$$

$$\langle \mathrm{Rm}(U,V)W, Z\rangle = \langle \mathrm{Rm}(W,Z)U, V\rangle, \tag{7.66c}$$

$$\langle \mathrm{Rm}(U,V)W, Z\rangle + \langle \mathrm{Rm}(V,W)U, Z\rangle + \langle \mathrm{Rm}(W,U)V, Z\rangle = 0. \tag{7.66d}$$

Proof. The antisymmetry identity (7.66a) follows directly from (7.62).

Identity (7.66d) follows directly from the first Bianchi identity (7.64).

Since Rm is bilinear in the last two components, the antisymmetry identity (7.66b) follows from showing that

$$\langle \mathrm{Rm}(U,V)X, X\rangle = 0 \tag{7.67}$$

for all $U, V, X \in T_xM$; indeed, for then

$$0 = \langle \mathrm{Rm}(U,V)W + Z, W + Z\rangle = \langle \mathrm{Rm}(U,V)W, Z\rangle + \langle \mathrm{Rm}(U,V)Z, W\rangle.$$

To see (7.67), we calculate that

$$\begin{aligned} 0 &= \frac{1}{2}\big(U(V|X|^2) - V(U|X|^2) - [U,V](|X|^2)\big) \\ &= \langle \nabla_U(\nabla_V X), X\rangle + \langle \nabla_V X, \nabla_U X\rangle - \langle \nabla_V(\nabla_U X), X\rangle \\ &\quad - \langle \nabla_U X, \nabla_V X\rangle - \langle \nabla_{[U,V]}X, X\rangle \\ &= \langle \mathrm{Rm}(U,V)X, X\rangle. \end{aligned}$$

The symmetry identity (7.66c) follows from adding up the following four cyclic permutations of the first Bianchi identity:

$$\begin{aligned} 0 &= \mathrm{Rm}(U,V,W,Z) + \mathrm{Rm}(V,W,U,Z) + \mathrm{Rm}(W,U,V,Z), \\ 0 &= \mathrm{Rm}(V,W,Z,U) + \mathrm{Rm}(W,Z,V,U) + \mathrm{Rm}(Z,V,W,U), \\ 0 &= \mathrm{Rm}(W,Z,U,V) + \mathrm{Rm}(Z,U,W,V) + \mathrm{Rm}(U,W,Z,V), \\ 0 &= \mathrm{Rm}(Z,U,V,W) + \mathrm{Rm}(U,V,Z,W) + \mathrm{Rm}(V,Z,U,W). \end{aligned}$$

Indeed, the sum of these equations yields

$$0 = 2\,\mathrm{Rm}(W,U,V,Z) + 2\,\mathrm{Rm}(Z,V,W,U). \qquad \square$$

7.7.2. Sectional curvature. It is natural and convenient to consider scalar functions associated to the Riemann curvature tensor Rm. Let $x \in M^n$. Let P be a 2-dimensional vector subspace of T_xM. The **sectional curvature** of P is defined as

$$\mathrm{sect}(P) := \frac{\mathrm{Rm}(V,W)W \cdot V}{|V|^2|W|^2 - (V \cdot W)^2}, \tag{7.68}$$

where V and W are any tangent vectors that span P. The claim is that the right-hand side is independent of the choice of V and W. We show this now.

Note that the denominator in (7.68) is equal to the area of the parallelogram spanned by U and V.

In particular, if e_1, e_2 is an orthonormal basis for the plane P, then

$$\operatorname{sect}(P) = \operatorname{Rm}(e_2, e_1)e_1 \cdot e_2. \tag{7.69}$$

Observe that if we write $V = V^1 e_1 + V^2 e_2$ and $W = W^1 e_1 + W^2 e_2$, then we compute that

$$\begin{aligned} |V|^2|W|^2 - (V \cdot W)^2 &= \left((V^1)^2 + (V^2)^2\right)\left((W^1)^2 + (W^2)^2\right) \\ &\quad - \left(V^1 W^1 + V^2 W^2\right)^2 \\ &= (V^1)^2(W^2)^2 + (V^2)^2(W^1)^2 - 2V^1W^1V^2W^2 \\ &= (V^1W^2 - V^2W^1)^2. \end{aligned}$$

We also compute that

$$\begin{aligned} &\operatorname{Rm}(V, W)W \cdot V \\ &\quad = \operatorname{Rm}\left(V^1 e_1 + V^2 e_2, W^1 e_1 + W^2 e_2\right)(W^1 e_1 + W^2 e_2) \cdot (V^1 e_1 + V^2 e_2) \\ &\quad = V^1W^2W^1V^2 \operatorname{Rm}(e_1, e_2)e_1 \cdot e_2 + V^1W^2W^2V^1 \operatorname{Rm}(e_1, e_2)e_2 \cdot e_1 \\ &\qquad + V^2W^1W^1V^2 \operatorname{Rm}(e_2, e_1)e_1 \cdot e_2 + V^2W^1W^2V^1 \operatorname{Rm}(e_2, e_1)e_2 \cdot e_1 \\ &\quad = \operatorname{Rm}(e_1, e_2)e_2 \cdot e_1 \begin{pmatrix} -V^1W^2V^2W^1 + (V^1)^2(W^2)^2 \\ +(V^2)^2(W^1)^2 - V^2W^1V^1W^2 \end{pmatrix} \\ &\quad = (V^1W^2 - V^2W^1)^2 \operatorname{Rm}(e_1, e_2)e_2. \end{aligned}$$

We conclude from all of this that $\operatorname{sect}(P)$ is independent of the choice of V and W spanning the plane P.

Let $\operatorname{Gr}(2, T_xM)$ denote the **Grassmannian** of 2-planes in T_xM (see Example 6.16). Then the sectional curvatures at $x \in M^n$ define a function

$$\operatorname{sect} : \operatorname{Gr}(2, T_xM) \to \mathbb{R}.$$

The Riemann curvature tensor at x is determined by the function sect; see equation (1.12) in Cheeger and Ebin [**CE08**].

When $n = 2$, the only 2-plane at $x \in M^2$ is $P = T_xM$. In this case, we call $\operatorname{sect}(P)$ the **Gauss curvature** of g at x, and we denote it by $K(x)$. By Gauss's Theorema Egregium (Theorem 4.43 and in particular (4.188)), this agrees with the Gauss curvature for surfaces in $\mathbb{R}^3$ defined by (4.93). Namely, if M^2 is a surface immersed in $\mathbb{R}^3$ with its induced Riemannian metric, then

$$K = \kappa_1\kappa_2 = \mathrm{II}(e_1, e_1)\,\mathrm{II}(e_2, e_2) = \langle R(e_1, e_2)e_2, e_1\rangle = \operatorname{sect}(P), \tag{7.70}$$

where e_1, e_2 are orthonormal eigenvectors of the Weingarten map L associated to the principal curvatures κ_1, κ_2.

7.7.3. The Riemann curvature operator acting on 2-forms. The Riemann curvature tensor also defines the **Riemann curvature operator**

$$\mathrm{Rm} : \Lambda^2 M \to \Lambda^2 M$$

by

$$\mathrm{Rm}\,(\alpha)_{ij} := \sum_{k,\ell,p,q=1}^{n} g^{kp} g^{\ell q} R_{ijk\ell}\, \alpha_{qp}. \tag{7.71}$$

By the symmetry $R_{ijkl} = R_{klij}$, we see that the Riemann curvature operator is self-adjoint with respect to the inner product on each $\Lambda^2_x M$:

$$\langle \mathrm{Rm}(\alpha), \beta \rangle = \langle \alpha, \mathrm{Rm}(\beta) \rangle .$$

Hence Rm has $\frac{n(n-1)}{2} = \dim(\Lambda^2_x M)$ real eigenvalues, counting multiplicity.

We say that a 2-form is **primitive** if it is the wedge product of two 1-forms. The sectional curvatures can be expressed in terms of the curvature operator acting on primitive 2-forms: If e_1, e_2 are orthonormal tangent vectors and if ω^1, ω^2 are their metrically dual 1-forms, then

$$\mathrm{sect}\,\big(\mathrm{span}\{e_1, e_2\}\big) = \big\langle\, \mathrm{Rm}(\omega^1 \wedge \omega^2), \omega^1 \wedge \omega^2 \big\rangle .$$

7.7.4. The Ricci tensor. Let (M^n, g) be a Riemannian manifold. The Riemann curvature tensor is somewhat complicated, being a 4-tensor. To simplify it, we can trace over two of its components. Namely, we define the **Ricci tensor** Ric to be the symmetric 2-tensor where $\mathrm{Ric}_x : T_x M \times T_x M \to \mathbb{R}$ is defined by

$$\mathrm{Ric}(V, W) := \sum_{i=1}^{n} \mathrm{Rm}(V, e_i, e_i, W), \tag{7.72}$$

where $\{e_i\}_{i=1}^n$ is any orthonormal basis for $T_x M$; see Exercises 7.2–7.4.

Given a symmetric 2-tensor α on M^n, its trace at a point p is an average in the following sense:

$$\mathrm{trace}_g(\alpha) = \frac{1}{\omega_n} \int_{S_p M} \alpha\,(V, V)\, d\mu_{S_p M}(V), \tag{7.73}$$

where $S_p M$ is the unit $(n-1)$-sphere in $(T_p M, g_p)$, $n\omega_n$ is its volume, and $d\mu_{S_p M}$ is its volume form.

Thus, for any unit vector $U \in T_x M$, we have that $\frac{1}{n-1}\,\mathrm{Ric}\,(U, U)$ is equal to the average of the sectional curvatures of planes containing the vector U. Specifically, we can parametrize the space of planes containing U by the $(n-2)$-sphere:

$$S_U := \{V \in T_x M \colon \langle V, U \rangle = 0,\ |V| = 1\}.$$

Given $V \in S_U$, define the 2-plane $\Pi_V := \operatorname{span}(V, U)$. Then

$$\frac{1}{n-1} \operatorname{Ric}(U,U) = \frac{1}{\operatorname{Vol}_{n-2}(S_U)} \int_{S_U} \operatorname{sect}(\Pi_V)\, d\mu_{S_U}(V), \tag{7.74}$$

where $d\sigma$ is the volume form of the Riemannian metric on $d\mu_{S_U}$ induced from its inclusion into $(T_x M, g(x))$.

7.7.5. The scalar curvature. The **scalar curvature** function is defined by

$$R := \sum_{j=1}^{n} \operatorname{Ric}(e_j, e_j) = \sum_{i \neq j} \langle \operatorname{Rm}(e_i, e_j) e_j, e_i \rangle, \tag{7.75}$$

where $\{e_i\}_{i=1}^n$ is an orthonormal frame. This definition is independent of the choice of $\{e_i\}_{i=1}^n$. When $n = 2$, we see that the scalar curvature is equal to twice the Gauss curvature: $R = 2K$. In all dimensions $n \geq 2$, $\frac{1}{n} R(p)$ is equal to the average of $\operatorname{Ric}(U,U)$ over all unit vectors $U \in T_p M$:

$$\frac{1}{n} R(p) = \frac{1}{\operatorname{Vol}_{n-1}(S_p M)} \int_{S_p M} \operatorname{Ric}(U,U)\, d\mu_{S_p M}(U). \tag{7.76}$$

As for Rm and Ric, we have the diffeomorphism invariance of the scalar curvature, which says that if $\varphi : N^n \to (M^n, g)$ is a local diffeomorphism, then

$$R_{\varphi^* g} = R_g \circ \varphi. \tag{7.77}$$

7.7.6. The contracted second Bianchi identity. Now that we have defined the Ricci tensor and the scalar curvature, we may return to the second Bianchi identity and consider its traces. By tracing (7.65) (letting $W = Z = e_i$ and summing, where $\{e_i\}_{i=1}^n$ is an orthonormal frame), we have that

$$(\nabla_U \operatorname{Ric})(V) - (\nabla_V \operatorname{Ric})(U) + \sum_{i=1}^{n} (\nabla_{e_i} \operatorname{Rm})(U, V, e_i) = 0. \tag{7.78}$$

By taking a second trace (now taking $U = e_j$ and dotting with e_j), we obtain

$$\left(\nabla_{e_j} \operatorname{Ric}\right)(V, e_j) - (\nabla_V \operatorname{Ric})(e_j, e_j) + \sum_{i=1}^{n} (\nabla_{e_i} \operatorname{Rm})(e_j, V, e_i, e_j) = 0,$$

which says that

$$\operatorname{div}(2 \operatorname{Ric} - Rg) = 0; \qquad \text{i.e.,} \qquad \operatorname{div}(\operatorname{Ric}) = \frac{1}{2} dR. \tag{7.79}$$

That is, the **Einstein tensor** $E := \operatorname{Ric} - \frac{1}{2} Rg$ is divergence-free. Equation (7.79) is known as the **contracted second Bianchi identity**.

7.7.7. The Riemann curvature tensor in low dimensions. In low dimensions, the Riemann curvature tensor takes special forms. In particular, we have the following result.

Proposition 7.37 (Riemann curvature tensor in dimensions 2 and 3).

(1) *If* $n = 2$, *then*

$$\mathrm{Rm}(U, V, W, Z) = \frac{1}{2} R\big(g(U, Z)g(V, W) - g(U, W)g(V, Z)\big). \tag{7.80}$$

(2) *If* $n = 3$, *then*

$$\begin{aligned}\mathrm{Rm}(U, V, W, Z) = {}& \mathrm{Ric}(U, Z)g(V, W) + g(U, Z)\,\mathrm{Ric}(V, W) \\ & - \mathrm{Ric}(U, W)g(V, Z) - g(U, W)\,\mathrm{Ric}(V, Z) \\ & - \frac{1}{2} R\big(g(U, Z)g(V, W) - g(U, W)g(V, Z)\big).\end{aligned} \tag{7.81}$$

Proof.

(1) Suppose $n = 2$. Let $\{e_1, e_2\}$ be an orthonormal frame at a point and denote $R_{ijkl} := \mathrm{Rm}(e_i, e_j, e_k, e_l)$. By the antisymmetries of the Riemann curvature tensor, the only nonzero components are

$$R_{1221} = R_{2112} = -R_{1212} = -R_{2121}.$$

Thus, the only nonzero components of the Ricci tensor are

$$R_{11} = R_{22} = R_{1221},$$

so that the scalar curvature is given by

$$R = 2R_{1221}.$$

Thus, with $U = U^1 e_1 + U^2 e_2$, etc., we have

$$\begin{aligned}&\mathrm{Rm}(U, V, W, Z) \\ &\quad = R_{1221}(U^1 V^2 W^2 Z^1 + U^2 V^1 W^1 Z^2 - U^1 V^2 W^1 Z^2 - U^2 V^1 W^2 Z^1).\end{aligned}$$

Formula (7.80) now follows from $R_{1221} = \frac{1}{2} R$ and that

$$\begin{aligned}&U^1 V^2 W^2 Z^1 + U^2 V^1 W^1 Z^2 - U^1 V^2 W^1 Z^2 - U^2 V^1 W^2 Z^1 \\ &\quad = g(U, Z)g(V, W) - g(U, W)g(V, Z).\end{aligned}$$

(2) Formula (7.81) follows from Lemma 7.38 below. □

7.8. Curvature decomposition

In this section we will see how the Riemann curvature tensor may be decomposed into simpler components.

7.8.1. The Weyl tensor. The **Weyl tensor** Wy, which is a covariant 4-tensor, is defined by the decomposition

$$\begin{aligned}\operatorname{Rm}(U,V,W,Z) =: & \frac{1}{n-2}\big(\operatorname{Ric}(U,Z)g(V,W)+g(U,Z)\operatorname{Ric}(V,W)\big)\\ & -\frac{1}{n-2}\big(\operatorname{Ric}(U,W)g(V,Z)+g(U,W)\operatorname{Ric}(V,Z)\big)\\ & -\frac{R}{(n-1)(n-2)}\big(g(U,Z)g(V,W)-g(U,W)g(V,Z)\big)\\ & +\operatorname{Wy}(U,V,W,Z). \end{aligned} \tag{7.82}$$

In terms of the trace-free Ricci tensor

$$\overset{\circ}{\operatorname{Ric}} := \operatorname{Ric} - \frac{1}{n}Rg \tag{7.83}$$

in place of Ric, the curvature decomposition above may be expressed as

$$\begin{aligned}\operatorname{Rm}(U,V,W,Z) = & \frac{1}{n-2}\Big(\overset{\circ}{\operatorname{Ric}}(U,Z)g(V,W)+g(U,Z)\overset{\circ}{\operatorname{Ric}}(V,W)\Big)\\ & -\frac{1}{n-2}\Big(\overset{\circ}{\operatorname{Ric}}(U,W)g(V,Z)+g(U,W)\overset{\circ}{\operatorname{Ric}}(V,Z)\Big)\\ & +\frac{R}{n(n-1)}\big(g(U,Z)g(V,W)-g(U,W)g(V,Z)\big)\\ & +\operatorname{Wy}(U,V,W,Z). \end{aligned} \tag{7.84}$$

Like the Riemann curvature tensor, the Weyl tensor has the following symmetries:

$$\operatorname{Wy}(U,V,W,Z)=\operatorname{Wy}(W,Z,U,V)=-\operatorname{Wy}(V,U,W,Z)=-\operatorname{Wy}(U,V,Z,W), \tag{7.85a}$$

$$0=\operatorname{Wy}(W,Z,U,V)+\operatorname{Wy}(Z,U,W,V)+\operatorname{Wy}(U,W,Z,V). \tag{7.85b}$$

Furthermore,

$$0=\big(\operatorname{trace}_{1,4}(\operatorname{Wy})\big)(V,W)=\sum_{i=1}^{n}\operatorname{Wy}(e_i,V,W,e_i). \tag{7.86}$$

This implies that all of the traces of the Weyl tensor are zero; that is, $\operatorname{trace}_{a,b}(\operatorname{Wy})=0$ for $1\leq a\neq b\leq 4$.

Let Wy_g denote the Weyl tensor of g. A standard result is that the Weyl tensor is conformally invariant; namely, for any function u on M^n, we have that

$$\operatorname{Wy}_{e^{2u}g}=e^{2u}\operatorname{Wy}_g\,. \tag{7.87}$$

The extra symmetries of Wy cause it to vanish in low dimensions; in particular, we have the following.

Lemma 7.38. *If $n = 3$, then the Weyl tensor vanishes:*

$$\mathrm{Wy} = 0.$$

Proof. It suffices to prove that all of the components

$$W_{ijkl} := \mathrm{Wy}\,(e_i, e_j, e_k, e_l)$$

vanish. By the antisymmetries of the Weyl tensor, we only need to check the cases where $i \neq j$ and $k \neq l$.

Consider a general Weyl tensor component W_{ijkl}. Since $1 \leq i, j, k, l \leq 3$, $i \neq j$, and $k \neq l$, the only possible nonvanishing components of Wy are of the forms

$$W_{ijji} = W_{jiij} = -W_{ijij} = -W_{jiji},$$
$$W_{ijki} = W_{jiik} = -W_{ijik} = -W_{jiki},$$

where $\{i, j, k\} = \{1, 2, 3\}$; i.e., i, j, k are distinct. Since Wy is totally trace-free, we have that

$$0 = W_{ijki} + W_{jjkj} + W_{kjkk} = W_{ijki}$$

for i, j, k distinct. We also see from (7.85b) that

$$W_{ijji} = -W_{kjjk} = W_{kiik} = -W_{jiij} = -W_{ijji},$$

which proves that $W_{ijji} = 0$. We conclude that the Weyl tensor vanishes. ☐

7.8.2. Irreducible components of curvature tensors. One can understand the Weyl tensor and the trace-free Ricci tensor in terms of decomposing the Riemann curvature tensor into its irreducible components, where we consider the space of curvature tensors as a representation space. Let (M^n, g) be a Riemannian manifold. Given a vector bundle E over M^n, let $S^2(E)$ denote the symmetric tensor product bundle of E (a.k.a. second symmetric power of E), which is the vector subbundle of $E \otimes E$ consisting of symmetric tensors.

By the symmetries (see (7.66)), we see that the Riemann curvature $(4,0)$-tensor is a section of the vector bundle $S^2(\Lambda^2 M)$. Moreover, by the first Bianchi identity (7.64), Rm is a section of the subbundle $\ker(b)$, the kernel of the vector bundle homomorphism (called the **Bianchi map**):

$$b : S^2(\Lambda^2 M) \to \Lambda^3 M \otimes T^*M$$

defined by

$$b\,(\Omega)\,(X, Y, Z, W) := \frac{1}{3}\,(\Omega\,(X, Y, Z, W) + \Omega\,(Y, Z, X, W) + \Omega\,(Z, X, Y, W))$$

for $\Omega \in S^2(\Lambda^2 M)$. We will call

$$\mathrm{C}\, M := \ker(b) \subset S^2(\Lambda^2 M) \tag{7.88}$$

the **bundle of curvature tensors**.

Let $x \in M^n$ and let g_x^* denote the dual metric on the cotangent space $T_x^* M$ (see §8.3.2 below). The orthogonal group $\mathrm{O}\,(T_x^* M, g_x^*)$ acts on the fiber $\mathrm{C}_x\, M$, where

$$\times : \mathrm{O}\,(T_x^* M, g_x^*) \times \mathrm{C}_x\, M \to \mathrm{C}_x\, M$$

is defined to be the 'diagonal action', which is uniquely characterized by the properties that for $A \in \mathrm{O}\,(T_x^* M, g_x^*)$, the map $\Omega \mapsto A \times \Omega$ (from $\mathrm{C}_x\, M$ to itself) is linear and

$$A \times (\alpha \otimes \beta \otimes \gamma \otimes \delta) := A\alpha \otimes A\beta \otimes A\gamma \otimes A\delta$$

for all $\alpha, \beta, \gamma, \delta \in T_x^* M$. It is easy to check that this action defines a **representation** of the group $\mathrm{O}\,(T_x^* M, g_x^*)$ on the vector space $\mathrm{C}_x\, M$; that is, the action defines a group homomorphism $\rho : \mathrm{O}\,(T_x^* M, g_x^*) \to \mathrm{GL}(\mathrm{C}_x\, M)$.

Recall that, given a representation $\rho : G \to \mathrm{GL}(V)$ of a group on a vector space V, if W is a vector subspace that is invariant under the action of G, then we call $G \to \mathrm{GL}(W)$ a **subrepresentation** and we call W a **subrepresentation space**. We say that the representation ρ is **irreducible** if any subrepresentation space is either $\{0\}$ or V.

As an $\mathrm{O}\,(T_x^* M, g_x^*)$ representation space, $\mathrm{C}_x\, M$ has a natural decomposition into its irreducible components. This yields a corresponding decomposition of any element of $\mathrm{C}_x\, M$, including the Riemann curvature tensor. Define the **Kulkarni–Nomizu product** (cf. [**Bes08**, Definition 1.110])

$$\odot : S^2(T^* M) \times S^2(T^* M) \to \mathrm{C}\, M$$

by

$$(\alpha \odot \beta)_{ijk\ell} := \alpha_{i\ell}\beta_{jk} + \alpha_{jk}\beta_{i\ell} - \alpha_{ik}\beta_{j\ell} - \alpha_{j\ell}\beta_{ik}. \tag{7.89}$$

We claim that the irreducible decomposition of $\mathrm{C}_x\, M$ as an $\mathrm{O}\,(T_x^* M, g_x^*)$-module is given by the fibers of the vector bundle decomposition:

$$\mathrm{C}\, M = (\mathbb{R} g \odot g) \oplus \left(S_0^2 M \odot g\right) \oplus \mathrm{W}\, M. \tag{7.90}$$

Here,

$$\mathbb{R} g \odot g := \{a g \odot g \colon a \in \mathbb{R}\}, \tag{7.91}$$

$S_0^2 M$ is the bundle of symmetric trace-free 2-tensors,

$$S_0^2 M \odot g := \{\omega \odot g \colon \omega \in S_0^2 M\}, \tag{7.92}$$

and

$$\mathrm{W} M := \ker(b) \cap \ker(c) \tag{7.93}$$

is called the **bundle of Weyl curvature tensors**, where

$$c : S^2(\Lambda^2 M) \to S^2 M$$

is the contraction map defined by

$$c(\Omega)(X,Y) := \sum_{i=1}^{n} \Omega(e_i, X, Y, e_i)$$

for $\Omega \in S^2(\Lambda^2 M)$, where $\{e_i\}_{i=1}^n$ is an orthonormal frame. By the symmetries of curvature tensors, we see that if $\Omega \in \mathrm{W}M$, then all of the traces of Ω vanish:

$$0 = g^{ij}\Omega_{ijkl} = g^{ij}\Omega_{ikjl} = g^{ij}\Omega_{iklj} = g^{ij}\Omega_{kijl} = g^{ij}\Omega_{kilj} = g^{ij}\Omega_{klij}.$$

Firstly, observe that (7.90) is an orthogonal decomposition with respect to the natural fiber metric on $\mathrm{C}M$. Indeed, that $(\mathbb{R}g \odot g)_x \oplus \left(S_0^2 M \odot g\right)_x$ and $(\mathrm{W}M)_x$ are orthogonal follows from Weyl tensors being trace-free. The orthogonality of $(\mathbb{R}g \odot g)_x$ and $(S_0^2 M \odot g)_x$ follows from the tensors in $(S_0^2 M)_x$ being trace-free. We have shown that the three subspaces are mutually orthogonal.

Secondly, we show that for each $x \in M^n$ the fibers of the three subbundles in (7.90) span $(\mathrm{C}M)_x$. Let $\Omega \in \mathrm{C}M$ and define

$$R^\Omega := g^{il} g^{jk} \Omega_{ijkl}, \quad \overset{\circ}{\mathrm{Ric}}{}^{\Omega}_{jk} := g^{il}\Omega_{ijkl} - \frac{1}{n} R^\Omega g_{jk},$$

and

$$\mathrm{Wy}^\Omega := \Omega - \frac{R^\Omega}{2n(n-1)} g \odot g - \frac{1}{n-2} \overset{\circ}{\mathrm{Ric}}{}^{\Omega} \odot g.$$

It suffices to show that $\mathrm{Wy}^\Omega \in \mathrm{W}M$. This is an easy calculation and we leave it as an exercise.

It is also easy to see that $(\mathbb{R}g \odot g)_x$ and $(S_0^2 M \odot g)_x$ are each invariant under the action of $\mathrm{O}(T_x^* M, g_x^*)$ and hence are each subrepresentation spaces. For example, if $A \in \mathrm{O}(T_x^* M, g_x^*)$, then

$$A \times g = A \times \left(\sum_i \omega^i \otimes \omega^i\right) = A_j^i A_k^i \omega^j \otimes \omega^k = \delta_{jk} \omega^i \otimes \omega^k = g,$$

where the matrix $A = (A_j^i)$ defined by $A_j^i \omega^j := A(\omega^i)$ satisfies $A^{\mathrm{T}} = A^{-1}$. Since the trace of a square matrix is invariant under conjugation, the action of $\mathrm{O}(T_x^* M, g_x^*)$ on $(S^2 M)_x$ preserves the trace.

Since the $\mathrm{O}(T_x^* M, g_x^*)$ action preserves the inner product on $(\mathrm{C}M)_x$, by the orthogonality of the decomposition in (7.90), we see that $(\mathrm{W}M)_x$ is also a subrepresentation space.

Since the subrepresentation space $(\mathbb{R}g \odot g)_x$ is 1-dimensional, it is irreducible. That the invariant subspace $(S_0^2 M \odot g)_x$ is irreducible follows from $(S_0^2 M)_x$ being irreducible. This, in turn, is because by taking, say, the element $\omega^1 \otimes \omega^1 - \omega^2 \otimes \omega^2$, we obtain elements of the form $\omega^i \otimes \omega^i - \omega^j \otimes \omega^j$ by the action, which span the trace-free 2-tensors that are diagonal with respect to $\{e_i\}_{i=1}^n$.

Finally, one can also show that the subspace of Weyl curvature tensors, which are totally trace-free, is irreducible.

The irreducible decomposition of the fibers of $\mathrm{C}\, M$ yields a corresponding irreducible decomposition of the Riemann curvature tensor:

$$\mathrm{Rm} = f g \odot g + (h \odot g) + \mathrm{Wy}, \tag{7.94}$$

where $f \in C^\infty(M^n)$, $h \in C^\infty\left(S_0^2 M\right)$, and $\mathrm{Wy} \in C^\infty(\mathrm{W}\, M)$. We will see below that Wy is the same as the Weyl tensor defined by (7.82) earlier. Since Wy is trace-free, taking the contraction c of this formula yields

$$\mathrm{Ric} = 2(n-1) f g + (n-2) h.$$

By taking a second contraction, while using that h is trace-free, we obtain

$$R = 2n(n-1) f.$$

Therefore, for $n \geq 3$, the **irreducible decomposition of** Rm is given by

$$\mathrm{Rm} = \frac{R}{2n(n-1)} g \odot g + \frac{1}{n-2} \overset{\circ}{\mathrm{Ric}} \odot g + \mathrm{Wy}, \tag{7.95}$$

where $\overset{\circ}{\mathrm{Ric}} := \mathrm{Ric} - \frac{R}{n} g$ is the traceless Ricci tensor and where $\mathrm{Wy} := \mathrm{Wy}_g$ is the Weyl tensor, which we take to be defined by formula (7.95) and is equivalent to the definition given by (7.82).

In local coordinates, we have by (7.89) that

$$(g \odot g)_{ijk\ell} = 2\left(g_{i\ell} g_{jk} - g_{ik} g_{j\ell}\right),$$

similarly for $\overset{\circ}{\mathrm{Ric}} \odot g$, and we thus obtain

$$\begin{aligned} R_{ijk\ell} = {} & \frac{R}{n(n-1)}\left(g_{i\ell} g_{jk} - g_{ik} g_{j\ell}\right) \\ & + \frac{1}{n-2}\left(\overset{\circ}{R}_{i\ell} g_{jk} + g_{i\ell} \overset{\circ}{R}_{jk} - \overset{\circ}{R}_{ik} g_{j\ell} - g_{ik} \overset{\circ}{R}_{j\ell}\right) + W_{ijk\ell}, \end{aligned} \tag{7.96}$$

where $W_{ijk\ell}$ are the components of Wy.

7.8.3. Concepts from conformal geometry.

The **Weyl–Schouten tensor** (or **Schouten tensor**) is the symmetric 2-tensor defined by

$$S := \mathrm{Ric} - \frac{1}{2(n-1)} R g. \tag{7.97}$$

By (7.95), we may express the Riemann curvature tensor as

$$\text{Rm} = \frac{1}{n-2} S \odot g + \text{Wy}, \tag{7.98}$$

where $\odot$ is the Kulkarni–Nomizu product defined by (7.89).

The **Cotton tensor** is defined by

$$\begin{aligned} C_{ijk} &:= \nabla_i S_{jk} - \nabla_j S_{ik} \\ &= \nabla_i R_{jk} - \nabla_j R_{ik} - \frac{1}{2(n-1)} \left(\nabla_i R g_{jk} - \nabla_j R g_{ik} \right). \end{aligned} \tag{7.99}$$

The Cotton tensor C is a 3-tensor that is antisymmetric in its first two components. We can think of C as the exterior covariant derivative of the Schouten tensor S, where S is considered as a 1-form with values in the tangent bundle TM.[1]

For any $n \geq 3$, one may compute that the divergence of the Weyl tensor satisfies

$$\sum_{\ell=1}^{n} \nabla^{\ell} W_{ijk\ell} = \frac{n-3}{n-2} C_{ijk}. \tag{7.100}$$

In particular, if $n \geq 4$, then the Weyl tensor vanishes for g if and only if the Cotton tensor vanishes for g.

On the other hand, recall in dimension 3 that the Weyl tensor always vanishes. In dimension 3, the Cotton tensor has the following conformal invariance property, which we leave to the reader to verify.

Lemma 7.39. *If $n = 3$ and $\widetilde{g} = \mathrm{e}^{2u} g$ for some function u, then*

$$\widetilde{C}_{ijk} = \mathrm{e}^{3u} C_{ijk},$$

where $\widetilde{C}$ denotes the Cotton tensor of $\widetilde{g}$.

Definition 7.40. A Riemannian manifold (M^n, g) is said to be **locally conformally flat** if for every point $p \in M^n$, there exist a neighborhood U of p and a function $u : U \to \mathbb{R}$ such that $\mathrm{e}^{2u} g$ is a flat metric on U.

We remark that in this case, for every $p \in M^n$ there exists a local coordinate system $\{x^i\}_{i=1}^n$ defined in a neighborhood V of p such that

$$g_{ij} = v\, \delta_{ij}$$

for some function $v : V \to \mathbb{R}$.

Every Riemannian surface (M^2, g) is locally conformally flat. Indeed, by (8.98) below, if $\widetilde{g} = \mathrm{e}^{2u} g$, then their Gauss curvatures are related by

$$\widetilde{K} = \mathrm{e}^{-2u}(K - \Delta u).$$

[1]Here we use the isomorphism between $T^*M \otimes T^*M$ and $T^*M \otimes TM$ induced by the metric g.

By the local solvability of the **Poisson equation** $\Delta u = K$, for any point $p \in M^2$, there exists a function u defined in some neighborhood U of p such that $\widetilde{K} = 0$ on U. Thus, g is locally conformally flat.

In higher dimensions, we have the following characterization of locally conformally flat.

Theorem 7.41 (Weyl, Schouten). *Let (M^n, g) be a Riemannian manifold. Then g is locally conformally flat if and only if*

(1) *for $n \geq 4$ the Weyl tensor vanishes,*

(2) *for $n = 3$ the Cotton tensor vanishes.*

To prove this theorem, we will need the following formula for the Riemann curvature tensor under a conformal change of the metric.

Lemma 7.42. *If $\widetilde{g} = \mathrm{e}^{2u} g$ for some function u, then the Riemann curvature tensors of g and $\widetilde{g}$ are related by*

$$\widetilde{R}_{ijk\ell} = R_{ijk\ell} - a_{i\ell} g_{jk} - a_{jk} g_{i\ell} + a_{ik} g_{j\ell} + a_{j\ell} g_{ik}, \tag{7.101}$$

where the 2-tensor a is defined by

$$a_{ij} := \nabla_i \nabla_j u - \nabla_i u \nabla_j u + \frac{1}{2} |\nabla u|^2 g_{ij}. \tag{7.102}$$

That is,

$$\mathrm{e}^{-2u}\, \widetilde{\mathrm{Rm}} = \mathrm{Rm} - a \odot g. \tag{7.103}$$

In particular, we infer from this that

$$\widetilde{\mathrm{Wy}} = \mathrm{e}^{2u}\, \mathrm{Wy},$$

which is (7.87).

Proof of Theorem 7.41. By the conformal invariance of the Weyl tensor (7.87), we have that if (M^n, g) is locally conformally flat, then the Weyl tensor vanishes. When $n = 3$, the fact that local conformal flatness implies that the Cotton tensor vanishes follows from Lemma 7.39.

Conversely, suppose that the Weyl tensor of g vanishes. We want to show that g is locally conformally flat. Consider the equation that the metric $\tilde{g} = \mathrm{e}^{2u} g$ is flat. Then, by (7.103) and (7.98), this equation is equivalent to

$$\begin{aligned} 0 &= \mathrm{e}^{-2u}\, \widetilde{\mathrm{Rm}} \\ &= \mathrm{Rm} - a \odot g \\ &= \left(\frac{1}{n-2} \left(\mathrm{Ric} - \frac{1}{2(n-1)} R g \right) - a \right) \odot g, \end{aligned} \tag{7.104}$$

where a is defined by (7.102). Since the map $\odot : S^2M \to \mathrm{C}\,M$ to the bundle of curvature tensors defined by $\odot(h) := h \odot g$ is injective, equation (7.104) is equivalent to the following equation for u:

$$\nabla^2 u - \nabla u \otimes \nabla u + \frac{1}{2}|\nabla u|^2 g = a = \frac{1}{n-2}S, \tag{7.105}$$

where S is the Schouten tensor.

Theorem 7.41 is now a consequence of (1) the following lemma, which gives a necessary and sufficient condition for when (7.105) is locally solvable for all $n \geq 3$, (2) the fact by (7.100) that $\mathrm{Wy} = 0$ implies $C = 0$ for $n \geq 4$, and (3) $\mathrm{Wy} = 0$ for $n = 3$. □

Lemma 7.43. *Let $n \geq 3$ and assume that the Weyl tensor of g vanishes. Then equation* (7.105) *is locally solvable if and only if the Schouten tensor S satisfies the following integrability condition:*

$$\nabla_i S_{jk} = \nabla_j S_{ik}, \tag{7.106}$$

that is, if and only if the Cotton tensor C_{ijk} vanishes.

Proof. Equation (7.105) is locally solvable if and only if locally there exists a 1-form X such that

$$\nabla_i X_j = \sigma_{ij} := \frac{1}{n-2}S_{ij} + X_i X_j - \frac{1}{2}|X|^2 g_{ij}. \tag{7.107}$$

This is because if X is a solution to (7.107), then by the symmetry of the right-hand side of (7.107), we have that $\nabla_i X_j = \nabla_j X_i$, so that locally there exists a function u such that $X = du$. We rewrite (7.107) as

$$\frac{\partial}{\partial x^i} X_j = \widetilde{\sigma}_{ij}, \tag{7.108}$$

where

$$\widetilde{\sigma}_{ij} := \frac{1}{n-2}S_{ij} + X_i X_j - \frac{1}{2}|X|^2 g_{ij} + \sum_{k=1}^{n} \Gamma_{ij}^k X_k. \tag{7.109}$$

Let $p \in M^n$ be any point and assume that the local coordinates $\{x^i\}_{i=1}^n$ used to express the equations above are defined in a neighborhood of p. Then the Frobenius theorem (Theorem 4.49) says that a necessary and sufficient condition to locally solve (7.108) with $X(p) = X_0$ for any $X_0 \in T_pM$ is the integrability condition:

$$\frac{\partial}{\partial x^k}\widetilde{\sigma}_{ij} = \frac{\partial}{\partial x^j}\widetilde{\sigma}_{ik}.$$

We may rewrite this integrability condition as

(7.110)

$$\begin{aligned}\nabla_k\sigma_{ij} - \nabla_j\sigma_{ik} &= -\sum_{\ell=1}^{n} R_{kij\ell}X^\ell \\ &= -\frac{1}{n-2}\sum_{\ell=1}^{n}\left(S_{k\ell}\,g_{ij} + S_{ij}\,g_{k\ell} - S_{kj}\,g_{i\ell} - S_{i\ell}\,g_{kj}\right)X^\ell,\end{aligned}$$

where for the last equality we used $W_{kij\ell} = 0$. By the definition (7.107) of σ_{ij}, (7.110) is equivalent to the equation

$$\begin{aligned}&\frac{1}{n-2}(\nabla_k S_{ij} - \nabla_j S_{ik}) \\ &\quad = -X_i\nabla_k X_j - X_j\nabla_k X_i + \sum_{\ell=1}^{n} X^\ell\nabla_k X_\ell\, g_{ij} \\ &\qquad + X_i\nabla_j X_k + X_k\nabla_j X_i - \sum_{\ell=1}^{n} X^\ell\nabla_j X_\ell\, g_{ik} - \sum_{\ell=1}^{n} R_{kij\ell}X^\ell \\ &\quad = \sum_{\ell=1}^{n} X^\ell\begin{pmatrix}\left(\nabla_k X_\ell - \frac{1}{n-2}S_{k\ell}\right)g_{ij} + \left(\nabla_j X_i - \frac{1}{n-2}S_{ji}\right)g_{k\ell} \\ -\left(\nabla_j X_\ell - \frac{1}{n-2}S_{j\ell}\right)g_{ik} - \left(\nabla_k X_i - \frac{1}{n-2}S_{ki}\right)g_{j\ell}\end{pmatrix} \\ &\quad = \sum_{\ell=1}^{n} X^\ell\begin{pmatrix}\left(X_k X_\ell - \frac{1}{2}|X|^2 g_{k\ell}\right)g_{ij} + \left(X_i X_j - \frac{1}{2}|X|^2 g_{ij}\right)g_{k\ell} \\ -\left(X_j X_\ell - \frac{1}{2}|X|^2 g_{j\ell}\right)g_{ik} - \left(X_k X_i - \frac{1}{2}|X|^2 g_{ki}\right)g_{j\ell}\end{pmatrix} \\ &\quad = 0,\end{aligned}$$

where we also used (7.110) and $\nabla_j X_k = \nabla_k X_j$. This completes the proof of both the lemma and Theorem 7.41. □

Since, for any constant sectional curvature manifold the Weyl tensor is equal to zero, the following is an immediate consequence of Theorem 7.41.

Corollary 7.44. *If (M^n, g) is a Riemannian manifold with constant sectional curvature, then (M^n, g) is locally conformally flat.*

We leave it to the reader to verify the following.

Corollary 7.45.

(1) *Let (N^m, g_N) and (P^k, g_P) be Riemannian manifolds satisfying*

$$\operatorname{sect}(g_N) \equiv C \quad \text{and} \quad \operatorname{sect}(g_P) \equiv -C,$$

where $C \in \mathbb{R}$. Then the Riemannian product $(N^m \times P^k, g_N + g_P)$ is locally conformally flat.

(2) *If* (N^m, g_N) *satisfies* $\operatorname{sect}(g_N) \equiv C$, *then the Riemannian product* $(N^m \times \mathbb{R}, g_N + dt^2)$ *is locally conformally flat.*

Finally, we remark that the **Bach tensor** is defined by

$$B_{ij} := \sum_{k,\ell=1}^{n} S_{k\ell} W_{ik\ell j} + \Delta S_{ij} - \sum_{k=1}^{n} \nabla^k \nabla_i S_{jk}. \tag{7.111}$$

The Bach tensor is trace-free and when $n = 4$ it is conformally invariant; see Bach [**Bac21**].

7.9. Constant sectional curvature manifolds

The condition of having constant sectional curvature is one of the strongest and most natural constraints one can place on a Riemannian manifold. Such manifolds exhibit a high degree of symmetry and have been central in the study of differential geometry. They serve as fundamental examples, benchmarks, and sources of geometric intuition. In this section, we discuss some of the properties of these special manifolds.

We have the following consequence of a result of Cartan, Ambrose, and Hicks (see [**CE08**, Theorems 1.42 and 1.43] for the proof).

Theorem 7.46. *Let* (M_1^n, g_1) *and* (M_2^n, g_2) *be simply connected complete Riemannian manifolds with constant sectional curvature* $K \in \mathbb{R}$. *Then* (M_1^n, g_1) *and* (M_2^n, g_2) *are isometric.*

Moreover, for any points $x_1 \in M_1^n$ *and* $x_2 \in M_2^n$ *and linear isometry* $L : (T_{x_1} M_1, g_1(x)) \to (T_{x_2} M_2, g_2(x))$, *there exists a unique isometry* $I : (M_1^n, g_1) \to (M_2^n, g_2)$ *such that* $I(x_1) = x_2$ *and* $dI_{x_1} = L$.

In this section we discuss models for simply connected complete Riemannian manifolds with constant sectional curvatures. The problem of the classification of nonsimply connected constant sectional curvature manifolds can be formulated as a problem regarding fundamental groups, and as such, there are group-theoretic approaches. See Wolf [**Wol11**].

7.9.1. Euclidean space. The standard model for the complete simply connected space form $\mathbb{E}^n$ of curvature $K = 0$ is Euclidean space $\mathbb{R}^n$ with the metric

$$g_0 = g_{\mathbb{E}} := \sum_{i=1}^{n} dx^i \otimes dx^i. \tag{7.112}$$

For Euclidean coordinates $\{x^i\}_{i=1}^n$, we have that the coordinate tangent vectors are $\frac{\partial}{\partial x^i} = e_i$, where $\{e_i\}$ is the standard Euclidean basis. The components of the Euclidean metric are $(g_0)_{ij} = g_{\mathbb{E}}(e_j, e_i) = \delta_{ij}$.

Recall from (4.51) that the **Euclidean covariant derivative** D on $\mathbb{E}^n$ is given by

$$D_{V_p}W := (V_p(W^1), \dots, V_p(W^n))$$

for a vector V_p at p and a vector field W defined in a neighborhood of p, where the W^i are the components of W. Euclidean space is flat: its Riemann curvature tensor defined by

$$R^{\mathbb{E}}(X,Y)Z = D_X(D_Y Z) - D_Y(D_X Z) - D_{[X,Y]}Z = 0 \tag{7.113}$$

for vector fields X, Y, Z is zero.

Any affine subspace of $\mathbb{E}^n$ is totally geodesic. A path $\gamma : \mathbb{R} \to \mathbb{E}^n$ is a unit-speed geodesic if and only if there exist $v \in \mathbb{R}^n$ and a unit vector u such that $\gamma(t) = v + tu$.

7.9.2. Isometries of Euclidean space. Let $\mathrm{O}(n)$ denote the Lie group of orthogonal transformations of $\mathbb{E}^n$; that is, a linear transformation $A : \mathbb{R}^n \to \mathbb{R}^n$ satisfies $A \in \mathrm{O}(n)$ if and only if $g_{\mathbb{E}}(Av, Aw) = g_{\mathbb{E}}(v, w)$ for all $v, w \in \mathbb{R}^n$.[2] Note that the $n \times n$ matrix $\mathrm{A} := (\mathrm{A}_{ij})$, where $\mathrm{A}_{ij} = g_{\mathbb{E}}(A(e_i), e_j)$, is an orthogonal matrix; i.e., $\mathrm{A}\mathrm{A}^{\mathrm{T}} = \mathrm{A}^{\mathrm{T}}\mathrm{A} = I_n$, the $n \times n$ identity matrix.

A diffeomorphism $T : \mathbb{E}^n \to \mathbb{E}^n$ is an isometry if and only if there exists an orthogonal transformation $A : \mathbb{R}^n \to \mathbb{R}^n$ and a translation $\tau : \mathbb{R}^n \to \mathbb{R}$ such that $T = \tau \circ A$. The isometries of $\mathbb{E}^n$ are called rigid motions.

Definition 7.47. An n-**dimensional crystallographic group** Γ is a discrete, uniform group of isometries of $\mathbb{E}^n$. That is, Γ is a subgroup of $\mathrm{Isom}(\mathbb{E}^n)$ which is a discrete subset and such that $\mathbb{E}^n/\Gamma$ is compact.

When $n = 2$, crystallographic groups are also known as *wallpaper groups*.

We say that a Riemannian manifold is **flat** if it has constant sectional curvature zero. If Γ is an n-dimensional crystallographic group, then $\mathbb{E}^n/\Gamma$ with the quotient metric induced by $g_{\mathbb{E}}$ is a closed (i.e., compact without boundary) flat Riemannian manifold. Conversely, a closed flat Riemannian n-manifold is the quotient of $\mathbb{E}^n$ by an n-dimensional crystallographic group.

Theorem 7.48 (Bieberbach [**Bie11, Bie12**]). *If M^n is a closed flat Riemannian n-manifold, then a finite cover of M^n is a flat torus, which is the quotient of $\mathbb{E}^n$ by n linearly independent translations.*

7.9.3. The sphere. For $K > 0$, the standard model for the simply connected space form of curvature K is the n-dimensional sphere of radius $r = K^{-1/2}$:

$$S^n(K^{-1/2}) = \{\mathbf{x} \in \mathbb{R}^{n+1} : \|\mathbf{x}\| = K^{-1/2}\}. \tag{7.114}$$

[2] Lie groups and the orthogonal group $\mathrm{O}(n)$ in particular will be discussed in Chapter 10.

The Riemannian metric g_K on $M_K^n = S^n(K^{-1/2}) = S^n(r)$ is the metric induced from the Euclidean metric on $\mathbb{R}^{n+1}$.

Given $v_1, v_2 \in S^n(r)$,

$$d_{S^n(r)}(v_1, v_2) = r\cos^{-1}(r^{-2}g_{\mathbb{E}}(v_1, v_2)). \tag{7.115}$$

We also have that

$$d_{\mathbb{E}}(v_1, v_2) = \|v_1 - v_2\| = 2r\sin\left(\frac{d_{S^n(r)}(v_1, v_2)}{2r}\right). \tag{7.116}$$

Recall that the second fundamental form of $S^n(r)$, as a hypersurface in $\mathbb{E}^{n+1}$, is given by $\mathrm{II}(v, w) = r^{-1}g_K(v, w)$, so that $S^n(r)$ is totally umbilical.

The Levi-Civita connection of g_K is given by

$$\nabla_v W = D_v W + r^{-1}g_K(v, W_{\mathbf{x}})\mathbf{x} \tag{7.117}$$

for $v \in T_{\mathbf{x}}S^n(r)$, $W \in \Gamma(TS^n(r))$, and $\mathbf{x} \in S^n(r)$.

Given $1 \le k \le n-1$, the k-dimensional complete totally geodesic submanifolds of $S^n(r)$ are the k-dimensional **equatorial submanifolds**, which can be defined by one of the following equivalent ways:

(1) an isometric embedding of $S^k(r)$ into $S^n(r)$,

(2) the image under an $(n+1)$-dimensional orthogonal transformation of the embedding of $S^k(r)$ into $S^n(r)$ defined by $(x_1, \ldots, x_{k+1}) \mapsto (x_1, \ldots, x_{k+1}, 0, \ldots, 0)$,

(3) the set of vectors in $S^n(r)$ perpendicular to vectors in an $(n-k)$-dimensional linear subspace of $\mathbb{R}^{n+1}$.

In particular, the entire geodesics are the (infinitely covered) great circles, i.e., the 1-dimensional equatorial submanifolds.

The diameter of $S^n(r)$ is equal to πr. In particular, by the Bonnet–Myers theorem (Theorem 7.68 below), $S^n(r)$ attains the maximum diameter among all complete Riemannian manifolds with the Ricci curvature lower bound $(n-1)r^{-2}$.

A **spherical space form** is defined to be an isometric quotient of $S^n(1)$ that is an n-manifold.

Suppose that $n = 2k$ is even. Then $\Gamma \subset \mathrm{O}(2k+1)$, where each element of Γ is fixed-point-free. Since the orthogonal group is odd-dimensional, this implies that $\Gamma \cong \mathbb{Z}_2$ and that Γ consists of $\pm\,\mathrm{id}_{S^n(1)}$. Hence S^n/Γ is the $2k$-dimensional real projective space.

There is a classification of 3-dimensional spherical space forms. In particular, the fundamental group is isomorphic to one of the following types of finite groups: cyclic (lens space), binary dihedral group, binary tetrahedral group, binary octahedral group, or binary icosahedral group (see Wolf [**Wol11**, Chapters 4–7]).

7.9.4. Hyperbolic space.

7.9.4.1. *The standard models of hyperbolic space.* For negative curvature, we normalize K to be equal to -1. **Hyperbolic n-space** is the Riemannian manifold defined, up to isometry, by any one of the following models:

(1) (**Poincaré ball model**) $\mathbb{D}^n := (D^n, g_{\mathbb{D}})$, where

$$D^n = \{\mathbf{x} \in \mathbb{R}^n : \|\mathbf{x}\| < 1\}$$

is the unit n-disk, and

$$g_{\mathbb{D}}(\mathbf{x}) = \frac{4}{(1 - \|\mathbf{x}\|^2)^2} g_{\mathbb{E}}. \tag{7.118}$$

See Figure 7.9.1.

(2) (**Upper half-space model**) $\mathbb{U}^n := (U^n, g_{\mathbb{U}})$, where

$$U^n = \{\mathbf{x} \in \mathbb{R}^n : x^n > 0\}$$

is the upper half n-space, and

$$g_{\mathbb{U}} := \frac{1}{(x^n)^2} g_{\mathbb{E}}. \tag{7.119}$$

See Figure 7.9.2.

(3) (**Hyperboloid model**) Let $\mathbb{E}^{n,1}$ denote Minkowski $(n+1)$-space, i.e., $\mathbb{R}^{n+1}$ with the Minkowski metric

$$g_{\mathbb{E}^{n,1}} := dx_1^2 + \cdots + dx_n^2 - dx_{n+1}^2, \tag{7.120}$$

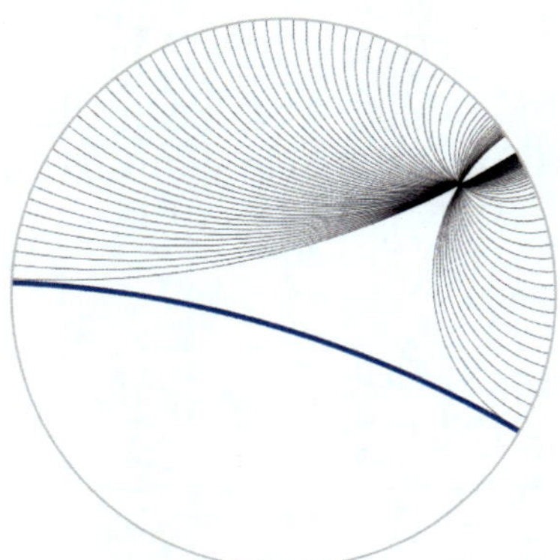

Figure 7.9.1. The Poincaré ball model of hyperbolic space. Credit: By Trevorgoodchild at English Wikipedia – Transferred from en.wikipedia to Commons., Public Domain, `https://commons.wikimedia.org/w/index.php?curid=36685779`

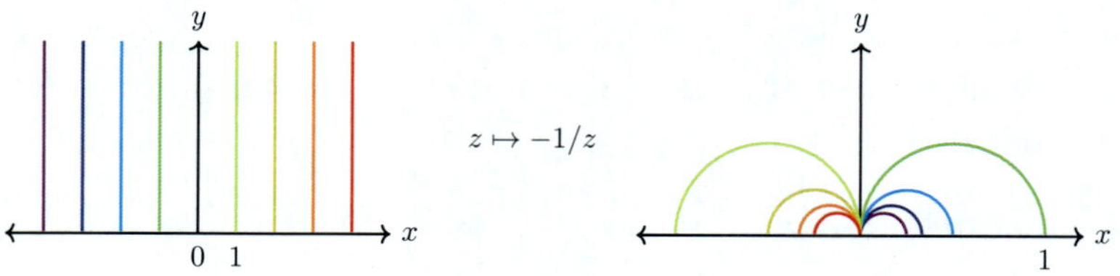

Figure 7.9.2. The isometry $z \mapsto -1/z$ of $\mathbb{U}^2$ maps vertical ray geodesics to semicircle geodesics.

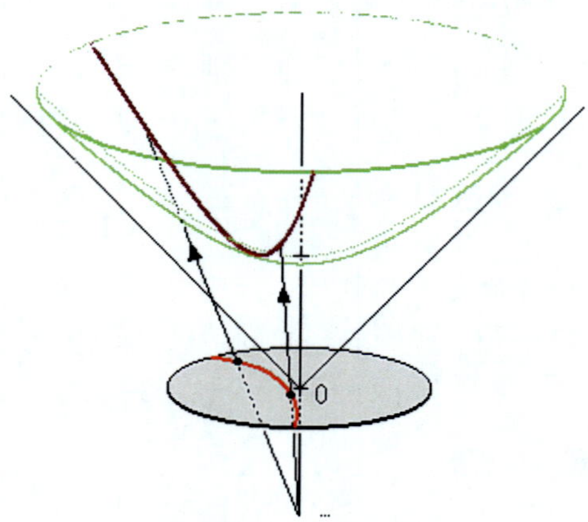

Figure 7.9.3. The hyperboloid model of hyperbolic space. Source: Wikimedia Commons. Credit: By Selfstudier – Own work, CC0, `https://commons.wikimedia.org/w/index.php?curid=18419030`

which is a Lorentzian metric.[3] The hyperboloid model $\mathbf{H}^n := (H^n, g_{\mathbf{H}})$ is the submanifold

$$H^n := \left\{\mathbf{x} \in \mathbb{R}^{n+1} : x_1^2 + \cdots + x_n^2 - x_{n+1}^2 = -1 \text{ and } x_{n+1} > 0\right\} \tag{7.121}$$

with the metric $g_{\mathbf{H}}$ induced by $g_{\mathbb{E}^{n,1}}$. See Figure 7.9.3. Note that although the ambient metric is indefinite, the induced metric is positive-definite.

7.9.4.2. *Isometries between models of hyperbolic space.* All of the three models above of hyperbolic n-space are isometric.

Lemma 7.49. *The diffeomorphism $\alpha : (H^3, g_H) \to (D^3, g_{\mathbb{D}})$ defined by*

$$\alpha(x) \doteqdot \frac{(x_1, x_2, x_3)}{1 + x_4} = \frac{x - x_4 e_4}{1 + x_4} \tag{7.122}$$

is an isometry.

[3]A **pseudo-Riemannian manifold** is defined like a Riemannian manifold except that the metric tensor is only assumed to be nondegenerate instead of positive-definite. For an $(n+1)$-dimensional manifold, a **Lorentzian metric** is defined as the case where the signature of the metric is $(n, 1)$.

Proof. Let $n \geq 2$. The hyperboloid model is

$$H^n = \left\{\boldsymbol{x} \in \mathbb{R}^{n+1} \colon x_1^2 + \cdots + x_n^2 - x_{n+1}^2 = -1 \text{ and } x_{n+1} > 0\right\}$$

with the metric g_H induced by $(\mathbb{R}^{n+1}, g_{\mathbb{E}^{n,1}})$, where

$$g_{\mathbb{E}^{n,1}} = dx_1^2 + \cdots + dx_n^2 - dx_{n+1}^2.$$

Given $y = (y_1, \ldots, y_{n+1}) \in \mathbb{R}^{n+1}$, let $\overline{y} = y - y_{n+1}e_{n+1}$. We shall show that for all $n \geq 2$, the map $\alpha : (H^n, g_H) \to (D^n, g_{\mathbb{D}})$ defined by

$$\alpha(\boldsymbol{x}) \doteqdot \frac{\boldsymbol{x} - x_{n+1}e_{n+1}}{1 + x_{n+1}} = \frac{\overline{\boldsymbol{x}}}{1 + x_{n+1}}, \tag{7.123}$$

where $\boldsymbol{x} = (x_1, \ldots, x_{n+1})$, is an isometry.

We have that α is a bijective smooth map with inverse $\alpha^{-1} : D^n \to H^n$ given by the smooth map

$$\alpha^{-1}(\boldsymbol{y}) = ((1 + \|\boldsymbol{y}\|)\boldsymbol{y}, \|\boldsymbol{y}\|) .$$

Indeed, we easily see that $\alpha(\alpha^{-1}(\boldsymbol{y})) = \boldsymbol{y}$ for $\boldsymbol{y} \in D^n$.

Observe that for any $\boldsymbol{x} \in H^n$ and $V = (V^1, \ldots, V^{n+1}) \in T_{\boldsymbol{x}}$ we have

$$\begin{aligned}\alpha_*(V) &= \frac{\overline{V}}{1 + x_{n+1}} - \frac{\overline{\boldsymbol{x}}}{(1 + x_{n+1})^2}V^{n+1} \\ &= \frac{(1 + x_{n+1})\overline{V} - V^{n+1}\overline{\boldsymbol{x}}}{(1 + x_{n+1})^2}.\end{aligned}$$

Moreover,

$$1 - \|\alpha(\boldsymbol{x})\|^2 = 1 - \frac{x_1^2 + \cdots + x_n^2}{(1 + x_{n+1})^2} = \frac{2}{1 + x_{n+1}}.$$

Let $W = (W^1, \ldots, W^{n+1}) \in T_{\boldsymbol{x}}$. We then have that

$$\begin{aligned}&(1 + x_{n+1})^2 g_{\mathbb{D}}(\alpha_*(V), \alpha_*(W)) \\ &\quad = \left((1 + x_{n+1})\overline{V} - V^{n+1}\overline{\boldsymbol{x}}\right) \cdot \left((1 + x_{n+1})\overline{W} - W^{n+1}\overline{\boldsymbol{x}}\right) \\ &\quad = (1 + x_{n+1})^2\overline{V} \cdot \overline{W} + V^{n+1}W^{n+1}\|\overline{\boldsymbol{x}}\|^2 \\ &\qquad - (1 + x_{n+1})\left(W^{n+1}\overline{V} \cdot \overline{\boldsymbol{x}} + V^{n+1}\overline{W} \cdot \overline{\boldsymbol{x}}\right).\end{aligned}$$

By applying $\|\overline{\boldsymbol{x}}\|^2 = -1 + x_{n+1}^2$, $\overline{V} \cdot \overline{\boldsymbol{x}} = V^{n+1}x_{n+1}$, and $\overline{W} \cdot \overline{\boldsymbol{x}} = W^{n+1}x_{n+1}$ to this equation, we obtain

$$g_{\mathbb{D}}(\alpha_*(V), \alpha_*(W)) = \overline{V} \cdot \overline{W} - V^{n+1}W^{n+1} = g_H(V, W).$$

That is, $\alpha^* g_{\mathbb{D}} = g_H$. This proves that α is an isometry.

Alternatively, we can show that α is an isometry as follows. We compute that

$$\begin{aligned} g_{\mathbb{D}}(\alpha_*(V), \alpha_*(V)) &= \frac{4}{1 - \|\alpha(\boldsymbol{x})\|^2} \|\alpha_*(V)\|^2 \\ &= (1 + x_{n+1})^{-2} \big\|(1 + x_{n+1})\overline{V} - V^{n+1}\overline{\boldsymbol{x}}\big\|^2. \end{aligned}$$

The right-hand side may be rewritten via

$$\begin{aligned} &\big\|(1 + x_{n+1})\overline{V} - V^{n+1}\overline{\boldsymbol{x}}\big\|^2 \\ &\quad = (1 + x_{n+1})^2 \left\|\overline{V}\right\|^2 - 2(1 + x_{n+1})\overline{V} \cdot \overline{\boldsymbol{x}}\, V^{n+1} + (V^{n+1})^2 \left\|\overline{\boldsymbol{x}}\right\|^2 \\ &\quad = (1 + x_{n+1})^2 \left\|\overline{V}\right\|^2 - 2(1 + x_{n+1})(V^{n+1})^2 x_{n+1} \\ &\qquad + (V^{n+1})^2 \left((x_{n+1})^2 - 1\right) \\ &\quad = (1 + x_{n+1})^2 \left(\left\|\overline{V}\right\|^2 - (V^{n+1})^2\right) \\ &\quad = (1 + x_{n+1})^2 g_H(V, V). \end{aligned}$$

Thus

$$g_{\mathbb{D}}(\alpha_*(V), \alpha_*(V)) = g_H(V, V)$$

for all $V \in T_{\boldsymbol{x}}H^n$. By polarization, we conclude that

$$g_{\mathbb{D}}(\alpha_*(V), \alpha_*(W)) = g_H(V, W)$$

for all $V, W \in T_{\boldsymbol{x}}H^n$.

Remark. By polarization, we mean obtaining an inner product from its norm via the formula

$$\langle V, W\rangle = \frac{1}{4}\left(\|V + W\|^2 - \|V - W\|^2\right). \qquad \square$$

Lemma 7.50. *The diffeomorphism* $\beta : \left(D^3, g_{\mathbb{D}}\right) \to \left(\mathbb{H}^3, g_{\mathbb{H}}\right)$ *defined by*

$$\beta\left(x\right) := 2\frac{x + e_3}{\left|x + e_3\right|^2} - e_3 \tag{7.124}$$

is an isometry.

Proof. Define $\beta : (D^n, g_{\mathbb{D}}) \to (\mathbb{H}^n, g_{\mathbb{H}})$ by

$$\beta\left(x\right) := 2\frac{x + e_n}{\left\|x + e_n\right\|^2} - e_n.$$

Observe that

$$\beta\left(x\right)_n = \frac{1 - |x|^2}{\left\|x + e_n\right\|^2}.$$

Let $V \in T_x D^n$. Then

$$\beta_*(V) = 2\frac{V}{\|x + e_n\|^2} - 4\frac{x + e_n}{\|x + e_n\|^4} V \cdot (x + e_n).$$

Thus, for $V, W \in T_xD^n$, we compute that

$$\begin{aligned} & g_{\mathbb{H}}(\beta_*(V), \beta_*(W)) \\ & = \frac{1}{\beta(x)_n^2} \beta_*(V) \cdot \beta_*(W) \\ & = \frac{4}{(1-|x|^2)^2} \left(V - 2\frac{V \cdot (x+e_n)}{\|x+e_n\|^2}(x+e_n) \right) \\ & \qquad \cdot \left(W - 2\frac{W \cdot (x+e_n)}{\|x+e_n\|^2}(x+e_n) \right) \\ & = \frac{4}{(1-\|x\|^2)^2} V \cdot W \\ & = g_{\mathbb{D}}(V, W). \end{aligned}$$

Therefore β is an isometry. □

Remark 7.51. Complete constant sectional curvature -1 Riemannian manifolds are called **hyperbolic manifolds**. The study of hyperbolic manifolds is a rich area. In dimension 3, they play an important role in Thurston's geometrization conjecture. See the books by Thurston [**Thu22**] and Kapovich [**Kap01**].

7.9.5. Simply connected space forms in spherical coordinates. Let (M_K^n, g_K) denote the simply connected space form of constant sectional curvature K, where $K \in \mathbb{R}$. When $K \leq 0$, we may take the underlying differentiable manifold M_K^n to be $\mathbb{R}^n$, and when $K > 0$, we may take M_K^n to be the 1-point compactification of the open Euclidean n-ball of radius $\frac{\pi}{\sqrt{K}}$, which is diffeomorphic to the n-sphere.

With the choices above for M_K^n, the Riemannian metric g_K may be defined as

$$g_K = dr^2 + \operatorname{sn}_K(r)^2 \, g_{S^{n-1}}, \tag{7.125}$$

where sn_K is the **modulated sine function** defined by

$$\operatorname{sn}_K(r) := \begin{cases} \frac{1}{\sqrt{K}} \sin\left(\sqrt{K} r\right) & \text{if } K > 0, \\ r & \text{if } K = 0, \\ \frac{1}{\sqrt{|K|}} \sinh\left(\sqrt{|K|} r\right) & \text{if } K < 0. \end{cases} \tag{7.126}$$

Observe that $\lim_{r\searrow 0}\frac{\mathrm{sn}_K(r)}{r}=1$. The **modulated cosine function** is defined by

$$\mathrm{cn}_K(r) := \begin{cases} \cos\left(\sqrt{K}r\right) & \text{if } K>0,\\ 1 & \text{if } K=0,\\ \cosh\left(\sqrt{|K|}r\right) & \text{if } K<0. \end{cases} \tag{7.127}$$

We will encounter the modulated cosine function when we consider the mean curvatures of geodesic spheres in space forms (see (11.98) below). Observe that

$$K\,\mathrm{sn}_K^2(r)+\mathrm{cn}_K^2(r)=1 \tag{7.128}$$

and that

$$\mathrm{sn}_K'(r)=\mathrm{cn}_K(r),\quad \mathrm{cn}_K'(r)=-K\,\mathrm{sn}_K(r),\quad \mathrm{sn}_K''(r)=-K\,\mathrm{sn}_K(r). \tag{7.129}$$

7.9.6. Isometries of simply connected space forms. Recall that the isometries of Euclidean n-space $\mathbb{E}^n$ are the rigid motions and the isometry group of the unit n-sphere $S^n(1)$ is $\mathrm{O}(n+1)$.[4]

To describe the isometry group of hyperbolic n-space, we consider the hyperboloid model $\mathbf{H}^n=(H^n,g_{\mathbf{H}})$, where

$$H^n := \{\mathbf{x}\in\mathbb{R}^{n+1} : g_{\mathbb{E}^{n,1}}(\mathbf{x},\mathbf{x})=-1,\ x^{n+1}>0\}\subset\mathbb{E}^{n,1}$$

is defined by (7.121) and $g_{\mathbf{H}}$ is the metric induced by $g_{\mathbb{E}^{n,1}}$. The isometry group of $\mathbf{H}^n$ is the Lorentz group $\mathrm{SO}(n,1)$ of special orthogonal transformations of $\mathbb{E}^{n,1}$, i.e., the Lie group of linear transformations $A:\mathbb{R}^{n+1}\to\mathbb{R}^{n+1}$ satisfying $\det(A)=1$ and $g_{\mathbb{E}^{n,1}}(Av,Aw)=g_{\mathbb{E}^{n,1}}(v,w)$ for all $v,w\in\mathbb{R}^{n+1}$. We remark that $A\in\mathrm{SO}(n,1)$ if and only if the $(n+1)\times(n+1)$ real matrix $\mathrm{A}:=(\mathrm{A}_{ij})$, where $\mathrm{A}_{ij}=g_{\mathbb{E}}(A(e_i),e_j)$ satisfies $\det(\mathrm{A})=1$ and

$$\mathrm{A}^T I_{n,1}\mathrm{A}=I_{n,1}, \tag{7.130}$$

where $I_{n,1}:=I_n\oplus(-I_1)$.

As special cases, we have

$$\mathrm{SO}(2,1)\cong\mathrm{SL}(2,\mathbb{R})$$

and

$$\mathrm{SO}(3,1)\cong\mathrm{SL}(2,\mathbb{C}).$$

[4]See §10.4.1 below for a discussion of the classical Lie groups, including $\mathrm{O}(n+1)$, $\mathrm{SO}(n,1)$, $\mathrm{SL}(n,\mathbb{R})$, and $\mathrm{SL}(n,\mathbb{C})$.

7.9.7. Mostow rigidity theorem. Rigidity theorems play a pivotal role in understanding the geometry and topology of manifolds. They often provide constraints on how flexible certain geometric structures can be. The Mostow rigidity theorem is a paramount result in this direction, particularly for hyperbolic manifolds. It asserts a deep connection between the algebraic structure of the fundamental group and the geometry of the manifold. Specifically, the theorem states the following; see Mostow [**Mos68**] (for an alternative proof, see Besson, Courtois, and Gallot [**BCG96**]).

Theorem 7.52. *Let (M_1^n, g_1) and (M_2^n, g_2) be finite-volume complete hyperbolic manifolds, where $n \geq 3$. If $\phi : \pi_1(M_1^n) \to \pi_1(M_2^n)$ is a group isomorphism, then there exists a unique isometry $f : (M_1^n, g_1) \to (M_2^n, g_2)$ whose induced map on fundamental groups satisfies $f_* = \phi$.*

As a consequence, we have the following.

Corollary 7.53. *Let (M_1^n, g_1) and (M_2^n, g_2) be finite-volume complete hyperbolic manifolds, where $n \geq 3$. Then the following are equivalent:*

(1) *(M_1^n, g_1) and (M_2^n, g_2) are isometric Riemannian manifolds.*
(2) *M_1^n and M_2^n are diffeomorphic smooth manifolds.*
(3) *M_1^n and M_2^n are homotopy equivalent as topological spaces.*

7.10. Second variation of arc length and stable geodesics

Variational principles lie at the heart of many geometric problems. One of the fundamental objects of interest is the arc length of curves, especially when it comes to understanding how this length changes under perturbations. Building on the first variation, which provides insight into the direction in which a curve should be varied to increase or decrease its length, the second variation delves deeper into the nature of this change. It determines whether the perturbations result in a maximum, minimum, or neither. In this section, we present a detailed exploration of the second variation of the arc length, starting with its formal definition.

7.10.1. Second variation of arc length.

7.10.1.1. *The second variation of the arc-length formula.* Suppose that we have a 2-parameter family of paths $\gamma_{u,v} : [a, b] \to M^n$, $u \in H$, and $v \in I$, where H and I are open intervals in $\mathbb{R}$. Assume that $\gamma_{u,v}(t)$ is smooth in u, v and piecewise smooth in t. Define the map

$$\Gamma : [a, b] \times H \times I \to M^n$$

by

$$\Gamma(t, u, v) := \gamma_{u,v}(t)$$

and let $T := d\Gamma(\frac{\partial}{\partial t})$, $U := d\Gamma(\frac{\partial}{\partial u})$, and $V := d\Gamma(\frac{\partial}{\partial v})$. By definition, Γ is continuous and there is a partition $a = t_0 < t_1 < \cdots < t_k = b$ such that Γ is smooth on each $[t_{i-1}, t_i] \times H \times I$, $1 \leq i \leq k$.

The **second variation of the arc-length formula** is given by the following.

Proposition 7.54 (Second variation of the arc length). *Suppose $0 \in H$ and $0 \in I$. If the path $\gamma_{0,0}(t)$ is parametrized by arc length, then*

(7.131)
$$\begin{aligned}\frac{\partial^2}{\partial u \partial v}\bigg|_{(u,v)=(0,0)} & \mathrm{L}\left(\gamma_{u,v}\right) \\ &= \int_a^b \left(\langle \nabla_T U, \nabla_T V\rangle - \langle \nabla_T U, T\rangle \langle \nabla_T V, T\rangle - \langle \mathrm{Rm}\left(U, T\right) T, V\rangle\right) dt \\ &\quad - \int_a^b \langle \nabla_U V, \nabla_T T\rangle \, dt + \langle \nabla_U V, T\rangle\big|_a^b ,\end{aligned}$$

where $\nabla_T U := \nabla^\Gamma_{\frac{\partial}{\partial t}} U$, $\nabla_T V := \nabla^\Gamma_{\frac{\partial}{\partial t}} V$, and $\nabla_U V := \nabla^\Gamma_{\frac{\partial}{\partial u}} V$ are covariant derivatives along the map Γ.

Proof. By the first variation of the arc-length formula (7.45), we have

$$\frac{\partial}{\partial v} \mathrm{L}(\gamma_{u,v}) = \int_a^b \left\langle \frac{T}{|T|}, \nabla^\Gamma_{\frac{\partial}{\partial t}} V \right\rangle dt.$$

By taking the partial derivative of this with respect to u, we compute that

$$\begin{aligned}\frac{\partial^2}{\partial u \partial v}\bigg|_{(u,v)=(0,0)} & \mathrm{L}\left(\gamma_{u,v}\right) \\ &= \int_a^b \frac{\partial}{\partial u}\bigg|_{u=0} \left\langle \frac{T}{|T|}, \nabla^\Gamma_{\frac{\partial}{\partial t}} V \right\rangle dt \\ &= \int_a^b \left(\left\langle \frac{T}{|T|}, \nabla^\Gamma_{\frac{\partial}{\partial u}} (\nabla^\Gamma_{\frac{\partial}{\partial t}} V) \right\rangle + \left\langle \nabla^\Gamma_{\frac{\partial}{\partial u}} \left(\frac{T}{|T|}\right), \nabla^\Gamma_{\frac{\partial}{\partial t}} V \right\rangle \right) dt \\ &= \int_a^b \left(\left\langle \frac{T}{|T|}, \mathrm{Rm}\left(U, T\right) V \right\rangle + \left\langle \frac{T}{|T|}, \nabla^\Gamma_{\frac{\partial}{\partial t}} (\nabla^\Gamma_{\frac{\partial}{\partial u}} V) \right\rangle \right) dt \\ &\quad + \int_a^b \frac{1}{|T|} \left\langle \nabla^\Gamma_{\frac{\partial}{\partial u}} T, \nabla^\Gamma_{\frac{\partial}{\partial t}} V \right\rangle dt - \int_a^b \frac{1}{|T|^3} \left\langle \nabla^\Gamma_{\frac{\partial}{\partial u}} T, T \right\rangle \left\langle T, \nabla^\Gamma_{\frac{\partial}{\partial t}} V \right\rangle dt\end{aligned}$$

The result for when Γ is smooth now follows from $|T| = 1$ and the integration by parts formula

$$\int_a^b \left\langle T, \nabla^\Gamma_{\frac{\partial}{\partial t}} (\nabla^\Gamma_{\frac{\partial}{\partial u}} V) \right\rangle dt = \left\langle T, \nabla^\Gamma_{\frac{\partial}{\partial u}} V \right\rangle \bigg|_a^b - \int_a^b \left\langle \nabla^\Gamma_{\frac{\partial}{\partial t}} T, \nabla^\Gamma_{\frac{\partial}{\partial u}} V \right\rangle dt.$$

In the above, we have used that $\nabla^{\Gamma}_{\frac{\partial}{\partial v}} T = \nabla^{\Gamma}_{\frac{\partial}{\partial t}} V$ by (7.34), that

$$\operatorname{Rm}(U,T)\,V = \nabla^{\Gamma}_{\frac{\partial}{\partial u}}\left(\nabla^{\Gamma}_{\frac{\partial}{\partial t}} V\right) - \nabla^{\Gamma}_{\frac{\partial}{\partial t}}\left(\nabla^{\Gamma}_{\frac{\partial}{\partial u}} V\right) - \nabla^{\Gamma}_{\left[\frac{\partial}{\partial u},\frac{\partial}{\partial t}\right]} V,$$

and that $\left[\frac{\partial}{\partial u},\frac{\partial}{\partial t}\right] = \vec{0}$.

In the piecewise smooth case we simply apply the second variation of the arc-length formula to each of the intervals $[t_{i-1}, t_i]$ and observe that the boundary terms $\langle \nabla_U V, T\rangle|_{t_{i-1}}^{t_i}$ cancel except for the endpoints of $[a,b]$. □

Corollary 7.55 (Second variation of arc length for fixed endpoints). *If $\gamma_v(t)$, where $t \in [a,b]$ and $v \in I$, is a smooth 1-parameter family of piecewise smooth paths with fixed endpoints, i.e., $\gamma_v(a) \equiv \gamma_0(a)$ and $\gamma_v(b) \equiv \gamma_0(b)$ for $v \in I$, and such that $\gamma_0(t)$ is a geodesic parametrized by arc length, then*

$$\frac{\partial^2}{\partial v^2}\bigg|_{v=0} \operatorname{L}(\gamma_v) = \int_a^b \left(\left| (\nabla_T V)^{\perp} \right|^2 - \langle \operatorname{Rm}(V,T)\,T, V\rangle \right) dt, \tag{7.132}$$

where $(\nabla_T V)^{\perp}$ denotes the orthogonal projection of $\nabla_T V$ onto $T^{\perp}$; that is, $(\nabla_T V)^{\perp} := \nabla_T V - \langle \nabla_T V, T\rangle\, T$.

7.10.1.2. *Stable geodesics.* A geodesic segment $\gamma : [a,b] \to M^n$ is **stable** if the second variation of the arc length, with respect to those variation vector fields which vanish at the endpoints, is nonnegative. For example, all minimal geodesic segments are stable.

A **geodesic loop** (a.k.a. **closed geodesic**) is a smooth map $\gamma : S^1 \to M^n$ which is a geodesic. A geodesic loop is called stable if the second variation of the arc length is nonnegative for arbitrary piecewise smooth variation vector fields. By Proposition 7.54 and Corollary 7.55, in the case of a manifold with nonpositive curvature we have the following.

Corollary 7.56 (Stability of geodesics in manifolds with nonpositive curvature). *If (M^n, g) has nonpositive sectional curvature and the paths γ_v are smooth and either (1) have fixed endpoints or (2) are closed loops, then*

$$\frac{\partial^2}{\partial v^2}\bigg|_{v=0} \operatorname{L}(\gamma_v) \geq 0.$$

That is, the geodesic γ_0 is stable.

On the other hand, great circles in spheres are unstable; see Exercise 7.5.

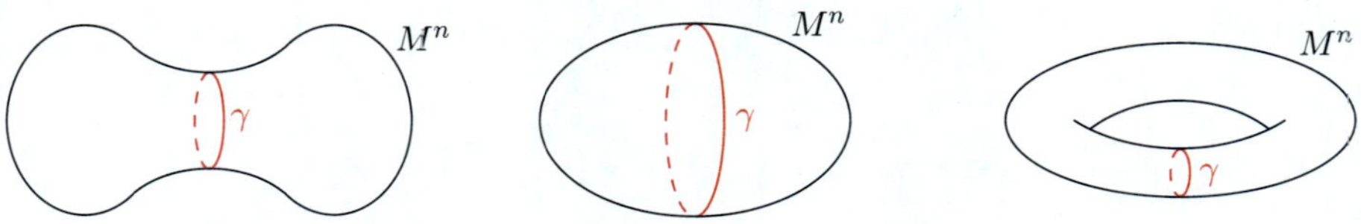

Figure 7.10.1. Left: (Strongly) stable geodesic in a negatively curved region in a manifold. Center: Unstable geodesic in a positively curved manifold. Right: (Weakly) stable geodesic in a flat torus.

7.11. Jacobi fields, conjugate points, and the index form

Fix a point p on a Riemannian manifold (M^n, g). To explore the geometry of (M^n, g), we may consider all geodesics emanating from p. To see how these geodesics spread out from one another, we study their variations.

7.11.1. Definition of a Jacobi field. Let $\gamma : [a, b] \to M^n$ be a constant-speed geodesic and let $\gamma_v : [a, b] \to M^n$, $v \in I \ni 0$, be a variation of γ such that each γ_v is a constant-speed geodesic and $\gamma_0 = \gamma$. We call the variation

$$J(t) := \left.\frac{\partial}{\partial v}\right|_{v=0} \gamma_v(t),$$

which is a vector field along γ, a **Jacobi field**. That is, a Jacobi field is the variation vector field of a 1-parameter family of geodesics, i.e., a variation where each path in the family is a geodesic.

7.11.2. Jacobi fields via the exponential map. Let (M^n, g) be a complete Riemannian manifold. Given $p \in M^n$ and $V, W \in T_pM$, define the 1-parameter family of geodesics $\gamma_u : [0, \infty) \to M^n$, $u \in \mathbb{R}$, by

$$\gamma_u(t) := \exp_p\big(t(V + uW)\big) = \gamma_{V+uW}(t). \tag{7.133}$$

We have $\gamma_u(0) = p$. The associated Jacobi field J_W along $\gamma_0 = \gamma_V$ is given by

$$J_W(t) := \left.\frac{\partial}{\partial u}\right|_{u=0} \gamma_{V+uW}(t) = (d\exp_p)_{tV}(tW). \tag{7.134}$$

The initial data are given by $J_W(0) = \vec{0}$ and $\nabla_V J_W(0) = W$. See Exercise 7.6 for a generalization of this to Jacobi fields that are not initially zero.

7.11.3. Jacobi fields on simply connected space forms.

Example 7.57 (Jacobi fields on model spaces). *Let (M^n_K, g_K) be the simply connected model space of constant sectional curvature $K \in \mathbb{R}$. Fix any point $p \in M^n_K$. Recall from* (7.125) *that we can write the Riemannian metric on M^n_K as $g_K = dt^2 + \operatorname{sn}_K(t)^2 g_{S^{n-1}}$, where sn_K is the modulated sine function defined by* (7.126) *and where $t = 0$ corresponds to the point $p \in M^n_K$. Let*

$\gamma : [0,\infty) \to M_K^n$ be a unit-speed geodesic emanating from p. Let $V \in T_p M_K$ be such that $\langle V, \gamma'(0)\rangle = 0$. Parallel transport V to obtain a vector field $V(t)$ along $\gamma(t)$. Then

$$J(t) := \operatorname{sn}_K(t)\, V(t) \tag{7.135}$$

is a Jacobi field along $\gamma(t)$ with $\langle J(t), \gamma'(t)\rangle \equiv 0$, $J(0) = \vec{0}$, and $\nabla_T J(0) = V$. Observe that, by (7.129), we have

$$\nabla_T(\nabla_T J)(t) = \operatorname{sn}_K''(t)\, V(t) = -K \operatorname{sn}_K(t)\, V(t) = -K\, J(t), \tag{7.136}$$

where $T(t) := \gamma'(t)$.

We can generalize this example to rotationally symmetric metrics; see Exercise 7.7.

7.11.4. The Jacobi equation for Jacobi fields on manifolds. Equation (7.136) and (7.214) are special cases of the following basic result which holds for any Riemannian manifold.

Lemma 7.58. *If $J(t)$ is a Jacobi field along a constant-speed geodesic $\gamma(t)$, then $J(t)$ satisfies the* ***Jacobi equation****:*

$$\nabla_T(\nabla_T J) + \operatorname{Rm}(J,T)\, T = \vec{0}, \tag{7.137}$$

where $T = T(t) := \gamma'(t)$ and we have used the notation of (7.37).

Proof. Let γ_v, $v \in I \in 0$, be a variation of γ such that each γ_v is a constant-speed geodesic, $\gamma_0 = \gamma$, and $\frac{\partial}{\partial v}\big|_{v=0} \gamma_v(t) = J(t)$. Let $\Gamma(t,v) := \gamma_v(t)$. Recall that ∇^Γ denotes the covariant derivative along the map Γ. By differentiating the geodesic equation $\nabla^\Gamma_{\frac{\partial}{\partial t}} T = \nabla_T T = \vec{0}$ and commuting covariant derivatives, we obtain

$$\begin{aligned}
\vec{0} &= \nabla^\Gamma_{\frac{\partial}{\partial v}} \big(\nabla^\Gamma_{\frac{\partial}{\partial t}} T\big) \\
&= \nabla^\Gamma_{\frac{\partial}{\partial t}} \big(\nabla^\Gamma_{\frac{\partial}{\partial v}} T\big) + \operatorname{Rm}(J,T)\, T \\
&= \nabla^\Gamma_{\frac{\partial}{\partial t}} \big(\nabla^\Gamma_{\frac{\partial}{\partial t}} J\big) + \operatorname{Rm}(J,T)\, T.
\end{aligned}$$

We remark that in the simplified notation of (7.37) for ∇^Γ, this computation may be expressed as

$$\vec{0} = \nabla_J(\nabla_T T) = \nabla_T(\nabla_J T) + \operatorname{Rm}(J,T)T = \nabla_T(\nabla_T J) + \operatorname{Rm}(J,T)T. \quad \square$$

Since the Jacobi equation is a linear second-order system of ordinary differential equations, we have the following.

Corollary 7.59 (Jacobi fields are determined by their initial data). *For any unit-speed geodesic $\gamma : [a,b] \to M^n$ and $V_0, V_1 \in T_{\gamma(a)}M$, there exists a unique Jacobi field $J(t)$ along the $\gamma(t)$ with initial data:*

$$J(a) = V_0 \quad \text{and} \quad \nabla_T J(a) = V_1, \tag{7.138}$$

which is equal to the unique solution to the Jacobi equation (7.137) *satisfying* (7.138).

Proof. Exercise 7.6 provides existence. Uniqueness follows from Lemma 7.58 and the standard uniqueness theorem for solutions to second-order linear ordinary differential equations with given initial data. □

Since γ is a geodesic and by the Jacobi equation (7.137), we have that

$$T(T\langle J,T\rangle) = T\langle \nabla_T J, T\rangle = \langle \nabla_T(\nabla_T J), T\rangle = -\langle R(J,T)T,T\rangle = 0.$$

Hence, the tangential component of the Jacobi field $\langle J,T\rangle(t)$ is a linear function of t; i.e., there exist constants c_1 and c_2 such that

$$\langle J,T\rangle(t) = c_1 t + c_2. \tag{7.139}$$

In particular, if $J(t)$ is a Jacobi field along a geodesic $\gamma : [a,b] \to M^n$ with $\langle J,T\rangle(a) = \vec{0}$ and $\langle J,T\rangle(b) = \vec{0}$, then $\langle J,T\rangle(t) \equiv \vec{0}$. For example, if $J(a) = \vec{0}$ and $J(b) = \vec{0}$, then $\langle J,T\rangle(t) \equiv \vec{0}$.

7.11.5. Conjugate points. Consider the unit n-sphere S^n and fix a point $\mathbf{x} \in S^n$. Let $\gamma_{\mathbf{V}}$ be a geodesic emanating from $\mathbf{x}$, where $\mathbf{V} \in T_{\mathbf{x}}S^n$ is a unit vector. Observe that $\gamma_{\mathbf{V}}$ stops minimizing after distance π at the antipodal point $-\mathbf{x}$ and at which point $\gamma_{\mathbf{V}}|_{[0,\pi]}$ traces out a semicircle. The 'obvious' reason for why $\gamma_{\mathbf{V}}$ stops minimizing is that the opposite direction geodesic $\gamma_{-\mathbf{V}}(t)$ meets $\gamma_{\mathbf{V}}(t)$ at $t = \pi$ and at which point they have the same length. So if we travel along $\gamma_{\mathbf{V}}(t)$ a little past $t = \pi$, then it is more efficient to travel along $\gamma_{-\mathbf{V}}(t)$ (a little before $t = \pi$) to get to the same point. This is an example of a *cut point* as we will discuss in §7.13.

However, there is a second reason for why $\gamma_{\mathbf{V}}$ stops minimizing, which is that all semicircles minimize length between $\mathbf{x}$ and its antipodal point $-\mathbf{x}$. In particular, for any family of unit tangent vectors $\mathbf{V}(s) \in T_{\mathbf{x}}S^n$ with $\mathbf{V}(0) = \mathbf{V}$, we have that each $\gamma_{\mathbf{V}(s)}|_{[0,\pi]}$ is a minimal geodesic from $\mathbf{x}$ to $-\mathbf{x}$. This causes the derivative of the exponential map $\exp_{\mathbf{x}}$ to be singular at $\pi V \in T_{\mathbf{x}}S^n$.

Definition 7.60. A point $x \in M^n$ is a **conjugate point** to $p \in M^n$ if x is a singular value of $\exp_p : T_pM \to M^n$. That is, $x = \exp_p(V)$, for some $V \in T_pM \setminus \{\vec{0}\}$, where

$$(d\exp_p)_V : T_V(T_pM) \to T_{\exp_p(V)}M$$

is singular; i.e., $(d\exp_p)_V$ has nontrivial kernel. We say that x is a conjugate point to p **along** γ_V (or along $\gamma_{V/|V|}$).

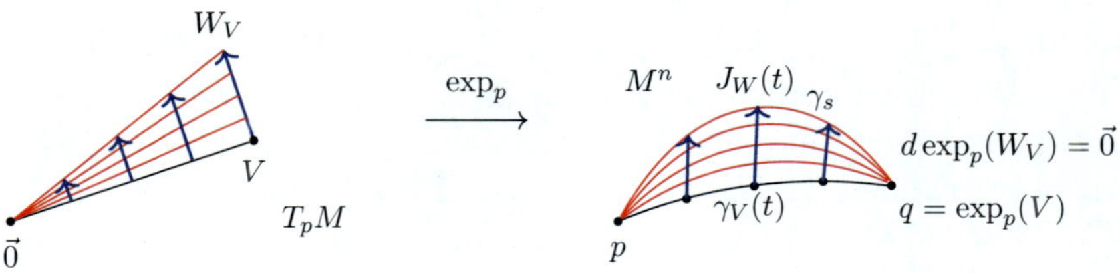

Figure 7.11.1. The point q is a conjugate point to p along $\gamma_V : [0,1] \to M^n$.

Lemma 7.61 (Conjugate points and Jacobi fields). *Let $\gamma_V : [0,1] \to M^n$ be a constant-speed geodesic. Then $x = \gamma_V(1)$ is a conjugate point to $p = \gamma_V(0)$ along γ_V if and only if there exists a nontrivial (i.e., not identically equal to zero) Jacobi field along γ_V vanishing at the endpoints of γ_V.*

Proof. We have that $x = \gamma_V(1) = \exp_p(V)$ is conjugate to p along γ_V if and only if there exists $W \in T_pM \setminus \{\vec{0}\}$ such that

$$\vec{0} = (d\exp_p)_V(W).$$

On the other hand, the Jacobi field J_W along γ_V satisfies

$$J_W(1) = (d\exp_p)_V(W).$$

Since $W \neq \vec{0}$, we have that J_W is nontrivial. The lemma follows since J_W vanishes at the endpoints if and only if $J_W(1) = \vec{0}$. □

An immediate consequence of the lemma is the following.

Corollary 7.62. *A point x is conjugate to p along a geodesic γ if and only if p is conjugate to x along the geodesic $-\gamma(t) := \gamma(-t)$.*

Proof. The corollary follows from Lemma 7.61 because the condition of being a Jacobi field $J(t)$ vanishing at the endpoints of $\gamma(t)$ is symmetric in these endpoints. □

Lemma 7.63. *Given a geodesic $\gamma : [a,b] \to M^n$ without conjugate points and vectors $X \in T_{\gamma(a)}M$ and $Y \in T_{\gamma(b)}M$, there exists a unique Jacobi field J along γ with $J(a) = X$ and $J(b) = Y$.*

Moreover, if $\langle X, T\rangle = \langle Y, T\rangle = 0$, then $\langle J(t), T(t)\rangle \equiv 0$ for $t \in [a,b]$.

Proof. The uniqueness of J follows from the difference of two such Jacobi fields being a Jacobi field that vanishes at its endpoints and from the geodesic

γ having no conjugate points. To prove the existence of J, given a vector $X \in T_{\gamma(a)}M$, define the affine map $A : T^{\perp}_{\gamma(a)}M \to T^{\perp}_{\gamma(b)}M$ by

$$A(Z) := J_Z(b),$$

where $J_Z(t)$ is the Jacobi field with $J_Z(a) = X$ and $\nabla_T J_Z(a) = Z$. Since γ has no conjugate points, the map A is injective. Since A is an affine map between vector spaces of the same dimension, this implies that A is surjective. Existence of a desired Jacobi field follows. Finally, that $\langle J(t), T(t)\rangle = 0$ when $\langle X, T\rangle = \langle Y, T\rangle = 0$ is a consequence of (7.139). $\square$

7.11.6. The index form. Motivated by the second variation of arc-length formula, we consider the following.

7.11.6.1. *Definition of the index form.*

Definition 7.64. Let $\gamma : [a, b] \to M^n$ be a unit-speed geodesic and let U and V be piecewise smooth vector fields along γ. The **index form** of γ is defined by

$$I(U, V) := \int_a^b \left(\langle \nabla_T U, \nabla_T V\rangle - \langle \mathrm{Rm}(U, T) T, V\rangle \right) dt. \tag{7.140}$$

The index form is a symmetric bilinear form on the space of piecewise smooth vector fields along γ.

If $\gamma_{u,v}$ is a 1-parameter family of paths, where $\gamma_{0,0}$ is a unit-speed geodesic and $\langle U, T\rangle \equiv 0$ or $\langle V, T\rangle \equiv 0$, then the second variation of the arc-length formula (7.131) may be expressed in terms of the index form as

$$\left.\frac{\partial^2}{\partial u \partial v}\right|_{(u,V)=(0,0)} \mathrm{L}(\gamma_{u,v}) = I(U, V) + \left.\langle \nabla_U V, T\rangle\right|_a^b, \tag{7.141}$$

where $U = \left.\frac{\partial}{\partial u}\right|_{(u,v)=(0,0)} \mathrm{L}(\gamma_{u,v})$ and $V = \left.\frac{\partial}{\partial v}\right|_{(u,v)=(0,0)} \mathrm{L}(\gamma_{u,v})$. In particular, if $\gamma_{u,v}$ is a 1-parameter family of paths with fixed endpoints and if $\gamma_{0,0}$ is a unit-speed geodesic, then

$$\left.\frac{\partial^2}{\partial u \partial v} \mathrm{L}(\gamma_{u,v})\right|_{(u,v)=(0,0)} = I(U, V). \tag{7.142}$$

The index form represents the Hessian of the length functional at a critical point (i.e., geodesic) on any space of paths with fixed endpoints. In the context of Morse theory, the number of negative eigenvalues of the Hessian of a function at a critical point is called the 'index'; see §9.1 below. Observe that a geodesic is stable if and only if the index form is positive semidefinite, reflecting that the second variation of the arc length of the geodesic is nonnegative in all directions.

7.11.6.2. *The index lemma.* Normal variations associated to families of geodesic paths, i.e., Jacobi fields, along a geodesic without conjugate points minimize the second variation among all normal variations with given endpoint values.

Lemma 7.65 (Index lemma)**.** *Suppose that $\gamma : [a,b] \to M^n$ is a unit-speed geodesic without conjugate points. Let $A \in T_{\gamma(a)}M$ and $B \in T_{\gamma(b)}M$ be tangent vectors. Let $\mathfrak{X}_{A,B}$ denote the space of all piecewise smooth vector fields V along γ with $V(a) = A$ and $V(b) = B$. Then the unique Jacobi field J in $\mathfrak{X}_{A,B}$ is the unique critical point of the function $V \mapsto I(V,V)$ and J minimizes the index form:*

$$I(J,J) \leq I(V,V) = \int_a^b \Big(|\nabla_T V|^2 - \langle \mathrm{Rm}\,(V,T)\,T, V\rangle \Big)\, dt \tag{7.143}$$

for all $V \in \mathfrak{X}_{A,B}$, with equality if and only if $V \equiv J$.

Proof. Let V and W be piecewise smooth vector fields along γ, where $V \in \mathfrak{X}_{A,B}$ and where W vanishes at the endpoints. Then the first derivative of the index form is given by

$$\begin{aligned} \frac{1}{2}\frac{d}{du}\bigg|_{u=0} I(V+uW, V+uW) &= \frac{1}{2}(I(V,W) + I(W,V)) \\ &= \int_a^b (\langle \nabla_T V, \nabla_T W\rangle - \langle \mathrm{Rm}\,(V,T)\,T, W\rangle)\, dt \\ &= -\int_a^b \big\langle \big(\nabla_T(\nabla_T V) + \mathrm{Rm}\,(V,T)\,T\big), W\big\rangle\, dt, \end{aligned} \tag{7.144}$$

where for the second equality we integrated by parts and used that W vanishes at the endpoints. Hence, the unique critical point of the function $V \mapsto I(V,V)$ on the space $\mathfrak{X}_{A,B}$ is the unique Jacobi field J in $\mathfrak{X}_{A,B}$ (see Lemma 7.63).

Furthermore, since I is a bilinear form, we have that the second derivative of the index form at the Jacobi field J is given by

$$\frac{1}{2}\frac{d^2}{du^2}\bigg|_{u=0} I(J+uW, J+uW) = I(W,W)$$

since $I(J+uW, J+uW) = I(J,J) + 2uI(J,W) + u^2 I(W,W)$.

Claim 7.66. *If $W \not\equiv 0$ and W vanishes at the endpoints, then $I(W,W) > 0$.*

Claim 7.66 implies Lemma 7.65. Indeed, let $Z \in \mathfrak{X}_{A,B}$ and define

$$f(u) := I(J + u(Z-J), J + u(Z-J)) \quad \text{for } u \in [0,1],$$

so that $f(0) = I(J,J)$ and $f(1) = I(Z,Z)$. We compute that $f(0) = I(J,J)$, $f'(0) = 0$ by (7.144) and since J is a Jacobi field, and $f''(u) = I(Z-J, Z-J)$.

Thus, by the fundamental theorem of calculus, we have that

$$I(Z,Z) - I(J,J) = f(1) - f(0) = \int_0^1 \int_0^u f''(t)\, dt\, du > 0,$$

where the strict inequality follows from Claim 7.66, provided $Z \not\equiv J$.

We now prove Claim 7.66 by computing $I(W,W)$ with respect to a basis of Jacobi fields. Let $\{J_i(t)\}_{i=1}^n$ be Jacobi fields along $\gamma(t)$ such that $J_i(a) = 0$ and $\{J_i(b)\}_{i=1}^n$ is a basis of $T_{\gamma(b)}M$. Since γ has no conjugate points, we have that $\{J_i(t)\}_{i=1}^n$ is a basis of $T_{\gamma(t)}M$ for each $t \in (a,b]$. Thus, since $W(a) = 0$, there exist piecewise smooth functions $f_i : [a,b] \to \mathbb{R}$ such that

$$W(t) = \sum_{i=1}^n f_i(t) J_i(t)$$

for $t \in [a,b]$. Note that $W(b) = 0$ implies that $f_i(b) = 0$ for $1 \leq i \leq n$. We have

$$\nabla_T W(t) = \sum_{i=1}^n \left(f_i'(t) J_i(t) + f_i(t) \nabla_T J_i(t) \right). \tag{7.145}$$

Now, integrating by parts and using that $f_i(b) = 0$, we obtain

$$\begin{aligned}\int_a^b \left| \sum_{i=1}^n f_i \nabla_T J_i \right|^2 dt &= \sum_{i,j=1}^n \int_a^b f_i f_j \langle \nabla_T J_i, \nabla_T J_j \rangle\, dt \\ &= - \sum_{i,j=1}^n \int_a^b (f_i' f_j + f_i f_j') \langle \nabla_T J_i, J_j \rangle\, dt \\ &\quad - \sum_{i,j=1}^n \int_a^b f_i f_j \langle \nabla_T (\nabla_T J_i), J_j \rangle\, dt.\end{aligned}$$

Since each J_i is a Jacobi field, by the Jacobi equation we have that the integrand of the last line is equal to

$$- \sum_{i,j=1}^n f_i f_j \langle \nabla_T (\nabla_T J_i), J_j \rangle = \sum_{i,j=1}^n f_i f_j \langle R(J_i, T)T, J_j \rangle = \langle R(W,T)T, W \rangle.$$

By these equations and expanding the norm squared of the quantity in (7.145), we obtain that

(7.146)

$$\begin{aligned} I(W,W) &= \int_a^b \left(|\nabla_T W|^2 - \langle \mathrm{Rm}\,(W,T)\,T, W \rangle \right) dt \\ &= \int_a^b \left| \sum_{i=1}^n f_i'(t) J_i(t) \right|^2 dt + \sum_{i,j=1}^n \int_a^b (f_i f_j' - f_i' f_j) \langle \nabla_T J_i, J_j \rangle\, dt. \end{aligned}$$

On the other hand, we can see that the last term on the right-hand side vanishes as follows. Since J_i and J_j are Jacobi fields, we have that

$$\begin{aligned} T\big(\langle \nabla_T J_i, J_j\rangle - \langle \nabla_T J_j, J_i\rangle\big) &= \langle \nabla_T(\nabla_T J_i), J_j\rangle - \langle \nabla_T(\nabla_T J_j), J_i\rangle \\ &= -\langle R(J_i,T)T, J_j\rangle + \langle R(J_j,T)T, J_i\rangle \\ &= 0. \end{aligned}$$

By this and $J_i(a) = J_j(a) = 0$, we obtain that

$$\langle \nabla_T J_i, J_j\rangle - \langle \nabla_T J_j, J_i\rangle \equiv 0$$

along γ. We conclude from (7.146) that

$$I(W,W) = \int_a^b \left| \sum_{i=1}^n f_i'(t) J_i(t) \right|^2 dt,$$

which is positive since $W \not\equiv 0$. □

7.11.7. Geodesics are not minimizing beyond their first conjugate point.

Lemma 7.67 (Geodesics do not minimize past their first conjugate point). *Let $\gamma : [0,\infty) \to M^n$ be a geodesic. If $\gamma(t_0)$ is the first conjugate point to $\gamma(0)$ along γ, then $\gamma|_{[0,t_1]}$ is not minimal for all $t_1 > t_0$.*

Proof. By our hypothesis and by Lemma 7.61, there exists a nontrivial Jacobi field J along $\gamma|_{[0,t_0]}$ with $J(0) = \vec{0}$ and $J(t_0) = \vec{0}$. We have that $\langle J, T\rangle \equiv 0$. Extend J to a piecewise smooth vector field $\overline{J}$ along $\gamma|_{[0,t_1]}$ by defining $\overline{J}(t) := \vec{0}$ for $t_0 \leq t \leq t_1$. Since J is a Jacobi field along $\gamma|_{[0,t_0]}$, with $J(0) = \vec{0}$ and $J(t_0) = \vec{0}$, by Exercise 7.8 we have that

$$0 = I_{[0,t_0]}(J,J) = -\int_0^{t_0} \langle \nabla_T(\nabla_T J) + \mathrm{Rm}\,(J,T)\,T, J\rangle\, dt = I_{[0,t_1]}(\overline{J},\overline{J}),$$

where the subscripts of I denote the intervals on which the indices and vector fields are defined.

Let $\delta \in (0, t_1 - t_0)$ be sufficiently small so that $\gamma(t_2)$ and $\gamma(t_3)$ are not conjugate to each other for all $t_2, t_3 \in [t_0 - \delta, t_0 + \delta]$. Define the vector field V along $\gamma|_{[t_0-\delta,t_0+\delta]}$ to be the unique Jacobi field satisfying

$$V(t_0 - \delta) := J(t_0 - \delta), \qquad V(t_0 + \delta) := \vec{0}.$$

We have that $\langle V, T\rangle \equiv 0$. We extend V to the piecewise smooth 'comparison' vector field $\overline{V}$ along $\gamma|_{[0,t_1]}$ defined by $\overline{V}|_{[0,t_0-\delta]} = J|_{[0,t_0-\delta]}$ and $\overline{V}|_{[t_0+\delta,t_1]} = \vec{0}$; see Figure 7.11.2. So $\overline{V}$ agrees with J (and $\overline{J}$) on $\gamma|_{[0,t_0-\delta]}$, while as we will see below, $\overline{V}$ is lengthwise more 'efficient' than $\overline{J}$ in the sense of having a smaller index.

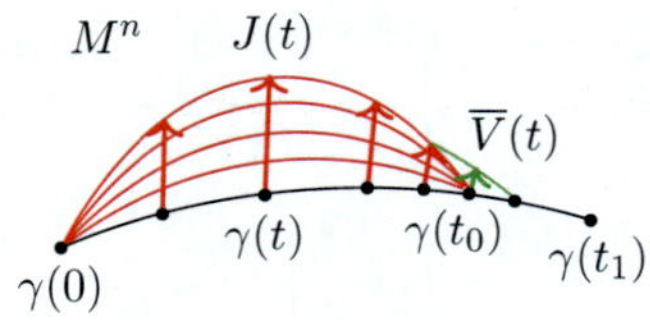

Figure 7.11.2. The vector field $\overline{V}(t)$ along $\gamma|_{[0,t_1]}(t)$. We have $\overline{V}(t) = J(t)$ for $t \leq t_0 - \delta$, $\overline{V}(t) = V(t)$ for $t_0 - \delta \leq t \leq t_0 + \delta$, and $\overline{V}(t) = \vec{0}$ for $t \geq t_0 + \delta$.

Now, since $\overline{J}|_{[t_0-\delta,t_0+\delta]}$ is not smooth, it is not a Jacobi field along $\gamma|_{[t_0-\delta,t_0+\delta]}$. Hence, by the index lemma, we have that

$$I_{[t_0-\delta,t_0+\delta]}(\overline{J},\overline{J}) > I_{[t_0-\delta,t_0+\delta]}(\overline{V},\overline{V}).$$

Since $\overline{J}$ and $\overline{V}$ agree on the intervals $[0, t_0 - \delta]$ and $[t_0 + \delta, t_1]$, it follows that

$$I_{[0,t_1]}(\overline{V},\overline{V}) < I_{[0,t_1]}(\overline{J},\overline{J}) = 0.$$

It now follows from the second variation of the arc-length formula (7.142) that there exists a variation of $\gamma|_{[0,t_1]}$ that decreases the arc length, proving that the geodesic segment $\gamma|_{[0,t_1]}$ is not minimal. □

7.11.8. Myers's theorem. A prevailing philosophy in Riemannian geometry posits that manifolds with positive curvature exhibit topologically 'simpler' structures compared to those with negative curvature.

The **diameter** of a Riemannian manifold (M^n, g) is defined by

$$\text{diam}\,(g) := \sup_{x,y\in M^n} d(x,y), \tag{7.147}$$

where d is the Riemannian distance associated to g.

The **Bonnet–Myers theorem** (a.k.a. **Myers's theorem**) is the following.

Theorem 7.68 (Bonnet and Myers [**Mye41**]). *If (M^n, g) is a complete Riemannian manifold with* Ric $\geq (n-1)\,kg$, *where k is a positive real number, then*

$$\text{diam}\,(g) \leq \frac{\pi}{\sqrt{k}}. \tag{7.148}$$

In particular, M^n is compact and $\pi_1(M^n) < \infty$.

Proof. Choose any point $p \in M^n$ and let $\gamma : [0, L] \to M^n$ be a unit-speed minimal geodesic emanating from p, where L is a positive number. The diameter estimate will follow from proving that $L \leq \frac{\pi}{\sqrt{k}}$.

Let $T(t) := \gamma'(t)$ be the unit tangent vector field along γ and let $\{e_i\}_{i=1}^{n-1}$ be orthonormal vectors at p perpendicular to $T(0)$. Parallel transport each vector e_i along γ to obtain a parallel vector field $e_i(t)$ along $\gamma(t)$ with $e_i(0) = e_i$. Then $\{e_1(t), \ldots, e_{n-1}(t), T(t)\}$ is a parallel orthonormal frame field along $\gamma(t)$. Modeling on the Jacobi fields for the n-sphere of constant sectional curvature $\frac{\pi^2}{L^2}$ (in which case γ would be a great semicircle), we define vector fields along $\gamma(t)$ by

$$V_i(t) := \sin\left(\frac{\pi t}{L}\right) e_i(t)$$

for $0 \leq t \leq L$. Observe that $|\nabla_T V_i|^2 = \frac{\pi^2}{L^2}\cos^2\left(\frac{\pi t}{L}\right)$ and that $V_i(0) = \vec{0}$ and $V_i(L) = \vec{0}$. Using that the integrals of sine squared and cosine squared are equal, we compute that the sum of the index forms (7.143) of the vector fields V_i is equal to

$$\begin{aligned}
\sum_{i=1}^{n-1} I(V_i, V_i) &= \int_0^L \sum_{i=1}^{n-1} \Big(|\nabla_T V_i|^2 - \langle \operatorname{Rm}(V_i, T)\, T, V_i \rangle \Big)\, dt \\
&= \int_0^L \sin^2\left(\frac{\pi t}{L}\right) \left((n-1)\frac{\pi^2}{L^2} - \operatorname{Ric}(T,T) \right) dt \\
&\leq (n-1)\left(\frac{\pi^2}{L^2} - k\right) \int_0^L \sin^2\left(\frac{\pi t}{L}\right) dt \\
&= (n-1)\left(\frac{\pi^2}{L^2} - k\right)\frac{L}{2},
\end{aligned} \tag{7.149}$$

where for the inequality we used the hypothesis $\operatorname{Ric}(T,T) \geq (n-1)k$.

Since $\gamma : [0, L] \to M^n$ is a minimal geodesics, either there are no conjugate points to p along γ or $\gamma(L)$ is the first conjugate point to p along γ. In either case, we may apply the index lemma (Lemma 7.65) (and Exercise 7.8) to obtain

$$I(V_i, V_i) \geq I(J, J) = 0 \quad \text{for all } i = 1, \ldots, n-1, \tag{7.150}$$

where $J(t) \equiv 0$ is the trivial Jacobi field. Applying these inequalities to (7.149) yields $\frac{\pi^2}{L^2} - k \geq 0$, so that $L \leq \frac{\pi}{\sqrt{k}}$. Since γ is an arbitrary minimal geodesic of length L, this implies that $\operatorname{diam}(g) \leq \frac{\pi}{\sqrt{k}}$, which is (7.148). Since the metric g is complete, we obtain that M^n is compact.

Now let $\pi : \widetilde{M}^n \to M^n$ be the universal covering space of M^n, so that in particular π is a local diffeomorphism. Let $\widetilde{g} := \pi^*(g)$ be the pull-back metric on $\widetilde{M}^n$. By definition, $\pi : \left(\widetilde{M}^n, \widetilde{g}\right) \to (M^n, g)$ is a local isometry. This implies that $\widetilde{\operatorname{Ric}} \geq (n-1)\widetilde{g}$ on $\widetilde{M}^n$. Thus, we may apply what we have just proved above to obtain that $\operatorname{diam}(\widetilde{g}) \leq \frac{\pi}{\sqrt{k}}$. Since $\widetilde{g}$ is complete (this follows from Theorem 7.33(2)), we also have that $\widetilde{M}^n$ is compact. It follows that $\pi_1(M^n) < \infty$. Indeed, were $\pi_1(M^n) = \infty$, then for any evenly covered neighborhood U in M^n (that is, $\pi^{-1}(U)$ is a disjoint union of open subsets, each of which is diffeomorphic to U), we would have that $\mathrm{Vol}_{\widetilde{g}}(\pi^{-1}(U)) = \infty$, which is a contradiction to $\widetilde{M}^n$ being compact. □

Corollary 7.69. *If M^n is a smooth closed manifold with infinite fundamental group, then M^n does not admit a Riemannian metric with positive Ricci curvature.*

In particular, if M^n is a smooth closed manifold of the same homotopy type as $X \times S^1$, where X is a topological space, then M^n does not admit a Riemannian metric with positive Ricci curvature.

Proof. Suppose that (M^n, g) is a smooth closed Riemannian manifold with positive Ricci curvature. Since M^n is compact, there exists a positive constant k such that $\mathrm{Ric} \geq (n-1)kg$. By Myers's theorem, $\pi_1(M^n) < \infty$. This implies the first statement of the corollary.

The second statement follows since if M^n is a smooth closed manifold of the same homotopy type as $X \times S^1$, then

$$\pi_1(M^n) \cong \pi_1(X \times S^1) \cong \pi_1(X) \times \pi_1(S^1) \cong \pi_1(X) \times \mathbb{Z},$$

which is infinite. □

7.12. The Rauch comparison theorem and its applications

The Rauch comparison theorem relates bounds on sectional curvatures of Riemannian manifolds to the behavior of their geodesics, providing insights into how geodesics diverge or converge based on curvature constraints.

7.12.1. Rauch comparison theorem.

By applying standard ODE comparison theory to the Jacobi equation, one can obtain the following.

Theorem 7.70 (Rauch comparison theorem [**Rau51**]). *Let (M_1^n, g_1) and (M_2^n, g_2) be Riemannian manifolds and let $\gamma_1 : [0, L] \to M_1^n$ and $\gamma_2 : [0, L] \to M_2^n$ be unit-speed geodesics. Assume that γ_2 has no conjugate points and that for all $t \in [0, L]$ and for all $V_1 \in (\gamma_1'(t))^\perp$ and $V_2 \in (\gamma_2'(t))^\perp$ we have that*

$$\mathrm{sect}\left(V_1 \wedge \gamma_1'(t)\right) \leq \mathrm{sect}\left(V_2 \wedge \gamma_2'(t)\right),$$

where $\mathrm{sect}(V \wedge W)$ denotes the sectional curvature of the plane spanned by V and W. If J_1 and J_2 are Jacobi fields along γ_1 and γ_2 with $J_1(0)$ and $J_2(0)$ tangent to γ_1 and γ_2, respectively, and if

$$\begin{aligned} |J_1(0)| &= |J_2(0)|, \\ \langle \nabla_{\gamma_1'} J_1(0), \gamma_1'(0) \rangle &= \langle \nabla_{\gamma_2'} J_2(0), \gamma_2'(0) \rangle, \\ |\nabla_{\gamma_1'} J_1(0)| &= |\nabla_{\gamma_2'} J_2(0)|, \end{aligned}$$

then

$$|J_1(t)| \geq |J_2(t)| \tag{7.151}$$

for all $t \in [0, L]$.

For a proof of this, see Cheeger and Ebin [**CE08**, Theorem 1.28] (we give a proof of a corollary below).

The Rauch comparison theorem says that if the sectional curvatures of one Riemannian manifold are less than or equal to the sectional curvatures of another Riemannian manifold, then for Jacobi fields that start out the same way, the lengths of the Jacobi field in the one manifold are greater than or equal to the lengths of the Jacobi field in the other manifold.

By considering the special case where

$$|J_1(0)| = |J_2(0)| = 0 \quad \text{and} \quad \left\langle \nabla_{\gamma_1'} J_1(0), \gamma_1'(0) \right\rangle = \left\langle \nabla_{\gamma_2'} J_2(0), \gamma_2'(0) \right\rangle = 0$$

and assuming that either g_1 or g_2 has constant sectional curvature, by the explicit formula (7.135) for Jacobi fields on model spaces we have the following.

Corollary 7.71 (Rauch upper comparison with constant curvature). *Let* (M^n, g) *be a Riemannian manifold with sectional curvatures less than or equal to* $k \in \mathbb{R}$ *and let* $\gamma : [0, L] \to M^n$ *be a unit-speed geodesic, where* $L \leq \frac{\pi}{\sqrt{k}}$ *if* $k > 0$. *If* $J(t)$ *is a Jacobi field along* $\gamma(t)$ *with* $J(0) = \vec{0}$ *and* $\nabla_{\gamma'} J(0) =: V$ *is a vector orthogonal to* $\gamma'(0)$, *then*

$$|J(t)| \geq |V| \operatorname{sn}_k(t) \tag{7.152}$$

for $t \in [0, L]$.

Proof. Although the corollary is a special case of Theorem 7.70, since we did not prove the theorem we give a proof of the corollary. We compute that

$$\frac{d}{dt}|J(t)| = \frac{1}{|J(t)|}\langle \nabla_T J, J \rangle,$$

where $T := \gamma'$, and

$$\frac{d^2}{dt^2}|J(t)| = \frac{1}{|J(t)|}\langle \nabla_T(\nabla_T J), J \rangle + \frac{1}{|J(t)|}|\nabla_T J|^2 - \frac{1}{|J(t)|^3}\langle \nabla_T J, J \rangle^2.$$

By applying the Jacobi equation (7.137), we obtain

$$\begin{aligned}\frac{d^2}{dt^2}|J(t)| &= -\frac{1}{|J(t)|}\langle \operatorname{Rm}(J, T)\, T, J \rangle + \frac{1}{|J(t)|}\left|\nabla_T J - \frac{1}{|J(t)|^2}\langle \nabla_T J, J \rangle J\right|^2 \\ &\geq -k|J(t)|,\end{aligned}$$

where the last inequality holds since J and T are orthogonal and the sectional curvatures are less than or equal to k. Now, since $J(0) = \vec{0}$ and

$\nabla_T J(0) = V$, we have that $|J(0)| = 0$ and $\frac{d}{dt}\big|_{t=0} |J(t)| = |V|$. In comparison, we have that the function $f(t) := |V| \operatorname{sn}_k(t)$ satisfies $\frac{d^2}{dt^2} f(t) = -kf(t)$, $f(0) = 0$, and $\frac{d}{dt}\big|_{t=0} f(t) = |V|$. Therefore, we may apply the ODE comparison theorem to conclude that $|J(t)| \geq |V| \operatorname{sn}_k(t)$ for all $t \in [0, L]$. □

7.12.2. Cartan–Hadamard theorem. When $k = 0$, the Rauch comparison theorem implies the following.

Theorem 7.72 (Cartan–Hadamard theorem [**Had98**], [**Car51**]). *If (M^n, g) is a complete Riemannian manifold with nonpositive sectional curvature, then for any $p \in M^n$, the exponential map $\exp_p : T_pM \to M^n$ is a covering map. In particular, the universal cover of M^n is diffeomorphic to Euclidean space $\mathbb{R}^n$.*

Proof. Given $p \in M^n$, let $\gamma : [0, \infty) \to M^n$ be a unit-speed geodesic with $\gamma(0) = p$, and let J be a Jacobi field along γ with $J(0) = \vec{0}$, J perpendicular to γ, and $\left|\nabla_{\gamma'} J(0)\right| > 0$. By Corollary 7.71 with $k = 0$ we have that

$$|J(t)| \geq \left|\nabla_{\gamma'} J(0)\right| t$$

for all $t \in [0, \infty)$. This implies that there are no conjugate points to p along γ and therefore $(d\exp_p)_V$ is nonsingular for all $V \in T_pM$. From this, one can deduce that the exponential map is a covering map (see the proof of Theorem 1.33 in [**CE08**] for details). Since the universal covering $(\widetilde{M}^n, \widetilde{g})$ with the lifted metric also has nonpositive sectional curvature, we obtain that given $\widetilde{p} \in \widetilde{M}^n$, the exponential map $\exp_{\widetilde{p}} : T_{\widetilde{p}}\widetilde{M} \to \widetilde{M}^n$ is a smooth and nonsingular covering map onto a simply connected manifold. This implies that $\exp_{\widetilde{p}}$ is a diffeomorphism. □

Remark 7.73. Let (M^n, g) be a complete Riemannian manifold with nonpositive sectional curvature. By covering space theory, we also have that the higher homotopy groups of M^n vanish: $\pi_i(M^n) = 0$ for $i \geq 2$. By definition, we say that M^n is a $K(\Pi, 1)$ space, where $\Pi = \pi_1(M^n)$, a.k.a. an Eilenberg–Mac Lane space.

7.12.3. The components of the metric in geodesic normal coordinates. Let $p \in M^n$. Given an orthonormal frame $\{e_i\}_{i=1}^n$ at p, let $\left\{X^i\right\}_{i=1}^n$ denote the Euclidean coordinates on the tangent space T_pM defined by

$$V =: \sum_{i=1}^n V^i e_i.$$

Correspondingly, **geodesic normal coordinates** $\{x^i\}_{i=1}^n$ centered at p are defined via the inverse of the exponential map by

$$x^i := X^i \circ \exp_p^{-1} : M^n \setminus \operatorname{Cut}(p) \to \mathbb{R}. \tag{7.153}$$

The components of the metric are defined by

$$g_{ij} := g(\partial_i, \partial_j). \tag{7.154}$$

Since $\{e_i\}_{i=1}^n$ is orthonormal, we have that

$$g_{ij}(p) = \delta_{ij}. \tag{7.155}$$

We now proceed to compute the expansion of g_{ij} near p.

For each $V \in T_pM$, consider the constant-speed geodesic γ_V emanating from p with $\gamma_V'(0) = V$. By definition, $\gamma_V(t) = \exp_p(tV)$, and hence

$$\gamma_V^i(t) := x^i \circ \gamma_V(t) = tV^i. \tag{7.156}$$

This in turn implies that

$$\frac{d\gamma_V^i}{dt}(t) = V^i \quad \text{and} \quad \frac{d^2\gamma_V^i}{dt^2}(t) = 0. \tag{7.157}$$

Using these equations, we see that the geodesic equation (7.41) for γ_V says that

$$\sum_{r,s=1}^n V^r V^s \Gamma_{rs}^k(\gamma_V(t)) = 0 \tag{7.158}$$

for all k. In particular, by taking $t = 0$, we obtain

$$\sum_{r,s=1}^n V^r V^s \Gamma_{rs}^k(p) = 0 \tag{7.159}$$

for all $V \in T_pM$.

Note that by (7.156), we also have that

$$dx^i\left(\frac{d\gamma_V}{dt}(t)\right) = V^i, \tag{7.160}$$

so that

$$\frac{d\gamma_V}{dt}(t) = \sum_{i=1}^n V^i \frac{\partial}{\partial x^i}(\gamma_V(t)). \tag{7.161}$$

The vanishing of the Christoffel symbols Γ_{ij}^k at p follows from first taking $V = e_i$ in (7.159) to obtain that $\Gamma_{ii}^k(p) = 0$ for all i and, secondly, by using this and taking $V = e_i + e_j$ in (7.159), so that $V^r = \delta_{ri} + \delta_{rj}$ and we may conclude that

$$\Gamma_{ij}^k(p) = 0. \tag{7.162}$$

Furthermore, by writing the tensor equation $\nabla g = 0$ (of the metric being parallel) in local coordinates, we obtain (cf. (4.217))

$$\partial_l g_{ij} = \sum_{r=1}^{n} \Gamma^r_{li} g_{rj} + \sum_{r=1}^{n} \Gamma^r_{lj} g_{ir}. \tag{7.163}$$

In particular, at the center p of geodesic normal coordinates,

$$\partial_l g_{ij}(p) = 0 \tag{7.164}$$

for all i, j, l.

By the formula (4.139) for the Christoffel symbols,

$$\Gamma^l_{jk} = \frac{1}{2} \sum_{m=1}^{n} g^{lm}(\partial_j g_{km} + \partial_k g_{jm} - \partial_m g_{jk}). \tag{7.165}$$

Differentiating this yields, at the center p of geodesic normal coordinates, that

$$\partial_i \Gamma^l_{jk}(p) = \frac{1}{2}(\partial_i \partial_j g_{kl} + \partial_i \partial_k g_{jl} - \partial_i \partial_l g_{jk}), \tag{7.166}$$

where we used (7.164) and the standard formula $\partial_i g^{lm} = -g^{lr} g^{ls} \partial_i g_{rs}$, which vanishes at p.

Recall from (4.185) that in any local coordinates, the components of the Riemann $(3,1)$-tensor are given by

$$R^l_{ijk} = \partial_i \Gamma^l_{jk} - \partial_j \Gamma^l_{ik} + \sum_{r=1}^{n} \Gamma^r_{jk} \Gamma^l_{ir} - \sum_{r=1}^{n} \Gamma^r_{ik} \Gamma^l_{jr}. \tag{7.167}$$

By this and (7.166), we calculate that at the center p of geodesic normal coordinates,

$$\begin{aligned} R_{ijkl}(p) = R^l_{ijk}(p) &= \partial_i \Gamma^l_{jk}(p) - \partial_j \Gamma^l_{ik}(p) \\ &= \frac{1}{2}(\partial_i \partial_j g_{kl} + \partial_i \partial_k g_{jl} - \partial_i \partial_l g_{jk}) \\ &\quad - \frac{1}{2}(\partial_j \partial_i g_{kl} + \partial_j \partial_k g_{il} - \partial_j \partial_l g_{ik}) \\ &= \frac{1}{2}(\partial_i \partial_k g_{jl} - \partial_i \partial_l g_{jk} - \partial_j \partial_k g_{il} + \partial_j \partial_l g_{ik}). \end{aligned} \tag{7.168}$$

We remark that we may use (7.168) to calculate that

$$R_{ijkl} + R_{jkil} + R_{kijl} = 0 \tag{7.169}$$

at the center of geodesic normal coordinates. Since this formula consists of the components of, and hence is equivalent to, the tensor identity

$$\mathrm{Rm}(U, V, W, X) + \mathrm{Rm}(V, W, U, X) + \mathrm{Rm}(W, U, V, X) = 0,$$

which by definition is independent of the coordinates, we have provided another proof of the first Bianchi identity (cf. Theorem 4.37).

Now, define

$$y_i(x) := \sum_{j=1}^n g_{ij}(x) x^j. \tag{7.170}$$

Differentiating this equation with respect to the coordinate vectors of geodesic normal coordinates, we obtain

$$\sum_{j=1}^n \partial_k g_{ij}(x) x^j = \frac{\partial y_i}{\partial x^k}(x) - g_{ik}(x). \tag{7.171}$$

Claim 7.74. *On $M^n \setminus \operatorname{Cut}(p)$,*

$$y_i(x) = x^i. \tag{7.172}$$

Note that the claim implies that

$$\sum_{j=1}^n \partial_k g_{ij}(x) x^j = \delta_{ik} - g_{ik}(x). \tag{7.173}$$

Furthermore, since the right-hand side is symmetric in i and k, we also obtain the identity

$$\sum_{j=1}^n \partial_k g_{ij}(x) x^j = \sum_{j=1}^n \partial_i g_{kj}(x) x^j. \tag{7.174}$$

Proof of the claim. Let $V \in T_pM$. Using (7.161), we compute that

$$\frac{d}{dt} y_i(\gamma_V(t)) = dy_i\left(\frac{d\gamma_V}{dt}\right) = \sum_{k=1}^n V^k \frac{\partial y_i}{\partial x^k}(\gamma_V(t)).$$

On the other hand,

$$\frac{d}{dt} x^i(\gamma_V(t)) = \frac{d}{dt}(tV^i) = V^i = \sum_{k=1}^n V^k \frac{\partial x^i}{\partial x^k}(\gamma_V(t)).$$

Thus,

$$\frac{d}{dt}(y_i - x^i)(\gamma_V(t)) = \sum_{k=1}^n V^k \frac{\partial (y_i - x^i)}{\partial x^k}(\gamma_V(t)). \tag{7.175}$$

Since $(y_i - x^i)(\gamma_V(0)) = 0$, by the uniqueness of solutions to ODEs, we obtain that $(y_i - x^i)(\gamma_V(t)) \equiv 0$. This completes the proof of Claim 7.74.

Recall from (7.158) that

$$\sum_{r,s=1}^n V^r V^s \Gamma_{rs}^k(\gamma_V(t)) = 0 \tag{7.176}$$

and by (7.174) (where we divided by t) that

$$\sum_{s=1}^{n} \partial_k g_{rs}(\gamma_V(t))V^s = \sum_{s=1}^{n} \partial_r g_{ks}(\gamma_V(t))V^s. \tag{7.177}$$

By substituting (7.165) into (7.176), while using (7.177) times V^r, we obtain that

$$\sum_{r,s=1}^{n} V^r V^s \partial_r g_{sk}(\gamma_V(t)) = 0 \tag{7.178}$$

for each k. This in turn implies that

$$\sum_{r,s=1}^{n} V^r V^s \partial_k g_{rs}(\gamma_V(t)) = 0. \tag{7.179}$$

Now, differentiating (7.178) with respect to t yields

$$\sum_{q,r,s=1}^{n} V^r V^s V^q \partial_q \partial_r g_{sk}(\gamma_V(t)) = 0.$$

Now, considering this as a function of V at $t = 0$ and differentiating in the direction of e_m, we obtain at p that

$$\sum_{q,s=1}^{n} V^s V^q \partial_q \partial_m g_{sk} + \sum_{q,r=1}^{n} V^r V^q \partial_q \partial_r g_{mk} + \sum_{r,s=1}^{n} V^r V^s \partial_m \partial_r g_{sk} = 0$$

for all $V \in T_pM$ and each m and k. In other words,

$$\sum_{q,s=1}^{n} V^s V^q \big(\partial_q \partial_m g_{sk} + \partial_q \partial_s g_{mk} + \partial_m \partial_s g_{qk}\big) = 0,$$

where the expression in parentheses is symmetric in q and s. Therefore, at p,

$$\partial_q \partial_m g_{sk} + \partial_q \partial_s g_{mk} + \partial_m \partial_s g_{qk} = 0 \tag{7.180}$$

for all q, m, s, k.

Similarly, by differentiating (7.179), we have that

$$\sum_{q,r,s=1}^{n} V^r V^s V^q \partial_q \partial_k g_{rs}(\gamma_V(t)) = 0. \tag{7.181}$$

This then yields

$$\sum_{q,s=1}^{n} V^s V^q \partial_q \partial_k g_{ms} + \sum_{q,r=1}^{n} V^r V^q \partial_q \partial_k g_{rm} + \sum_{r,s=1}^{n} V^r V^s \partial_m \partial_k g_{rs} = 0,$$

so that at p,

$$\partial_q \partial_k g_{ms} + \partial_s \partial_k g_{qm} + \partial_m \partial_k g_{qs} = 0$$

for all q, m, s, k. By switching the indices m and k, we may rewrite this as

$$\partial_q \partial_m g_{ks} + \partial_s \partial_m g_{qk} + \partial_k \partial_m g_{qs} = 0. \tag{7.182}$$

Now, by combining (7.180) and (7.182), we obtain at p that

$$\partial_q \partial_s g_{mk} = \partial_k \partial_m g_{qs} = -\partial_q \partial_m g_{sk} - \partial_m \partial_s g_{qk}. \tag{7.183}$$

In particular,

$$V^q V^s \partial_q \partial_s g_{mk} = -2 V^q V^s \partial_q \partial_m g_{sk}. \tag{7.184}$$

By applying (7.183) to (7.168), we have the following formula for the components of Rm at p:

$$\begin{aligned} R_{ijkl} &= \frac{1}{2}(\partial_i \partial_k g_{jl} + \partial_j \partial_l g_{ik} - \partial_i \partial_l g_{jk} - \partial_j \partial_k g_{il}) \\ &= \partial_i \partial_k g_{jl} - \partial_j \partial_k g_{il}. \end{aligned} \tag{7.185}$$

By this and (7.184), we obtain

$$\begin{aligned} V^j V^k R_{ijkl} &= V^j V^k \partial_i \partial_k g_{jl} - V^j V^k \partial_j \partial_k g_{il} \\ &= -\frac{3}{2} V^j V^k \partial_j \partial_k g_{il}. \end{aligned} \tag{7.186}$$

The expansion of the metric, near p and in geodesic normal coordinates centered at p, is given by the following (cf. Lemma 3.4 on p. 210 of [**SY94**]).

Lemma 7.75. *In geodesic normal coordinates $\{x^i\}_{i=1}^n$ centered at a point p, we have*

$$\begin{aligned} g_{ij}(x) = \delta_{ij} &- \frac{1}{3} \sum_{r,s=1}^{n} R_{irsj}(p) x^r x^s - \frac{1}{6} \sum_{r,s,t=1}^{n} \nabla_t R_{irsj}(p) x^r x^s x^t \\ &+ \sum_{r,s,t,u=1}^{n} \left(-\frac{1}{20} \nabla_t \nabla_u R_{irsj} + \frac{2}{45} \sum_{m=1}^{n} R_{irsm} R_{jtum} \right) x^r x^s x^t x^u + \mathrm{O}(r^5), \end{aligned}$$

where $x^r := x^r(x)$.

Proof (Sketch). We calculate the terms of the Taylor expansion for g_{ij} in ascending order of degree.

(0) The degree 0 term is $g_{ij}(p) = \delta_{ij}$ by (7.155).

(1) The degree 1 term is

$$\sum_{l=1}^{n} \partial_l g_{ij}(p) x^l = 0,$$

where the vanishing is by (7.164).

(2) Using (7.186), we calculate that the degree 2 term is

$$\frac{1}{2}\sum_{k,l=1}^{n}\partial_k\partial_l g_{ij}(p)x^kx^l = -\frac{1}{3}\sum_{k,l=1}^{n}R_{iklj}(p)x^kx^l.$$

(≥ 3) The higher-degree terms are calculated similarly, where the calculations are straightforward but more involved. □

7.13. Cut locus and injectivity domain and range

In this section, we consider (M^n, g) to be a complete Riemannian manifold and delve into concepts related to when geodesics stop minimizing.

7.13.1. Geodesic rays.

Definition 7.76. A unit-speed geodesic $\gamma : [0,\infty) \to M^n$ is a **geodesic ray** if γ is minimal on each finite subinterval of $[0,\infty)$; equivalently, $\gamma|_{[0,T]}$ is minimal for all $T > 0$.

Note that for any compact Riemannian manifold, there do not exist geodesic rays. This is simply because geodesics stop minimizing when or before their lengths reach the diameter of the manifold.

On the other hand, for any complete noncompact Riemannian manifold, there always exists at least one geodesic ray. To see this we can argue as follows. Choose any $p \in M^n$ and let x_i be a sequence of points with $d_i := d(x_i, p) \to \infty$. Since g is a complete metric, for each i there exists a minimal unit-speed geodesic $\gamma_{V_i} : [0, d_i] \to M^n$ from p to x_i, with $\gamma'_{V_i}(0) = V_i$, where $V_i \in T_pM$ is a unit vector. Since the unit sphere $S_pM \subset T_pM$ is compact, there exists a subsequence such that

$$\lim_{i\to\infty} V_i =: V_\infty \in S_pM \tag{7.187}$$

exists. The existence of a ray is now a consequence of Exercise 7.9.

Observe that for Euclidean space $\mathbb{R}^n$, for any $\mathbf{x} \in \mathbb{R}^n$ and $\mathbf{V} \in S^{n-1}$, the path $\gamma_{\mathbf{V}}(t) := \mathbf{x} + t\mathbf{V}$ is a geodesic ray.

7.13.2. Cut points and the cut loci in M^n and in T_pM. Let $p \in M^n$ and let $\gamma : [0,\infty) \to M^n$ be a unit-speed geodesic with $\gamma(0) = p$. Then either γ is a geodesic ray or there exists a unique $t_\gamma \in (0,\infty)$ such that $d(\gamma(t), p) = t$ for $t \leq t_\gamma$, and $d(\gamma(t), p) < t$ for $t > t_\gamma$. That is, γ *first stops minimizing at* $t = t_\gamma$.

Definition 7.77. We say that $\gamma(t_\gamma)$ is a **cut point to p along γ**.

If γ is a ray, then define $t_\gamma := \infty$ with the understanding in this instance that $t \leq t_\gamma = \infty$ above actually means $t < \infty$.

If $\gamma(t_0)$ is a conjugate point to p along γ, then $t_0 \geq t_\gamma$. This is because, by Lemma 7.67, any geodesic stops minimizing at the time of or before its first conjugate point.

Definition 7.78. The **cut locus** $\operatorname{Cut}(p)$ **of** p **in** M^n is defined to be the set of all cut points of p.

Now let

$$\overline{D}_p := \left\{V \in T_pM : d\left(\exp_p(V), p\right) = |V|\right\}, \tag{7.188}$$

which is a closed subset of T_pM. This is the set of all tangent vectors V for which $\gamma_V|_{[0,1]}$ is a minimal geodesic from p to $\exp_p(V)$. In other words,

$$\overline{D}_p = \left\{V \in T_pM : d\left(\gamma_V(1), \gamma_V(0)\right) = \mathrm{L}(\gamma_V|_{[0,1]})\right\},$$

where L denotes the length. Denote $t_V := t_{\gamma_V}$. We also have

$$\overline{D}_p = \left\{tV : V \in S_pM,\ t \in [0, t_V]\right\}. \tag{7.189}$$

The following is Proposition 5.4 in [**CE08**].

Lemma 7.79. *The function $V \mapsto t_V$ is continuous on S_pM.*

Definition 7.80. The set

$$C_p := \partial(\overline{D}_p) = \left\{t_V V : V \in S_pM \text{ and } t_V < \infty\right\} \tag{7.190}$$

is called the **cut locus of** p **in the tangent space** T_pM. Here, ∂ denotes the boundary.

By Lemma 7.79, C_p is a closed subset of T_pM. By definition, we have $V \in C_p$ if and only if the geodesic $\gamma_V(t)$ first stops minimizing at $t = 1$. It is easy to see that

$$\operatorname{Cut}(p) = \exp_p(C_p).$$

We leave it to the reader to show that $\operatorname{Cut}(p)$ is a closed subset of M^n.

7.13.3. Characterization of cut points.

Lemma 7.81. *A point $\gamma(r)$ is a cut point to p along γ if and only if r is the smallest positive number such that either $\gamma(r)$ is a conjugate point to p along γ or there exist two distinct minimal geodesics joining p and $\gamma(r)$.*

Proof. See Lemma 5.2 in Cheeger and Ebin [**CE08**]. □

7.13.4. The injectivity domain and range.

7.13.4. The injectivity domain and range. Define

$$D_p := \operatorname{int}\left(\overline{D}_p\right) = \overline{D}_p \setminus C_p \subset T_pM \tag{7.191}$$

to be the **injectivity domain** of $\exp_p$. We have that

$$D_p = \left\{tV : V \in S_pM,\ t \in [0, t_V)\right\}.$$

See Figure 7.13.1.

We claim that

$$\exp_p|_{D_p} : D_p \to M^n \setminus \operatorname{Cut}(p)$$

is a diffeomorphism. Firstly, to see that $\exp_p|_{D_p}$ is injective, suppose that $V, W \in \overline{D}_p \setminus \{\vec{0}\}$ satisfy $\exp_p(V) = \exp_p(W) =: q$. Then $\gamma_V|_{[0,1]}$ and $\gamma_W|_{[0,1]}$ are both minimal geodesics from p to $q \neq p$. By Lemma 7.81, q is a cut point of p and $V, W \in C_p$. This proves that $\exp_p|_{D_p}$ is injective. On the other hand, for each $V \in D_p$, we have that $d(\exp_p)|_V$ is nonsingular since otherwise $\exp_p(V)$ would be a conjugate point of p. Hence $\exp_p|_{D_p}$ is a smooth embedding with image equal to the open subset $\operatorname{Cut}(p)$ of M^n.

Let

$$\mathcal{D}_p := \exp_p(D_p) = M^n \setminus \operatorname{Cut}(p) \subset M^n, \tag{7.192}$$

which we call the **injectivity range** of $\exp_p$. By the remarks above, $\mathcal{D}_p$ is diffeomorphic to D_p.

Recall that $d_p(x) = d(x, p)$. If $V \in D_p$, then

$$d_p(\exp_p(V)) = d(\exp_p(V), p) = |V| =: \rho(V). \tag{7.193}$$

Since $\exp_p|_{D_p}$ is a diffeomorphism and $V \mapsto |V|$ is smooth on $T_pM \setminus \{\vec{0}\}$, we have the following result.

Lemma 7.82. *For any $p \in M^n$, the distance function d_p is smooth on the set $\mathcal{D}_p \setminus \{p\} = M^n \setminus (\operatorname{Cut}(p) \setminus \{p\})$.*

On $T_pM \setminus \{\vec{0}\}$, let $\frac{\partial}{\partial r}$ be the unit radial vector field defined by

$$\left.\frac{\partial}{\partial r}\right|_V = \frac{V}{|V|}. \tag{7.194}$$

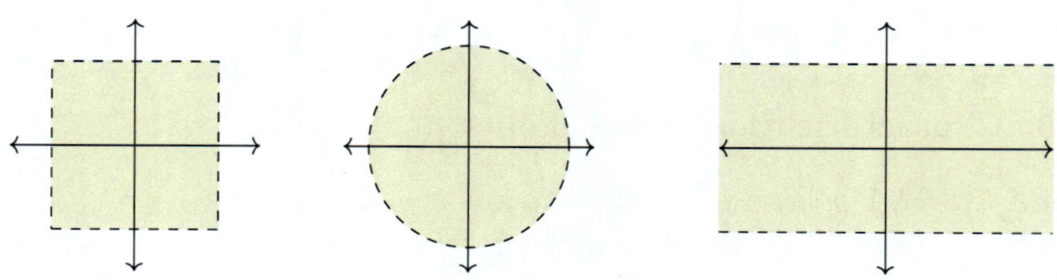

Figure 7.13.1. The injectivity domains of (left) a flat torus, (center) a round sphere, (right) a cylinder.

In other words, this is the unit outward normal vector field to the spheres of positive radii in T_pM.

Given $p \in M^n$ and $r > 0$, the **geodesic sphere** of radius r and centered at p is defined by

$$S_r(p) := \{x \in M^n \colon d(x,p) = r\}. \tag{7.195}$$

Corollary 7.83. *Let $p \in M^n$, let $V \in T_pM$ be a unit vector, and let $r \in (0, t_V)$. Then the geodesic sphere $S_r(p)$ is a smooth $(n-1)$-dimensional submanifold in a neighborhood of $x := \gamma_V(r)$ and*

$$\gamma_V'(r) = (\nabla d_p)_x \tag{7.196}$$

is a unit vector perpendicular to $T_xS_r(p)$; i.e., $(\nabla d_p)_x$ is a unit normal vector to $S_r(p)$ at x.

Proof. By differentiating (7.193), we obtain that

$$d(d_p)\big((d\exp_p)_{rV}(W)\big) = d\rho_{rV}(W) = \frac{\langle W, V\rangle}{|V|}.$$

On the other hand, by the Gauss lemma (Lemma 7.29), we have that

$$\langle (d\exp_p)_{rV}(W), (d\exp_p)_{rV}(V)\rangle = \langle W, V\rangle.$$

Therefore,

$$\langle (d\exp_p)_{rV}(W), (\nabla d_p)_x\rangle = \frac{\langle (d\exp_p)_{rV}(W), (d\exp_p)_{rV}(V)\rangle}{|V|}.$$

Since this holds for any $W \in T_pM$ and since $(d\exp_p)_{rV}$ is an isomorphism by our hypothesis, we obtain

$$(\nabla d_p)_x = (d\exp_p)_{rV}\left(\frac{V}{|V|}\right) = (d\exp_p)_{rV}\left(\frac{\partial}{\partial r}\right) = \gamma_V'(r).$$

The lemma follows since $\gamma_V'(r)$ is a unit vector and $(\nabla d_p)_x$ is normal to the level set $S_r(p)$ of the function d_p. $\square$

7.13.5. Examples.

Example 7.84. *Let (M_K^n, g_K) be an n-sphere of constant sectional curvature $K > 0$. Let $\gamma : \left[0, \frac{\pi}{\sqrt{K}}\right] \to M_K^n$ be a great semicircle. Let $V \in T_{\gamma(0)}M_K$ be a nonzero vector satisfying $\langle V, \gamma'(0)\rangle = 0$ and parallel transport V to obtain a vector field $V(t)$ along $\gamma(t)$. As in Example* 7.57,

$$J(t) := \sin\left(\sqrt{K}t\right)V(t)$$

is a Jacobi field along $\gamma(t)$ which vanishes at the endpoints. Hence, with respect to the standard model of (M_K^n, g_K), any two antipodal points on the sphere are conjugate to each other. Note that the diameter of (M_K^n, g_K) is equal to $\frac{\pi}{\sqrt{K}}$ and that geodesics stop minimizing exactly at distance $\frac{\pi}{\sqrt{K}}$.

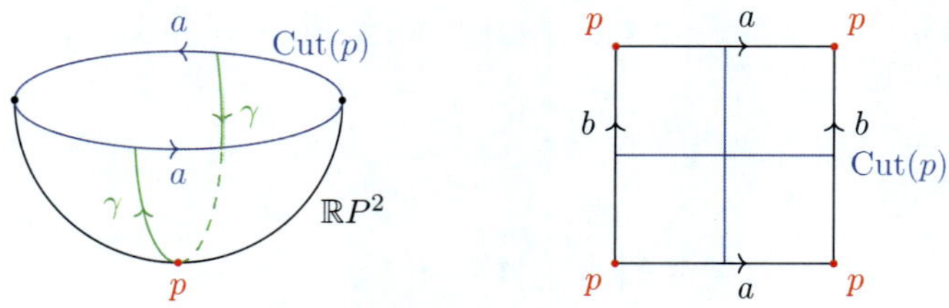

Figure 7.13.2. Left: The cut locus on $\mathbb{R}P^2$. Right: The cut locus on the 2-torus.

Example 7.85. *For the projective plane $\mathbb{R}P^2$ with the constant curvature 1-metric, for any point $p \in \mathbb{R}P^2$ and geodesic γ emanating from p, the first conjugate point along γ is at distance π, whereas the first cut point along γ is at distance $\pi/2$. See Figure* 7.13.2.

Example 7.86. *Let S^1 denote the unit circle. For the flat square torus $S^1 \times S^1$, the cut locus of a point $p = (p_1, p_2)$ is the union of two intersecting circles:*

$$\mathrm{Cut}(p) = (\{-p_1\} \times S^1) \cup (S^1 \times \{-p_2\}),$$

where $-p_1$ and $-p_2$ are antipodal to p_1 and p_2, respectively.

7.13.6. The cut locus has measure zero. Without resorting to considering a Riemannian manifold (M^n, g) and defining its Riemannian measure, we can define what it means for a set to have measure zero in a smooth manifold.

Definition 7.87. A subset A of a smooth manifold M^n has **measure zero** if for each set of local coordinates $(\mathcal{U}, \varphi_{\mathcal{U}})$, the subset $\varphi_{\mathcal{U}}(A \cap \mathcal{U})$ of $\mathbb{R}^n$ has Lebesgue measure zero.

Let $\{(\mathcal{U}, \varphi_{\mathcal{U}})\}_{\mathcal{U} \in \mathfrak{U}}$ be a collection of coordinate charts that cover A. By Lee [**Lee13**, Lemma 6.6], A has measure zero if and only if $\varphi_{\mathcal{U}}(A \cap \mathcal{U})$ of $\mathbb{R}^n$ has Lebesgue measure zero for $\mathcal{U} \in \mathfrak{U}$. We remark that if A has measure zero, then $M^n \setminus A$ is dense in M^n; see [**Lee13**, Proposition 6.8].

The following is Theorem 6.9 in [**Lee13**].

Lemma 7.88. *Let $N^n \to M^n$ be a smooth map between smooth manifolds of the same dimension. If a subset A of N^n has measure zero, then its image $f(A)$ has measure zero in M^n.*

Recall that the cut locus in the tangent space is defined by

$$C_p = \{t_V V : V \in S_p M \text{ and } t_V < \infty\},$$

where t_V is the first time at which $\gamma_V(t)$ stops minimizing. Since the function $t \mapsto t_V$ is measurable, we have that C_p has measure zero in T_pM ((radial) graphs of measurable functions have measure zero). Since $\exp_p : T_pM \to M$ is a smooth map and since $\operatorname{Cut}(p) = \exp_p(C_p)$, by Lemma 7.88 we conclude the following.

Proposition 7.89. *For any $p \in M^n$, the cut locus $\operatorname{Cut}(p)$ has measure zero in M^n.*

7.13.7. Gradient of the distance function inside the cut locus.

Lemma 7.90. *Let $p \in M^n$. If $x \in M^n \setminus (\operatorname{Cut}(p) \setminus \{p\})$, then*

$$\nabla d_p(x) = \gamma' \left(d_p(x) \right), \tag{7.197}$$

where $\gamma : [0, d_p(x)] \to M^n$ is the unique unit-speed minimal geodesic from p to x.

Proof. Let $\gamma_v : [0, L] \to M^n$, $v \in I \ni 0$, be a variation of $\gamma = \gamma_0$ with $\gamma_v(0) = p$ and for $v \in I$. We have $\mathrm{L}(\gamma_v) \geq d(p, \gamma_v(L))$, while $\mathrm{L}(\gamma_0) = d(p, \gamma_0(L))$ since γ_0 is a minimal geodesic. That is, the function $v \mapsto \mathrm{L}(\gamma_v)$ supports the function $v \mapsto (p, \gamma_v(L))$ from above at $v = 0$. Hence, since ∇d_p exists at $x = \gamma_0(L)$, we have that

$$\langle \nabla d_p, V \rangle = \left. \frac{\partial}{\partial v} \right|_{v=0} \mathrm{L}(\gamma_v),$$

where $V := \left. \frac{\partial}{\partial v} \right|_{v=0} \gamma_v(L)$. On the other hand, by the first variation formula (7.48),

$$\left. \frac{\partial}{\partial v} \right|_{v=0} \mathrm{L}(\gamma_v) = \langle \gamma_0'(L), V \rangle .$$

Therefore $\langle \nabla d_p, V \rangle = \langle \gamma_0'(L), V \rangle$. Since we may take $V \in T_xM$ to be arbitrary, (7.197) follows. □

7.13.7.1. *Synge's theorem.* Next we consider the case of positive curvature. With the idea of minimizing in a homotopy class of an element of $\pi_1(M^n)$ and some simple linear algebra, one also obtains:

Theorem 7.91 (Synge [**Syn36**]). *If (M^n, g) is an even-dimensional, orientable, closed Riemannian manifold with positive sectional curvature, then M^n is simply connected.*

Proof. Suppose, for a contradiction, that the manifold M^n were not simply connected. Then there would exist a nontrivial free (i.e., not based) homotopy class Θ of constant-speed loops. Regarding this free homotopy class, we will combine the following with a second variation argument to obtain a contradiction.

Claim 7.92. *There exists a smooth closed constant-speed geodesic* $\gamma : [0, 1] \to M^n$ *representing the free homotopy class* Θ *with minimal length among all constant-speed loops representing* Θ.

Proof of Claim 7.92. Let $\Lambda := \inf_{\gamma\in\Theta} \mathrm{L}(\gamma)$. Since M^n is compact and Θ is nontrivial, it is easy to see that $\Lambda > 0$. Let $\gamma_i : [0,1] \to M^n$ be a sequence of constant-speed loops in Θ with $2\Lambda > \mathrm{L}(\gamma_i) \to \Lambda$. Let $k := \left\lceil \frac{2\Lambda}{\mathrm{inj}(M)} \right\rceil$. Then

$$\mathrm{L}\left(\gamma_i|_{[\frac{j-1}{k},\frac{j}{k}]} \right) = \frac{\mathrm{L}(\gamma_i)}{k} < \mathrm{inj}(M^n)$$

for $1 \leq j \leq k$. Thus, without loss of generality, we may assume that each piece $\gamma_i|_{[\frac{j-1}{k},\frac{j}{k}]}$ is a constant-speed minimal geodesic (the speeds depend on i and j in general and are bounded by 2Λ); that is, each γ_i is a 'broken geodesic' with corners at the times $\frac{j}{k}$. Therefore, by the compactness of M^n and its unit tangent bundle SM, there exists a subsequence of γ_i that converges to a broken geodesic $\gamma : [0,1] \to M^n$ with $\mathrm{L}(\gamma) = \Lambda$ and $\gamma \in \Theta$. Now, by applying the first variation of arc-length formula, it follows that γ is a smooth geodesic loop. This proves the claim. □

Continuing with our 'contradiction hypothesis', we argue as follows. Now fix a point $p \in \gamma$ and consider parallel transport around γ to obtain a linear isometry $\iota : T_pM \to T_pM$ of the even-dimensional vector space T_pM. We have that $\iota(\gamma'(0)) = \gamma'(0)$ because γ is a smooth geodesic loop. Since M^n is orientable,

$$\iota : \gamma'(0)^\perp \to \gamma'(0)^\perp$$

is a special orthogonal transformation of an odd-dimensional real vector space with an inner product. The eigenvalues of an orthogonal transformation are complex numbers with length 1 arising in conjugate pairs. Since the dimension of $\gamma'(0)^\perp$ is odd and the determinant of ι is equal to 1, we conclude that 1 is an eigenvalue of ι. Hence there exists a smooth parallel unit vector field $S(t)$ along the loop $\gamma(t)$ that is normal to $\gamma'(t)$. The second variation of the arc-length formula (7.132) for a variation γ_v with $\frac{\partial}{\partial v}\gamma_v\big|_{v=0} = V$ now implies that

$$\frac{\partial^2}{\partial v^2}\bigg|_{v=0} \mathrm{L}(\gamma_v) = -\int_0^1 \langle \mathrm{Rm}(V,T)\,T, V\rangle \, dt < 0,$$

where the inequality is because the sectional curvatures are positive. However, this contradicts the fact that γ has minimal length in its free homotopy class. Hence M^n is simply connected and the proof of the theorem is complete. □

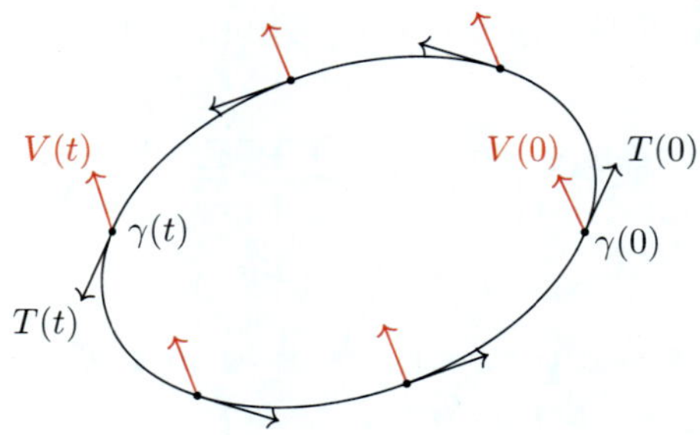

Figure 7.13.3. A stable geodesic loop γ in an even-dimensional manifold and a parallel vector field $V(t)$ along $\gamma(t)$ that is orthogonal to $T(t) = \gamma'(t)$.

7.13.8. Injectivity radius. Let (M^n, g) be a Riemannian manifold. The **injectivity radius** $\operatorname{inj}(p)$ of a point $p \in M^n$ is defined to be the largest radius r such that $B_r(\vec{0})$ is contained in the injectivity domain D_p in T_pM. That is, $\operatorname{inj}(p)$ is the largest radius r for which $\exp_p : T_pM \to M^n$ is injective when restricted to $B_r(\vec{0})$. It is easy to see that such a largest r exists.

It is also easy to see that the injectivity radius function $p \mapsto \operatorname{inj}(p)$ is continuous. The injectivity radius of (M^n, g) is defined as

$$\operatorname{inj}(g) = \inf_{p \in M^n} \operatorname{inj}(p). \tag{7.198}$$

If M^n is a closed manifold, then $\operatorname{inj}(g) > 0$.

Example 7.93. *Let (M^2, g) be a flat 2-torus $S^1(r_1) \times S^1(r_2)$, where $S^1(r_i)$ denotes the circle of radius r_i for $i = 1, 2$. Then,*

$$\operatorname{inj}(p) = \pi \min\{r_1, r_2\} \tag{7.199}$$

for each $p \in M^2$.

Theorem 7.94 (Klingenberg [**Kli61**]). *Let (M^n, g) be a complete Riemannian manifold.*

(1) *If M^n is closed and* $\operatorname{sect}(g) \leq K$, *then*

$$\operatorname{inj}(g) \geq \min\left\{\frac{\pi}{\sqrt{K}}, \frac{1}{2}\textit{length of the shortest closed geodesic}\right\}.$$

(2) *If M^n is simply connected and $0 < \frac{1}{4}K < \operatorname{sect}(g) \leq K$, then*

$$\operatorname{inj}(M^n, g) \geq \frac{\pi}{\sqrt{K}}.$$

(3) *If M^n is closed, even-dimensional, orientable, and $0 < \operatorname{sect}(g) \leq K$, then*

$$\operatorname{inj}(g) \geq \frac{\pi}{\sqrt{K}}.$$

For an exposition of the proof of Theorem 7.94, see Cheeger and Ebin [**CE08**, Corollary 5.7].

By a result of Cheeger, Gromov, and Taylor [**CGT82**], in the presence of a local curvature bound, a volume ratio lower bound implies an injectivity radius estimate.

Proposition 7.95. *For any constant $\kappa > 0$ and dimension $n \geq 2$, there exists a real number $\iota_0 > 0$ depending only on κ and n such that if a complete Riemannian manifold (M^n, g), point $p \in M^n$, and radius $r \in (0, \infty)$ satisfy the local curvature bound $|\mathrm{Rm}| \leq r^{-2}$ in $B_r(p)$ and the volume ratio lower bound $\frac{\mathrm{Vol}(B_r(p))}{r^n} \geq \kappa$, then*

$$\operatorname{inj}(p) \geq \iota_0 r.$$

In [**CGT82**] they also proved the following injectivity radius decay estimate.

Proposition 7.96. *Let (M^n, g) be a complete Riemannian manifold with the absolute values of its sectional curvatures less than or equal to a constant K_0 and let $o \in M^n$ be a point with injectivity radius $\operatorname{inj}(o) \geq \iota_0$, where ι_0 is a positive constant. Then there exist constants $c = c(n, K_0) > 0$ and $C = C(n, K_0) < \infty$ such that for any $x \in M^n$ we have that*

$$\operatorname{inj}(x) \geq c \min\{\iota_0, 1\}^n \, \mathrm{e}^{-Cd(x,o)}. \tag{7.200}$$

7.14. Statement of the Cheeger–Gromov compactness theorem

A foundational result in the study of sequences of Riemannian manifolds is the Cheeger–Gromov compactness theorem. Key prerequisites for this theorem include:

- **Control over exponential maps:** Through the Rauch comparison theorem (Theorem 7.70), we gain C^0 control of the exponential maps of Riemannian manifolds under sectional curvature bounds.
- **Control of embedding radii:** By imposing lower bounds on the injectivity radii, we ensure control over the radii of balls where exponential maps serve as embeddings.
- **Volume control:** The Bishop–Gromov volume comparison theorem provides a mechanism to control the volume.

With these controls in place, we can establish compactness results for sequences of Riemannian manifolds.

Definition 7.97. We say that a sequence of complete pointed Riemannian manifolds $\{(M_i^n, g_i, x_i)\}_{i=1}^\infty$ **converges in the C^∞ pointed Cheeger–Gromov sense** to a complete pointed Riemannian manifold $(M_\infty^n, g_\infty, x_\infty)$ as $i \to \infty$ provided the following hold. There exist

(1) an exhaustion $\{U_i\}_{i=1}^\infty$ of M_∞^n by open subsets with $x_\infty \in U_i$ and
(2) a sequence of diffeomorphisms $\Phi_i : U_i \to V_i := \Phi_i(U_i) \subset M_i^n$ with $\Phi_i(x_\infty) = x_i$

such that $\left(U_i, \Phi_i^*(g_i|_{V_i})\right)$ converges in C^∞ to (M_∞^n, g_∞) on compact subsets of M_∞^n.

Theorem 7.98 (Cheeger and Gromov). *Let $\{(M_i^n, g_i, x_i)\}_{i=1}^\infty$ be a sequence of complete pointed Riemannian manifolds that satisfy the following:*

(1)
$$\left|\nabla_{g_i}^k \operatorname{Rm}_{g_i}\right| \leq C_k \quad \text{on } M_i^n$$
for all $k \geq 0$ and $i \geq 1$, where $\{C_k\}$ is a sequence of constants independent of i, and

(2)
$$\operatorname{inj}_{g_i}(x_i) \geq \iota_0$$
for some constant $\iota_0 > 0$.

Then, after passing to a subsequence, $\{(M_i^n, g_i, x_i)\}_{i=1}^\infty$ converges in the C^∞ pointed Cheeger–Gromov sense to a complete pointed Riemannian manifold $(M_\infty^n, g_\infty, x_\infty)$ as $i \to \infty$.

See Cheeger [**Che70**], Gromov [**Gro07**], Greene and Wu [**GW88**], Peters [**Pet87**], and Hamilton [**Ham95a**]. The C^∞ version of the Cheeger–Gromov compactness theorem, which we stated above, is Theorem 2.3 in [**Ham95a**].

The Cheeger–Gromov compactness theorem has important applications to Einstein manifolds and Ricci flow (see, e.g., [**CCG⁺07**, Chapters 3 and 4] for an exposition).

7.15. Holonomy

Let (M^n, g) be a Riemannian manifold. The (Riemannian) holonomy of its Levi-Civita connection measures the nontriviality of parallel translation along loops.

7.15.1. Parallel transport along loops.

Given a path $\gamma : [0,1] \to M^n$, let

$$P_\gamma^\parallel : T_{\gamma(0)}M \to T_{\gamma(1)}M \tag{7.201}$$

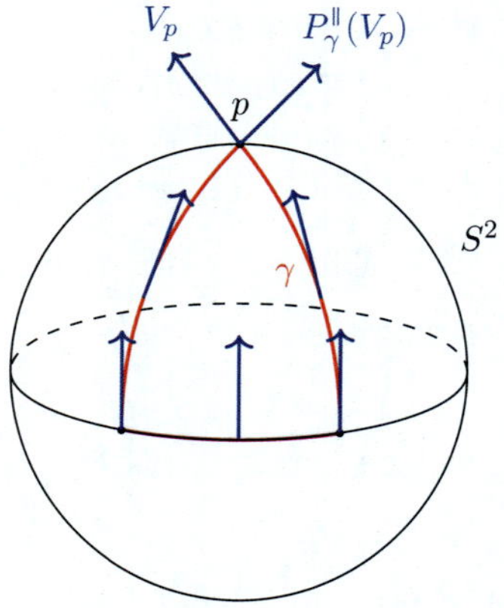

Figure 7.15.1. Parallel transport of a vector along a triangular loop γ in the 2-sphere maps V_p to $P^{\|}_{\gamma}(V_p)$. The angle of the geodesic triangle at p is equal to the rotation angle of parallel transport.

denote parallel transport along γ. That is, given $V_0 \in T_{\gamma(0)}M$, by Lemma 7.20 we may extend V_0 to a parallel vector field $V(t)$ along $\gamma(t)$, and we define $P^{\|}_{\gamma}(V_0) := V(1)$. Note that since parallel transport preserves the inner product of vector fields, we have that

$$P^{\|}_{\gamma} : \left(T_{\gamma(0)}M, g_{\gamma(0)}\right) \to \left(T_{\gamma(1)}M, g_{\gamma(1)}\right)$$

is an isometry. Here, g_x denotes the metric at x, which is an inner product on T_xM.

Now let β be a piecewise smooth loop based at a point $p \in M^n$. Then parallel transport $P^{\|}_{\beta} : T_pM \to T_pM$ is an orthogonal transformation; i.e., $P^{\|}_{\beta} \in \mathrm{O}(T_pM, g_p)$; see Figure 7.15.1. Given two piecewise smooth loops α, β based at p, let $\alpha \smile \beta : [0,1] \to M^n$ denote their concatenation defined by

$$\alpha \smile \beta(t) := \begin{cases} \alpha(2t) & \text{if } t \in [0, 1/2], \\ \beta(1-2t) & \text{if } t \in [1/2, 1]. \end{cases} \tag{7.202}$$

We have the composition rule:

$$P^{\|}_{\alpha \smile \beta} = P^{\|}_{\beta} \circ P^{\|}_{\alpha}. \tag{7.203}$$

Define $\alpha^{-1}(t) := \alpha(1-t)$. Then $P^{\|}_{(\alpha^{-1})} = \left(P^{\|}_{\alpha}\right)^{-1}$.

7.15.2. The holonomy group. The **holonomy group** at $p \in M^n$ is defined to be

$$\mathrm{Hol}_p := \{P^{\|}_{\alpha} : \alpha \in \mathrm{Loop}_p\}, \tag{7.204}$$

where Loop_p denotes the space of piecewise smooth loops based at p. Since parallel translations of pairs of vector fields along a loop preserve their inner product, we have that $\mathrm{Hol}_p \in \mathrm{O}(T_pM, g_p)$.

If, in the definition above, we restrict α to be a contractible (i.e., a null-homotopic) loop based at p, then we obtain the **restricted holonomy group** at p, which is denoted by Hol_p^0.[5]

From the concatenation properties above, we see that both the holonomy group and the restricted holonomy group at p are both subgroups of $\mathrm{O}(T_pM, g_p)$. One can prove that Hol_p^0 is the identity component of Hol_p and is a Lie subgroup of $\mathrm{O}(T_pM, g_p)$ (see Exercise 7.16). In the case of an oriented Riemannian manifold, the holonomy group at p is a subgroup of $\mathrm{SO}(T_pM, g_p)$ because parallel transport preserves both the inner product and the orientation.

Assume that M^n is connected. Then any two holonomy groups are conjugate in the following sense. Let $p, q \in M^n$ and let $\gamma : [0,1] \to M^n$ be a path from p to q. We have

$$\mathrm{Hol}_q = P_\alpha^\parallel \, \mathrm{Hol}_p \left(P_\alpha^\parallel\right)^{-1}. \tag{7.205}$$

By identifying (T_pM, g_p) with Euclidean space $(\mathbb{R}^n, g_{\mathbb{E}})$ by an isometry, we may consider $\mathrm{Hol}_p < \mathrm{O}(n, \mathbb{R})$, which is well-defined up to conjugation in $\mathrm{O}(n, \mathbb{R})$.

If α and β are loops based at p, with β contractible, then $\alpha \smile \beta \smile \alpha^{-1}$ is contractible. Hence, Hol_p^0 is a normal subgroup of Hol_p.

7.15.3. The monodromy group.

From the definition of the restricted holonomy group, we see that there is a surjective homomorphism (i.e., epimorphism) of the fundamental group:

$$\Phi : \pi_1(M^n) \to \mathrm{Hol}_p / \mathrm{Hol}_p^0, \tag{7.206}$$

called the **monodromy representation**. The group $\mathrm{Hol}_p / \mathrm{Hol}_p^0$ is called the **monodromy group**. In particular, if M^n is simply connected, then $\mathrm{Hol}_p^0 = \mathrm{Hol}_p$, and the monodromy group is trivial.

If (M^n, g) is flat, i.e., $\mathrm{Rm} \equiv 0$, then $\mathrm{Hol}_p^0 = 0$. So, in this case, the monodromy group and the holonomy group are the same.

7.15.4. Infinitesimal holonomy and curvature.

Let $p \in M^n$. The **holonomy Lie algebra** $\mathfrak{hol}_p$ is the Lie algebra of the holonomy group Hol_p (equivalently, the restricted holonomy group Hol_p^0).[6]

[5] A loop based at p is **null-homotopic** if it is homotopic to the constant loop; see Munkres [**Mun00**, Chapter 8].

[6] See §10.4.5 below for the definition of Lie algebra.

Let V, W be vectors in T_pM. Let $\Gamma : (-\varepsilon_0, \varepsilon_0) \times (-\varepsilon_0, \varepsilon_0) \to M^n$, where $\varepsilon_0 > 0$, be a smooth map such that $\Gamma(0,0) = p$,

$$d\Gamma(1,0) = V, \qquad \text{and} \qquad d\Gamma(0,1) = W.$$

Given $\varepsilon \in (0, \varepsilon_0)$, define $\mathbf{c}_\varepsilon : [0,1] \to (-\varepsilon_0, \varepsilon_0) \times (-\varepsilon_0, \varepsilon_0)$ to be a parametrization of the square with vertices $(0,0)$, $(\varepsilon, 0)$, $(\varepsilon, \varepsilon)$, $(0, \varepsilon)$, oriented in that order. Let $P^{\|}_{\gamma_\varepsilon} : T_pM \to T_pM$ be parallel transport along $\gamma_\varepsilon := \Gamma \circ \mathbf{c}_\varepsilon$. Then

$$\operatorname{Rm}(V, W)X = \lim_{\varepsilon \to 0} \frac{P^{\|}_{\gamma_\varepsilon}(X) - X}{\varepsilon^2}. \tag{7.207}$$

That is, $\operatorname{Rm}(V, W) : T_pM \to T_pM$ is given by

$$\operatorname{Rm}(V, W) = \lim_{\varepsilon \to 0} \frac{P^{\|}_{\gamma_\varepsilon} - \mathrm{id}_{T_pM}}{\varepsilon^2} \in \mathfrak{hol}_p. \tag{7.208}$$

We learned the following basic argument from Deane Yang. Let $p \in M^n$ and suppose that $\gamma : [0,1] \to M^n$ is a null-homotopic loop, and suppose that $H : [0,1] \times [0,1] \to M$ is a smooth null-homotopy for γ:

$$H(s,0) = p = H(0,t) = H(1,t), \qquad H(s,1) = \gamma(s).$$

Define coordinate vector fields along the map H by (see §7.3.1)

$$T(t,s) := dH_{(t,s)}(\partial_t), \qquad S(t,s) := dH_{(t,s)}(\partial_s).$$

Let $V_p \in T_pM$. Firstly, parallel transport V_p along $\gamma(\bar{s}) = H(\bar{s}, 1)$, $0 \leq \bar{s} \leq s$, to obtain $V(s,1)$. Secondly, parallel transport the vector $V(s,1)$ along $\bar{t} \mapsto H(s, \bar{t})$, $1 \geq \bar{t} \geq t$, to obtain $V(s,t)$, and hence we have defined a vector field V along the map H. We see that

$$V(0,t) = V_p, \qquad V(1,t) = P^{\|}_\gamma V_p.$$

Let $W_p \in T_pM$. Extend W_p to a vector field along the map H by parallel transport along the curves $\bar{t} \mapsto H(s, \bar{t})$, $0 \leq \bar{t} \leq t$, each of which emanates from $H(s,0) = p$; that is, $W(s,t)$ is defined by

$$\nabla^H_{\partial_t} W(s,t) = 0, \qquad W(s,0) = W_p.$$

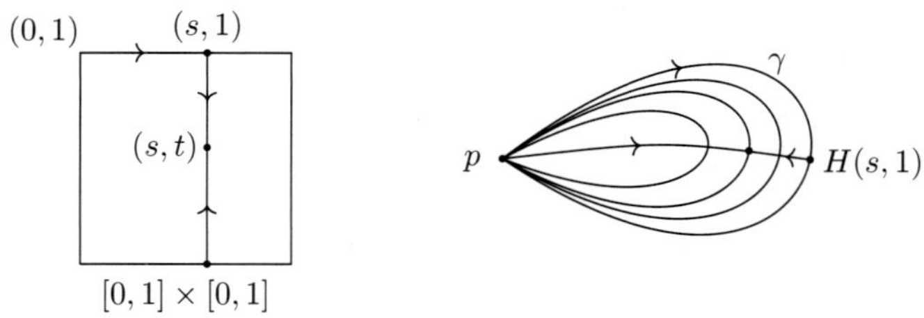

Figure 7.15.2. Left: The unit square. Left and right: The paths used to define by parallel transport of the vector fields V and W along the map H.

Hence, by using the fundamental theorem of calculus and by using $\nabla^H_{\partial_s} W(s,0) = \vec{0}$, $\nabla^H_{\partial_s} V(s,1) = \vec{0}$, $\nabla^H_{\partial_t} W(s,t) = \vec{0}$, and $\nabla^H_{\partial_t} V(s,t) = 0$, we compute that

$$
\begin{aligned}
(7.209) \qquad \langle W_p, P^{\|}_\gamma V_p - V_p\rangle &= \langle W(1,0), V(1,0)\rangle - \langle W(0,0), V(0,0)\rangle \\
&= \int_0^1 \partial_s \langle V(s,0), W(s,0)\rangle\, ds \\
&= \int_0^1 \langle \nabla^H_{\partial_s} V(s,0), W(s,0)\rangle\, ds \\
&= -\int_0^1 \left(\int_0^1 \partial_t \langle \nabla^H_{\partial_s} V(s,t), W(s,t)\rangle\, dt \right) ds \\
&= -\int_0^1 \int_0^1 \langle \nabla^H_{\partial_t} (\nabla^H_{\partial_s} V), W\rangle(s,t)\, dt\, ds \\
&= \int_0^1 \int_0^1 \langle \mathrm{Rm}(S,T)V, W\rangle(s,t)\, dt\, ds.
\end{aligned}
$$

Note that to obtain the last equality we used that

$$\mathrm{Rm}(S,T)V = \nabla^H_{\partial_s} (\nabla^H_{\partial_t} V) - \nabla^H_{\partial_t} (\nabla^H_{\partial_s} V) = -\nabla^H_{\partial_t} (\nabla^H_{\partial_s} V).$$

Define the map $P^{\|}_{s,t} : T_pM \to T_{H(s,t)}M$ by parallel transport along the path $\bar{t} \mapsto H(s,\bar{t})$, $0 \leq \bar{t} \leq t$ (note that $H(s,0) = p$). Observe that

$$(P^{\|}_{s,t})^{-1}(W(s,t)) = W_p.$$

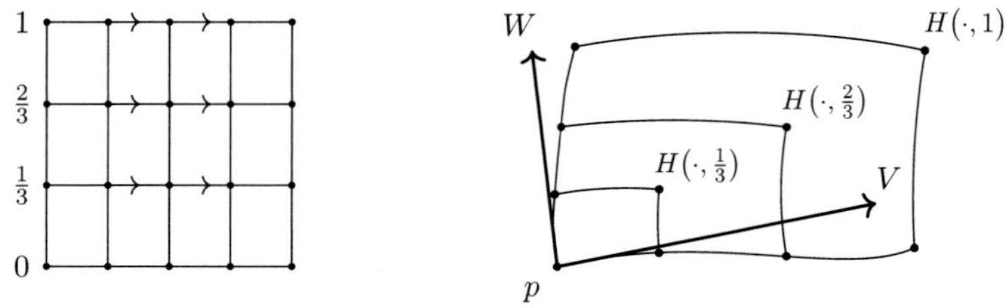

Figure 7.15.3. A null-homotopy using vectors V and W.

By (7.209), we have

$$(7.210)\quad \langle W_p, P^{\|}_\gamma V_p - V_p\rangle = \int_0^1\int_0^1 \langle (P^{\|}_{s,t})^{-1}\big(\mathrm{Rm}(S,T)V(s,t)\big), W_p\rangle\, dt\, ds.$$

Define $\widehat{P}^{\|}_{s,t} : T_pM \to T_{H(s,t)}M$ by first parallel transporting along the path $\bar{s} \mapsto H(\bar{s},1)$, $0 \leq \bar{s} \leq s$, and secondly parallel transporting along the path $\bar{t} \mapsto H(s,\bar{t})$, $1 \geq \bar{t} \geq t$. Observe that

$$V(s,t) = \widehat{P}^{\|}_{s,t}(V_p).$$

Therefore, (7.210) may be rewritten as

$$\langle W_p, P^{\|}_\gamma V_p - V_p\rangle = \int_0^1\int_0^1 \langle \big((P^{\|}_{s,t})^{-1} \circ \mathrm{Rm}(S,T) \circ \widehat{P}^{\|}_{s,t}\big)(V_p), W_p\rangle\, ds\, dt.$$

We conclude that

$$(7.211)\qquad P^{\|}_\gamma - \mathrm{id}_{T_pM} = \int_0^1\int_0^1 (P^{\|}_{s,t})^{-1} \circ \mathrm{Rm}(S,T) \circ \widehat{P}^{\|}_{s,t}\, ds\, dt.$$

We can now prove (7.207) by a homotopy as in Figure 7.15.3.

The **Ambrose–Singer theorem** says that the curvatures at a point p generate the holonomy Lie algebra at that point:

Theorem 7.99 (Ambrose and Singer [**AS53**]). *Let (M^n, g) be a Riemannian manifold and let $p \in M^n$. The holonomy Lie algebra $\mathfrak{hol}_p$ is generated by the elements in $\mathfrak{so}(T_pM, g_p)$ of the form $\mathrm{Rm}(V,W) : T_pM \to T_pM$, where $V, W \in T_pM$.*

7.15.5. The de Rham splitting theorem.

A fundamental question in Riemannian geometry is to determine conditions under which manifolds split as Riemannian products. The de Rham splitting theorem addresses this question.

Let $p \in M^n$ and let E be a vector subspace of T_pM. We say that E is a **holonomy invariant subspace** if

$$\mathrm{Hol}^0_p(E) := \left\{P^{\|}_\alpha(E) \colon \alpha \in \mathrm{Loop}_p\right\} \subset E.$$

It is easy to see that if $E \subset T_pM$ is a holonomy invariant subspace, then

$$E^\perp := \left\{ V \in T_pM \colon \langle V, W \rangle = 0 \text{ for all } W \in E \right\}$$

is also a holonomy invariant subspace.

We say that an invariant subspace is **irreducible** if it is not a nontrivial direct sum of invariant subspaces. We may decompose T_pM into irreducible holonomy invariant subspaces:

$$T_pM = (E_1)_p \oplus \cdots \oplus (E_k)_p \, .$$

By locally applying parallel translation along paths, we have in any simply connected neighborhood $\mathcal{U}$ of p that the restriction of TM to $\mathcal{U}$ has the bundle decomposition

$$TM|_{\mathcal{U}} = E_1 \oplus \cdots \oplus E_k, \tag{7.212}$$

where $E_i \to \mathcal{U}$ are subbundles invariant under parallel translation such that for each $x \in \mathcal{U}$,

$$T_xM = (E_1)_x \oplus \cdots \oplus (E_k)_x$$

is the decomposition of T_xM into its irreducible holonomy invariant subspaces. We call the local splitting (7.212) the **irreducible holonomy decomposition** of TM over $\mathcal{U}$.

We say that a connected, simply connected Riemannian manifold is **irreducible** if it is not the product of two Riemannian manifolds of positive dimension. This is equivalent to the irreducible holonomy decomposition of TM over M^n being TM itself.

We have the following global version of the holonomy splitting theorem for Riemannian manifolds.

Theorem 7.100 (De Rham holonomy splitting). *Let (M^n, g) be a complete simply connected Riemannian manifold. Suppose that $TM = E_1 \oplus \cdots \oplus E_k$ is the irreducible holonomy decomposition of TM and let $r_i := \dim E_i$. Then (M^n, g) splits as a Riemannian product of simply connected irreducible Riemannian manifolds:*

$$(M^n, g) = (N_1^{r_1}, h_1) \times \cdots \times (N_k^{r_k}, h_k), \tag{7.213}$$

where $E_i = \pi_i^*(TN_i)$ and $\pi_i : N_1^{r_1} \times \cdots \times N_k^{r_k} \to N_i^{r_i}$ is the projection map, for $1 \leq i \leq k$.

7.15.6. The Berger holonomy group classification. Understanding holonomy is crucial as it is a fundamental invariant that reflects the geometries of manifolds. Addressing the desire to classify holonomy groups, Berger's theorem provides a comprehensive solution.

Definition 7.101. We say that a Riemannian manifold (M^n, g) is **locally symmetric** if $\nabla \operatorname{Rm} \equiv 0$; i.e., the covariant derivative of the Riemann curvature tensor vanishes.

Definition 7.102. A complete Riemannian manifold (M^n, g) is called a **symmetric space** if for every point $p \in M^n$ there exists an isometry $I : (M^n, g) \to (M^n, g)$ such that $I(p) = p$ and $dI_p = -\operatorname{id}_{T_pM}$.

Theorem 7.103. *Let (M^n, g) be a simply connected Riemannian manifold. Then (M^n, g) is a symmetric space if and only if (M^n, g) is locally symmetric.*

The **rank** of a symmetric space is the dimension of the maximal flat submanifold.

We have the following **Berger holonomy classification theorem** [**Ber55**]. For the definitions of the classical groups $\mathrm{SO}(n)$, $\mathrm{U}(m)$, $\mathrm{SU}(m)$, and $\mathrm{Sp}(m)$ see §10.4.1 below. The group G_2 is an exceptional simple Lie group. $\mathrm{Spin}(n)$ is the double cover of $\mathrm{SO}(n)$.

Theorem 7.104. *If (M^n, g) is a simply connected irreducible complete Riemannian manifold, then either (M^n, g) is isometric to a symmetric space of rank ≥ 2 or the holonomy group is isomorphic to one of the following:*

- $\mathrm{SO}(n)$, *(generic) orientable Riemannian,*
- $\mathrm{U}(n/2)$, *(generic) Kähler,*
- $\mathrm{SU}(n/2)$, *Ricci-flat Kähler,*
- $\mathrm{Sp}(n/4)\,\mathrm{Sp}(1)$, *quaternion-Kähler (Einstein),*
- $\mathrm{Sp}(n/4)$, *hyperkähler (Kähler and Ricci-flat),*
- G_2 $(n = 7)$, *special holonomy Ricci-flat,*
- $\mathrm{Spin}(7)$ $(n = 8)$, *special holonomy Ricci-flat,*
- $\mathrm{Spin}(9)$ $(n = 16)$, *symmetric.*

A key idea in the proof of the theorem is that if an irreducible Riemannian manifold is not a symmetric space of rank ≥ 2, then its holonomy group acts transitively on the unit sphere of each tangent space. See Simons [**Sim62**] for an alternate proof of Berger's theorem. Later, Alekseevskiĭ [**Ale68**] and Brown and Gray [**BG72**] proved that $\mathrm{Spin}(9)$ falls into the symmetric space case.

See the books by Salamon [**Sal89**] and Joyce [**Joy00, Joy07**] for Riemannian manifolds with special holonomy, with Joyce emphasizing G_2 and $\mathrm{Spin}(7)$.

7.16. Notes and commentary

More thorough treatments of Riemannian geometry are given for example in the classic books by Hicks [**Hic65**], Cheeger and Ebin [**CE08**], Spivak [**Spi79a, Spi79b, Spi79c, Spi79d, Spi79e**], and Petersen [**Pet16**]. For detailed treatments of the basic theory, see also the standard texts by do Carmo [**dC92**] and Lee [**Lee18**]. A highly influential book is by Gromov [**Gro07**]. See also the well-known books by Gallot, Hulin, and Lafontaine [**GHL04**], Sakai [**Sak96**], Jost [**Jos17**], and Chavel [**Cha06**]. We have not discussed generalizations of Riemannian geometry in this book. For a treatment of metric geometry, see the book by Burago, Burago, and Ivanov [**BBI01**].

7.17. Exercises

Exercise 7.1 (The difference of two connections is a tensor). *Let ∇ and ∇' be two linear connections on the tangent bundle TM of a smooth manifold M^n (see Lemma* 4.28*, which gives the Levi-Civita connection as an example of a linear connection). Prove that $\nabla - \nabla'$ is a tensor. Note that the key point is to show that*

$$(\nabla - \nabla')_V(fW) = f(\nabla - \nabla')_V W$$

for any vector fields V, W and function f.

Exercise 7.2. *Prove that the definition* (7.72) *of the Ricci tensor* Ric_x *is independent of the choice of orthonormal basis $\{e_i\}_{i=1}^n$ for T_xM, $x \in M^n$.*

Exercise 7.3. *Prove that* $\mathrm{Ric}_x : T_xM \times T_xM \to \mathbb{R}$ *is a bilinear form.*

Exercise 7.4 (The Ricci tensor is symmetric). *Prove that* $\mathrm{Ric}_x(V, W) = \mathrm{Ric}_x(W, V)$ *for all $V, W \in T_xM$, $x \in M^n$.*

Exercise 7.5. *Show that any great circle (which is a closed geodesic) in a constant curvature sphere is unstable. This is in contrast to Corollary* 7.56.

Exercise 7.6. *Consider p, V, and W as in* §7.11.2 *and now let $Z \in T_pM$. Choose any path $c(u)$ so that $c'(0) = Z$. (The case above is where $Z = \vec{0}$ and $c(u) \equiv p$.) Parallel transport V and W along $c(u)$ to obtain vector fields $V(u)$ and $W(u)$ along $c(u)$. Define $\gamma_u : [0, \infty) \to M^n$, $u \in \mathbb{R}$, by*

$$\gamma_u(t) := \exp_{c(u)}\big(t(V(u) + uW(u))\big).$$

Show that $J_{Z,W}(t) := \frac{\partial}{\partial u}\big|_{u=0} \gamma_u(t)$ is a Jacobi field along the geodesic γ_0 satisfying $J_{Z,W}(0) = Z$ and $\nabla_T J_{Z,W}(0) = W$. Note that $J_W = J_{\vec{0},W}$.

Exercise 7.7. *Consider a rotationally symmetric Riemannian manifold* (M^n, g)*, where* $M^n = (a,b) \times S^{n-1}$ *and the metric is of the form*

$$g = dt^2 + f(t)^2 g_{S^{n-1}}$$

for some positive function $f : (a,b) \to \mathbb{R}$*. Let* $\gamma : (a,b) \to M^n$ *be a geodesic of the form* $\gamma(t) = (t, \theta)$*, where* $\theta \in S^{n-1}$ *is fixed. Observe that* $J(t) = f(t)V$*, where* $V \in TS^{n-1} \cong T(\{t\} \times S^{n-1})$*, is a Jacobi field along* $\gamma(t)$*. Show that* $J(t)$ *satisfies the equation*

$$\nabla_T(\nabla_T J) + \mathrm{Rm}\,(J,T)\,T = \vec{0} \tag{7.214}$$

on M^n*, where* $T(t) = \gamma'(t)$*. Hint: Show that formula* (8.146b) *implies that* $\mathrm{Rm}(E,T)T = -\frac{f''(t)}{f(t)}E$ *for any unit vector* E *perpendicular to* T*.*

Exercise 7.8. *Let* $\gamma : [a,b] \to M^n$ *be a unit-speed geodesic whose first conjugate point is at* $t = b$*. Prove that if* J *is a Jacobi field with* $J(a) = \vec{0}$ *and* $J(b) = \vec{0}$*, then* $I(J,J) = 0$ *minimizes the index form* I *defined by* (7.140) *among all piecewise smooth vector fields* V *along* γ *with* $V(a) = \vec{0}$ *and* $V(b) = \vec{0}$*.*

Exercise 7.9. *Prove that* $\gamma_{V_\infty} : [0,\infty) \to M^n$ *is a ray, where* V_∞ *is given by* (7.187)*.*

Exercise 7.10. *Let* $\{x^1, \ldots, x^m, y^1, \ldots, y^m\}$ *denote the Euclidean coordinates on* $\mathbb{R}^{2m}$*. Define*

$$V = \sum_{i=1}^{m} dx^i \otimes dy^i - \sum_{j=1}^{m} dy^j \otimes dx^j. \tag{7.215}$$

Prove that V *is a Killing vector field on* $\mathbb{R}^{2m}$*. Show that* V *generates the circle action given by* $(\mathbf{x}, \mathbf{y}) \mapsto \big(\cos(t)\mathbf{x}, \sin(t)\mathbf{y}\big)$*.*

Exercise 7.11. *Prove that if* V *is a Killing vector field and if* $\gamma(s)$ *is a unit-speed geodesic, then*

$$\frac{d}{ds}\langle V_{\gamma(s)}, \gamma'(s)\rangle \equiv 0. \tag{7.216}$$

Exercise 7.12. *Let* V *be a Killing vector field on a Riemannian manifold* (M^n, g)*. Prove that* $V(R)$*,* $\mathcal{L}_V \mathrm{Ric}$*, and* $\mathcal{L}_V \mathrm{Rm}$ *all vanish.*

Exercise 7.13. *Prove that for any vector field* V *on a Riemannian manifold* (M^n, g)*,*

$$\Delta V + \mathrm{Ric}(U) = \mathrm{div}(\mathcal{L}_V g) - \frac{1}{2}\nabla\,\mathrm{tr}(\mathcal{L}_V g). \tag{7.217}$$

In particular, if V *is a Killing vector field, i.e.,* $\mathcal{L}_V g = 0$*, then*

$$\Delta V + \mathrm{Ric}(V) = 0. \tag{7.218}$$

Exercise 7.14 (Bochner). *Let V be a Killing vector field on a Riemannian manifold (M^n, g).*

(1) *Prove that if (M^n, g) has negative Ricci curvature, then V vanishes.*
(2) *Prove that if (M^n, g) has nonpositive Ricci curvature, then V is parallel.*

Exercise 7.15. *Let V and W be Killing vector fields. Prove that their Lie bracket $[V, W]$ is a Killing vector field.*

Exercise 7.16. *Prove that the restricted holonomy group Hol_p^0 defined in* §7.15.2 *is equal to the identity component of Hol_p and is a Lie subgroup of $\mathrm{O}(T_pM)$.*

Chapter 8

Differential Forms and the Method of Moving Frames on Manifolds

In this chapter, we begin by considering orientations on smooth manifolds. We then review differential forms and integration, including Stokes's theorem. The meat of the chapter is devoted to Cartan's method of moving frames, which provides a convenient way to discuss and calculate the curvatures of Riemannian manifolds. We revisit the Hessian and Laplacian and discuss conformally equivalent metrics, hyperbolic 2-space, and basic facts about the topology and geometry of surfaces, including the uniformization theorem. We conclude the chapter with a discussion of submanifolds of Riemannian manifolds using moving frames.

In the next chapter we will apply the methods of this chapter to prove the Gauss–Bonnet formula for closed surfaces.

8.1. Orientations on manifolds

The notion of an orientation on a manifold is useful for topological reasons and for defining integration on manifolds. Roughly speaking, an orientation on a manifold is a continuous choice of orientations on its tangent spaces. So we first discuss orientations on vector spaces.

8.1.1. Orientations on vector spaces. Let $\mathbb{V}$ be a finite-dimensional vector space. Let $T : \mathbb{V} \to \mathbb{V}$ be a linear transformation. Its **determinant** $\det T$ is defined as follows. Let $\{b_i\}_{i=1}^n$, where $n = \dim \mathbb{V}$, be a basis for $\mathbb{V}$.

In this context, whenever we refer to a 'basis', we mean an 'ordered basis'. The components of T are defined by

$$T(b_i) =: \sum_{j=1}^{n} T_i^j b_j. \tag{8.1}$$

We define

$$\det T := \det(T_i^j). \tag{8.2}$$

One can easily prove that $\det T$ is independent of the choice of basis $\{b_i\}_{i=1}^n$.

We say that two bases of $\mathbb{V}$ are **equivalently oriented** if the linear transformation taking one to the other has positive determinant. That is, bases $\{b_i\}_{i=1}^n$ and $\{b_i'\}_{i=1}^n$ are equivalently oriented if the unique linear transformation $T : \mathbb{V} \to \mathbb{V}$ defined by $T(b_i) = b_i'$ for $1 \leq i \leq n$ satisfies $\det T > 0$.[1] This produces an equivalence relation on the set of all bases of $\mathbb{V}$ with exactly two equivalence classes. We call an equivalence class of bases an **orientation class** of $\mathbb{V}$. A choice of one of the two orientation classes of $\mathbb{V}$ is called an **orientation** on $\mathbb{V}$. If two orientations on $\mathbb{V}$ are not equal, we say that they are **opposite**.

Example 8.1. *Let $n \geq 2$. An orientation $\mathcal{O}$ on $\mathbb{R}^n$ is defined by the equivalence class of the Euclidean basis $\{\mathbf{e}_1, \mathbf{e}_2, \ldots, \mathbf{e}_n\}$ of $\mathbb{R}^n$. The opposite orientation $\mathcal{O}'$ on $\mathbb{R}^n$ is the equivalence class of $\{\mathbf{e}_2, \mathbf{e}_1, \ldots, \mathbf{e}_n\}$. Let $A \in \mathrm{GL}(n)$; then $\{A(\mathbf{e}_1), A(\mathbf{e}_2), \ldots, A(\mathbf{e}_n)\}$ is in $\mathcal{O}$ if and only if $\det(A) > 0$. In other words, $\{A(\mathbf{e}_1), A(\mathbf{e}_2), \ldots, A(\mathbf{e}_n)\}$ is in $\mathcal{O}'$ if and only if $\det(A) < 0$.*

The highest exterior power $\Lambda^n\mathbb{V}^*$ is a vector space isomorphic to $\mathbb{R}$. An orientation on $\mathbb{V}$ is equivalent to choosing one of the two components of $\Lambda^n\mathbb{V}^* \setminus \{0\}$. Hence, an orientation on $\mathbb{V}$ is determined by the choice of a nonzero n-form on $\mathbb{V}$. In particular, an element ω of $\Lambda^n\mathbb{V}^* \setminus \{0\}$ defines an orientation on $\mathbb{V}$ by a basis $\{b_i\}_{i=1}^n$ being positively oriented if and only if $\omega(b_1, \ldots, b_n) > 0$. For example, for the basis $\{b_i\}_{i=1}^n$ we may define the nonzero n-form $\omega := b_1^* \wedge \cdots \wedge b_n^*$, which satisfies $\omega(b_1, \ldots, b_n) = 1$. If $\{\omega^j\}_{j=1}^n$ is a basis for $\mathbb{V}^*$ and if $A : \mathbb{V}^* \to \mathbb{V}^*$ is a linear isomorphism, then

$$A(\omega^1) \wedge \cdots \wedge A(\omega^n) = \det(A)\omega^1 \wedge \cdots \wedge \omega^n. \tag{8.3}$$

So the element $A(\omega^1) \wedge \cdots \wedge A(\omega^n)$ of $\Lambda^n\mathbb{V}^* \setminus \{0\}$ defines the same orientation as $\omega^1 \wedge \cdots \wedge \omega^n$ if and only if $\det(A) > 0$.

8.1.2. Orientations on topological manifolds.

An **orientation** $\mathcal{O}$ on a topological manifold M^n, which possibly has nonempty boundary, is a *continuous choice* of orientation classes $\mathcal{O}_p$ for each T_pM, $p \in M^n$. By a

[1] Note that T is a vector space isomorphism.

continuous choice, we mean that for each $p \in M^n$ there exists an open neighborhood $\mathcal{U}$ of p and a continuous frame field $\{B_i\}_{i=1}^n$ such that $\{B_i(x)\}_{i=1}^n$ is in the orientation class $\mathcal{O}_x$ for all $x \in \mathcal{U}$. We call a basis of T_pM in the orientation class $\mathcal{O}_p$ a **positively oriented** basis.

If a manifold admits an orientation, then we say that it is **orientable**. We call a manifold with an orientation an **oriented manifold**. If a manifold is connected and orientable, then it has exactly two orientations. If a manifold is not orientable, then we say that is **nonorientable**.

If $\mathcal{O}$ is an orientation for a manifold M^n, then we denote the other orientation for M^n by $-\mathcal{O}$ and we call it the **opposite orientation**.

Note that if two orientations $\mathcal{O}_1$ and $\mathcal{O}_2$ on M^n satisfy $(\mathcal{O}_1)_p = (\mathcal{O}_2)_p$ for some point $p \in M^n$, then $\mathcal{O}_1$ and $\mathcal{O}_2$ are the same orientation.

If U^n is an open subset of an oriented manifold M^n, then the orientation $\mathcal{O}$ on M^n restricts to an orientation on U^n, which we denote by $\mathcal{O}|_U$. Hence, we have the following:

Lemma 8.2. *If U^n is an open submanifold of a manifold M^n and if U^n is nonorientable, then M^n is nonorientable.*

In §8.1.6 below we discuss examples of nonorientable manifolds.

8.1.3. Connected sums of manifolds. Let M_1^n and M_2^n be smooth manifolds. A **connected sum** $M_1^n \# M_2^n$ is obtained by removing smooth n-balls $B_1 \subset M_1^n$ and $B_2 \subset M_2^n$ and gluing together the two boundaries ∂B_1 and ∂B_2 by a diffeomorphism. In this way we can define $M_1^n \# M_2^n$ as a smooth manifold.

If M_1^n and M_2^n are connected and oriented, then one can define the connected sum so that the resulting smooth manifold $M_1^n \# M_2^n$ is unique up to diffeomorphism (see Kervaire and Milnor [**KM63**]). In this case, one of the requirements is that the diffeomorphism used to glue ∂B_1 and ∂B_2 together is orientation reversing. However, without further restrictions on this diffeomorphism, the diffeomorphism type of the resulting manifold is not unique. In particular, the first instance of showing this was Milnor's exotic 7-sphere (see [**Mil56**]). On the other hand, in low dimensions, the connected sum operation in the smooth category is unique in general.

8.1.4. The induced orientation on the boundary of a manifold. Let M^n be an oriented manifold with nonempty boundary ∂M. Then the orientation on M^n induces an orientation on ∂M as follows. Given $x \in \partial M$, let N be an outward pointing tangent vector at x. We say that a basis $\{b_i\}_{i=1}^{n-1}$ of vectors tangent to ∂M at x is positively oriented if $\{N, b_1, \ldots, b_{n-1}\}$ is positively oriented in M^n. We call this orientation on ∂M the **induced orientation**.

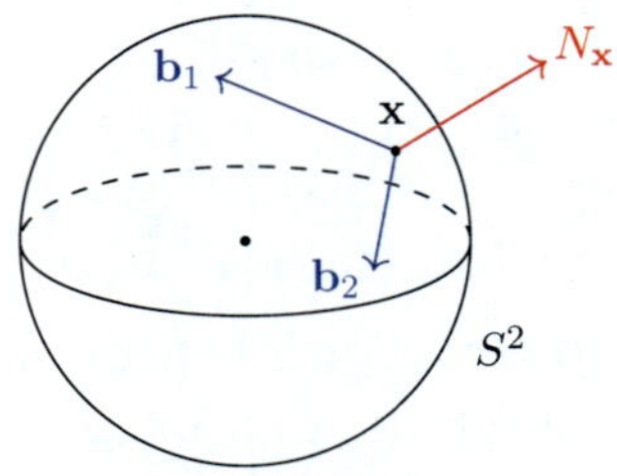

Figure 8.1.1. A positively oriented frame $\{\mathbf{b}_1, \mathbf{b}_2\}$ on S^2.

Example 8.3. *An n-sphere is orientable. Consider the unit n-sphere S^n in $\mathbb{R}^{n+1}$. Let N denote the outward unit normal to S^n. That is, $N_\mathbf{x} = \mathbf{x}$ for $\mathbf{x} \in S^n$. We define a basis $\{\mathbf{b}_i\}_{i=1}^n$ of tangent vectors to S^n at a point $\mathbf{x} \in S^n$ to be positively oriented if $\{N_\mathbf{x}, \mathbf{b}_1, \ldots, \mathbf{b}_n\}$ is positively oriented on $\mathbb{R}^{n+1}$. This orientation on S^n is equal to the orientation induced on the boundary of the closed unit ball $\bar{B}^{n+1}$, where the orientation on $\bar{B}^{n+1}$ is in turn induced from the standard orientation on $\mathbb{R}^{n+1}$. See Figure* 8.1.1.

For example, the induced orientation on S^1 is the one where the counterclockwise pointing unit tangent vector field, considered as a 1*-dimensional frame field, is positively oriented.*

Example 8.4. *Let $U^n := \{\mathbf{x} \in \mathbb{R}^n : x^n \geq 0\}$ be the upper half-space, which has a natural orientation as a subset of $\mathbb{R}^n$. Then U^n is a smooth manifold with boundary equal to the hyperplane $\partial U = \{\mathbf{x} \in \mathbb{R}^n : x^n = 0\}$. The unit outward normal vector field to ∂U is equal to $N = -\mathbf{e}_n$. Therefore $\{(-1)^n \mathbf{e}_1, \mathbf{e}_2, \ldots, \mathbf{e}_{n-1}\}$ is a positively oriented basis for the induced orientation on ∂U because $\{-\mathbf{e}_n, (-1)^n \mathbf{e}_1, \mathbf{e}_2, \ldots, \mathbf{e}_{n-1}\}$ is a positively oriented basis for $\mathbb{R}^n$.*

8.1.5. Orientation-preserving and orientation-reversing maps.

Let M_1^n and M_2^n be oriented manifolds. If $f : M_1^n \to M_2^n$ is an immersion, then we say that f is **orientation preserving** if df maps positively oriented bases to positively oriented bases. It is easy to see that the composition of orientation-preserving maps is orientation preserving.

If an immersion f is not orientation preserving, then we say that f is **orientation reversing**. In this case, df maps positively oriented bases to negatively oriented bases and vice versa.

Example 8.5. *For the unit n-sphere S^n, any reflection is orientation reversing. For example, the map $\rho_i : S^n \to S^n$ defined by*

$$\rho_i(x^1, \ldots, x^{n+1}) := (x^1, \ldots, x^{i-1}, -x^i, x^{i+1}, x^{n+1}) \tag{8.4}$$

is orientation reversing.

On the other hand, consider the antipodal map $\alpha : S^n \to S^n$ *defined by*

$$\alpha(x^1, \dots, x^{n+1}) := (-x^1, \dots, -x^{n+1}). \tag{8.5}$$

This is the restriction of a linear map with determinant equal to $(-1)^{n+1}$. *Hence,* α *is orientation reversing if* n *is even and it is orientation preserving if* n *is odd.*

8.1.6. Nonorientable manifolds. Observe the following sufficient condition for a manifold to be nonorientable. Let $\gamma : [a, b] \to M^n$ be a continuous loop; i.e., $\gamma(a) = \gamma(b) =: p$. Let $b_i : [a, b] \to TM$, $1 \leq i \leq n$, be continuous maps with the property that $\{b_i(t)\}_{i=1}^n$ is a basis for $T_{\gamma(t)}M$ for all $t \in [a, b]$. If $\{b_i(a)\}_{i=1}^n$ and $\{b_i(b)\}_{i=1}^n$ have opposite orientations as bases for T_pM, then M^n is nonorientable.

Example 8.6. *A Möbius strip* M^2 *is nonorientable; see Figure* 8.1.2. *We can construct* M^2 *as the quotient space of the unit square* $[0, 1] \times (0, 1)$ *by the identification* $(0, x) \sim (1, 1 - x)$.[2] *Let* $[\mathbf{x}] \in M^2$ *denote the equivalence class of* $\mathbf{x} \in [0, 1] \times (0, 1)$. *Define the continuous loop* $\gamma : [0, 1] \to M^2$ *by*

$$\gamma(t) = [(t, 0.5)].$$

Let $\pi : [0, 1] \times (0, 1) \to M^2$ *be the quotient map. Define* $e_i : [0, 1] \to TM$ *by*

$$e_i(t) := d\pi\big(\mathbf{e}_i(t, 0.5)\big) \quad \textit{for } i = 1, 2,$$

where $\{\mathbf{e}_1, \mathbf{e}_2\}$ *is the standard Euclidean basis. Then* $\{e_1(0), e_2(0)\}$ *and* $\{e_1(1), e_2(1)\} = \{e_1(0), -e_2(0)\}$ *are bases for* $T_{[(0,0.5)]}M$ *with opposite orientations. Therefore, the Möbius strip* M^2 *is nonorientable.*

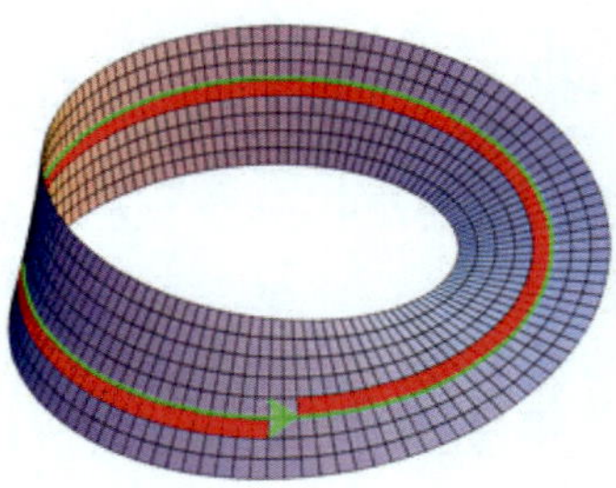

Figure 8.1.2. A Möbius strip. The green curve is a loop. Its unit tangent vector field is a smooth vector field along the loop. The thickened red strip represents a continuous choice of a side of the loop for a unit normal vector field along the loop. After traversing the loop once, the direction of the normal becomes the opposite of its original direction. Credit: By Jean-Luc W – Own work, CC BY-SA 3.0, `https://commons.wikimedia.org/w/index.php?curid=6515695`

[2]See Definition 2.5 for the definition of quotient space.

Since the Möbius strip M^2 is nonorientable and since M^2 can be considered as an open submanifold of the real projective plane $\mathbb{R}P^2$,[3] by Lemma 8.2 we have that $\mathbb{R}P^2$ is nonorientable. We also remark that any connected sum of real projective planes $\#^k \mathbb{R}P^2$ is nonorientable; this is because it contains an embedded Möbius strip.

8.2. Differential forms and integration

We assume that the reader is familiar with the theory of integration of differential forms on manifolds, so we only give a cursory review of this theory. The reader may consult, e.g., Spivak [**Spi65**], Lee [**Lee13**] (second edition), and Guillemin and Haine [**GH19**] amongst many references.

8.2.1. Differential forms on manifolds. Let M^n be a smooth manifold. A **differential k-form** α assigns to each $x \in M^n$ a multilinear map

$$\alpha_x : (T_x M)^k \to \mathbb{R} \tag{8.6}$$

(i.e., α is a tensor) such that α is **alternating**; i.e., for any permutation $\sigma \in S_k$ in the symmetric group of degree k, we have

$$\alpha_x^\sigma(V_1, \ldots, V_k) := \alpha_x(V_{\sigma(1)}, \ldots, V_{\sigma(k)}) = \operatorname{sgn}(\sigma)\alpha_x(V_1, \ldots, V_k), \tag{8.7}$$

where $\operatorname{sgn}(\sigma)$ denotes the sign of the permutation σ. Recall that $|S_k| = k!$. Given a permutation, it can be expressed as a product of transpositions (swaps of two elements), and the number of such transpositions is either always even or always odd. The sign of the permutation is 1 or -1 if it is even or odd, respectively. For each permutation σ, the map $\alpha_x \mapsto \alpha_x^\sigma$ is linear. If τ is a permutation, then

$$\alpha_x^{\tau \circ \sigma} = (\alpha_x^\sigma)^\tau. \tag{8.8}$$

We call an n-form on an n-dimensional manifold a **top-degree form**.

Given $x \in M^n$, let $\bigotimes^k T_x^* M$ be the vector space of k-tensors at x. Let $\Lambda^k T_x^* M$ be the vector space of alternating k-tensors at x, i.e., k-forms at x. We define the **alternating map**

$$\operatorname{Alt} : \bigotimes^k T_x^* M \to \bigotimes^k T_x^* M$$

by

$$\operatorname{Alt}(\beta_x) := \frac{1}{k!} \sum_{\sigma \in S_k} \operatorname{sgn}(\sigma)\beta_x^\sigma. \tag{8.9}$$

Note by (8.7) that the image of the map Alt is equal to $\Lambda^k T_x^* M$ and that Alt restricted to $\Lambda^k T_x^* M$ is equal to the identity map.

[3]Topologically, the real projective plane $\mathbb{R}P^2$ can be constructed by identifying the points on the boundary of a disk by the antipodal map. By taking a collar of the quotient of this boundary in $\mathbb{R}P^2$, we obtain an embedded Möbius strip. See Figure 7.13.2.

The alternating map is a linear map and has the properties that

$$\mathrm{Alt}(\beta_x^\sigma) = \big(\mathrm{Alt}(\beta_x)\big)^\sigma = \mathrm{sgn}(\sigma)\,\mathrm{Alt}(\beta_x). \tag{8.10}$$

The **wedge product** of a k-form α and an ℓ-form β is defined by (cf. [**Lee13**, (14.3)])

$$\alpha \wedge \beta := \frac{(k+\ell)!}{k!\ell!}\,\mathrm{Alt}(\alpha \otimes \beta). \tag{8.11}$$

For example, if α and β are 1-forms, then it follows that

$$\alpha \wedge \beta = \alpha \otimes \beta - \beta \otimes \alpha.$$

Moreover, if γ is a 1-form, then

$$\begin{aligned} \alpha \wedge \beta \wedge \gamma = {} & \alpha \otimes \beta \otimes \gamma - \alpha \otimes \gamma \otimes \beta + \gamma \otimes \alpha \otimes \beta \\ & - \gamma \otimes \beta \otimes \alpha + \beta \otimes \gamma \otimes \alpha - \beta \otimes \alpha \otimes \gamma. \end{aligned} \tag{8.12}$$

The wedge product has the properties (see [**Lee13**, Proposition 14.11]) that it is bilinear, associative, and for any k-form α and ℓ-form β,

$$\alpha \wedge \beta = (-1)^{k\ell} \beta \wedge \alpha. \tag{8.13}$$

In particular, if α and β are 1-forms, then $\alpha \wedge \beta = -\beta \wedge \alpha$. Moreover, for any integer $1 \leq m \leq n$, 1-forms $\alpha^1, \ldots, \alpha^m$, and tangent vectors $V_1, \ldots, V_m$, we have

$$(\alpha^1 \wedge \cdots \wedge \alpha^m)(V_1, \ldots, V_m) = \det(\alpha^j(V_i)). \tag{8.14}$$

Moreover, for any $\sigma \in S_m$,

$$\alpha^{\sigma(1)} \wedge \cdots \wedge \alpha^{\sigma(m)} = \mathrm{sgn}(\sigma) \alpha^1 \wedge \cdots \wedge \alpha^m. \tag{8.15}$$

Given local coordinates $(\mathcal{U}, \{x^i\}_{i=1}^n)$, for any $x \in \mathcal{U}$, we have that the collection

$$\{dx^{i_1} \wedge \cdots \wedge dx^{i_k}\}_{1 \leq i_1 < \cdots < i_k \leq n}$$

is a basis for $\Lambda^k T_x^* M$. In particular, $\dim(\Lambda^k T_x^* M) = \binom{n}{k}$. So the top-degree exterior algebra satisfies $\Lambda^n T_x^* M \cong \mathbb{R}$.

Given a tangent vector $V \in T_x M$, the associated **interior product**

$$\iota_V : \bigotimes{}^k T_x^* M \to \bigotimes{}^{k-1} T_x^* M \tag{8.16}$$

is defined by

$$(\iota_V \beta_x)(V_1, \ldots, V_{k-1}) := \beta_x(V, V_1, \ldots, V_{k-1}). \tag{8.17}$$

Another standard notation for the interior product is $V \lrcorner \beta_x$.

For each V, the ι_V is a linear map. Moreover, $V \mapsto \iota_V$ is linear. If β_x is a k-tensor and γ_x is an ℓ-tensor, then

$$\iota_V(\beta_x \otimes \gamma_x) = (\iota_V \beta_x) \otimes \gamma_x \tag{8.18}$$

if $k \geq 1$. The case $k = 0$ corresponds to $\iota_V(c\gamma_x) = c\iota_V \gamma_x$.

8.2.2. Integration of differential forms. Given a map $F : N^m \to M^n$ and a k-tensor β on M^n, its **pull-back** to N^m by f is defined by

$$(F^*\beta)(V_1, \ldots, V_k) := \beta\big(dF(V_1), \ldots, dF(V_k)\big).$$

Now let M^n and N^n be smooth manifolds of the same dimension and let $F : M^n \to N^n$ be a smooth map. The pull-back on the top-degree exterior algebra has the following nice property. Let $(\mathcal{U}, \{x^i\}_{i=1}^n)$ and $(\mathcal{V}, \{y^j\}_{j=1}^n)$ be local coordinates on M^n and N^n, respectively. Let $u : \mathcal{V} \to \mathbb{R}$ be a continuous function. Then, on $F^{-1}(\mathcal{V}) \cap \mathcal{U}$ we have (see Proposition 14.20 of [**Lee13**])

$$F^*(u\,dy^1 \wedge \cdots \wedge dy^n) = (u \circ F) \det\left(\frac{\partial F^j}{dx^i}\right) dx^1 \wedge \cdots \wedge dx^n, \tag{8.19}$$

where $F^j := y^j \circ F$.

The basis for defining integration of differential forms on manifolds is the compatibility of (8.19) with the change of variables formulas for Euclidean integration. Let M^n be an oriented smooth manifold and let ω be a continuous n-form on M^n with compact support $\operatorname{supp}(\omega) := \operatorname{cl}(\{x \in M^n : \omega_x \neq 0\})$, where cl denotes the closure. The integral $\int_{M^n} \omega$ is defined as follows.

Let $\mathcal{C} := \{(\mathcal{U}_\alpha, \mathbf{x}_\alpha = \{x_\alpha^i\}_{i=1}^n)\}_{\alpha \in A}$ be a finite collection of positively oriented coordinate charts covering $\operatorname{supp}(\omega)$. Let $\{\phi_\alpha\}_{\alpha \in A}$ be a partition of unity subordinate to $\mathcal{C}$, so that by definition $\operatorname{supp}(\phi_\alpha) \subset \mathcal{U}_\alpha$ and $\sum_{\alpha \in A} \phi_\alpha = 1$. We define

$$\int_{M^n} \omega := \sum_{\alpha \in A} \int_{\mathcal{U}_\alpha} \phi_\alpha \omega := \sum_{\alpha \in A} \int_{\mathbf{x}_\alpha(\mathcal{U}_\alpha)} (\mathbf{x}_\alpha^{-1})^*(\phi_\alpha \omega), \tag{8.20}$$

where the integrals in the rightmost sum are over open subsets of $\mathbb{R}^n$. The integration of an n-form on $\mathbb{R}^n$ is defined by

$$\int_{\mathbb{R}^n} f\,dx^1 \wedge \cdots \wedge dx^n := \int_{\mathbb{R}^n} f, \tag{8.21}$$

where the right-hand side is the usual Riemann integral of a continuous function with compact support. Using (8.19) and the usual change of variables formula for the integration of functions on $\mathbb{R}^n$, one can show that the definition (8.20) of the integration of top-degree differential forms on manifolds is well-defined.

We have the following naturality property for integration of differential forms. If $F : M^n \to N^n$ is an orientation-preserving diffeomorphism and if ω is an n-form on N^n, then

$$\int_{M^n} F^*\omega = \int_{N^n} \omega. \tag{8.22}$$

See [**Lee13**, Chapter 16] for a detailed presentation of integration on manifolds.

8.2.3. Exterior derivative. Let M^n be a smooth manifold. Let $\Omega^k(M^n)$, where $1 \leq k \leq n$, denote the space of smooth k-forms on M^n. Let ω be a k-form on M^n. Let $(\mathcal{U}, \{x^i\}_{i=1}^n)$ be a coordinate chart for M^n. Let $I = (i_1, \ldots, i_k)$ be a multi-k-index and denote the k-form

$$dx^I := dx^{i_1} \wedge \cdots \wedge dx^{i_k} \tag{8.23}$$

and the function

$$\omega_I := \omega(\partial_{i_1}, \ldots, \partial_{i_k}), \tag{8.24}$$

where $\partial_{i_j} := \frac{\partial}{\partial x^{i_j}}$. We may express ω in local coordinates as

$$\omega = \sum_I \omega_I \, dx^I. \tag{8.25}$$

The **exterior derivative** is defined by (cf. [**Lee13**, (14.19)])

$$d\omega = \sum_I d\omega_I \wedge dx^I. \tag{8.26}$$

Here, the 1-form $d\omega_I$ is defined by $d\omega_I(V) = V(\omega_I)$, where $V(\omega_I)$ denotes the directional derivative of ω_I along V. This definition of $d\omega$ is independent of the choice of local coordinates $\{x^i\}_{i=1}^n$.

For example, if $\omega = \sum_{i=1}^n \omega_i \, dx^i$ is a 1-form, then

$$d\omega = \sum_{i,j=1}^n \frac{\partial \omega_i}{\partial x^j} dx^j \wedge dx^i = \sum_{i<j} \left(\frac{\partial \omega_i}{\partial x^j} - \frac{\partial \omega_j}{\partial x^i} \right) dx^j \wedge dx^i.$$

The exterior derivative $d : \Omega^k(M^n) \to \Omega^{k+1}(M^n)$ is a linear map and has the following properties (see [**Lee13**, Proposition 14.23]):

(1) (Acting on functions) If f is a function, then

$$df(V) := V(f) \tag{8.27}$$

for any tangent vector V.

(2) (Product rule) If ω and η are differential forms, then

$$d(\omega \wedge \eta) = d\omega \wedge \eta + (-1)^k \omega \wedge d\eta, \tag{8.28}$$

where k is the degree of ω.

(3) (Coboundary operator)

$$d^2 := d \circ d = 0 : \Omega^k(M^n) \to \Omega^{k+2}(M^n). \tag{8.29}$$

That is, the coboundary of the coboundary is equal to zero.

(4) (Naturality) If $F : M^n \to N^m$ is a smooth map and ω is a differential form on N^m, then

$$d(F^*\omega) = F^*(d\omega). \tag{8.30}$$

The exterior derivative d is the unique collection of operators from $\Omega^k(M^n)$ to $\Omega^{k+1}(M^n)$ satisfying properties (1)–(3).

8.2.4. Properties of the exterior derivative. We have the following formula for the exterior derivative of a k-form ω in terms of directional derivatives and the Lie bracket (see [**Lee13**, Proposition 14.32]):

$$\begin{aligned}(d\omega)(V_0, \dots, V_k) & \\ &= \sum_{j=0}^{k} (-1)^j \, V_j\big(\omega(V_0, \dots, \hat{V}_j, \dots, V_k)\big) \\ &\quad + \sum_{0 \le i < j \le k} (-1)^{i+j} \, \omega\big([V_i, V_j], V_0, \dots, \hat{V}_i, \dots, \hat{V}_j, \dots, V_k\big).\end{aligned} \tag{8.31}$$

By the product rule for covariant differentiation and the fact that the Levi-Civita connection ∇ is torsion-free, we can express $d\omega$ in terms of covariant differentiation as

$$(d\omega)(V_0, \dots, V_k) = \sum_{j=0}^{k} (-1)^j \, (\nabla_{V_j}\omega)\big(V_0, \dots, \hat{V}_j, \dots, V_k\big); \tag{8.32}$$

see Exercise 8.6. The right-hand side of (8.32) is also referred to as the **exterior covariant derivative**. So, for a Riemannian manifold, the exterior derivative and the exterior covariant derivative are one and the same.

For example, if ω is a 1-form, then by (8.31) and (8.32) we have that

$$\begin{aligned}(d\omega)(V_0, V_1) &= V_0\big(\omega(V_1)\big) - V_1\big(\omega(V_0)\big) - \omega\big([V_0, V_1]\big) \\ &= \left(\nabla_{V_0}\omega\right)(V_1) - \left(\nabla_{V_1}\omega\right)(V_0).\end{aligned} \tag{8.33}$$

The Lie derivative of a differential form satisfies the formula

$$\mathcal{L}_V\omega = \iota_V(d\omega) + d(\iota_V\omega) \tag{8.34}$$

for any vector field V and differential form ω, where ι denotes the interior product. See Theorem 14.35 of [**Lee13**]; Lee calls (8.34) "Cartan's magic formula".

8.2.5. Stokes's theorem. The generalization of the fundamental theorem of calculus to manifolds is the following (cf. [**Lee13**, Theorem 16.11]).

Theorem 8.7 (Stokes's theorem). *Let M^n be an oriented compact manifold with boundary ∂M, where ∂M has the induced orientation. If ω is a differential $(n-1)$-form on M^n, then*

$$\int_{M^n} d\omega = \int_{\partial M} \omega. \tag{8.35}$$

On the right-hand side, ω actually denotes $i^*\omega$, where $i : \partial M \to M^n$ is the inclusion map.

Example 8.8. *If M^n is the closed interval $[a,b]$ in $\mathbb{R}$ and f is a 0-form on $[a,b]$, i.e., f is function, then Stokes's theorem $\int_{[a,b]} df = \int_{\partial[a,b]} f$ says that*

$$\int_a^b f'(x)\,dx = f(b) - f(a),$$

which is the fundamental theorem of calculus.

8.2.6. The volume form. Let (M^n, g) be an oriented Riemannian manifold. The volume form $d\mu$ is a globally defined n-form on M^n defined as follows. Let $x \in M^n$. Let $\{e_i\}_{i=1}^n$ be a positively oriented orthonormal frame at x, so that $g(e_i, e_j) = \delta_{ij}$ for $1 \leq i, j \leq n$. Let $\{\omega^j\}_{j=1}^n$ be the dual orthonormal coframe of 1-forms, so that $\omega^j(e_i) = \delta_{ij}$ for $1 \leq i, j \leq n$. The **volume form** of (M^n, g) is defined by

$$d\mu := \omega^1 \wedge \omega^2 \wedge \cdots \wedge \omega^n. \tag{8.36}$$

We have (cf. [**Lee13**, Proposition 15.29])

$$d\mu(e_1, \ldots, e_n) = 1. \tag{8.37}$$

Suppose that $\{x^i\}_{i=1}^n$ are positively oriented local coordinates; i.e., the local frame field $\left\{\frac{\partial}{\partial x^i}\right\}_{i=1}^n$ is positively oriented. Then the volume form may be expressed as

$$d\mu = \sqrt{\det(g_{ij})}\, dx^1 \wedge dx^2 \wedge \cdots \wedge dx^n. \tag{8.38}$$

To see this, define the matrix $(A_i^\alpha)_{i,\alpha=1}^n$ by $\frac{\partial}{\partial x^i} = \sum_{\alpha=1}^n A_i^\alpha e_\alpha$. We compute that

$$d\mu\left(\frac{\partial}{\partial x^1}, \ldots, \frac{\partial}{\partial x^n}\right) = \det(A_i^\alpha)\, d\mu(e_1, \ldots, e_n) = \det(A_i^\alpha).$$

On the other hand,

$$g_{ij} = \sum_{\alpha=1}^n A_i^\alpha A_j^\alpha,$$

so that $\det(g_{ij}) = (\det(A_i^\alpha))^2$. This proves (8.38).

8.2.7. The divergence theorem. In this subsection we derive the divergence theorem from Stokes's theorem. The proof is an instructive exercise in calculating with the exterior, covariant, and Lie derivatives and the interior product.

Let β be a p-form on an oriented Riemannian manifold (M^n, g) with boundary. Recall from (8.34) the general formula for the Lie derivative acting on differential forms:

$$\mathcal{L}_V \beta = \iota_V(d\beta) + d(\iota_V \beta). \tag{8.39}$$

Now assume that β is an n-form. Since $d\beta = 0$, we have

$$\mathcal{L}_V \beta = d(\iota_V \beta). \tag{8.40}$$

Hence,

$$\int_{M^n} \mathcal{L}_V \beta = \int_{\partial M} \iota_V \beta. \tag{8.41}$$

Taking, $\beta = d\mu$, we have that

$$\mathcal{L}_V(d\mu) = d(\iota_V d\mu) = \operatorname{div}(V) d\mu, \tag{8.42}$$

where $\operatorname{div}(V) = \sum_{i=1}^n \langle \nabla_{e_i} V, e_i \rangle$ is the divergence of the vector field V as defined by (5.53). Let $V =: \sum_{i=1}^n V^i e_i$, where $\{e_i\}_{i=1}^n$ is a positively oriented orthonormal frame. To see the last equality in (8.42), using (8.32) we compute that

$$\begin{aligned} d(\iota_V d\mu)(e_1, \ldots, e_n) &= \sum_{i=1}^n (-1)^{i-1} (\nabla_{e_i}(\iota_V d\mu))(e_1, \ldots, \hat{e}_i, \ldots, e_n) \\ &= \sum_{i=1}^n (-1)^{i-1} e_i\big((\iota_V d\mu)(e_1, \ldots, \hat{e}_i, \ldots, e_n)\big) \\ &\quad - \sum_{i=1}^n \sum_{j \neq i} (-1)^{i-1} (\iota_V d\mu)(e_1, \ldots, \nabla_{e_i} e_j, \ldots, \hat{e}_i, \ldots, e_n) \\ &= \sum_{i=1}^n e_i(V_i) + \sum_{i \neq j} V^j \langle \nabla_{e_i} e_j, e_i \rangle = \operatorname{div}(V), \end{aligned}$$

where to obtain the last line we used that $\langle \nabla_{e_i} e_j, e_j \rangle = 0$ and that

$$\begin{aligned} V^j d\mu & \left(e_j, e_1, \ldots, \sum_{k=1}^n \langle \nabla_{e_i} e_j, e_k \rangle e_k, \ldots, \hat{e}_i, \ldots, e_n \right) \\ &= V^j \langle \nabla_{e_i} e_j, e_i \rangle \, d\mu(e_j, e_1, \ldots, e_i, \ldots, \hat{e}_i, \ldots, e_n) \\ &= (-1)^i V^j \langle \nabla_{e_i} e_j, e_i \rangle, \end{aligned}$$

where the e_k and the e_i are in the $(j+1)$-st slot.

Let $x \in \partial M$ be a boundary point. Let $\{e_i\}_{i=1}^{n-1}$ be an orthonormal frame on ∂M at x that is positively oriented with respect to the induced orientation on ∂M; that is, $\{N, e_1, \ldots, e_{n-1}\}$ is positively oriented, where N is the unit outward normal to ∂M. The **area form** $d\sigma$ of ∂M at x is defined by

$$d\sigma := e_1^* \wedge \cdots \wedge e_{n-1}^*. \tag{8.43}$$

On ∂M we have

$$i^*(\iota_V d\mu) = V \cdot N\, d\sigma, \tag{8.44}$$

where $i : \partial M \to M^n$ is the inclusion map. To see (8.44), we calculate that

$$\begin{aligned} i^*(\iota_V d\mu)(e_1, \ldots, e_{n-1}) &= d\mu(V, e_1, \ldots, e_{n-1}) \\ &= \langle V, N \rangle\, d\mu(N, e_1, \ldots, e_{n-1}) \\ &= \langle V, N \rangle. \end{aligned}$$

Therefore, we obtain the following from (8.41); cf. [**Lee13**, Theorem 16.32].

Corollary 8.9 (Divergence theorem). *Let (M^n, g) be an oriented compact Riemannian manifold with boundary ∂M, where ∂M has the induced orientation. If V is a vector field on M^n, then*

$$\int_{M^n} \operatorname{div}(V)\, d\mu = \int_{\partial M} V \cdot N d\sigma, \tag{8.45}$$

where $d\sigma$ is the volume $(n-1)$-form of ∂M.

This is a generalization of the Euclidean divergence theorem (Proposition 5.12). For applications of the divergence theorem, see §11.1.4 below on integration by parts, which is a very useful technique in geometric analysis.

Let $f : M^n \to \mathbb{R}$ be a function and take the vector field $V = \nabla f$ to be equal to the gradient of f in the divergence theorem. We then obtain

$$\int_{M^n} \Delta f\, d\mu = \int_{\partial M} \nabla f \cdot N\, d\sigma = \int_{\partial M} N(f)\, d\sigma, \tag{8.46}$$

where $\Delta f := \operatorname{div}(\nabla f)$ is the **Laplacian** of f. In particular, if M^n is a closed manifold, then

$$\int_{M^n} \Delta f\, d\mu = 0. \tag{8.47}$$

8.3. Moving frames

On a manifold M^n, local coordinates $\{x^i\}_{i=1}^n$ are used to define a smooth structure; but they are not specially adapted to a Riemannian metric g on M^n. For example, let $\mathcal{U}$ be an open subset of $\mathbb{R}^n$ and let $g_{ij} : \mathcal{U} \to \mathbb{R}$ be *any* smooth functions with the property that for each $\mathbf{x} \in \mathcal{U}$, the matrix $(g_{ij}(\mathbf{x}))_{i,j=1}^n$ is positive-definite. Then $g := \sum_{i,j=1}^n g_{ij} dx^i \otimes dx^j$ is a Riemannian metric on $\mathcal{U}$. In fact, given local coordinates $\{x^i\}_{i=1}^n$, knowing the metric g is the same as knowing its components g_{ij}.

The search for a more well-adapted way to express the metric g motivates Cartan's method of moving frames, which we discuss in this section; cf. Hicks [**Hic65**]. In short, instead of using local coordinates to express geometric invariants, we will use an orthonormal frame field defined on an open subset.

8.3.1. Orthonormal frame and coframe fields.

Let $\{e_i\}_{i=1}^n$ be an **orthonormal frame field** on an open subset $\mathcal{V}$ of M^n; that is, each e_i is a vector field on $\mathcal{V}$ and for each $x \in \mathcal{V}$, $\{e_i(x)\}_{i=1}^n$ is an orthonormal basis for T_xM. In other words,

$$g(e_i, e_j) = \delta_{ij}.$$

We also call $\{e_i\}_{i=1}^n$ a **moving frame**.

Note that moving frames are generally only locally defined. For example, the **hairy ball theorem** says that any even-dimensional sphere does not admit a continuous everywhere nonvanishing vector field (see Poincaré [**Poi85**] and Brouwer [**Bro11**]; for an exposition, see Milnor [**Mil78**]). See Figure 8.3.1 for a picture of a vector field on the 2-sphere that has only one zero.

Example 8.10. *Recall from* (7.119) *that the upper half-plane model for* 2*-dimensional hyperbolic space is* $U^2 = \{\mathbf{x} =: (x^1, x^2) \in \mathbb{R}^2 : x^2 > 0\}$ *with the Riemannian metric* $g_{\mathbb{U}} = \frac{1}{(x^2)^2} g_{\mathbb{E}}$. *A global orthonormal frame field for* $(U^2, g_{\mathbb{U}})$ *is given by*

$$e_1(\mathbf{x}) = x^2 \mathbf{e}_1, \quad e_2(\mathbf{x}) = x^2 \mathbf{e}_2. \tag{8.48}$$

Note that for any tangent vector at (x^1, x^2) *which is unit with respect to* $g_{\mathbb{U}}$, *its length with respect to the Euclidean metric* $g_{\mathbb{E}}$ *is equal to* $(x^2)^{-1}$. *See Figure* 8.3.2.

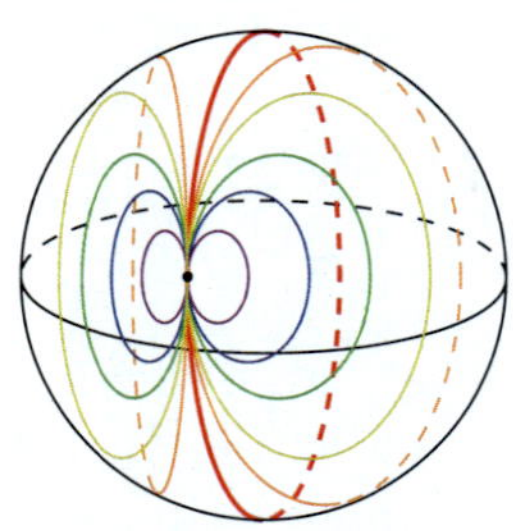

Figure 8.3.1. The flow lines of a vector field on S^2 with a single zero.

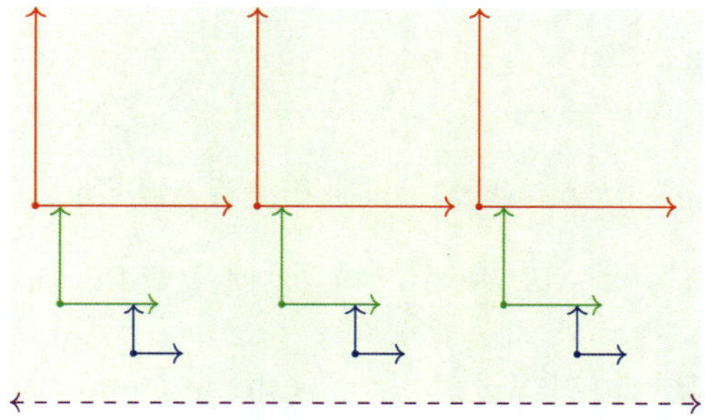

Figure 8.3.2. A moving frame on the upper half-plane model of 2-dimensional hyperbolic space.

Let $\{\omega^j\}_{j=1}^n$ be the **dual orthonormal coframe field** on $\mathcal{V}$; that is,

$$\omega^j(e_i) = \delta_{ij} \quad \text{for } 1 \le i, j \le n. \tag{8.49}$$

Observe that the metric may be expressed as

$$g = \sum_{j=1}^n \omega^j \otimes \omega^j. \tag{8.50}$$

This formula holds simply because

$$g(e_i, e_k) = \delta_{ik} = \sum_{j=1}^n \omega^j(e_i)\omega^j(e_k) = \sum_{j=1}^n (\omega^j \otimes \omega^j)(e_i, e_k).$$

So the metric g may be recovered from the orthonormal frame field on the open subset on which it is defined. On the other hand, g cannot be recovered from local coordinates without knowing the components g_{ij}.

8.3.2. The inner product on the cotangent bundle. We remark, in the language of §7.1.5, that $(e_i)^\flat = \omega^i$ since $(e_i)^\flat(e_j) = g(e_i, e_j) = \delta_{ij} = \omega^i(e_j)$ for all $1 \le j \le n$. That is, ω^i is the metrically dual cotangent vector of e_i for each i.

Note that if $V = \sum_{i=1}^n V^i e_i$, where $V^i \in \mathbb{R}$, then $V^\flat = \sum_{i=1}^n V^i \omega^i$. And if $\alpha = \sum_{i=1}^n \alpha_i \omega^i$, then $\alpha^\natural = \sum_{i=1}^n \alpha_i e_i$. So we have $V^\flat = \alpha$ if and only if $\alpha^\natural = V$. In particular, $(\omega^i)^\natural = e_i$. In other words, the linear transformations

$$\cdot^\flat : T_xM \to T_x^*M \quad \text{and} \quad \cdot^\natural : T_x^*M \to T_xM$$

are inverses of each other.

The **dual metric** (inner product) g^* on T_x^*M is defined by

$$g^*(\alpha, \beta) := g(\alpha^\natural, \beta^\natural). \tag{8.51}$$

Observe that

$$g^*(\omega^i, \omega^j) = \delta_{ij}. \tag{8.52}$$

Thus, $\{\omega^j(x)\}_{j=1}^n$ is an orthonormal basis for T_x^*M for each $x \in \mathcal{V}$.

Remark 8.11 (The dual metric in terms of local coordinates). Suppose that $(\mathcal{U}, \mathbf{x})$, where $\mathbf{x} = (x^1, \dots, x^n)$, are local coordinates on M^n. Then $\{dx^j\}_{j=1}^n$ is the dual basis of $\left\{\frac{\partial}{\partial x^i}\right\}_{i=1}^n$. We have that

$$\left(\frac{\partial}{\partial x^i}\right)^\flat \left(\frac{\partial}{\partial x^j}\right) = g\left(\frac{\partial}{\partial x^i}, \frac{\partial}{\partial x^j}\right) = g_{ij}.$$

And

$$\delta_{ij} = dx^j\left(\frac{\partial}{\partial x^i}\right) = g\left((dx^j)^\natural, \frac{\partial}{\partial x^i}\right).$$

This implies that

$$(dx^j)^\natural = \sum_{k=1}^n (g^{-1})_{jk} \frac{\partial}{\partial x^k}. \tag{8.53}$$

From this it is easy to derive that

$$g^*(dx^i, dx^j) = (g^{-1})_{ij} =: g^{ij}.$$

That is, the g^{ij} are the components of the inner product g^* on the cotangent space with respect to the basis $\{dx^j\}_{j=1}^n$. Hence, if $\alpha = \sum_{i=1}^n \alpha_i dx^i$ and $\beta = \sum_{i=1}^n \beta_j dx^j$, then

$$g^*(\alpha, \beta) = \sum_{i,j=1}^n g^{ij} \alpha_i \beta_j. \tag{8.54}$$

8.3.3. Connection 1-forms.

Just as we can express the Riemannian metric g in terms of an orthonormal (co)frame field $\{\omega^j\}_{j=1}^n$, we can express the Levi-Civita connection in terms of an orthonormal frame field $\{e_i\}_{i=1}^n$. Given a tangent vector $V \in T_xM$, $x \in \mathcal{V}$, we have that $\nabla_V e_i \in T_xM$. Hence this tangent vector is a linear combination of the $e_j(x)$.

We define the **connection 1-forms** ω_i^j, $1 \le i, j \le n$, as the components of the Levi-Civita connection with respect to $\{e_i\}_{i=1}^n$; that is,

$$\nabla_V e_i =: \sum_{j=1}^n \omega_i^j(V) e_j, \tag{8.55}$$

for all $1 \le i \le n$ and all $V \in T_xM$, $x \in \mathcal{V}$. In other words,

$$\omega_i^j(V) = g(\nabla_V e_i, e_j). \tag{8.56}$$

Observe that

$$\begin{aligned} 0 &= V\big(g(e_i, e_j)\big) \\ &= g(\nabla_V e_i, e_j) + g(e_i, \nabla_V e_j) \\ &= \omega_i^j(V) + \omega_j^i(V). \end{aligned}$$

Therefore, as differential 1-forms on $\mathcal{V}$, we have that

$$\omega_i^j = -\omega_j^i. \tag{8.57}$$

Example 8.12 (Euclidean space). *For Euclidean space $(\mathbb{R}^n, g_{\mathbb{E}})$, the standard basis $\{\mathbf{e}_i\}_{i=1}^n$ is a globally defined orthonormal frame. Since $\nabla_V \mathbf{e}_i = \mathbf{0}$ for any $V \in T\mathbb{R}^n$, we have that the associated connection 1-forms vanish: $\omega_i^j \equiv 0$ for $1 \leq i, j \leq n$.*

Recall from §6.4 that a vector field V on M^n is simply a section of the tangent bundle TM; that is, $V : M^n \to TM$, where $V(x) \in T_x M$. Dually, a differential 1-form α on M^n is a section of the cotangent bundle T^*M; that is, $\alpha : M^n \to T^*M$, where $\alpha(x) \in T_x^*M$; see §3.3.2. Recall that covariant differentiation of 1-forms is defined via the product rule and covariant differentiation of vector fields. That is,

$$(\nabla_V \alpha)(W) := V(\alpha(W)) - \alpha(\nabla_V W). \tag{8.58}$$

The motivation for this definition is that it is equivalent to the product rule:

$$V(\alpha(W)) = (\nabla_V \alpha)(W) + \alpha(\nabla_V W), \tag{8.59}$$

where we think of the contraction $\alpha(W)$ as a product.

Since $\omega^j(e_i) = \delta_j^i$, we have

$$(\nabla_V \omega^j)(e_i) = V\big(\omega^j(e_i)\big) - \omega^j(\nabla_V e_i) = -\omega_i^j(V). \tag{8.60}$$

Therefore, the covariant derivatives of the dual orthonormal coframe vector fields ω^j are given in terms of the connection 1-forms by

$$\nabla_V \omega^j = -\sum_{i=1}^n \omega_i^j(V)\omega^i. \tag{8.61}$$

To understand covariant differentiation of vector fields is equivalent to understanding covariant differentiation of 1-forms. The advantage of considering 1-forms is that we have formulated the exterior derivative as an operator on differential forms.

We also remark that the discussion above defines the covariant derivative on the cotangent bundle T^*M. With covariant differentiation defined on both TM and T^*M, we can use the product rule to define covariant differentiation on any tensor bundle, i.e., for tensors of type (r, s).

8.3.4. Parallel translation in terms of the connection 1-forms. Let $\{e_i\}_{i=1}^n$ be an orthonormal frame field defined on an open subset $\mathcal{U}$ of M^n. Let $\gamma : (a,b) \to \mathcal{U}$ be a smooth path. Set $T(t) := \gamma'(t)$. Let $V(t)$ be a vector field along $\gamma(t)$, so that $V(t) \in T_{\gamma(t)}M$. The components of $V(t)$ are defined by

$$V(t) = \sum_{i=1}^n V^i(t) e_i(\gamma(t)).$$

We compute that

$$\nabla_{T(t)} V(t) = \sum_{i=1}^n \frac{dV^i}{dt}(t) e_i(\gamma(t)) + \sum_{i,j=1}^n V^i(t) \omega_i^j(T(t)) e_j(\gamma(t)) \tag{8.62}$$

$$= \sum_{i=1}^n \left(\frac{dV^i}{dt}(t) + \sum_{j=1}^n V^j(t) \omega_j^i(T(t)) \right) e_i(\gamma(t)).$$

Thus, we have that $V(t)$ is parallel along $\gamma(t)$ if and only if

$$\frac{dV^i}{dt}(t) + \sum_{j=1}^n V^j(t) \omega_j^i(T(t)) = 0 \quad \text{for all } 1 \leq i \leq n. \tag{8.63}$$

In particular, the equation for the path $\gamma(t)$ to be a geodesic is

$$\gamma''(t)^i + \sum_{j=1}^n \gamma'(t)^j \omega_j^i(\gamma'(t)) = 0 \quad \text{for all } 1 \leq i \leq n, \tag{8.64}$$

where $\gamma'(t) =: \sum_{i=1}^n \gamma'(t)^i e_i$.

8.3.5. Curvature 2-forms. Similarly to expressing the Levi-Civita connection in terms of our orthonormal frame field, we can express the Riemann curvature tensor Rm in terms of the orthonormal frame field. As we will see in the following sections, this provides a convenient way to compute the curvature tensors of Riemannian manifolds. The **curvature 2-forms** Ω_i^j, $1 \leq i,j \leq n$, on $\mathcal{V}$ are defined by

$$\sum_{j=1}^n \Omega_i^j(V,W) e_j := \mathrm{Rm}(V,W) e_i. \tag{8.65}$$

Observe that

$$\Omega_i^j(V,W) = g\big(\mathrm{Rm}(V,W)e_i, e_j\big). \tag{8.66}$$

Moreover,

$$\Omega_i^j = \sum_{k,l=1}^n R_{ijkl}\, \omega^k \wedge \omega^l. \tag{8.67}$$

Note also that

$$\Omega_i^j(e_k, e_\ell) = g\big(\operatorname{Rm}(e_k, e_\ell)e_i, e_j\big) = R_{k\ell ij}. \tag{8.68}$$

By the antisymmetry of the Riemann curvature 4-tensor in its *first* two components, we have that each Ω_i^j is a 2-form. By the antisymmetry of the Riemann curvature 4-tensor in its *last* two components, we have that

$$\Omega_i^j = -\Omega_j^i \tag{8.69}$$

for all $1 \leq i, j \leq n$.

Of course, we can recover the Riemann curvature tensor from the curvature 2-forms by the formula

$$\operatorname{Rm}(V, W)U = \sum_{i,j=1}^{n} \Omega_i^j(V, W)U^i e_j,$$

where $U =: \sum_{i=1}^n U^i e_i$.

8.4. Cartan structure equations

The Cartan structure equations relate an orthonormal coframe with its connection 1-forms and curvature 2-forms via the exterior derivative and the wedge product. They provide a convenient way to calculate the curvatures of Riemannian metrics.

8.4.1. The first Cartan structure equation. Recall that, for a differentiable manifold M^n, the exterior derivative is a first-order linear differential operator acting on differential forms. In particular, if α is a 1-form on M^n, then $d\alpha$ is a 2-form on M^n. By (8.33), we have that the exterior derivative is the antisymmetrization of the covariant derivative:

$$(d\alpha)(V_0, V_1) = (\nabla_{V_0}\alpha)(V_1) - (\nabla_{V_1}\alpha)(V_0). \tag{8.70}$$

In particular, taking the exterior derivative of the dual orthonormal coframe vector fields yields

$$\begin{aligned} d\omega^i(V_0, V_1) &= \left(\nabla_{V_0}\omega^i\right)(V_1) - \left(\nabla_{V_1}\omega^i\right)(V_0) \\ &= -\omega_j^i(V_0)\omega^j(V_1) + \omega_j^i(V_1)\omega^j(V_0) \\ &= \left(\sum_{j=1}^{n} \omega^j \wedge \omega_j^i\right)(V_0, V_1). \end{aligned}$$

Thus, we obtain the **first Cartan structure equation**:

$$d\omega^i = \sum_{j=1}^{n} \omega^j \wedge \omega_j^i. \tag{8.71}$$

This provides equations satisfied by the connection 1-forms ω_j^i along with the conormal basis of 1-forms $\{\omega^j\}_{j=1}^n$.

8.4.2. Revisiting the proof of the first Cartan structure equation. Let $\mathbb{V}$ be an n-dimensional vector space. Let $\{e_i\}_{i=1}^n$ be a (not necessarily orthonormal) basis for $\mathbb{V}$ and let $\{\omega^j\}_{j=1}^n$ be its dual basis for $\mathbb{V}^*$. Define $\mathbb{V}^* \otimes \mathbb{V}$ as the span of $\{\omega^j \otimes e_i\}_{i,j=1}^n$; that is,

$$\mathbb{V}^* \otimes \mathbb{V} := \left\{ \sum_{i,j=1}^{n} a_j^i \omega^j \otimes e_i : a_j^i \in \mathbb{R} \right\}. \tag{8.72}$$

Given $\alpha \in \mathbb{V}^*$ and $W \in \mathbb{V}$, the tensor $\alpha \otimes W \in \mathbb{V}^* \otimes \mathbb{V}$ may be identified with the linear map $\alpha \otimes W : \mathbb{V} \to \mathbb{V}$ defined by

$$(\alpha \otimes W)(V) := \alpha(V) W. \tag{8.73}$$

This identification induces an isomorphism from $\mathbb{V}^* \otimes \mathbb{V}$ to the space $\mathrm{Lin}(\mathbb{V}, \mathbb{V})$ of linear maps from $\mathbb{V}$ to itself.

With this in mind, we can re-express the derivation the first Cartan structure equation as follows. Denote $(\nabla e_i)(V) := \nabla_V e_i$. By (8.55), we have that the covariant derivative of the vector field e_i on $\mathcal{V}$ satisfies

$$\nabla e_i = \sum_{j=1}^{n} \omega_i^j \otimes e_j. \tag{8.74}$$

We have $(\nabla e_i)_X \in T_X^* M \otimes T_X M$ for all $X \in \mathcal{V}$. Denote $(\nabla \omega^i)(V) := \nabla_V \omega^i$, so that

$$(\nabla \omega^i)(V, W) := \big((\nabla \omega^i)(V)\big)(W) = (\nabla_V \omega^i)(W). \tag{8.75}$$

Note that $(\nabla \omega^i)_X : T_X M \times T_X M \to \mathbb{R}$ is a bilinear form. By (8.61), we have

$$\nabla \omega^i = -\sum_{j=1}^{n} \omega_j^i \otimes \omega^j. \tag{8.76}$$

Furthermore, by (8.70) and (8.76),

$$\begin{aligned} d\omega^i(V,W) &= (\nabla\omega^i)(V,W) - (\nabla\omega^i)(W,V) \\ &= -\sum_{j=1}^{n}(\omega^i_j \otimes \omega^j)(V,W) + \sum_{j=1}^{n}(\omega^i_j \otimes \omega^j)(W,V) \\ &= \left(\sum_{j=1}^{n}\omega^j \wedge \omega^i_j\right)(V,W). \end{aligned}$$

We have proved (8.71) again.

The connection 1-forms ω^j_i can be expressed in terms of the exterior derivatives of the coframe; see Exercise 8.12.

8.4.3. The second Cartan structure equation. Just as the connection 1-forms ω^j_i satisfy a nice system of equations, the first Cartan structure equation, so do the curvature 2-forms Ω^j_i. To this end, we consider the second covariant derivatives of the orthonormal frame vector fields. Using (4.156), for $x \in \mathcal{U}$ define $\nabla^2 e_i : T_xM \times T_xM \to T_xM$ by

$$(\nabla^2 e_i)(V,W) := \nabla^2_{V,W}e_i = \nabla_V(\nabla_W e_i) - \nabla_{\nabla_V W}e_i.$$

Since $(\nabla^2 e_i)_x$ is a bilinear map for each $x \in \mathcal{U}$, we can think of it as an element of $T^*_xM \otimes T^*_xM \otimes T_xM$. We remark, given the definition of ∇e_i above, that $\nabla^2 e_i$ is equal to the second covariant derivative $\nabla^2 e_i := \nabla(\nabla e_i)$ of the vector field e_i because of the following calculation using the product rule:

$$\begin{aligned} &(\nabla^2 e_i)(V,W,U) \\ &\quad = \big(\nabla_V(\nabla e_i)\big)(W,U) \\ &\quad = V\big((\nabla e_i)(W,U)\big) - (\nabla e_i)(\nabla_V W,U) - (\nabla e_i)(W,\nabla_V U) \\ &\quad = V\big((\nabla_W e_i)(U)\big) - (\nabla_{\nabla_V W}e_i)(U) - (\nabla_W e_i)(\nabla_V U) \\ &\quad = \big(\nabla_V(\nabla_W e_i)\big)(U) + (\nabla_W e_i)(\nabla_V U) \\ &\qquad - (\nabla_{\nabla_V W}e_i)(U) - (\nabla_W e_i)(\nabla_V U) \\ &\quad = \big(\nabla_V(\nabla_W e_i)\big)(U) - (\nabla_{\nabla_V W}e_i)(U) \\ &\quad = (\nabla^2_{V,W}e_i)(U). \end{aligned}$$

We compute that

$$\begin{aligned}
\nabla^2_{V,W} e_i &= \nabla_V(\nabla_W e_i) - \nabla_{\nabla_V W} e_i \\
&= \nabla_V\left(\sum_{j=1}^n \omega_i^j(W) e_j\right) - \nabla_{\nabla_V W} e_i \\
&= \sum_{j=1}^n (\nabla_V \omega_i^j)(W) e_j + \sum_{j=1}^n \omega_i^j(\nabla_V W) e_j \\
&\quad + \sum_{j=1}^n \omega_i^j(W) \nabla_V e_j - \nabla_{\nabla_V W} e_i \\
&= \sum_{j=1}^n (\nabla_V \omega_i^j)(W) e_j + \sum_{j=1}^n \omega_i^j(W) \sum_{k=1}^n \omega_j^k(V) e_k \\
&= \sum_{j=1}^n \left((\nabla_V \omega_i^j)(W) + \sum_{k=1}^n \omega_i^k(W) \omega_k^j(V) \right) e_j.
\end{aligned}$$

By simply switching V and W, we obtain

$$\nabla^2_{W,V} e_i = \sum_{j=1}^n \left((\nabla_W \omega_i^j)(V) + \sum_{k=1}^n \omega_i^k(V) \omega_k^j(W) \right) e_j.$$

Thus, by (4.157), the curvature 2-forms (which measure the noncommutativity of the operation of taking two covariant derivatives) are given by

$$\begin{aligned}
&\Omega_i^j(V,W) \\
&\quad = g\big(\operatorname{Rm}(V,W)e_i, e_j\big) \\
&\quad = g(\nabla^2_{V,W} e_i, e_j) - g(\nabla^2_{W,V} e_i, e_j) \\
&\quad = (\nabla_V \omega_i^j)(W) + \sum_{k=1}^n \omega_i^k(W)\omega_k^j(V) - (\nabla_W \omega_i^j)(V) - \sum_{k=1}^n \omega_i^k(V)\omega_k^j(W) \\
&\quad = (\nabla_V \omega_i^j)(W) - (\nabla_W \omega_i^j)(V) - \left(\sum_{k=1}^n \omega_i^k(V)\omega_k^j(W) - \omega_i^k(W)\omega_k^j(V) \right) \\
&\quad = d\omega_i^j(V,W) - \sum_{k=1}^n (\omega_i^k \wedge \omega_k^j)(V,W).
\end{aligned}$$

We have derived the **second Cartan structure equation**:

$$\Omega_i^j = d\omega_i^j - \sum_{k=1}^n \omega_i^k \wedge \omega_k^j. \tag{8.77}$$

Observe that if $n = 2$, then we have that

$$\Omega_2^1 = d\omega_2^1 - \sum_{k=1}^{2} \omega_2^k \wedge \omega_k^1 = d\omega_2^1. \tag{8.78}$$

Thus, the curvature 2-form Ω_2^1 is exact on the domain of the orthonormal frame field $\{e_1, e_2\}$. Recall from (4.188) that the Gauss curvature of g is given by $K = g\big(\mathrm{Rm}(e_2, e_1)e_1, e_2\big)$. So, by (8.66),

$$\Omega_1^2(e_2, e_1) = g\big(\mathrm{Rm}(e_2, e_1)e_1, e_2\big) = K.$$

The Gauss curvature of g is given by

$$K = \Omega_2^1(e_1, e_2) = d\omega_2^1(e_1, e_2). \tag{8.79}$$

For computing the Gauss curvature of a rotationally symmetric surface, see Exercise 8.14.

In §8.7 below we will use the second Cartan structure equation to calculate the curvature 2-forms for the upper half-plane model of hyperbolic 2-space.

8.5. The Hessian and Laplacian

The Hessian and Laplacian, which we first encountered for submanifolds in §5.4, are both fundamental operators in the study of differential geometry and geometric analysis. In this section, after giving their invariant definitions for smooth manifolds, we express them in terms of both local coordinates and moving frames.

8.5.1. Definition of the Hessian of a function. Let (M^n, g) be a Riemannian manifold. Let $u : M^n \to \mathbb{R}$ be a C^2 function. The **covariant Hessian** of u is the symmetric 2-tensor $\nabla^2 u$ defined by

$$(\nabla^2 u)(V, W) := \nabla^2_{V,W} u := V(W(u)) - (\nabla_V W)(u) \tag{8.80}$$

for vector fields V and W on M^n. It is easy to see from this definition that

$$\nabla^2_{fV,W} u = f \nabla^2_{V,W} u \tag{8.81}$$

for any function f. We also compute that

$$\begin{aligned}
\nabla^2_{V,W} u - \nabla^2_{W,V} u &= V(W(u)) - (\nabla_V W)(u) - W(V(u)) + (\nabla_W V)(u) \\
&= [V, W](u) - (\nabla_V W)(u) + (\nabla_W V)(u) \\
&= 0,
\end{aligned}$$

where the last equality is because ∇ is torsion-free. We have shown that the Hessian of u is symmetric:

$$\nabla^2_{V,W} u = \nabla^2_{W,V} u. \tag{8.82}$$

Hence

$$\nabla^2_{V,fW} u = \nabla^2_{fW,V} u = f\nabla^2_{W,V} u = f\nabla^2_{V,W} u. \tag{8.83}$$

By Corollary 4.39, this proves that the Hessian $\nabla^2 u$ is a symmetric 2-tensor.

Remark 8.13. In abstract, we have shown that the covariant Hessian is a symmetric 2-tensor for the following reasons. Let $A(V,W) := \nabla^2_{V,W} u$. Then A is bilinear in V and W. By (8.82), $A(V,W) = A(W,V)$, so that A is symmetric. By (8.81) and (8.83), we have $A(fV,W) = A(V,fW) = fA(V,W)$, so that A is a 2-tensor.

8.5.2. The Hessian in local coordinates. Let $(\mathcal{U}, \{x^i\}_{i=1}^n)$ be local coordinates on M^n. We define

$$\nabla_i \nabla_j u := \nabla^2_{\partial_i,\partial_j} u = (\nabla^2 u)(\partial_i, \partial_j). \tag{8.84}$$

Recall from (4.138) that the Christoffel symbols are defined by $\nabla_{\partial_i}\partial_j = \sum_{k=1}^n \Gamma_{ij}^k \partial_k$. We compute using (8.80) that

$$\begin{aligned} \nabla_i \nabla_j u &= \partial_i(\partial_j(u)) - (\nabla_{\partial_i}\partial_j)(u) \\ &= \frac{\partial^2 u}{\partial x^i \partial x^j} - \sum_{k=1}^n \Gamma_{ij}^k \frac{\partial u}{\partial x^k}. \end{aligned} \tag{8.85}$$

8.5.3. The definition of the Laplacian. The **Laplacian** of a function $u : M^n \to \mathbb{R}$ is defined by

$$\Delta u := \operatorname{trace}_g(\nabla^2 u). \tag{8.86}$$

With respect to local coordinates $\{x^i\}$, the Laplacian is given by

$$\Delta u = \sum_{i,j=1}^n g^{ij} \nabla_i \nabla_j u = \sum_{i,j=1}^n g^{ij} \left(\frac{\partial^2 u}{\partial x^i \partial^j} - \sum_{k=1}^n \Gamma_{ij}^k \frac{\partial u}{\partial x^k} \right); \tag{8.87}$$

cf. (5.59).

For another version of the Laplacian in local coordinates, see Exercise 8.17; for the Laplacian of a rotationally symmetric surface, see Exercise 8.18.

With respect to a local orthonormal frame field $\{e_i\}_{i=1}^n$ for (M^n, g), we have

$$\Delta u = \sum_{i=1}^n \nabla^2_{e_i,e_i} u = \sum_{i=1}^n \left(e_i(e_i(u)) - (\nabla_{e_i} e_i)(u) \right). \tag{8.88}$$

Recall from (8.55) that the connection 1-forms are defined by $\nabla_V e_i = \sum_{j=1}^n \omega_i^j(V) e_j$. Thus,

$$\Delta u = \sum_{i=1}^n e_i(e_i(u)) - \sum_{i,j=1}^n \omega_i^j(e_i) e_j(u). \tag{8.89}$$

When $\dim(M) = n = 2$, by (8.89), we have that

$$\Delta u = e_1(e_1(u)) + e_2(e_2(u)) - \omega_1^2(e_1)e_2(u) - \omega_2^1(e_2)e_1(u), \tag{8.90}$$

where $\omega_2^1 = -\omega_1^2$.

8.6. Conformally equivalent metrics

A Riemannian metric provides, at each point of a manifold, a measure of distance and angle between tangent vectors. The notion of conformality considers metrics that may have different measures of distance but the same measures of angles.

8.6.1. Definitions of conformality. Let $u : M^n \to \mathbb{R}$. Consider the Riemannian metric

$$\widetilde{g} = e^{2u} g, \tag{8.91}$$

where we call u the conformal factor. The metrics $\widetilde{g}$ and g share the property that for any tangent vectors V and W at the same point, the angles between V and W are the same with respect to g and $\widetilde{g}$. (Exercise: Prove this.) For this reason, we say that $\widetilde{g}$ is **pointwise conformal to** g.

Now suppose that $\phi : N^n \to M^n$ is an immersion, i.e., a local diffeomorphism. Let h be a Riemannian metric on N^n and let g be a Riemannian metric on M^n. We say that $\phi : (N^n, h) \to (M^n, g)$ is a **conformal map** if $\phi^* g$ is pointwise conformal to h; that is, there exists $u : N^n \to \mathbb{R}$ such that $\phi^* g = e^{2u} h$. Note that if $\widetilde{g}$ is pointwise conformal to g on M^n, then the identity map $\mathrm{id} : (M^n, \widetilde{g}) \to (M^n, g)$ is a conformal map.

8.6.2. Stereographic projection is a conformal map. Let (S^n, g_{S^n}) be the unit n-sphere. Recall that the inverse of stereographic projection $F : \mathbb{R}^n \to S^n \setminus \{\mathbf{e}_{n+1}\} \subset S^n$ is given by (4.37). In (4.38) we computed that

$$(g_{S^n})_{ij}(F(\mathbf{u})) := g_{S^n}\left(\frac{\partial F}{\partial u^i}(\mathbf{u}), \frac{\partial F}{\partial u^j}(\mathbf{u})\right) = \frac{16}{(\|\mathbf{u}\|^2 + 4)^2}\delta_{ij}.$$

In other words,

$$(F^* g_{S^n})_{ij}(\mathbf{u}) := (F^* g_{S^n})_{\mathbf{u}}(\mathbf{e}_i, \mathbf{e}_j) = \frac{16}{(\|\mathbf{u}\|^2 + 4)^2}\delta_{ij}. \tag{8.92}$$

We write this invariantly as

$$(F^* g_{S^n})_{\mathbf{u}} = \frac{16}{(\|\mathbf{u}\|^2 + 4)^2}(g_{\mathbb{E}})_{\mathbf{u}}. \tag{8.93}$$

Thus, the embedding $F : (\mathbb{R}^n, g_{\mathbb{E}}) \to (S^n, g_{S^n})$ is a conformal map.

8.6.3. Conformal equivalence and the uniformization theorem.

Definition 8.14. A **Riemann surface** $\mathcal{S} = (M^2, \mathbf{A})$ is a smooth 2-dimensional manifold M^2 with a (not maximal) atlas $\mathbf{A} := \{(\mathcal{U}, \varphi_{\mathcal{U}})\}_{\mathcal{U}\in\mathfrak{U}}$ of charts with the property that each transition function

$$\varphi_{\mathcal{U}\mathcal{V}} := \varphi_{\mathcal{U}} \circ \varphi_{\mathcal{V}}^{-1} : \varphi_{\mathcal{V}}(\mathcal{U} \cap \mathcal{V}) \to \varphi_{\mathcal{U}}(\mathcal{U} \cap \mathcal{V})$$

is a holomorphic map, where we consider the domain and range of $\varphi_{\mathcal{U}\mathcal{V}}$ as contained in $\mathbb{C} \cong \mathbb{R}^2$. We call $\mathbf{A}$ a **holomorphic atlas** and each chart in $\mathbf{A}$ a **holomorphic chart**.

We remark that a Riemann surface is the same as a 1-dimensional complex manifold; see Wells [**Wel08**]. Examples of Riemann surfaces will be given in the course of our discussion.

Definition 8.15. Let $\mathcal{S}_1 = (M_1^2, \mathbf{A}_1)$ and $\mathcal{S}_2 = (M_2^2, \mathbf{A}_2)$ be Riemann surfaces. A map $f : M_1^2 \to M_2^2$ is **holomorphic** if for every pair of charts $(\mathcal{U}_1, \varphi_{\mathcal{U},1}) \in \mathbf{A}_1$ and $(\mathcal{U}_2, \varphi_{\mathcal{U},2}) \in \mathbf{A}_2$, the map $\varphi_{\mathcal{U},2} \circ f \circ \varphi_{\mathcal{U},1}^{-1}$ is a holomorphic map between (possibly vacuous) open subsets of $\mathbb{C}$.

Two Riemann surfaces are **biholomorphic** (or **isomorphic**) if there exists a bijective holomorphic map between the two Riemann surfaces whose inverse is also a holomorphic map. This map is called a **biholomorphism**. It is easy to see that the relation on the set of all Riemann surfaces of being biholomorphic is an equivalence relation.

The uniformization theorem in complex analysis, due to Poincaré and Köbe, is the following theorem. For proofs of the uniformization theorem, see the books by Springer [**Spr57**], Siegel [**Sie69**], Ahlfors [**Ahl10**], Farkas and Kra [**FK92**], and Schlag [**Sch14a**]. See the articles by Abikoff [**Abi81**] and Gray [**Gra94**] for historical surveys; a modern presentation of the historical proofs is in the book by de Saint-Gervais [**dSG16**].

Theorem 8.16. *If $\mathcal{S}$ is a simply connected Riemann surface, then $\mathcal{S}$ is biholomorphic to the Riemann sphere S^2, the complex plane $\mathbb{C}$, or the open unit disk D^2, each of which is a Riemann surface.*

Definition 8.17. Let M^n be a smooth manifold. On the space of metrics on M^n, we can define an equivalence relation by $g_1 \sim g_2$ if g_1 is pointwise conformal to g_2. Let

$$[g] := \left\{ e^{2u} g \colon u \in C^\infty(M^n) \right\}$$

denote the **pointwise conformal class** of a Riemannian metric g on M^n; this is the equivalence class of g under $\sim$.

Let $\mathcal{S} = (M^2, \mathbf{A})$ be a Riemann surface. For each $x \in M^2$, the Riemann surface structure $\mathcal{S}$ induces an inner product on $T_x M$ up to scaling. Namely, any holomorphic chart $(\mathcal{U}, \varphi_{\mathcal{U}})$ containing x defines an inner product on T_xM by pulling back the Euclidean inner product on $\mathbb{C} = \mathbb{R}^2$. The inner produce thus defined is well-defined up to scaling. That is, the inner products on T_xM defined by any two holomorphic charts containing x differ only by a multiplicative constant. From this we conclude that a Riemann surface $\mathcal{S}$ defines a conformal class $[g]$ of Riemannian metrics on M^2. For this reason, we often denote a Riemann surface by $(M^2, [g])$.

We say that Riemannian manifolds (N^n, h) and (M^n, g) are **conformally equivalent** if there exists a conformal diffeomorphism ϕ mapping (N^n, h) to (M^n, g). In other words, we have the equality $[\phi^* g] = [h]$ of conformal classes.

The uniformization theorem in complex analysis (Theorem 8.16) is equivalent to the following result; see Exercise 8.23.

Theorem 8.18. *Any simply connected Riemannian surface (M^2, g) is conformally equivalent to either the unit sphere S^2, the Euclidean plane $\mathbb{R}^2$, or the open unit disk D^2.*

Since S^2 is not diffeomorphic to either $\mathbb{R}^2$ or D^2, the following is an immediate consequence of Theorem 8.18.

Theorem 8.19. *For any surfaces N^2 and M^2, both diffeomorphic to the 2-sphere, if h and g are any Riemannian metrics on N^2 and M^2, respectively, then (N^2, h) and (M^2, g) are conformally equivalent.*

Since S^2 minus any point is conformally equivalent to $\mathbb{R}^2$ (for example, by stereographic projection from that point), we obtain the following.

Corollary 8.20. *If (M^2, g) is a Riemannian surface, where M^2 is diffeomorphic to the 2-sphere, then for any $p \in M^2$ we have that $(M^2 \setminus \{p\}, g)$ is conformally equivalent to $(\mathbb{R}^2, g_{\mathbb{E}})$.*

8.6.4. Using moving frames to calculate curvatures of pointwise conformal metrics. In this subsection, we will use the second Cartan structure equation to compute the curvatures of conformally related metrics on surfaces. Let (M^2, g) be a 2-dimensional Riemannian manifold. Recall from (8.79) that

$$K = \Omega_1^2(e_2, e_1) = d\omega_1^2(e_2, e_1). \tag{8.94}$$

The fact that $\Omega_1^2 = d\omega_1^2$ is an exact 2-form on the open subset $\mathcal{U}$ of M^2, on which the orthonormal frame field $\{e_1, e_2\}$ is defined, is crucial to the proof of the Gauss–Bonnet formula in the next chapter.

Now suppose that $u : M^2 \to \mathbb{R}$ is a function. Consider the Riemannian metric $\widetilde{g} = e^{2u} g$. Let $\{e_1, e_2\}$ be an orthonormal frame field for g on an open subset $\mathcal{U}$ of M^2.

We have that $\{\widetilde{e}_1, \widetilde{e}_2\} = \{\mathrm{e}^{-u}e_1, \mathrm{e}^{-u}e_2\}$ is an orthonormal frame field for g on $\mathcal{U}$. Its dual orthonormal coframe field is $\{\widetilde{\omega}^1, \widetilde{\omega}^2\} = \{\mathrm{e}^u\omega^1, \mathrm{e}^u\omega^2\}$.

We have

$$\begin{aligned}
d\widetilde{\omega}^i &= \mathrm{e}^u\big(du \wedge \omega^i + d\omega^i\big)\\
&= \mathrm{e}^u\left(\sum_{j=1}^2 e_j(u)\omega^j \wedge \omega^i + \sum_{j=1}^2 \omega^j \wedge \omega^i_j\right)\\
&= \mathrm{e}^u\sum_{j=1}^2 \omega^j \wedge \big(e_j(u)\omega^i - e_i(u)\omega^j + \omega^i_j\big)\\
&= \sum_{j=1}^2 \widetilde{\omega}^j \wedge \big(\omega^i_j + e_j(u)\omega^i - e_i(u)\omega^j\big),
\end{aligned}$$

where the third equality is because $\omega^j \wedge \omega^j = 0$. Thus, if we define

$$\widetilde{\omega}^i_j := \omega^i_j + e_j(u)\omega^i - e_i(u)\omega^j, \tag{8.95}$$

then

$$d\widetilde{\omega}^i = \sum_{j=1}^2 \widetilde{\omega}^j \wedge \widetilde{\omega}^i_j. \tag{8.96}$$

By Exercise 8.12, we have that the connection 1-forms $\widetilde{\omega}^i_j$ of $\{\widetilde{e}_1, \widetilde{e}_2\}$ are uniquely determined by equation (8.96) and the antisymmetry $\widetilde{\omega}^i_j = -\widetilde{\omega}^j_i$, so that they are given by (8.95).

Let the 2-forms $\widetilde{\Omega}^i_j$, $1 \le i, j \le 2$, be the curvature 2-forms of g with respect to the orthonormal frame field $\{\widetilde{e}_1, \widetilde{e}_2\}$. Then

$$\begin{aligned}
\widetilde{\Omega}^2_1 &= d\widetilde{\omega}^2_1\\
&= d\left(\omega^2_1 + e_1(u)\omega^2 - e_2(u)\omega^1\right)\\
&= d\omega^2_1 + e_1\big(e_1(u)\big)\,\omega^1 \wedge \omega^2 - e_2\big(e_2(u)\big)\,\omega^2 \wedge \omega^1\\
&\quad + e_1(u)\,d\omega^2 - e_2(u)\,d\omega^1\\
&= \Omega^2_1 + \big(e_1\big(e_1(u)\big) + e_2\big(e_2(u)\big)\big)\,\omega^1 \wedge \omega^2\\
&\quad + e_1(u)\omega^1 \wedge \omega^2_1 - e_2(u)\omega^2 \wedge \omega^1_2\\
&= \Omega^2_1 + \big(e_1\big(e_1(u)\big) + e_2\big(e_2(u)\big)\big)\,\omega^1 \wedge \omega^2 + du \wedge \omega^2_1.
\end{aligned}$$

On the other hand, regarding the Laplacian of the conformal factor u, by (8.90) we have

$$\Delta u = e_1(e_1(u)) + e_2(e_2(u)) + (du \wedge \omega^2_1)(e_1, e_2).$$

Therefore, we conclude that

$$\widetilde{\Omega}_1^2 = \Omega_1^2 + (\Delta u)\omega^1 \wedge \omega^2. \tag{8.97}$$

By applying this 2-form to (e_2, e_1), we obtain

$$K - \Delta u = \widetilde{\Omega}_1^2(e_2, e_1) = \mathrm{e}^{2u}\widetilde{K},$$

where $\widetilde{K}$ is the Gauss curvature of $\widetilde{g}$. Thus, we have the following.

Lemma 8.21. *If g and $\widetilde{g} = \mathrm{e}^{2u}g$ are metrics on a 2-dimensional manifold, then their Gauss curvatures are related by*

$$\widetilde{K} = \mathrm{e}^{-2u}(K - \Delta u). \tag{8.98}$$

Observe that if $\mathrm{e}^{2u} = c$ is a constant, so that $\widetilde{g} = cg$, then $\widetilde{K} = c^{-1}K$. For example, if $\bar{g}$ is the standard metric on the 2-sphere S^2, that is, $(S^2, \bar{g})$ is a *unit* 2-sphere, then the Riemannian manifold $(S^2, r^2\bar{g})$ is a 2-sphere of radius r and has constant Gauss curvature equal to r^{-2}.

For a generalization of Lemma 8.21 to higher dimensions, see Exercise 8.20.

8.6.5. The Laplacian of a conformally related metric.

In dimension 2, the Laplace operator changes conformally under a conformal change of the metric.

Lemma 8.22 (Laplacian with respect to a conformal metric on a surface)**.** *Assume that $\widetilde{g} = \mathrm{e}^{2u}g$ on M^2 and let $\widetilde{\Delta}$ denote the Laplacian with respect to $\widetilde{g}$. Then, for any function $f : M^2 \to \mathbb{R}$ we have*

$$\widetilde{\Delta} f = \mathrm{e}^{-2u}\Delta f. \tag{8.99}$$

Proof. We have that $\widetilde{e}_i := \mathrm{e}^{-u}e_i$, $i = 1, 2$, is an orthonormal frame field for $\widetilde{g}$ and that $\widetilde{\omega}^j := \mathrm{e}^u\omega^j$, $j = 1, 2$, is its dual orthonormal coframe field for $\widetilde{g}$. By equation (8.95), we have that

$$\widetilde{\omega}_1^2 = \omega_1^2 + e_1(u)\omega^2 - e_2(u)\omega^1, \tag{8.100}$$

where the $\widetilde{\omega}_i^j$ are the connection 1-forms of the orthonormal frame field $\{\widetilde{e}_1, \widetilde{e}_2\}$ for $\widetilde{g}$. Hence, using (8.90), we calculate that

$$\begin{aligned}
\widetilde{\Delta} f &= \widetilde{e}_1(\widetilde{e}_1(f)) + \widetilde{e}_2(\widetilde{e}_2(f)) - \widetilde{\omega}_1^2(\widetilde{e}_1)\widetilde{e}_2(f) - \widetilde{\omega}_2^1(\widetilde{e}_2)\widetilde{e}_1(f) \\
&= \mathrm{e}^{-2u}\big(e_1(e_1(f)) + e_2(e_2(f)) - e_1(u)e_1(f) - e_2(u)e_2(f)\big) \\
&\quad + \mathrm{e}^{-2u}\big(- \omega_1^2(e_1)e_2(f) - \omega_2^1(e_2)e_1(f) + e_2(u)e_2(f) + e_1(u)e_1(f)\big) \\
&= \mathrm{e}^{-2u}\Delta u.
\end{aligned}$$

□

By applying Lemma 8.22 to Lemma 8.21, we have the following.

Corollary 8.23. *If g and $\widetilde{g} = e^{2u} g$ are metrics on a 2-dimensional manifold, then*

$$\widetilde{K} = e^{-2u} K - \widetilde{\Delta} u, \tag{8.101}$$

where $\widetilde{\Delta}$ is the Laplacian with respect to $\widetilde{g}$.

8.7. Hyperbolic 2-space

Continuing with Example 8.10, consider the upper half-plane U^2 with the Riemannian metric

$$g_{\mathbb{U}} = \frac{1}{(x^2)^2} \big(dx^1 \otimes dx^1 + dx^2 \otimes dx^2\big). \tag{8.102}$$

We have the orthonormal coframe field defined on all of $\mathbb{R}^2_+$ by

$$\omega^1 = \frac{1}{x^2} dx^1, \quad \omega^2 = \frac{1}{x^2} dx^2, \tag{8.103}$$

which is dual to the orthonormal frame field $e_1 = x^2 \frac{\partial}{\partial x^1}$, $e_2 = x^2 \frac{\partial}{\partial x^2}$ given by (8.48). Clearly,

$$g = \omega^1 \otimes \omega^1 + \omega^2 \otimes \omega^2. \tag{8.104}$$

Since

$$d\omega^1 = -\frac{1}{(x^2)^2} dx^2 \wedge dx^1 = -\omega^2 \wedge \omega^1$$

and $d\omega^2 = 0$, by (8.143) below in Exercise (8.12) we have that

$$\omega_1^2(e_j) = \frac{1}{2} \left(d\omega^1 \left(e_j, e_2\right) + d\omega^j \left(e_1, e_2\right)\right).$$

Hence

$$\omega_1^2(e_1) = d\omega^1 \left(e_1, e_2\right) = \omega^2(e_2)\omega^1(e_1) = 1$$

and $\omega_1^2(e_2) = 0$. Therefore, we have the connection 1-form formula:

$$\omega_1^2 = \omega^1. \tag{8.105}$$

The curvature 2-form Ω_1^2 is thus calculated as

$$\Omega_1^2 = d\omega_1^2 = d\omega^1 = -\omega^2 \wedge \omega^1. \tag{8.106}$$

Therefore, the Gauss curvature $K_{\mathbb{U}}$ of $g_{\mathbb{U}}$ is given by

$$K_{\mathbb{U}} = \Omega_1^2(e_2, e_1) = -1. \tag{8.107}$$

8.8. Uniformization of geometric surfaces

In differential geometry, there is the idea of finding 'nice' or 'canonical' Riemannian metrics on smooth manifolds. Perhaps the most ideal metrics are those with constant curvature. In this section, after a preliminary discussion of the classification of closed topological surfaces, we state the differential geometric version of the uniformization theorem of complex analysis.

8.8.1. Constructing closed surfaces. One can start with a 2-dimensional disk, subdivide the boundary into arcs, and identify pairs of arcs to obtain closed surfaces. For example, a torus may be obtained from taking a disk, subdividing its boundary circle into 4 arcs, and identifying the arcs suitably in pairs; see Figures 8.8.1 and 8.8.2. A double-holed torus, i.e., a surface of genus 2, may be obtained from taking a disk, subdividing its boundary circle into 8 arcs, and identifying the arcs suitably in pairs; see Figures 8.8.3 and 8.8.4.

In each of these examples, we are considering the surface topologically as a *CW-complex*, a.k.a. *cell complex*. A 2-dimensional (finite) **CW-complex**

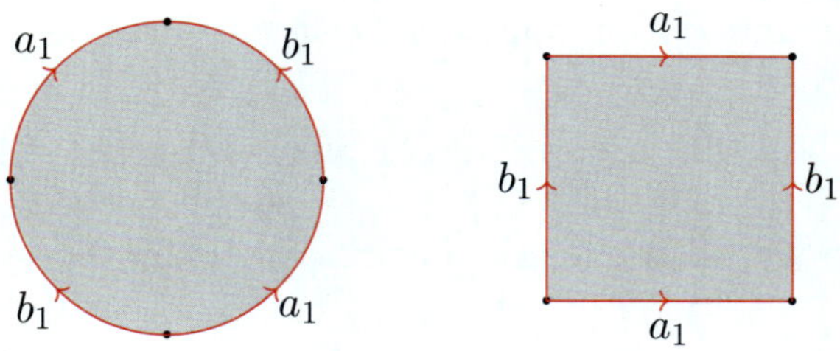

Figure 8.8.1. Left: A disk with 4 sides identified. Right: A square with its sides identified in the same manner.

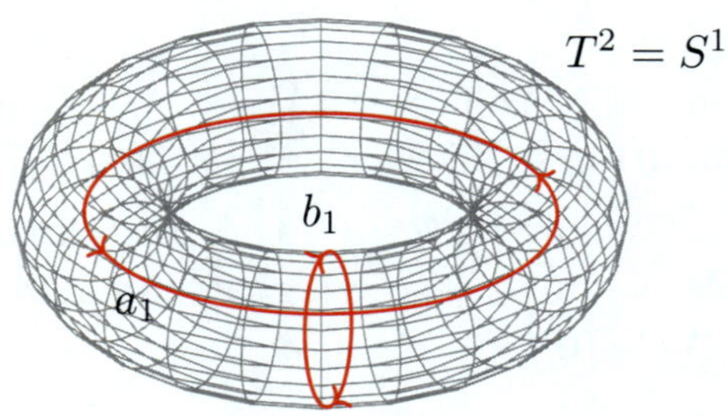

Figure 8.8.2. A cellular decomposition of a closed surface of genus 1, i.e., a torus.

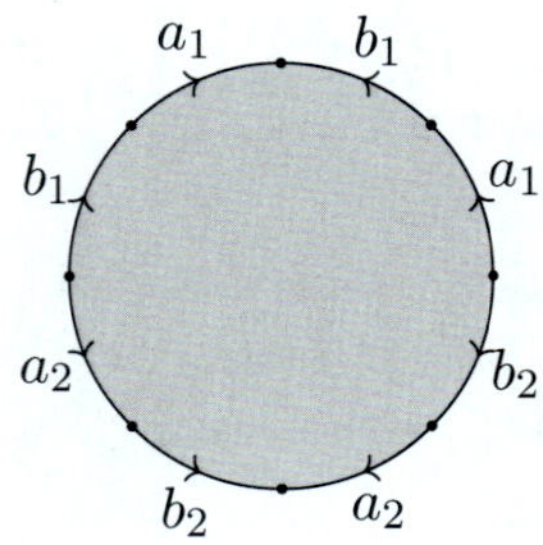

Figure 8.8.3. A disk with 8 sides identified yields a closed surface S_2 of genus 2.

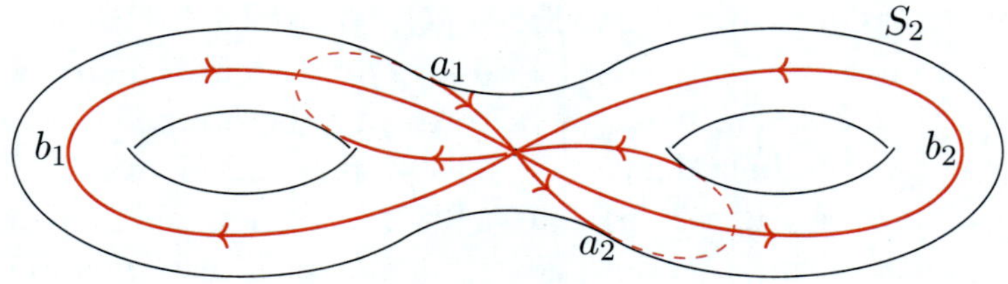

Figure 8.8.4. A cellular decomposition of a closed surface S_2 of genus 2.

is defined as follows.[4] Let D^0 denote a point, let D^1 denote a closed interval, and let D^2 denote a closed disk. We call each D^k a k-cell for $k = 0, 1, 2$.

0. Let X_0 be a finite disjoint union of 0-cells (with the discrete topology), called the 0-skeleton.

1. Let $\{D^1_\alpha\}$ be a finite disjoint collection of 1-cells. For each α, let $g_\alpha : \partial D^1_\alpha \to X_0$ be a map. Let X_1 be the quotient space of the disjoint union of the D^1_α's and X_0 under the identifications by the maps g_α, called the 1-skeleton.

2. Let $\{D^2_\beta\}$ be a finite disjoint collection of 2-cells. For each β, let $h_\beta : \partial D^2_\beta \to X_1$ be a continuous map. Let X_2 be the quotient space of the disjoint union of the D^2_β's and X_1 under the identifications by the maps h_β, called the 2-skeleton.

If X is a topological space homeomorphic to X_2, then we say that this is a **cellular decomposition** of X (as a 2-dimensional CW-complex).

Higher-dimensional CW-complexes are constructed similarly, involving the attachment of cells of progressively higher dimensions, as detailed in G. Whitehead's book [**Whi78**].

8.8.2. Triangulated surface. A **triangulation** of a closed topological surface M^2 is finite collection $\{T_\alpha\}$ of closed subsets, called **triangles**, with the following properties:

(i) The union of the T_α is M^2.

(ii) There is a homeomorphism ϕ_α from T_α to a triangle T'_α in the Euclidean plane.

(iii) Call the preimages of the vertices and edges of T'_α under ϕ_α the **vertices** and **edges** of T_α, respectively. We require that if $\alpha \neq \beta$, then $T_\alpha \cap T_\beta$ is equal to either a single vertex or a single edge.

See Figures 8.8.5 and 8.8.6.

[4]In the terminology 'CW-complex', the 'C' stands for 'closure-finite' and the 'W' stands for 'weak topology'.

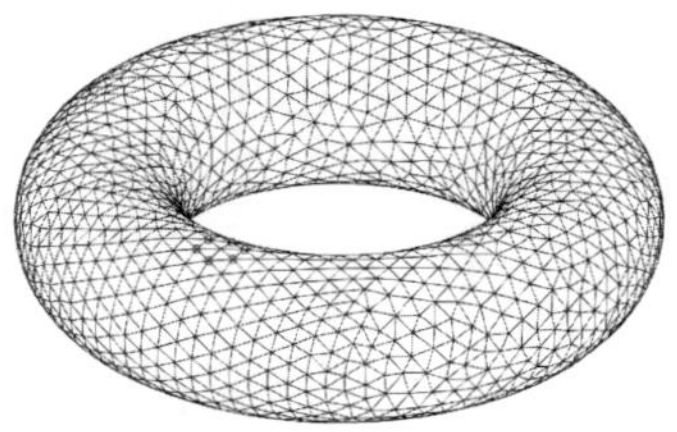

Figure 8.8.5. A triangulated torus with many triangles. Credit: By Ag2gaeh – Own work, CC BY-SA 3.0, `https://commons.wikimedia.org/w/index.php?curid=30856793`

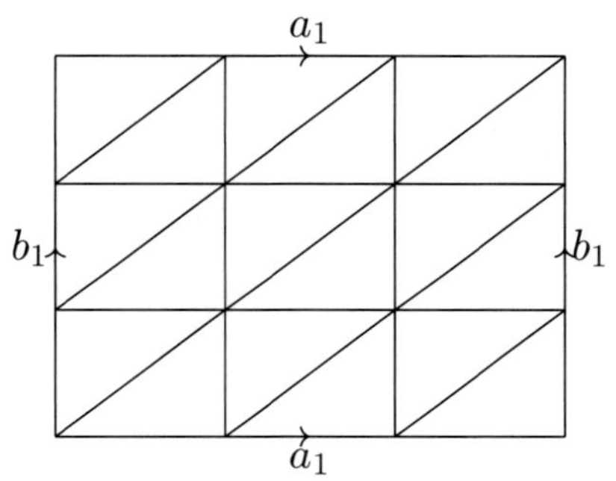

Figure 8.8.6. A simpler triangulation of a torus.

For the following classical result of Radó, see Moise's book [**Moi77**].

Theorem 8.24 (Radó). *For any closed topological surface M^2, there exists a triangulation of M^2.*

8.8.3. The Euler characteristic. Given a triangulated closed topological surface M^2, let V, E, F denote the number of vertices, edges, triangles (faces) of the triangulation, respectively. The **Euler characteristic** is defined by

$$\chi(M^2) := V - E + F. \tag{8.108}$$

One can compute the Euler characteristic in various ways. For the torus, we can have for example the following:

(1) For the triangulation of the torus T^2 in Figure 8.8.6, we have $V = 9$, $E = 27$, and $F = 18$. Hence $\chi(T^2) = 0$.

(2) For the cell-decomposition of a torus in Figure 8.8.1, we have $V = 1$, $E = 2$, and $F = 1$ (see (9.7) below for the formula for the Euler characteristic of a CW-complex). Again we see that $\chi(T^2) = 0$.

8.8.4. Classification of topological surfaces. By a topological surface we mean a 2-dimensional topological manifold, i.e., a C^0 manifold. See Figure 8.8.7.

The classification of closed surfaces in the topological and smooth categories are given as follows. See Theorem 5.1 in Chapter 1 of Massey's book [**Mas77**]. This result has been attributed to Brahana, Dehn, and Heegaard.

Theorem 8.25.

(1) *Any orientable closed smooth surface is diffeomorphic to either a sphere or a connected sum of* $\mathbf{g}$ *tori, where* $\mathbf{g} \geq 1$.

(2) *Any nonorientable closed smooth surface is diffeomorphic to a connected sum of real projective planes.*

For the following result, see Munkres [**Mun56**].

Theorem 8.26. *Let* M_1^2 *and* M_2^2 *be closed smooth surfaces. Let* $\mathcal{T}_1$ *and* $\mathcal{T}_2$ *be the induced topologies of* M_1^2 *and* M_2^2*, respectively. If the topological spaces* $(M_1^2, \mathcal{T}_1)$ *and* $(M_2^2, \mathcal{T}_2)$ *are homeomorphic, then* M_1^2 *and* M_2^2 *are diffeomorphic as smooth surfaces.*

Noncompact surfaces are more topologically complicated. Firstly, any simply connected noncompact topological surface M^2 is homeomorphic to

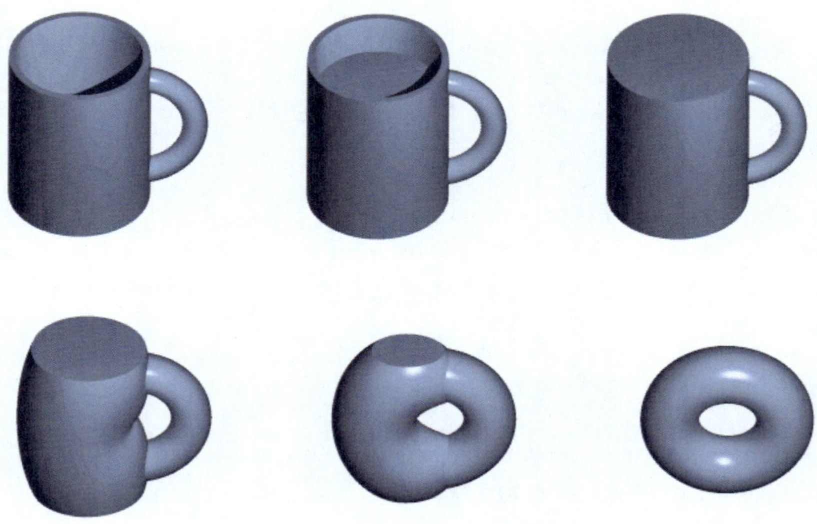

Figure 8.8.7. From left to right: A coffee cup morphing into a doughnut. (Stills from an animation.) Credit: By Kieff – This file was derived from: Mug and Torus morph.gif, Public Domain, `https://commons.wikimedia.org/w/index.php?curid=2358985`

$\mathbb{R}^2$. More generally, Ian Richards [**Ric63**] has given a classification of noncompact triangulable surfaces; see the references therein for earlier developments.

8.8.5. The geometric uniformization of Riemann surfaces. Suppose that g_0 is a Riemannian metric on a 2-dimensional manifold M^2. Let $g = \mathrm{e}^{2u} g_0$. By (8.98),

$$K = \mathrm{e}^{-2u}(K_0 - \Delta_0 u), \tag{8.109}$$

where Δ_0 denotes the Laplacian with respect to g_0 and where K_0 is the Gauss curvature of the metric g_0.

One can pose the following question. Does there exist a function u on M^2 such that the metric $g = \mathrm{e}^{2u} g_0$ conformal to g_0 satisfies $K \equiv k$, where $k \in \mathbb{R}$? This constant curvature k equation is equivalent to

$$\Delta_0 u - K_0 = -k\mathrm{e}^{2u}. \tag{8.110}$$

Observe that by the Gauss–Bonnet formula (see Theorem 9.15 below), k has the same sign as the Euler characteristic $\chi(M^2)$. The answer to this question is yes and this is called the differential geometric version of the uniformization theorem in complex analysis (Theorem 8.18). In particular, this result can be deduced from the existence of isothermal coordinates (see Chern [**Che55**] and Jost [**Jos06**]) and the uniformization theorem. In Chapter 14 we will discuss the heat flow approach of Hamilton, which gives a new proof of this classical result.

Theorem 8.27 (Differential geometric version of uniformization theorem). *For any Riemannian metric g_0 on a closed surface M^2, there exists a positive function v such that the new metric vg_0 has constant Gauss curvature.*

We can think of the differential geometric version of the uniformization theorem as saying that in the conformal class of any Riemannian metric on a smooth closed surface there exists a 'best' metric. Consider an oriented closed surface. For a surface of genus greater than or equal to 2, within any pointwise conformal class there exists a unique metric of constant curvature -1. For a surface of genus 1, in any pointwise conformal class the zero curvature metric is unique, up to scaling the metric by constant multiples. On the other hand, for genus 0, i.e., when M^2 is diffeomorphic to S^2, given a metric of constant curvature 1, by pulling this metric back by any conformal diffeomorphism that is not an isometry, we obtain a distinct constant curvature 1-metric in the same pointwise conformal class.

In the next chapter we will state and prove the famous Gauss–Bonnet formula. A consequence of this formula is that the sign of the constant Gauss curvature in Theorem 8.27 is the same as the sign of the Euler characteristic of M^2.

8.9. Submanifolds of Riemannian manifolds

Generalizing the notion of submanifolds of Euclidean space discussed in Part I, we have the notion of submanifolds of Riemannian manifolds. Using moving frames adapted to hypersurfaces in Riemannian manifolds, we discuss the second fundamental form and the Gauss and Codazzi equations.

8.9.1. Embedded submanifolds. Let Σ^m and M^n be smooth manifolds. As in Chapter 2, we say that a map $F : \Sigma^m \to M^n$ is a **smooth embedding** if F is both a topological embedding and a smooth immersion. We call the image $F(\Sigma^m)$ an **embedded submanifold** of M^n. We will often assume that $\Sigma^m \subset M^n$ and take F to be the inclusion map.

Let Σ^m be an embedded submanifold of M^n. Then, for each $x \in \Sigma^m$, we have the orthogonal decomposition

$$T_x M = T_x \Sigma \oplus N_x \Sigma \tag{8.111}$$

of the ambient tangent space into the tangent and normal spaces of the submanifold. Note that the normal space of a submanifold of a smooth manifold is defined in the same way as for a submanifold of a Euclidean space (see (3.64)).

8.9.2. The first and second fundamental forms. Henceforth we assume that M^n is endowed with a Riemannian metric $\overline{g}$. The **first fundamental form** g of Σ^m is the induced metric

$$g(V, W) := \overline{g}(V, W) \qquad \text{for } V, W \in T\Sigma. \tag{8.112}$$

The Levi-Civita connection ∇ of g satisfies the formula

$$\nabla_V W = (\overline{\nabla}_V W)^\top \tag{8.113}$$

for $V \in T\Sigma$ and $W \in C^\infty(T\Sigma)$, where $\cdot^\top : T_x M \to T_x \Sigma$ denotes the tangential projection.

The **second fundamental form** II of Σ^m is defined by

$$B(V, W) := (\overline{\nabla}_V W)^\perp \qquad \text{for } V, W \in T\Sigma, \tag{8.114}$$

where $\cdot^\perp : T_x M \to N_x \Sigma$ denotes the normal projection; cf. (4.115) for submanifolds of Euclidean spaces.

Both the first and second fundamental forms are symmetric 2-tensors.

8.9.3. Moving frames adapted to a submanifold. Let Σ^m be an embedded submanifold of a Riemannian manifold $(M^n, \overline{g})$. Let $x \in \Sigma^m$ and let $\{e_i\}_{i=1}^n$ be an orthonormal frame field defined on an open subset $\mathcal{U}$ of M^n containing x. Assume that $e_i \in T_y \Sigma$ for all $1 \leq i \leq m$ and $y \in \Sigma^m \cap \mathcal{U}$, so that $e_j \in N_y \Sigma$ for all $m + 1 \leq j \leq n$. In this case we say that the frame $\{e_i\}_{i=1}^n$ is **adapted to** the submanifold Σ^m.

Let $\{\omega^j\}_{j=1}^n$ be the orthonormal coframe field on $\mathcal{U}$ dual to $\{e_i\}_{i=1}^n$ and let $\{\overline{\omega}_i^j\}_{i,j=1}^n$ denote the associated connection 1-forms defined by

$$\overline{\nabla}_X e_i =: \sum_{j=1}^n \overline{\omega}_i^j e_j \tag{8.115}$$

for $1 \leq i \leq n$. Note that $\{e_i\}_{i=1}^m$ restricts to a local orthonormal frame field on Σ^m. Its associated connection 1-forms $\{\omega_i^j\}_{i,j=1}^m$ on $\Sigma^m \cap \mathcal{U}$ are defined by

$$\nabla_X e_i =: \sum_{j=1}^m \omega_i^j e_j \tag{8.116}$$

for $1 \leq i \leq m$. We have on $\Sigma^m \cap \mathcal{U}$ that

$$\overline{\omega}_i^j = \omega_i^j \qquad \text{for } 1 \leq i, j \leq m. \tag{8.117}$$

For $1 \leq i, j \leq m$,

$$B_{ji} := B(e_j, e_i) = (\overline{\nabla}_{e_j} e_i)^\perp = \sum_{k=m+1}^n \overline{\omega}_i^k(e_j) e_k \in N_x\Sigma. \tag{8.118}$$

If we define the components B_{ji}^k, $1 \leq i, j \leq m$ and $m+1 \leq k \leq n$, of the second fundamental form by

$$B(e_j, e_i) := \sum_{k=m+1}^n B_{ji}^k e_k, \tag{8.119}$$

then we have that

$$B_{ij}^k = B_{ji}^k = \overline{\omega}_i^k(e_j). \tag{8.120}$$

By (8.71) and (8.77), associated to the local orthonormal frame field $\{e_i\}_{i=1}^n$ on M^n we have the Cartan structure equations

$$d\omega^i = \sum_{j=1}^n \omega^j \wedge \overline{\omega}_j^i, \tag{8.121a}$$

$$\overline{\Omega}_i^j = d\overline{\omega}_i^j - \sum_{k=1}^n \overline{\omega}_i^k \wedge \overline{\omega}_k^j \tag{8.121b}$$

for $1 \leq i, j \leq n$. Associated to the local orthonormal frame field $\{e_i\}_{i=1}^m$ on Σ^m we have the Cartan structure equations

$$d\omega^i = \sum_{j=1}^{m} \omega^j \wedge \omega_j^i, \tag{8.122a}$$

$$\Omega_i^j = d\omega_i^j - \sum_{k=1}^{m} \omega_i^k \wedge \omega_k^j \tag{8.122b}$$

for $1 \leq i, j \leq m$.

For $1 \leq i, j \leq m$ we have that

$$\Omega_i^j = \overline{\Omega}_i^j + \sum_{k=m+1}^{n} \overline{\omega}_i^k \wedge \overline{\omega}_k^j. \tag{8.123}$$

8.9.4. The Gauss and Codazzi equations on hypersurfaces. We now specialize to hypersurfaces, so that $m = n - 1$. In this case $N := e_n$ is a choice of a local unit normal vector field to Σ^{n-1}. If $M^n = \mathbb{R}^n$, then

$$B(V, W) = -\operatorname{II}(V, W)N \tag{8.124}$$

for $V, W \in T_x\Sigma$, where the $B(V, W)$ on the left-hand side is according to definition (8.114) and the $\operatorname{II}(V, W)$ on the right-hand side is according to definition (4.78). In general, on a Riemannian manifold (M^n, g), we also define the second fundamental form II of a hypersurface Σ^{n-1} as a scalar function by (8.124); that is,

$$\operatorname{II}(V, W) = -\overline{g}(\overline{\nabla}_V W, N) = \overline{g}(\overline{\nabla}_V N, W). \tag{8.125}$$

By (8.123) we have that for $1 \leq i, j \leq n - 1$,

$$\Omega_i^j = \overline{\Omega}_i^j + \overline{\omega}_i^n \wedge \overline{\omega}_n^j. \tag{8.126}$$

We may assume that the local orthonormal frame field $\{e_i\}_{i=1}^n$ satisfies $\nabla_{e_n} e_n = 0$; that is, the integral curves to e_n are geodesics in $(\mathcal{U}, \overline{g})$. This implies that $\overline{\omega}_i^n(e_n) = -\overline{\omega}_n^i(e_n) = 0$ for all i. Therefore

$$\overline{\omega}_i^n = -\sum_{k=1}^{n-1} \operatorname{II}(e_i, e_k)\omega^k, \tag{8.127}$$

where $\operatorname{II}(e_i, e_k)$ is considered as a scalar. Therefore (8.126) implies that

$$\Omega_i^j = \overline{\Omega}_i^j - \sum_{k,\ell=1}^{n-1} \operatorname{II}(e_i, e_k) \operatorname{II}(e_j, e_\ell)\omega^k \wedge \omega^\ell \tag{8.128}$$

for $1 \leq i, j \leq n - 1$. This implies that for $1 \leq i, j, k, \ell \leq n - 1$,

$$\Omega_i^j(e_k, e_\ell) = \overline{\Omega}_i^j(e_k, e_\ell) - \operatorname{II}(e_i, e_k) \operatorname{II}(e_j, e_\ell) + \operatorname{II}(e_i, e_\ell) \operatorname{II}(e_j, e_k). \tag{8.129}$$

By (8.68), this says that

$$R_{k\ell ij} = \overline{R}_{k\ell ij} - \mathrm{II}_{ik}\,\mathrm{II}_{j\ell} + \mathrm{II}_{i\ell}\,\mathrm{II}_{jk}\,. \tag{8.130}$$

This equation is the **Gauss equation** for a hypersurface in a Riemannian manifold. By tracing this equation, we obtain

$$g^{kj} R_{k\ell ij} = \overline{g}^{kj}\overline{R}_{k\ell ij} + g^{kj}(-\,\mathrm{II}_{ik}\,\mathrm{II}_{j\ell} + \mathrm{II}_{i\ell}\,\mathrm{II}_{jk}), \tag{8.131}$$

where we sum over $1 \leq j, k \leq n-1$. This says that

$$R_{\ell i} = \overline{R}_{\ell i} - \overline{R}_{n\ell in} - g^{kj}\,\mathrm{II}_{ik}\,\mathrm{II}_{j\ell} + \mathrm{II}_{i\ell}\,H, \tag{8.132}$$

where H is the mean curvature and which is called the **traced Gauss equation**. By tracing again (i.e., by multiplying by $g^{\ell i} = \overline{g}^{\ell i}$ and summing over $1 \leq i, \ell \leq n-1$), we obtain

$$R = \overline{R} - 2\overline{R}_{nn} - |\mathrm{II}|^2 + H^2, \tag{8.133}$$

which is the **double traced Gauss equation**. In the discussion above, barred curvature quantities denote those associated to $\overline{g}$.

The Codazzi equation for hypersurfaces arises as follows. By (8.121b), we have that

$$\overline{\Omega}^j_n = d\overline{\omega}^j_n - \sum_{k=1}^n \overline{\omega}^k_n \wedge \overline{\omega}^j_k.$$

The 2-tensor

$$\mathrm{II} := \sum_{j=1}^n \overline{\omega}^j_n \otimes \omega^j \tag{8.134}$$

is equal to the second fundamental form. Recall from (8.33) that

$$d\overline{\omega}^j_n(V, W) = (\nabla_V \overline{\omega}^j_n)(W) - (\nabla_W \overline{\omega}^j_n)(V). \tag{8.135}$$

We compute that (suppressing the summation signs for summing over k)

$$\begin{aligned}(\nabla_U \mathrm{II})(V, e_j) &= \big(\nabla_U(\overline{\omega}^k_n \otimes \omega^k)\big)(V, e_j)\\ &= (\nabla_U \overline{\omega}^k_n)(V)\omega^k(e_j) + \overline{\omega}^k_n(V)(\nabla_U \omega^k)(e_j)\\ &= (\nabla_U \overline{\omega}^j_n)(V) - \overline{\omega}^k_n(V)\overline{\omega}^k_j(U)\end{aligned} \tag{8.136}$$

where we used (8.60) for the last equality. Therefore,

$$\begin{aligned}&(\nabla_U \mathrm{II})(V, e_j) - (\nabla_V \mathrm{II})(U, e_j)\\ &= (\nabla_U \overline{\omega}^j_n)(V) - (\nabla_V \overline{\omega}^j_n)(U) - \overline{\omega}^k_n(V)\overline{\omega}^k_j(U) + \overline{\omega}^k_n(U)\overline{\omega}^k_j(V)\\ &= d\overline{\omega}^j_n(U, V) + (\overline{\omega}^k_n \wedge \overline{\omega}^k_j)(U, V)\\ &= \overline{\Omega}^j_n(U, V).\end{aligned}$$

We conclude the **Codazzi equation** for a hypersurface in a Riemannian manifold:

$$(\nabla_U \mathrm{II})(V, e_j) - (\nabla_V \mathrm{II})(U, e_j) = \overline{\Omega}^j_n(U, V). \tag{8.137}$$

8.9.5. Level sets. Let (M^n, g) be a Riemannian manifold. Suppose that $f : M^n \to \mathbb{R}$ is a smooth function and that $c \in \mathbb{R}$ is a **regular value** of f; that is, df_x is surjective for all $x \in \Sigma^{n-1} := f^{-1}(c)$. Then by the implicit function theorem (see §2.2.4), Σ^{n-1} is a smooth hypersurface in M^n, $|\nabla f|(x) \neq 0$ for all $x \in \Sigma^{n-1}$, and

$$N := \frac{\nabla f}{|\nabla f|} \tag{8.138}$$

is a choice of smooth unit normal vector field to Σ^{n-1}. Therefore, the second fundamental form II of Σ^{n-1} is given by

$$\begin{aligned} \mathrm{II}(V, W) &= \langle \nabla_V N, W \rangle \\ &= \left\langle \nabla_V \left(\frac{\nabla f}{|\nabla f|} \right), W \right\rangle \\ &= \frac{1}{|\nabla f|} \langle \nabla_V \nabla f, W \rangle \\ &= \frac{1}{|\nabla f|} \nabla^2 f(V, W) \end{aligned} \tag{8.139}$$

for $V, W \in T\Sigma$; here, we have used for the third equality that $\langle \nabla f, W \rangle = 0$.

8.10. Notes and commentary

Hicks [**Hic65**] has an excellent discussion of moving frames in Riemannian geometry; this includes the use of moving frames to derive the Gauss and Codazzi equations for hypersurfaces of Riemannian manifolds. For further reading, in the context of exterior differential systems, see Ivey and Landsberg [**IL16**].

8.11. Exercises

Exercise 8.1. *Show that there is a covering map from the cylinder* $S^1 \times (0, 1)$ *to the Möbius strip* M^2 *which is a double cover. If you like, you can consider* $S^1 \times (0, 1)$ *as* $[0, 2] \times (0, 1)$ *with the identification* $(0, y) \sim (2, y)$.

Exercise 8.2. *Prove that the Klein bottle is a nonorientable surface (see, e.g., Massey* [**Mas77**] *for the definition of the Klein bottle).*

Exercise 8.3. *Prove that the Cartesian product of two orientable manifolds is orientable.*
Hint: Show that orientations for two manifolds naturally determine an orientation for their product.

Exercise 8.4. *Prove that any odd-dimensional real projective space* $\mathbb{R}P^{2k-1}$*, where* k *is a positive integer, is an orientable manifold.*
Hint: The antipodal map $\mathbf{x} \to -\mathbf{x}$ *is orientation preserving as a map from* $\mathbb{R}^{2k}$ *to itself and hence from* S^{2k-1} *to itself.*

Exercise 8.5. *Let* M^n *be a smooth manifold and suppose that* Γ *is a discrete group of diffeomorphisms acting properly discontinuously on* M^n*. Prove that if* M^n *is oriented and* Γ *acts by orientation-preserving diffeomorphisms, then the quotient manifold* M^n/Γ *is naturally oriented.*

Exercise 8.6. *Prove* (8.32).
Hint: The product rule says that

$$V_j\big(\omega(V_0,\dots,\hat{V}_j,\dots,V_k)\big) = (\nabla_{V_j}\omega)(V_0,\dots,\hat{V}_j,\dots,V_k) + \sum_{i\neq j}\omega(V_0,\dots,\nabla_{V_j}V_i\dots,\hat{V}_j,\dots,V_k), \tag{8.140}$$

where $\nabla_{V_j}V_i$ *is where* V_i *normally would be and we may have either* $i<j$ *or* $i>j$.

Exercise 8.7 (Wedge product rule for Lie derivative). *If* α *and* β *are differential forms, then*

$$\mathcal{L}_V(\alpha\wedge\beta) = (\mathcal{L}_V\alpha)\wedge\beta + \alpha\wedge(\mathcal{L}_V\beta).$$

Exercise 8.8. *Prove that if* ω *is a differential form, then for any vector fields* W *and* V,

$$\mathcal{L}_W(\iota_V\omega) = \iota_V(\mathcal{L}_W\omega) + \iota_{[W,V]}\omega. \tag{8.141}$$

Exercise 8.9. *Let* $R := \prod_{i=1}^n [a_i,b_i]$ *be a hyperrectangle in* $\mathbb{R}^n$ *and let* $F : R \to \mathbb{R}$ *be a continuously differentiable function. Work out what Stokes's theorem says when we apply it to the differential form*

$$\omega := F\,dx^1\wedge\cdots\wedge\widehat{dx^i}\wedge\cdots\wedge dx^n,$$

where $1\le i\le n$ *and* $\widehat{\cdot}$ *denotes omission.*

Exercise 8.10. *Using Stokes's theorem (Theorem* 8.7*), prove Green's theorem in the plane, Stokes's theorem for surfaces in* $\mathbb{R}^3$*, and the divergence theorem in* $\mathbb{R}^{n+1}$ *(Proposition* 5.12*).*

Exercise 8.11. *Let* f *and* h *be functions on a compact Riemannian manifold* (M^n,g)*. What does the divergence theorem say when we apply it to the vector field* $f\nabla h - h\nabla f$*?*

A consequence of the identity that you just proved is that the Laplacian is a self-adjoint operator on closed manifolds. Can you figure out what is meant by this?

Exercise 8.12 (Formula for the connection 1-forms). *Prove that the first Cartan structure equation* (8.71) *implies that*

$$d\omega^k(e_i, e_j) = \omega_i^k(e_j) - \omega_j^k(e_i). \tag{8.142}$$

Using this, derive the following formula for the connection 1*-forms:*

$$\omega_i^k(e_j) = \frac{1}{2}\left(d\omega^i(e_j, e_k) + d\omega^j(e_i, e_k) - d\omega^k(e_j, e_i)\right). \tag{8.143}$$

Note that this shows that the first Cartan structure equation and the antisymmetry of ω_i^k in i and k uniquely determine these connection 1*-forms.*

Exercise 8.13. *Use* (8.143) *to calculate the connection* 1*-forms and curvature* 2*-forms of a metric $g := \mathrm{e}^{2u} g_{\mathbb{E}}$ defined on an open subset $\mathcal{U}$ of $\mathbb{R}^2$, where $u : \mathcal{U} \to \mathbb{R}$ is a smooth function.*

Exercise 8.14. *Consider the Riemannian metric $g := ds^2 + f(s)^2 d\theta^2$ on an annulus $(a, b) \times (\mathbb{R}/2\pi\mathbb{Z})$. Using the method of moving frames, show that the Gauss curvature of g is given by*

$$K(s, \theta) = -\frac{f''(s)}{f(s)}. \tag{8.144}$$

Hint: First compute ω_2^1 and then show that

$$\Omega_2^1 = -f''(s)\, ds \wedge d\theta.$$

Exercise 8.15. *The previous exercise generalizes to higher dimensions. Let $g := ds^2 + f(s)^2 g_{S^{n-1}}$, where $g_{S^{n-1}}$ is the unit-sphere metric on S^{n-1}, be a rotationally symmetric Riemannian metric on an annulus $M^n := (a, b) \times S^{n-1}$. Let $\{\bar{e}_i\}_{i=1}^{n-1}$ be an orthonormal frame field on an open subset $\mathcal{U}$ of S^{n-1} and let $e_n := \frac{\partial}{\partial s}$. Then $\{e_i\}_{i=1}^{n}$ is an orthonormal frame field on $(a, b) \times \mathcal{U}$, where*

$$e_i := f(s)^{-1} (\iota_s)_* (\bar{e}_i)$$

for $1 \le i \le n-1$ and where $\iota_s : S^{n-1} \to \{s\} \times S^{n-1}$ is the inclusion map. Let $\{\bar{\omega}^j\}_{j=1}^{n-1}$ be the orthonormal coframe field on $\mathcal{U}$ dual to $\{\bar{e}_i\}_{i=1}^{n-1}$. Then $\{\omega^j\}_{j=1}^{n}$ is the dual orthonormal coframe field on $(a, b) \times \mathcal{U}$, where $\omega^n := ds$ and

$$\omega^j := f(s) (\iota_s^{-1})^* (\bar{\omega}^j)$$

for $1 \leq j \leq n-1$*. Note that since* $g_{S^{n-1}}$ *has constant sectional curvature equal to* 1*, we have that*

$$d\bar{\omega}^i = \sum_{j=1}^{n-1} \bar{\omega}^j \wedge \bar{\omega}^i_j,$$

$$d\bar{\omega}^i_j - \sum_{k=1}^{n-1} \bar{\omega}^k_i \wedge \bar{\omega}^j_k = (\mathrm{Rm}_{g_{S^{n-1}}})^j_i = \bar{\omega}^i \wedge \bar{\omega}^j.$$

(1) *Show that the connection* 1*-forms of* g *are given by the following: For* $1 \leq i, j \leq n-1$*,*

$$\omega^i_j = \bar{\omega}^i_j, \qquad \omega^i_n = \frac{f'(s)}{f(s)}\omega^i. \tag{8.145}$$

(2) *Show that the curvature* 2*-forms of* g *are given by the following: For* $1 \leq i, j \leq n-1$*,*

$$(\mathrm{Rm})^i_j = \frac{1 - f'(s)^2}{f(s)^2}\omega^i \wedge \omega^j, \tag{8.146a}$$

$$(\mathrm{Rm})^i_n = -\frac{f''(s)}{f(s)}\omega^i \wedge \omega^n. \tag{8.146b}$$

This shows that the set of 2*-forms* $\{\omega^i \wedge \omega^j\}_{1 \leq i < j \leq n}$ *is an orthogonal basis of eigenvectors for the Riemann curvature operator* Rm.

For curvature calculations of warped products and doubly warped products, see Petersen [**Pet16**, §4.2 and §4.3].

Exercise 8.16.

(1) *Continuing with the previous exercise, show that if a* 2*-plane* Π *in a tangent space* T_xM *contains the radial vector* $\frac{\partial}{\partial s}$*, then its sectional curvature is given by*

$$K_{\mathrm{rad}} := \mathrm{sect}(\Pi) = -\frac{f''(s)}{f(s)}. \tag{8.147}$$

On the other hand, if a 2*-plane* Π′ *is tangent to a slice* $\{s\} \times S^{n-1}$*, then we have*

$$K_{\mathrm{sph}} := \mathrm{sect}(\Pi') = \frac{1 - f'(s)^2}{f(s)^2}. \tag{8.148}$$

We call K_{rad} *the* ***radial sectional curvature*** *of* g *and we call* K_{sph} *the* ***spherical sectional curvature*** *of* g*.*

(2) *Consider the cases where* $f(s)$ *is equal to* $\sin(s)$*,* s*, and* $\sinh(s)$*.*

Exercise 8.17. *Denote* $|g| := \det(g_{ij})$. *Show that*

$$\Delta u = \frac{1}{\sqrt{|g|}} \sum_{i,j=1}^{n} \frac{\partial}{\partial x^i} \left(\sqrt{|g|}\, g^{ij} \frac{\partial u}{\partial x^j} \right). \tag{8.149}$$

Exercise 8.18. *Let* $g = ds^2 + f(s)^2 d\theta^2$. *Show that the Laplacian with respect to* g *is given by*

$$\Delta u = \frac{\partial^2 u}{\partial s^2} + \frac{1}{f(s)^2} \frac{\partial^2 u}{\partial \theta^2} + \frac{f'(s)}{f(s)} \frac{\partial u}{\partial s}. \tag{8.150}$$

Hint: Show that $\omega_2^1 = -\frac{f'(s)}{f(s)} \omega^2$.

Exercise 8.19. *Use* (8.149) *to give a shorter proof of Lemma* 8.22. *Then generalize the lemma to higher dimensions.*

Exercise 8.20. *Let* (M^n, g) *be a Riemannian manifold and let* $\tilde{g} := \mathrm{e}^{2u} g$. *Let* $\{\omega^i\}_{i=1}^n$ *be a local orthonormal coframe field for* g. *Then* $\{\tilde{\omega}^i\}_{i=1}^n$, *where* $\tilde{\omega}^i := \mathrm{e}^u \omega^i$, *is a local orthonormal coframe field for* $\tilde{g}$. *Let* $\{e_i\}_{i=1}^n$ *and* $\{\tilde{e}_i\}_{i=1}^n$ *denote the orthonormal frame fields dual to* $\{\omega^i\}$ *and* $\{\tilde{\omega}^i\}$, *respectively, so that* $\tilde{e}_i = \mathrm{e}^{-u} e_i$. *Show that the associated Riemann curvature 2-forms* $\mathrm{Rm}_i^{\,j}$ *and* $\widetilde{\mathrm{Rm}}_i^j$ *are related by the formula*

$$\begin{aligned} \widetilde{\mathrm{Rm}}_i^j &= \mathrm{Rm}_i^{\,j} + \nabla^2_{e_k, e_i} u\, \omega^k \wedge \omega^j - \nabla^2_{e_k, e_j} u\, \omega^k \wedge \omega^i \\ &\quad + \|\nabla u\|^2 \omega^i \wedge \omega^j + du \wedge \left(e_j(u)\, \omega^i - e_i(u)\, \omega^j \right), \end{aligned} \tag{8.151}$$

where ∇ *denotes covariant differentiation with respect to* g.

Exercise 8.21. *Use Lemma* 8.21 *to give an alternate derivation of the formula for* $K_{\mathbb{U}}$.

Exercise 8.22. *Let* $U^n := \{\mathbf{x} \in \mathbb{R}^n : x^n > 0\}$ *be the* n*-dimensional upper half-space. Define the Riemannian metric* $g_{\mathbb{U}} := (x^n)^{-2} g_{\mathbb{E}}$, *where* $g_{\mathbb{E}}$ *is the Euclidean metric on* $\mathbb{R}^n$. *Prove that* $g_{\mathbb{U}}$ *has constant* -1 *sectional curvature.*

Exercise 8.23. *Prove that Theorem* 8.16 *and Theorem* 8.18 *are equivalent.*

Chapter 9

The Gauss–Bonnet and Poincaré–Hopf Theorems

In this chapter, building upon the framework of moving frames developed in the previous chapter, we present a proof of the Gauss–Bonnet formula for 2-dimensional closed oriented Riemannian manifolds and orbifolds using this technique. To pave the way for this proof, we will delve into foundational topics such as Morse theory, intersection numbers, degrees of maps, integration on surfaces, and the Poincaré–Hopf theorem.

9.1. Morse functions and the Morse lemma

It is natural to consider a closed manifold as being slowly built up by a family of sublevel sets of a given function on the manifold as the 'height' goes from the minimum to the maximum of the function. For example, imagine a clear glass container shaped like a torus with a tiny hole at the top; slowly fill this container with dyed water and see how the topology of the part of the container that is filled up with water changes. For this to work well to track the topological changes that occur when the heights of the sublevel sets pass critical values of the function, we require the function to satisfy a nondegeneracy condition at its critical points. Such functions, called *Morse functions*, which we will discuss in this section, are useful for the study of the topology of manifolds. See Figure 9.1.1.

9.1.1. Nondegenerate critical points and Morse functions. Let M^n be a smooth manifold and let $f : M^n \to \mathbb{R}$ be a smooth function. Recall

Figure 9.1.1. The level sets of a Morse function on a 2-dimensional torus. Credit: Public Domain, `https://commons.wikimedia.org/w/index.php?curid=1520502`

that a point p in M^n is a **critical point** of f if $df_p = 0$. We call $f(p)$ a **critical value** of f.

Let p be a critical point of f and let $\{x^i\}_{i=1}^n$ be local coordinates in a neighborhood of p. We say that p is a **nondegenerate critical point** if the Hessian matrix $\left(\frac{\partial^2 f}{\partial x^i \partial x^j}(p)\right)$ is nonsingular. This definition is independent of the choice of local coordinates. Indeed, equivalently, the covariant Hessian $\nabla^2 f$ (a symmetric 2-tensor) has a trivial nullspace; i.e., if $V \in T_pM$ is such that $\nabla^2 f(V, W) = 0$ for all $W \in T_pM$, then $V = 0$. The aforestated equivalence follows from $\nabla_i \nabla_j f = \partial_i \partial_j f - \Gamma_{ij}^k \partial_k f$ and $\partial_k f(p) = 0$.

Definition 9.1. We say that a smooth function $f : M^n \to \mathbb{R}$ is a **Morse function** if each of its critical points is nondegenerate.

The study of Morse functions on smooth manifolds is called **Morse theory**.

Since $\left(\frac{\partial^2 f}{\partial x^i \partial x^j}(p)\right)$ is symmetric, it is diagonalizable with n real eigenvalues, counting multiplicity. The condition of $\left(\frac{\partial^2 f}{\partial x^i \partial x^j}(p)\right)$ being nonsingular is equivalent to the eigenvalues being all nonzero. The **Morse index** $\operatorname{index}_p(f)$ of f at p is defined to be the number of negative eigenvalues. The Morse index of f at p encapsulates the essential qualitative behavior of f near p.

Example 9.2. *For a torus T^2 embedded in $\mathbb{R}^3$ as in Figure* 9.1.1, *the 'height function' $f : T^2 \to \mathbb{R}$ defined by $f(x, y, z) := z$ is a Morse function. From the figure, we see that f has* 4 *critical points, with Morse indices* 0, 1, 1, *and* 2, *in increasing order of the associated critical values.*

For the proof of the following, see Milnor's book on the h-cobordism theorem [**Mil65**, Theorem 2.7].

Lemma 9.3. *For any closed smooth manifold M^n, there exists a Morse function on M^n.*

This result lays the foundation for employing Morse functions in the study of the differential topology of manifolds.

9.1.2. The inverse and implicit function theorems.

Recall:

Theorem 9.4 (Inverse function theorem). *Let $f : M^n \to N^m$ be a smooth function between manifolds. If $p \in M^n$ is such that $df_p : T_pM \to T_{f(p)}N$ is an injection, then there exists an open neighborhood U of p such that $f(U)$ is a smooth n-dimensional submanifold of N^m and $f|_U : U \to f(U)$ is a diffeomorphism. Note that if $m = n$, then $f(U)$ is an open subset of N^m.*

We have the following version of the implicit function theorem (cf. Theorem 2.11).

Theorem 9.5 (Implicit function theorem). *Let $f : M^n \to N^m$ be a smooth function between manifolds. If $p \in M^n$ is such that $df_p : T_pM \to T_{f(p)}N$ is a surjection, then there exists an open neighborhood U of p such that $f^{-1}(q) \cap U$, where $q = f(p)$, is an $(n-m)$-dimensional submanifold of M^n.*

The inverse and implicit function theorems are both special cases of the following **constant rank theorem** (where the constant ranks are n and m, respectively); see the Constant-Rank Level Set Theorem 5.12 in [**Lee13**]. This says that level sets of constant rank maps are submanifolds.

Theorem 9.6. *Let $F : M^n \to N^m$ be a smooth map between smooth manifolds. If F has constant rank k, then for every $y \in N^m$, the level set $F^{-1}(y)$ is a properly embedded $(n-k)$-dimensional submanifold of M^n.*

9.1.3. The Morse lemma.

In a neighborhood of any nondegenerate critical point p of a function f, there are coordinates that put f into a canonical form (see Milnor's Morse theory book [**Mil63**, Lemma 2.2]). This shows that the Morse index characterizes the topological nature of the behavior of f near p. See Figure 9.1.2.

Proposition 9.7 (Morse lemma). *Let M^n be a smooth manifold and let f be a smooth function on M^n. If p is a nondegenerate critical point of f of index k, then there exist local coordinates $\mathbf{x} = \{x^i\}_{i=1}^n$ in a neighborhood $\mathcal{U}$ such that $\mathbf{x}(p) = \mathbf{0}$ and in $\mathcal{U}$ we have*

$$f = f(p) - \sum_{i=1}^{k} (x^i)^2 + \sum_{j=k+1}^{n} (x^j)^2. \tag{9.1}$$

In particular, in $\mathcal{U}$ we have that $\frac{\partial f}{\partial x^i}$ equals $-2x^i$ if $i \leq k$ and $\frac{\partial f}{\partial x^i}$ equals $2x^i$ if $i \geq k+1$.

Figure 9.1.2. Level sets of a function with a critical point of saddle type. The Morse lemma (Proposition 9.7) implies that, by a suitable change of coordinates, the contours can be straightened out to a family of standard hyperbolas together with the lines $y = \pm x$. Credit: By Sam Derbyshire – This W3C-unspecified diagram was created with Mathematica., CC BY-SA 3.0, `https://commons.wikimedia.org/w/index.php?curid=18447127`

Example 9.8. *Suppose that $n = 2$ and $p \in M^2$ is a nondegenerate critical point of a function f with $f(p) = 0$. Then there exist local coordinates $\{x^1, x^2\}$ in a neighborhood of p such that:*

(1) *(Local minimum)* $f = +(x^1)^2 + (x^2)^2$ *if* $\operatorname{index}_p(f) = 0$.

(2) *(Saddle point)* $f = -(x^1)^2 + (x^2)^2$ *if* $\operatorname{index}_p(f) = 1$.

(3) *(Local maximum)* $f = -(x^1)^2 - (x^2)^2$ *if* $\operatorname{index}_p(f) = 2$.

Proof of Proposition 9.7. Let $\mathbf{y} = \{y^\alpha\}_{\alpha=1}^n$ be local coordinates in some neighborhood $\mathcal{V}$ of p with $\mathbf{y}(p) = \mathbf{0}$. We first express f in a nice form with respect to these coordinates. Namely, we claim that there exist smooth functions g_α on $\mathcal{V}$ such that

$$f - f(p) = \sum_{\alpha=1}^n y^\alpha g_\alpha . \tag{9.2}$$

Indeed, by the fundamental theorem of calculus and the chain rule, we have for $\bar{f} := f \circ \mathbf{y}^{-1}$ and $\mathbf{v} \in \mathbf{y}(\mathcal{V}) \subset \mathbb{R}^n$ that

$$\bar{f}(\mathbf{v}) - \bar{f}(\mathbf{0}) = \int_0^1 \frac{d}{dt}\bar{f}(t\mathbf{v})\,dt = \mathbf{v} \cdot \int_0^1 \nabla_{\mathbb{R}^n} \bar{f}(t\mathbf{v})\,dt.$$

Taking $\mathbf{v} = \mathbf{y}(q) = (y^1(q), \ldots, y^n(q))$, where $q \in \mathcal{V}$, this says that

$$f(q) - f(p) = \sum_{\alpha=1}^n y^\alpha(q) g_\alpha(q),$$

where

$$g_\alpha(q) = \int_0^1 \frac{\partial f}{\partial y^\alpha}\big(\mathbf{y}^{-1}(t\mathbf{y}(q))\big)\,dt$$

are smooth functions. This proves the claim.

Now observe that by differentiating (9.2), we have

$$g_\alpha(p) = \frac{\partial f}{\partial y^\alpha}(p) = 0$$

since p is a critical point of f. We may apply (9.2) again, now to each g_α instead of f, to obtain

$$g_\alpha = \sum_{\beta=1}^n y^\beta h_{\alpha\beta}$$

for some smooth functions $h_{\alpha\beta}$ on $\mathcal{V}$. Thus

$$f - f(p) = \sum_{\alpha,\beta=1}^n y^\alpha y^\beta h_{\alpha\beta} \quad \text{on } \mathcal{V},$$

where by symmetrizing we may assume that $h_{\alpha\beta} = h_{\beta\alpha}$. Observe that

$$(2h_{\alpha\beta}(p)) = \left(\frac{\partial^2 f}{\partial y^\alpha y^\beta}(p)\right)$$

is a nonsingular matrix by hypothesis.

Now we find the desired coordinates by an n-step induction. Let $1 \leq k \leq n$ and suppose by induction that we have found coordinates $\{z^i\}_{i=1}^n$ in a neighborhood $\mathcal{V}_k$ of p such that

$$f - f(p) = \sum_{i=1}^{k-1} \pm(z^i)^2 + \sum_{i,j=k}^n z^i z^j H_{ij} \tag{9.3}$$

for some smooth functions $\{H_{ij}\}_{i,j=k}^n$ satisfying $H_{ij} = H_{ji}$ and $H_{kk} \neq 0$ in a smaller neighborhood $\mathcal{V}'_k$.

We define $\{w^i\}_{i=1}^n$ by only changing the k-th coordinate appropriately:

$$\begin{aligned} w^i &= z^i \quad \text{for } i \neq k, \\ w^k &= \sqrt{|H_{kk}|}\left(z^k + \sum_{i=k+1}^n z^i \frac{H_{ik}}{H_{kk}}\right). \end{aligned}$$

Observe that

$$\begin{aligned} \frac{\partial w^j}{\partial z^i} &= \delta_i^j,\, j \neq k, & \frac{\partial w^k}{\partial z^i} &= 0,\, i < k, \\ \frac{\partial w^k}{\partial z^k} &= \sqrt{|H_{kk}|}, & \frac{\partial w^k}{\partial z^i} &= \sqrt{|H_{kk}|}\frac{H_{ik}}{H_{kk}},\, i > k. \end{aligned}$$

In particular, $\det\left(\frac{\partial w^j}{\partial z^i}\right) = \sqrt{|H_{kk}|} \neq 0$. Thus, by the inverse function theorem (Theorem 9.4), $\{w^i\}_{i=1}^n$ are smooth coordinates on some smaller neighborhood.

Clearly,

$$\sum_{i=1}^{k-1} \pm(z^i)^2 + \sum_{i,j=k+1}^{n} z^i z^j H_{ij} = \sum_{i=1}^{k-1} \pm(w^i)^2 + \sum_{i,j=k+1}^{n} w^i w^j H_{ij}.$$

We compute that

$$\begin{aligned} \frac{H_{kk}}{|H_{kk}|} w^k w^k &= H_{kk}\left(z^k + \sum_{i=k+1}^{n} z^i \frac{H_{ik}}{H_{kk}}\right)^2 \\ &= H_{kk} z^k z^k + 2\sum_{i=k+1}^{n} H_{ik} z^i z^k + \sum_{i,j=k+1}^{n} \frac{H_{ik}H_{jk}}{H_{kk}} z^i z^j. \end{aligned}$$

Therefore,

$$f - f(p) = \sum_{i=1}^{k} \pm(w^i)^2 + \sum_{i,j=k+1}^{n} w^i w^j G_{ij} \tag{9.4}$$

for some smooth functions $\{G_{ij}\}_{i,j=k+1}^n$ satisfying $G_{ij} = G_{ji}$; namely

$$G_{ij} = H_{ij} - \frac{H_{ik}H_{jk}}{H_{kk}}.$$

Since $(G_{ij}(p))_{i,j=k+1}^n$ is nonsingular, we may assume by a linear change of coordinates, that $G_{(k+1)(k+1)} \neq 0$ at p and hence in a neighborhood of p. This completes the proof by induction on k. □

9.1.4. Homotopy types of manifolds.

A well-known application of Morse theory is the following; see, e.g., [**Mil63**, Theorem 3.5].

Theorem 9.9. *If M^n is a smooth manifold and if $f : M^n \to \mathbb{R}$ is a Morse function with the property that each sublevel set $M(a)$ is compact (which is true if M^n is closed), then M^n has the homotopy type of a finite CW-complex with an i-cell for each critical point of index i.*

For example, the height function of a 2-torus, pictured in Figure 9.1.1, has critical points of index 0, 1, 1, and 2 and hence the 2-torus has the homotopy type of a finite CW-complex with one 0-cell, two 1-cells, and one 2-cell.

9.1.5. Vector fields with isolated zeroes. By the Morse lemma (Proposition 9.7), we see that for any Morse function, its critical points are isolated. Hence, if M^n is a closed manifold, then the set of critical points is finite. In particular, if g is a Riemannian metric on a closed manifold M^n, then the gradient vector field ∇f of a Morse function f has a finite zero-set. Since a Morse function exists on any closed manifold by Lemma 9.3, this proves the existence of a smooth vector field with a finite zero-set on any closed manifold.[1]

An alternative way to prove the existence of a vector field with isolated zeroes on any closed manifold is to view the vector field as a section of the tangent bundle and use the fact that we can perturb this section so that it intersects the zero-section transversally. Namely, the zero-section is the map $Z : M^n \to TM$ defined by $Z(x) := \vec{0}_x$, where $\vec{0}_x$ denotes the zero vector in T_xM. Of course, $Z(M^n)$ is a submanifold of TM diffeomorphic to M^n. Let

$$H_x := dZ_x(T_xM) = T_{\vec{0}_x}(Z(M^n)) \subset T_{\vec{0}_x}(TM),$$

which we call a **horizontal space**.

Now let $V : M^n \to TM$ be a smooth vector field. Then $V(M^n)$ is a submanifold of TM diffeomorphic to M^n. Suppose that $x \in M^n$ is a zero of V; that is, $V_x = \vec{0}_x$. We say that V intersects the zero-section Z transversally at x if H_x and $T_{\vec{0}_x}(V(M^n))$ span $T_{\vec{0}_x}(TM)$; since the dimensions of $V(M^n)$ and $Z(M^n)$ are complementary, this is equivalent to $H_x \cap T_{\vec{0}_x}(V(M^n)) = \{\vec{0}\}$. A general result about transversality (the **transversality theorem**; see [**GP10**, p. 68 in Chapter 2, §3]) tells us that the vector field V can be perturbed so that it intersects the zero-section Z transversally. Thus, we may assume that the zeroes of V are isolated.

9.2. Intersection number and degree

In this section, we define the intersection number of a map f, between manifolds of the same dimension and at a regular value of f.

9.2.1. Intersection numbers of maps. Let M_1^n and M_2^n be manifolds of the same dimension, and let $f : M_1^n \to M_2^n$. We say that $y \in M_2^n$ is a **regular value** of f if for every $x \in f^{-1}(y)$ we have $df_x(T_xM_1) = T_yM_2$. In other words, $df_x : T_xM_1 \to T_yM_2$ is a vector space isomorphism.

[1] To obtain such a vector field, one does not need to introduce a Riemannian metric on the smooth manifold M^n; see Milnor [**Mil65**, §3], where **gradient-like vector fields** are discussed.

Now assume that the manifolds M_1^n and M_2^n are closed and oriented. Given a regular value y of f, we have that $f^{-1}(y)$ consists of a finite number of points. For each point $x \in f^{-1}(y)$, its **orientation number** is $+1$ (respectively, -1) if $df_x : T_xM_1 \to T_yM_2$ is orientation preserving (respectively, orientation reversing).

The (oriented) **intersection number**

$$I(f, \{y\}) := \sum_{x \in f^{-1}(y)} (\text{orientation number of } x)$$

of f and $\{y\}$ is the sum of the orientation numbers of the preimages of y. The intersection number $I(f, \{y\})$ is an integer independent of the choice of the regular value y (see, e.g., Guillemin and Pollack [**GP10**, Chapter 3, §3]). The **degree** of f is defined to be this integer:

$$\deg(f) := I(f, \{y\}).$$

Note that, by the following theorem of Sard (see [**Mil97**, §2 and §3]), the set of critical values of f has measure zero; i.e., the set of regular values of f has full measure.

Theorem 9.10 (Sard). *Let $f : M^n \to N^m$ be a smooth map. We say that $x \in M^n$ is a critical point of f if df_x has rank less than m. Let C denote the set of critical points of f in M^n. Then the set of critical values $f(C)$ has measure zero in N^m.*

Example 9.11. *Let $M_1^n = M_2^n = S^1$ be the unit circle. Let k be an integer and define the map $f : S^1 \to S^1$ by $f(z) := z^k$ for $z \in S^1 \subset \mathbb{C}$. Assume that $k \neq 0$. Then every $y \in S^1$ is a regular value of f and the intersection number $I(f, \{y\})$ is equal to k. Thus, the degree of the map f is equal to k. On the other hand, if $k = 0$, then the only critical value of f is the point $1 \in S^1 \subset \mathbb{C}$. So the degree in this case is equal to 0.*

Example 9.12. *Let $f : [0, 2\pi] \to [0, 2\pi]$ be a smooth function with $f(0) = 0$ and $f(2\pi) = 2\pi$. Define $\bar{f} : S^1 \to S^1$ by $\bar{f}([x]) = [f(x)]$ for $x \in [0, 2\pi]$, where $S^1 = \mathbb{R}/2\pi\mathbb{Z}$. Assume that f is such that the map $\bar{f}$ is smooth. Then* $\deg(f) = 1$. *See Figure 9.2.1 for an example.*

9.2.2. Index of a vector field. Let V be a vector field on an oriented manifold M^n and suppose that $p \in M^n$ is an isolated zero of V. Choose any positively oriented local coordinates $\mathbf{x} = \{x^i\}_{i=1}^n$ defined on a neighborhood $\mathcal{U}$ of p with $\mathbf{x}(p) = \mathbf{0}$ such that V has no zeroes in $\mathcal{U}$. Choose $\varepsilon > 0$ so that the $(n-1)$-sphere S_ε of radius ε is contained in $\mathbf{x}(\mathcal{U})$. Let $V =: \sum_{i=1}^n V^i \frac{\partial}{\partial x^i}$.

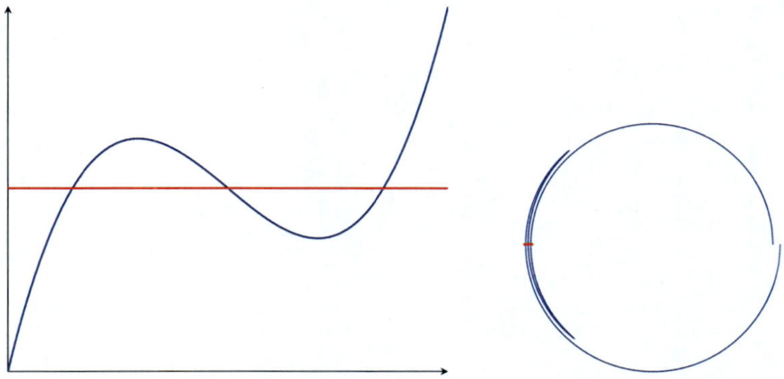

Figure 9.2.1. Left: The graph of a smooth function $f : [0, 2\pi] \to [0, 2\pi]$ with $f(0) = 0$ and $f(2\pi) = 2\pi$. For $y = \pi$, we have $\deg(\bar{f}) = I(\bar{f}, \{[y]\}) = 1 + (-1) + 1 = 1$, where $[y] \in S^1$ denotes the equivalence class of y. Right: Visualizing the surjective map $\bar{f}$.

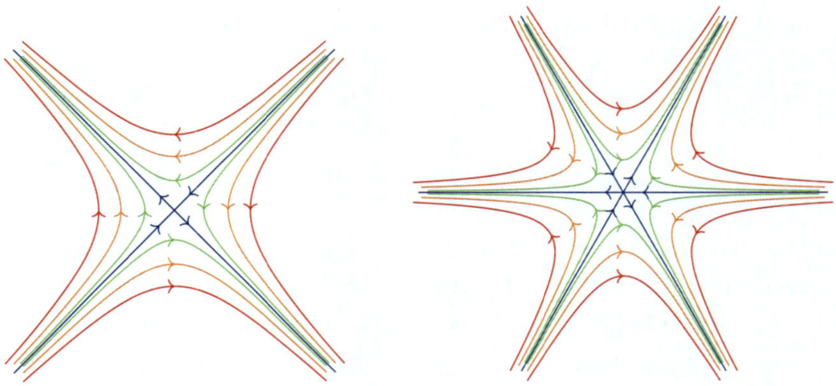

Figure 9.2.2. Left: The integral curves of a vector field with index -1. Right: The integral curves of a vector field with index -2.

Define the map

$$f : S_\varepsilon \to S_1$$

by

$$f(\mathbf{x}) = \frac{\mathbf{V}}{\|\mathbf{V}\|}, \tag{9.5}$$

where $\mathbf{V} := (V^1, \ldots, V^n) \in \mathbb{R}^n$. The map f is well-defined since V has no zeroes in $\mathcal{U}$.

Definition 9.13. The **index** of V at p is defined to be the degree of the map f given by (9.5), where S_ε and S_1 are endowed with the orientations induced by M^n and the local coordinates $\{x^i\}_{i=1}^n$; see [**Mil97**, §6]. The index is independent of the choice of $\{x^i\}_{i=1}^n$. See Figures 9.2.2 and 9.2.3.

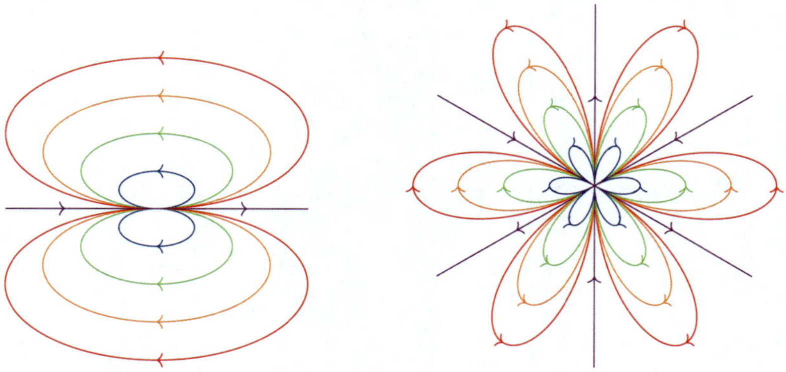

Figure 9.2.3. Left: The integral curves of a vector field with index 2. Right: The integral curves of a vector field with index 4.

9.2.3. The Poincaré–Hopf theorem. The Euler characteristic χ is a fundamental topological invariant. It has many different but equivalent definitions. For example, the most general definition holds for topological spaces. Given a topological space S, we define its **Euler characteristic** by

$$\chi(S) = \sum_{i=0}^{\infty}(-1)^i b_i(S), \tag{9.6}$$

where $b_i(S)$ denotes the rank of the i-th singular homology group $H_i(S;\mathbb{R})$ and is called the **i-th Betti number** of S.[2] The Euler characteristic of S is well-defined provided $b_i(S)$ vanishes for all but a finite number of i.

For a finite CW-complex K, its Euler characteristic is equal to

$$\chi(K) = \sum_{i\geq 0}(-1)^i k_i(K), \tag{9.7}$$

where $k_i(K)$ denotes the number of i-cells of K (this follows from cellular homology being isomorphic to singular homology for CW-complexes; see Hatcher [**Hat02**, Theorem 2.44]).

The Euler characteristic $\chi(M^n)$ of a manifold M^n is equal to the oriented intersection number in its tangent bundle TM of a smooth section (i.e., vector field) and the zero-section provided they meet transversally (see Chapter 3 of [**GP10**] for why this intersection number is well-defined; note that all sections of a vector bundle are homotopic to each other by a straight line homotopy).

This fact, about intersection numbers of sections of TM, is equivalent to the following **Poincaré–Hopf theorem** (see [**Mil97**, §6] or [**GP10**, p. 134 in Chapter 3, §5]).

[2]For the definition of homology, see, e.g., Hatcher [**Hat02**, Chapter 2].

Theorem 9.14 (Poincaré and Hopf). *The sum of the indices of a vector field on M^n with isolated zeroes is equal to $\chi(M^n)$.*

Assuming the fact that the sum of the indices of a vector field on M^n with isolated zeroes is independent of the vector field, we give a proof of this theorem using Morse theory.[3]

Let f be a Morse function on M^n. Recall that, by definition, this means that the critical points of f are all nondegenerate; that is, the Hessians of f at its critical points are nonsingular. By the Morse lemma (Proposition 9.7), if p is a critical point of f, then there exists a coordinate system $\mathbf{x} = (x^1, \ldots, x^n)$ on a neighborhood $\mathcal{U}$ of p such that $\mathbf{x}(p) = \mathbf{0}$ and

$$f = f(p) - \sum_{i=1}^{I} (x^i)^2 + \sum_{j=I+1}^{n} (x^j)^2 \quad \text{in } \mathcal{U},$$

where $0 \leq I \leq n$. The integer $\operatorname{index}_p(f) := I$ is equal to the Morse index of f at p. Since $\frac{\partial f}{\partial x^i}$ is equal to $-2x^i$ if $i \leq I$ and equal to $2x^i$ if $i > I$, we see that the critical points of a Morse function are isolated. Since M^n is compact, there are thus only finitely many critical points of f. At a critical point p of f, the index of the vector field ∇f at p is equal to $(-1)^{\operatorname{index}_p(f)}$ since this is the degree of the map $(x^1, \ldots, x^n) \mapsto (-x^1, \ldots, -x^I, x^{I+1}, \ldots, x^n)$ from S^{n-1} to itself:

$$\operatorname{index}_p(\nabla f) = (-1)^{\operatorname{index}_p(f)}.$$

See Figure 9.2.4 for the integral curves of f near minimum and maximum points.

Let $f_{\min} = c_1 < c_2 < \cdots < c_m = f_{\max}$ denote the critical values of a Morse function f on M^n. Given $a \in \mathbb{R}$, let

$$M(a) := \{x \in M^n \colon f(x) \leq a\} \tag{9.8}$$

be a **sublevel set** of the function f. If $a \in (c_1, c_2)$, then $M(a)$ is diffeomorphic to the disjoint union of $|f^{-1}(c_1)|$ closed n-balls and hence homotopy equivalent to the disjoint union of $|f^{-1}(c_1)|$ 0-cells. This is because each critical point with critical value c_1 is a minimum point and hence has index equal to 0. As we increase a, for each critical point p of f with index I, the passing of a through the associated critical value corresponds homotopically for $M(a)$ to the *attaching* of an I-cell; cf. Theorem 9.9. (E.g., a local minimum corresponds to a 0-cell, a local maximum corresponds to an n-cell, a saddle point on a surface ($n = 2$) corresponds to a 1-cell.) See Figures 9.2.5 and 9.2.6.

[3] Actually, a standard proof of the Poincaré–Hopf theorem is, thinking of a vector field as a section of the tangent bundle, to homotope it to a vector field whose zeroes are each either $+1$ or -1 and then to triangulate the manifold and show directly from this that the sum of the indices is equal to the Euler characteristic. This bypasses the need to show that the sum of the indices is independent of the vector field.

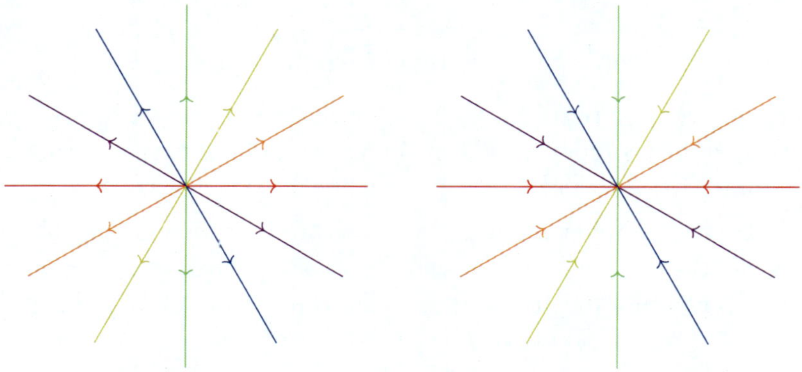

Figure 9.2.4. Left: The integral curves to ∇f near a minimum point of a Morse function f in canonical form. Right: Near a maximum point of f. The integral curves near a saddle point are pictured on the left side of Figure 9.2.2.

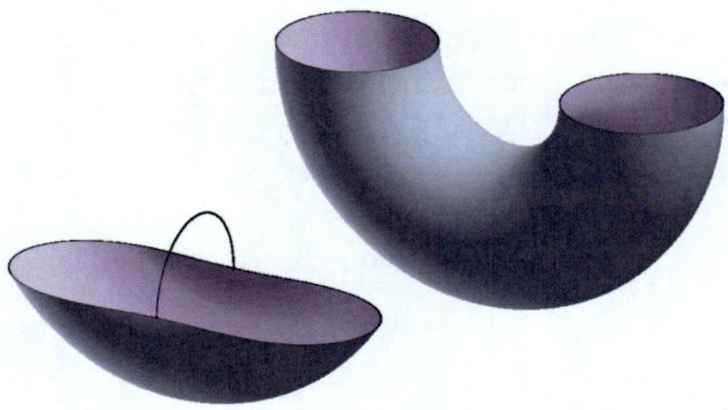

Figure 9.2.5. The sublevel set on the right is homotopic to the topological space obtained by attaching a 1-cell to the sublevel set on the left. Credit: Public Domain, `https://commons.wikimedia.org/w/index.php?curid=1520470`

Hence, the Euler characteristic of M^n, which is equal to the alternating sum $\sum_{i=0}^{n}(-1)^i$(number of i-cells), is given by

$$
\begin{aligned}
(9.9)\qquad \chi(M^n) &= \sum_{i=0}^{n}(-1)^i\big(\text{number of critical points of } f \text{ of index } i\big)\\
&= \sum_{\text{critical points } p \text{ of } f} (-1)^{\operatorname{index}_p(f)}\\
&= \sum_{\text{zeroes } p \text{ of } \nabla f} \operatorname{index}_p(\nabla f).
\end{aligned}
$$

We remark that by cellular homology, we have the inequality

$$
(9.10)\qquad b_i(M^n) \le k_i(M^n) = c_i(f) := \#\text{ of critical points of } f \text{ of index } i.
$$

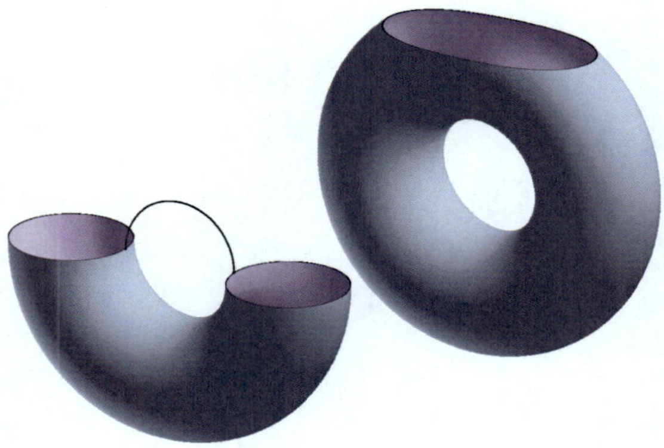

Figure 9.2.6. Again, the sublevel set on the right is homotopic to the topological space obtained by attaching a 1-cell to the sublevel set on the left. Credit: Public Domain, `https://commons.wikimedia.org/w/index.php?curid=1520472`

The **Morse inequalities** say that for every $0 \leq j \leq n$,

$$\sum_{i=0}^{j}(-1)^{j-i}c_i(f) \geq \sum_{i=0}^{j}(-1)^{j-i}b_i(M^n); \tag{9.11}$$

see Milnor [**Mil63**, §5] for the proof of this. Note that from this, we successively deduce that $c_0(f) - b_0(M^n) \geq 0$, $c_1(f) - b_1(M^n) \geq 0$, ..., $c_n(f) - b_n(M^n) \geq 0$. Thus, the Morse inequalities strengthen the inequalities in (9.10).

In [**Wit82**], Witten gave an innovative proof of the Morse inequalities using ideas from theoretical physics.

9.3. Statement of the Gauss–Bonnet formula

Consider the **total Gauss curvature** of a closed Riemannian surface (M^2, g); that is, the integral

$$\int_{M^2} K\, d\mu, \tag{9.12}$$

where K is the Gauss curvature defined in §7.7.2.

For example, if g_r is the 2-sphere of radius $r > 0$, then its Gauss curvature is equal to $K_r = \frac{1}{r^2}$ and its area is equal to $\mathrm{Area}(g_r) = 4\pi r^2$. Thus, the total Gauss curvature is equal to

$$\int_{S^2} K_r\, d\mu_{g_r} = \frac{1}{r^2} \cdot 4\pi r^2 = 4\pi. \tag{9.13}$$

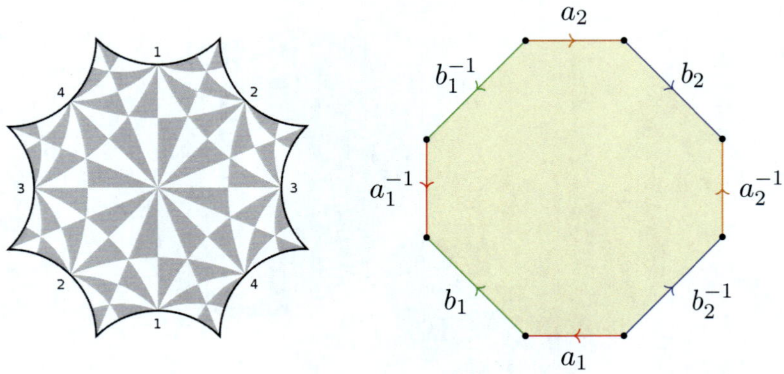

Figure 9.3.1. Left: An octagonal fundamental domain in the Poincaré ball model $\mathbb{D}^2$ of a hyperbolic surface of genus 2. Credit: By Parcly Taxel – Own work, FAL, `https://commons.wikimedia.org/w/index.php?curid=87326215`. Right: The boundary identifications for the octagon.

On the other hand, the torus T^2 has many flat metrics. If g is such a metric on T^2, then we have that

$$\int_{T^2} K\, d\mu = 0. \tag{9.14}$$

A particular Riemannian surface (S_2, g) of genus 2 may be obtained by taking a regular hyperbolic octagon with interior angles equal to $\pi/8$ by identifying the edges according to Figure 8.8.3. The result yields a metric g on S_2 with constant Gauss curvature -1, called the **Bolza surface**. See Figure 9.3.1. Using hyperbolic trigonometry, one can compute that the area of the Bolza surface is equal to 4π. Therefore,

$$\int_{S_2} K\, d\mu = -4\pi. \tag{9.15}$$

We have the following completely general result regarding the total Gauss curvature of a closed oriented Riemannian surface. See Figure 9.3.2.

Theorem 9.15 (Gauss–Bonnet formula [**Gau65**], [**Bon48**]). *Let (M^2, g) be a closed oriented Riemannian surface. Then*

$$\int_{M^2} K\, d\mu = 2\pi\chi(M^2). \tag{9.16}$$

Example 9.16. *Consider the rotationally symmetric metric $g := ds^2 + f(s)^2 d\theta^2$ defined on the annulus $M^2 := (0, L) \times (\mathbb{R}/2\pi\mathbb{Z})$. Recall from (8.144) that*

$$K(s, \theta) = -\frac{f''(s)}{f(s)}.$$

Figure 9.3.2. Left: A double-holed torus has total Gauss curvature -4π. Right: A triple-holed torus has total Gauss curvature -8π. Credit: By Oleg Alexandrov – This diagram was created with MATLAB., Public Domain, `https://commons.wikimedia.org/w/index.php?curid=2697455`

Thus,

$$\int_{M^2} K\,d\mu = 2\pi \int_0^L -\frac{f''(s)}{f(s)} f(s)\,ds = 2\pi(f'(0) - f'(L)) \tag{9.17}$$

by the fundamental theorem of calculus. Suppose that M^2 compactifies to a 2-sphere S^2 and that the Riemannian metric g extends smoothly over the points where $s = 0$ and $s = L$. Then $f'(0) = 1$ and $f'(L) = -1$, which implies that $\int_{S^2} K\,d\mu = 4\pi$, in agreement with the Gauss–Bonnet formula.

9.4. Proof of the Gauss–Bonnet formula

Let M^2 be a closed oriented surface of genus $\mathbf{g}$. Its Euler characteristic is equal to

$$\chi(M^2) = b_0(M^2) - b_1(M^2) + b_2(M^2) = 1 - 2g + 1 = 2 - 2\mathbf{g}.$$

Let V be a vector field on M^2. Recall from §9.1.5 that we may assume that V has **isolated zeros**; that is, the subset

$$\mathcal{Z}(V) := \left\{x \in M^2 \colon V(x) = \vec{0}\right\} \subset M^2$$

is a discrete set; i.e., each point in $\mathcal{Z}$ is an isolated point. Since M^2 is closed, this is equivalent to $\mathcal{Z}(V)$ being a finite set.

Let $I(V)$ denote the index of the vector field V. By the Poincaré–Hopf theorem (Theorem 9.14),

$$I(V) = \chi(M^2). \tag{9.18}$$

Let

$$\mathcal{V} := M^2 \setminus \mathcal{Z}(V), \tag{9.19}$$

which is an open subset of M^2 since $\mathcal{V}$ is equal to M^2 minus a finite number of points.

Now let g be a Riemannian metric on M^2. Define the unit vector field e_1 on $\mathcal{V}$ by

$$e_1 := \frac{V}{|V|}, \tag{9.20}$$

where $|V| := \sqrt{g(V,V)}$ is the norm of V with respect to g.

Since (M^2, g) is an oriented Riemannian surface, we have the almost complex structure $J : TM \to TM$ defined by $J(\vec{0}) = \vec{0}$ and for $V \neq \vec{0}$, $J(V)$ is such that $\{V, J(V)\}$ is a positively oriented orthogonal frame with $|J(V)| = |V|$. Then J is a linear transformation with $J^2 = -\operatorname{id}_{T_xM}$ on each tangent space. We consider J as a counterclockwise rotation by 90 degrees; see Figure 9.4.1. Define

$$e_2 := J(e_1). \tag{9.21}$$

Then $\{e_1, e_2\}$ is a positively oriented orthonormal frame field on $\mathcal{V} \subset M^2$.

Let ω_i^j and Rm_i^j denote the connection 1-forms and curvature 2-forms, respectively, associated to this orthonormal frame field. Using (8.78), we see that the Gauss–Bonnet integrand may be expressed as

$$K\, d\mu = \mathrm{Rm}_2^1(e_1, e_2)\omega^1 \wedge \omega^2 = \mathrm{Rm}_2^1 = d\omega_2^1, \tag{9.22}$$

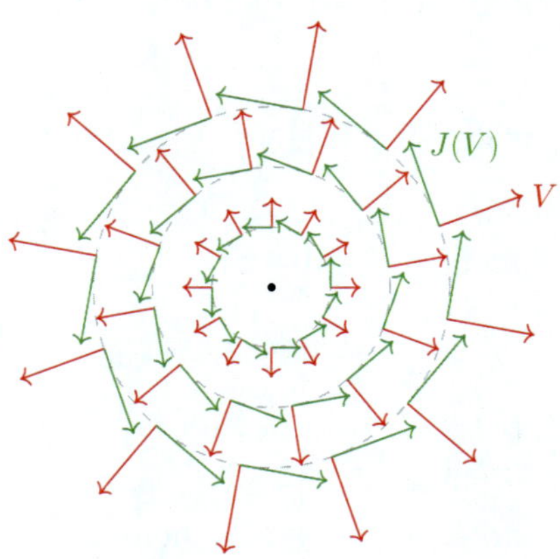

Figure 9.4.1. A vector field V and its counterclockwise 90-degree rotation $J(V)$.

since for any 2-form β on our Riemannian surface, we have

$$\beta = \beta(e_1, e_2)\omega^1 \wedge \omega^2.$$

(This last fact follows from $\omega^1 \wedge \omega^2 = \omega^1 \otimes \omega^2 - \omega^2 \otimes \omega^1$.)

If we are on a 2-torus T^2, so that $\chi(T^2) = 0$, then there exists a nowhere-zero vector field V on T^2. Hence, in this case, we may take $\mathcal{V} = T^2$. As a consequence, by Stokes's theorem (Theorem 8.7),

$$\int_{T^2} K\, d\mu = \int_{T^2} d\omega_2^1 = 0. \tag{9.23}$$

In general, the set $\mathcal{Z}(V)$ is nonempty. Since $\mathcal{Z}(V)$ is finite, we may write it as $\mathcal{Z}(V) = \{p_1, \ldots, p_k\}$ for some positive integer k. Let $r_0 > 0$ be one-half the minimum distance between distinct points in $\mathcal{Z}(V)$. For any positive real number ϵ less than both r_0 and the injectivity radius of g, we have that $M^2 \backslash \bigcup_{i=1}^k B_\epsilon(p_i)$ is a compact surface with boundary consisting of k disjoint smooth topological circles. Clearly,

$$\int_{M^2} K\, d\mu = \lim_{\epsilon \to 0} \int_{M^2 \backslash \bigcup_{i=1}^k B_\epsilon(p_i)} K\, d\mu. \tag{9.24}$$

By Stokes's theorem,

$$\begin{aligned}\int_{M^2 \backslash \bigcup_{i=1}^k B_\epsilon(p_i)} K\, d\mu &= \int_{M^2 \backslash \bigcup_{i=1}^k B_\epsilon(p_i)} d\omega_2^1 \\ &= \int_{\bigcup_{i=1}^k \partial B_\epsilon(p_i)} \omega_2^1,\end{aligned} \tag{9.25}$$

where the boundary $\bigcup_{i=1}^k \partial B_\epsilon(p_i)$ has the usual orientation induced from $M^2 \setminus \bigcup_{i=1}^k B_\epsilon(p_i)$.

We now evaluate each of the integrals contained in the sum (9.25), with the integral $\int_{\partial B_\epsilon(p_i)} \omega_2^1$ of the indicated connection 1-form being carried out over a small geodesic circle. Let $1 \leq i \leq k$. Let N denote the unit *inward* normal to the embedded circle $\partial B_\epsilon(p_i)$; i.e., N points into $B_\epsilon(p_i)$ or in other words N is the unit *outward* normal to the manifold with boundary $M^2 \backslash \bigcup_{i=1}^k \partial B_\epsilon(p_i)$. Let T be the clockwise pointing unit tangent vector field to $\partial B_\epsilon(p_i)$. Then $\{N, T\}$ is a positively oriented orthonormal frame field along $\partial B_\epsilon(p_i)$. So, by the definition of the induced orientation in §8.1.4, T is positively oriented on $\partial B_\epsilon(p_i)$.

Let L denote the length of $\partial B_\epsilon(p_i)$ and let $\gamma : [0, L] \to M^2$ be a unit-speed clockwise parametrization of $\partial B_\epsilon(p_i)$, so that $\gamma'(s) = T_{\gamma(s)}$. We express the unit tangent T and normal N vector fields in terms of the orthonormal frame field $\{e_1, e_2\}$. Choose $\theta_0 \in \mathbb{R}$ so that

$$T(0) := T_{\gamma(0)} = \cos(\theta_0) e_1(\gamma(0)) + \sin(\theta_0) e_2(\gamma(0)). \tag{9.26}$$

Then there exists a unique smooth function $\theta : [0, L] \to \mathbb{R}$ satisfying

$$T(s) := T_{\gamma(s)} = \cos(\theta(s))e_1(\gamma(s)) + \sin(\theta(s))e_2(\gamma(s)) \tag{9.27}$$

for $s \in [0, L]$ and $\theta(0) = \theta_0$. Note that θ represents a smooth choice of angle between the vector fields T and e_1 along γ; that is, θ is the angle *from T to e_1* in the clockwise direction. We then have that

$$N(s) := N_{\gamma(s)} = \sin(\theta(s))e_1(\gamma(s)) - \cos(\theta(s))e_2(\gamma(s)). \tag{9.28}$$

Note that

$$\begin{pmatrix} N \\ T \end{pmatrix} = \begin{pmatrix} \sin(\theta) & -\cos(\theta) \\ \cos(\theta) & \sin(\theta) \end{pmatrix} \begin{pmatrix} e_1 \\ e_2 \end{pmatrix}.$$

Example 9.17. *Let $T^2 = \mathbb{R}^2/\Gamma$, where Γ is a lattice, be a flat 2-torus. We have that the standard Euclidean orthonormal frame field $\{\mathbf{e}_1, \mathbf{e}_2\}$ projects by $\pi : \mathbb{R}^2 \to T^2$ to a well-defined global orthonormal frame field on T^2, which we denote by $\{e_1, e_2\}$. Let $p \in T^2$ and choose $\varepsilon > 0$ small enough so that the closed ball $\overline{B_\varepsilon(p)}$ is embedded in T^2; i.e., ε is less than the injectivity radius of T^2. Parametrize $\partial B_\varepsilon(p)$ by the map $\gamma : [0, 2\pi\varepsilon] \to T^2$ defined by $\gamma(s) := \pi(\widetilde{p} + \varepsilon e^{-is/\varepsilon})$, where $\widetilde{p}$ is any lift of p; i.e., $\pi(\widetilde{p}) = p$. Then the function $\theta : [0, 2\pi\varepsilon] \to \mathbb{R}$ with $\theta(0) = -\frac{\pi}{2}$ satisfies* (9.27) *with*

$$T(s) := T(\gamma(s)) = -\sin(s/\varepsilon)e_1 - \cos(s/\varepsilon)e_2.$$

Therefore, $\theta(s) = -\frac{s}{\varepsilon} - \frac{\pi}{2}$.

Note that $V := e_1$ is a vector field globally defined on T^2 with no zeroes. But, for the sake of argument, we can consider the example of V with an isolated zero at the point $p := [(0, 0)] \in T^2$, defined so that in a neighborhood $\mathcal{U}$ of p we have

$$V([(x, y)]) := (x^2 + y^2)e_1.$$

In this case, we have $\frac{V}{\|V\|} = e_1$ in $\mathcal{U}$, which is the same choice as in (9.20). *Observe that* $\operatorname{index}_p(V) = 0$ *and $\theta(2\pi\varepsilon) - \theta(0) = -2\pi$. In summary, this is an example of the following general result:*

Lemma 9.18. *For $\epsilon > 0$ sufficiently small, we have that*

$$\theta(L) - \theta(0) = 2\pi\left(\operatorname{index}_{p_i}(V) - 1\right). \tag{9.29}$$

See Figure 9.4.2.

Proof. Let $\epsilon > 0$ be sufficiently small as in the discussion above. Choose, for convenience, orientation-preserving geodesic normal coordinates $\varphi = \{x^1, x^2\} : \mathcal{D}_{p_i} \setminus \{p_i\} \to \mathbb{R}^2$ centered at p_i for each $1 \leq i \leq k$. Then, $\varphi(\partial B_\epsilon(p_i)) = S^1(\epsilon)$ is equal to the circle of radius ϵ centered at the origin $\mathbf{0}$ in $\mathbb{R}^2$. Since φ is an embedding, we have that the vector field $\varphi_*(V)$, defined in a neighborhood of $\mathbf{0}$ in $\mathbb{R}^2$, has a zero at $\mathbf{0}$ and

$$\operatorname{index}_{p_i}(V) = \operatorname{index}_{\mathbf{0}}(\varphi_*(V)). \tag{9.30}$$

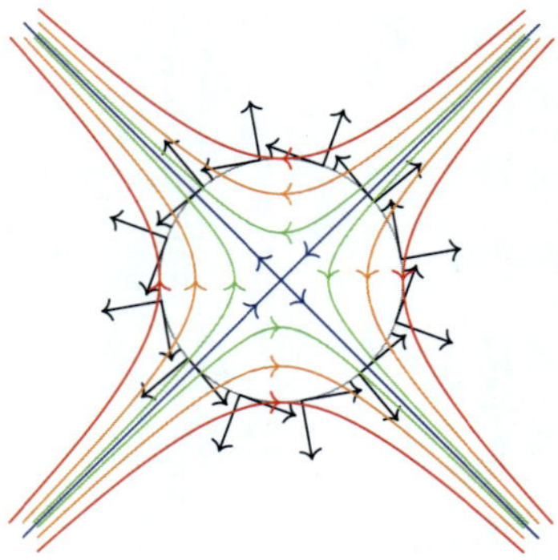

Figure 9.4.2. The unit tangent and normal fields along a small circle and the flow lines of a vector field with index -1. The change in angle $\theta(L) - \theta(0)$ is equal to -4π.

We have that $\varphi_*(T)$ is a nonzero vector field that is tangential along $S^1(\epsilon)$. Thus, the relative change in the angle from $\mathbf{e}_1$ to the Euclidean coordinate vector $\varphi_*(T)$ (in the clockwise direction) is equal to 2π as one traverses the circle $S^1(\epsilon)$ in the clockwise direction. By (9.30) and the definition of index, we have that the relative change in the angle from the Euclidean coordinate vector $\mathbf{e}_1$ to $\varphi_*(V)$ (in the clockwise direction) is equal to $2\pi \operatorname{index}_{p_i}(V)$. Therefore, the relative change in angle from $\varphi_*(T)$ to $\varphi_*(V)$ as one traverses $S^1(\epsilon)$ in the clockwise direction is equal to $2\pi(\operatorname{index}_{p_i}(V) - 1)$.

On the other hand, $\theta(L) - \theta(0)$, which is equal to the relative change in angle from T and V, is an integer multiple of 2π. Since φ is almost an isometry for $\epsilon > 0$ sufficiently small, the change in angle between $\varphi_*(T)$ and $\varphi_*(V)$ must be exactly equal to $\theta(L) - \theta(0)$. This completes the proof of the lemma. □

Let $\gamma : [0, L] \to \mathbb{R}$ be a unit-speed path in a Riemannian surface (M^2, g), let $T(s) := \gamma'(s)$ be its unit tangent vector field, and let $N(s)$ be a choice of unit normal vector field along $\gamma(s)$. The **geodesic curvature** of γ (with respect to the choice N) is defined by

$$\kappa(s) := \langle \nabla_{T(x)} T(s), N(s) \rangle. \tag{9.31}$$

This is a direct generalization of the definition (4.8) of the curvature of a plane curve.

Lemma 9.19. *The connection* 1*-form* ω_1^2 *along the embedded circle* $\partial B_\epsilon(p_i)$ *is given by*

$$\omega_2^1(T) = \kappa + T(\theta), \tag{9.32}$$

where κ *is the geodesic curvature of* $\partial B_\epsilon(p_i)$*, defined with respect to the unit normal field* N*.*

Proof. Using (9.27), we compute that

$$\begin{aligned}\nabla_{T(s)}T(s) &= \nabla_{T(s)}\big(\cos(\theta(s))e_1(\gamma(s)) + \sin(\theta(s))e_2(\gamma(s))\big)\\ &= \cos(\theta(s))\nabla_{T(s)}e_1(\gamma(s)) + \sin(\theta(s))\nabla_{T(s)}e_2(\gamma(s))\\ &\quad - \sin(\theta(s))\frac{d\theta}{ds}(s)e_1(\gamma(s)) + \cos(\theta(s))\frac{d\theta}{ds}(s)e_2(\gamma(s))\\ &= \cos(\theta(s))\omega_1^2(T(s))e_2(\gamma(s)) + \sin(\theta(s))\omega_2^1(T(s))e_1(\gamma(s))\\ &\quad - \frac{d\theta}{ds}(s)N(s)\\ &= \omega_2^1(T(s))N(s) - \frac{d\theta}{ds}(s)N(s),\end{aligned}$$

where $N(s) := N(\gamma(s))$. The curvature of $\gamma(s)$ with respect to its unit normal vector field $N(s)$ is

$$\kappa(s) := \kappa(\gamma(s)) = \nabla_{T(s)}T(s)\cdot N(s).$$

We conclude that

$$\kappa(s) = \omega_2^1(T(s)) - \frac{d\theta}{ds}(s). \tag{9.33}$$

This proves (9.32) and hence the lemma. □

Proof of the Gauss–Bonnet theorem (Theorem 9.15). We calculate that

$$\begin{aligned}\int_{\partial B_\epsilon(p_i)}\omega_2^1 &= \int_{\partial B_\epsilon(p_i)}\omega_2^1(T)\,ds\\ &= \int_{\partial B_\epsilon(p_i)}(\kappa + T(\theta))\,ds\\ &= \int_{\partial B_\epsilon(p_i)}\kappa\,ds + \theta(L) - \theta(0)\\ &= \int_{\partial B_\epsilon(p_i)}\kappa\,ds + 2\pi(\mathrm{index}_{p_i}(V) - 1),\end{aligned} \tag{9.34}$$

where the second equality is by (9.32) and where the last equality is by (9.29).

By substituting (9.34) into (9.25), we obtain

$$\int_{M^2\setminus\bigcup_{i=1}^k B_\epsilon(p_i)} K\,d\mu = \sum_{i=1}^k \int_{\partial B_\epsilon(p_i)} \omega_2^1 \tag{9.35}$$

$$= \sum_{i=1}^k \int_{\partial B_\epsilon(p_i)} \kappa\,ds + 2\pi \sum_{i=1}^k (\mathrm{index}_{p_i}(V) - 1)$$

$$= \sum_{i=1}^k \left(\int_{\partial B_\epsilon(p_i)} \kappa\,ds - 2\pi \right) + 2\pi \sum_{i=1}^k \mathrm{index}_{p_i}(V)$$

$$= \sum_{i=1}^k \left(\int_{\partial B_\epsilon(p_i)} \kappa\,ds - 2\pi \right) + 2\pi\chi(M^2),$$

where the last equality is by the Poincaré–Hopf theorem.

Finally, we have for each i that

$$\lim_{\epsilon\to 0} \int_{\partial B_\epsilon(p_i)} \kappa\,ds = 2\pi. \tag{9.36}$$

This is because (9.36) is true on the Euclidean plane and Riemannian surfaces are locally Euclidean geometrically. We conclude that

$$\int_{M^2} K\,d\mu = \lim_{\epsilon\to 0} \int_{M^2\setminus\bigcup_{i=1}^k B_\epsilon(p_i)} K\,d\mu$$
$$= 2\pi\chi(M^2).$$

This completes the proof of the Gauss–Bonnet formula. □

The Gauss–Bonnet formula has a generalization to manifolds of even dimensions, due to Allendoerfer and Weil. An intrinsic proof of this theorem is due to Chern [**Che44**]. We will discuss the Chern–Gauss–Bonnet formula in the special case of dimension 4 in §10.8.

9.5. The Gauss–Bonnet formula on surfaces with boundary

A **Morse function on a manifold** M^n **with boundary** ∂M is a smooth function $f : M^n \to [a, b]$, where $a < b$, such that either $f^{-1}(\{a\}) = \partial M$ or $f^{-1}(\{b\}) = \partial M$ and such that each critical point of f is in the interior $M^n \setminus \partial M$ of M^n and is nondegenerate. By [**Mil65**, Theorem 2.5], every compact manifold M^n with boundary admits a Morse function.[4]

Now assume that (M^n, g) is a compact Riemannian manifold with boundary ∂M and let $f : M^n \to [a, b]$ be a Morse function on M^n. Without loss

[4]Milnor proves the existence of Morse functions more generally on **triads** $(W; V_0, V_1)$, where the boundary of the manifold W is the disjoint union of V_0 and V_1. A manifold M^n with boundary ∂M is the special case $(M^n; \emptyset, \partial M)$ of a triad.

of generality, we may assume that $f^{-1}(\{b\}) = \partial M$. Let $V = \nabla f$ be the gradient of f. Then V is outward pointing on ∂M and the zeroes of V are isolated (and hence finite) and in the interior of M^n. By the Poincaré–Hopf theorem for manifolds with boundary (see [**Mil97**, p. 35]), we have that the sum of the indices of V is equal to $\chi(M^n)$.

The Gauss–Bonnet formula for compact surfaces with boundary says the following.

Theorem 9.20. *Let (M^2, g) be a compact oriented Riemannian surface with boundary ∂M and let N denote the unit outward pointing normal field to ∂M. Then*

$$\int_{M^2} K\, d\mu = 2\pi\chi(M^2) + \int_{\partial M} \kappa ds, \tag{9.37}$$

where κ denotes the geodesic curvature of ∂M as defined by (9.31).

The proof of Theorem 9.20 is left as Exercise 9.3.

Example 9.21. *Suppose that M^2 is a smooth topological disk embedded in $\mathbb{R}^2$. Since $K \equiv 0$ and $\chi(M^2) = 1$, we have that*

$$\int_{\partial M} \kappa ds = -2\pi. \tag{9.38}$$

Here, our sign convention for κ is the opposite of the usual one for closed embedded curves in the plane since it is defined with respect to the outward unit normal instead of the inward unit normal.

9.6. The Gauss–Bonnet formula on orbifolds

An important generalization of manifold is the concept of *orbifold*. The Gauss–Bonnet formulas extends from smooth surfaces to smooth orbifolds.

9.6.1. 2-dimensional orbifolds with isolated singularities. The local models for smooth manifolds are Euclidean spaces (and their open subsets). So, in dimension 2, the local model is the plane $\mathbb{R}^2$. In order to define 2-dimensional orbifolds, we enlarge the class of local models by adding the quotient spaces $\mathbb{R}^2/\mathbb{Z}_r$, where $r \geq 2$ and where the cyclic group $\mathbb{Z}_r$ acts on $\mathbb{R}^2 \cong \mathbb{C}$ by multiplication by $e^{2\pi i/r}$. We may replace $\mathbb{R}^2$ by the open unit ball B_1.

Intuitively, a 2-dimensional smooth closed **orbifold** is a topological closed surface O^2 with a smooth structure on the complement of a finite set of points S, called the set of **singular points**. At any singular point $x \in O^2$, a neighborhood $\mathcal{U}$ of x is modeled on $\mathbb{R}^2/\mathbb{Z}_r$ for some $r \geq 2$ depending on x, where the point x corresponds to the vertex $[\mathbf{0}]$ of the model $\mathbb{R}^2/\mathbb{Z}_r$. We call r the **order** of the singular point x. The map $\phi : \mathbb{R}^2 \to \mathbb{R}^2/\mathbb{Z}_r \cong \mathcal{U}$

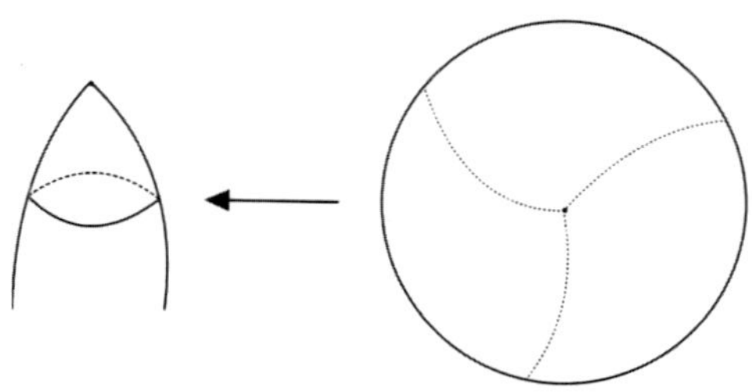

Figure 9.6.1. An orbifold coordinate chart. Pictured is the case $r = 3$.

is called an **orbifold coordinate chart** (although the direction of the map is opposite of that of a manifold coordinate chart), where $\cong$ indicates that the map is a bijection; see Figure 9.6.1. For simplicity, we define the orbifold O^2 by a finite number of orbifold coordinate charts such that each singular point is associated to exactly one orbifold coordinate chart.

Let $\phi_1 : \mathbb{R}^2 \to \mathbb{R}^2/\mathbb{Z}_{r_1} \cong \mathcal{U}_1$ and $\phi_2 : \mathbb{R}^2 \to \mathbb{R}^2/\mathbb{Z}_{r_2} \cong \mathcal{U}_2$ be orbifold coordinate charts. Then $\mathcal{U}_1 \cap \mathcal{U}_2$ is an open subset of $M^2 \setminus S$. Thus, it makes sense to require that the transition function of a pair of orbifold coordinate charts is a smooth map from an open subset of $(\mathbb{R}^2/\mathbb{Z}_{r_1}) \setminus [\mathbf{0}]$ to an open subset of $(\mathbb{R}^2/\mathbb{Z}_{r_2}) \setminus [\mathbf{0}]$ in the manifold sense. See Cooper, Hodgson, and Kerckhoff [**CHK00**] for further discussion of orbifolds, including their formal definition.

We say that a function or tensor on an orbifold O^2 is smooth near a conical singularity if its local lift is smooth. For example, if $f : O^2 \to \mathbb{R}$ is a function, then we say that f is smooth if for every coordinate chart $\phi : \mathbb{R}^2 \to \mathbb{R}^2/\mathbb{Z}_r \cong \mathcal{U}$, the function $f \circ \phi : \mathbb{R}^2 \to \mathbb{R}$ is smooth.

The tangent space at a smooth point $x \in O^2 \setminus S$ is defined in the same way as for a smooth surface. Making the analogous definition at a singular point $x \in S$ yields a tangent space $T_x O$ that is 'isomorphic' to $T_x M/\mathbb{Z}_r$, where $r \geq 2$ is the order of x.

However, we will usually pass to the local lift. In particular, a smooth Riemannian metric g on O^2 may be defined in the usual manifold way on $O^2 \setminus S$ with the extra condition that for an orbifold coordinate chart $\phi : \mathbb{R}^2 \to \mathbb{R}^2/\mathbb{Z}_r \cong \mathcal{U}$, the Riemannian metric $\phi^*(g|_{\mathcal{U}\setminus\{x\}})$ on $\mathbb{R}^2 \setminus \{\mathbf{0}\}$ extends smoothly to $\mathbb{R}^2$.

9.6.2. The 2-dimensional orbifold Euler characteristic. Let (O^2, g) be a 2-dimensional closed oriented Riemannian orbifold with isolated singular points. Let $\{y_1, \ldots, y_k\}$ be the set of singular points of O^2. Let $r_1, \ldots, r_k$ be the orders of $y_1, \ldots, y_k$, respectively. That is, a neighborhood of y_i is modeled on $\mathbb{R}^2/\mathbb{Z}_{r_i}$. Then its **orbifold Euler characteristic** is defined by

$$\chi_{\text{orb}}(O^2) := \chi(O^2) + \sum_{i=1}^{k} \left(\frac{1}{r_i} - 1 \right). \tag{9.39}$$

For example:

(1) (Teardrop) If O^2 is a topological 2-sphere with a sole singular point of order r_1, then

$$\chi_{\text{orb}}(O^2) = 1 + \frac{1}{r_1}.$$

(2) (Football) If O^2 is a topological 2-sphere with two singular points of orders r_1 and r_2, then

$$\chi_{\text{orb}}(O^2) = \frac{1}{r_1} + \frac{1}{r_2}.$$

Example 9.22. *As in Example* 9.16, *let* $g := ds^2 + f(s)^2 d\theta^2$ *be a rotationally symmetric metric defined on* $M^2 := (0, L) \times (\mathbb{R}/2\pi\mathbb{Z})$. *Recall from* (9.17) *that*

$$\int_{M^2} K \, d\mu = 2\pi(f'(0) - f'(L)).$$

Suppose that M^2 *compactifies to a topological 2-sphere and that the Riemannian metric* g *extends smoothly over the points, where* $s = 0$ *and* $s = L$ *with respect to a smooth orbifold structure* O^2 *and where* $s = 0$ *and* $s = L$ *correspond to singular points of orders* r_1 *and* r_2, *respectively. Then*

$$f'(0) = \frac{1}{r_1} \quad \text{and} \quad f'(L) = -\frac{1}{r_2}, \tag{9.40}$$

which implies that

$$\int_{O^2} K \, d\mu = 2\pi \left(\frac{1}{r_1} + \frac{1}{r_2} \right) = 2\pi \chi_{\text{orb}}(O^2).$$

In the next subsection, we will see that this formula is true in general.

Remark 9.23. To intuitively see why (9.40) is true, for any integer $r > 1$ consider the cone metric $g_r := ds^2 + s^2 d\theta^2$, where $\theta \in \mathbb{R}/\frac{2\pi}{r}\mathbb{Z}$. This represents the flat orbifold metric on $\mathbb{R}^2/\mathbb{Z}_r$. Changing variables by $\widetilde{\theta} := r\theta$, we have that $\widetilde{\theta} \in \mathbb{R}/2\pi\mathbb{Z}$ and $g_r = ds^2 + \left(\frac{s}{r}\right)^2 d\widetilde{\theta}^2$.

9.6.3. Orbifold index of a vector field. Let V be a smooth vector field on O^2. Since V is smooth, by definition, if $\phi : \mathbb{R}^2 \to \mathbb{R}^2/\mathbb{Z}_r \cong \mathcal{U}$ is an orbifold coordinate chart for a singular point $x \in O^2$, then the vector field $\widetilde{V} := \phi^*(V|_{\mathcal{U}\setminus\{x\}})$ on $\mathbb{R}^2 \setminus \{\mathbf{0}\}$ extends smoothly to $\mathbb{R}^2$. Since $r \geq 2$, this implies that $V_x = \vec{0}$. That is, any smooth vector field V vanishes at each singular point.

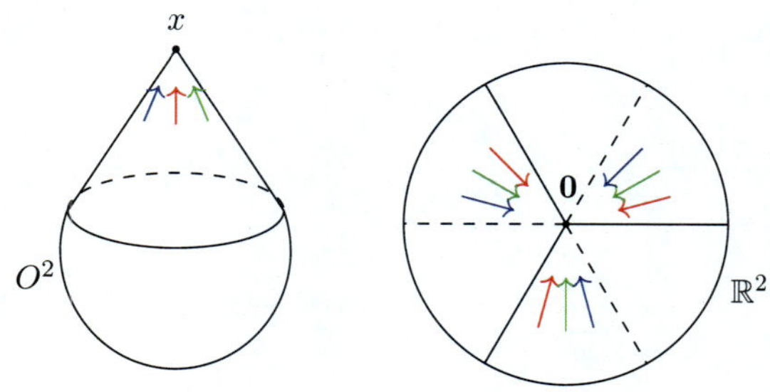

Figure 9.6.2. Left: A vector field V near a singular point x of an orbifold O^2. Right: Its local lift $\widetilde{V}$ to $\mathbb{R}^2$. In this example, $\mathrm{index}_{\mathbf{0}}(\widetilde{V}) = 1$, $r = 3$, and $\mathrm{index}_x^{\mathrm{orb}}(V) = \frac{1}{3}$.

As in the manifold case, for any smooth closed 2-dimensional orbifold O^2, one can prove the existence of a smooth vector field V on O^2 whose zeroes are isolated. Let V be such a vector field. Let $x \in O^2$ be a zero of V. If x is a smooth point, then the index of V at x is defined in the usual manifold way:

$$\mathrm{index}_x^{\mathrm{orb}}(V) := \mathrm{index}_x(V).$$

On the other hand, if x is a singular point of order r, then we define

$$\mathrm{index}_x^{\mathrm{orb}}(V) := \frac{1}{r}\,\mathrm{index}_{\mathbf{0}}(\widetilde{V}).$$

See Figure 9.6.2. By the definition (9.39) of $\chi_{\mathrm{orb}}(O^2)$ and by the Poincaré–Hopf theorem, we have

$$\chi_{\mathrm{orb}}(O^2) = \mathrm{index}^{\mathrm{orb}}(V) := \sum_{x \in S} \mathrm{index}_x^{\mathrm{orb}}(V). \tag{9.41}$$

9.6.4. The Gauss–Bonnet formula for closed orbifolds. The Gauss–Bonnet formula, which is proved in a way similar to that for smooth surfaces, says the following for orbifolds.

Theorem 9.24 (Gauss–Bonnet formula for orbifolds). *If (O^2, g) is a closed Riemannian orbifold with isolated singularities, then*

$$\int_{O^2} K\, d\mu = 2\pi\, \chi_{\mathrm{orb}}(O^2). \tag{9.42}$$

Proof. As in our discussion above, let V be a smooth vector field on O^2 with isolated zeroes. Let $\{p_i\}_{i=1}^{k+\ell} = \{y_1, \dots, y_k, z_1, \dots, z_\ell\}$ be the set of zeros of V, which contains the set of singular points $\{y_1, \dots, y_k\}$ of O^2. By (9.35),

$$\int_{O^2 \setminus \bigcup_{i=1}^{k+\ell} B_\epsilon(p_i)} K\,d\mu = \sum_{i=1}^{k+\ell}\left(\int_{\partial B_\epsilon(p_i)} \kappa\,ds - 2\pi\right) + 2\pi \sum_{i=1}^{k} \operatorname{index}_{p_i}(V)$$
$$= \sum_{i=1}^{k+\ell}\left(\int_{\partial B_\epsilon(p_i)} \kappa\,ds - 2\pi\right) + 2\pi\chi(O^2).$$

For zeros of V that are the singular points, in lieu of (9.36), we have

$$\lim_{\epsilon\to 0}\int_{\partial B_\epsilon(y_i)} \kappa\,ds = \frac{2\pi}{r_i}, \tag{9.43}$$

where r_i is the order of the singular point y_i. The theorem follows from the two equations above, (9.36), and the definition (9.39) of the orbifold Euler characteristic:

$$\int_{O^2} K\,d\mu = \lim_{\epsilon\to 0}\int_{O^2 \setminus \bigcup_{i=1}^{k+\ell} B_\epsilon(p_i)} K\,d\mu$$
$$= \sum_{i=1}^{k}\left(\frac{2\pi}{r_i} - 2\pi\right) + 2\pi\chi(O^2)$$
$$= 2\pi\chi_{\mathrm{orb}}(O^2).$$

□

9.7. Notes and commentary

Most books on differential geometry discuss the Gauss–Bonnet formula; see for example [**Hic65**], [**dC16**], [**Spi79c**], and [**Lee18**].

Classic references for Morse theory are the books by Milnor [**Mil63**, **Mil65**]. A historically and mathematically significant application of Morse theory is the Bott periodicity theorem (see Bott [**Bot59**] as well as [**Mil63**]).

We can describe the result, in the case of the unitary group, as follows.[5] The inclusion map $i_k : \mathrm{U}(n) \to \mathrm{U}(n+1)$ induces the map on fundamental groups $(i_k)_* : \pi_k(\mathrm{U}(n)) \to \pi_k(\mathrm{U}(n+1))$, which is an isomorphism for $k \leq 2n+1$. This group is called the **k-th stable homotopy group of the unitary group** and is denoted by $\pi_k(\mathrm{U})$.

The **Bott periodicity theorem** for the unitary group says that $\pi_k(\mathrm{U}) \cong \pi_{k+2}(\mathrm{U})$ for $k \geq 0$. Furthermore, if k is even, then this group is trivial and if k is odd, then this group is an infinite cyclic group, i.e., isomorphic to $\mathbb{Z}$.

[5] The result also applies to the orthogonal and symplectic groups.

9.8. Exercises

Exercise 9.1. *For the sake of conceptual clarity, work out the proof of Proposition* 9.7 *in the case where* $M^1 = \mathbb{R}$, *so that* $n = 1$.

Exercise 9.2. *Let* k *be a positive integer. Show that the index of the vector field defined by* $V(z) := z^k$ *for* $z \in \mathbb{C}$ *is equal to* k. *Show that the index of the vector field* $W(z) := \bar{z}^k$ *is equal to* $-k$.

Exercise 9.3. *Prove Theorem* 9.20.
Hint: Let T *be the unit tangent vector field to each component of* ∂M, *so that* $\{N, T\}$ *is positively oriented. If you like (this is not necessary), you may assume that the vector field* V *is a positive multiple of* N *on* ∂M. *Let* $e_1 = V/\|V\|$ *on* M^2 *minus the zero set of* V.

Exercise 9.4. *Let* (M^2, g) *be a closed oriented Riemannian surface. Prove that there exists a point* $x \in M^2$ *such that the Gauss curvature* $K(x)$ *has the same sign as* $\chi(M^2)$.

Exercise 9.5. *Extend the result of Example* 9.16 *by removing the assumptions that* $f'(0) = 1$ *and* $f'(L) = -1$. *Namely, verify the Gauss–Bonnet formula for surfaces with boundary (Theorem* 9.20*) in this case.*

Exercise 9.6. *Let* T *be a region in an oriented Riemannian surface* (M^2, g) *whose piecewise smooth boundary consists of three geodesic segments and where* T *is homeomorphic to a disk; i.e.,* T *is a* ***geodesic triangle***. *Let* $0 < \alpha_1, \alpha_2, \alpha_3 < \pi$ *be the interior angles of* T. *Prove that*

$$\int_T K\, d\mu = \alpha_1 + \alpha_2 + \alpha_3 - \pi. \tag{9.44}$$

For example, if (M^2, g) *is the unit* 2*-sphere, then the left-hand side of* (9.44) *is equal to the area of* T. *On the other hand, if* (M^2, g) *is the hyperbolic disk, then the left-hand side of* (9.44) *is equal to minus the area of* T.

Exercise 9.7. *The following generalizes the previous exercise. Let* (M^2, g) *be a compact oriented Riemannian surface with piecewise smooth boundary* ∂M. *Let* θ_{total} *be the sum of the exterior angles at the corner points of* ∂M. *Prove that*

$$\int_{M^2} K\, d\mu + \int_{\partial M} \kappa\, ds + \theta_{\text{total}} = 2\pi\chi(M^2). \tag{9.45}$$

Exercise 9.8. *Let* (M^2, g) *be a closed oriented Riemannian surface. Suppose that* $\mathcal{T}$ *is a triangulation of* M^2 *by geodesic triangles; i.e., the edges of the triangles of* $\mathcal{T}$ *are geodesic segments. Show that the Gauss–Bonnet formula for* (M^2, g) *follows from applying Exercise* 9.6 *to each triangle of* $\mathcal{T}$.
Hint: What is the relation between the number of edges E *and the number of vertices* V *of a triangulation?*

Exercise 9.9. *Let S be a closed oriented topological surface. We say that a metric (in the sense of metric spaces) d on S is a* **polyhedral metric** *on S if the following holds. There exists a triangulation $\mathcal{T}$ of S such that each triangle t of $\mathcal{T}$, with the restricted metric $d|_t$, is isometric to a Euclidean triangle. The* **discrete curvature** *$K(v)$ of a vertex v is defined to be 2π minus the sum of the Euclidean interior angles at v. Let V denote the set of vertices of $\mathcal{T}$. Prove the discrete Gauss–Bonnet formula:*

$$\sum_{v \in V} K(v) = 2\pi\chi(S). \tag{9.46}$$

Chapter 10

Bundles and the Chern–Gauss–Bonnet Formula

Figure 10.0.1. Shiing-Shen Chern. Credit: By George M. Bergman, CC BY-SA 4.0, `https://commons.wikimedia.org/w/index.php?curid=73958062`

So far, we have encountered the tangent bundle and the cotangent bundle. These are examples of special types of *fiber bundles* called *vector bundles.* We have also considered the unit tangent bundle, which is an example of a fiber bundle that is not a vector bundle. Connected via group actions to fiber bundles and vector bundles are *principal bundles.* In this chapter we will learn about all of these structures.

We further provide proofs of Liouville's theorem for the geodesic flow on the unit tangent bundle and the Chern–Gauss–Bonnet formula, and we discuss both index theorems on 4-manifolds and Einstein manifolds.

10.1. The tangent bundle revisited

Let M^n be a smooth manifold. Recall from §3.2 and §6.4.2 that the tangent bundle TM is the disjoint union of the tangent spaces of M^n; that is,

$$TM := \bigcup_{x \in M^n} \{x\} \times T_x M. \tag{10.1}$$

See Figure 10.1.1. So far, we have considered this object only as a set, and not as a topological space or manifold. We will now show that we can naturally put the latter two structures on the tangent bundle TM of the manifold M^n.

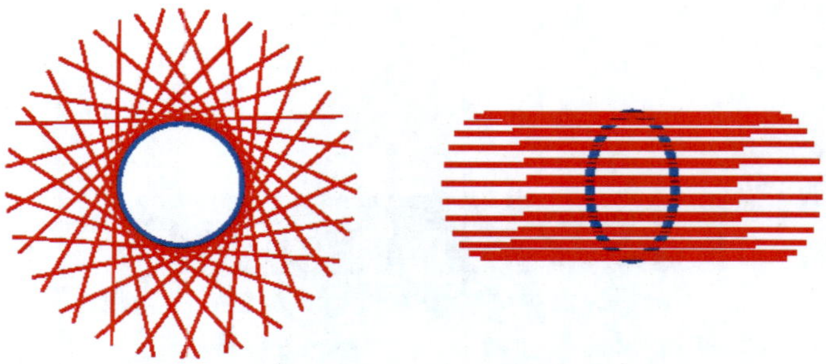

Figure 10.1.1. Picturing the tangent bundle of S^1. Credit: By Oleg Alexandrov – self-made, with en:Matlab then tweaked with en:Inkscape, Public Domain, `https://commons.wikimedia.org/w/index.php?curid=2233661`

10.1.1. The tangent bundle as a smooth manifold.

Let $\pi : TM \to M^n$ be the projection map defined by $\pi(x, V) := x$. Let $\mathbf{A}$ be an atlas for M^n (see Definition 6.2). Let $(\mathcal{U}, \varphi_{\mathcal{U}}) \in \mathbf{A}$ be a coordinate chart of M^n, so that $\varphi_{\mathcal{U}} : \mathcal{U} \to \mathbb{R}^n$. Denote $\mathbf{x} = (x^1, \ldots, x^n) := \varphi_{\mathcal{U}}$ as in (6.23). We define an associated coordinate chart $(\mathcal{U}^*, \varphi_{\mathcal{U}^*})$ on TM as follows:

$$\varphi_{\mathcal{U}^*} : \mathcal{U}^* := \pi^{-1}(\mathcal{U}) \to \varphi_{\mathcal{U}}(\mathcal{U}) \times \mathbb{R}^n \subset \mathbb{R}^{2n} \tag{10.2}$$

by

$$\varphi_{\mathcal{U}^*}(x, V) := (\varphi_{\mathcal{U}}(x), V^1, \ldots, V^n), \tag{10.3}$$

where the functions $V^i : \mathcal{U} \to \mathbb{R}$ are defined by

$$V =: \sum_{i=1}^n V^i \frac{\partial}{\partial x^i}. \tag{10.4}$$

Observe that $d\varphi_{\mathcal{U}}\big(\frac{\partial}{\partial x^i}\big) = \mathbf{e}_i$, which implies

$$d\varphi_{\mathcal{U}}(V) = \sum_{i=1}^{n} V^i \mathbf{e}_i = (V^1, \dots, V^n)$$

and hence

$$\varphi_{\mathcal{U}^*}(x, V) = (\varphi_{\mathcal{U}}(x), (d\varphi_{\mathcal{U}})_x(V)). \tag{10.5}$$

For simplicity, we write this as

$$\varphi_{\mathcal{U}^*} = (\varphi_{\mathcal{U}}, d\varphi_{\mathcal{U}}). \tag{10.6}$$

We will use the charts $\{(\mathcal{U}^*, \varphi_{\mathcal{U}^*})\}_{(\mathcal{U},\varphi_{\mathcal{U}})\in\mathbf{A}}$ to define both the topological and manifold structures of TM.

Firstly, the topological structure on TM is defined as follows. We choose as a basis for the topology of TM the sets of the form $\varphi_{\mathcal{U}^*}^{-1}(\mathcal{S})$, where $\mathcal{S}$ is an open subset of $\varphi_{\mathcal{U}}(\mathcal{U}) \times \mathbb{R}^n$ and $(\mathcal{U}, \varphi_{\mathcal{U}}) \in \mathbf{A}$. We leave it to the reader to check that this collection indeed defines a topological basis.

Secondly, we define the manifold structure on TM as follows. By the definition of the topology of TM, we have that each map $\varphi_{\mathcal{U}^*} : \mathcal{U}^* \to \varphi_{\mathcal{U}}(\mathcal{U}) \times \mathbb{R}^n$ is a homeomorphism onto an open subset of $\mathbb{R}^{2n}$. We can now define the atlas $\mathbf{A}^*$ on TM as the collection $\{(\mathcal{U}^*, \varphi_{\mathcal{U}^*})\}_{(\mathcal{U},\varphi_{\mathcal{U}})\in\mathbf{A}}$. Observe that if $(\mathcal{U}, \varphi_{\mathcal{U}})$ and $(\mathcal{V}, \varphi_{\mathcal{V}})$ are charts on M^n, then we have the associated transition function for TM given by

$$\varphi_{\mathcal{V}^*} \circ \varphi_{\mathcal{U}^*}^{-1} : \varphi_{\mathcal{U}}(\mathcal{U} \cap \mathcal{V}) \times \mathbb{R}^n \to \varphi_{\mathcal{V}}(\mathcal{U} \cap \mathcal{V}) \times \mathbb{R}^n. \tag{10.7}$$

It is easy to see that this defines an atlas on TM, so that TM is a $2n$-dimensional smooth manifold.

Observe that by (10.5), we have that

$$\varphi_{\mathcal{V}^*} \circ \varphi_{\mathcal{U}^*}^{-1}(\mathbf{x}, \mathbf{V}) = \big((\varphi_{\mathcal{V}} \circ \varphi_{\mathcal{U}}^{-1})(\mathbf{x}), d(\varphi_{\mathcal{V}} \circ \varphi_{\mathcal{U}}^{-1})(\mathbf{V})\big). \tag{10.8}$$

If M^n is oriented, then its tangent bundle has a naturally defined orientation; see Exercise 10.4.

10.1.2. The tangent bundle of the n-sphere. By Remark 6.10 regarding Exercise 2.11, for the n-sphere S^n, we have two charts (given by the stereographic projections from the north and south poles)

$$\varphi_{\mathcal{U}} = G : S^n \setminus \{\mathbf{N}\} \to \mathbb{R}^n, \qquad \varphi_{\mathcal{V}} = H : S^n \setminus \{\mathbf{S}\} \to \mathbb{R}^n,$$

and the transition function is $\varphi_{\mathcal{V}} \circ \varphi_{\mathcal{U}}^{-1} = H \circ G^{-1} = 4\iota$, where $\iota(\mathbf{x}) = \frac{\mathbf{x}}{\|\mathbf{x}\|^2}$ is the inversion map. Let S^n_+ and S^n_- denote the closed upper and lower hemispheres, respectively. Then $\varphi_{\mathcal{U}}(S^n_-) = \overline{B}_2(\mathbf{0})$ and $\varphi_{\mathcal{U}}(S^n_+) = \overline{B}_2(\mathbf{0})$. The n-sphere S^n is the union of two copies of the closed ball $\overline{B}_2(\mathbf{0})$ glued together along their boundaries $\partial\overline{B}_2(\mathbf{0}) = S^{n-1}_2$, each the $(n-1)$-sphere of radius 2, by the map $4\iota|_{S_2^{n-1}} = \mathrm{id}_{S_2^{n-1}}$. On the other hand, the tangent bundle TS^n is the union of two copies of the trivial bundles $\overline{B}_2(\mathbf{0}) \times \mathbb{R}^n$ glued together along their boundaries $S_2^{n-1} \times \mathbb{R}^n$ by the map $(4\iota|_{S_2^{n-1}}, 4(d\iota)|_{S_2^{n-1}})$. We evaluate this map as

$$(\mathbf{x}, \mathbf{V}) \mapsto \left(\mathbf{x}, \mathbf{V} - \frac{\langle \mathbf{x}, \mathbf{V}\rangle}{2}\mathbf{x}\right) = \left(\mathbf{x}, \mathbf{V} - 2\left\langle \frac{\mathbf{x}}{2}, \mathbf{V}\right\rangle \frac{\mathbf{x}}{2}\right) \tag{10.9}$$

for $\mathbf{x} \in S_2^{n-1}$ and $\mathbf{V} \in \mathbb{R}^n$. Note that $\frac{\mathbf{x}}{2}$ is a unit vector, so that the map

$$\mathbf{V} \mapsto \mathbf{V} - 2\left\langle \frac{\mathbf{x}}{2}, \mathbf{V}\right\rangle \frac{\mathbf{x}}{2} \tag{10.10}$$

comprising the second component of (10.9) is equal to the reflection about the hyperplane in $\mathbb{R}^n$ perpendicular to $\mathbf{x}$. Note that for each $\mathbf{x}$ in S_2^{n-1} (which corresponds to the equator in S^n), the linear map in (10.10) is orientation reversing. So, to see that S^n is orientable, if we choose one orientation on one copy of $\overline{B}_2(\mathbf{0})$, then we choose the other orientation on the other copy of $\overline{B}_2(\mathbf{0})$ and then the gluing will be orientation preserving.

For example, in the case of the 2-sphere S^2, if we let $\rho(x^1, x^2) = (-x^1, x^2)$ be the reflection about the x^2-axis, then the composition of the reflection about the line perpendicular to $e^{i\theta}$, where $\theta \in \mathbb{R}$, with the map ρ is given by rotation by the angle -2θ. In other words, the tangent bundle S^2 is obtained from two copies of the trivial bundle $\overline{B}_2(\mathbf{0}) \times \mathbb{R}^2$, with opposite orientations, by identifying the two boundaries $S_2^1 \times \mathbb{R}^2$ by the map

$$(e^{i\theta}, \mathbf{V}) \mapsto (e^{i\theta}, e^{i(-2\theta)}\mathbf{V}). \tag{10.11}$$

As we will see later, the factor of -2 in the exponent in the second factor is related to the fact that the Euler characteristic of S^2 is equal to 2.

10.1.3. Vector fields as sections of the tangent bundle. Let V be a smooth vector field on M^n; that is, $V : M^n \to TM$ is a smooth map such that $\pi \circ V = \mathrm{id}_{M^n}$, where $\pi : TM \to M^n$ is the projection map. Observe that V is an embedding and so its image $V(M^n)$ is a smooth n-dimensional submanifold of TM. We call the map V a **section of the tangent bundle**.

10.2. The geodesic flow on the unit tangent bundle

In this section, we consider the geodesic flow of a complete Riemannian manifold.

10.2.1. The Sasaki metric on the tangent bundle. Let (M^n, g) be a complete Riemannian manifold. We define a natural metric on TM as follows. Let $v \in TM$ and let $p = \pi(v)$, where $\pi : TM \to M^n$ denotes the projection map. We also denote v as (p, v).

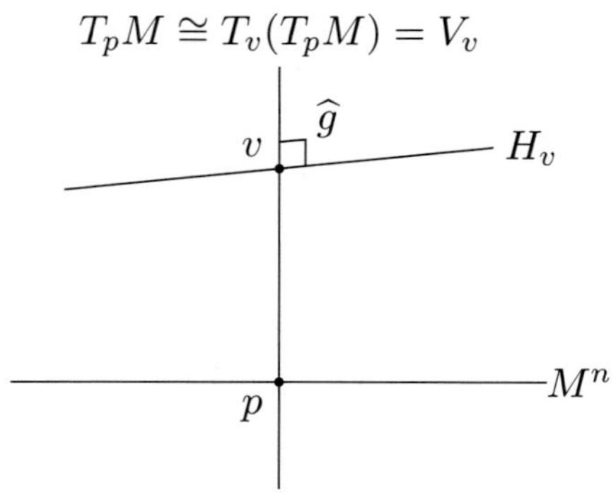

Figure 10.2.1. The Sasaki metric $\widehat{g}$ on TM. Given $v \in TM$, the vertical space V_v and horizontal space H_v are perpendicular with respect to $\widehat{g}_v$.

The tangent space at v to the submanifold (fiber) $T_pM \subset TM$ is called the **vertical subspace** V_v of $T_v(TM)$. We have a natural vector space isomorphism

$$i_v : V_v \to T_pM$$

since we can naturally identify $T_v(T_pM)$ with T_pM. The Euclidean metric $g_p := g(p)$ on T_pM gives a metric $g_{V_v} := i_v^*(g_p)$ on V_v. So,

$$(V_v, g_{V_v}) = (T_pM, g_p) \qquad \text{(isometric)}.$$

Given $w \in T_pM$, we call $i_v^{-1}(w)$ the **vertical lift** of w to V_v.

The Levi-Civita connection ∇ of (M^n, g) defines a **horizontal subspace** H_v of $T_v(TM)$. The horizontal subspaces are characterized by the property that for any path $\alpha(t)$ in M^n and parallel vector field $X(t)$ along $\alpha(t)$, $X'(t) \in T_{X(t)}(TM)$ is in $H_{X(t)}$. We have

$$T_v(TM) = V_v + H_v.$$

The derivative $(d\pi)_v : T_v(TM) \to T_pM$ restricts to a vector space isomorphism

$$\pi_v := (d\pi)_v|_{H_v} : H_v \to T_pM.$$

Define the metric $g_{H_v} = \pi_v^*(g_p)$ on H_v. Given $w \in T_pM$, we call $\pi_v^{-1}(w)$ the **horizontal lift** of w to H_v.

The **Sasaki metric** $\widehat{g}$ on the tangent bundle is the unique Riemannian metric on TM such that for each $v \in TM$,

$$\widehat{g}_v|_{V_v} = g_{V_v}, \quad \widehat{g}_v|_{H_v} = g_{H_v}, \quad V_v \perp_{\widehat{g}_v} H_v;$$

see Figure 10.2.1 and Sasaki [**Sas58**].

10.2.2. Liouville's theorem. Recall that the **unit tangent bundle** SM consists of all vectors $v \in TM$ of norm $|v| = 1$. Let $\overline{g}$ be the Riemannian metric on SM induced by its inclusion into $(TM, \widehat{g})$. We call $\overline{g}$ the **Sasaki metric on** the unit tangent bundle SM. We have the decomposition $T_vSM = H_v \oplus \overline{V}_v$, where

$$\overline{V}_v := v^\perp = \{w \in T_pM : g_p(w,v) = 0\} \subset T_pM \cong V_v.$$

See Figure 10.2.2. We have $\overline{V}_v \perp_{\overline{g}_v} H_v$. Let $\overline{\pi} : SM \to M$ denote the projection map. Observe that $\overline{V}_v = \ker(\overline{\pi}_*)$, where $\overline{\pi}_* = d\overline{\pi}_v : T_vSM \to T_pM$.

Recall that $n\omega_n$ is the volume of the Euclidean $(n-1)$-sphere.

Definition 10.1. The **Liouville measure** Θ on SM is defined to be $(n\omega_n)^{-1}$ times the Riemannian measure of $(SM, \overline{g})$. The $(n-1)$-dimensional volume of each sphere S_pM is equal to 1 with respect to the Liouville measure Θ.

For any $v \in SM$, denote by $\gamma_v : \mathbb{R} \to M$ the arc-length geodesic with $\gamma_v'(0) = v$. The **geodesic flow** is the $\mathbb{R}$-action on SM,

$$GF_t : SM \to SM,$$

defined by

$$GF_t(v) = \gamma_v'(t).$$

See Figures 10.2.3 and 10.2.4. Observe that $\{GF_t\}_{t\in\mathbb{R}}$ is a 1-parameter group of diffeomorphisms of M^n.

Let W be the horizontal vector field on SM defined by horizontal lifting:

$$W_u := \pi_u^{-1}(u) \in H_u \subset T_u(SM) \qquad \text{for } u \in SM. \tag{10.12}$$

The geodesic flow GF_t is generated by the vector field W; that is,

$$\frac{d}{dt}GF_t(v) = \pi^{-1}_{\gamma_v'(t)}\big(\gamma_v'(t)\big) = W_{GF_t(v)}, \qquad GF_0(v) = v. \tag{10.13}$$

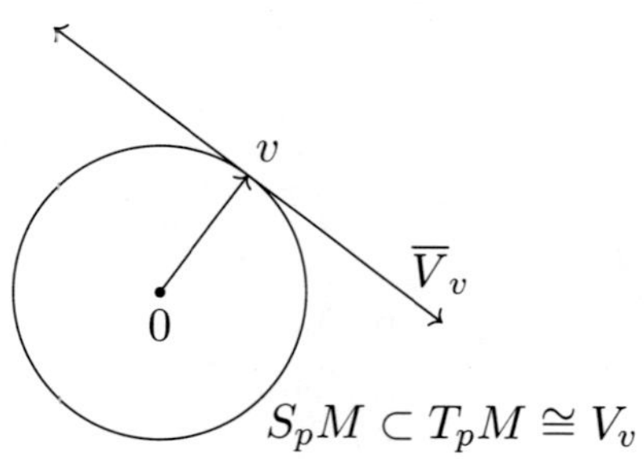

Figure 10.2.2. The hyperplane $\overline{V}_v$ in V_v for $v \in S_pM$.

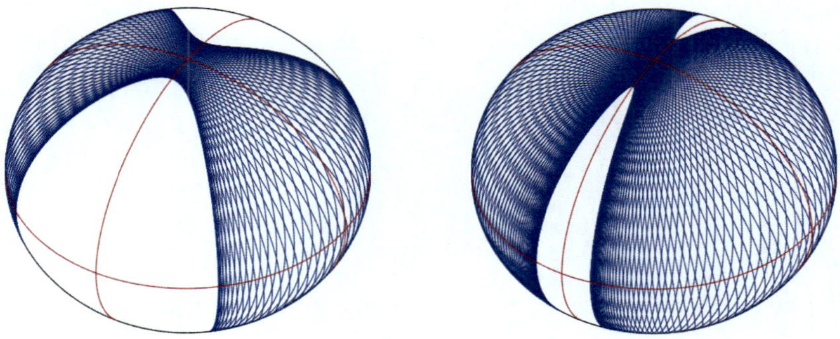

Figure 10.2.3. Geodesics on an ellipsoid. Credit: By Cffk – Own work, CC BY-SA 3.0, `https://commons.wikimedia.org/w/index.php?curid=27525009`

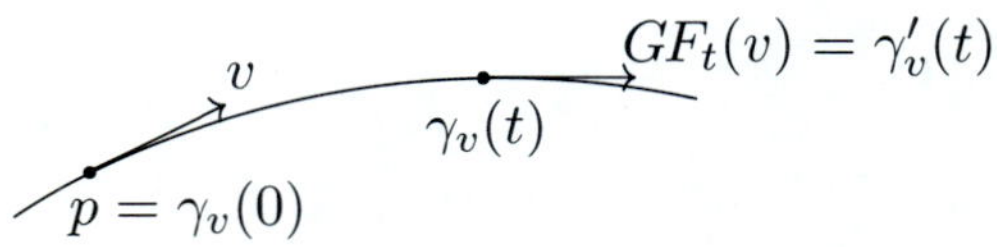

Figure 10.2.4. The geodesic flow $v \mapsto GF_t(v)$.

This is because $\gamma_v'(t)$ is parallel along and tangent to $\gamma_v(t)$, so $\frac{d}{dt}GF_t(v)$ is a horizontal vector projecting to $\gamma_v'(t)$.

Example 10.2. *Consider the geodesic flow on $\mathbb{R}^n$. Firstly, we have a naturally defined diffeomorphism*

$$S\mathbb{R}^n \cong \mathbb{R}^n \times S^{n-1} = \left\{ \begin{pmatrix} \mathbf{x} \\ V \end{pmatrix} : \mathbf{x}, V \in \mathbb{R}^n,\ \|V\| = 1 \right\}.$$

The Sasaki metric $\overline{g}$ on $S\mathbb{R}^n$ is the same as the Riemannian metric induced by the inclusion of $S\mathbb{R}^n$ into the Euclidean space $T\mathbb{R}^n \cong \mathbb{R}^{2n}$. The geodesic flow $GF_t : S\mathbb{R}^n \to S\mathbb{R}^n$ is given by

$$GF_t \begin{pmatrix} \mathbf{x} \\ V \end{pmatrix} = \begin{pmatrix} \mathbf{x} + tV \\ V \end{pmatrix}.$$

Therefore, its derivative $d(GF_t)_{(\mathbf{x},V)} : T_{(\mathbf{x},V)}(S\mathbb{R}^n) \to T_{(\mathbf{x}+tV,V)}(S\mathbb{R}^n)$ is given by the matrix

$$d(GF_t)_{(\mathbf{x},V)} = \begin{pmatrix} \mathrm{id}_{\mathbb{R}^n} & t\,\mathrm{id}_{T_V(S^{n-1})} \\ 0 & \mathrm{id}_{T_V(S^{n-1})} \end{pmatrix}.$$

Observe that this matrix has determinant equal to 1. Thus, the map $GF_t : (S\mathbb{R}^n, \overline{g}) \to (S\mathbb{R}^n, \overline{g})$ is measure preserving.

The geodesic flow is measure preserving in general; see, e.g., Besse [**Bes78**, Proposition 1.56].

Lemma 10.3 (Liouville's theorem). *For any $t \in \mathbb{R}$, the geodesic flow map $GF_t : (SM, \overline{g}) \to (SM, \overline{g})$ preserves the Liouville measure. Equivalently, $GF_t^*(d\mu_{\overline{g}}) = d\mu_{\overline{g}}$.*

Hence, if B is a measurable subset of SM and $f : SM \to \mathbb{R}$ is an integrable function, then

$$\int_B f(v)\, d\mu_{\overline{g}}(v) = \int_{GF_{-t}(B)} f\big(GF_t(v)\big)\, d\mu_{\overline{g}}(v). \tag{10.14}$$

Proof. Since GF_t is generated by W, we just need to show that

$$\mathcal{L}_W d\mu_{\overline{g}} = 0.$$

For simplicity, we assume that M^n is oriented. Let $(U, \{x^i\}_{i=1}^n)$ be positively oriented local coordinates on M^n. For any point $p \in U$ and $v \in T_pM$, let $v =: \sum_{i=1}^n v^i \partial_{x^i}$, where $\partial_{x^i} := \frac{\partial}{\partial x^i}$ denote the coordinate vectors. We have that

$$\big(TM|_U = \pi^{-1}(U), \{x^1, \ldots, x^n, v^1, \ldots, v^n\}\big)$$

are local coordinates on TM. Let $g_{ij} = g(\partial_{x^i}, \partial_{x^j})$. The Riemannian volume form on M^n is

$$d\mu_g = \sqrt{\det g_{ij}}\, dx^1 \wedge \cdots \wedge dx^n. \tag{10.15}$$

We denote the coordinate vectors on $TM|_U$ by

$$\widehat{\partial}_{x^1}, \ldots, \widehat{\partial}_{x^n}, \widehat{\partial}_{v^1}, \ldots, \widehat{\partial}_{v^n},$$

where we put hats on the ∂ to remind us that these coordinate vectors are on TM and to avoid redundant notation. The vertical lift of $w = \sum_{i=1}^n w^i \partial_{x^i} \in T_pM$ to V_v is

$$w_v^V = i_v^{-1}(w) = \sum_{i=1}^n w^i \widehat{\partial}_{v^i}. \tag{10.16}$$

We prove below that the horizontal lift of w to H_v is given by

$$w_v^H = \pi_v^{-1}(w) = \sum_i w^i \left(\widehat{\partial}_{x^i} - \sum_{j,k} v^j \Gamma_{ij}^k \widehat{\partial}_{v^k} \right), \tag{10.17}$$

where Γ_{ij}^k are the Christoffel symbols. In particular, the horizontal lift of ∂_{x^i} is

$$(\partial_{x^i})_v^H = \widehat{\partial}_{x^i} - \sum_{j,k} v^j \Gamma_{ij}^k \widehat{\partial}_{v^k}. \tag{10.18}$$

It is easy to see that $d\pi(w_v^H) = w$.

Parallel translate $v \in T_pM$ along $\gamma_w(t)$ to get a path $t \mapsto v(t)$ in TM with $v(0) = v$. Then

$$w_v^H = v'(0) \in T_v(TM).$$

Write $v(t) = \sum_i v^i(t)(\partial_{x^i})_{\gamma_w(t)}$. Since $v(t)$ is parallel along $\gamma_w(t)$, we have by calculating on M that

$$\begin{aligned} 0 &= \sum_i w^i \nabla_{\partial_{x^i}} \left(\sum_j v^j(t)\partial_{x^j} \right) \Bigg|_{t=0} \\ &= \sum_j (v^j)'(0)\,\partial_{x^j} + \sum_{i,j} w^i v^j \Gamma_{ij}^k \,\partial_{x^k}, \end{aligned}$$

by the product rule and the definition of the Christoffel symbols. Thus,

$$(v^k)'(0) = -\sum_{i,j} w^i v^j \Gamma_{ij}^k(p). \tag{10.19}$$

So, by calculating on TM, we have

$$\begin{aligned} w_v^H = v'(0) &= \sum_k (v^k)'(0)\,\widehat{\partial}_{v^k} + \sum_i w^i \widehat{\partial}_{x^i} \\ &= -\sum_{i,j,k} w^i v^j \Gamma_{ij}^k(p)\,\widehat{\partial}_{v^k} + \sum_i w^i \widehat{\partial}_{x^i}. \end{aligned}$$

This proves (10.17). To see the equality on the top line, we note that if we consider instead the path $\widetilde{v}(t) := \sum_i v^i(\partial_{x^i})_{\gamma_w(t)}$ (where the v^i are constant) in TM, then the local coordinates of $\widetilde{v}(t)$ are

$$\left(x^1(\gamma_w(t)), \ldots, x^n(\gamma_w(t)), v^1, \ldots, v^n\right),$$

so we have $\widetilde{v}'(0) = \sum_i w^i \widehat{\partial}_{x^i}$. Therefore the equality on the top line follows from the product rule.

We proceed to calculate the volume form $d\mu_{\widehat{g}}$ of $(TM, \widehat{g})$ in local coordinates. The components of the Sasaki metric $\widehat{g}$ on TM are given by the following:

(1)

$$\widehat{g}\big(\widehat{\partial}_{v^i}, \widehat{\partial}_{v^j}\big) = g_{ij}, \tag{10.20}$$

since $\widehat{\partial}_{v^i}$ corresponds to ∂_{x^i} under the natural isomorphism i_v between V_v and T_pM.

(2)

$$\begin{aligned} \widehat{g}\big(\widehat{\partial}_{x^i}, \widehat{\partial}_{x^j}\big) &= \widehat{g}\left((\partial_{x^i})^H + \sum_{k,\ell} v^k \Gamma_{ik}^\ell \,\widehat{\partial}_{v^\ell} \,,\, (\partial_{x^j})^H + \sum_{p,q} v^p \Gamma_{jp}^q \,\widehat{\partial}_{v^q} \right) \\ &= g_{ij} + \sum_{k,\ell,p,q} v^k v^p \Gamma_{ik}^\ell \Gamma_{jp}^q g_{\ell q}, \end{aligned} \tag{10.21}$$

where we used that $H_v \perp_{\widehat{g}_v} V_v$.

(3)

$$\widehat{g}(\widehat{\partial}_{v^i}, \widehat{\partial}_{x^j}) = \widehat{g}\left(\widehat{\partial}_{v^i}, (\partial_{x^j})^H + \sum_{p,q} v^p \Gamma^q_{jp} \widehat{\partial}_{v^q}\right) = \sum_{p,q} v^p \Gamma^q_{jp} g_{iq}. \tag{10.22}$$

Defining

$$h_{ij} := \sum_{p,q} v^p \Gamma^q_{jp} g_{iq}, \tag{10.23}$$

we have

$$\widehat{g}(\widehat{\partial}_{v^i}, \widehat{\partial}_{x^j}) = h_{ij} \tag{10.24}$$

and

$$\widehat{g}(\widehat{\partial}_{x^i}, \widehat{\partial}_{x^j}) = g_{ij} + \sum_{k,\ell} h_{ki} h_{\ell j} g^{k\ell}. \tag{10.25}$$

This implies

$$\det \widehat{g} = (\det g)^2. \tag{10.26}$$

To see this last formula, we observe the block matrix identity

$$\begin{pmatrix} A & B \\ C & D \end{pmatrix} \begin{pmatrix} I & -A^{-1}B \\ 0 & I \end{pmatrix} = \begin{pmatrix} A & 0 \\ C & D - CA^{-1}B \end{pmatrix}. \tag{10.27}$$

Thus,

$$\det \begin{pmatrix} A & B \\ C & D \end{pmatrix} = \det(A) \det(D - CA^{-1}B). \tag{10.28}$$

As a special case, we have

$$\det \begin{pmatrix} G & H \\ H^{\mathrm{T}} & G + H^{\mathrm{T}} G^{-1} H \end{pmatrix} = \big(\det(G)\big)^2, \tag{10.29}$$

and (10.26) follows from applying this to $G = (g_{ij})$ and $H = (h_{ij})$. Hence, the volume form of the Sasaki metric on TM is given by

$$d\mu_{\widehat{g}} = \det(g_{ab})\, \widehat{d}x^1 \wedge \cdots \wedge \widehat{d}x^n \wedge \widehat{d}v^1 \wedge \cdots \wedge \widehat{d}v^n, \tag{10.30}$$

where $\widehat{d}$ is the (exterior) derivative on TM.

The volume form of the metric $\overline{g}$ on SM is given in local coordinates by

$$(d\mu_{\overline{g}})_v = \hat{\iota}_v (d\mu_{\widehat{g}})_v = \det(g_{ab})\, \hat{\iota}_v \big(\widehat{d}x^1 \wedge \cdots \wedge \widehat{d}x^n \wedge \widehat{d}v^1 \wedge \cdots \wedge \widehat{d}v^n\big) \tag{10.31}$$

at $v \in SM$, where $\hat{\iota}_v$ denotes the interior product on TM with respect to the vertical vector field $v \mapsto v$.

Next, we calculate the components of the vector field W generating the geodesic flow. Given $v \in S_pM$, set

$$GF_t(v) = \gamma_v'(t) =: \sum_i v^i(t)\partial_{x^i}(t),$$

where $\partial_{x^i}(t) := \partial_{x^i}\big(\gamma_v(t)\big)$ and $v^i(0) = v^i$. Since γ_v is a geodesic, we have

$$(v^i)'(t) + \sum_{j,k} \Gamma^i_{jk}(t) v^j(t) v^k(t) = 0,$$

where $\Gamma^i_{jk}(t) := \Gamma^i_{jk}\big(\gamma_v(t)\big)$. Thus, the vector field W at the point $GF_t(v)$ is given by

$$\begin{aligned} W_{GF_t(v)} &= \pi^{-1}_{\gamma_v'(t)}\left(\sum_i v^i(t)\partial_{x^i}(t)\right) \\ &= \sum_i v^i(t)\left(\widehat{\partial}_{x^i} - \sum_{j,k} v^j(t)\Gamma^k_{ij}(t)\widehat{\partial}_{v^k}\right). \end{aligned} \tag{10.32}$$

In particular, for $v \in S_pM$,

$$W_v = \sum_i v^i\left(\widehat{\partial}_{x^i} - \sum_{j,k} v^j\Gamma^k_{ij}(p)\widehat{\partial}_{v^k}\right). \tag{10.33}$$

The components of W, defined by (on $SM \subset TM$)

$$W_v = \sum_i W_v^{\widehat{i}}\widehat{\partial}_{x^i} + \sum_k W_v^k\widehat{\partial}_{v^k}, \tag{10.34}$$

are given by

$$W_v^{\widehat{i}} = v^i, \tag{10.35a}$$

$$W_v^k = -\sum_{i,j} v^i v^j \Gamma^k_{ij}. \tag{10.35b}$$

In particular, if $p = \pi(v)$ is the center of geodesic normal coordinates, then $W_v = \sum_i v^i\widehat{\partial}_{x^i}$.

We first show that the geodesic flow preserves the volume form $d\mu_{\widehat{g}}$ of $(TM, \widehat{g})$. Recall Cartan's formula $\mathcal{L}_W = \widehat{d}\widehat{\iota}_W + \widehat{\iota}_W\widehat{d}$ for the Lie derivative acting on differential forms; see (8.34). Since $d\mu_{\widehat{g}}$ is a top-dimensional form, we compute using (10.35) that at all $v \in TM$,

$$
\begin{aligned}
&\mathcal{L}_W(d\mu_{\widehat{g}})_v \\
&= \widehat{d}\hat{\iota}_W(d\mu_{\widehat{g}})_v \\
&= (-1)^{i-1}\widehat{d}\big(\det(g_{ab})v^i\big) \wedge \widehat{d}x^1 \wedge \cdots \widehat{\widehat{d}x^i} \cdots \wedge \widehat{d}x^n \wedge \widehat{d}v^1 \wedge \cdots \wedge \widehat{d}v^n \\
&\quad + (-1)^{n+k}\widehat{d}\big(\det(g_{ab})v^i v^j \Gamma_{ij}^k\big) \wedge \widehat{d}x^1 \wedge \cdots \wedge \widehat{d}x^n \wedge \widehat{d}v^1 \wedge \cdots \widehat{\widehat{d}v^k} \cdots \wedge \widehat{d}v^n,
\end{aligned}
$$

where $\widehat{\widehat{d}x^i}$ indicates that we omit $\widehat{d}x^i$, and likewise for $\widehat{\widehat{d}v^k}$. Note that the terms $\det(g_{ab})$ and Γ_{ij}^k depend only on $p = \pi(v)$. We may assume that the coordinates $\{x^i\}$ are geodesic normal coordinates centered at the point p at which we are computing. We claim that the formula above yields

$$
\mathcal{L}_W(d\mu_{\widehat{g}}) = 0. \tag{10.36}
$$

Indeed, this is because $\widehat{d}\det(g_{ab})_v = 0$ (by $\partial_{x^i} g_{jk}(p) = 0$), $\widehat{d}v^i \wedge \widehat{d}v^i = 0$, $\Gamma_{ij}^k(p) = 0$, $\widehat{d}x^j \wedge \widehat{d}x^j = 0$, and $\widehat{\partial}_{v^k}\Gamma_{ij}^k = 0$.

Finally, from this we can also show that $\mathcal{L}_W(d\mu_{\overline{g}}) = 0$ on SM. To see this, using (8.141) we first calculate on TM that

$$
\begin{aligned}
\mathcal{L}_W(\hat{\iota}_I d\mu_{\widehat{g}}) &= \hat{\iota}_I(\mathcal{L}_W d\mu_{\widehat{g}}) + \hat{\iota}_{[W,I]} d\mu_{\widehat{g}} \\
&= \hat{\iota}_{[W,I]} d\mu_{\widehat{g}},
\end{aligned}
$$

where I is the vertical vector field on TM defined by $I_v = v$. Using $I_v = \sum_j v^j \widehat{\partial}_{v^j}$, we calculate that

$$
\begin{aligned}
[W, I] &= WI - IW \\
&= \sum_{k,j} W_v^k \widehat{\partial}_{v^k}(v^j)\widehat{\partial}_{v^j} - \sum_{k,j} v^j \widehat{\partial}_{v^j}\big(W_v^i \widehat{\partial}_{x^i} + W_v^i \widehat{\partial}_{v^i}\big) \\
&= \sum_j W_v^j \widehat{\partial}_{v^j} - \sum_i v^i \widehat{\partial}_{x^i} - \sum_{k,j} v^j \widehat{\partial}_{v^j}(W_v^i)\widehat{\partial}_{v^i} \\
&= -\sum_i v^i \widehat{\partial}_{x^i} + \sum_{j,k} v^j v^k \Gamma_{jk}^i \widehat{\partial}_{v^i} \\
&= -W.
\end{aligned}
$$

Thus, we have that

$$
\mathcal{L}_W(\hat{\iota}_I d\mu_{\widehat{g}}) = \hat{\iota}_{[W,I]} d\mu_{\widehat{g}} = -\hat{\iota}_W d\mu_{\widehat{g}}. \tag{10.37}
$$

Recall that $(\hat{\iota}_I d\mu_{\widehat{g}})|_{SM} = d\mu_{\overline{g}}$. By restricting to SM, we obtain

$$
\mathcal{L}_W d\mu_{\overline{g}} = \mathcal{L}_W(\hat{\iota}_I d\mu_{\widehat{g}})|_{SM} = -(\hat{\iota}_W d\mu_{\widehat{g}})|_{SM} = 0. \tag{10.38}
$$

This completes the proof of Lemma 10.3. □

Remark 10.4. The proof of Liouville's theorem that we gave used local coordinates. One can give a proof using moving frames.

10.3. Vector bundles

The notion of a vector bundle is a natural generalization of the notion of the tangent bundle of a manifold.

10.3.1. Definition of vector bundle. A **(topological) vector bundle** of rank k, where k is a positive integer, consists of topological spaces E and B and a continuous surjective map

$$\pi : E \to B$$

satisfying the following requirements. For each point $x \in B$, the **fiber** $E_x := \pi^{-1}(\{x\})$ is endowed with a real-dimension k vector space structure. For each $x \in B$ there exists an open neighborhood U of x and a homeomorphism

$$\varphi_U : \pi^{-1}(U) \to U \times \mathbb{R}^k$$

such that:

(1) $p_1 \circ \varphi_U = \pi|_U$, where $p_1 : U \times \mathbb{R}^k \to U$ is the projection onto the first factor.

(2) $\varphi_U|_{E_y} : E_y \to \{y\} \times \mathbb{R}^k$ is a vector space isomorphism for each $y \in U$, where $\{y\} \times \mathbb{R}^k$ has the vector space structure obtained by identifying it with $\mathbb{R}^k$.

We call E the **total space**, B the **base space**, π the **projection map**, and (U, φ_U) a **local trivialization** of the vector bundle. We usually denote a vector bundle by $\pi : E \to B$ and sometimes just by E for short.

When the base and entire spaces are smooth manifolds, we will assume that all of the maps are smooth, and we call $\pi : E \to B$ a **smooth vector bundle**.

Given two local trivializations (U, φ_U) and (V, φ_V), the associated **transition function** is the map

$$\varphi_{VU} := \varphi_V \circ (\varphi_U)^{-1} : (U \cap V) \times \mathbb{R}^k \to (U \cap V) \times \mathbb{R}^k. \tag{10.39}$$

Observe that $\varphi_{UV} = (\varphi_{VU})^{-1}$ and if (W, φ_W) is a third local trivialization, then we have

$$\varphi_{WV} \circ \varphi_{VU} = \varphi_{WU}. \tag{10.40}$$

This last equation is called the **cocycle condition** and is often written as

$$\varphi_{WV} \circ \varphi_{VU} \circ \varphi_{UW} = \mathrm{id}\,. \tag{10.41}$$

Since $\varphi_{VU} : \{x\} \times \mathbb{R}^k \to \{x\} \times \mathbb{R}^k$ is a vector space isomorphism for each $x \in U \cap V$, we may consider φ_{VU} as a map

$$\varphi_{VU} : U \cap V \to \mathrm{GL}(k, \mathbb{R}). \tag{10.42}$$

The transition functions have the following properties.

Lemma 10.5.

(1) $\varphi_{UU}(x) = \mathrm{id}_k$ *for* $x \in U$, *where* $\mathrm{id}_k : \mathbb{R}^k \to \mathbb{R}^k$ *is the identity map.*

(2) $\varphi_{UW}(x) \cdot \varphi_{WV}(x) \cdot \varphi_{VU}(x) = \mathrm{id}_k$ *for* $x \in U \cap V \cap W$.

In particular, $\varphi_{UV}(x) \cdot \varphi_{VU}(x) = \mathrm{id}_k$ *for* $x \in U \cap V$.

10.3.2. Examples of vector bundles.

Example 10.6.

(1) *If* M^n *is a smooth manifold, then its tangent bundle* $\pi : TM \to M^n$ *is a rank* n *smooth vector bundle, and we define its smooth vector bundle structure as follows.*

Let $\mathbf{A} := \{(\mathcal{U}, \phi_{\mathcal{U}})\}_{\mathcal{U} \in \mathfrak{U}}$ *be an atlas on* M^n, *where we denote*

$$\mathbf{x} = (x^1, \ldots, x^n) := \phi_{\mathcal{U}}$$

as in (6.23). *For each local coordinate chart* $(\mathcal{U}, \phi_{\mathcal{U}})$, *the corresponding local trivialization of* TM,

$$\varphi_{\mathcal{U}} : \pi^{-1}(\mathcal{U}) \to \mathcal{U} \times \mathbb{R}^n, \tag{10.43}$$

is defined by

$$\varphi_{\mathcal{U}}(x, V) := (x, V^1, \ldots, V^n), \tag{10.44}$$

where the functions $V^i : \mathcal{U} \to \mathbb{R}$ *are defined by* $V =: \sum_{i=1}^n V^i \frac{\partial}{\partial x^i}$.

Note that this definition is similar to that of $\varphi_{\mathcal{U}*}$ *in* §10.1.1, *where a smooth manifold structure on* TM *is defined by choosing* $\varphi_{\mathcal{U}*}$ *to be local coordinate charts for* TM. *With respect to the smooth manifold structure on* TM, *the maps* $\varphi_{\mathcal{U}}$ *defined by* (10.44) *are diffeomorphisms.*

(2) *Similarly, the cotangent bundle* $\pi : T^*M \to M^n$ *is a rank* n *smooth vector bundle.*

(3) *If* M^n *is an embedded submanifold of* $\mathbb{R}^N$, *then its normal bundle* $\pi : NM \to M^n$ *is a rank* $N - n$ *smooth vector bundle, and we can see this as follows. Recall that for every* $\mathbf{x} \in M^n$ *there exists a neighborhood* $\mathcal{U}$ *of* $\mathbf{x}$ *in* $\mathbb{R}^N$ *and a smooth map* $f : \mathcal{U} \to \mathbb{R}^{N-n}$ *such that* $f^{-1}(\mathbf{0}) = M^n \cap \mathcal{U}$. *The components of* f *are given by* $f = (f^1, \ldots, f^{N-n})$, *where each* $f^i : \mathcal{U} \to \mathbb{R}$. *Then the set of vectors* $\{(\nabla f^i)_{\mathbf{x}}\}_{i=1}^{N-n}$ *forms a basis for the normal space* $N_{\mathbf{x}}M$ *for each* $\mathbf{x} \in M^n \cap \mathcal{U}$. *Correspondingly, the local trivialization of* NM,

$$\varphi_{\mathcal{U}} : \pi^{-1}(M^n \cap \mathcal{U}) \to (M^n \cap \mathcal{U}) \times \mathbb{R}^{N-n}, \tag{10.45}$$

is defined by

$$\varphi_{\mathcal{U}}(\mathbf{x}, W) := (\mathbf{x}, W^1, \ldots, W^{N-n}), \tag{10.46}$$

where the functions $W^i : M^n \cap \mathcal{U} \to \mathbb{R}$ *are defined by* $W =: \sum_{i=1}^{N-n} W^i \nabla f^i$.

If $k = 1$, then we call a vector bundle a **line bundle**. If $k = 2$, then we call a vector bundle a **plane bundle**.

Example 10.7 (Tautological line bundle of $\mathbb{R}P^n$). *Let $\mathbb{R}P^n$ be the n-dimensional real projective space as defined in Example* 6.13. *The **tautological line bundle** $\pi : E \to \mathbb{R}P^n$ has its fibers defined by*

$$E_{[\mathbf{x}]} := \mathbb{R}\mathbf{x} \quad \text{for } [\mathbf{x}] \in \mathbb{R}P^n.$$

As a topological space, we define the entire space by

$$E := \big\{([\mathbf{x}], a\mathbf{x}) : \mathbf{x} \in \mathbb{R}^{n+1} \setminus \{\mathbf{0}\},\ a \in \mathbb{R}\big\} \subset \mathbb{R}P^n \times \mathbb{R}^{n+1}.$$

Then, by defining the projection map π as the projection onto the first factor, we obtain a rank 1 *vector bundle over $\mathbb{R}P^n$.*

Example 10.8 (Tautological vector bundle of $\mathrm{Gr}(k, n)$). *Let $\mathrm{Gr}(k, n)$ be the Grassmannian of k-dimensional vector subspaces of $\mathbb{R}^n$ as defined by Example* 6.16. *The **tautological vector bundle** $\pi : E \to \mathrm{Gr}(k, n)$ has its fibers defined by*

$$E_S := S \subset \mathbb{R}^n \quad \text{for } S \in \mathrm{Gr}(k, n).$$

As a topological space, we define the entire space by

$$E := \big\{(S, v) : v \in S,\ S \in \mathrm{Gr}(k, n)\big\} \subset \mathrm{Gr}(k, n) \times \mathbb{R}^n.$$

Then, by defining the projection map π as the projection onto the first factor, we obtain a rank k vector bundle over $\mathrm{Gr}(k, n)$.

10.3.3. How to patch together local products to form a vector bundle. Conversely, suppose that we have a topological space B, an open cover $\{U_\alpha\}_{\alpha \in A}$ of B, and for each pair of sets U_α, U_β in the open cover we have a continuous map

$$g_{\alpha\beta} := \varphi_{U_\beta U_\alpha} : U_\alpha \cap U_\beta \to \mathrm{GL}(k, \mathbb{R}) \tag{10.47}$$

satisfying conditions (1) and (2) in Lemma 10.5. Then we can construct a vector bundle $\pi : E \to B$ with transition functions given by the maps $\varphi_{U_\beta U_\alpha}$, $\alpha, \beta \in A$. Indeed, we define

$$\widetilde{E} := \bigsqcup_{\alpha \in A} U_\alpha \times \mathbb{R}^k \tag{10.48}$$

and we make the identifications

$$U_\alpha \times \mathbb{R}^k \ni (x, v) \sim \big(x, \varphi_{U_\beta U_\alpha}(v)\big) \in U_\beta \times \mathbb{R}^k \tag{10.49}$$

for $x \in U_\alpha \cap U_\beta$ and $v \in \mathbb{R}^k$. Then the quotient space $E := \widetilde{E}/\sim$, together with the naturally defined projection map $\pi : E \to B$, is the desired vector bundle. See Figure 10.3.1.

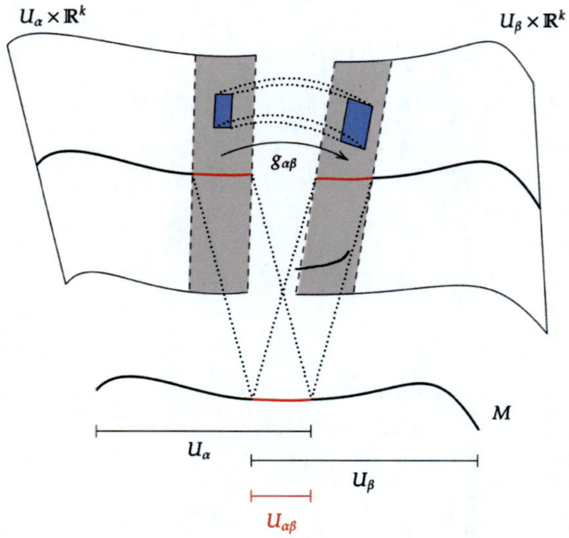

Figure 10.3.1. A transition function $g_{\alpha\beta} := \varphi_{U_\beta U_\alpha} : (U_\alpha \cap U_\beta) \times \mathbb{R}^k \to (U_\alpha \cap U_\beta) \times \mathbb{R}^k$; $U_{\alpha\beta} := U_\alpha \cap U_\beta$. (In the text, we use B for the base space instead of the M used in the figure.) Credit: By Tazerenix – Own work, CC BY-SA 4.0, `https://commons.wikimedia.org/w/index.php?curid=110091887`

If the base $B = M^n$ is a smooth manifold, then we require that the maps $g_{\alpha\beta}$ in (10.47) be smooth. In this case, $\pi : E = \widetilde{E}/\sim \to M^n$ has the structure of a smooth vector bundle.

10.3.4. Trivial vector bundles and parallelizable manifolds. Given any topological space B,

$$\pi : B \times \mathbb{R}^k \to B,$$

where π is the projection map onto the first factor, is a vector bundle, called the **trivial vector bundle** of rank k.

Definition 10.9. If a tangent bundle of a manifold is isomorphic to a trivial vector bundle, then we say that the manifold is **parallelizable**.

Every parallelizable manifold is orientable (see Exercise 10.6).

Every closed orientable 3-manifold is parallelizable (see Milnor and Stasheff [**MS74**, Problem 12-B]).

A sphere S^n is parallelizable if and only if $n = 1$, 3, or 7 (see Bott and Milnor [**BM58**] and Kervaire [**Ker58**] and Hatcher's K-theory book [**Hat**, Theorem 2.16]).

10.3.5. Orientations on vector bundles. An **orientation** on a vector bundle $\pi : E \to B$ is an orientation on each fiber E_x, $x \in B$, with the

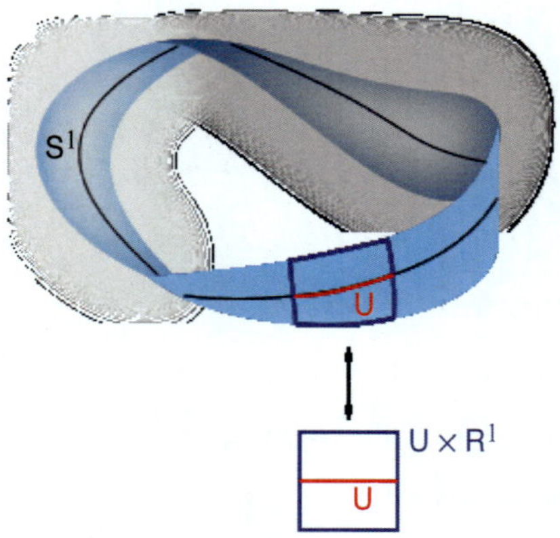

Figure 10.3.2. The Möbius strip as a line bundle over S^1. Credit: By IkamusumeFan – Own work, CC BY-SA 4.0, `https://commons.wikimedia.org/w/index.php?curid=41691537`

following property.[1] For every $x \in B$ there exists an open neighborhood U of x and a local trivialization $\varphi_U : \pi^{-1}(U) \to U \times \mathbb{R}^k$ such that the map φ_U restricted to each fiber is orientation preserving, where $\mathbb{R}^k$ is endowed with the standard orientation. In this case we say that the vector bundle is **oriented**. If a vector bundle admits an orientation, then we say that it is **orientable**; otherwise we say that it is **nonorientable**.

Example 10.10.

(1) *If M^n is an oriented manifold, then its tangent bundle TM has a natural orientation induced by the orientation on M^n.*

(2) *The Möbius strip, considered as a rank* 1 *vector bundle over S^1, is nonorientable. See Figure* 10.3.2.

10.3.6. Bundle metrics.

Definition 10.11. Let $\pi : E \to B$ be a rank k vector bundle. A **bundle metric** $\langle \cdot, \cdot \rangle$ on E is a continuous choice of inner products on the fibers E_x for $x \in B$.

The prototypical example of a bundle metric is the tangent bundle TM of a Riemannian manifold (M^n, g). Here, the inner product on each T_xM is given by g_x. The Riemannian metric g also induces natural bundle metrics on tensor bundles of M^n.

[1] The notion of orientations on vector spaces is defined in §8.1.1.

10.3.7. Operations on vector bundles. Let $\pi_1 : E_1 \to B$ and $\pi_2 : E_2 \to B$ be vector bundles with the same base space. Their **Whitney sum** $\pi : E_1 \oplus E_2 \to B$ is the vector bundle with fibers $(E_1)_x \oplus (E_2)_x$ for $x \in B$. As a topological space, we define the entire space by

$$E_1 \oplus E_2 := \{(v_1, v_2) \in E_1 \times E_2 : \pi_1(v_1) = \pi_2(v_2)\}.$$

The projection map is given by

$$\pi(v_1, v_2) := \pi_1(v_1) = \pi_2(v_2).$$

We have $\mathrm{rank}(E_1 \oplus E_2) = \mathrm{rank}(E_1) + \mathrm{rank}(E_2)$.

The **tensor product bundle** $\pi : E_1 \otimes E_2 \to B$ is the vector bundle with fibers $(E_1)_x \otimes (E_2)_x$ for $x \in B$, so that the entire space as a set is

$$E := \bigsqcup_{x \in B} (E_1)_x \otimes (E_2)_x.$$

The projection map π is the obvious map. We may choose an open covering $\{U_\alpha\}_{\alpha \in A}$ of B and local trivializations $(U_\alpha, \varphi^1_{U_\alpha})$ and $(U_\alpha, \varphi^2_{U_\alpha})$ of E_1 and E_2, respectively. Define maps

$$\varphi^1_{U_\alpha} \otimes \varphi^2_{U_\alpha} : \pi^{-1}(U_\alpha) \to U_\alpha \times (\mathbb{R}^{k_1} \otimes \mathbb{R}^{k_2}),$$

where k_1 and k_2 are the ranks of E_1 and E_2, respectively. These maps, in conjunction with the open cover and local trivializations, induce a topology on E, making it a vector bundle of rank $k_1 k_2$.

Example 10.12. *Let M^n be a smooth manifold. Given nonnegative integers r and s, not both zero, we have the tensor bundle on M^n defined by*

$$\bigotimes^{r,s} M = \left(\bigotimes^r T^*M\right) \otimes \left(\bigotimes^s TM\right). \tag{10.50}$$

We also have the smooth vector bundle of alternating covariant tensors $\Lambda^k M := \Lambda^k T^*M$ and its contravariant version $\Lambda^k TM$.

In general, for a rank r vector bundle E over a topological space B, we have the tensor bundles

$$\bigotimes^{k,\ell} E = \left(\bigotimes^k E^*\right) \otimes \left(\bigotimes^\ell E\right)$$

and alternating tensor bundles $\Lambda^k E$ and $\Lambda^k E^*$. We define

$$\Lambda^*(E) := \bigoplus_{k=0}^r \Lambda^k E. \tag{10.51}$$

10.3.8. Vector bundle morphisms. Let $\pi_1 : E_1 \to B_1$ and $\pi_2 : E_2 \to B_2$ be vector bundles. A **vector bundle homomorphism** from E_1 to E_2 consists of a pair of maps $f : B_1 \to B_2$ and $\phi : E_1 \to E_2$ with the following properties:

(1) $\pi_2 \circ \phi = f \circ \pi_1 : E_1 \to B_2$. This implies that $\phi((E_1)_x) \subset (E_2)_{f(x)}$ for all $x \in B_1$.

(2) The restricted map $\phi : (E_1)_x \to (E_2)_{f(x)}$ is a linear map of vector spaces.

We say that a vector bundle homomorphism $\phi : E_1 \to E_2$ is a **vector bundle isomorphism** if ϕ is a bijection and if its inverse $\phi^{-1} : E_2 \to E_1$ is a vector bundle homomorphism. If there exists a vector bundle isomorphism between two vector bundles E_1 and E_2, then we say that the two vector bundles are **isomorphic**; we denote this by $E_1 \cong E_2$.

Given a topological space B and a positive integer k, let $\mathrm{Vect}^k(B)$ denote the set of isomorphism classes of rank k vector bundles over B.

Let $\mathrm{Vect}^k_+(B)$ denote the set of isomorphism classes of oriented rank k vector bundles over B, where two oriented vector bundles are defined to be isomorphic if there exists an orientation-preserving isomorphism between them.

10.3.9. Pull-back vector bundles. Let $\pi : E \to B$ be a vector bundle and let $f : A \to B$ be a continuous map between topological spaces. Then we can define the **pull-back vector bundle** $\pi' : f^*(E) \to A$ as follows. The fibers are $(f^*(E))_x = E_{f(x)}$ for $x \in A$. As a topological space, we define the entire space by

$$f^*(E) := \{(x, v) : v \in E_{f(x)}\} \subset A \times E.$$

Then, by defining the projection map π' as the projection onto the first factor, we obtain vector bundle $f^*(E)$ over A. The rank of the pull-back bundle $f^*(E)$ is the same as the rank of the original bundle E.

Example 10.13. *Let $f : N^m \to M^n$ be a smooth map between smooth manifolds. Then $f^*(TM)$ is a rank n smooth vector bundle over N^m.*

The pull-back operation has the following naturality properties:

(1) $(f \circ g)^*(E) \cong g^*(f^*(E))$.

(2) $f^*(E_1 \oplus E_2) \cong f^*(E_1) \oplus f^*(E_2)$.

(3) $f^*(E_1 \otimes E_2) \cong f^*(E_1) \otimes f^*(E_2)$.

Let $\pi : E \to B$ be a vector bundle and let A be a paracompact topological space. If maps $f_0, f_1 : A \to B$ are homotopic, then the pull-back bundles $f_0^*(E)$ and $f_1^*(E)$ over A are isomorphic (see [**Hat**, Theorem 1.6]).

10.3.10. Sections of vector bundles. A **section** of a vector bundle $\pi : E \to B$ is a map $\sigma : B \to E$ satisfying $\pi \circ \sigma = \mathrm{id}_B$.

The **zero-section** $\vec{0}$ of a vector bundle $\pi : E \to B$ is the map $\vec{0} : B \to E$ defined by $\vec{0}(x) = \vec{0}_x$ for $x \in B$, where $\vec{0}_x$ is the zero-vector for the vector space E_x.

Observe that for the Möbius strip M^2 and for any continuous section $\sigma : S^1 \to M^2$, its image $\sigma(S^1)$ intersects the image of the zero-section $\vec{0}(S^1)$.

Definition 10.14. Suppose that M^n is a smooth manifold. A section $\sigma : M^n \to E$ of a smooth vector bundle $\pi : E \to M^n$ is a **smooth section** if σ is a smooth map. We denote the module (often viewed as a sheaf) of smooth sections by $C^\infty(E)$.[2]

A smooth vector field is the same as a smooth section of the tangent bundle TM. A smooth 1-form is the same as a smooth section of the cotangent bundle T^*M. More generally, a smooth section of the tensor bundle $\bigotimes^{r,s} M$ in (10.50) is called a smooth tensor of type (r, s).

10.3.11. The clutching construction. The set of oriented plane bundles over S^2, up to isomorphism, may be identified with $\mathbb{Z}$. One can describe these plane bundles by the **clutching construction**.

More generally, consider the n-sphere S^n as the union of two hemispheres $S^n = S^n_+ \cup S^n_-$, where $S^n_+ \cap S^n_- =: S^n_0 \cong S^{n-1}$ is the equator. Let k be a positive integer and let $f : S^n_0 \to \mathrm{GL}_+(k, \mathbb{R})$ be a continuous map, called the **clutching map**, where $\mathrm{GL}_+(k, \mathbb{R}) := \{\mathrm{A} \in \mathrm{GL}(k, \mathbb{R}) \colon \det(\mathrm{A}) > 0\}$. We define the rank k bundle $\pi : E \to S^n$ by

$$E := (S^n_+ \times \mathbb{R}^k) \sqcup (S^n_- \times \mathbb{R}^k) / \sim,$$

where $\sim$ is the identification defined by

$$(x, v) \sim (x, f(x)(v))$$

for $x \in S^n_0$ and $v \in \mathbb{R}^k$.

For example, by (10.11), the tangent bundle TS^2 may be obtained by the clutching construction with clutching map $f : S^2_0 \to \mathrm{SO}(2) \subset \mathrm{GL}(2, \mathbb{R})$ defined by $f(z) = z^{-2}$.

10.3.12. Classifying vector bundles. The vector bundle $\pi : E \to S^n$ defined in §10.3.11 is determined up to isomorphism by the homotopy class of the clutching map $f : S^n_0 \to \mathrm{GL}_+(k, \mathbb{R})$ (see [**Hat**, Proposition 1.14]). Note that $\mathrm{GL}_+(k, \mathbb{R})$ has the same homotopy type as $\mathrm{SO}(k)$ (which is a deformation retract).

In particular, if $n = k = 2$, then we have that S^n_0 is diffeomorphic to S^1 and $\mathrm{GL}_+(k, \mathbb{R})$ has the same homotopy type as $\mathrm{SO}(2) \cong S^1$. Using the natural orientations on these spaces, we see that plane bundles over S^2 are classified by the degree of the map f, which takes values in $\mathbb{Z}$.

We also see that there exist exactly two line bundles over S^1 up to vector bundle isomorphism: $S^1 \times \mathbb{R}$ and the Möbius strip. As the entire space

[2] $C^\infty(E)$ is a module over $C^\infty(M^n)$.

of a vector bundle, we define the Möbius strip to be the quotient space $M^2 := [0, 2\pi] \times \mathbb{R}/\sim$, where $(0, y) \sim (2\pi, -y)$. Let $S^1 := [0, 2\pi]/\{0 \sim 2\pi\}$. Then the projection map $\pi : M^2 \to S^1$ is a rank 1 vector bundle.

10.4. Lie groups

Roughly speaking, a Lie group is a set that has the structure of both a smooth manifold and a group, where the structures are compatible. We begin our discussion with the classical groups, which are the most commonly used examples. Then we consider Lie groups in general, and we consider Lie groups endowed with Riemannian metrics compatible with the group structure.

10.4.1. The classical groups. In this subsection, we consider certain subgroups of the group of linear transformations of a vector space, which preserve a bilinear or sesquilinear form on the vector space.[3] We have already encountered many of these groups in our discussion up to this point.

Let $\mathbb{F}$ be either the real field $\mathbb{R}$, complex field $\mathbb{C}$, or quaternion division ring $\mathbb{H}$.[4] Consider an n-dimensional vector space $\mathbb{V}$ over $\mathbb{F}$, where for simplicity we usually take $\mathbb{V}$ to be $\mathbb{F}^n$.

The **general linear group** $\mathrm{GL}(n, \mathbb{F})$ of degree n over $\mathbb{F}$ is defined to be the set of invertible $n \times n$ matrices with entries in $\mathbb{F}$ together with the operation of matrix multiplication. This is the same as the **automorphism group** $\mathrm{Aut}(\mathbb{F}^n)$ of $\mathbb{F}^n$, i.e., the group of invertible $\mathbb{F}$-linear transformations of $\mathbb{F}^n$, with group multiplication given by composition of linear transformations.

A **bilinear form** β on $\mathbb{V}$ is a map $\beta : \mathbb{V} \times \mathbb{V} \to \mathbb{F}$ satisfying

$$\begin{aligned} \beta(u + cv, w) &= \beta(u, w) + c\beta(v, w), \\ \beta(u, v + cw) &= \beta(u, v) + c\beta(u, w) \end{aligned}$$

for $u, v, w \in \mathbb{V}$ and $c \in \mathbb{F}$. A bilinear form β is **nondegenerate** if $v \in \mathbb{V}$ satisfying $b(v, w) = 0$ for all $w \in \mathbb{V}$ implies that $v = 0$.

The **classical groups** (named after Herman Weyl's book of the same title [**Wey97**]) are the automorphism groups of vector spaces endowed with nondegenerate bilinear or sesquilinear forms. These forms are assumed to be symmetric, skew-symmetric, or Hermitian symmetric (if $\mathbb{F} = \mathbb{C}$). Automorphism groups that preserve a volume form, such as $\mathrm{SL}(n; \mathbb{F})$ described below, are also often included in this list.

[3]By a sesquilinear form, we usually mean a Hermitian form.

[4]$\mathbb{C}$ and $\mathbb{H}$ are the only finite-dimensional division rings containing $\mathbb{R}$ as a proper subring.

10.4.1.1. *The orthogonal group.* Given nonnegative integers p and q satisfying $p+q=n$, consider the nondegenerate symmetric bilinear form on $\mathbb{F}^n$ defined by

$$\beta(v,w) := v^1w^1 + \cdots + v^pw^p - v^{p+1}w^{p+1} - \cdots - v^nw^n, \tag{10.52}$$

where $v=(v^1,\ldots,v^n)$ and $w=(w^1,\ldots,w^n)$. With β given by (10.52), we denote

$$\langle v,w\rangle_{p,q} := \beta(v,w). \tag{10.53}$$

The associated **orthogonal group** is defined by

$$\mathrm{O}(p,q,\mathbb{F}) := \{\mathrm{A}\in \mathrm{GL}(n,\mathbb{F}) : \langle \mathrm{A}(v),\mathrm{A}(w)\rangle_{p,q} = \langle v,w\rangle_{p,q}\}. \tag{10.54}$$

For example, the orthogonal group associated to the Euclidean inner product on $\mathbb{F}^n$, where $\mathbb{F}=\mathbb{R}$ or $\mathbb{F}=\mathbb{C}$,[5] is usually denoted by $\mathrm{O}(n,\mathbb{F}) := \mathrm{O}(n,0,\mathbb{F})$, with the further abbreviation $\mathrm{O}(n) := \mathrm{O}(n,\mathbb{R})$. The **Lorentz group** is $\mathrm{O}(3,1)$.

For a proof that $\mathrm{O}(n)$ is a smooth $\frac{n(n-1)}{2}$-dimensional manifold, see Lee [**Lee13**, Example 7.27]. It follows that $\mathrm{O}(n)$ is a Lie group.

10.4.1.2. *The symplectic group.* Let $\mathbb{F}$ be either $\mathbb{R}$ or $\mathbb{C}$. Define a blinear form ω_n on $\mathbb{F}^{2n}$ by

$$\begin{aligned}\omega_n(v,w) := {} & v^1w^{n+1} + v^2w^{n+2} + \cdots + v^nw^{2n} \\ & - w^1v^{n+1} - w^2v^{n+2} - \cdots - w^nv^{2n}.\end{aligned} \tag{10.55}$$

The associated **symplectic group** is defined by

$$\mathrm{Sp}(n,\mathbb{F}) := \{\mathrm{A}\in \mathrm{GL}(2n,\mathbb{F}) : \omega_n(\mathrm{A}(v),\mathrm{A}(w)) = \omega_n(v,w)\}. \tag{10.56}$$

We use the abbreviation $\mathrm{Sp}(n) := \mathrm{Sp}(n,\mathbb{R})$ for the real symplectic group.

See [**Lee13**, Problem 22-3] for the proof that $\mathrm{Sp}(n)$ is a Lie group.

10.4.1.3. *The unitary group.* Define the Hermitian symmetric sesquilinear form H on $\mathbb{C}^n$ by

$$H(v,w) := v^1\overline{w}^1 + \cdots + v^n\overline{w}^n. \tag{10.57}$$

The associated **unitary group** is defined by

$$\mathrm{U}(n) := \{\mathrm{A}\in \mathrm{GL}(n,\mathbb{C}) : H(\mathrm{A}(v),\mathrm{A}(w)) = H(v,w)\}. \tag{10.58}$$

$\mathrm{U}(n)$ is an n^2-dimensional Lie group.

[5] Note that the Euclidean inner product on $\mathbb{C}$ is complex bilinear (not Hermitian bilinear).

10.4.1.4. *Special linear, orthogonal, and unitary groups.* The **special linear group** is defined by

$$\mathrm{SL}(n;\mathbb{F}) := \{\mathrm{A} \in \mathrm{GL}(n,\mathbb{F}) : \det(\mathrm{A}) = 1\}. \tag{10.59}$$

$\mathrm{SL}(n,\mathbb{R})$ is an (n^2-1)-dimensional Lie group.

The **special orthogonal group** is defined by

$$\mathrm{SO}(p,q,\mathbb{F}) := \{\mathrm{A} \in \mathrm{O}(p,q,\mathbb{F}) : \det(\mathrm{A}) = 1\} \subset \mathrm{GL}(n,\mathbb{F}) \tag{10.60}$$

for $\mathbb{F} = \mathbb{R}$ or $\mathbb{F} = \mathbb{C}$. In particular,

$$\mathrm{SO}(n) := \mathrm{SO}(n,0,\mathbb{R}) = \{\mathrm{A} \in \mathrm{O}(n) : \det(\mathrm{A}) = 1\} \subset \mathrm{GL}(n,\mathbb{R}). \tag{10.61}$$

$\mathrm{SO}(n)$ is an $\frac{n(n-1)}{2}$-dimensional Lie group.

The **special unitary group** is defined by

$$\mathrm{SU}(n) := \{\mathrm{A} \in \mathrm{U}(n) : \det(\mathrm{A}) = 1\} \subset \mathrm{GL}(n,\mathbb{C}). \tag{10.62}$$

$\mathrm{SU}(n)$ is an (n^2-1)-dimensional Lie group.

A natural generalization of classical groups discussed in the previous section consists of the class of spaces called Lie groups. We give the precise definition below.

10.4.2. Topological groups.

10.4.2. Topological groups. A **topological group** is a set G endowed with both a topological structure and a group structure such that these two structures are compatible. More precisely, it is a set G endowed with a group operation $\cdot$ and a topology such that the map $\mu : G \times G \to G$ defined by $\mu(\sigma,\tau) = \sigma \cdot \tau$ and the map $\iota : G \to G$ defined by $\iota(\sigma) = \sigma^{-1}$ are both continuous.

Given $\sigma \in G$, we define its associated left multiplication $L_\sigma : G \to G$ by

$$L_\sigma(\tau) = \sigma \cdot \tau$$

and its associated right multiplication $R_\sigma : G \to G$ by

$$R_\sigma(\tau) = \tau \cdot \sigma.$$

Let e denote the identity element of G. We have that $L_e : G \to G$ is equal to the identity map id_G. Thus,

$$L_\sigma \circ L_{\sigma^{-1}} = L_{\sigma^{-1}} \circ L_\sigma = L_e = \mathrm{id}_G,$$

so that $L_{\sigma^{-1}} = (L_\sigma)^{-1}$.

Since μ is continuous, we have that L_σ is continuous for each $\sigma \in G$. Thus, $L_\sigma : G \to G$ is a homeomorphism for each $\sigma \in G$.

Example 10.15. *The unit circle S^1 in the complex plane, which consists of complex numbers of absolute value* 1, *is a topological group. The group operation is multiplication, where $\mu(z, w) = zw$ represents complex multiplication. This operation makes S^1 into a group, and the topology on S^1 induced from the complex plane is compatible with the group operation.*

Example 10.16. *Let $(S, \mathcal{T})$ be a topological space and let* Homeo(S) *denote the group of homeomorphisms from $(S, \mathcal{T})$ to itself. The* ***compact-open topology*** *on* Homeo(S) *is defined by the collection of sets*

$$B(K, U) := \{f \in \mathrm{Homeo}(S) : f(K) \subset U\},$$

where $K \subset S$ is a compact set and $U \subset S$ is an open set, being a subbase for the compact-open topology. That is, the compact-open topology on Homeo(S) *is the smallest topology containing the collection $B(K, U)$.*

If $(S, \mathcal{T})$ is compact and Hausdorff, then Homeo(S) *is a topological group.*

10.4.3. Definition of Lie group. A **Lie group** is a set G endowed with both a smooth manifold structure and a group structure, such that these two structures are compatible. Specifically, let $\cdot$ represent the group multiplication; the map $\mu : G \times G \to G$, defined by $\mu(\sigma, \tau) = \sigma \cdot \tau^{-1}$, is a C^∞ (smooth) function.

For example, the classical groups are Lie groups. Generally, that the classical groups are manifolds follows from the implicit function theorem. The group multiplications are compositions when the elements are considered as automorphisms, or matrix multiplications when the elements are considered as matrices.

10.4.4. Left-invariant vector fields. Given $\sigma \in G$, the **left and right multiplication maps** $L_\sigma : G \to G$ and $R_\sigma : G \to G$ are defined by

$$L_\sigma(\tau) = \sigma \cdot \tau \quad \text{and} \quad R_\sigma(\tau) = \tau \cdot \sigma \quad \text{for all } \tau \in G, \tag{10.63}$$

respectively. Note that for each $\sigma \in G$, L_σ and R_σ are diffeomorphisms with inverses $L_{\sigma^{-1}}$ and $R_{\sigma^{-1}}$, respectively.

Definition 10.17. A vector field V on a Lie group $(G, \cdot)$ is said to be **left-invariant** if $(L_\sigma)_*(V) = V$ holds for all $\sigma \in G$. That is, $dL_\sigma(V_\tau) = V_{\sigma\cdot\tau}$ for all $\sigma, \tau \in G$.

If V is a left-invariant vector field, then we have that

$$V_\sigma = dL_\sigma(V_e) \qquad \text{for all } \sigma \in G. \tag{10.64}$$

Conversely, if $v \in T_eG$, the tangent space at the identity element, then $V_\sigma := dL_\sigma(v)$, $\sigma \in G$, defines a left-invariant vector field V on G satisfying

$V_e = v$. So we see that the set of left-invariant vector fields on a Lie group G is a finite-dimensional real vector space isomorphic to T_eG.

10.4.5. Lie algebra associated to a Lie group. A **Lie algebra** $\mathfrak{l}$ is a vector space $\mathfrak{l}$ together with a map from $\mathfrak{l}\times\mathfrak{l}$ to $\mathfrak{l}$, denoted by $(X, Y) \mapsto [X, Y]$; that is, it is bilinear and skew-symmetric and satisfies the **Jacobi identity**:

$$[[X,Y],Z] + [[Y,Z],X] + [[Z,X],Y] = 0 \tag{10.65}$$

for all $X, Y, Z \in \mathfrak{l}$.

We define the **Lie algebra associated to** G to be the vector space $\mathfrak{g}$ of left-invariant vector fields on G together with the Lie bracket operation. The reason for why $\mathfrak{g}$ is a Lie algebra is that the Lie bracket of left-invariant vector fields is left-invariant by $dL_\sigma([V, W]) = [dL_\sigma(V), dL_\sigma(W)]$, and the Jacobi identity holds by (4.59). As we have seen above, we may identify the Lie algebra $\mathfrak{g}$ with T_eG.

Example 10.18 (Lie algebras of classical groups).

(1) *Let* $\mathfrak{gl}(n, \mathbb{F})$ *denote the vector space of all* $n\times n$ *matrices with entries in* $\mathbb{F}$*, i.e., the vector space of endomorphisms of* $\mathbb{F}^n$*. Then* $\mathfrak{gl}(n, \mathbb{F})$ *is the Lie algebra of the Lie group* $\mathrm{GL}(n, \mathbb{F})$.

(2) *Let*
$$\mathfrak{so}(n) := \{A \in \mathfrak{gl}(n, \mathbb{R}) : A^{\mathrm{T}} = -A\},$$
where A^{T} *denotes the transpose of* A*. Then* $\mathfrak{so}(n)$ *is the Lie algebra of both of the Lie groups* $\mathrm{O}(n)$ *and* $\mathrm{SO}(n)$.

(3) *The Lie algebra of* $\mathrm{U}(n)$ *is*
$$\mathfrak{u}(n) := \{A \in \mathfrak{gl}(n, \mathbb{C}) : \overline{A}^{\mathrm{T}} = -A\},$$
where $\overline{A}^{\mathrm{T}}$ *denotes the conjugate transpose of* A.

(4) *The Lie algebra of* $\mathrm{SU}(n)$ *is*
$$\mathfrak{su}(n) := \{A \in \mathfrak{u}(n) : \mathrm{trace}(A) = 0\}.$$

(5) *The Lie algebra of* $\mathrm{Sp}(n)$ *is*
$$\mathfrak{sp}(n) := \{A \in \mathfrak{gl}(2n, \mathbb{R}) : A^{\mathrm{T}}\Omega A = \Omega\},$$
where $\Omega := \begin{pmatrix} 0 & I_n \\ -I_n & 0 \end{pmatrix}$.

10.4.6. Orientability. Any Lie group is orientable. Indeed, given an n-dimensional Lie group G, choose any basis $\{b_i\}_{i=1}^n$ for T_eG. Then the associated left-invariant vector fields $\{B_i\}_{i=1}^n$, where $(B_i)_e = b_i$, define an orientation on G. In fact, this argument shows that G is a parallelizable manifold; i.e., its tangent bundle TG is isomorphic to a trivial bundle.

10.4.7. Adjoint representations. The noncommutativity of a Lie group is reflected by conjugation. The **group of automorphisms** $\mathrm{Aut}(G)$ of a Lie group G is simply the group of automorphisms of G as a group. One can easily show that any automorphism is necessarily a diffeomorphism. The **automorphism group** $\mathrm{Aut}(\mathfrak{g})$ of the associated Lie algebra is the group of vector space isomorphisms of $\mathfrak{g}$ that preserve the Lie bracket.

Define the group homomorphism $\Psi : G \to \mathrm{Aut}(G)$ by

$$\Psi_g(h) := \Psi(g)(h) := ghg^{-1}. \tag{10.66}$$

Given $g \in G$, define $\mathrm{Ad}_g : \mathfrak{g} \to \mathfrak{g}$ by

$$\mathrm{Ad}_g := (d\Psi_g)_e : T_eG \to T_eG, \tag{10.67}$$

where $\mathfrak{g}$ is identified with T_eG. We leave it as an exercise that Ad_g preserves the Lie algebra structure of $\mathfrak{g}$. Hence $\mathrm{Ad}_g \in \mathrm{Aut}(\mathfrak{g})$. The group homomorphism

$$\mathrm{Ad} : G \to \mathrm{Aut}(\mathfrak{g})$$

defined by $\mathrm{Ad}(g) := \mathrm{Ad}_g$ is called the **adjoint representation** of a Lie group G on its Lie algebra $\mathfrak{g}$.

10.4.8. The exponential map of a Lie group. The **exponential map** for a Lie group, not to be confused with the exponential map of a Riemannian manifold, is defined as follows. Given $A \in \mathfrak{g}$, let $\gamma : \mathbb{R} \to G$ be the integral curve to A, as a left-invariant vector field on G, with $\gamma(0) = e$ the identity. The exponential map $\exp : \mathfrak{g} \to G$ is defined by

$$\exp(A) := \gamma(1). \tag{10.68}$$

In the case of a Lie group of matrices, we have that

$$\exp(A) = \sum_{k=1}^{\infty} \frac{1}{k!} A^k. \tag{10.69}$$

The reason for this is that

$$\gamma(t) := \exp(tA) = \sum_{k=1}^{\infty} \frac{1}{k!} (tA)^k$$

satisfies $\gamma'(0) = A$ and $\exp((t+s)A) = \exp(tA) \cdot \exp(sA)$ for all $s, t \in \mathbb{R}$.

10.4.9. Left-invariant Riemannian metrics on Lie groups.

Definition 10.19. A Riemannian metric g on a Lie group G is called **left-invariant** if the left multiplication map L_σ is an isometry of (G, g) for every $\sigma \in G$; i.e., $(L_\sigma)^* g = g$. In other words, this condition is equivalent to

$$g_{\sigma\cdot\tau}(dL_\sigma(V), dL_\sigma(W)) = g_\tau(V, W) \quad \text{for all } V, W \in T_\tau G,\ \sigma, \tau \in G.$$

Similarly, we say that g is **right-invariant** if $(R_\sigma)^* g = g$ for every $\sigma \in G$.

These definitions extend in the obvious ways to the notions of left- and right-invariant tensors and differential forms.

Observe that $g_\sigma(dL_\sigma(V), dL_\sigma(W)) = g_e(V, W)$ for $V, W \in T_eG$. Conversely, if $\langle \cdot, \cdot \rangle$ is an inner product on T_eG, we have that

$$g_\sigma(V, W) := \langle dL_{\sigma^{-1}}(V), dL_{\sigma^{-1}}(W) \rangle \quad \text{for } V, W \in T_\sigma G$$

defines a left-invariant metric on G. Thus, the set of left-invariant metrics on G is bijective with the set of inner products on its Lie algebra $\mathfrak{g}$.

The following lemma provides explicit formulas for the Levi-Civita connection and the Riemann curvature tensor on a Lie group G endowed with a left-invariant metric g, expressed in terms of the Lie brackets of the associated left-invariant vector fields.

Lemma 10.20. *Let g be a left-invariant metric on a Lie group $(G, \cdot)$. Suppose that X, Y, Z, and W are left-invariant vector fields. Then:*

(1) *The Levi-Civita connection is given by*

$$\langle \nabla_X Y, Z \rangle = \frac{1}{2} \left(\langle [X, Y], Z \rangle - \langle [X, Z], Y \rangle - \langle [Y, Z], X \rangle \right). \tag{10.70}$$

Observe that $\nabla_X Y$ is a left-invariant vector field.

(2) *The Riemann curvature tensor is given by*

$$\langle \mathrm{Rm}(X, Y)Z, W \rangle = \langle \nabla_X Z, \nabla_Y W \rangle - \langle \nabla_Y Z, \nabla_X W \rangle - \left\langle \nabla_{[X,Y]} Z, W \right\rangle. \tag{10.71}$$

In particular,

$$\langle \mathrm{Rm}(X, Y)Y, X \rangle = \langle \nabla_X Y, \nabla_Y X \rangle - \langle \nabla_Y Y, \nabla_X X \rangle - \left\langle \nabla_{[X,Y]} Y, X \right\rangle. \tag{10.72}$$

Observe that $\mathrm{Rm}(X, Y)Z$ *is a left-invariant vector field.*

Proof.

(1) The inner products of left-invariant vector fields are constant on G. Consequently, we have $X \langle Y, Z \rangle = 0$, $Y \langle X, Z \rangle = 0$, and $Z \langle X, Y \rangle = 0$. Formula (10.70) for the Levi-Civita connection follows from this observation and (4.132).

(2) By the definition (7.61) of the Riemann curvature tensor and the product rule (7.13), we have

$$\begin{aligned} \langle \mathrm{Rm}(X, Y)Z, W \rangle &= \langle \nabla_X (\nabla_Y Z), W \rangle - \langle \nabla_Y (\nabla_X Z), W \rangle - \left\langle \nabla_{[X,Y]} Z, W \right\rangle \\ &= X \langle \nabla_Y Z, W \rangle - \langle \nabla_Y Z, \nabla_X W \rangle - Y \langle \nabla_X Z, W \rangle \\ &\quad + \langle \nabla_X Z, \nabla_Y W \rangle - \left\langle \nabla_{[X,Y]} Z, W \right\rangle. \end{aligned} \tag{10.73}$$

Since $\langle \nabla_Y Z, W\rangle$ and $\langle \nabla_X Z, W\rangle$ are constant functions, we have

$$X \langle \nabla_Y Z, W\rangle = Y \langle \nabla_X Z, W\rangle = 0. \tag{10.74}$$

Formula (10.71) for the Riemann curvature tensor now follows from applying (10.74) to (10.73).

□

10.4.10. Unimodular Lie groups.

10.4.10.1. *Definition of a unimodular Lie group.* Let G be an n-dimensional Lie group. Let ω_e be a nonzero n-form at the identity e of G. Then ω_e extends to a left-invariant n-form ω on G. The measure on G associated to ω is called a **Haar measure**. We say that G is **unimodular** if ω is right-invariant. This definition is independent of the choice of ω_e. We remark that Milnor [**Mil76**] has classified 3-dimensional unimodular Lie groups.

Let G be an n-dimensional Lie group. Let $\{F_i\}_{i=1}^n$ be a basis of left-invariant vector fields on G. The **structure constants** c_{ij}^k of its associated Lie algebra $\mathfrak{g}$ are defined by

$$[F_i, F_j] =: \sum_{k=1}^n c_{ij}^k F_k.$$

10.4.10.2. *The unimodular Lie group* SU(2) *and Berger spheres.* Now consider the 3-dimensional unimodular Lie group $G = \mathrm{SU}(2)$. This is the same as Spin(3), the double cover of SO(3). We have that SU(2) is diffeomorphic to S^3 and SO(3) is diffeomorphic to $\mathbb{R}P^3$. We may choose $\{F_i\}_{i=1}^3$ so that $[F_i, F_j] = -2F_k$, where (i, j, k) is a cyclic permutation of $(1, 2, 3)$. On SU(2), consider the left-invariant metric

$$g := \sum_{i=1}^3 g_i F_i^* \otimes F_i^*. \tag{10.75}$$

Then, by (10.70), we have for any ℓ and for i, j, k as above,

$$\langle \nabla_{F_i} F_j, F_\ell\rangle = \frac{1}{2}\left(\langle [F_i, F_j], F_\ell\rangle - \langle [F_i, F_\ell], F_j\rangle - \langle [F_j, F_\ell], F_i\rangle\right),$$

so that

$$\nabla_{F_i} F_j = \left(-1 - \frac{g_j}{g_k} + \frac{g_i}{g_k}\right) F_k.$$

It is also easy to see that

$$\nabla_{F_i} F_i = 0.$$

By (10.72), we have for i, j, k as above,

$$\begin{aligned} R_{ijji} &:= \langle \operatorname{Rm}(F_i, F_j)F_j, F_i \rangle \\ &= \langle \nabla_{F_i} F_j, \nabla_{F_j} F_i \rangle - \langle \nabla_{F_j} F_j, \nabla_{F_i} F_i \rangle - \left\langle \nabla_{[F_i, F_j]} F_j, F_i \right\rangle \\ &= g_k \left(-1 - \frac{g_j}{g_k} + \frac{g_i}{g_k} \right) \left(1 - \frac{g_j}{g_k} + \frac{g_i}{g_k} \right) + 2g_i \left(1 - \frac{g_k}{g_i} + \frac{g_j}{g_i} \right) \\ &= g_k \left(-3 + \left(\frac{g_i - g_j}{g_k} \right)^2 \right) + 2g_i + 2g_j. \end{aligned}$$

Now we choose $g_1 = g_2 = 1$ and $g_3 = \varepsilon$, where $\varepsilon > 0$. These choices yield left-invariant metrics on $\mathrm{SU}(2)$, which are called **Berger spheres**. We then have that

$$\begin{aligned} R_{1221} &= 4 - 3\varepsilon, \\ R_{3113} &= \varepsilon^2, \\ R_{2332} &= \varepsilon^2. \end{aligned}$$

So the principal sectional curvatures of a Berger sphere are given by

$$\begin{aligned} K(F_1 \wedge F_2) &= 4 - 3\varepsilon, \\ K(F_3 \wedge F_1) &= \varepsilon, \\ K(F_2 \wedge F_3) &= \varepsilon. \end{aligned}$$

10.5. Fiber bundles

Fiber bundles are generalizations of vector bundles where the fibers no longer are required to have a linear structure.

10.5.1. Fibrations. The notion of fibration is a topological generalization of fiber bundle, which will be discussed beginning in the next subsection. When looking at fiber bundles (defined below) from a topological perspective such as homotopy theory, it is natural to consider fibrations.[6]

A **fibration** is a continuous map between topological spaces $\pi : E \to B$ which satisfies the following **homotopy lifting property**: For any topological space X, any continuous map $f : X \times [0,1] \to B$, and any lifting $\widetilde{f_0} : X \to E$ of the map $f \circ i_0$ (i.e., $f \circ i_0 = \pi \circ \widetilde{f_0}$), there exists a continuous map

$$\widetilde{f} : X \times [0,1] \to E$$

[6]In this subsection we assume that the reader is familiar with homotopy groups; see G. W. Whitehead's book for background [**Whi78**].

lifting f (i.e., $f = \pi \circ \widetilde{f}$) and satisfying $\widetilde{f}_0 = \widetilde{f} \circ i_0$. Here, $i_0 : X \to X \times [0,1]$ denotes the inclusion map defined by $i_0(x) = (x, 0)$. This homotopy lifting property may be depicted by the following commutative diagram:

$$\begin{array}{ccc} X & \xrightarrow{\widetilde{f}_0} & E \\ {\scriptstyle i_0}\downarrow & \overset{\widetilde{f}}{\nearrow} & \downarrow{\scriptstyle \pi} \\ X \times [0,1] & \xrightarrow{f} & B \end{array}$$

The **fibration long exact sequence of homotopy groups** says that for a fibration $\pi : E \to B$, basepoint $b_0 \in B$, and fiber $F = \pi^{-1}(b_0)$, we have the following long exact sequence of homotopy groups:

(10.76)
$$\cdots \to \pi_k(F) \xrightarrow{i_*} \pi_k(E) \xrightarrow{\pi_*} \pi_k(B) \to \pi_{k-1}(F) \xrightarrow{i_*} \cdots \to \pi_0(F) \xrightarrow{i_*} \pi_0(E),$$

where $i : F \to E$ is the inclusion map (see Husemoller [**Hus94**, Theorem 5.3 in Chapter 1]).

10.5.2. Definition of fiber bundle. A **fiber bundle** is defined to be a triple (E, B, F) of topological spaces together with a continuous surjective map $\pi : E \to B$ satisfying the following property: For each $x \in B$ there exists an open neighborhood U and a homeomorphism, called a **local trivialization**,

$$\varphi_U : \pi^{-1}(U) \to U \times F$$

such that $p_1 \circ \varphi_U = \pi|_U : \pi^{-1}(U) \to U$.

In particular, for each $x \in B$, we have that $E_x := \pi^{-1}(\{x\})$ is homeomorphic to F. We call E the **total space**, B the **base space**, π the **projection map**, F the **fiber space**, and E_x the **fiber over** x. We typically denote a fiber bundle by $F \to E \xrightarrow{\pi} B$. We call a system of local trivializations $\{(U_\alpha, \varphi_{U_\alpha})\}_{\alpha \in A}$ satisfying $\bigcup_{\alpha \in A} U_\alpha = B$ a **fiber bundle atlas**.

A **section** of a fiber bundle is a continuous map $\sigma : B \to E$ such that $\pi \circ \sigma = \mathrm{id}_B$. That is, a section is a continuous assignment, to each point in the base, of an element in the fiber over that point.

Given two fiber bundles $F_i \to E_i \xrightarrow{\pi_i} B_i$, $i = 1, 2$, a **fiber bundle map** is a pair of continuous maps $\epsilon : E_1 \to E_2$ and $\beta : B_1 \to B_2$ such that

$$\beta \circ \pi_1 = \pi_2 \circ \epsilon. \tag{10.77}$$

If $B_1 = B_2$, i.e., the fiber bundles have the same base space, then we require that β is the identity map. In this case, the fiber bundle map is called a **fiber bundle isomorphism** if ϵ is a homeomorphism. Moreover, we say that the fiber bundles are **isomorphic**.

If the base space B is a paracompact topological space (as we will always assume), then a fiber bundle $\pi : E \to B$ is a fibration (see Husemoller [**Hus94**, Theorem 5.2 in Chapter 1]).

10.5.3. Examples of fiber bundles. Given two topological spaces F and B, the **product fiber bundle** is $F \to B \times F \xrightarrow{p_1} B$, where p_1 is the projection onto the first factor. Here we have a global trivialization and we may choose a fiber bundle atlas consisting of only one fiber bundle chart, which is the identity map of the total space. We say that a fiber bundle is **trivial** if it is isomorphic to a product fiber bundle.

The Möbius strip in Example 2.16 may be considered as a fiber bundle over $S^1 = [0, 2\pi]/(0 \sim 2\pi)$ with fiber $\left[-\frac{1}{2}, \frac{1}{2}\right]$.

The Hopf fibration, which is given in Example 6.34, is a fiber bundle over S^2 with fiber space S^1. The total space of the Hopf bundle is homeomorphic to S^3. On the other hand, the unit tangent bundle of S^2 has its total space homeomorphic to $\mathbb{R}P^3$.

Given a vector bundle $\pi : E \to B$ with a bundle metric (as in Definition 10.11), the associated **unit sphere bundle** has total space

$$S(E) := \{v \in E \colon \langle v, v\rangle = 1\}. \tag{10.78}$$

The projection map is the restricted projection map $\pi := \pi|_{S(E)} : S(E) \to B$. It is easy to see that the unit sphere bundle $\pi : S(E) \to B$ is a fiber bundle.

As a special case, consider a Riemannian manifold (M^n, g). For this manifold, we have the unit tangent bundle SM, which was discussed in §10.2.2.

10.5.4. Structure groups and G atlases. Let $F \to E \xrightarrow{\pi} B$ be a fiber bundle. Let G be a topological group with the identity element denoted by e and suppose that we are given a continuous action on the left by G on the fiber space F. That is, we have a continuous map $\alpha : G \times F \to F$ such that $g \cdot x := \alpha(g, x)$ satisfies the following:

(1) $e \cdot x = x$ for all $x \in F$.

(2) $g \cdot (h \cdot x) = (gh) \cdot x$ for all $g, h \in G$ and $x \in F$.

Observe that $g^{-1} \cdot (g \cdot x) = g \cdot (g^{-1} \cdot x) = x$ for all $g \in G$ and $x \in F$. For each $g \in G$, define the **left multiplication map** as

$$L_g : F \to F \quad \text{by } L_g(x) := g \cdot x. \tag{10.79}$$

We see that $L_{g^{-1}} = (L_g)^{-1}$, so that L_g is a homeomorphism of F for each $g \in G$. So G acts on the left on F by homeomorphisms.

Given a fiber bundle $F \to E \xrightarrow{\pi} B$, a covering of local trivializations $\{(U_\alpha, \varphi_\alpha)\}_{\alpha \in A}$ is called a G **atlas** for the fiber bundle if for all $\alpha, \beta \in A$ we have that the transition function

$$\varphi_\alpha \circ \varphi_\beta^{-1} : (U_\alpha \cap U_\beta) \times F \to (U_\alpha \cap U_\beta) \times F$$

is given by

$$\varphi_\alpha \circ \varphi_\beta^{-1}(x, v) = (x, L_{\mathrm{g}_{\alpha\beta}(x)}(v)), \tag{10.80}$$

where $\mathrm{g}_{\alpha\beta} : U_\alpha \cap U_\beta \to G$ is a continuous map. Here G is the **structure group** of the G atlas (and of the bundle) and each map $\mathrm{g}_{\alpha\beta}$ is a G**-transition function**.

Observe that we have the properties

$$\mathrm{g}_{\alpha\alpha} = 1, \quad \mathrm{g}_{\beta\alpha} = \mathrm{g}_{\alpha\beta}^{-1}, \quad \text{and} \quad \mathrm{g}_{\alpha\beta}\mathrm{g}_{\beta\gamma}\mathrm{g}_{\gamma\alpha} = 1. \tag{10.81}$$

The last equation is called the **cocycle condition**. We will usually assume that the G-action on F is faithful; i.e., if $g \cdot x = x$ for all $x \in F$, then $g = e$.

Conversely, if we are given a topological group G acting continuously on a topological space F, an open cover $\{U_\alpha\}_{\alpha \in A}$ of a topological space B, and continuous maps $g_{\alpha\beta} : U_\alpha \cap U_\beta \to G$ satisfying (10.81), then by starting with an initial trivialization $T_\alpha = \mathrm{id} : U_\alpha \times F \to U_\alpha \times F$, we can use the $g_{\alpha\beta}$ as transition functions to construct a fiber bundle $F \to E \xrightarrow{\pi} B$. Here, we may define the total space by

$$E := \bigsqcup_{\alpha \in A} (U_\alpha \times F) / \sim,$$

where

$$(x, f) \in U_\beta \times F \sim (x, L_{g_{\alpha\beta}(x)}(f)) \in U_\alpha \times F.$$

We leave it as an exercise to define the local trivializations $\{(U_\alpha, \varphi_\alpha)\}_{\alpha \in A}$ and show using the cocycle condition (10.81) that a well-defined fiber bundle atlas results.

Example 10.21 (Fiber bundles and their structure groups).

(1) *For a product fiber bundle $B \times F \to B$ we may take an atlas to consist of the single trivialization* $\mathrm{id} : B \times F \to B \times F$ *and the structure group to be $G = \{1\}$.*

(2) *The Möbius band $\mathbb{R} \to E \to S^1$, where $S^1 := \{\mathrm{e}^{it} : t \in \mathbb{R}\}$, may be defined by identifying the products $(S^1 - \{1\}) \times \mathbb{R}$ and $(S^1 - \{-1\}) \times \mathbb{R}$ via*

$$(\mathrm{e}^{it}, v) \sim (\mathrm{e}^{it}, v) \quad \textit{for } 0 < t < \pi,$$
$$(\mathrm{e}^{it}, v) \sim (\mathrm{e}^{it}, -v) \quad \textit{for } -\pi < t < 0.$$

Here, the structure group is $G = \mathbb{Z}_2$.

Let H be a subgroup of G, with the induced action on F. We can **reduce** the structure group G of a fiber bundle to H if there exists an H atlas.

Example 10.22. *If $\pi : E \to B$ is a rank k vector bundle, then any vector bundle atlas for E is a G atlas for $G = \mathrm{GL}(k, \mathbb{R})$, where $\mathrm{GL}(k, \mathbb{R})$ acts in the usual way on $\mathbb{R}^k$.*

By the Gram–Schmidt orthonormalization process, we have that $\mathrm{O}(k)$ is a smooth deformation retract of $\mathrm{GL}(k, \mathbb{R})$. Thus, the structure group of any rank k vector bundle can be reduced to $\mathrm{O}(k)$.

In particular, if M^n is a smooth manifold, then any vector bundle atlas for the tangent bundle TM is a $\mathrm{GL}(n, \mathbb{R})$ atlas and the structure group can be reduced to $\mathrm{O}(n)$.

10.5.5. Associated bundles. Let $F \to E \xrightarrow{\pi} B$ be a fiber bundle and let G be a structure group for E. Let F' be a topological space and assume that G acts on the left on F'. Let $\{(U_\alpha, \varphi_\alpha : \pi^{-1}(U_\alpha) \to U_\alpha \times F)\}_{\alpha \in A}$ be a fiber bundle atlas for E, used to describe the bundle's local-to-global structure. We define the **associated fiber bundle** $F' \to E' \xrightarrow{\pi'} B$ as follows. Let

$$E' := \Big(\bigsqcup_{\alpha \in A} U_\alpha \times F' \Big) \Big/ \sim,$$

where

$$(x, f') \in U_\beta \times F' \sim (x, L_{g_{\alpha\beta}(x)}(f')) \in U_\alpha \times F'$$

and $g_{\alpha\beta} : U_\alpha \cap U_\beta \to G$ satisfies (10.80). One easily checks that $\sim$ is an equivalence relation. The projection map is the unique map such that the following diagram commutes:

$$\begin{array}{ccc} U_\alpha \times F' & \longrightarrow & E' \\ {\scriptstyle p_1}\big\downarrow & & \big\downarrow{\scriptstyle \pi'} \\ U_\alpha & \hookrightarrow & B \end{array}$$

where the map $U_\alpha \times F' \to E'$ is the quotient map associated to $\sim$. Since the cocycle condition for the $g_{\alpha\beta}$ holds, we obtain a well-defined fiber bundle.

10.6. Principal bundles

The group of automorphisms of a vector space V, denoted $\mathrm{GL}(V)$, acts freely and transitively on the set of frames of V. For a given vector bundle, by bundling the sets of frames across its vector space fibers, we construct a fiber bundle upon which the Lie group $\mathrm{GL}(V)$ acts. This idea naturally extends to the concept of a principal fiber bundle, which is the focus of our discussion in this section.

A **principal bundle** (a.k.a. **principal fiber bundle**) is a fiber bundle $\pi : P \to B$ and a topological group G with a right action $\rho : P \times G \to P$ that preserves the fibers of P and such that G acts freely and transitively on each P_x, $x \in B$. That is:

(1) For each $g \in G$, the **right multiplication map**

$$R_g : P \to P$$

defined by $R_g(y) := y \cdot g := \rho(y, g)$ satisfies $R_g(P_x) \subset P_x$ for each $x \in B$.

(2) $(y \cdot g) \cdot h = y \cdot (gh)$ for all $g, h \in G$ and $y \in P$.

(3) For any $y_1, y_2 \in P_x$ there exists $g \in G$ such that $y_1 \cdot g = y_2$.

(4) If $g(y) = y$ for some $y \in P$, then $g = e$, where e is the identity element.

We denote the principal bundle with its projection map and base space by the sequence $G \to P \xrightarrow{\pi} B$.

When the base and entire spaces are smooth manifolds and the fiber is a Lie group, we will assume that all of the maps are smooth, and we call $G \to P \xrightarrow{\pi} B$ a **smooth principal bundle**.

Given two principal bundles with the same fiber and the same base space $G \to P_i \xrightarrow{\pi_i} B$, $i = 1, 2$, a **principal bundle isomorphism** is a homeomorphism $\phi : P_1 \to P_2$ such that $\pi_2 \circ \phi = \phi_1$ and

$$\phi(x \cdot g) = \phi(x) \cdot g$$

for $x \in P_1$ and $g \in G$. That is, ϕ is a fiber bundle isomorphism that is G-**equivariant**. In this case, we say that the principal bundles are **isomorphic**. We say that a principal bundle is **trivial** if it is isomorphic to the Cartesian product of the base with the fiber.

Observe that a (smooth) principal bundle $G \to P \xrightarrow{\pi} B$ is trivial if and only if it admits a (smooth) continuous section. For suppose $\sigma : B \to P$ is a (smooth) continuous section. Then the map $\iota : B \times G \to P$, defined by $\iota(b, g) = \sigma(b) \cdot g$, is a (smooth) principal bundle isomorphism.

Example 10.23 (Frame bundle of a smooth manifold). *Let M^n be a smooth manifold. A **frame** $\mathcal{F}$ at a point $x \in M^n$ is an ordered n-tuple of tangent vectors $(v_1, \ldots, v_n)$ that form a basis for T_xM. The **frame bundle** FM is the set of all frames on M^n with the projection map $p : FM \to M^n$ sending a frame to its basepoint. To see that FM is a principal $\mathrm{GL}(n, \mathbb{R})$-bundle over M^n, let $\{\mathbf{e}_i\}_{i=1}^n$ be the standard basis for $\mathbb{R}^n$. There is a one-to-one correspondence between frames at a point x and vector space isomorphisms from $\mathbb{R}^n$ to T_xM, defined by mapping $\mathcal{F} = (v_1, \ldots, v_n)$ to the isomorphism $\mathcal{F} : \mathbb{R}^n \to T_xM$ defined by*

$$\mathcal{F}(\mathbf{e}_i) = v_i \,. \tag{10.82}$$

Then the right $\mathrm{GL}(n,\mathbb{R})$ *action on* FM *is defined by*

$$R_g(\mathcal{F}) = \mathcal{F} \circ g \quad \textit{for } g \in \mathrm{GL}(n,\mathbb{R}).$$

We leave it to the reader to define a natural topology and smooth manifold structure on FM *and to further verify that* FM *is a principal bundle.*

Example 10.24 (Orthonormal frame bundle of a Riemannian manifold). *Let* (M^n, g) *be a Riemannian manifold. An* ***orthonormal frame*** $\mathcal{O}$ *at a point* $x \in M^n$ *is an ordered* n*-tuple of tangent vectors* $(v_1, \ldots, v_n)$ *that form an orthonormal basis for* T_xM*. The* ***orthonormal frame bundle*** OM *is the set of all orthonormal frames on* M^n *with the projection map* $p : OM \to M^n$ *sending an orthonormal frame to its basepoint. To see that* OM *is a principal* $\mathrm{O}(n)$*-bundle over* M^n*, we observe that there is a one-to-one correspondence between orthonormal frames at a point* x *and linear isometries from* $(\mathbb{R}^n, g_{\mathbb{E}}) \to (T_xM, g(x))$*, defined by mapping the frame* $\mathcal{O} = (v_1, \ldots, v_n)$ *to the linear isometry defined by* $\mathcal{O} : \mathbf{e}_i \mapsto v_i$*. The right* $\mathrm{O}(n)$ *action on* OM *is defined by*

$$R_g(\mathcal{O}) = \mathcal{O} \circ g \quad \textit{for } g \in \mathrm{O}(n).$$

Observe that a moving frame $\{e_i\}_{i=1}^n$ *on an open set* $\mathcal{U}$ *(see §8.3) is equivalent to a section of* OM *restricted to* $\mathcal{U}$*.*

10.7. Connections on vector bundles and principal bundles

Connections provide a means to differentiate objects in the context of geometry and topology. One of the most classical examples is the Levi-Civita connection on the tangent bundle, which differentiates vector fields along curves on a manifold. This concept, however, has broader applicability. Just as we can differentiate vector fields, we can generalize this operation to differentiate sections of more general vector bundles. Furthermore, the idea of a connection is not limited to vector bundles; it can be further generalized to principal bundles, which play a central role in the theory of fiber bundles and gauge theories in physics. In this section, we aim to introduce and explore these generalizations.

10.7.1. Connections on vector bundles.

Definition 10.25. A **linear connection** D on a vector bundle E over a smooth manifold M^n is an $\mathbb{R}$-bilinear map

$$D : TM \times C^\infty(E) \to C^\infty(E),$$

satisfying the product rule

$$D_V(f\sigma) = f \cdot D_V\sigma + V(f)\,\sigma$$

for $V \in TM$, $f \in C^\infty(M)$, and $\sigma \in C^\infty(E)$, where we used the notation $D_V\sigma := D(V,\sigma)$ and where $C^\infty(E)$ denotes the vector space of smooth sections of E.

For example, the Levi-Civita connection ∇ on the tangent bundle TM of a Riemannian manifold (M^n, g) is a linear connection.

Suppose that D_1 and D_2 are linear connections on a vector bundle E over a smooth manifold M^n. Then $D_1 - D_2 : TM \times C^\infty(E) \to C^\infty(E)$ is an $\mathbb{R}$-bilinear map and

$$(D_1 - D_2)_V(f\sigma) = f \cdot (D_1)_V\sigma + V(f)\,\sigma - \big(f \cdot (D_1)_V\sigma + V(f)\,\sigma\big) = f(D_1 - D_2)_V(\sigma). \tag{10.83}$$

This implies for $x \in M^n$ and $V \in T_xM$ that $\big((D_1 - D_2)_V\sigma\big)_x$ depends only on V and $\sigma(x)$; cf. Corollary 4.39. That is, $D_1 - D_2$ is a tensor field; in particular, $D_1 - D_2$ is a section of the vector bundle $TM^* \otimes E^* \otimes E$.

10.7.2. Connections on principal bundles over manifolds. Let M^n be a smooth manifold and let $\pi : P \to M^n$ be a smooth principal bundle with Lie group G. Given a point $y \in P$, it is contained in the fiber $P_{\pi(y)}$, which we may identify with G via the map $\rho_y : G \to P_{\pi(y)}$ defined by

$$\rho_y(g) := y \cdot g. \tag{10.84}$$

We define the **vertical space** at y by

$$V_y := T_y P_{\pi(y)}. \tag{10.85}$$

Note that $y \mapsto V_y$ is a $\dim(G)$-dimensional distribution on P.

A **connection** Γ on P is a smooth assignment to each $y \in P$ of a vector subspace H_y of T_yP such that:

(1)

$$T_yP = V_y \oplus H_y; \tag{10.86}$$

that is, T_yP is the direct sum of V_y and H_y, so that each $p \in T_yP$ may be uniquely written as $v + h$, where $v \in V_y$ and $h \in H_y$.

(2) For each $y \in P$ and $g \in G$,

$$H_{y \cdot g} = dR_g(H_y), \tag{10.87}$$

where R_g denotes right mulitplication by g.

Thus, $y \mapsto H_y$ is an n-dimensional distribution on P that is invariant under the right action of G on P. We call H_y the **horizontal subspace** at y.

Let $\mathfrak{g}$ denote the Lie algebra of G, which we identify with the tangent space T_eG at the identity. Given $A \in \mathfrak{g}$, the **fundamental vector field**

A^* on P associated to A is defined as follows. Given $y \in Y$, consider the derivative $(d\rho_y)_e : T_eG \to V_y = T_yP_{\pi(y)}$ of the map $\rho_y : G \to P_{\pi(y)}$ defined by (10.84). We define

$$A^*_y := (d\rho_y)_e(A). \tag{10.88}$$

Observe that, given $y \in P$, the map $\iota_y : \mathfrak{g} \to V_y$, defined by

$$\iota_y(A) = A^*_y, \tag{10.89}$$

is a vector space isomorphism. Note that

$$[A, B]^* = [A^*, B^*]. \tag{10.90}$$

The assignment of a Lie algebra element to its associated fundamental vector field has the following property. For $A \in \mathfrak{g}$, we have that

$$dR_g(A^*) = (\mathrm{Ad}_{g^{-1}}(A))^*. \tag{10.91}$$

To see this, we compute that

$$dR_g(A^*_{y\cdot g^{-1}})_y = dR_g\big(d(\rho_{y\cdot g^{-1}})_e(A)\big) = d(R_g \circ \rho_{y\cdot g^{-1}})_e(A).$$

Now,

$$(R_g \circ \rho_{y\cdot g^{-1}})(h) = y \cdot (g^{-1}hg) = \rho_y(g^{-1}hg) \in P_{\pi(y)}.$$

Thus,

$$d(R_g \circ \rho_{y\cdot g^{-1}})_e(A) = \frac{d}{dt}\bigg|_{t=0} \rho_y(g^{-1}\mathrm{e}^{tA}g) = (d\rho_y)_e(\mathrm{Ad}_{g^{-1}}(A)) = (\mathrm{Ad}_{g^{-1}}(A))^*_y,$$

and (10.91) follows.

A connection Γ on P may be equivalently described as follows. The **connection form** ω on P associated to the connection Γ is the $\mathfrak{g}$-valued 1-form on P defined for $X \in T_yP$ by

$$\omega(X) := A, \tag{10.92}$$

where $A \in \mathfrak{g}$ is the unique Lie algebra element such that

$$A^*_y = v_y(X), \tag{10.93}$$

where $v_y(X)$ is the vertical component of X. Evidently, $v_y(X) = \vec{0}$ if and only if $X \in H_y$ is a horizontal vector. Also, observe that since A^* is a vertical vector field on P, we have that

$$\omega(A^*_y) = A \quad \text{for all } A \in \mathfrak{g},\ y \in P. \tag{10.94}$$

In general, a $\mathfrak{g}$-valued 1-form on P is a **connection 1-form** on P if it satisfies (10.94) and

$$R^*_g\omega = \mathrm{Ad}(g^{-1}) \circ \omega \quad \text{for all } g \in G. \tag{10.95}$$

That is, for all $g \in G$ and all tangent vectors X on P, we have

$$(R^*_g\omega)(X) = \mathrm{Ad}(g^{-1})(\omega(X)). \tag{10.96}$$

To see this, we note that by (10.86), we just need to prove (10.96) for the cases where X is horizontal and where X is vertical.

(1) Let $X \in H_y$. Then the right-hand side vanishes: $\mathrm{Ad}(g^{-1})(\omega(X)) = \vec{0}$ since $\omega(X) = \vec{0}$. On the other hand, since $X \in H_y$, we have $dR_g(X) \in H_{y\cdot g}$. Thus, $(R_g^*\omega)(X) = \omega(dR_g(X)) = \vec{0}$. This proves (10.96) in this case.

(2) Let $X \in V_y$. We may assume that $X = A_y^*$, where $A \in \mathfrak{g}$. We compute using (10.91) that

$$\begin{aligned}(R_g^*\omega)_y(A^*) &= \omega_{y\cdot g}(dR_g(A^*)) = \omega_{y\cdot g}\big((\mathrm{Ad}_{g^{-1}}(A))^*\big)\\ &= \mathrm{Ad}_{g^{-1}}(A) = \mathrm{Ad}_{g^{-1}}(\omega_y(A^*))\end{aligned}$$

and (10.96) follows in this case also.

Conversely, suppose that ω is a $\mathfrak{g}$-valued 1-form on P satisfying (10.94) and (10.95). We then define

$$H_y := \{X \in T_yP : \omega(X) = \vec{0}\} \tag{10.97}$$

for all $y \in P$. We claim that the H_y are the horizontal spaces for a connection Γ on P. Firstly, let $X \in H_y$. We compute that

$$\omega_{y\cdot g}(dR_g(X)) = \mathrm{Ad}(g^{-1})(\omega(X)) = \mathrm{Ad}(g^{-1})(\vec{0}) = \vec{0}.$$

Hence, $dR_g(X) \in H_{y\cdot g}$. Thus, $dR_g(H_y) \subset H_{y\cdot g}$. Similarly, $dR_{g^{-1}}(H_{y\cdot g}) \subset H_y$. Since $dR_{g^{-1}} \circ dR_g = \mathrm{id}_{H_y}$, we see that $dR_g : H_y \to H_{y\cdot g}$ is a vector space isomorphism. Secondly, $\omega : T_yP \to \mathfrak{g}$ is surjective by (10.94). Thus,

$$\dim(H_y) = \dim(\ker(\omega_y)) = \dim(P) - \dim(\mathfrak{g}) = n$$

and hence $T_yP = V_y \oplus H_y$. We conclude that the H_y are the horizontal spaces for a connection Γ on P.

10.7.3. The curvature form.

10.7.3. The curvature form. Let ω be a connection 1-form on a principal bundle P. For each $y \in P$, associated to the direct sum decomposition $T_yP = V_y \oplus H_y$, let $\pi_H : T_yP \to H_y$ be the horizontal projection. The **curvature form** of the connection 1-form ω is defined by

$$\Omega := d\omega \circ \pi_H; \tag{10.98}$$

that is,

$$\Omega(X, Y) = d\omega\big(\pi_H(X), \pi_H(Y)\big). \tag{10.99}$$

Clearly, Ω is a $\mathfrak{g}$-valued 2-form on P.

A $\mathfrak{g}$-valued differential k-form α on P is a **tensorial form** if

$$R_g^*\alpha = \mathrm{Ad}(g^{-1}) \circ \alpha \quad \text{for all } g \in G \tag{10.100}$$

and $\alpha(X_1, \ldots, X_k) = 0$ whenever one of the X_i's is vertical.

By (10.99), $\Omega(X,Y)=0$ if X or Y is vertical. We also compute using (10.95) that

$$\begin{aligned}(R_g^*\Omega)(X,Y) &= \Omega\big(dR_g(X),dR_g(Y)\big)\\ &= d\omega\big(dR_g(\pi_H(X)),dR_g(\pi_H(Y))\big)\\ &= (R_g^*(d\omega))(\pi_H(X),\pi_H(Y))\\ &= d(R_g^*\omega)(\pi_H(X),\pi_H(Y))\\ &= d\big(\operatorname{Ad}(g^{-1})\circ\omega\big)(\pi_H(X),\pi_H(Y))\\ &= \operatorname{Ad}(g^{-1})\big(d\omega(\pi_H(X),\pi_H(Y))\big)\\ &= \operatorname{Ad}(g^{-1})(\Omega(X,Y)).\end{aligned}$$

This proves that Ω is a tensorial $\mathfrak{g}$-valued 2-form.

Proposition 10.26. *For $X,Y\in TP$,*

$$\Omega(X,Y)=d\omega(X,Y)+[\omega(X),\omega(Y)]. \tag{10.101}$$

Proof. Observe that the expressions on both sides of (10.101) are bilinear. Thus, it suffices to consider the cases where each of X and Y is vertical or horizontal.

(1) If X and Y are both horizontal, then

$$d\omega(X,Y)+[\omega(X),\omega(Y)]=d\omega(X,Y)=d\omega(\pi_H(X),\pi_H(Y))=\Omega(X,Y).$$

(2) Suppose that $X=A^*$ and $Y=B^*$ are both vertical, where $A,B\in\mathfrak{g}$. We compute using (8.33) and (10.94) that

$$\begin{aligned}d\omega(A^*,B^*) &= A^*(\omega(B^*))-B^*(\omega(A^*))-\omega([A^*,B^*])\\ &= A^*(B)-B^*(A)-\omega([A,B]^*)\\ &= -[A,B]\\ &= -[\omega(A^*),\omega(B^*)],\end{aligned}$$

where $A^*(B)$ and $B^*(A)$ are zero since they are directional derivatives of constant $\mathfrak{g}$-valued functions and where we used (10.90). On the other hand,

$$\Omega(A^*,B^*)=d\omega(\pi_H(A^*),\pi_H(B^*))=\vec{0},$$

so that (10.101) is proved in this case.

(3) Since both sides of (10.101) are antisymmetric in X and Y, for the final case it suffices to assume that $X=A^*$ is vertical and Y is horizontal.

Firstly,

$$\Omega(A^*, Y) = \vec{0} = [\omega(A^*), \omega(Y)]$$

since Y is horizontal. We compute that

$$\begin{aligned} d\omega(A^*, Y) &= A^*(\omega(Y)) - Y(\omega(B^*)) - \omega([A^*, Y]) \\ &= -\omega([A^*, Y]) \end{aligned}$$

since $\omega(Y) = 0$ and $\omega(B^*) = B$ is constant. To prove (10.101) it suffices to show that $[A^*, Y]$ is horizontal. We compute that

$$[A^*, Y] = \lim_{t \to 0} \frac{1}{t}\big(Y - dR_{e^{tA}}(Y)\big)$$

is horizontal since Y and $dR_{e^{tA}}(Y)$ are horizontal. □

10.8. The Chern–Gauss–Bonnet formula

The Gauss–Bonnet formula for even-dimensional closed Riemannian manifolds was originally proved by Allendoerfer and Weil [**All40, AW43**]. However, their proof was *extrinsic* and relied on embedding a Riemannian polyhedron into Euclidean space. Chern gave an *intrinsic* proof of the higher-dimensional Gauss–Bonnet formula for even-dimensional closed Riemannian manifolds.

In this section we discuss Chern's proof, closely following his original paper [**Che44**]. One of the key ideas of the proof is to consider the unit tangent bundle of a Riemannian manifold, which is a fiber bundle with its fiber diffeomorphic to the $(n-1)$-sphere. In the case where $n = 2$, this is not necessary and the geometry can be done directly on the manifold M^n. However, when $n := 2m \geq 4$, it is necessary to consider fiber bundles.

Chern's proof is remarkable and demonstrates deep insight and a mastery of intricate calculations using Cartan's method of moving frames. Moreover, Chern's method has had a profound influence on geometry, topology, and analysis and in particular has led to his subsequent discovery of Chern classes of complex vector bundles.

10.8.1. The Gauss–Bonnet formula in dimension 2 revisited.

A general fact is that on any smooth closed manifold M^n there exists a smooth vector field W on M^n with a single zero at a point which we call p_1 such that

$$\text{index}_{p_1}(W) = \chi(M^n), \tag{10.102}$$

where χ denotes the Euler characteristic defined by (9.6); see Alexandroff and Hopf [**AH74**, p. 550].

Now assume that (M^2, g) is a closed oriented Riemannian surface. Let

$$\Omega := K\, d\mu,$$

which is a globally defined 2-form on M^2. Let $\pi : SM \to M^2$ denote the projection map. Consider the pulled-back 2-form $\pi^*(\Omega)$ on the unit tangent bundle SM. By taking $e_1 := W/|W|$, where W is a vector field such as in the previous paragraph, we have a positively oriented orthonormal frame field $\{e_1, e_2\}$ defined on $\mathcal{U} := M^2 \setminus \{p_1\}$, where $e_2 := J(e_1)$ and

$$J : TM \to TM \tag{10.103}$$

is the almost complex structure.[7] By (8.78), we have that

$$\Omega = K\, d\mu = \mathrm{Rm}_2^1 = d\omega_2^1,$$

where we note that the 1-form ω_2^1 is defined on $\mathcal{U}$.

Let

$$\Pi := \pi^*\Omega,$$

which is a 2-form defined on SM. On $SM|_{\mathcal{U}}$ we have

$$\Pi := \pi^*(\Omega) = \pi^*(d\omega_2^1) = d(\pi^*\omega_2^1) \tag{10.104}$$

since π^* and d commute (by (8.30)). We consider the unit vector field e_1 as a section $e_1 : M^2 \to SM$ of the unit tangent bundle, which is a 3-manifold. Consider the 2-dimensional submanifold

$$N^2 := e_1(\mathcal{U}) \tag{10.105}$$

of SM. Define $\overline{\pi} := \pi|_{N^2} : N^2 \to M^2$, which is a diffeomorphism. We have that

$$\pi^*(\Omega)|_{N^2} = \overline{\pi}^*(\Omega). \tag{10.106}$$

Therefore,

$$\int_{N^2} \pi^*(\Omega) = \int_{N^2} \overline{\pi}^*(\Omega) = \int_{M^2} \Omega = \int_{M^2} K\, d\mu, \tag{10.107}$$

where the second equality is a consequence of the naturality of the integration of differential forms (see (8.22)). We now proceed to calculate the left-hand side.

The closure $\overline{N^2}$ of N^2 is not necessarily a submanifold with boundary in SM. However, as a homology 1-cycle, its 'boundary' ∂N is equal to $\chi(M^2)$

[7] Given any tangent vector W, there exists a unique tangent vector $J(W)$ satisfying $|J(W)| = |W|$, such that $\{W, J(W)\}$ is orthogonal and positively oriented if $W \neq 0$, and $J(0) = 0$. Here, J is the almost complex structure defined by (10.103).

times the fiber $S_{p_1}M$. From this we see that we should have by Stokes's theorem that

(10.108)
$$\int_{N^2} \pi^*(\Omega) = \int_{N^2} d(\pi^*\omega_2^1) = \int_{\partial N} \pi^*\omega_2^1 = \chi(M^2)\,\mathrm{L}(S_{p_1}M) = 2\pi\chi(M^2).$$

To make this proof rigorous, we consider $\mathcal{U}_\varepsilon := M^2 \setminus B_\varepsilon(p_1)$, where $\varepsilon > 0$ is less than the injectivity radius of g. Then $\mathcal{U}_\varepsilon$ is a compact 2-dimensional submanifold of M^2 with boundary $\partial\mathcal{U}_\varepsilon = \partial B_\varepsilon(p_1)$ diffeomorphic to a circle. We have that

$$N_\varepsilon^2 := e_1(\mathcal{U}_\varepsilon) \subset N^2$$

is a compact 2-dimensional submanifold of SM with its boundary diffeomorphic to a circle. Note that $\pi(\partial N_\varepsilon) = \partial B_\varepsilon(p_1)$. We define the map

$$f : \partial B_\varepsilon(p_1) \to S_{p_1}M \tag{10.109}$$

by defining $f(x)$ to be equal to the parallel translation of the vector $e_1(x)$ to the point p_1 along the unique minimal geodesic from x to p_1. With respect to the induced orientations on $\partial B_\varepsilon(p_1)$ and $S_{p_1}M$, by the definition of index we have that

$$\deg(f) = \mathrm{index}_{p_1}(W) = \chi(M^2). \tag{10.110}$$

It is in this sense that we have the heuristic result that $\partial N = \chi(M^2)S_{p_1}M$ used in (10.108). Now, by Stokes's theorem,

$$\int_{N_\varepsilon^2} \pi^*(\Omega) = \int_{\partial N_\varepsilon} \pi^*\omega_2^1. \tag{10.111}$$

Let $\gamma : [0, L] \to \partial B_\varepsilon(p_1)$ be a unit-speed counterclockwise parametrization of the embedded circle $\partial B_\varepsilon(p_1)$. Let $\widehat{\pi} := \pi|_{\partial N_\varepsilon} : \partial N_\varepsilon \to \partial B_\varepsilon(p_1)$, which is a diffeomorphism. Thus,

$$\int_{\partial N_\varepsilon} \pi^*\omega_2^1 = \int_{\partial N_\varepsilon} \widehat{\pi}^*\omega_2^1 = \int_{\partial B_\varepsilon(p_1)} \omega_2^1 \to 2\pi\chi(M^2) \tag{10.112}$$

as $\varepsilon \to 0$. So this proves the Gauss–Bonnet formula in essentially the same way as in the previous chapter. However, in dimension 4, we will have to stick to the unit tangent bundle SM to prove the Chern–Gauss–Bonnet formula.

10.8.2. The unit tangent bundle. Let (M^{2m}, g) be an even-dimensional oriented closed Riemannian manifold, where $m \geq 2$. Let

$$SM := \{V \in TM : |V| = 1\}$$

denote the unit tangent bundle, which is a $(4m-1)$-dimensional oriented smooth closed manifold (the orientation on M^{2m} induces one on TM, which in turn induces an orientation on SM). Recall from (10.102) that there exists a smooth vector field W on M^{2m} with a single isolated zero $p_1 \in M^{2m}$ and

$\text{index}_{p_1}(W) = \chi(M^{2m})$.[8] Then, we have the associated unit vector field $U : M^{2m} \setminus \{p_1\} \to SM$ defined by

$$U_x := \frac{W_x}{|W_x|} \quad \text{for } x \in M^{2m} \setminus \{p_1\}. \tag{10.113}$$

We have that the image

$$N^{2m} := U\big(M^{2m} \setminus \{p_1\}\big)$$

is an (incomplete) $2m$-dimensional smooth submanifold of SM. Note that we have that

$$N^{2m} = \lim_{\varepsilon \to 0} U\big(M^{2m} \setminus B_\varepsilon(p_1)\big) \tag{10.114}$$

is the limit of smooth manifolds with boundary. We will explain below in what sense we have $\partial N = \chi(M^{2m}) S_{p_1} M$, that is, the unit tangent space at p_1 with multiplicity $\chi(M^{2m})$.

10.8.3. The idea of transgression. Let $\{e_i\}_{i=1}^{2m}$ be a positively oriented orthonormal frame field on an open set $\mathcal{U} \subset M^{2m} \setminus \{p_1\}$, with the dual orthonormal coframe field $\{\omega^j\}_{j=1}^{2m}$. Let $\{\omega_i^j\}_{i,j=1}^{2m}$ and $\{\Omega_i^j\}_{i,j=1}^{2m} := \{\mathrm{Rm}_i^j\}_{i,j=1}^{2m}$ denote the associated connection 1-forms and curvature 2-forms on $\mathcal{U}$, respectively.

The **Pfaffian curvature** $2m$**-form** on M^{2m} is defined by

$$\mathcal{P}(\Omega) := (-1)^{m-1} \frac{1}{2^{2m} m! \pi^m} \sum_{\sigma \in S_{2m}} \operatorname{sgn}(\sigma) \Omega_{\sigma(1)}^{\sigma(2)} \wedge \Omega_{\sigma(3)}^{\sigma(4)} \wedge \cdots \wedge \Omega_{\sigma(2m-1)}^{\sigma(2m)}. \tag{10.115}$$

It is well known (see Exercise 10.12 below) that $\mathcal{P}(\Omega)$ is independent of the choice of the positively oriented orthonormal frame $\{e_i\}_{i=1}^{2m}$. Therefore, $\mathcal{P}(\Omega)$ is a globally defined (top degree) $2m$-form on M^{2m}. Because of this, the differential form $\mathcal{P}(\Omega)$ is **intrinsically** defined.

Remark 10.27. In comparison, the Pfaffian of a $2m \times 2m$ real matrix $\mathrm{A} = (a_{ij})_{i,j=1}^{2m}$ is defined by

$$\mathrm{Pf}(\mathrm{A}) := \frac{1}{2^{2m} m!} \sum_{\sigma \in S_{2m}} \operatorname{sgn}(\sigma) a_{\sigma(1)\sigma(2)} \cdots a_{\sigma(2m-1)\sigma(2m)}. \tag{10.116}$$

See Spivak [**Spi79e**, §13.3] for a discussion.

[8] In fact, we only need a smooth vector field with isolated (equivalently, finitely many) zeroes. The proof of the Chern–Gauss–Bonnet formula is essentially unchanged in this case.

Theorem 10.28 (Chern–Gauss–Bonnet formula). *If (M^{2m}, g) is an even-dimensional oriented closed Riemannian manifold, then*

$$\int_{M^{2m}} \mathcal{P}(\Omega) = \chi(M^{2m}), \tag{10.117}$$

where the Pfaffian curvature form $\mathcal{P}(\Omega)$ is defined by (10.115) *and χ denotes the Euler characteristic defined by* (9.6).

We now proceed to prove this theorem. Let $\pi : SM \to M^{2m}$ be the projection map. Chern's fundamental result hinges on showing that the $2m$-form $\pi^*\mathcal{P}(\Omega)$ is an exact differential form on the bundle SM restricted to $M^{2m} \setminus \{p_1\}$.

Firstly, write in component form

$$V =: \sum_{i=1}^{2m} V^i e_i$$

for $V \in SM|_{\mathcal{U}}$. In this way the local frame defines component functions

$$c^i : SM|_{\mathcal{U}} \to \mathbb{R}$$

where $c^i(V) := V^i$. On the unit tangent bundle $SM|_{\mathcal{U}}$ define the 1-forms $\{\theta^i\}_{i=1}^{2m}$ by

$$\theta^i := dc^i + \sum_{j=1}^{2m} c^j \pi^* \omega_j^i; \tag{10.118}$$

that is,

$$(\theta^i)_V := (dc^i)_V + \sum_{j=1}^{2m} V^j (\pi^* \omega_j^i)_V. \tag{10.119}$$

Since

$$\sum_{i=1}^{2m} (c^i)^2 = 1, \tag{10.120}$$

we have $\sum_{i=1}^{2m} c^i \, dc^i = 0$, which implies that the 1-forms $\{\theta^i\}_{i=1}^{2m}$ satisfy the linear relation

$$\sum_{i=1}^{2m} c^i \theta^i = 0 \tag{10.121}$$

at each point in SM, since $\pi^*\omega_j^i$ is antisymmetric in i and j.

Lemma 10.29. *The* 1*-forms θ^i on the unit tangent bundle SM are vertical in the sense that they vanish when applied to horizontal vectors. That is, θ^i is the vertical projection of the exact* 1*-form dc^i.*

Proof. Firstly, any horizontal vector on SM can be described as follows. Let $V(t)$ be a parallel unit vector field along a path $\gamma(t)$ in M^n. Then $V(t)$ is a path in SM with $\pi(V(t)) = \gamma(t)$. We have that $V'(t) \in T_{V(t)}SM$ is a horizontal vector field along the path $V(t)$ in SM. Secondly, we compute that

$$\begin{aligned}\theta^i(V'(t)) &= dc^i(V'(t)) + \sum_{j=1}^{2m} V^j(t)\pi^*\omega_j^i(V'(t)) \\ &= \frac{dV^i}{dt}(t) + \sum_{j=1}^{2m} V^j(t)\omega_j^i(\gamma'(t)) \\ &= 0,\end{aligned} \tag{10.122}$$

where the last equality is by (8.63) and since $V(t)$ is parallel along $\gamma(t)$.

That θ^i is the vertical projection of the 1-form dc^i follows from $\pi^*\omega_j^i$ being horizontal in the sense that it vanishes when applied to any horizontal vector. □

Remark 10.30. Since the fibers of SM are $(n-1)$-dimensional, the vertical vector fields $\{\theta^i\}_{i=1}^n$ are linearly dependent; formula (10.121) exhibits that coefficients for the linear dependence are given by $\{c^i\}_{i=1}^n$.

Suppose that $\{\overline{e}_i\}_{i=1}^n$ is another positively oriented orthonormal frame field on $\mathcal{U}$. Then

$$\overline{e}_i := \sum_{j=1}^n a_i^j e_j, \tag{10.123}$$

where $\mathrm{A} := (a_i^j)_{i,j=1}^n : \mathcal{U} \to \mathrm{SO}(n)$; i.e., $\mathrm{A}^{-1} = \mathrm{A}^{\mathrm{T}}$ and $\det(\mathrm{A}) = 1$. Moreover, $e_i = \sum_{j=1}^n a_j^i \,\overline{e}_j$ and the dual 1-forms are related by

$$\overline{\omega}^j = \sum_{i=1}^n a_j^i \omega^i$$

and $\omega^i = \sum_{i=1}^n a_j^i \overline{\omega}^j$. Recall that from the first Cartan structure equation $d\omega^i = \sum_{j=1}^n \omega^j \wedge \omega_j^i$ (see (8.71)) we have that the connection 1-forms for $\{\overline{e}_i\}_{i=1}^n$ satisfy

$$\overline{\omega}_i^j = \sum_{k,\ell=1}^n a_i^k a_j^\ell \omega_k^\ell - \sum_{k=1}^n a_i^k da_j^k \tag{10.124}$$

(note that $\sum_{k=1}^n a_i^k da_j^k = -\sum_{k=1}^n a_j^k da_i^k$). Using this, one calculates that the component functions, the 1-forms $\overline{\theta}^i$, and the curvature 2-forms for

$\{\overline{e}_i\}_{i=1}^n$ satisfy

(10.125)
$$\overline{c}^i = \sum_{j=1}^n a_i^j c^j$$

(where $a_j^i := a_j^i \circ \pi : SM|_{\mathcal{U}} \to \mathbb{R}$),

(10.126)
$$\overline{\theta}^i = \sum_{j=1}^n a_i^j \theta^j,$$

and

(10.127)
$$\overline{\Omega}_i^j = \sum_{k,\ell=1}^n a_i^k a_j^\ell \Omega_k^\ell,$$

respectively, where henceforth we will often use the Einstein summation convention. Let $\mathrm{A} := (a_i^j)_{i,j=1}^n$ and $\Omega := (\Omega_i^j)_{i,j=1}^n$. We may write (10.127) as

$$\overline{\Omega} = \mathrm{A}\Omega\mathrm{A}^{\mathrm{T}}.$$

By taking the exterior derivative of (10.118), we obtain

(10.128)
$$\begin{aligned} d\theta^i &= \sum_{j=1}^n \left(dc^j \wedge \pi^*\omega_j^i + c^j \pi^* d\omega_j^i\right) \\ &= \sum_{j=1}^n \left(\left(\theta^j - \sum_{k=1}^n c^k \pi^* \omega_k^j\right) \wedge \pi^*\omega_j^i + c^j \pi^* \left(\Omega_j^i + \sum_{k=1}^n \omega_j^k \wedge \omega_k^i\right)\right) \\ &= \sum_{j=1}^n \left(\theta^j \wedge \pi^*\omega_j^i + c^j \pi^* \Omega_j^i\right), \end{aligned}$$

where the second equality follows from the second Cartan structure equation (8.77). This equation holds as 2-forms on $SM|_{\mathcal{U}}$.

Denote $\Omega_i^j := \pi^*\Omega_i^j$, where we pulled back the 2-form Ω_i^j on M^n to the unit tangent bundle SM, while using the same notation for simplicity. As we will see below, motivated by the Cartan structure equations, we define two sets of differential forms for $0 \leq k \leq m-1$ by

(10.129)
$$\Phi_k = \sum_{\sigma \in S_{2m}} \operatorname{sgn}(\sigma) c^{\sigma(1)} \theta^{\sigma(2)} \wedge \cdots \wedge \theta^{\sigma(2m-2k)} \wedge \Omega_{\sigma(2m-2k+1)}^{\sigma(2m-2k+2)} \wedge \cdots \wedge \Omega_{\sigma(2m-1)}^{\sigma(2m)}$$

and

(10.130)

$$
\begin{aligned}
\Psi_k
&= \sum_{\sigma \in S_{2m}} \operatorname{sgn}(\sigma) \Omega^{\sigma(2)}_{\sigma(1)} \wedge \theta^{\sigma(3)} \wedge \cdots \wedge \theta^{\sigma(2m-2k)} \wedge \Omega^{\sigma(2m-2k+2)}_{\sigma(2m-2k+1)} \wedge \cdots \wedge \Omega^{\sigma(2m)}_{\sigma(2m-1)} \\
&= \sum_{\sigma \in S_{2m}} \operatorname{sgn}(\sigma) \theta^{\sigma(1)} \wedge \cdots \wedge \theta^{\sigma(2m-2k-2)} \wedge \Omega^{\sigma(2m-2k)}_{\sigma(2m-2k-1)} \wedge \cdots \wedge \Omega^{\sigma(2m)}_{\sigma(2m-1)}.
\end{aligned}
$$

The forms Φ_k are smooth $(2m-1)$-forms defined on all of SM and the forms Ψ_k are smooth $2m$-forms defined on all of SM. This is because, by using (10.125), (10.126), and (10.127), one can show that each of these forms is independent of the choice of the positively oriented local orthonormal frame field $\{e_i\}_{i=1}^{2m}$; see Exercise 10.13. Observe that

$$
(-1)^{m-1} \frac{1}{2^{2m} m! \pi^m} \Psi_{m-1} = \pi^* \mathcal{P}(\Omega) \tag{10.131}
$$

is the Pfaffian curvature form.

Lemma 10.31. *For each $0 \leq k \leq m-1$ we have*

$$
d\Phi_k = \Psi_{k-1} + \frac{2m-2k-1}{2(k+1)} \Psi_k, \tag{10.132}
$$

where $\Psi_{-1} := 0$.

Proof. To present expressions more concisely, we will omit the wedge product symbol $\wedge$ and abbreviate the summation sign as $\sum_\sigma := \sum_{\sigma \in S_{2m}}$ in our notation from now on. Let $\mathrm{s}(\sigma) := \operatorname{sgn}(\sigma)$ and $\sigma_i := \sigma(i)$. Taking the exterior derivative of (10.129) yields

$$
\begin{aligned}
d\Phi_k = &\sum_\sigma \mathrm{s}(\sigma) dc^{\sigma_1} \theta^{\sigma_2} \cdots \theta^{\sigma_{2m-2k}} \Omega^{\sigma_{2m-2k+2}}_{\sigma_{2m-2k+1}} \cdots \Omega^{\sigma_{2m}}_{\sigma_{2m-1}} \\
&+ \sum_\sigma \mathrm{s}(\sigma)(2m-2k-1) c^{\sigma_1} d\theta^{\sigma_2} \theta^{\sigma_3} \cdots \theta^{\sigma_{2m-2k}} \Omega^{\sigma_{2m-2k+2}}_{\sigma_{2m-2k+1}} \cdots \Omega^{\sigma_{2m}}_{\sigma_{2m-1}} \\
&- \sum_\sigma \mathrm{s}(\sigma) k c^{\sigma_1} \theta^{\sigma_2} \cdots \theta^{\sigma_{2m-2k}} d\Omega^{\sigma_{2m-2k+2}}_{\sigma_{2m-2k+1}} \Omega^{\sigma_{2m-2k+4}}_{\sigma_{2m-2k+3}} \cdots \Omega^{\sigma_{2m}}_{\sigma_{2m-1}}.
\end{aligned}
$$

Because of the independence of this formula for $d\Phi_k$ from the choice of the positively oriented local orthonormal frame field $\{e_i\}_{i=1}^{2m}$, given any point $x \in M^{2m}$, we may assume that $\{e_i\}_{i=1}^{2m}$ is defined in a neighborhood of x and satisfies $\omega_i^j(x) = 0$ for all $1 \leq i, j \leq 2m$. In the following, all of our calculations are carried out at x and are done with respect to such a frame.[9] By the second Bianchi identity, we have that

$$
d\Omega_i^j = \omega_i^k \Omega_k^j - \omega_k^j \Omega_i^k = 0
$$

[9] This is a key simplification for the subsequent calculations. If we were to compute using a general frame instead, the resulting computations would be significantly more complex.

at x. By (10.118), we have $dc^{\sigma_1} = \theta^{\sigma_1}$ and by (10.128) we have $d\theta^{\sigma_2} = c^j \Omega_j^{\sigma_2}$. Thus,

$$d\Phi_k = \sum_\sigma s(\sigma)\theta^{\sigma_1}\theta^{\sigma_2}\cdots\theta^{\sigma_{2m-2k}}\Omega^{\sigma_{2m-2k+2}}_{\sigma_{2m-2k+1}}\cdots\Omega^{\sigma_{2m}}_{\sigma_{2m-1}}$$
$$+\sum_{\sigma,j} s(\sigma)(2m-2k-1)c^{\sigma_1}c^j\Omega_j^{\sigma_2}\theta^{\sigma_3}\cdots\theta^{\sigma_{2m-2k}}\Omega^{\sigma_{2m-2k+2}}_{\sigma_{2m-2k+1}}\cdots\Omega^{\sigma_{2m}}_{\sigma_{2m-1}}.$$

Since the first term on the right-hand side is equal to Ψ_{k-1}, Lemma 10.31 is a consequence of the following.

Claim 10.32.

(10.133)
$$\Psi_k = 2(k+1)\sum_{\sigma\in S_{2m},j\neq 2} s(\sigma)c^{\sigma_1}c^{\sigma_j}\Omega^{\sigma_2}_{\sigma_j}\theta^{\sigma_3}\cdots\theta^{\sigma_{2m-2k}}\Omega^{\sigma_{2m-2k+2}}_{\sigma_{2m-2k+1}}\cdots\Omega^{\sigma_{2m}}_{\sigma_{2m-1}}.$$

To prove this claim, the idea is to use the identities $\sum_j (c^{\sigma_j})^2 = 1$ and $\sum_j c^{\sigma_j}\theta^{\sigma_j} = 0$ to show that the right-hand side of (10.133) is equal to

$$\Psi_k = \sum_\sigma s(\sigma)\Omega^{\sigma_2}_{\sigma_1}\theta^{\sigma_3}\cdots\theta^{\sigma_{2m-2k}}\Omega^{\sigma_{2m-2k+2}}_{\sigma_{2m-2k+1}}\cdots\Omega^{\sigma_{2m}}_{\sigma_{2m-1}}.$$

Note that for $j\neq 2$, we have $j=1$, $j\in\{3,\dots,2m-2k\}$, or $j\in\{2m-2k+1,\dots,2m\}$. Denoting the summand on the right-hand side of (10.133) by $S_{\sigma,j}$, we have that the right-hand side is equal to $2(k+1)$ times

$$\sum_\sigma S_{\sigma,1} + \sum_{\sigma,3\le j\le 2m-2k} S_{\sigma,j} + \sum_{\sigma,2m-2k+1\le j\le 2m} S_{\sigma,j}. \tag{10.134}$$

The last term vanishes (see Exercise 10.14 below), so this expression is equal to

$$P_k + 2(m-k-1)\Sigma_k,$$

where

$$P_k := \sum_{\sigma\in S_{2m}} s(\sigma)(c^{\sigma_1})^2\Omega^{\sigma_2}_{\sigma_1}\theta^{\sigma_3}\cdots\theta^{\sigma_{2m-2k}}\Omega^{\sigma_{2m-2k+2}}_{\sigma_{2m-2k+1}}\cdots\Omega^{\sigma_{2m}}_{\sigma_{2m-1}} \tag{10.135}$$

and

$$\Sigma_k := \sum_{\sigma\in S_{2m}} s(\sigma)c^{\sigma_1}c^{\sigma_3}\Omega^{\sigma_2}_{\sigma_3}\theta^{\sigma_3}\cdots\theta^{\sigma_{2m-2k}}\Omega^{\sigma_{2m-2k+2}}_{\sigma_{2m-2k+1}}\cdots\Omega^{\sigma_{2m}}_{\sigma_{2m-1}}. \tag{10.136}$$

Thus Claim 10.32 follows from showing that

$$\Psi_k = 2(k+1)\big(P_k + 2(m-k-1)\Sigma_k\big). \tag{10.137}$$

To see this, let

$$T_k := \sum_{\sigma\in S_{2m}} s(\sigma)(c^{\sigma_3})^2\Omega^{\sigma_2}_{\sigma_1}\theta^{\sigma_3}\cdots\theta^{\sigma_{2m-2k}}\Omega^{\sigma_{2m-2k+2}}_{\sigma_{2m-2k+1}}\cdots\Omega^{\sigma_{2m}}_{\sigma_{2m-1}}. \tag{10.138}$$

We observe, using (10.120), that

$$\begin{aligned}P_k &= \sum_{\sigma\in S_{2m}} \mathrm{s}(\sigma)\Big(1-\sum_{j\neq 1}(c^{\sigma_j})^2\Big)\Omega^{\sigma_2}_{\sigma_1}\,\theta^{\sigma_3}\cdots\theta^{\sigma_{2m-2k}}\,\Omega^{\sigma_{2m-2k+2}}_{\sigma_{2m-2k+1}}\cdots\Omega^{\sigma_{2m}}_{\sigma_{2m-1}}\\ &= \Psi_k - P_k - (2m-2k-2)T_k - 2kP_k,\end{aligned}$$

where the last three terms on the right-hand side correspond to $j=2$, $3\le j\le 2m-2k$, and $2m-2k+1\le j\le 2m$, respectively. Thus,

$$\Psi_k = 2(k+1)P_k + 2(m-k-1)T_k. \tag{10.139}$$

On the other hand, using (10.121), we compute that

$$\begin{aligned}\Sigma_k &= -\sum_{\sigma,j\neq 3} \mathrm{s}(\sigma)\,c^{\sigma_1}\,\Omega^{\sigma_2}_{\sigma_3}\,c^{\sigma_j}\,\theta^{\sigma_j}\,\theta^{\sigma_4}\cdots\theta^{\sigma_{2m-2k}}\,\Omega^{\sigma_{2m-2k+2}}_{\sigma_{2m-2k+1}}\cdots\Omega^{\sigma_{2m}}_{\sigma_{2m-1}}\\ &= T_k - (2k+1)\Sigma_k,\end{aligned}$$

where, on the right-hand side of the second line, the first term corresponds to $j=1$, the second term corresponds to the $2k+1$ indices $j=2$ and $2m-2k+1\le j\le 2m$, and for $4\le j\le 2m-2k$ the terms vanish since $\theta^{\sigma_j}\wedge\theta^{\sigma_j}=0$. Thus,

$$T_k = 2(k+1)\Sigma_k. \tag{10.140}$$

Equation (10.137) follows directly from (10.139) and (10.140). This completes the proof of Claim 10.32 and consequently that of Lemma 10.31. □

Lemma 10.33. *For each $0\le k\le m-1$ we have*

$$\Psi_k = \sum_{j=0}^{k}\frac{(-1)^j\,2^{j+1}(k+1)k\cdots(k-j+1)}{(2m-2k+2j-1)(2m-2k+2j-3)\cdots(2m-2k-1)}d\Phi_{k-j}, \tag{10.141}$$

which is an exact form.

Proof. We prove the formulas by induction on k. The base case holds since formula (10.141) for $k=0$, which says that $\Psi_0=\frac{2}{2m-1}d\Phi_0$, follows from (10.132) with $k=0$. For the inductive hypothesis, assume that (10.141) holds for some k with $0\le k\le m-2$. Then (10.132) implies that

$$\begin{aligned}&\Psi_{k+1}-\frac{2(k+2)}{2m-2k-3}d\Phi_{k+1}\\ &= -\frac{2(k+2)}{2m-2k-3}\Psi_k\\ &= \sum_{j=0}^{k}\frac{(-1)^{j+1}\,2^{j+2}(k+2)(k+1)\cdots(k-j+1)}{(2m-2k+2j-1)(2m-2k+2j-3)\cdots(2m-2k-3)}d\Phi_{k-j}.\end{aligned}$$

Setting $\bar{\jmath} := j+1$, we obtain

$$\begin{aligned}\Psi_{k+1} &- \frac{2(k+2)}{2m-2k-3} d\Phi_{k+1} \\ &= \sum_{\bar{\jmath}=1}^{k+1} \frac{(-1)^{\bar{\jmath}} 2^{\bar{\jmath}+1}((k+1)+1)(k+1)\cdots((k+1)-\bar{\jmath}+1)}{(2m-2(k+1)+2\bar{\jmath}-1)\cdots(2m-2(k+1)-1)} d\Phi_{k+1-\bar{\jmath}}.\end{aligned}$$

This proves (10.141) for k replaced by $k+1$, which is the desired inductive conclusion. Therefore the lemma follows from induction. □

We are now in position to complete the proof of the Chern–Gauss–Bonnet formula. By (10.131) and (10.141) with $k = m-1$, we have that

$$\begin{aligned}\mathcal{P}(\Omega) &= \frac{(-1)^{m-1}}{2^{2m} m! \pi^m} \sum_{j=0}^{m-1} \frac{(-1)^j 2^{j+1} m(m-1)\cdots(m-j)}{(2j+1)(2j-1)\cdots 1} d\Phi_{m-1-j} \\ &= \frac{1}{\pi^m} \sum_{j=0}^{m-1} \frac{(-1)^{m-1+j}}{(m-1-j)! 2^{2m-j-1}(2j+1)(2j-1)\cdots 1} d\Phi_{m-1-j} \\ &= \frac{1}{\pi^m} \sum_{i=0}^{m-1} \frac{(-1)^i}{i! 2^{m+i}(2m-2i-1)(2m-2i-3)\cdots 1} d\Phi_i.\end{aligned}$$

That is,

$$\pi^* \mathcal{P}(\Omega) = d\Pi, \tag{10.142}$$

where

$$\Pi := \frac{1}{\pi^m} \sum_{i=0}^{m-1} \frac{(-1)^i}{i! 2^{m+i}(2m-2i-1)(2m-2i-3)\cdots 1} \Phi_i. \tag{10.143}$$

Let

$$N_\varepsilon^{2m} := U\big(M^{2m} \setminus B_\varepsilon(p_1)\big) \subset SM. \tag{10.144}$$

By Stokes's theorem, we have that

$$\begin{aligned}\int_{M^{2m}} \mathcal{P}(\Omega) &= \int_{N^{2m}} \pi^* \mathcal{P}(\Omega) \\ &= \lim_{\varepsilon \to 0} \int_{N_\varepsilon^{2m}} \pi^* \mathcal{P}(\Omega) \\ &= \lim_{\varepsilon \to 0} \int_{\partial N_\varepsilon} \Pi.\end{aligned} \tag{10.145}$$

We will compute the right-hand side of (10.145) with the help of the following.

Claim 10.34. *For each* $1 \le i \le m-1$,

$$\lim_{\varepsilon \to 0} \int_{\partial N_\varepsilon} \Phi_i = 0. \tag{10.146}$$

To see this, recall from (10.129) that

$$\Phi_i = \sum_\sigma \mathrm{s}(\sigma) c^{\sigma_1} \theta^{\sigma_2} \cdots \theta^{\sigma_{2m-2i}} \pi^* \Omega^{\sigma_{2m-2i+2}}_{\sigma_{2m-2i+1}} \cdots \pi^* \Omega^{\sigma_{2m}}_{\sigma_{2m-1}}, \tag{10.147}$$

where we reinserted the pull-back π^* before each $\Omega^{\sigma_{2j}}_{\sigma_{2j-1}}$ in our notation. When $i \ge 1$, we have that the expression (10.147) for Φ_i contains at least one curvature 2-form. Since each curvature 2-form $\pi^* \Omega^{\sigma_{2j}}_{\sigma_{2j-1}}$ vanishes when applied to vertical vectors, and in particular to vectors tangent to $S_{p_1} M$, Claim 10.34 follows.

By Claim 10.34, we have that

$$\begin{aligned} \int_{M^{2m}} \mathcal{P}(\Omega) &= \frac{1}{2^m \pi^m (2m-1)(2m-3)\cdots 1} \lim_{\varepsilon \to 0} \int_{\partial N_\varepsilon} \Phi_0 \\ &= \frac{(m-1)!}{2\pi^m (2m-1)!} \lim_{\varepsilon \to 0} \int_{\partial N_\varepsilon} \Phi_0. \end{aligned} \tag{10.148}$$

By (10.129) with $k = 0$, we have that the global $(2m-1)$-form Φ_0 on SM is given locally on $SM|_{\mathcal{U}}$ by

$$\begin{aligned} \Phi_0 &= \sum_{\sigma \in S_{2m}} \operatorname{sgn}(\sigma) c^{\sigma(1)} \theta^{\sigma(2)} \wedge \cdots \wedge \theta^{\sigma(2m)} \\ &= (2m-1)! \sum_{j=1}^{2m} (-1)^{j-1} c^j \theta^1 \wedge \cdots \widehat{\theta^j} \cdots \wedge \theta^{2m}, \end{aligned}$$

where $\widehat{\theta^j}$ denotes that we omit θ^j from the wedge product. We also have that the volume $(2m-1)$-form of the unit sphere $S_{p_1} M$ in $T_{p_1} M$ is given by

$$d\mu_{S_{p_1} M} = \sum_{j=1}^{2m} (-1)^{j-1} c^j \theta^1 \wedge \cdots \widehat{\theta^j} \cdots \wedge \theta^{2m},$$

where the (vertical) differential 1-forms θ^i on the right-hand side are restricted to $S_{p_1} M$. Therefore, on $S_{p_1} M$ we have that $\Phi_0 = (2m-1)! d\mu_{S_{p_1} M}$, so that

$$\int_{S_{p_1} M} \Phi_0 = (2m-1)! \operatorname{Vol}(S_{p_1} M) = \frac{(2m-1)! 2\pi^m}{(m-1)!}. \tag{10.149}$$

The main remaining step in the proof of the Chern–Gauss–Bonnet formula is the following claim. This claim is natural in the sense that, intuitively, as $\varepsilon \to 0$, ∂N_ε limits to $S_{p_1} M$ with multiplicity $\chi(M^{2m})$.

Claim 10.35. *We have that*

$$\lim_{\varepsilon\to 0}\int_{\partial N_\varepsilon}\Phi_0=\mathrm{index}_{p_1}(W)\int_{S_{p_1}M}\Phi_0=\chi(M^{2m})\int_{S_{p_1}M}\Phi_0, \tag{10.150}$$

where ∂N_ε is defined by (10.144)*, the orientation on $S_{p_1}M$ is induced by N^{2m}, and the last equality is a consequence of the Poincaré–Hopf theorem.*

Proof of Claim 10.35. Recall that W is a smooth vector field on M^{2m} with a single isolated zero at the point p_1 and that $U=\frac{W}{|W|}$ is a unit vector field on $M^{2m}\setminus\{p_1\}$. Assume that $\varepsilon<\mathrm{inj}(g)$. Then N_ε^{2m} is a $2m$-dimensional smooth submanifold (with boundary) of SM. We have that its boundary $\partial N_\varepsilon=U(\partial B_\varepsilon(p_1))=U(S_\varepsilon(p_1))$ is diffeomorphic to S^{2m-1}. Define the map

$$\rho_\varepsilon:\partial N_\varepsilon\to S_{p_1}M$$

as follows. Let $U_x\in\partial N_\varepsilon$, so that $x=\pi(U)\in S_\varepsilon(p_1)$. Since $\varepsilon<\mathrm{inj}(g)$, there exists a unique minimal geodesic $\gamma:[0,\varepsilon]\to M^{2m}$ from x to p_1. Let $\rho_\varepsilon(U_x)$ be the parallel transport to p_1 of U_x along γ, so that $\rho_\varepsilon(U_x)\in S_{p_1}M$.

By the Definition 9.13 of the index of a vector field, it is easy to see that the degree of the map ρ_ε is equal to the index of W at p_1, which in turn is equal to the Euler characteristic $\chi(M^{2m})$. We also have that

$$\lim_{\varepsilon\to 0}\int_{\partial N_\varepsilon}\Phi_0=\lim_{\varepsilon\to 0}\int_{\partial N_\varepsilon}\rho_\varepsilon^*(\Phi_0). \tag{10.151}$$

The claim now follows from the change of variable (pull-back naturality) formula

$$\int_{\partial N_\varepsilon}\rho_\varepsilon^*(\Phi_0)=\deg(\rho_\varepsilon)\int_{S_{p_1}M}\Phi_0=\chi(M^{2m})\int_{S_{p_1}M}\Phi_0. \tag{10.152}$$

□

Finally, from (10.148), Claim 10.35, and (10.149), we conclude that

$$\int_{M^{2m}}\mathcal{P}(\Omega)=\frac{(m-1)!}{2\pi^m(2m-1)!}\chi(M^{2m})\int_{S_{p_1}M}\Phi_0=\chi(M^{2m}).$$

This completes the proof of the Chern–Gauss–Bonnet theorem (Theorem 10.28).

10.9. Index theorems on 4-manifolds

In this section, for 4-dimensional Riemannian manifolds, we survey the Hirzebruch signature theorem (see [**Hir95**]) and re-express the Chern–Gauss–Bonnet formula.

10.9.1. (Anti-)self-dual decomposition of 2-forms. Let (M^4, g) be an oriented closed Riemannian 4-manifold. By (11.200), we have that the Hodge star operator $* : \Lambda^2 T^*M \to \Lambda^2 T^*M$ satisfies

$$\langle *\gamma, \eta \rangle \, d\mu = *\gamma \wedge *\eta = *\eta \wedge *\gamma = \langle *\eta, \gamma \rangle$$

for $\gamma, \eta \in \Lambda^2 T_x^* M$, so that $*$ is a self-adjoint operator acting on 2-forms on M^4. By (11.201), we have

$$*^2 = \mathrm{id}_{\Lambda^2 T^* M} \, .$$

Therefore, the eigenvalues λ of $*$ are real and satisfy $\lambda^2 = 1$, so that they are equal to $+1$ and -1. Define the eigen-subbundles associated to the $+1$ and -1 eigenvalues to be $\Lambda_+^2 T^*M$ and $\Lambda_-^2 T^*M$, respectively. We thus have the orthogonal decomposition of the vector space of 2-forms on M^4:

$$\Lambda^2 T^*M = \Lambda_+^2 T^*M \oplus \Lambda_-^2 T^*M \, . \tag{10.153}$$

Let $\{e_1, e_2, e_3, e_4\}$ be a positively oriented orthonormal frame at a point $x \in M^4$, and let $\{\omega^1, \omega^2, \omega^3, \omega^4\}$ denote its dual orthonormal coframe. We have that

$$*(\omega^a \wedge \omega^b) = \omega^c \wedge \omega^d$$

whenever a, b, c, d is an even permutation of $1, 2, 3, 4$. Thus, we have that a basis of 2-forms for $\Lambda_+^2 T_x^* M$ is given by

(10.154)
$$\alpha_1 := \omega^1 \wedge \omega^2 + \omega^3 \wedge \omega^4, \quad \alpha_2 := \omega^1 \wedge \omega^3 + \omega^4 \wedge \omega^2, \quad \alpha_3 := \omega^1 \wedge \omega^4 + \omega^2 \wedge \omega^3$$

and that a basis of 2-forms for $\Lambda_-^2 T_x^* M$ is given by

(10.155)
$$\alpha_4 := \omega^1 \wedge \omega^2 - \omega^3 \wedge \omega^4, \quad \alpha_5 := \omega^1 \wedge \omega^3 - \omega^4 \wedge \omega^2, \quad \alpha_6 := \omega^1 \wedge \omega^4 - \omega^2 \wedge \omega^3 .$$

It is easy to see that all of the six 2-forms above are orthogonal to each other and have the same nonzero length $\sqrt{2}$. Hence, $\{\alpha_1, \alpha_2, \alpha_3\}$ forms a basis for $\Lambda_+^2 T_x^* M$, and $\{\alpha_4, \alpha_5, \alpha_6\}$ forms an orthogonal basis for $\Lambda_-^2 T_x^* M$.

10.9.2. Block decomposition of the curvature operator. Using the orthogonal decomposition (10.153), we can decompose the curvature operator $\mathrm{Rm} : \Lambda^2 T^*M \to \Lambda^2 T^*M$, which is self-adjoint, as

$$\mathrm{Rm} = \begin{pmatrix} A & B \\ B^{\mathrm{T}} & C \end{pmatrix}, \tag{10.156}$$

where

$$A : \Lambda_+^2 T^*M \to \Lambda_+^2 T^*M, \; B : \Lambda_+^2 T^*M \to \Lambda_-^2 T^*M, \; C : \Lambda_-^2 T^*M \to \Lambda_-^2 T^*M$$

and $B^{\mathrm{T}} : \Lambda_-^2 T^*M \to \Lambda_+^2 T^*M$ denotes the transpose of B. Here, A and C are self-adjoint.

Recall by (7.84) that in any dimension n, the curvature 4-tensor Rm has the decomposition

$$\mathrm{Rm} = \frac{R}{2n(n-1)} g \odot g + \frac{1}{n-2} \overset{\circ}{\mathrm{Rc}} \odot g + \mathrm{Wy}, \tag{10.157}$$

where

$$(\alpha \odot \beta)_{ijk\ell} := \alpha_{i\ell}\beta_{jk} + \alpha_{jk}\beta_{i\ell} - \alpha_{ik}\beta_{j\ell} - \alpha_{j\ell}\beta_{ik}$$

for symmetric 2-tensors α and β. By (10.157), since $n = 4$, we have that

$$\mathrm{Rm} = \frac{R}{24} g \odot g + \frac{1}{2} \overset{\circ}{\mathrm{Rc}} \odot g + \mathrm{Wy}\,. \tag{10.158}$$

Observe for any 2-form α that

$$(g \odot g)(\alpha)_{ij} = \frac{1}{2}(g \odot g)_{ijkl}\alpha_{lk} = (g_{i\ell}g_{jk} - g_{ik}g_{j\ell})\alpha_{lk} = 2\alpha_{ij}\,,$$

so that $g \odot g : \Lambda^2 T^*M \to \Lambda^2 T^*M$ is given by $g \odot g = 2\,\mathrm{id}$. In local coordinates, (10.158) says for $n = 4$ that

$$R_{ijk\ell} = -\frac{R}{12}\left(g_{i\ell}g_{jk} - g_{ik}g_{j\ell}\right) + \frac{1}{2}\left(\overset{\circ}{R}_{i\ell}g_{jk} + g_{i\ell}\overset{\circ}{R}_{jk} - \overset{\circ}{R}_{ik}g_{j\ell} - g_{ik}\overset{\circ}{R}_{j\ell}\right) + W_{ijk\ell}\,, \tag{10.159}$$

where $\overset{\circ}{R}_{ij}$ denote the components of the trace-free Ricci tensor $\overset{\circ}{\mathrm{Rc}}$.

As an endomorphism of $\Lambda^2 T^*M$, the Weyl tensor commutes with the Hodge star operator:

$$\mathrm{Wy} \circ * = * \circ \mathrm{Wy} : \Lambda^2 T^*M \to \Lambda^2 T^*M\,,$$

which follows from the identities of the Weyl tensor such as $W_{1212} = W_{3434}$. This implies that $\mathrm{Wy}(\Lambda^2_\pm T^*M) \subset \Lambda^2_\pm T^*M$. Thus, the Weyl tensor Wy may be written in block form as

$$\mathrm{Wy} = \begin{pmatrix} \mathrm{Wy}_+ & 0 \\ 0 & \mathrm{Wy}_- \end{pmatrix}, \tag{10.160}$$

where $\mathrm{Wy}_\pm : \Lambda^2_\pm T^*M \to \Lambda^2_\pm T^*M$ denote the self-dual and anti-self-dual parts of the Weyl tensor, respectively.

To simplify the calculations, we assume that the positively oriented orthonormal frame $\{e_1, e_2, e_3, e_4\}$ consists of eigenvectors for the Ricci tensor. We compute that

$$\left(\overset{\circ}{\mathrm{Rc}} \odot g\right)(\alpha)_{ij} = \overset{\circ}{R}_{ik}\alpha_{kj} - \overset{\circ}{R}_{jk}\alpha_{ki} = \left(\overset{\circ}{R}_{ii} + \overset{\circ}{R}_{jj}\right)\alpha_{ij} \tag{10.161}$$

for any 2-form α. Let a, b, c, d be an even permutation of $1, 2, 3, 4$. Let $\alpha_\pm = \omega^a \wedge \omega^b \pm \omega^c \wedge \omega^d$. Then $\overset{\circ}{\mathrm{Rc}} \odot g : \Lambda^2_\pm T^*M \to \Lambda^2_\mp T^*M$ satisfies

$$\left(\overset{\circ}{\mathrm{Rc}} \odot g\right)(\alpha_\pm) = \left(\overset{\circ}{R}_{aa} + \overset{\circ}{R}_{bb}\right)\alpha_\mp \tag{10.162}$$

for any self-dual 2-form α_+ and for any anti-self-dual 2-form α_-. For example,

$$\big(\overset{\circ}{\mathrm{Rc}} \odot g\big)(\alpha_1) = \big(\overset{\circ}{R}_{11} + \overset{\circ}{R}_{22}\big)\,\alpha_4,$$

and similarly for the other α_i, where $\alpha_1, \ldots, \alpha_6$ are defined by (10.154) and (10.155). Hence, $\overset{\circ}{\mathrm{Rc}} \odot g : \Lambda^2_\pm T^*M \to \Lambda^2_\mp T^*M$ is equal to B, B^{T}, respectively.

Thus:

Lemma 10.36. *The curvature operator* $\mathrm{Rm} : \Lambda^2 T^*M \to \Lambda^2 T^*M$ *on a* 4*-manifold, which satisfies* (10.158)*, is given in block form* (10.156) *by*

$$A = \mathrm{Wy}_+ + \frac{R}{12}\,\mathrm{id}_{\Lambda^2_+ T^*M}, \qquad C = \mathrm{Wy}_- + \frac{R}{12}\,\mathrm{id}_{\Lambda^2_- T^*M}, \tag{10.163}$$

and B *and* B^{T} *are given by* (10.162).

10.9.3. The Pfaffian curvature form on a 4-manifold. In this section we rewrite the Chern–Gauss–Bonnet formula when the dimension of our manifold is equal to 4. Let (M^4, g) be an oriented closed Riemannian 4-manifold. From (10.115), the global Pfaffian curvature form $\mathcal{P}(\Omega)$ on M^4 is given by

$$\mathcal{P}(\Omega) = \frac{1}{32\pi^2} \sum_{\sigma \in S_4} \mathrm{s}(\sigma)\, \Omega_{\sigma_1}^{\sigma_2} \wedge \Omega_{\sigma_3}^{\sigma_4} \tag{10.164}$$

with respect to a positively oriented local orthonormal frame field $\{e_i\}_{i=1}^4$ on an open subset $\mathcal{U}$ of M^4. By (8.66) we have that

$$\Omega_i^j(e_k, e_l) = R_{klij} = R_{ijkl}.$$

Using this, we may rewrite on $\mathcal{U}$ the Pfaffian curvature form as

$$\mathcal{P}(\Omega) = \frac{1}{32\pi^2} \sum_{\sigma,\tau \in S_4} A_{\sigma,\tau}\, d\mu, \tag{10.165}$$

where $d\mu = \omega^1 \wedge \omega^2 \wedge \omega^3 \wedge \omega^4$ is the volume form of g, where $\{\omega^j\}_{j=1}^4$ denotes the coframe dual to $\{e_i\}_{i=1}^4$, and the functions $A_{\sigma,\tau}$ on $\mathcal{U}$ are defined by

$$A_{\sigma,\tau} := \mathrm{s}(\sigma)\,\mathrm{s}(\tau)\, R_{\sigma_1\sigma_2\tau_1\tau_2} R_{\sigma_3\sigma_4\tau_3\tau_4}. \tag{10.166}$$

Here, we have also used that if α is a 2-form, then $\alpha = \sum_{i,j} \alpha(e_i, e_j)\omega^i \wedge \omega^j$.

Since the $\{e_1, e_2, e_3, e_4\}$ are eigenvectors for the Ricci tensor, by (10.159) we have that $R_{1234} = W_{1234}$, where W_{ijkl} denotes the components of the Weyl tensor Wy.

For $\sigma, \tau \in S_4$, the symmetric group of degree 4, define the integer

$$b(\sigma, \tau) := \big|\{\sigma_1, \sigma_2\} \cap \{\tau_1, \tau_2\}\big| \in \{0, 1, 2\},$$

where $|X|$ denotes the cardinality of a set X. Each pair (σ, τ) satisfies one of the following three cases:

(1) $b(\sigma, \tau) = 0$. Then

$$A_{\sigma,\tau} = (R_{\sigma_1\sigma_2\tau_1\tau_2})^2 = (W_{\sigma_1\sigma_2\tau_1\tau_2})^2.$$

Since for each $\sigma \in S_4$ there are four τ's in S_4 with $b(\sigma, \tau) = 0$, we thus have

$$\sum_{\sigma,\tau\in S_4\,:\,b(\sigma,\tau)=2} A_{\sigma,\tau} = 32\big((W_{1234})^2 + (W_{1342})^2 + (W_{1423})^2\big). \tag{10.167}$$

(2) $b(\sigma, \tau) = 1$. We have

$$\begin{aligned}\sum_{\sigma,\tau\in S_4\,:\,b(\sigma,\tau)=1} A_{\sigma,\tau} &= 64\big(R_{1213}R_{3442} + R_{1214}R_{3423} + R_{1314}R_{4223}\big)\\ &\quad + 64\big(R_{2123}R_{4314} + R_{2124}R_{4331} + R_{3132}R_{2441}\big)\\ &= 64\big((W_{1213})^2 + (W_{1214})^2 + (W_{1314})^2\big)\\ &\quad + 64\big((W_{2123})^2 + (W_{2124})^2 + (W_{3132})^2\big),\end{aligned} \tag{10.168}$$

since $R_{ijik} = W_{ijik}$ for i, j, k distinct and since $W_{3442} = -W_{3112}$, etc.

(3) $b(\sigma, \tau) = 2$. Then

$$A_{\sigma,\tau} = R_{\sigma_1\sigma_2\sigma_1\sigma_2}R_{\sigma_3\sigma_4\sigma_3\sigma_4}.$$

Thus

$$\sum_{\sigma,\tau\in S_4\,:\,b(\sigma,\tau)=2} A_{\sigma,\tau} = 32\big(R_{1212}R_{3434} + R_{1313}R_{2424} + R_{1414}R_{2323}\big). \tag{10.169}$$

Next, we observe that

$$\begin{aligned}|\,\mathrm{Wy}\,|^2 &= \sum_{i,j,k,l=1}^{4} (W_{ijkl})^2 \quad \text{(there are 144 terms)}\\ &= 8\big((W_{1234})^2 + (W_{1342})^2 + (W_{1423})^2\big)\\ &\quad + 16\big((W_{1213})^2 + (W_{1214})^2 + (W_{1314})^2\big)\\ &\quad + 16\big((W_{2123})^2 + (W_{2124})^2 + (W_{3132})^2\big)\\ &\quad + 8\big((W_{1212})^2 + (W_{1313})^2 + (W_{1414})^2\big),\end{aligned} \tag{10.170}$$

where, to simplify the last line above, we used the identities

$$W_{1212} = W_{3434}, \quad W_{1313} = W_{2424}, \quad W_{1414} = W_{2323}.$$

Therefore, by combining all of the calculations above, we obtain

(10.171)

$$
\begin{aligned}
8\pi^2 \frac{\mathcal{P}(\Omega)}{d\mu} &= \frac{1}{4} \sum_{\sigma,\tau \in S_4} A_{\sigma,\tau} \\
&= 8\big((W_{1234})^2 + (W_{1342})^2 + (W_{1423})^2\big) \\
&\quad + 16\big((W_{1213})^2 + (W_{1214})^2 + (W_{1314})^2\big) \\
&\quad + 16\big((W_{2123})^2 + (W_{2124})^2 + (W_{3132})^2\big) \\
&\quad + 8\big(R_{1212}R_{3434} + R_{1313}R_{2424} + R_{1414}R_{2323}\big) \\
&= |\operatorname{Wy}|^2 - 8\big(W_{1212}W_{3434} + W_{1313}W_{2424} + W_{1414}W_{2323}\big) \\
&\quad + 8\big(R_{1212}R_{3434} + R_{1313}R_{2424} + R_{1414}R_{2323}\big)\,,
\end{aligned}
$$

where the first equality follows from the calculations for $A_{\sigma,\tau}$ and where the second equality follows from the calculation of the norm-square of the Weyl tensor $|\operatorname{Wy}|^2$.

Let $\overset{\circ}{\operatorname{Ric}} := \operatorname{Ric} - \frac{1}{4}Rg$ denote the trace-free Ricci tensor. Using by (7.84) that

$$
2R_{1212} = \frac{R}{6} - \overset{\circ}{R}_{11} - \overset{\circ}{R}_{22} + 2W_{1212}\,,
$$

etc., we compute that

$$
\begin{aligned}
&4\big(R_{1212}R_{3434} + R_{1313}R_{2424} + R_{1414}R_{2323}\big) \\
&- 4\big(W_{1212}W_{3434} + W_{1313}W_{2424} + W_{1414}W_{2323}\big) \\
&= \big(\tfrac{R}{6} - \overset{\circ}{R}_{11} - \overset{\circ}{R}_{22}\big)\big(\tfrac{R}{6} - \overset{\circ}{R}_{33} - \overset{\circ}{R}_{44}\big) + \big(\tfrac{R}{6} - \overset{\circ}{R}_{11} - \overset{\circ}{R}_{33}\big)\big(\tfrac{R}{6} - \overset{\circ}{R}_{22} - \overset{\circ}{R}_{44}\big) \\
&\quad + \big(\tfrac{R}{6} - \overset{\circ}{R}_{11} - \overset{\circ}{R}_{44}\big)\big(\tfrac{R}{6} - \overset{\circ}{R}_{22} - \overset{\circ}{R}_{33}\big) \\
&= \frac{R^2}{12} - \big(\overset{\circ}{R}_{11} + \overset{\circ}{R}_{22}\big)^2 - \big(\overset{\circ}{R}_{11} + \overset{\circ}{R}_{33}\big)^2 - \big(\overset{\circ}{R}_{11} + \overset{\circ}{R}_{44}\big)^2 \\
&= \frac{R^2}{12} - \big|\overset{\circ}{\operatorname{Ric}}\big|^2,
\end{aligned}
$$

with the first equality above because of cancellations following from

$$
\Big(\frac{R}{6} - \overset{\circ}{R}_{11} - \overset{\circ}{R}_{22}\Big)W_{3434} + \Big(\frac{R}{6} - \overset{\circ}{R}_{33} - \overset{\circ}{R}_{44}\Big)W_{1212} = \frac{R}{3}W_{1212}\,,
$$

etc., where we used that $W_{1212} = W_{3434}$ and that $\overset{\circ}{\operatorname{Ric}}$ and Wy are trace-free (see (7.86)). We obtain that

$$
8\pi^2\mathcal{P}(\Omega) = \Big(|\operatorname{Wy}|^2 + \frac{R^2}{24} - \frac{1}{2}\big|\overset{\circ}{\operatorname{Ric}}\big|^2\Big)\,d\mu. \tag{10.172}
$$

We conclude from (10.172) and $|\,\mathrm{Wy}_+|^2 + |\,\mathrm{Wy}_-|^2 = |\,\mathrm{Wy}\,|^2$, where Wy_+ and Wy_- are defined by (10.160), that in dimension 4 the Chern–Gauss–Bonnet formula (10.117) may be expressed as follows.

Theorem 10.37 (Chern–Gauss–Bonnet formula on a 4-manifold).

$$\chi(M^4) = \frac{1}{8\pi^2}\int_{M^4}\left(\frac{R^2}{24} + |\,\mathrm{Wy}_+|^2 + |\,\mathrm{Wy}_-|^2 - \frac{|\overset{\circ}{\mathrm{Ric}}|^2}{2}\right) d\mu. \tag{10.173}$$

10.9.4. Hirzebruch signature theorem. Another fundamental 'index theorem' is the Hirzebruch signature theorem for $4k$-dimensional oriented closed manifolds. We discuss the case $k = 1$.

Let M^4 be an oriented closed 4-manifold. Now we consider the signature

$$\tau(M^4) := b^2_+(M^4) - b^2_-(M^4), \tag{10.174}$$

where $b^2_+(M^4)$ and $b^2_-(M^4)$ denote the dimensions of the spaces of self-dual and anti-self-dual harmonic 2-forms on (M^4, g), respectively. Note that the quantities $b^2_+(M^4)$ and $b^2_-(M^4)$ represent the dimensions of the maximal positive and negative definite subspaces of the second cohomology group $H^2(M^4;\mathbb{R})$, with respect to the cup product as a bilinear form.[10]

By the **Hirzebruch signature theorem** (see [**Hir95**], [**MS74**, Theorem 19.4]), we have

$$\tau(M^4) = \frac{1}{3}p_1(TM), \tag{10.175}$$

where $p_1(TM)$ is the first Pontryagin number of the tangent bundle of M^4. We have

$$p_1(TM) = \int_{M^4} p_1,$$

where p_1 is the first Pontryagin form, which is the 4-form on M^4 defined by

$$\begin{aligned} p_1 &:= \sum_{i,j=1}^{4} \Omega_i^j \wedge \Omega_j^i \\ &= \operatorname{sgn}(\tau) R_{ij\tau_1\tau_2} R_{ji\tau_3\tau_4} d\mu, \end{aligned} \tag{10.176}$$

where for the second equality we used (8.67). Using 'Chern–Weil theory', one can prove that (see Walschap [**Wal04**, Chapter 6, §5])

$$p_1 = \frac{1}{4\pi^2}\left(|\,\mathrm{Wy}_+|^2 - |\,\mathrm{Wy}_-|^2\right) d\mu. \tag{10.177}$$

Therefore,

$$\tau(M^4) = \frac{1}{12\pi^2}\int_M \left(|\,\mathrm{Wy}_+|^2 - |\,\mathrm{Wy}_-|^2\right) d\mu. \tag{10.178}$$

[10] See Hatcher [**Hat02**, §3.2] for the definition of cup product.

Like the Chern–Gauss–Bonnet formula, the Hirzebruch signature theorem relates a topological invariant, in this case the signature of a manifold, to the integral of a curvature invariant, in this case the Weyl tensor. As such, it provides another important link between topology and geometry.

10.10. Einstein manifolds

We define a Riemannian manifold (M^n, g) to be an **Einstein manifold** (g is called an **Einstein metric**) if

$$\mathrm{Ric} = \frac{1}{n} R g. \tag{10.179}$$

We remark that if a Riemannian metric g satisfies $\mathrm{Ric} = fg$ for some function f, then $R = \mathrm{trace}(\mathrm{Ric}) = \mathrm{trace}(fg) = nf$, so that $f = \frac{1}{n}R$. Thus the Einstein metric equation is equivalent to the condition that the Ricci tensor is equal to a function times the metric.

10.10.1. Einstein manifolds have constant scalar curvature. Recall that the contracted second Bianchi identity (7.79) says that

$$\mathrm{div}(\mathrm{Ric}) = \frac{1}{2} dR.$$

By combining this with the Einstein metric equation, we obtain

$$\frac{1}{2} dR = \mathrm{div}\left(\frac{1}{n} R g\right) = \frac{1}{n} dR. \tag{10.180}$$

Proposition 10.38. *If (M^n, g) is a connected Einstein manifold and if $n \geq 3$, then the scalar curvature R is a constant function.*

Proof. Since $n \neq 2$, (10.180) implies that $dR \equiv 0$ on M^n. Since M^n is connected, we conclude that R is constant. □

Corollary 10.39. *If (M^n, g), where $n \geq 3$, is a connected Riemannian manifold with the property that for each $x \in M^n$, the sectional curvatures of all planes in $T_x M$ are equal, then the sectional curvatures are also independent of x.*

Proof. The hypothesis implies that there exists a function $f : M^n \to \mathbb{R}$ such that

$$\mathrm{Rm} = f g \odot g.$$

By tracing this equation, we obtain $\mathrm{Ric} = 2(n-1) f g$. Hence, $f = \frac{1}{2n(n-1)} R$ is equal to a constant by Proposition 10.38. The corollary follows. □

10.10.2. The Hitchin–Thorpe inequality. The **Hitchin–Thorpe inequality** says the following (see Thorpe [**Tho69**] and Hitchin [**Hit74**]).

Theorem 10.40. *If (M^4, g) is a closed oriented 4-dimensional Einstein manifold, then*

$$\chi(M^4) \geq \frac{3}{2}|\tau(M^4)|. \tag{10.181}$$

Moreover, equality holds if and only if $R \equiv 0$ and $\mathrm{Wy}_+ \equiv 0$ or $\mathrm{Wy}_- \equiv 0$.

Proof. Since an Einstein metric satisfies $\overset{\circ}{\mathrm{Rc}} = 0$, by the Chern–Gauss–Bonnet formula (10.173) we have that

$$\begin{aligned}\chi(M^4) &= \frac{1}{8\pi^2}\int_{M^4}\left(\frac{R^2}{24} + |\,\mathrm{Wy}_+\,|^2 + |\,\mathrm{Wy}_-\,|^2\right)d\mu \\ &\geq \frac{1}{8\pi^2}\int_{M^4}\left(|\,\mathrm{Wy}_+\,|^2 + |\,\mathrm{Wy}_-\,|^2\right)d\mu \\ &\geq \frac{1}{8\pi^2}\int_{M^4}\left||\,\mathrm{Wy}_+\,|^2 - |\,\mathrm{Wy}_-\,|^2\right|d\mu \\ &= \frac{3}{2}|\tau(M^4)|,\end{aligned} \tag{10.182}$$

where the last equality is by (10.178). It is easy to see that equality in (10.182) holds if and only if $R \equiv 0$ and $\mathrm{Wy}_+ \equiv 0$ or $\mathrm{Wy}_- \equiv 0$. □

10.11. Notes and commentary

An extensive treatment of differential geometry, including bundles, is presented in the two-volume book by Kobayashi and Nomizu [**KN96a, KN96b**]. Our discussion of connections on principal bundles closely follows [**KN96a**, Chapter 2, §1]. A book on fiber bundles is by Husemoller [**Hus94**]. An introduction to characteristic classes is by Milnor and Stasheff [**MS74**]. A book on Einstein manifolds is by Besse [**Bes08**].

A significant area of study in differential geometry focuses on spin manifolds and the Dirac operator. See the book by Lawson and Michelsohn [**LM89**]. Lichnerowicz [**Lic63**] proved a notable Bochner-type formula, which implies that for a closed spin Riemannian manifold with positive scalar curvature, any harmonic spinor must vanish identically. By the Atiyah–Singer index theorem, this implies that any such manifold must have trivial $\widehat{A}$-genus.

In broad terms, the Atiyah–Singer index theorem says that for an elliptic partial differential operator on a closed manifold, the analytical index is equal to the topological index. The special cases of this theorem include the Chern–Gauss–Bonnet formula, the Hirzebruch signature formula, and the Riemann–Roch formula, among others. See Rosenberg [**Ros97**] for an introduction and Roe [**Roe98**] and Berline, Getzler, and Vergne [**BGV04**] for more advanced texts.

10.12. Exercises

Exercise 10.1. *Let M^n and N^m be smooth manifolds. Let $p_1 : M^n \times N^m \to M^n$ and $p_2 : M^n \times N^m \to N^m$ be the projection maps. Show that the tangent bundle $T(M \times N)$ of $M^n \times N^m$ is naturally isomorphic to $p_1^*TM \oplus p_2^*TN$.*

Exercise 10.2. *Let $\pi : \widetilde{M}^n \to M^n$ be a smooth covering map. Show that the tangent bundle of the cover $T\widetilde{M}$ is naturally isomorphic to the pull-back bundle $\pi^*(TM)$.*

Exercise 10.3. *Following up on the previous exercise, show that if $T\widetilde{M}$ is not parallelizable, then TM is not parallelizable.*

Exercise 10.4. *Show that if M^n is an oriented smooth manifold, then its tangent bundle TM has a (natural) orientation as a $2n$-dimensional manifold. This in turn induces an orientation on the unit tangent bundle SM.*

Exercise 10.5. *Following up on the previous exercise, show that even if M^n is nonorientable, TM has an orientation as a $2n$-dimensional manifold. Hint: If $\mathbb{V}$ is a finite-dimensional vector space, then $\mathbb{V} \times \mathbb{V}$ has a natural orientation.*

Exercise 10.6. *Prove that every parallelizable manifold (see Definition* 10.9*) is orientable.*

Exercise 10.7. *Prove that the Whitney sum of two vector bundles over the same base space, defined in* §10.3.7, *is a vector bundle.*

Exercise 10.8. *Prove that the tensor product of two vector bundles over the same base space, defined in* §10.3.7, *is a vector bundle.*

Exercise 10.9. *Let $\pi : E \to M$ be a vector bundle with a bundle metric. Show that the unit sphere bundle $\pi : S(E) \to M$ has a G atlas with $G = \mathrm{O}(n)$.*

Exercise 10.10. *Let $V \to E \xrightarrow{\pi} B$ be a vector bundle. Show that we have the principal bundle*

$$\mathrm{GL}(V) \to P_{\mathrm{GL}}(E) \xrightarrow{\pi} B,$$

where $\mathrm{GL}(V)$ is the general linear group of V and where the total space $P_{\mathrm{GL}}(E)$ is the set of ordered bases (i.e., frames) for the fibers of E.

Exercise 10.11. *Now let* $V \to E \overset{\pi}{\to} B$ *be a vector bundle with a fiber metric. Show that we have the principal bundle*

$$\mathrm{O}(V) \to P_{\mathrm{O}}(E) \overset{\pi}{\to} B,$$

where $\mathrm{O}(V)$ *is the orthogonal group of* V *and where the total space* $P_{\mathrm{O}}(E)$ *is the set of ordered orthonormal bases for the fibers of* E*.*

Exercise 10.12. *Show that the Pfaffian curvature form* $\mathcal{P}(\Omega)$ *defined by* (10.115) *is independent of the choice of orthonormal frame* $\{e_i\}_{i=1}^{2m}$*.*
Hint: Show that $\mathcal{P}(\overline{\Omega}) = \det(\mathrm{A})\mathcal{P}(\Omega)$*. Cf. Spivak* [**Spi79e**, §13.3].

Exercise 10.13. *Show that* Φ_k *and* Ψ_k*, given by* (10.129) *and* (10.130)*, respectively, are well-defined differential forms on* SM*.*

Exercise 10.14. *Prove that the sum* $\sum_{\sigma, 2m-2k+1 \leq j \leq 2m} S_{\sigma,j}$ *in* (10.134) *vanishes.*

Exercise 10.15 (Hamilton; see [**Ham82**, Lemma 11.6] and [**Hui85**, Lemma 4.3])**.** *Let* (M^n, g) *be a Riemannian manifold. Define the covariant* 3*-tensor* E *by*

(10.183)
$$E_{ijk} = \frac{n-2}{2(n-1)(n+2)}\left(\nabla_j R g_{ik} + \nabla_k R g_{ij}\right) + \frac{n}{(n-1)(n+2)} \nabla_i R g_{jk}.$$

Define the covariant 3*-tensor* F *by* $\nabla_i R_{jk} = E_{ijk} + F_{ijk}$*. Show that*

$$\langle E, F \rangle = 0 \tag{10.184}$$

and

$$|E|^2 = \frac{3n-2}{2(n-1)(n+2)} |\nabla R|^2. \tag{10.185}$$

Conclude that

$$\left| \nabla \operatorname{Ric} - \frac{1}{n} \nabla R \otimes g \right|^2 \geq \frac{(n-2)^2}{2n(n-1)(n+2)} |\nabla R|^2. \tag{10.186}$$

Give another proof of Proposition 10.38 *using this formula.*

Part III

Elliptic and Parabolic Equations in Geometric Analysis

Chapter 11

Linear Elliptic and Parabolic Equations

The prototypical *elliptic* equation is the *Laplace equation*, which is a linear second-order elliptic (scalar) equation. The prototypical *parabolic* equation is the *heat equation*, which is a linear second-order parabolic (scalar) equation. In addition to these two equations, we will discuss eigenfunction equations for the Laplacian as well as the (Hodge) Laplace equation for differential forms, whose solutions are called harmonic forms.

In this chapter, we also discuss the fundamental Bishop–Gromov volume comparison theorem and the Cheeger–Gromoll splitting theorem. The techniques in this chapter are related to the Bochner technique.

11.1. The fundamental Bochner–Weitzenböck formula and integration by parts

The Laplace operator, a.k.a. the Laplacian, is central to the study of the geometry and topology of manifolds. The study of the Laplace operator on a Riemannian manifold includes considering the covariant derivative operator and its commutator with the Laplacian. This is where the curvature, especially the Ricci curvature, comes into play. Furthermore, a powerful tool is the *maximum principle*, which is just the first and second derivative tests from multivariable calculus.

11.1.1. The second derivative test on manifolds. Let (M^n, g) be a Riemannian manifold and let $u : M^n \to \mathbb{R}$ be a C^2 function. Recall from

(8.80) that the Hessian of u is the symmetric 2-tensor defined by

$$(\nabla^2 u)(V, W) = \nabla^2_{V,W} u = V(W(u)) - (\nabla_V W)(u),$$

and recall from (8.86) that the Laplacian acting on functions is defined by

$$\Delta u = \mathrm{trace}_g(\nabla^2 u). \tag{11.1}$$

Now suppose that $u : M^n \to \mathbb{R}$ attains its minimum at a point $x_0 \in M^n$. We claim that $(\nabla^2 u)_{x_0} \geq 0$ is positive semidefinite at x_0. To see this, we will now show that for any vector $V \in T_{x_0} M$ we have

$$(\nabla^2 u)_{x_0}(V, V) \geq 0. \tag{11.2}$$

Let $\epsilon > 0$ and consider the (unique) constant-speed geodesic $\gamma_V : (-\epsilon, \epsilon) \to M^n$ satisfying $\gamma_V(0) = x_0$ and $\gamma_V'(0) = V$. We have that the one-variable function $t \mapsto u(\gamma_V(t))$ attains its minimum value $u(\gamma_V(0)) = u(x_0)$ at $t = 0$. Therefore,

$$0 \leq \left.\frac{d^2}{dt^2}\right|_{t=0} u(\gamma_V(t)). \tag{11.3}$$

Define $T_t = \gamma_V'(t)$, which is a vector field along the geodesic γ_V. Note that $T_0 = V$. We have that

$$\frac{d^2}{dt^2} u(\gamma_V(t)) = T_t(T_t(u)) \tag{11.4}$$

is a second directional derivative of u. We remark that (11.4) holds for any geodesic γ_V, i.e., at any point of M^n, not necessarily a critical point of u.

Since γ_V is a constant-speed geodesic, we have $\nabla_{T_t} T_t = 0$ along $\gamma_V(t)$. Therefore,

$$\frac{d^2}{dt^2} u(\gamma_V(t)) = T_t(T_t(u)) - (\nabla_{T_t} T_t)(u) = (\nabla^2 u)(T_t, T_t),$$

and we conclude from (11.3) that

$$0 \leq \left.\frac{d^2}{dt^2}\right|_{t=0} u(\gamma_V(t)) = (\nabla^2 u)_{x_0}(V, V). \tag{11.5}$$

This proves (11.2).

By tracing (11.2), we have that at any minimum point x_0 of u,

$$\Delta u(x_0) \geq 0. \tag{11.6}$$

Similarly, if u attains its maximum at x_1, then $(\nabla^2 u)_{x_1}$ is negative semidefinite at x_1 and $\Delta u(x_1) \leq 0$. We call all of this the **second derivative test** or the **(elliptic) maximum principle**. This generalizes the second derivative test from multivariable calculus.

Of course, at any critical point x_2 of u, including at a minimum point or at a maximum point, we have the vanishing of the gradient:

$$\nabla u(x_2) = 0; \tag{11.7}$$

this is the **first derivative test**.

11.1.2. The Hessian acting on 1-forms. The Hessian operator not only acts on functions but also on tensors and differential forms. We consider the case of its action on 1-forms, which we will find useful for computing so-called *Bochner–Weitzenböck-type formulas* (see [**Boc46**], [**Wei23**]).

Let α be a 1-form on M^n. The Hessian of α is defined by

$$\nabla^2_{U,V}\alpha = \nabla_U(\nabla_V\alpha) - \nabla_{\nabla_U V}\alpha. \tag{11.8}$$

For example, if $u : M^n \to \mathbb{R}$ is a function, then

$$\nabla^2_{U,V}du = \nabla_U(\nabla_V du) - \nabla_{\nabla_U V}du. \tag{11.9}$$

Recall from (4.157) that the Riemann curvature tensor Rm may be expressed as

$$\mathrm{Rm}(U,V)W = \nabla^2_{U,V}W - \nabla^2_{V,U}W.$$

That is, Rm is the commutator of the Hessian acting on vector fields. Similarly, we can consider the commutator of the Hessian acting on 1-forms. We have the following.

Lemma 11.1. *If α is a C^2 1-form, then*

$$(\nabla^2_{U,V}\alpha)(W) - (\nabla^2_{V,U}\alpha)(W) = \alpha\big(\mathrm{Rm}(V,U)W\big). \tag{11.10}$$

In particular, if u is a C^3 function, then

$$(\nabla^2_{U,V}du)(W) - (\nabla^2_{V,U}du)(W) = du\big(\mathrm{Rm}(V,U)W\big). \tag{11.11}$$

Proof. The main idea is to use the product rule for covariant differentiation. In this way we can convert the covariant differentiation of 1-forms into the covariant differentiation of vector fields. In particular, we begin by calculating that

(11.12)

$$\begin{aligned}
&(\nabla^2_{U,V}\alpha)(W)\\
&\quad= \big(\nabla_U(\nabla_V\alpha)\big)(W) - (\nabla_{\nabla_U V}\alpha)(W)\\
&\quad= U\big((\nabla_V\alpha)(W)\big) - (\nabla_V\alpha)(\nabla_U W) - (\nabla_U V)(\alpha(W)) + \alpha(\nabla_{\nabla_U V}W)\\
&\quad= U\big(V(\alpha(W)) - \alpha(\nabla_V W)\big) - V\big(\alpha(\nabla_U W)\big) + \alpha\big(\nabla_V(\nabla_U W)\big)\\
&\qquad - (\nabla_U V)(\alpha(W)) + \alpha(\nabla_{\nabla_U V}W).
\end{aligned}$$

Rearranging the terms on the right-hand side, we obtain

$$(11.13)\qquad \begin{aligned}(\nabla^2_{U,V}\alpha)(W) &= \alpha\big(\nabla_V(\nabla_U W)\big) + \alpha(\nabla_{\nabla_U V}W)\\ &\quad + U\big(V(\alpha(W))\big) - (\nabla_U V)(\alpha(W))\\ &\quad - U\big(\alpha(\nabla_V W)\big) - V\big(\alpha(\nabla_U W)\big).\end{aligned}$$

By switching U and V we obtain

$$(11.14)\qquad \begin{aligned}(\nabla^2_{V,U}\alpha)(W) &= \alpha\big(\nabla_U(\nabla_V W)\big) + \alpha(\nabla_{\nabla_V U}W)\\ &\quad + V\big(U(\alpha(W))\big) - (\nabla_V U)(\alpha(W))\\ &\quad - V\big(\alpha(\nabla_U W)\big) - U\big(\alpha(\nabla_V W)\big).\end{aligned}$$

Since the terms on the second and third lines of both (11.13) and (11.14) cancel, we obtain

$$(11.15)\qquad \begin{aligned}(\nabla^2_{U,V}\alpha)(W) - (\nabla^2_{V,U}\alpha)(W) &= \alpha\big(\nabla_V(\nabla_U W)\big) + \alpha(\nabla_{\nabla_U V}W)\\ &\quad - \alpha\big(\nabla_U(\nabla_V W)\big) - \alpha(\nabla_{\nabla_V U}W)\\ &= \alpha\big(\operatorname{Rm}(V,U)W\big).\end{aligned}$$

This completes the proof of the lemma. □

11.1.3. The fundamental Bochner formula. Let (M^n, g) be a Riemannian manifold and let $u : M^n \to \mathbb{R}$. As a warm-up for calculating with the Laplacian, we compute that

$$(11.16)\qquad \frac{1}{2}\Delta(u^2) = \operatorname{div}(u\nabla u) = u\Delta u + |du|^2.$$

Recall that the inner product g_x on a tangent space T_xM induces an inner product g_x^* on the cotangent space T_x^*M. When we take norms and inner products of 1-forms, we use this inner product. Furthermore, this inner product extends to the fibers of tensor bundles. For example, if α and β are 2-tensors, then their inner product at a point $x \in M^n$ is given by

$$\langle \alpha, \beta\rangle := \sum_{i,j=1}^{n} \alpha(e_i, e_j)\beta(e_i, e_j),$$

where $\{e_i\}_{i=1}^n$ is an orthonormal frame at x (cf. (5.111)). With respect to local coordinates,

$$\langle \alpha, \beta\rangle = \sum_{i,j,k,\ell=1}^{n} g^{ik}g^{j\ell}\alpha_{ij}\beta_{k\ell};$$

cf. (5.110). The pointwise norm of α is defined by

$$(11.17)\qquad |\alpha| := \sqrt{\langle\alpha,\alpha\rangle};$$

cf. (5.112).

Lemma 11.2 (Commutator of Δ and d). *For any C^3 function u on a Riemannian manifold,*

$$\Delta(du) = d(\Delta u) + \operatorname{Ric}(du), \tag{11.18}$$

where we consider Ric *as a bundle map* $\operatorname{Ric} : T^*M \to T^*M$.

Proof. Let $\{e_i\}_{i=1}^n$ be a local orthonormal frame field on an open subset $\mathcal{U}$ of M^n. We compute on $\mathcal{U}$ that

$$\begin{aligned}(\Delta(du))(W) &= \sum_{i=1}^n (\nabla^2_{e_i,e_i} du)(W) \\ &= \sum_{i=1}^n (\nabla^2_{e_i,W} du)(e_i) \\ &= \sum_{i=1}^n (\nabla^2_{W,e_i} du)(e_i) + \sum_{i=1}^n du\big(\operatorname{Rm}(W,e_i)e_i\big) \\ &= W(\Delta u) + du(\operatorname{Ric}(W)) \\ &= (d(\Delta u))(W) + \operatorname{Ric}(\nabla u, W).\end{aligned} \tag{11.19}$$

The lemma follows immediately from this formula. Let us now break down the calculation above. The reason for why the second equality in (11.19) is true is that

$$(\nabla^2_{U,V} du)(W) = (\nabla^2 du)(U,V,W) = (\nabla^2 du)(U,W,V) = (\nabla^2_{U,W} du)(V)$$

for any tangent vectors U, V, and W, since the Hessian ∇du of the function u is a symmetric 2-tensor.[1] The third equality in (11.19) follows from the commutator formula for covariant derivatives acting on 1-forms:

$$(\nabla^2_{V,W}\alpha)(U) = (\nabla^2_{W,V}\alpha)(U) + \alpha\big(\operatorname{Rm}(W,V)U\big); \tag{11.20}$$

see (11.11). The fourth equality in (11.19) is because the trace and covariant differentiation commute:

$$\operatorname{trace}_{2,3}(\nabla^2 du)(W) = d\big(\operatorname{trace}(\nabla du)\big)(W) = W(\Delta u),$$

where $\operatorname{trace}_{2,3}$ denotes the trace on the second and third components. ☐

[1]Note also that if α is a symmetric 2-tensor, then $(\nabla_X \alpha)(Y,Z)$ is symmetric in Y and Z since it is equal to $X(\alpha(Y,Z)) - \alpha(\nabla_X Y, Z) - \alpha(Y, \nabla_X Z)$, which is symmetric in Y and Z. We use this with $\alpha = \nabla du$.

The **fundamental Bochner formula** (a.k.a. **fundamental Bochner–Weitzenböck formula**) says the following.

Proposition 11.3. *For any function u on (M^n, g), we have*

$$\frac{1}{2}\Delta|du|^2 = \langle du, d(\Delta u)\rangle + |\nabla du|^2 + \operatorname{Ric}(\nabla u, \nabla u).$$

In other words,

$$\frac{1}{2}\Delta|\nabla u|^2 = \langle \nabla u, \nabla\Delta u\rangle + |\nabla^2 u|^2 + \operatorname{Ric}(\nabla u, \nabla u). \tag{11.21}$$

Proof. We compute using the product rule for covariant differentiation

$$\begin{aligned}\Delta|du|^2 &= \operatorname{trace}\left(\nabla^2\langle du, du\rangle\right)\\ &= 2\operatorname{trace}\left(\langle\nabla^2 du, du\rangle + \langle\nabla du, \nabla du\rangle\right)\\ &= 2\langle\Delta(du), du\rangle + 2|\nabla du|^2\\ &= 2\langle d(\Delta u) + \operatorname{Ric}(du), du\rangle + 2|\nabla^2 u|^2,\end{aligned}$$

where the last equality is by Lemma 11.2; the proposition follows. □

In particular, Proposition 11.3 implies that if the Ricci curvature of (M^n, g) is nonnegative, then

$$\frac{1}{2}\Delta|\nabla u|^2 \geq \langle\nabla u, \nabla\Delta u\rangle + |\nabla^2 u|^2.$$

11.1.4. Integration by parts. Recall that in the case of closed manifolds, the divergence theorem (see Corollary 8.9) says the following.

Proposition 11.4 (Divergence theorem)**.** *If (M^n, g) is an oriented closed Riemannian manifold and if V is a vector field on M^n, then*

$$\int_{M^n} \operatorname{div}(V)\, d\mu = 0. \tag{11.22}$$

As we know from calculus, integration by parts is the fundamental theorem of calculus combined with the product rule for differentiation. Similarly, we have the following.

Lemma 11.5 (Integration by parts on manifolds)**.** *Let u and v be functions on a closed Riemannian manifold (M^n, g). Then*

$$\int_{M^n} \langle\nabla u, \nabla v\rangle\, d\mu = -\int_{M^n} u\Delta v\, d\mu. \tag{11.23}$$

Proof. The lemma follows from the divergence theorem (Proposition 11.4) and the product rule in the form

$$\operatorname{div}(u\nabla v) = \langle\nabla u, \nabla v\rangle + u\Delta v. \tag{11.24}$$

□

Since $\langle \nabla u, \nabla v \rangle$ is symmetric in u and v, we have the following.

Corollary 11.6. *For any functions u and v on a closed Riemannian manifold (M^n, g),*

$$\int_{M^n} u \Delta v \, d\mu = \int_{M^n} v \Delta u \, d\mu. \tag{11.25}$$

Lemma 11.7. *Every harmonic function on a closed Riemannian manifold is constant.*

Proof. Suppose that $\Delta u = 0$ on M^n. Since M^n is closed, by the divergence theorem, we have

$$0 = \int_{M^n} u \Delta u \, d\mu = - \int_{M^n} |\nabla u|^2 \, d\mu. \tag{11.26}$$

This implies $\nabla u \equiv \vec{0}$ on M^n, which in turn implies that u is constant. □

Given tensors γ and η of the same degree, their **L^2-inner product** is defined by

$$\langle\langle \gamma, \eta \rangle\rangle_{L^2} := \int_{M^n} \langle \gamma_x, \eta_x \rangle_x \, d\mu(x), \tag{11.27}$$

where $\langle \gamma_x, \eta_x \rangle_x$ denotes the pointwise inner product of γ and η at x defined by (5.110). The associated **L^2-norm** is defined by

$$\|\gamma\|_{L^2} := \left(\langle\langle \gamma, \gamma \rangle\rangle_{L^2} \right)^{1/2} = \left(\int_{M^n} |\gamma_x|_x^2 \, d\mu(x) \right)^{1/2}, \tag{11.28}$$

where $|\gamma_x|_x$ denotes the pointwise norm of γ at x defined by (5.112).

By Corollary 11.6, the Laplacian Δ acting on functions is a **formally self-adjoint operator** on L^2; that is, if u, v are C^2 functions on M^n, then

$$\langle\langle \Delta u, v \rangle\rangle_{L^2} = \langle\langle u, \Delta v \rangle\rangle_{L^2}. \tag{11.29}$$

11.2. Eigenvalues of the Laplacian on closed manifolds

We begin our study of the Laplace operator by considering some fundamental estimates for its first eigenvalue on a closed Riemannian manifold. The techniques we employ, which are ubiquitous in geometric analysis, are the Bochner technique, the maximum principle, and integration by parts.

A classic reference for eigenvalues of the Laplacian on manifolds is the book by Berger, Gauduchon, and Mazet [**BGM71**].

11.2.1. Definitions of eigenvalues and eigenfunctions of the Laplacian. Let (M^n, g) be a Riemannian manifold. A real number λ is an **eigenvalue of** (minus) **the Laplacian** $-\Delta$ if there exists a function $u : M^n \to \mathbb{R}$ that is not identically zero and satisfies

$$\Delta u + \lambda u = 0. \tag{11.30}$$

The function u is the associated **eigenfunction of the Laplacian**. Abusing terminology, we usually say that λ is an eigenvalue of the Laplacian Δ, instead of its minus.

Now suppose that the manifold M^n is closed. Multiplying (11.30) by the function u and integrating by parts (see (11.23)) yields

$$\begin{aligned} 0 &= \int_{M^n} (\Delta u + \lambda u) u \, d\mu \\ &= \int_{M^n} (-|\nabla u|^2 + \lambda u^2) \, d\mu. \end{aligned} \tag{11.31}$$

Since we have assumed that u is not identically zero, we have $\int_M u^2 \, d\mu > 0$. Therefore we have the eigenvalue formula

$$\lambda = \frac{\int_{M^n} |\nabla u|^2 \, d\mu}{\int_{M^n} u^2 \, d\mu} \geq 0. \tag{11.32}$$

Note that any nonzero constant function is an eigenfunction of the Laplacian with associated eigenvalue zero. On the other hand, if u is a nonconstant eigenfunction with associated eigenvalue λ, then $\lambda > 0$ since $\int_{M^n} |\nabla u|^2 \, d\mu > 0$.

11.2.2. Elementary examples of eigenvalues and eigenfunctions.

11.2.2.1. *Eigenvalues and eigenfunctions on the circle.*

Example 11.8. *Let (M^1, g) be the circle $S^1(r)$ of radius r (and hence length (circumference) $2\pi r$). Let $\theta \in [0, 2\pi]$ denote the angle parameter (coordinate) of (M^1, g). Then an arc-length parameter is given by $s := r\theta$ and $\frac{\partial}{\partial s} = \frac{1}{r}\frac{\partial}{\partial \theta}$ is a unit tangent vector field. The Laplacian is the operator $\Delta = \frac{\partial^2}{\partial s^2} = \frac{1}{r^2}\frac{\partial^2}{\partial \theta^2}$. A basis of eigenfunctions of the Laplacian is given by*

$$\cos\left(\frac{ks}{r}\right) = \cos(k\theta), \quad \sin\left(\frac{ks}{r}\right) = \cos(k\theta), \tag{11.33}$$

for positive integers k, together with the constant function 1. The identical eigenvalue of the Laplacian associated to the two eigenfunctions in (11.33) is equal to the constant $\frac{k^2}{r^2}$. The general eigenfunction with eigenvalue $\frac{k^2}{r^2}$ is given by the linear combination

$$u := a\cos\left(\frac{ks}{r}\right) + b\sin\left(\frac{ks}{r}\right), \tag{11.34}$$

where $a, b \in \mathbb{R}$. *Observe that for* u *defined by* (11.34), *we have*

$$\int_{M^1} \left| \frac{\partial u}{\partial s} \right|^2 ds = \frac{k^2}{r^2} \int_{M^1} u^2 \, ds.$$

This is a special case of (11.32).

11.2.2.2. *Eigenvalues and eigenfunctions on spheres.* Let S^n be the unit sphere in $\mathbb{R}^{n+1}$. Let $\alpha = (\alpha_1, \ldots, \alpha_{n+1})$ be a multi-index of nonnegative integers and let $|\alpha| := \alpha_1 + \cdots + \alpha_{n+1}$. A **homogeneous polynomial of degree** k is a function $P : \mathbb{R}^{n+1} \to \mathbb{R}$ of the form

$$P(\mathbf{x}) = \sum_{|\alpha|=k} c_\alpha \mathbf{x}^\alpha, \tag{11.35}$$

where $c_\alpha \in \mathbb{R}$ and $\mathbf{x}^\alpha := (x^1)^{\alpha_1} \cdots (x^{n+1})^{\alpha_{n+1}}$ for $\mathbf{x} = (x^1, \ldots, x^{n+1}) \in \mathbb{R}^{n+1}$. Observe that

$$P(c\mathbf{x}) = c^k P(\mathbf{x}) \tag{11.36}$$

for any $c \in \mathbb{R}$ and $\mathbf{x} \in \mathbb{R}^{n+1}$. Let $N : \mathbf{x} \mapsto N_{\mathbf{x}} := \frac{\mathbf{x}}{\|\mathbf{x}\|}$ be the unit radial vector field on $\mathbb{R}^{n+1} \setminus \{\mathbf{0}\}$. By (11.36), we have that the first and second directional derivatives of P in the direction $N_{\mathbf{x}}$ are given by

$$N_{\mathbf{x}}(P) = k P(\mathbf{x}) \quad \text{and} \quad N_{\mathbf{x}}(N(P)) = k(k-1) P(\mathbf{x}) \tag{11.37}$$

for $\mathbf{x} \in S^n$.

Let $\Delta_{\mathbb{E}}$ denote the Euclidean Laplacian of $\mathbb{R}^{n+1}$ and let Δ_{S^n} denote the Laplacian of S^n. Let D be the Euclidean covariant derivative on $\mathbb{R}^{n+1}$ and let ∇ be the induced covariant derivative on S^n. Let $f : \mathbb{R}^{n+1} \to \mathbb{R}$. Using (4.125), we compute that for $V, W \in TS^n$,

$$\begin{aligned} (\nabla_V df)(W) &= V(df(W)) - df(\nabla_V W) \\ &= V(W(f)) - (D_V W)(f) - \mathrm{II}(V, W) N(f) \\ &= (D_V df)(W) - \langle V, W \rangle N(f) \end{aligned} \tag{11.38}$$

since $\mathrm{II} = \mathrm{I} = \langle \cdot, \cdot \rangle$ for S^n. Let $\{e_i\}_{i=1}^n$ be a local orthonormal frame for S^n at a point $\mathbf{x} \in S^n$ and let $e_{n+1} := N$, which is the unit outward pointing normal to S^n. By tracing (11.38), we have that

$$\begin{aligned} \Delta_{S^n} f &= \sum_{i=1}^n (\nabla_{e_i} df)(e_i) \\ &= \sum_{i=1}^{n+1} (D_{e_i} df)(e_i) - (D_{e_{n+1}} df)(e_{n+1}) - \sum_{i=1}^n \langle e_i, e_i \rangle N(f) \\ &= \Delta_{\mathbb{R}^{n+1}} f - N(N(f)) - nN(f), \end{aligned}$$

where f on the left-hand side denotes the restriction of f to S^n. Denoting $\frac{\partial}{\partial r} := N$, this says that

$$\Delta_{S^n} f = \Delta_{\mathbb{R}^{n+1}} f - \frac{\partial^2 f}{\partial r^2} - n\frac{\partial f}{\partial r}. \tag{11.39}$$

Let P be a **harmonic polynomial** of degree k on $\mathbb{R}^{n+1}$, so that P is a homogeneous polynomial of degree k and $\Delta_{\mathbb{R}^{n+1}} P = 0$. By (11.37), we have that on S^n,

$$\Delta_{S^n} P = -k(n+k-1)P. \tag{11.40}$$

That is, P restricted to S^n is an eigenfunction of the Laplacian on S^n with the associated eigenvalue equal to $k(n+k-1)$. For example, any linear function $L(\mathbf{x}) := \sum_{i=1}^{n+1} c_i x^i$ is a harmonic polynomial of degree 1 and hence restricts to an eigenfunction of Δ_{S^n} with eigenvalue n. Since n is the smallest nonzero eigenvalue of the Laplacian of S^n, we call n the **first eigenvalue** of the Laplacian of S^n and denote it by $\lambda_1(\Delta_{S^n})$. More generally, we define the first eigenvalue $\lambda_1(\Delta)$ of the Laplacian of a Riemannian manifold (M^n, g) in the corresponding way (see §11.2.3.4 below).

Let $\mathcal{H}_k(S^n)$ denote the space of harmonic polynomials of degree k on $\mathbb{R}^{n+1}$ restricted to S^n. Then each function $P \in \mathcal{H}_k(S^n)$ is an eigenfunction of Δ_{S^n} with eigenvalue $k(n+k-1)$. By a standard theorem (see [**BGM71**, Chapter 3, §C]), we have the L^2 Hilbert space decomposition

$$L^2(S^n) = \bigoplus_{k=0}^{\infty} \mathcal{H}_k(S^n). \tag{11.41}$$

11.2.3. The spectrum of the Laplacian.

In this section we present some basic facts regarding the spectrum of the Laplacian of a closed Riemannian manifold. For more details, see [**BGM71**], [**Cha84**], and the references in Remark 11.9 below.

Let f be a smooth function on M^n. Define the **Sobolev norm** of f by[2]

$$\|f\|_{W^{k,p}} := \left(\sum_{j=0}^{k} \int_{M^n} |\nabla^j f|^p d\mu \right)^{1/p}. \tag{11.42}$$

The **Sobolev space** $W^{k,p}(M^n)$ is the Banach space completion of $C^\infty(M^n)$ with respect to the Sobolev norm $\|\cdot\|_{W^{k,p}}$.

[2]The 'W' in $W^{k,p}$ stands for 'Wladimir,' in honor of Sergei Wladimirovich Sobolev.

11.2.3.1. *The discreteness of the spectrum.* Let (M^n, g) be a closed Riemannian manifold. Consider the linear partial differential operator

$$-\Delta + I : W^{2,2}(M^n) \to L^2(M^n),$$

where I denotes the inclusion map. This linear operator is invertible and its inverse

$$(-\Delta + I)^{-1} : L^2(M^n) \to W^{2,2}(M^n)$$

is a bounded linear operator by elliptic regularity theory (see Remark 11.9 below) and hence

$$(-\Delta + I)^{-1} : L^2(M^n) \to W^{2,2}(M^n) \hookrightarrow L^2(M^n)$$

is a compact linear operator by the Rellich–Kondrachov compactness theorem (see Evans [**Eva10**]). This implies that the eigenvalues μ_i of $(-\Delta+I)^{-1}$ are discrete and

$$1 = \mu_0 > \mu_1 \geq \mu_2 \geq \cdots$$

and $\lim_{i\to\infty} \mu_i = 0$. Thus, the eigenvalues of the Laplacian are $\lambda_i = \mu_i^{-1} - 1$ and satisfy

$$0 = \lambda_0 < \lambda_1 \leq \lambda_2 \leq \cdots, \tag{11.43}$$

where $\lim_{i\to\infty} \lambda_i = \infty$.

Remark 11.9. A particularly nice treatment of compact operators is presented in the book by Riesz and Sz.-Nagy [**RSN90**]. Elliptic regularity theory is discussed in Gilbarg and Trudinger [**GT01**] and Evans [**Eva10**].

11.2.3.2. *Eigenfunction expansion in* $L^2(M^n)$. For each eigenvalue λ_i of the Laplacian, we may choose an associated eigenfunction ϕ_i such that the collection $\{\phi_i\}_{i=0}^{\infty}$ is a complete orthonormal basis for the Hilbert space $L^2(M^n)$ (see Berger, Gauduchon, and Mazet [**BGM71**]). For any $f \in L^2(M^n)$ we have that

$$f = \sum_{i=0}^{\infty} \langle\langle f, \phi_i \rangle\rangle_{L^2} \, \phi_i \tag{11.44}$$

and its L^2-norm is given by

$$\|f\|_{L^2}^2 = \sum_{i=0}^{\infty} \langle\langle f, \phi_i \rangle\rangle_{L^2}^2 . \tag{11.45}$$

11.2.3.3. *The Rayleigh principle.* Let V_i denote the span of $\phi_0, \ldots, \phi_i$ in $W^{1,2}(M^n)$. By the **Rayleigh principle**, we have for $i \geq 0$ that

$$\lambda_{i+1} = \inf_{f \perp \mathrm{V}_i} \frac{\int_{M^n} |\nabla f|^2 d\mu}{\int_{M^n} f^2 \, d\mu}, \tag{11.46}$$

where $f \perp \mathrm{V}_i$ means that $\langle\langle f, \phi \rangle\rangle_{L^2} = 0$ for all $\phi \in \mathrm{V}_i$; see Chavel [**Cha84**, §I.5]. The quotient on the right-hand side of (11.46) is called the **Rayleigh–Ritz quotient** (see [**Ray45**]).

11.2.3.4. *The first eigenvalue of the Laplacian.* Let (M^n, g) be a closed Riemannian manifold and let

$$E := \{\lambda \in \mathbb{R}_+ : \lambda \text{ is an eigenvalue of } -\Delta\} \tag{11.47}$$

be the set of positive eigenvalues. By (11.43), E is an infinite discrete set. In particular, E has a minimum element, which we call the **first eigenvalue of the Laplacian** and we denote this minimum by $\lambda_1(\Delta)$.

For example, as we have seen above, the Laplacian of the unit n-sphere satisfies $\lambda_1(\Delta_{S^n}) = n$.

11.2.3.5. *Scaling properties of eigenvalues.* Let $c > 0$. If λ is an eigenvalue for Δ_g, then $c^{-1}\lambda$ is an eigenvalue for Δ_{cg}, and the λ-eigenspace for Δ_g is equal to the $c^{-1}\lambda$-eigenspace for Δ_{cg}. In particular,

$$\lambda_1(\Delta_{cg}) = c^{-1}\lambda_1(\Delta_g). \tag{11.48}$$

11.2.4. Lichnerowicz theorem. For closed Riemannian manifolds with positive Ricci curvature, the first eigenvalue of the Laplacian has a sharp lower bound.

Theorem 11.10 (Lichnerowicz [**Lic77**]). *If (M^n, g) is a closed Riemannian manifold with* $\mathrm{Ric} \geq (n-1)\, Kg$, *where K is a positive real number, then*

$$\lambda_1(\Delta) \geq nK. \tag{11.49}$$

Proof. Suppose that $u \not\equiv 0$ is a function and λ is a positive real number satisfying the eigenfunction equation $\Delta u + \lambda u = 0$. Using the algebraic inequality (5.108), the fundamental Bochner formula (11.21), and the eigenfunction equation, we compute that

$$\begin{aligned} \frac{1}{2}\Delta |\nabla u|^2 &= \left|\nabla^2 u\right|^2 + \langle \nabla \Delta u, \nabla u \rangle + \mathrm{Ric}(\nabla u, \nabla u) \\ &\geq \frac{1}{n}(\Delta u)^2 - \lambda |\nabla u|^2 + (n-1)\, K\, |\nabla u|^2 \\ &= \frac{1}{n}\lambda^2 u^2 + ((n-1)\, K - \lambda)\, |\nabla u|^2 . \end{aligned} \tag{11.50}$$

Integrating this and applying the divergence theorem yields

$$0 \geq \frac{1}{n}\lambda^2 \int_M u^2 d\mu + ((n-1)K - \lambda) \int_M |\nabla u|^2 \, d\mu. \tag{11.51}$$

Dividing this by $\int_M u^2 d\mu > 0$ and using the eigenvalue equality (11.32) implies that

$$0 \geq \frac{1}{n}\lambda^2 + ((n-1)K - \lambda)\lambda. \tag{11.52}$$

Since $\lambda > 0$, we conclude that $\lambda \geq nK$ and the estimate (11.49) follows. □

Remark 11.11. Obata [**Oba62**] proved that equality in (11.49) holds if and only if (M^n, g) is isometric to the n-sphere of radius $1/\sqrt{K}$.

11.2.5. Lower bound for the first eigenvalue of the Laplacian. We note that the first eigenvalue of the circle $S^1(r)$ of radius r is equal to $\frac{1}{r^2}$. As a Riemannian manifold, the diameter of $S^1(r)$ is equal to πr. Therefore, we have for $S^1(r)$ that

$$\lambda_1(\Delta) = \frac{1}{r^2} = \frac{\pi^2}{\operatorname{diam}(S^1(r))^2}. \tag{11.53}$$

A beautiful first eigenvalue lower bound depending on a gradient estimate is the following.

Theorem 11.12 (Li and Yau [**LY80**]). *Let (M^n, g) be a closed Riemannian manifold with nonnegative Ricci curvature. Then*

$$\lambda_1(\Delta) \geq \frac{\pi^2}{4 \operatorname{diam}(g)^2}, \tag{11.54}$$

where $\operatorname{diam}(g)$ *is the diameter of* (M^n, g)*, defined by* (7.147).

Proof. Suppose that λ is a positive eigenvalue of the Laplacian with an associated eigenfunction $u : M^n \to \mathbb{R}$ which is not identically zero. That is, $\Delta u + \lambda u = 0$ and $u \not\equiv 0$. Define $f : M^n \to \mathbb{R}$ by

$$f := f_\varepsilon := |\nabla u|^2 + (\lambda + \varepsilon)\, u^2, \tag{11.55}$$

where $\varepsilon > 0$. The ε serves as a 'fudge factor', which we will let tend to zero and which is included to ensure that the argument works effectively. See the remark after this proof for a motivation for the definition of f. Using (11.21) and (11.16), we compute that

$$\frac{1}{2}\Delta f = \langle \nabla u, \nabla \Delta u \rangle + |\nabla^2 u|^2 + \operatorname{Ric}(\nabla u, \nabla u) + (\lambda + \varepsilon)\, u \Delta u + (\lambda + \varepsilon)\, |\nabla u|^2.$$

Therefore, using the eigenvalue equation, we obtain

$$\frac{1}{2}\Delta f = |\nabla^2 u|^2 - \lambda (\lambda + \varepsilon)\, u^2 + \varepsilon |\nabla u|^2 + \operatorname{Ric}(\nabla u, \nabla u). \tag{11.56}$$

Since Ric ≥ 0 by assumption, we further obtain

$$\frac{1}{2}\Delta f \geq |\nabla^2 u|^2 - \lambda(\lambda+\varepsilon)u^2 + \varepsilon|\nabla u|^2. \tag{11.57}$$

By the elliptic maximum principle (see §11.1.1), at a point $x_0 \in M^n$ where f attains its maximum, $\nabla f(x_0) = 0$ and $\Delta f(x_0) \leq 0$. Let $\{e_i\}_{i=1}^n$ be any orthonormal frame at x_0. We calculate at x_0 that for $1 \leq i \leq n$,

$$\begin{aligned} 0 = \frac{1}{2}e_i(f) &= \frac{1}{2}e_i\big(|\nabla u|^2 + (\lambda+\varepsilon)u^2\big) \\ &= (\nabla^2 u)(e_i, \nabla u) + (\lambda+\varepsilon)u e_i(u). \end{aligned}$$

Now choose our orthonormal frame $\{e_i\}_{i=1}^n$ so that

$$\nabla u(x_0) = |\nabla u|(x_0)e_1.$$

Then, at x_0, we have $e_i(u) = \nabla u(x_0) \cdot e_i = |\nabla u|(x_0)\delta_{1i}$. It follows that at x_0,

$$0 = (\nabla^2 u)(e_i, |\nabla u| e_1) + (\lambda+\varepsilon)u|\nabla u|\delta_{1i}.$$

Thus, if $|\nabla u|(x_0) \neq 0$, then

$$(\nabla^2 u)(e_i, e_1) + (\lambda+\varepsilon)u\delta_{1i} = 0 \tag{11.58}$$

for $1 \leq i \leq n$. In particular, taking $i = 1$, we have that

$$(\nabla^2 u)(e_1, e_1) + (\lambda+\varepsilon)u = 0. \tag{11.59}$$

Therefore, if $\nabla u(x_0) \neq \vec{0}$, then at x_0 we have[3]

$$|\nabla^2 u|^2 \geq \big((\nabla^2 u)(e_1, e_1)\big)^2 = (\lambda+\varepsilon)^2 u^2. \tag{11.60}$$

By applying this inequality to (11.57), we obtain at x_0 that

$$\begin{aligned} 0 \geq \frac{1}{2}\Delta f &\geq (\lambda+\varepsilon)^2 u^2 - \lambda(\lambda+\varepsilon)u^2 + \varepsilon|\nabla u|^2 \\ &= \varepsilon(\lambda+\varepsilon)u^2 + \varepsilon|\nabla u|^2, \end{aligned} \tag{11.61}$$

which is a contradiction to the assumption that $\nabla u(x_0) \neq \vec{0}$. Hence, $\nabla u(x_0) = \vec{0}$ at any point x_0 at which f attains its maximum.

Let $\mu := \max_{x \in M^n} |u(x)|$. Then

$$f(x) \leq f(x_0) = (\lambda+\varepsilon)u(x_0)^2 \leq (\lambda+\varepsilon)\mu^2$$

for all $x \in M^n$. Thus, for all $x \in M^n$,

$$|\nabla u|^2(x) \leq (\lambda+\varepsilon)\big(\mu^2 - u(x)^2\big).$$

[3]Observe that this is superior to the algebraic estimate $|\nabla^2 u|^2 \geq \frac{1}{n}(\Delta u)^2 = \frac{1}{n}\lambda^2 u^2$ implied by (5.108).

Since the inequality above is true for all $\varepsilon > 0$, by taking $\varepsilon \to 0$ we obtain that

$$|\nabla u|^2 \leq \lambda \left(\mu^2 - u^2\right)$$

on all of M^n.

Now u vanishes somewhere on M^n since it follows from the divergence theorem that

$$0 = \int_{M^n} (\Delta u + \lambda u)\, d\mu = \lambda \int_{M^n} u\, d\mu,$$

and $\lambda > 0$. Let γ be a minimal geodesic joining points x_1 and x_2, where $u(x_1) = 0$ and $|u(x_2)| = \mu$. Then, by the inequality above,

$$\begin{aligned} \operatorname{diam}(g)\sqrt{\lambda} &\geq \int_\gamma \sqrt{\lambda} ds \geq \int_\gamma \frac{|\nabla u|}{\sqrt{\mu^2 - u^2}} ds \\ &\geq \int_0^\mu \frac{du}{\sqrt{\mu^2 - u^2}} = \frac{\pi}{2}. \end{aligned}$$

In conclusion, we have proved that if $\Delta u + \lambda u = 0$, then

$$\lambda \geq \frac{\pi^2}{4\operatorname{diam}(g)^2}.$$

The theorem follows. $\square$

Remark 11.13. For the case of M^1 being the circle of radius r as in Example 11.8, we have for $\lambda = \frac{k^2}{r^2}$ and $\varepsilon = 0$ that f defined by (11.55) is given by

$$\begin{aligned} f &= \left|\frac{\partial u}{\partial s}\right|^2 + \frac{k^2}{r^2} u^2 \\ &= \left(-a\frac{k}{r} \sin\left(\frac{ks}{r}\right) + b\frac{k}{r} \cos\left(\frac{ks}{r}\right)\right)^2 \\ &\quad + \frac{k^2}{r^2}\left(a\cos\left(\frac{ks}{r}\right) + b\sin\left(\frac{ks}{r}\right)\right)^2 \\ &= \frac{k^2}{r^2}(a^2 + b^2), \end{aligned}$$

which is a constant function on M^1.

Circles, being intrinsically flat, are model cases for Riemannian manifolds in the study of eigenvalues of the Laplacian. We can describe the choice of f in definition (11.55) as — for the purpose of applying the maximum principle — following the principle of choosing quantities which are constant in the model cases.

Remark 11.14. Li and Yau [**LY80**] were able to improve their estimate (11.54) to $\lambda_1(\Delta) \geq \frac{\pi^2}{2\operatorname{diam}(g)^2}$. The sharp estimate $\lambda_1(\Delta) \geq \frac{\pi^2}{\operatorname{diam}(g)^2}$ was later proved by Zhong and Yang [**ZY84**]. See Li [**Li93**, §5] for an exposition of Zhong and Yang's improvement.

11.3. The geometry of geodesic spheres

In this section we consider geodesic spherical coordinates and the second fundamental forms and mean curvatures of geodesic spheres.

11.3.1. The second fundamental form of geodesic spheres and the Riccati equation.

11.3.1.1. *Mean curvature and second fundamental form of geodesic spheres.* Let (M^n, g) be a Riemannian manifold. Let $p \in M^n$. Define the **distance function** $d_p : M^n \to \mathbb{R}$ by

$$d_p(x) := d(x, p). \tag{11.62}$$

The **geodesic sphere** of radius r centered at p, where $r > 0$, is defined by

$$S_r(p) := \{x \in M^n : d_p(x) = r\}. \tag{11.63}$$

Recall from Lemma 7.82 that the distance function d_p is smooth on $\mathcal{D}_p \setminus \{p\}$, where $\mathcal{D}_p$ is the injectivity range of $\exp_p$ defined by (7.192). The gradient of a smooth function is normal to its level sets. Since ∇d_p is a unit vector by Lemma 7.90, we have that the gradient ∇d_p is a unit normal vector field to the smooth hypersurface $S_r(p) \cap \mathcal{D}_p$ in M^n, which is a level set of $d_p|_{\mathcal{D}_p \setminus \{p\}}$. Thus, for all $r > 0$ we have that the set $S_r(p) \cap \mathcal{D}_p$ is a smooth hypersurface in M^n. In other words, for each $x \in \mathcal{D}_p \setminus \{p\}$ and $r := d(x,p)$, the geodesic sphere $S_r(p)$, on which x lies, is a smooth hypersurface in a neighborhood of x; moreover, the gradient $\nabla d_p(x)$ is a unit normal vector to $S_r(p)$ at x.

Definition 11.15. Let $\gamma_V : [0, L] \to M^n$ be a minimal geodesic with $\gamma_V(0) = p$ and $\gamma_V'(0) = V$, where V is a unit tangent vector. Let $\{e_i\}_{i=1}^{n-1}$ be such that $\{e_1, \ldots, e_{n-1}, V\}$ is an orthonormal frame at p. Let $\{J_i(r)\}_{i=1}^{n-1}$ be the unique Jacobi fields along γ_V satisfying

$$J_i(0) = \vec{0} \quad \text{and} \quad \nabla_V J_i = e_i \in T_pM \quad \text{for } 1 \leq i \leq n-1. \tag{11.64}$$

We call $\{J_i(r)\}_{i=1}^{n-1}$ the **Jacobi fields associated to** $\{e_i\}_{i=1}^{n-1}$.

For each i, define the 1-parameter family of geodesics $\gamma_{i,s} : [0, L] \to M^n$ by

$$\gamma_{i,s}(r) := \exp_p\big(r(V + se_i)\big) = \gamma_{V+se_i}(r). \tag{11.65}$$

Then J_i is the variation of this family $\gamma_{i,s}$ at $s = 0$:

$$J_i(r) = \left.\frac{\partial}{\partial s}\right|_{s=0} \gamma_{V+se_i}(r) = (d\exp_p)_{rV}(re_i); \tag{11.66}$$

see (7.134).

Let II denote the second fundamental form of $S_r(p) \cap \mathcal{D}_p$ with respect to the choice of normal ∇d_p. Define the **components of the first and second fundamental forms in geodesic spherical coordinates** by

$$g_{ij}(r) := g(J_i(r), J_j(r)) \qquad \text{and} \qquad \mathrm{II}_{ij}(r) := \mathrm{II}(J_i(r), J_j(r)) \tag{11.67}$$

for $1 \leq i, j \leq n-1$.

By (8.125), we have that

$$\mathrm{II}_{ij} = -\langle \nabla d_p, \nabla_{J_i} J_j \rangle = \langle \nabla_{J_i}(\nabla d_p), J_j \rangle = \langle \nabla_{\nabla d_p} J_i, J_j \rangle, \tag{11.68}$$

where the last equality is by Lemma 7.19(iv) and because J_i and ∇d_p are the 'coordinate variations' of the map $F(r,s) := \gamma_{i,s}(r)$ from a rectangle in $\mathbb{R}^2$ to M^n (see §7.3.1). We compute that

$$\begin{aligned} \frac{d}{dr} g_{ij} &= \nabla d_p\big(g(J_i, J_j)\big) \\ &= g\big(\nabla_{\nabla d_p} J_i, J_j\big) + g\big(J_i, \nabla_{\nabla d_p} J_j\big) \\ &= 2\,\mathrm{II}_{ij}\,. \end{aligned} \tag{11.69}$$

Definition 11.16. The **Jacobian of the exponential map** is defined by

$$\mathrm{J}(r) := \mathrm{J}(V, r) := \sqrt{\det(g_{ij}(r))} = \sqrt{\det(\langle J_i(r), J_j(r)\rangle)}, \tag{11.70}$$

where the $J_i(r)$ satisfy (11.64) and where $V \in S_pM$ is a unit vector.

By (11.66) and (11.70),

$$\mathrm{J}(r) = \sqrt{\det(\langle (d\exp_p)_{rV}(re_i), (d\exp_p)_{rV}(re_j)\rangle)}\,. \tag{11.71}$$

We compute using (11.69) that

$$\frac{d}{dr} \ln \mathrm{J}(r) = \frac{1}{2} g^{ij} \frac{d}{dr} g_{ij} = g^{ij}\, \mathrm{II}_{ij} = H \tag{11.72}$$

is equal to the mean curvature. Here we used the standard formula that if $A(r)$ is a 1-parameter family of square matrices, then

$$\frac{d}{dr} \det A(r) = \operatorname{trace}\left(A(r)^{-1} \frac{dA}{dr}(r) \right) \det A(r).$$

11.3.1.2. *Expansions for the mean curvature and second fundamental form.*

Lemma 11.17. *Let $g_{ij}(r)$ and $\mathrm{II}_{ij}(r)$ be defined by (11.67). In geodesic spherical coordinates, as $r \to 0$, we have that the metric, second fundamental form, and mean curvature satisfy for $1 \leq i, j \leq n-1$,*

$$g_{ij}(r) = r^2 \delta_{ij} + \mathrm{O}(r^4), \tag{11.73}$$

$$\mathrm{II}_{ij}(r) = r\, \delta_{ij} + \mathrm{O}(r^3), \tag{11.74}$$

$$H(r) = \frac{n-1}{r} + \mathrm{O}(r), \tag{11.75}$$

respectively.

Proof. We first prove (11.73) by using Taylor's theorem. We have the following computations of the derivatives of $g_{ij}(r)$ at $r = 0$. Since $J_i(0) = \vec{0}$, we have $g_{ij}(0) = 0$,

$$\left.\frac{d}{dr}\right|_{r=0} g_{ij}(r) = \left(\langle \nabla_V J_i, J_j \rangle + \langle J_i, \nabla_V J_j \rangle\right)\big|_{r=0} = 0,$$

$$\begin{aligned}\left.\frac{d^2}{dr^2}\right|_{r=0} g_{ij}(r) &= \left(\langle \nabla_V(\nabla_V J_i), J_j \rangle + \langle J_i, \nabla_V(\nabla_V J_j) \rangle + 2\langle \nabla_V J_i, \nabla_V J_j \rangle\right)\big|_{r=0} \\ &= \left(- \langle \mathrm{Rm}(J_i, V)V, J_j \rangle - \langle \mathrm{Rm}(J_j, V)V, J_i \rangle\right)\big|_{r=0} + 2\langle e_i, e_j \rangle \\ &= 2\delta_{ij}, \end{aligned} \tag{11.76}$$

where we used the Jacobi equation (7.137) to obtain the second equality in (11.76), and (see Exercise 11.9)

$$\left.\frac{d^3}{dr^3}\right|_{r=0} g_{ij}(r) = 0. \tag{11.77}$$

Since $g_{ij}(r)$ is smooth at $r = 0$, by Taylor's theorem, we conclude (11.73).

Since $g_{ij}(r)$ is smooth at $r = 0$, we have that

$$\mathrm{II}_{ij}(r) = \frac{1}{2}\frac{d}{dr} g_{ij}(r)$$

is also smooth at $r = 0$ and $\mathrm{II}_{ij}(0) = 0$. For each integer $k \geq 1$,

$$\frac{d^k}{dr^k} \mathrm{II}_{ij}(0) = \frac{1}{2}\frac{d^{k+1}}{dr^{k+1}} g_{ij}(0). \tag{11.78}$$

In particular,

$$\frac{d}{dr} \mathrm{II}_{ij}(0) = \frac{1}{2}\frac{d^2}{dr^2} g_{ij}(0) = \delta_{ij} \quad \text{and} \quad \frac{d^2}{dr^2} \mathrm{II}_{ij}(0) = \frac{1}{2}\frac{d^3}{dr^3} g_{ij}(0) = 0. \tag{11.79}$$

Equation (11.74) follows.

Finally, (11.75) follows from (11.74) and

$$g^{ij}(r) = \frac{1}{r^2}\delta_{ij} + \mathrm{O}(1). \tag{11.80}$$

To see the last display, let $g^{ij}(r) =: \frac{1}{r^2}\delta_{ij} + a^{ij}(r)$. We calculate that

$$\delta_{ik} = g^{ij}(r) g_{jk}(r) = \left(\frac{1}{r^2}\delta_{ij} + a^{ij}(r)\right)\left(r^2\delta_{jk} + \mathrm{O}(r^4)\right).$$

Thus,

$$\frac{1}{r^2}\delta_{ij} + a^{ij}(r) = \delta_{ik}\left(\frac{1}{r^2}\delta_{jk} + \mathrm{O}(1)\right),$$

and (11.80) follows. □

11.3.1.3. *Geodesic spherical coordinates per se.* The components of the metric g_{ij} in geodesic spherical coordinates are defined using Jacobi fields by (11.67). However, we have not actually defined local coordinates. We can do this as follows. Let $p \in M^n$ and let $S_pM = \{v \in T_pM : |v| = 1\}$ be the unit tangent space at p. Let $\{\theta^i\}_{i=1}^{n-1}$ be local coordinates on S_pM. Then **geodesic spherical coordinates** centered at p are local coordinates $\{x^i\}_{i=1}^n$ defined on $B_{\mathrm{inj}(p)}(p) \setminus \{p\}$ by

$$x^i(x) = \theta^i\left(\frac{\exp_p^{-1}(x)}{|\exp_p^{-1}(x)|}\right), \qquad x^n(x) = |\exp_p^{-1}(x)| = d(x,p) \tag{11.81}$$

for $1 \leq i \leq n-1$. We denote $r := x^n$. Let $g_{ij} = g\big(\frac{\partial}{\partial x^i}, \frac{\partial}{\partial x^j}\big)$ for $1 \leq i,j \leq n$. By the Gauss lemma (Lemma 7.29), we have that

$$g_{nn} = 1 \quad \text{and} \quad g_{in} = 0 \ \text{ for } \ 1 \leq i \leq n-1.$$

Note that this implies that

$$g^{nn} = 1 \quad \text{and} \quad g^{in} = 0 \ \text{ for } \ 1 \leq i \leq n-1.$$

Let $\mathrm{II}_{ij} = \mathrm{II}\left(\frac{\partial}{\partial x^i}, \frac{\partial}{\partial x^j}\right)$, $1 \leq i,j \leq n-1$, denote the components of the second fundamental form in geodesic spherical coordinates. Since $\frac{\partial}{\partial x^n}$ is the unit (outward) normal to the geodesic spheres, we have that

$$\mathrm{II}_{ij} = -g\left(\nabla_{\frac{\partial}{\partial x^i}}\frac{\partial}{\partial x^j}, \frac{\partial}{\partial x^n}\right) = -\Gamma_{ij}^n. \tag{11.82}$$

Hence, the mean curvatures of the geodesic spheres may be expressed as

$$H = -\sum_{i,j=1}^{n-1} g^{ij}\Gamma_{ij}^n. \tag{11.83}$$

11.3.1.4. *The Laplacian and Hessian in geodesic spherical coordinates.* In geodesic spherical coordinates $\{x^a\}_{a=1}^n$ as defined by (11.81), the Laplacian acting on a function $f : M^n \to \mathbb{R}$ is given by

$$\begin{aligned}\Delta f &= \sum_{a,b=1}^{n} g^{ab}\left(\frac{\partial^2 f}{\partial x^a \partial x^b} - \sum_{c=1}^{n} \Gamma_{ab}^c \frac{\partial f}{\partial x^c}\right) \\ &= \frac{\partial^2 f}{\partial (x^n)^2} + \sum_{a,b=1}^{n-1} g^{ab}\left(\frac{\partial^2 f}{\partial x^a \partial x^b} - \Gamma_{ab}^n \frac{\partial f}{\partial x^n} - \sum_{c=1}^{n-1} \Gamma_{ab}^c \frac{\partial f}{\partial x^c}\right) \\ &= \frac{\partial^2 f}{\partial r^2} + H\frac{\partial f}{\partial r} + \Delta_{S_r(p)}(f|_{S_r(p)}) \\ &= \frac{\partial^2 f}{\partial r^2} + \left(\frac{\partial}{\partial r}\ln \mathrm{J}\right)\frac{\partial f}{\partial r} + \Delta_{S_r(p)}(f|_{S_r(p)})\end{aligned} \tag{11.84}$$

since $\Gamma_{nn}^a = 0$ for $a = 1, \ldots, n$ and where

$$\Delta_{S_r(p)} = \sum_{a,b=1}^{n-1} g^{ab}\left(\frac{\partial^2 f}{\partial x^a \partial x^b} - \sum_{c=1}^{n-1} \Gamma_{ab}^c \frac{\partial f}{\partial x^c}\right)$$

is the Laplacian of $S_r(p)$ with respect to the induced metric. Note that the standard formula for the Euclidean Laplacian in spherical coordinates is a special case of (11.84).

In particular, if $f := f \circ d_p$ is a function of the distance to p, then we obtain

$$\Delta f = f'' + f' \partial_r \ln \mathrm{J} = f'' + f' H. \tag{11.85}$$

In particular, the Laplacian of the distance function $r(\cdot) := d_p(\cdot)$ at a point x is equal to the mean curvature at x of the geodesic sphere on which x lies:

$$\Delta r = H \qquad \text{on } \mathcal{D}_p \setminus \{p\}. \tag{11.86}$$

Consequently, the square of the distance function satisfies

$$\Delta\left(\frac{r^2}{2}\right) = rH + 1 \qquad \text{on } \mathcal{D}_p, \tag{11.87}$$

where $rH(p) := n - 1$. For example, on Euclidean space $(\mathbb{R}^n, g_{\mathbb{E}})$ we have

$$\Delta_{\mathbb{E}} r = \frac{n-1}{r} \text{ on } \mathbb{R}^n \setminus \{\mathbf{0}\}, \qquad \Delta_{\mathbb{E}}\left(\frac{r^2}{2}\right) = n \text{ on } \mathbb{R}^n.$$

Now assume that $f = f(d_p)$ is a radial function. Then, from

$$\nabla_a \nabla_b f = \nabla^2 f(\partial_a, \partial_b) = \frac{\partial^2 f}{\partial x^a \partial x^b} - \sum_{c=1}^{n} \Gamma_{ab}^c \frac{\partial f}{\partial x^c},$$

we obtain that for $1 \leq i, j < n$,

$$\begin{aligned}\nabla_n \nabla_n f &= f'', \\ \nabla_i \nabla_n f &= 0, \\ \nabla_i \nabla_j f &= f' \, \mathrm{II}_{ij} .\end{aligned}$$

In particular, for the distance function $r = d_p$ we have for $1 \leq i, j < n$ and $1 \leq a \leq n$ that

$$\nabla_i \nabla_j r = \mathrm{II}_{ij}, \qquad \nabla_a \nabla_n r = 0 \qquad \text{on } \mathcal{D}_p \setminus \{p\}. \tag{11.88}$$

That is, the Hessian of the distance function at a point x is equal to the second fundamental form of the geodesic sphere on which x lies. Observe that by tracing (11.88), we recover (11.86).

11.3.1.5. *The Riccati equation.* Fix $p \in M^n$. We compute the radial derivative of the components of the second fundamental form of the geodesic sphere $S_r(p)$ in geodesic spherical coordinates. Let $T(r) := \gamma'(r)$, so that $T(0) = V$. We calculate that (cf. (11.76))

$$\begin{aligned}2\frac{d}{dr}\,\mathrm{II}_{ij}(r) &= \frac{d^2}{dr^2} g_{ij}(r) \\ &= \langle \nabla_T(\nabla_T J_i), J_j\rangle + \langle J_i, \nabla_T(\nabla_T J_j)\rangle + 2\langle \nabla_T J_i, \nabla_T J_j\rangle \\ &= -\langle \mathrm{Rm}(J_i, T)T, J_j\rangle - \langle \mathrm{Rm}(J_j, T)T, J_i\rangle + 2\langle \nabla_{J_i} T, \nabla_{J_j} T\rangle,\end{aligned}$$

where we used the Jacobi equation to obtain the last equality. Thus,

$$\frac{d}{dr}\,\mathrm{II}_{ij}(r) = \frac{d}{dr}\big(\mathrm{II}(J_i(r), J_j(r))\big) = -\langle \mathrm{Rm}(J_i, T)T, J_j\rangle + \langle \mathrm{L}(J_i), \mathrm{L}(J_j)\rangle, \tag{11.89}$$

where L is the Weingarten map. This is the **Riccati equation** (see Gray [**Gra04**, Chapter 3]).

The entries g^{ij} of the inverse matrix of the components of the metric $(g_{ij})_{i,j=1}^n$ satisfy

$$\frac{d}{dr} g^{ij} = -\sum_{k,\ell=1}^{n-1} g^{ik} g^{j\ell} \frac{d}{dr} g_{k\ell} = -2 \sum_{k,\ell=1}^{n-1} g^{ik} g^{j\ell}\, \mathrm{II}_{k\ell} \tag{11.90}$$

for $1 \le i, j \le n$. By using this and (11.89), we compute that

$$\begin{aligned}\frac{d}{dr}H &= \frac{d}{dr}\left(\sum_{i,j}^{n-1} g^{ij}\,\mathrm{II}_{ij}\right)\\ &= g^{ij}\frac{d}{dr}\,\mathrm{II}_{ij} + \mathrm{II}_{ij}\,\frac{d}{dr}g^{ij}\\ &= g^{ij}\big(-\langle \mathrm{Rm}(J_i,T)T,J_j\rangle + \langle \mathrm{L}(J_i),\mathrm{L}(J_j)\rangle\big)\\ &\quad - 2g^{ik}g^{j\ell}\,\mathrm{II}_{k\ell}\,\mathrm{II}_{ij}\\ &= -\operatorname{Ric}(T,T) - |\mathrm{II}|_g^2,\end{aligned}$$

where we used that $g^{ij}\langle \mathrm{L}(J_i),\mathrm{L}(J_j)\rangle = |\mathrm{II}|_g^2$. That is, we have the following result.

Lemma 11.18. *Given any $p \in M^n$, along any geodesic emanating from p the mean curvature function $r \mapsto H(r)$ of the geodesic spheres $S_r(p)$ satisfies*

$$\frac{d}{dr}H(r) = -\operatorname{Ric}(\gamma'(r),\gamma'(r)) - |\mathrm{II}|_g^2(r). \tag{11.91}$$

Assuming a Ricci curvature lower bound, by the Riccati equation (11.91) and the algebraic inequality $|\mathrm{II}|^2 \ge \frac{1}{n-1}H^2$, we have the following.

Corollary 11.19. *If $\operatorname{Ric} \ge (n-1)Kg$, then the mean curvature function of the geodesic spheres satisfies*

$$\frac{d}{dr}\left(\frac{H(r)}{n-1}\right) \le -K - \left(\frac{H(r)}{n-1}\right)^2. \tag{11.92}$$

Recall that by (11.75), we have that $H(r)$ satisfies the 'initial condition':

$$\lim_{r\to 0_+}\frac{rH(r)}{n-1} = 1. \tag{11.93}$$

11.3.2. Geodesic spheres in simply connected space forms and rotationally symmetric metrics.

Recall from Exercise 8.16 that if we have a rotationally symmetric metric of the form

$$g(r,v) = dr^2 + \phi(r)^2\, g_{S^{n-1}}(v) \tag{11.94}$$

for some 'warping' function $\phi(r)$, then the sectional curvatures of g are given by

$$K_{\mathrm{rad}}(r,v) = -\frac{\phi''(r)}{\phi(r)} \quad\text{and}\quad K_{\mathrm{sph}}(r,v) = \frac{1-(\phi'(r))^2}{\phi^2(r)}, \tag{11.95}$$

where $K_{\text{rad}}(r,v)$ is the sectional curvature of planes containing the radial vector $\frac{\partial}{\partial r}$ and $K_{\text{sph}}(r,v)$ is the sectional curvature of planes perpendicular to $\frac{\partial}{\partial r}$. Moreover, if $\phi : [0,\rho) \to [0,\infty)$, $\rho \in (0,\infty]$, with $\phi(r) > 0$ for $r > 0$, then the metric g defined by (11.94) extends smoothly over the origin if and only if $\phi(0) = 0$, $\phi'(0) = 1$, and $\frac{d^{2k}\phi}{dr^{2k}}(0) = 0$ for $k \geq 1$ (see Lemma 2.10 in [**CK04**]). We will assume these conditions, so that we can take the underlying manifold M^n to be equal to $[0,\rho) \times S^{n-1}$ with $\{0\} \times S^{n-1}$ identified to a point, which we call the origin and denote by o.

For example, we easily check that for any $K \in \mathbb{R}$, if we take the warping function to be $\phi(r) = \operatorname{sn}_K(r)$, where sn_K is the modulated sine function defined by (7.126), then $g = g_K$ is a constant sectional curvature K-metric, which extends smoothly to $M^n = M^n_K$.

Another way of deriving formula (11.95) for the radial sectional curvatures is to consider the geodesic spheres centered at o. Given an orthonormal basis of tangent vectors $\{e_i\}_{i=1}^{n-1}$ in TS^{n-1} with respect to the metric $g_{S^{n-1}}$, we have that

$$J_i(r) := e_i|_{\{r\}\times S^{n-1}}$$

are Jacobi fields along a geodesic $\gamma : r \mapsto (r,v)$ satisfying $J_i(0) = \vec{0}$ and $\nabla_{\gamma'(0)} J_i = e_i$. We have that

$$g_{ij}(r,v) = g(J_i(r), J_j(r)) = \phi^2(r)\, g_{S^{n-1}}(e_i,e_j) = \phi^2(r)\,\delta_{ij}. \tag{11.96}$$

Using (11.69), we compute that

$$\mathrm{II}_{ij}(r,v) = \frac{1}{2}\frac{d}{dr} g_{ij} = \phi(r)\phi'(r)\delta_{ij} = \frac{\phi'(r)}{\phi(r)} g_{ij}(r,v). \tag{11.97}$$

That is, the spheres $S_r(o)$ are **totally umbilical** with principal curvature $\kappa(r)$ equal to $\kappa(r) = \frac{\phi'(r)}{\phi(r)}$.[4] The intrinsic sectional curvatures of the hypersurface $S_r(o)$, since it has radius $\phi(r)$, are equal to

$$K_{\text{intrinsic}}(r) = \frac{1}{\phi^2(r)}.$$

From the Gauss equation (8.130), we have

$$K_{\text{sph}}(r,v) = K_{\text{intrinsic}}(r) - \kappa^2(r) = \frac{1}{\phi^2(r)} - \frac{(\phi')^2(r)}{\phi^2(r)},$$

which agrees with (11.95).

[4] A hypersurface is totally umbilical if at each point all of its principal curvatures are equal.

By tracing (11.97), we see that the mean curvature of $S_r(o)$ at (r, v) is given by $H(r,v) = (n-1)\frac{\phi'(r)}{\phi(r)}$. In particular, for the constant sectional curvature K-metric g_K, the mean curvature $H_K(r)$ of the geodesic sphere $(S_K)_r(o)$ is given by
(11.98)

$$H_K(r) := (n-1)\frac{\mathrm{cn}_K(r)}{\mathrm{sn}_K(r)} = \begin{cases} (n-1)\sqrt{K}\cot\left(\sqrt{K}r\right) & \text{if } K > 0, \\ \dfrac{n-1}{r} & \text{if } K = 0, \\ (n-1)\sqrt{|K|}\coth\left(\sqrt{|K|}r\right) & \text{if } K < 0, \end{cases}$$

where sn_K and cn_K are the modulated sine and cosine functions defined by (7.126) and (7.127), respectively. If $K \leq 0$, then we take the domain of H_K to be $(0,\infty)$. If $K > 0$, then we take the domain of H_K to be $(0, \pi/\sqrt{K})$.

Observe, as a special case of (11.75), that $H_K(r) = \frac{n-1}{r} + \mathrm{O}(r)$. Note that $H_K(r)$ is a solution to the equality case of (11.92); that is,

$$\frac{d}{dr}\left(\frac{H_K(r)}{n-1}\right) = -K - \left(\frac{H_K(r)}{n-1}\right)^2. \tag{11.99}$$

11.3.3. The geodesic sphere mean curvature comparison theorem. Given $x \in \mathcal{D}_p \setminus \{p\}$, define $H(x)$ to be the mean curvature of $S_r(p)$ at x, where $r := d(x,p)$. By ordinary differential equation comparison methods, we prove the following.

Theorem 11.20 (Mean curvature comparison for geodesic spheres). *Let (M^n, g) be a complete Riemannian manifold such that the Ricci curvature satisfies the lower bound* $\mathrm{Ric} \geq (n-1)Kg$ *for some $K \in \mathbb{R}$. Then, given any $p \in M^n$, we have that the mean curvature of the geodesic sphere $S_r(p)$ satisfies*

$$H(x) \leq H_K(r) \tag{11.100}$$

for all points $x \in S_r(p)$ at which the distance function d_p is smooth; i.e., $x \in \mathcal{D}_p \setminus \{p\}$.

Moreover (the 'equality case'), if $H(x) = H_K(r)$ for some $x \in \mathcal{D}_p \setminus \{p\}$, then, along the unique unit-speed minimal geodesic $\gamma : [0,r] \to M^n$ joining p and x, the ('radial') sectional curvatures of all 2-planes containing $\gamma'(s)$ are equal to K.

Proof. We compute from (11.92) and (11.99) that

(11.101)
$$\begin{aligned}
&\frac{d}{dr}\left(\operatorname{sn}_K^2(r)\left(\frac{H(r)-H_K(r)}{n-1}\right)\right)\\
&\quad\le 2\operatorname{sn}_K(r)\operatorname{cn}_K(r)\left(\frac{H(r)-H_K(r)}{n-1}\right)-\operatorname{sn}_K^2(r)\frac{H^2(r)-H_K^2(r)}{(n-1)^2}\\
&\quad=-\operatorname{sn}_K^2(r)\left(\frac{H(r)-H_K(r)}{n-1}\right)^2\\
&\quad\le 0,
\end{aligned}$$

where we used $\operatorname{cn}_K(r)=\frac{1}{n-1}\operatorname{sn}_K(r)H_K(r)$ to obtain the equality on the third line. Since $\lim_{r\to 0_+}\operatorname{sn}_K^2(r)(H(r)-H_K(r))=0$, the monotonicity formula (11.101) implies that

$$\operatorname{sn}_K^2(r)(H(r)-H_K(r))\le 0.$$

Inequality (11.100) follows.

Finally, we consider the equality case. Here, $H(x)=H_K(r)$ for some $x\in\mathcal{D}_p\setminus\{p\}$, which implies by (11.101) that $H(\gamma(s))=H_K(s)$ for all $s\in(0,r_x]$. By returning to the mean curvature evolution equation (11.91), we also conclude that $\mathrm{II}=\frac{1}{n-1}Hg$ along γ. Thus $\mathrm{II}=\frac{1}{n-1}H_K(s)g$ at $\gamma(s)$ for all $s\in[0,r_x]$. Now, the second fundamental form evolution equation (11.89) implies that the sectional curvatures of all 2-planes containing $\gamma'(s)$ are equal to K. The theorem follows. □

We remark that if $K>0$, then by Myers's theorem (Theorem 7.68), we have $\operatorname{diam}(M^n,g)\le\pi/\sqrt{K}$, and if $d(x,p)=\pi/\sqrt{K}$, then x is a conjugate point of p.

Corollary 11.21 (Jacobian comparison theorem)**.** *If* (M^n,g) *is a complete Riemannian manifold satisfying* $\operatorname{Ric}\ge(n-1)Kg$ *for some* $K\in\mathbb{R}$*, then*

$$\frac{d}{dr}\left(\frac{\mathrm{J}(r)}{\operatorname{sn}_K^{n-1}(r)}\right)\le 0. \tag{11.102}$$

Proof. Using (11.72) and (11.100), we compute that

(11.103)
$$\frac{d}{dr}\ln\left(\frac{\mathrm{J}(r)}{\operatorname{sn}_K^{n-1}(r)}\right)=\frac{d}{dr}\ln\mathrm{J}(r)-(n-1)\frac{\operatorname{cn}_K(r)}{\operatorname{sn}_K(r)}=H-H_K(r)\le 0.$$

□

11.3.4. The local Laplacian comparison theorem. Since the regular part $S_r(p) \cap \mathcal{D}_p$ of each geodesic sphere $S_r(p)$ is a smooth level set of the distance function d_p, by (8.139) and since $|\nabla d_p| = 1$, we have that

$$\mathrm{II}(x) = (\nabla^2 d_p)|_{TS_r(p)}(x) \tag{11.104}$$

for all $x \in S_r(p) \cap \mathcal{D}_p$, where $r = d_p(x)$. That is, the second fundamental form of a geodesic sphere is equal to the Hessian of the distance function restricted to its tangent spaces.

By tracing formula (11.104) and using that

$$(\nabla^2 d_p)(\nabla d_p, \nabla d_p) = \frac{1}{2}\langle \nabla |\nabla d_p|^2, \nabla d_p \rangle = 0,$$

we obtain that

$$H(x) = \Delta d_p(x), \tag{11.105}$$

where Δ is the Laplacian with respect to g acting on functions on M^n. So, the mean curvature of a geodesic sphere is equal to the Laplacian of the distance function. Thus, as a consequence of the 'mean curvature comparison' theorem (Theorem 11.20), we have the following **local Laplacian comparison theorem** (see [**Li12**]).

Theorem 11.22 (Local Laplacian comparison theorem). *Let (M^n, g) be a complete Riemannian manifold with* $\mathrm{Ric} \geq (n-1)Kg$ *for some $K \in \mathbb{R}$. For any $p \in M^n$ we have that the function d_p is smooth on $\mathcal{D}_p \setminus \{p\}$ and*

$$\Delta d_p(x) \leq H_K(d_p(x)) \qquad \text{for } x \in \mathcal{D}_p \setminus \{p\}, \tag{11.106}$$

where H_K is defined by (11.98).

11.3.5. The global Laplacian comparison theorem. The Laplacian comparison theorem holds on all of M^n in the sense of distributions as we prove in Theorem 11.24 below.

Definition 11.23. We say that an inequality $\Delta f \leq F$ on M^n holds **in the sense of distributions** if for any nonnegative C^∞ function φ on M^n with compact support, we have that

$$\int_{M^n} f \Delta \varphi \, d\mu \leq \int_{M^n} F \varphi \, d\mu. \tag{11.107}$$

Note that if f is C^2 and if $\Delta f \leq F$ on M^n holds in the usual sense, then (11.107) holds for all nonnegative C^∞ functions φ, and so $\Delta f \leq F$ holds in the sense of distributions.

Theorem 11.24 (Global Laplacian comparison theorem)**.** *Let (M^n, g) be a complete Riemannian manifold with* $\operatorname{Ric} \geq (n-1)Kg$ *for some $K \in \mathbb{R}$. For any $p \in M^n$ we have that the inequality*

$$\Delta d_p \leq H_K \circ d_p \tag{11.108}$$

holds in the sense of distributions on M^n.

Proof. For $r > 0$, let

$$C(r) := \left\{ v \in S_pM : \gamma_v(t) = \exp_p(tv),\ t \in [0, r],\ \text{is minimizing} \right\}. \tag{11.109}$$

This is the set of unit tangent vectors v at p such that the geodesic with initial velocity v minimizes up to at least distance r. Observe the set-monotonicity property that if $r_1 \leq r_2$, then $C(r_2) \subset C(r_1)$.

Let $\mathrm{J}(v, r)$ be the Jacobian of the exponential map (see (11.70) and (11.71)[5]) and $d\mu_{S_pM}(v)$ be the volume form of the metric on S_pM induced from the Euclidean metric g_p on T_pM. The cut locus of p has measure zero and, by the pull-back formula, we have that

$$\exp_p^* (d\mu) = \mathrm{J}\, d\mu_{S_pM} \wedge dr \tag{11.110}$$

inside the cut locus of p. Thus, for any integrable function f on a geodesic ball $B_{\overline{r}}(p)$, where $\overline{r} > 0$, we have that

$$\int_{B_{\overline{r}}(p)} f(x)\, d\mu\,(x) = \int_0^{\overline{r}} \left(\int_{C(r)} f\big(\exp_p(v, r)\big)\, \mathrm{J}\,(v, r)\, d\mu_{S_pM}(v) \right) dr.$$

In particular, for any nonnegative function $\varphi \in C^\infty(M^n)$ with compact support,

$$\begin{aligned} &\int_{M^n} \varphi(x)\, H_K(d_p(x))\, d\mu(x) \\ &\qquad = \int_0^\infty \int_{C(r)} \varphi\left(\exp_p(v, r)\right) H_K(r)\, \mathrm{J}(v, r)\, d\mu_{S_pM}(v)\, dr. \end{aligned}$$

Given a unit vector $v \in S_pM$, let r_v be the largest value of r such that $t \mapsto \gamma_v(t) = \exp_p(tv)$ minimizes up to $t = r$; cf. Definition 7.77. By Fubini's theorem, we have

$$\int_{M^n} \varphi(x)\, H_K(d_p(x))\, d\mu(x) = \int_{S_pM} \int_0^{r_v} \varphi(\exp_p(v, r)) H_K(r)\, \mathrm{J}(v, r) dr\, d\mu_{S_pM}(v).$$

Now for $0 < r < r_v$, by the geodesic sphere mean curvature comparison inequality (11.100) and by (11.72), we have that

$$H_K(r)\, \mathrm{J}(v, r) \geq H(v, r)\, \mathrm{J}(v, r) = \frac{\partial}{\partial r} \mathrm{J}(v, r).$$

[5]Namely, $\mathrm{J}(v, r)$ is equal to the Jacobian $\mathrm{J}(r)$ in (11.70) associated to the geodesic γ_v.

Hence,

$$\begin{aligned}\int_{M^n} \varphi(x) H_K(d_p(x))\, d\mu(x) &\geq \int_{S_pM}\int_0^{r_v} \varphi(\exp_p(v,r)) \frac{\partial}{\partial r} \mathrm{J}(v,r)\, dr\, d\mu_{S_pM}(v) \\ &= -\int_{S_pM}\int_0^{r_v} \frac{\partial}{\partial r}(\varphi\circ\exp_p)(v,r)\, \mathrm{J}(v,r)\, dr\, d\mu_{S_pM}(v) \\ &\quad + \int_{S_pM} \varphi(\exp_p(v,r_v))\, \mathrm{J}(v,r_v)\, d\mu_{S_pM}(v),\end{aligned}$$

where we integrated by parts and used $\lim_{r\to 0_+} \mathrm{J}(v,r) = 0$. Since the last line is nonnegative, we have by the Gauss lemma (Lemma 7.29) that

$$\int_{M^n} \varphi(x) H_K(d_p(x))\, d\mu(x) \geq -\int_{M^n} \langle \nabla\varphi, \nabla d_p\rangle\, d\mu = \int_{M^n} d_p \Delta\varphi\, d\mu,$$

where the last equality follows from the fact that the function d_p is Lipschitz on M^n and the divergence theorem holds for Lipschitz functions (see [**Eva10**], [**EG15**]). This completes the proof of Theorem 11.24. □

11.3.6. The Laplacian comparison theorem in the barrier sense. In addition to holding in the distributional sense, the Laplacian comparison theorem holds in the barrier sense.

Definition 11.25. We say that a continuous function $f : M^n \to \mathbb{R}$ and a function $h : M^n \to \mathbb{R}$ satisfy

$$\Delta f \geq h \quad \text{on } M^n \textbf{ in the barrier sense}$$

if for every $p \in M^n$ and $\varepsilon > 0$ there exists a C^2 function β (called a **lower barrier** of f at p) defined on a neighborhood U of p satisfying $\beta(p) = f(p)$, $\beta \leq f$ on U, and $\Delta\beta \geq h - \varepsilon$ on U.

The definition of $\Delta f \leq h$ in the barrier sense, and of a C^2 function β being an **upper barrier** of f at p, is defined analogously. Namely, the only changes to the definition above are that we now require $\beta \geq f$ and $\Delta\beta \leq h + \varepsilon$ on U.

Lemma 11.26 (Upper bound for Δd in terms of the Ricci curvature). *Let (M^n, g) be a complete Riemannian manifold, let $p \in M^n$, and let $x \in M^n \setminus \{p\}$. Let $\gamma : [0, r_x] \to M^n$ be any unit-speed minimal geodesic joining p to x and let $\zeta : [0, r_x] \to \mathbb{R}$, where $r_x := d_p(x)$, be any continuous piecewise smooth function satisfying $\zeta(0) = 0$ and $\zeta(r_x) = 1$. Then, we have that*

$$\Delta d_p(x) \leq \int_0^{r_x} \left((n-1)(\zeta'(s))^2 - \zeta^2 \operatorname{Ric}(\gamma'(s), \gamma'(s))\right) ds \tag{11.111}$$

in the barrier sense of Definition 11.25.

Proof. First we will construct an upper barrier for the distance function d_p in a neighborhood of $x \in M^n \setminus \{p\}$. Note that we allow x to be in the cut locus of p. Let

$$\varepsilon := \min\left\{\min_{s\in[0,r_x]}\left(\mathrm{inj}_g(\gamma(s))\right), \frac{1}{2}d_p(x)\right\} > 0,$$

and let $B_\varepsilon(\vec{0})$ denote the ball of radius ε centered at the origin in T_xM.

We extend the minimal geodesic $\gamma : [0, r_x] \to M^n$ to an n-parameter family of paths $\{\gamma^V\}_{V\in B_\varepsilon(\vec{0})}$ as follows. Given $V \in T_xM$, define the vector field $V(s)$ along $\gamma(s)$ by

$$V(s) \in T_{\gamma(s)}M \quad \text{for } s \in [0, r_x],$$

to be the parallel translation of V along γ, so that $V = V(r_x)$.

Define the continuous piecewise smooth vector field $\overline{V}$ along γ by

$$\overline{V}(s) = \zeta(s)\, V(s) \in T_{\gamma(s)}M, \tag{11.112}$$

where $\zeta : [0, r_x] \to \mathbb{R}$ is as in the hypothesis. Note that $\overline{V}(0) = \vec{0} \in T_pM$ and $\overline{V}(r_x) = V \in T_xM$. Define

$$\gamma^V(s) = \exp_{\gamma(s)}\left(\overline{V}(s)\right)$$

for $s \in [0, r_x]$ and $V \in B_\varepsilon(\vec{0})$. Then:

(1) $\gamma^{\vec{0}}(s) = \gamma(s)$ for $s \in [0, r_x]$.

(2) $\gamma^V(0) = p$.

(3) $\gamma^V(r_x) = \exp_x(V)$.

(4) $\frac{\partial}{\partial r}\big|_{r=0}\gamma^{rV}(s) = \overline{V}(s)$ for $s \in [0, r_x]$.

We have

$$\mathrm{L}\left(\gamma^V\right) \geq d_p(\exp_x(V)) \quad \text{for } V \in B_\varepsilon(\vec{0}),$$
$$\mathrm{L}\left(\gamma^{\vec{0}}\right) = d_p(x).$$

This implies that the C^∞ function $\beta : B_\varepsilon(x) \to \mathbb{R}$ defined by

$$\beta(y) = \mathrm{L}\left(\gamma^{\exp_x^{-1}(y)}\right)$$

is an upper barrier for d_p at x; that is,

$$\beta(y) \geq d_p(y) \quad \text{for } y \in B_\varepsilon(x),$$
$$\beta(x) = d_p(x).$$

(Note that $\exp_x^{-1} : B_\varepsilon(x) \to \mathbb{R}$ is well-defined by the definition of ε.) Thus, in the barrier sense of Definition 11.25, we have

$$\Delta d_p(x) \leq \Delta\beta(x). \tag{11.113}$$

Now we turn to calculating $\Delta\beta(x)$. Let $\{e_1, \ldots, e_{n-1}, \gamma'(r_x)\}$ be an orthonormal basis at x. Then the parallel translate

$$\{e_1(s), \ldots, e_{n-1}(s), \gamma'(s)\}$$

is an orthonormal basis of $T_{\gamma(s)}M$ for $s \in [0, r_x]$. By the second variation of arc-length formula (Proposition 7.54) in the direction $\overline{V}$, which holds for families of continuous piecewise smooth paths, we have that

$$\begin{aligned} \frac{\partial^2}{\partial t^2}\bigg|_{t=0} \mathrm{L}(\gamma^{tV}) &= \int_0^{r_x} \left(\left|\nabla_{\gamma'}\overline{V}(s)\right|^2 - \left\langle \mathrm{Rm}(\gamma'(s), \overline{V}(s))\,\overline{V}(s), \gamma'(s) \right\rangle \right) ds \\ &= \int_0^{r_x} \left((\zeta'(s))^2 |V(s)|^2 - \zeta^2 \left\langle \mathrm{Rm}(\gamma'(s), V(s))V(s), \gamma'(s) \right\rangle \right) ds \end{aligned} \tag{11.114}$$

for all $V \in B_\varepsilon(\vec{0})$.

We compute that

$$\begin{aligned} &\Delta\beta(x) \\ &= \sum_{i=1}^{n-1} \nabla\nabla\beta(e_i, e_i) + \nabla\nabla\beta\left(\gamma'(r_x), \gamma'(r_x)\right) \\ &= \sum_{i=1}^{n-1} \frac{\partial^2}{\partial t^2}\bigg|_{t=0} \beta\left(\exp_x(te_i)\right) + \frac{\partial^2}{\partial t^2}\bigg|_{t=0} \beta\left(\exp_x\left(t\gamma'(r_x)\right)\right) \\ &= \sum_{i=1}^{n-1} \frac{\partial^2}{\partial t^2}\bigg|_{t=0} \mathrm{L}(\gamma^{te_i}) + \frac{\partial^2}{\partial t^2}\bigg|_{t=0} (r_x + t) \\ &= \int_0^{r_x} \left((n-1)\left(\zeta'(s)\right)^2 - \zeta^2 \sum_{i=1}^{n-1} \left\langle \mathrm{Rm}\left(\gamma'(s), e_i(s)\right) e_i(s), \gamma'(s) \right\rangle \right) ds, \end{aligned}$$

where we used (11.114) to obtain the last equality. That (11.111) holds in the barrier sense follows from (11.113) and the formula

$$\sum_{i=1}^{n-1} \left\langle \mathrm{Rm}\left(\gamma'(s), e_i(s)\right) e_i(s), \gamma'(s) \right\rangle = \mathrm{Ric}\left(\gamma'(s), \gamma'(s)\right).$$

This completes the proof of the lemma. □

By choosing variations modeled on variations of geodesics in space forms, we obtain the following.

Theorem 11.27 (Laplacian comparison theorem in the barrier sense). *Let (M^n, g) be a complete Riemannian manifold with* $\mathrm{Ric} \geq (n-1)Kg$ *for some $K \in \mathbb{R}$. For any $p \in M^n$ we have on $M^n \setminus \{p\}$ that*

$$\Delta d_p \leq H_K \circ d_p \tag{11.115}$$

in the barrier sense, where H_K is defined by (11.98).

Proof. Let $x \in M^n \setminus \{p\}$ and let $r_x := d_p(x)$. We choose $\zeta(s) = \frac{\operatorname{sn}_K(s)}{\operatorname{sn}_K(r_x)}$, $s \in [0, r_x]$, in Lemma 11.26. Then, by using the hypothesis $\operatorname{Ric} \geq (n-1)K$ and the modulated (co)sine identities (7.128) and (7.129), we obtain from (11.111) that in the barrier sense,

$$\begin{aligned}
\Delta d_p(x) &\leq \frac{n-1}{\operatorname{sn}_K^2(r_x)} \int_0^{r_x} \left(\operatorname{cn}_K^2(s) - K \operatorname{sn}_K^2(s)\right) ds \\
&= \frac{n-1}{\operatorname{sn}_K^2(r_x)} \int_0^{r_x} \left(\operatorname{cn}_K(s) \operatorname{sn}_K'(s) + \operatorname{cn}_K'(s) \operatorname{sn}_K(s)\right) ds \\
&= \frac{n-1}{\operatorname{sn}_K^2(r_x)} \operatorname{cn}_K(r_x) \operatorname{sn}_K(r_x) \\
&= H_K(r_x).
\end{aligned}$$

□

11.3.7. The Bonnet–Myers theorem revisited. We can use geodesic sphere mean curvature comparison to give another proof of the Bonnet–Myers theorem (see Theorem 7.68).

Theorem 11.28 (Bonnet and Myers). *If (M^n, g) is a complete Riemannian manifold with $\operatorname{Ric} \geq (n-1)Kg$, where $K > 0$, then*

$$\operatorname{diam}(g) \leq \pi/\sqrt{K}. \tag{11.116}$$

In particular, M^n is compact and $\pi_1(M^n) < \infty$.

Proof. Choose any point $p \in M^n$ and suppose $\gamma : [0, L] \to M^n$ is a unit-speed minimal geodesic emanating from p. Then d_p is smooth on $\gamma\big((0, L)\big)$, and for every $r \in (0, L)$ the geodesic sphere $S_r(p)$ is smooth in a neighborhood of $\gamma(r)$, so that $H(r)$ is well-defined. By Theorem 11.20, we have

$$H(r) \leq (n-1)\sqrt{K} \cot\left(\sqrt{K} r\right)$$

along $\gamma|_{(0,L)}$. Since $\lim_{r \to (\pi/\sqrt{K})_-} \cot\left(\sqrt{K} r\right) = -\infty$, we have that $H(r)$ tends to $-\infty$ at or before $r = \pi/\sqrt{K}$. We conclude from this that $L \leq \pi/\sqrt{K}$, which implies (11.116). The rest of the proof is the same as in the proof of Theorem 7.68. □

11.3.8. The Hessian comparison theorem. The Laplacian comparison theorem holds under a Ricci curvature lower bound. Under the stronger hypothesis of a sectional curvature lower bound, we obtain the Hessian comparison theorem.

Theorem 11.29 (Local Hessian comparison theorem). *Let (M^n, g) be a complete Riemannian manifold with* sect $\geq K$*, where $K \in \mathbb{R}$. Then, for any $p \in M^n$, the distance function d_p satisfies*

$$\nabla^2 d_p = \mathrm{II} \leq \frac{1}{n-1}(H_K \circ d_p) g \tag{11.117}$$

on $\mathcal{D}_p \setminus \{p\}$, where H_K is defined by (11.98).

Proof. We will calculate using the geodesic spherical coordinates $\{x^i\}_{i=1}^n$ centered at p, as defined in §11.3.1.3. Our calculations will hold in $\mathcal{D}_p \setminus \{p\}$. Let $\mathrm{II}_{ij} = \mathrm{II}\left(\frac{\partial}{\partial x^i}, \frac{\partial}{\partial x^j}\right)$, for $1 \leq i, j \leq n$, be the components of the second fundamental form. Recall from (11.82) that $\mathrm{II}_{ij} = -\Gamma_{ij}^n$. Let V be a unit vector at p and let γ_V be the unit-speed geodesic emanating from p with $\gamma_V'(0) = V$. Since each $J_i := \frac{\partial}{\partial x^i}$ is a Jacobi field along the geodesic γ_V, we have by (11.89) that

$$\frac{\partial}{\partial r} \mathrm{II}_{ij}(r) = -\langle \mathrm{Rm}(J_i, \gamma_V')\gamma_V', J_j\rangle + \langle \mathrm{L}(J_i), \mathrm{L}(J_j)\rangle, \tag{11.118}$$

where L is the Weingarten map. Consider II as a tensor field along γ_V. The covariant derivative of II along γ_V satisfies

$$\left(\nabla^{\gamma_V}_{\frac{\partial}{\partial r}} \mathrm{II}\right)_{ij} = \frac{\partial}{\partial r} \mathrm{II}_{ij} - \Gamma_{ni}^k \mathrm{II}_{kj} - \Gamma_{nj}^k \mathrm{II}_{ik}, \tag{11.119}$$

where $\frac{\partial}{\partial r} = \frac{\partial}{\partial x^n} = \gamma_V'$. By substituting (11.118) into this equation and using that $\Gamma_{ni}^k = \mathrm{II}_i^k$ and the hypothesis sect $\geq K$, we obtain

$$\begin{aligned}\left(\nabla^{\gamma_V}_{\frac{\partial}{\partial r}} \mathrm{II}\right)_{ij} &= -\langle \mathrm{Rm}(J_i, \gamma_V')\gamma_V', J_j\rangle - \langle \mathrm{L}(J_i), \mathrm{L}(J_j)\rangle \\ &\leq -K g_{ij} - \sum_{k,\ell=1}^n \mathrm{II}_{ik}\, g^{k\ell}\, \mathrm{II}_{\ell j},\end{aligned} \tag{11.120}$$

where $g_{ij} := g\left(\frac{\partial}{\partial x^i}, \frac{\partial}{\partial x^j}\right)$. Using $\nabla^{\gamma_V}_{\frac{\partial}{\partial r}} g = 0$ and (11.99), we compute that

$$\left(\nabla^{\gamma_V}_{\frac{\partial}{\partial r}}\left(\mathrm{II} - \tfrac{H_K}{n-1} g\right)\right)_{ij} \leq \left(\tfrac{H_K}{n-1}\right)^2 - \sum_{k,\ell=1}^n \mathrm{II}_{ik}\, g^{k\ell}\, \mathrm{II}_{\ell j}. \tag{11.121}$$

From (11.74), we can easily derive that $\mathrm{II} - \frac{H_K}{n-1} g = \mathrm{O}(r)$. Hence, it follows from the ordinary differential inequality (11.121) that

$$\mathrm{II} - \tfrac{H_K}{n-1} g \leq 0.$$

This completes the proof of the theorem. □

As a consequence of the Hessian comparison theorem, we obtain the Rauch lower comparison theorem; cf. the Rauch upper comparison theorem (Corollary 7.71).

Corollary 11.30 (Rauch lower comparison with constant curvature). *Let (M^n, g) be a Riemannian manifold with sectional curvatures greater than or equal to $K \in \mathbb{R}$ and let $\gamma : [0, L] \to M^n$ be a unit-speed geodesic, where $L \leq \frac{\pi}{\sqrt{K}}$ if $K > 0$. If $J(r)$ is a Jacobi field along $\gamma(r)$ with $J(0) = \vec{0}$ and $\nabla_{\gamma'} J(0) =: W$ is a vector orthogonal to $\gamma'(0)$, then*

$$|J(r)| \leq |W| \operatorname{sn}_K(r) \tag{11.122}$$

for $r \in [0, L]$.

Proof. By hypothesis, $\langle J(r), \gamma'(r) \rangle = 0$. Thus, by (11.69) and the Hessian comparison theorem, we have that

$$\frac{1}{2} \frac{\partial}{\partial r} \langle J, J \rangle = \nabla^2 d_p (J, J) = \mathrm{II}\,(J, J) \leq \frac{1}{n-1} H_K \langle J, J \rangle . \tag{11.123}$$

From this we obtain that

$$\frac{\partial}{\partial r} \left(\frac{|J(r)|}{\operatorname{sn}_K(r)} \right) \leq 0.$$

This monotonicity formula, together with the initial conditions for J, yield the corollary. □

11.4. The Bishop–Gromov volume comparison theorems

The Bishop–Gromov volume comparison theorem is a pivotal result in Riemannian geometry that elucidates the relationship between the Ricci curvature of a Riemannian manifold and the growth of its volume, providing a powerful tool for understanding the large-scale geometric and topological properties of Riemannian manifolds with Ricci curvature lower bounds.

11.4.1. Bishop–Gromov volume comparison theorem. We will find the following calculus lemma useful to prove the comparison theorem.

Lemma 11.31. *If $f : (0, \rho) \to \mathbb{R}$ and $g : (0, \rho) \to \mathbb{R}$ are continuous functions satisfying $f \geq 0$, $g > 0$ and if the ratio $r \mapsto \frac{f(r)}{g(r)}$ is nonincreasing for $r > 0$, then*

$$r \longmapsto \frac{\int_0^r f(\bar{r}) d\bar{r}}{\int_0^r g(\bar{r}) d\bar{r}} \tag{11.124}$$

is nonincreasing for $r \in (0, \rho)$.

Proof. By hypothesis, for $\overline{r} \in (0,r)$ we have that $\frac{f(\overline{r})}{g(\overline{r})} \geq \frac{f(r)}{g(r)}$, so that $f(r)g(\overline{r}) \leq g(r)f(\overline{r})$. Thus,[6]

$$f(r)\int_0^r g(\overline{r})\,d\overline{r} \leq g(r)\int_0^r f(\overline{r})\,d\overline{r}.$$

We conclude that

$$\left(\int_0^r g(\overline{r})\,d\overline{r}\right)^2 \frac{d}{dr}\left(\frac{\int_0^r f(\overline{r})\,d\overline{r}}{\int_0^r g(\overline{r})\,d\overline{r}}\right) = f(r)\int_0^r g(\overline{r})\,d\overline{r} - g(r)\int_0^r f(\overline{r})\,d\overline{r} \leq 0$$

by (11.125) (see (11.125) in the footnote below). □

In fact, under the hypotheses of Lemma 11.31, the function

$$(r,s) \longmapsto \frac{\int_r^s f(\overline{r})\,d\overline{r}}{\int_r^s g(\overline{r})\,d\overline{r}} \tag{11.126}$$

is nonincreasing in r and nonincreasing in s for $0 < r < s < \rho$; see Exercise 11.11.

Now let B_r^K denote a ball of radius r in the simply connected model space (M_K^n, g_K) of constant sectional curvature $K \in \mathbb{R}$. Note that the volume $\mathrm{Vol}_{g_K}(B_r^K)$ is independent of the choice of center of B_r^K.

Theorem 11.32 (Bishop–Gromov volume comparison theorem [**BC01**], [**Gro07**]). *Suppose that (M^n, g) is a complete Riemannian manifold with $\mathrm{Ric} \geq (n-1)Kg$, where $K \in \mathbb{R}$. Then, for any $p \in M^n$, the volume ratio*

$$r \mapsto \frac{\mathrm{Vol}_g(B_r(p))}{\mathrm{Vol}_{g_K}(B_r^K)}$$

is a nonincreasing function of r. Hence, for all $0 < r_1 \leq r_2$, we have that

$$\frac{\mathrm{Vol}_g(B_{r_1}(p))}{\mathrm{Vol}_g(B_{r_2}(p))} \geq \frac{\mathrm{Vol}_{g_K}(B_{r_1}^K)}{\mathrm{Vol}_{g_K}(B_{r_2}^K)}. \tag{11.127}$$

Moreover,

$$\mathrm{Vol}_g(B_r(p)) \leq \mathrm{Vol}_{g_K}(B_r^K) \tag{11.128}$$

for all $r > 0$.

If equality holds in (11.128) for some $p \in M^n$ and $r > 0$, then $(B_r(p), g)$ is isometric to (B_r^K, g_K).[7]

[6] Alternatively, by the mean value theorem (see Rudin [**Rud76**, Theorem 5.9]), there exists $r_1 \in (0,r)$ such that

$$\frac{\int_0^r f(\overline{r})\,d\overline{r}}{\int_0^r g(\overline{r})\,d\overline{r}} = \frac{f(r_1)}{g(r_1)} \geq \frac{f(r)}{g(r)}. \tag{11.125}$$

[7] Cf. [**Esc87**].

Proof. Define

$$f(r) := \int_{C(r)} \mathrm{J}(v,r)\, d\mu_{S_pM}(v) = \int_{S_pM} \mathrm{J}(v,r)\chi_{C(r)}(v)\, d\mu_{S_pM}(v),$$

$$g(r) := \int_{S_pM} \mathrm{sn}_K(r)^{n-1} d\mu_{S_pM}(v) = n\omega_n\, \mathrm{sn}_K(r)^{n-1},$$

where ω_n is the volume of the unit Euclidean n-ball. Then $f(r)$ and $g(r)$ are continuous functions which satisfy

$$\mathrm{Vol}_g\left(B_r(p)\right) = \int_0^r f(\overline{r})\, d\overline{r},$$

$$\mathrm{Vol}_{g_K}(B_r^K) = \int_0^r g(\overline{r})\, d\overline{r}.$$

By (11.102) and the set-monotonicity of $C(r)$, for each $v \in S_pM$ we have that the function $r \mapsto \dfrac{\mathrm{J}(v,r)\chi_{C(r)}(v)}{s_K^{n-1}(r)}$ is nonincreasing on $(0, r_v)$. Thus, we have that $\frac{f(r)}{g(r)}$ is nonincreasing for $r > 0$. By Lemma 11.31, we conclude that the function $r \mapsto \frac{\mathrm{Vol}_g(B_r(p))}{\mathrm{Vol}_{g_K}(B_r^K)}$ is nonincreasing.

Now suppose that equality holds in (11.128) for some p and some $r > 0$:

$$\mathrm{Vol}_g(B_r(p)) = \mathrm{Vol}_{g_K}\left(B_r^K\right).$$

In this case, the monotonicity of $s \mapsto \dfrac{\mathrm{J}(v,s)\chi_{C(s)}(v)}{s_K^{n-1}(s)}$ for $s \in (0,r)$ and the fact that the limit as $s \to 0$ is equal to 1 imply that $\mathrm{J}(v,s) = s_K^{n-1}(s)$ for all $v \in S_pM$ and $s \in (0,r)$. This and (11.103) yield that $H(x) = H_K(r_x)$ for $x \in B_r(p)$. In turn, by Theorem 11.20, we have that the radial sectional curvatures of g are equal to K in $B_r(p)$. This implies that the sectional curvatures of all 2-planes at p are equal to K.

We now prove that $B_r(p)$ is isometric to B_r^K. Let V be a unit vector at p and let $\gamma_V : [0,r) \to M^n$ be the unit-speed geodesic emanating from p with $\gamma_V'(0) = V$. Let $\{e_1, \dots, e_{n-1}, V\}$ be an orthonormal frame at p. For each $1 \leq i \leq n-1$, let $J_i(s)$ be the unique Jacobi field along $\gamma_V(s)$ with $J_i(0) = \vec{0}$ and $\nabla_V J_i = e_i$. The parallel translates

$$\left\{e_1(s), \dots, e_{n-1}(s), \gamma_V'(s)\right\}$$

form an orthonormal basis of $T_{\gamma_V(s)}M$ for each $s \in [0,r)$. Let $T := T(s) := \gamma_V'(s)$, so that $T(0) = V$. By the Jacobi equation (7.137), we have that

$$\nabla_T(\nabla_T J_i) = -\operatorname{Rm}(J_i, T)T = -KJ_i. \tag{11.129}$$

From this and the ODE uniqueness theorem, we obtain that

$$J_i(s) = \mathrm{sn}_K(s)e_i(s).$$

Therefore, the metric g at the point $\gamma_V(s)$ is given by

$$g_{\gamma_V(s)} = \gamma_V'(s)^* \otimes \gamma_V'(s)^* + \sum_{i=1}^{n-1} e_i(s)^* \otimes e_i(s)^* \tag{11.130}$$
$$= T(s)^* \otimes T(s)^* + \operatorname{sn}_K^2(s) \sum_{i=1}^{n-1} J_i(s)^* \otimes J_i(s)^*$$

for each unit vector V at p and for all $s \in (0, r)$. This proves that $(B_r(p), g)$ is isometric to (B_r^K, g_K). □

11.4.2. Euclidean volume growth rigidity. We have the following volume characterization of Euclidean space.

Corollary 11.33. *Let (M^n, g) be a complete noncompact Riemannian manifold with nonnegative Ricci curvature. If we have*

$$\lim_{r\to\infty} \frac{\operatorname{Vol}(B_r(p))}{r^n} = \omega_n \tag{11.131}$$

for some $p \in M^n$, then (M^n, g) is isometric to Euclidean n-space.

Proof. For any Riemannian manifold, we have that $\lim_{r\to 0} \frac{\operatorname{Vol}(B_r(p))}{r^n} = \omega_n$. Thus, by (11.131) and the Bishop–Gromov volume comparison theorem, we have that $\frac{\operatorname{Vol}(B_r(p))}{r^n} \equiv \omega_n$ for all $r > 0$. By the equality case of Theorem 11.32, we have that $(B_r(p), g)$ is isometric to the Euclidean ball $(B_r(\mathbf{0}), g_{\mathbb{E}})$ for all $r > 0$. This implies that (M^n, g) is isometric to Euclidean n-space. □

11.4.3. Bishop–Gromov relative volume comparison theorem. Let $p \in M^n$ and $0 \le r_1 \le r_2$. Define the closed annulus

$$A_{r_1,r_2}(p) := \overline{B}_{r_2}(p) \setminus B_{r_1}(p).$$

Let $A^K_{r_1,r_2}$ denote an annulus in the simply connected model space (M^n_K, g_K) of constant sectional curvature $K \in \mathbb{R}$. Note that the volume $\operatorname{Vol}_{g_K}(A^K_{r_1,r_2})$ is independent of the choice of the center of $A^K_{r_1,r_2}$ since $A^K_{r_1,r_2}$ is well-defined up to isometry. The **Bishop–Gromov relative volume comparison theorem** says the following (see [**Gro07**]).

Theorem 11.34. *Let (M^n, g) be a complete Riemannian manifold with* $\operatorname{Ric} \ge (n-1)Kg$, *where $K \in \mathbb{R}$. If $0 \le r_1 < r_2 \le s_2$ and $r_1 \le s_1 < s_2$,*[8] *then*

$$\frac{\operatorname{Vol}_g(A_{s_1,s_2}(p))}{\operatorname{Vol}_{g_K}(A^K_{s_1,s_2})} \le \frac{\operatorname{Vol}_g(A_{r_1,r_2}(p))}{\operatorname{Vol}_{g_K}(A^K_{r_1,r_2})}. \tag{11.132}$$

[8] Equivalently, $0 \le r_1 < r_2$, $0 \le s_1 < s_2$, $r_1 \le s_1$, and $r_2 \le s_2$. In other words, the inner and outer radius of the annulus $A_{r_1,r_2}(p)$ are less than or equal to the inner and outer radius of $A_{s_1,s_2}(p)$, respectively.

Equivalently,

$$\frac{\mathrm{Vol}_g\left(A_{s_1,s_2}(p)\right)}{\mathrm{Vol}_g\left(A_{r_1,r_2}(p)\right)} \leq \frac{\mathrm{Vol}_{g_K}\left(A^K_{s_1,s_2}\right)}{\mathrm{Vol}_{g_K}\left(A^K_{r_1,r_2}\right)}. \tag{11.133}$$

In other words, the volume ratio function

$$(r,s) \longmapsto \frac{\mathrm{Vol}_g\left(A_{r,s}(p)\right)}{\mathrm{Vol}_{g_K}\left(A^K_{r,s}\right)} \tag{11.134}$$

is nonincreasing in r and nonincreasing in s.

Proof. Define

$$v_K(r) := n\omega_n \int_0^r \mathrm{sn}_K^{n-1}(t)\,dt = \mathrm{Vol}_{g_K}(B^K_r),$$

so that

$$\frac{dv_K}{dr}(r) = n\omega_n\,\mathrm{sn}_K^{n-1}(r). \tag{11.135}$$

Claim 11.35. *The function*

$$\Phi(r,s) := \frac{\int_r^s \mathrm{J}(v,t)\chi_{C(t)}(v)\,dt}{v_K(s)-v_K(r)} \tag{11.136}$$

is nonincreasing in r and nonincreasing in s.

The theorem follows from this claim, which we prove below. Indeed, by integrating $\Phi(r,s)$ over S_pM and using Fubini's theorem, we have that the function

$$(r,s) \longmapsto \frac{\mathrm{Vol}_g(A_{r,s}(p))}{v_K(s)-v_K(r)} = \frac{\mathrm{Vol}_g(A_{r,s}(p))}{\mathrm{Vol}_{g_K}\left(A^K_{r,s}\right)} \tag{11.137}$$

is nonincreasing in r and nonincreasing in s.

Proof of Claim 11.35. Using

$$v_K(s) - v_K(r) = n\omega_n \int_r^s \mathrm{sn}_K^{n-1}(t)\,dt$$

for the second equality below, we compute that

$$\begin{aligned}
\frac{\partial\Phi}{\partial r}(r,s) &= \frac{\frac{dv_K}{dr}(r)\int_r^s \mathrm{J}(v,t)\chi_{C(t)}(v)\,dt - \mathrm{J}(v,r)\chi_{C(r)}(v)(v_K(s)-v_K(r))}{(v_K(s)-v_K(r))^2} \\
&= n\omega_n \frac{\int_r^s \left(\mathrm{sn}_K^{n-1}(r)\,\mathrm{J}(v,t)\chi_{C(t)}(v) - \mathrm{sn}_K^{n-1}(t)\,\mathrm{J}(v,r)\chi_{C(r)}(v)\right)dt}{(v_K(s)-v_K(r))^2} \\
&\leq 0,
\end{aligned}$$

where the last inequality follows from $\frac{\mathrm{J}(v,t)}{\mathrm{sn}_K^{n-1}(t)} \leq \frac{\mathrm{J}(v,r)}{\mathrm{sn}_K^{n-1}(r)}$ for $t \geq r$ by (11.102) and since $\chi_{C(t)}(v) \leq \chi_{C(r)}(v)$ by the set-monotonicity of $C(r)$. We leave it to the reader to show that $\frac{\partial \Phi}{\partial s}(r,s) \leq 0$; see Exercise 11.8. This completes the proof of Claim 11.35.

Finally, as we have seen above, Theorem 11.34 follows from Claim 11.35. □

11.4.4. Yau's linear volume growth estimate. We note that complete noncompact hyperbolic manifolds can have finite volume; for example, hyperbolic manifolds can have cusp ends (see Ratcliffe [**Rat19**], Kapovich [**Kap01**]). In contrast, the following result of Yau says that complete noncompact Riemannian manifolds with nonnegative Ricci curvature must have at least linear volume growth (see [**Yau76**]).

Corollary 11.36. *Let (M^n, g) be a complete noncompact Riemannian manifold with nonnegative Ricci curvature. For any point $p \in M^n$, there exists a constant $C > 0$ such that for any $r \geq 3$,*

$$\mathrm{Vol}\,(B_r(p)) \geq Cr.$$

Proof. Let $x \in M^n$ be any point with $d(x,p) =: r \geq 2$. By the Bishop–Gromov relative volume comparison theorem with $K = 0$ (i.e., (11.133)), we have that

$$\frac{\mathrm{Vol}_g\,(A_{r-1,r+1}(p))}{\mathrm{Vol}_g\,(A_{0,r-1}(p))} \leq \frac{\mathrm{Vol}_{g_0}\left(A^0_{r-1,r+1}\right)}{\mathrm{Vol}_{g_0}\left(A^0_{0,r-1}\right)};$$

that is,

$$\frac{\mathrm{Vol}\,(B_{r+1}(x)) - \mathrm{Vol}\,(B_{r-1}(x))}{\mathrm{Vol}\,(B_{r-1}(x))} \leq \frac{(r+1)^n - (r-1)^n}{(r-1)^n} \leq \frac{C(n)}{r}. \tag{11.138}$$

Since $d(p,x) = r$, we have that $B_1(p) \subset B_{r+1}(x) \setminus B_{r-1}(x)$ and that $B_{r-1}(x) \subset B_{2r-1}(p)$. Thus (11.138) implies that

$$\mathrm{Vol}\,(B_{2r-1}(p)) \geq \mathrm{Vol}\,(B_{r-1}(x)) \geq \frac{r}{C(n)}\,\mathrm{Vol}\,(B_1(p)).$$

The corollary follows since this inequality holds for all $r \geq 2$. □

11.4.5. Cheng's diameter rigidity theorem. Recall from Myers's theorem that the hypothesis $\mathrm{Ric} \geq (n-1)Kg$ implies the diameter inequality $\mathrm{diam}(M^n, g) \leq \frac{\pi}{\sqrt{K}}$. We have the following fundamental result of Cheng [**Che75**], which says that in the diameter equality case, we have rigidity.

Theorem 11.37. *Let (M^n, g) be a complete Riemannian manifold satisfying $\mathrm{Ric} \geq (n-1)Kg$, where K is a positive real number. If $\mathrm{diam}(M^n, g) = \frac{\pi}{\sqrt{K}}$, then (M^n, g) is isometric to the n-sphere of radius $\frac{1}{\sqrt{K}}$, i.e., the geometry with constant sectional curvature K.*

Proof. We follow the proof in [**Shi83**]. Observe that our diameter hypothesis is the equality case of Myers's theorem. Since g is complete, there exist points p and q on M^n such that $d(p,q) = \frac{\pi}{\sqrt{K}}$. By the Bishop volume comparison theorem, we have that

$$\frac{\operatorname{Vol}_g\left(B_{\frac{\pi}{2\sqrt{K}}}(p)\right)}{\operatorname{Vol}_g(M^n)} = \frac{\operatorname{Vol}_g\left(B_{\frac{\pi}{2\sqrt{K}}}(p)\right)}{\operatorname{Vol}_g\left(B_{\frac{\pi}{\sqrt{K}}}(p)\right)} \geq \frac{\operatorname{Vol}_{g_K}\left(B^K_{\frac{\pi}{2\sqrt{K}}}\right)}{\operatorname{Vol}_{g_K}\left(B^K_{\frac{\pi}{\sqrt{K}}}\right)} = \frac{1}{2} \tag{11.139}$$

since the volume of a hemisphere is equal to one-half of the volume of a sphere. Since this argument applies equally well to balls centered at q, we obtain that

(11.140)

$$\operatorname{Vol}_g\left(B_{\frac{\pi}{2\sqrt{K}}}(p)\right) \geq \frac{1}{2}\operatorname{Vol}_g(M^n) \quad \text{and} \quad \operatorname{Vol}_g\left(B_{\frac{\pi}{2\sqrt{K}}}(q)\right) \geq \frac{1}{2}\operatorname{Vol}_g(M^n).$$

Since the open balls $B_{\frac{\pi}{2\sqrt{K}}}(p)$ and $B_{\frac{\pi}{2\sqrt{K}}}(q)$ are disjoint in M^n, so that their volumes add up to at most the volume of M^n, the inequalities in (11.140) must be equalities, so that

$$\frac{\operatorname{Vol}_g\left(B_{\frac{\pi}{2\sqrt{K}}}(x)\right)}{\operatorname{Vol}_g\left(B_{\frac{\pi}{\sqrt{K}}}(x)\right)} = \frac{\operatorname{Vol}_{g_K}\left(B^K_{\frac{\pi}{2\sqrt{K}}}\right)}{\operatorname{Vol}_{g_K}\left(B^K_{\frac{\pi}{\sqrt{K}}}\right)} = \frac{1}{2}$$

for $x = p$ and $x = q$. By the equality case of Theorem 11.32, we have that $(M^n, g) = \left(\overline{B}_{\frac{\pi}{\sqrt{K}}}(p), g\right)$ is isometric to the n-sphere $(M^n_K, g_K) = \left(\overline{B}^K_{\frac{\pi}{\sqrt{K}}}, g_K\right)$ of radius $\frac{1}{\sqrt{K}}$. □

11.4.6. Growth of fundamental groups. In this subsection we study the fundamental group of a complete Riemannian manifold (M^n, g) with nonnegative Ricci curvature. The idea is that applying the Bishop–Gromov volume comparison theorem on the universal covering of (M^n, g) yields a growth estimate for the fundamental group of M^n, which we will now describe.

Let $(G, \cdot)$ be a subgroup of the fundamental group $\pi_1(M^n)$ of a Riemannian manifold (M^n, g). Suppose that $(G, \cdot)$ is generated by a finite set $\mathbf{g} = \{\mathrm{g}_1, \ldots, \mathrm{g}_k\}$ of elements of G, called **generators**. In this case $(G, \cdot)$ is said to be **finitely generated**. The **growth function** $\gamma : \mathbb{N} \to \mathbb{N}$ associated to $(G, \cdot, \mathbf{g})$ is defined for positive integers r as the number $\gamma(r)$ of distinct elements which can be expressed as words of length at most r in the generators and their inverses.[9]

[9] A **word** is a written product of group elements and their inverses. The **length** of a word is the number of group elements in the written product. The word of length zero, a.k.a. the empty word, is the identity element.

For example, if $(G, \cdot)$ is a free abelian group with the basis $\{\mathrm{g}_1, \ldots, \mathrm{g}_k\}$, then the distinct words of length at most r are $\mathrm{g}_1^{i_1} \cdots \mathrm{g}_k^{i_k}$, where $\sum_{j=1}^{k} |i_j| \leq r$. As a crude estimate, it is easy to see in this case that $\gamma(r) \leq (2r+1)^k$. For example, if $k = 2$, then one can show that $\gamma(r) = 2r^2 + 2r + 1$, which of course is less than $(2r+1)^2$.

The following result of Milnor establishes that any finitely generated subgroup of the fundamental group of a complete Riemannian manifold exhibits at most polynomial growth, with the degree of the polynomial being no greater than the dimension of the manifold itself.

Theorem 11.38 (Milnor [**Mil68**]). *Let (M^n, g) be a complete Riemannian manifold with nonnegative Ricci curvature. Then, for any finitely generated subgroup $(G, \cdot, \mathbf{g})$ of $\pi_1(M^n)$, there exists a constant C such that the associated growth function γ satisfies*

$$\gamma(r) \leq C r^n. \tag{11.141}$$

Proof. We sketch the proof and leave it as Exercise 11.13 for the reader to fill in the details. Consider the universal cover $\pi : \widetilde{M^n} \to M^n$, let $\widetilde{g} = \pi^* g$ be the lifted metric, and fix a basepoint $\widetilde{p} \in \widetilde{M^n}$. We may identify $\pi_1(M^n)$ with the group of deck transformations of $\widetilde{M^n}$ (see Massey [**Mas77**, Chapter 5]). Let

$$\mu := \max_{1 \leq i \leq k} d_{\widetilde{g}}\big(\widetilde{p}, \mathrm{g}_i(\widetilde{p})\big) < \infty.$$

Choose $\varepsilon > 0$ so that $B_\varepsilon(\widetilde{p}) \cap B_\varepsilon(\mathrm{g}(\widetilde{p})) = \emptyset$ for all $\mathrm{g} \in \pi_1(M^n) \setminus \{1\}$. For example, we may take $\varepsilon := \frac{1}{2} \operatorname{inj}_g\big(\pi(\widetilde{p})\big)$.

The result follows from estimating the number of balls of the form $\mathrm{g}\big(B_\varepsilon(\widetilde{p})\big)$ that $B_{r\mu+\varepsilon}(\widetilde{p})$ contains, where $\mathrm{g} \in G$ is an element that can be expressed as a word in the finite set of generators $\mathbf{g}$ of length at most r, where r is any positive integer. This is accomplished by applying the Bishop–Gromov volume comparison theorem, with $K = 0$, i.e., under the hypothesis of nonnegative Ricci curvature (see (11.127)). $\square$

11.5. Splitting theorems

The Cheeger–Gromoll splitting theorem is a pillar for the understanding of complete Riemannian manifolds with nonnegative Ricci curvature and more generally manifolds with a Ricci curvature lower bound (see [**CG72**, Theorem 2]). In this section, we discuss this seminal result, which reveals profound connections between geometry and analysis.

11.5.1. The mean value inequality for subharmonic functions. The following mean value inequality, which follows from the Laplacian comparison theorem, has an application in the proof of the splitting theorem.

Proposition 11.39 (Mean value inequality for subharmonic functions). *If (M^n, g) is a complete Riemannian manifold with $\operatorname{Ric} \geq 0$ and if $f \leq 0$ is a Lipschitz function with $\Delta f \geq 0$ in the sense of distributions, then for all $p \in M^n$ and $0 < r < \operatorname{inj}(p)$,*

$$f(p) \leq \frac{1}{\omega_n r^n} \int_{B_r(p)} f d\mu,$$

where ω_n is the volume of the unit Euclidean n-ball.

Proof. Let $S_r(p) := \partial B_r(p)$, which is a smooth $(n-1)$-sphere embedded in M^n since $r < \operatorname{inj}(p)$ by hypothesis. By the divergence theorem, we have

$$\begin{aligned} 0 &\leq \frac{1}{r^{n-1}} \int_{B_r(p)} \Delta f \, d\mu \\ &= \frac{1}{r^{n-1}} \int_{S_r(p)} \langle \nabla f, N \rangle \, d\mu_{S_r(p)} \\ &= \int_{S_p M} \frac{\partial}{\partial r} (f(\exp_p(rV))) \frac{\mathrm{J}(V, r)}{r^{n-1}} d\mu_{S_p M}(V), \end{aligned} \tag{11.142}$$

where N denotes the unit outward normal to $S_r(p)$ and $\mathrm{J}(V, r)$ is the Jacobian of the exponential map as defined by (11.70). Since we have that $\frac{\partial}{\partial r} \frac{\mathrm{J}(V,r)}{r^{n-1}} \leq 0$ from the inequality $H = \frac{\partial}{\partial r} \ln \mathrm{J} \leq \frac{n-1}{r}$ and since $f \leq 0$, we obtain that

$$\begin{aligned} 0 &\leq \int_{S_p M} \left(\frac{\partial}{\partial r} (f(\exp_p(rV))) \frac{\mathrm{J}(V, r)}{r^{n-1}} + f(\exp_p(rV)) \frac{\partial}{\partial r} \frac{\mathrm{J}(V, r)}{r^{n-1}} \right) d\mu_{S_p M}(V) \\ &= \frac{d}{dr} \left(\int_{S_p M} f(\exp_p(rV)) \frac{\mathrm{J}(V, r)}{r^{n-1}} d\mu_{S_p M}(V) \right) \\ &= \frac{d}{dr} \left(\frac{1}{r^{n-1}} \int_{S_r(p)} f d\mu_{S_r(p)} \right). \end{aligned}$$

Since $\lim_{r \to 0} \frac{1}{r^{n-1}} \int_{S_r(p)} f d\mu_{S_r(p)} = n\omega_n f(p)$, where $n\omega_n$ is the volume of the unit Euclidean $(n-1)$-sphere, integrating the above inequality over $[0, s]$ yields

$$s^{n-1} f(p) \leq \frac{1}{n\omega_n} \int_{S_s(p)} f d\mu_{S_s(p)}.$$

Integrating this again, now over $[0, r]$, implies

$$f(p) \leq \frac{1}{\omega_n r^n} \int_{B_r(p)} f d\mu. \qquad \square$$

11.5.2. Complete vector fields. Let M^n be a differentiable manifold. A vector field V on M^n is **complete** if for any $p \in M^n$ there exists an integral curve $\gamma_p : \mathbb{R} \to M^n$ to V with $\gamma_p(0) = p$. In this case, V generates a 1-parameter group of diffeomorphisms $\{\varphi_t\}_{t\in\mathbb{R}}$ of M^n defined by $\varphi_t(p) = \gamma_p(t)$.

We have the following criterion for completeness of a vector field on a Riemannian manifold. We leave the proof of this result as Exercise 11.10.

Lemma 11.40. *Let (M^n, g) be a complete Riemannian manifold, and let $p \in M^n$. Suppose that $F : [0,\infty) \to [0,\infty)$ is a locally Lipschitz function with the property that any solution $u(t)$ to the ordinary differential equation $\frac{du}{dt} = F(u)$, with $u(0) \geq 0$, exists for all $t \geq 0$. If a vector field V on M^n satisfies $|V|(x) \leq F(d(x,p))$ for all $x \in M^n$, then V is complete.*

11.5.3. Busemann functions. Busemann functions, which are defined on noncompact manifolds, play a central role in the proof of the Cheeger–Gromoll splitting theorem. In this subsection, we discuss their properties.

A **geodesic line** is a unit-speed geodesic $\gamma : (-\infty,\infty) \to M^n$ with the property that for all $s_1, s_2 \in (-\infty,\infty)$,

$$d(\gamma(s_1), \gamma(s_2)) = |s_2 - s_1|. \tag{11.143}$$

A unit-speed geodesic $\gamma : [0,\infty) \to M^n$ is a **geodesic ray** if it satisfies (11.143) for all $s_1, s_2 \in [0,\infty)$.

Example 11.41. *Any unit-speed geodesic $\gamma : (-\infty,\infty) \to \mathbb{R}^n$ is a geodesic line. Any unit-speed geodesic $\gamma : [0,\infty) \to \mathbb{R}^n$ is a geodesic ray.*

Example 11.42. *Consider the flat 2-dimensional cylinder $S^1 \times \mathbb{R} = \mathbb{R}^2/\sim$, where $(x,y) \sim (x + 2\pi, y)$ for $x, y \in \mathbb{R}$. Let $\gamma : [0,\infty) \to S^1 \times \mathbb{R}$ be a unit-speed geodesics emanating from $[0,0] \in S^1 \times \mathbb{R}$. Then there exists a unit vector $(u,v) \in \mathbb{R}^2$ such that*

$$\gamma(s) = [(su, sv)].$$

We have that γ is a geodesic ray if and only if $u = 0$ (which implies that $v = \pm 1$).

Given a geodesic ray $\gamma : [0,\infty) \to M^n$, the **Busemann function**

$$b_\gamma : M^n \to \mathbb{R}$$

associated to γ is defined by

$$b_\gamma(x) := \lim_{s\to\infty} (s - d(\gamma(s), x)). \tag{11.144}$$

Intuitively, a Busemann function associated to a ray γ is the renormalized distance to a point at infinity in the direction of γ.

The Busemann function is well-defined since, given $x \in M^n$, the function on the right-hand side of (11.144),

$$s \mapsto s - d(\gamma(s), x),$$

is nondecreasing and bounded above by $d(x, \gamma(0))$; this follows from the triangle inequality.

Example 11.43. *For Euclidean space, Busemann functions are linear: If* $\mathbf{V}, \mathbf{W} \in \mathbb{R}^n$*, with* $\|\mathbf{V}\| = 1$*, then the Busemann function* b_γ *associated to the geodesic ray* $\gamma : [0, \infty) \to \mathbb{R}^n$ *defined by* $\gamma(s) := \mathbf{W} + s\mathbf{V}$ *is the linear function*

$$b_\gamma(\mathbf{x}) = \langle \mathbf{x} - \mathbf{W}, \mathbf{V} \rangle. \tag{11.145}$$

Example 11.44 (Busemann function of hyperbolic space)**.** *Let* $(D^n, g_{\mathbb{D}})$ *be the Poincaré ball model of hyperbolic space as defined by* (7.118)*; i.e.,* $g_{\mathbb{D}}(\mathbf{x}) = \frac{4}{(1-\|\mathbf{x}\|^2)^2} g_{\mathbb{E}}$*. Let* $\mathbf{v}$ *be a unit vector in* $\mathbb{R}^n$*. The unit-speed geodesic* $\gamma_{\mathbf{v}}$ *in* D *emanating from the origin* $\mathbf{0}$ *with initial velocity* $\mathbf{v}$ *is given by* $\gamma_{\mathbf{v}}(t) = \tanh\left(\frac{t}{2}\right)\mathbf{v}$*. The Busemann function associated to* $\gamma_{\mathbf{v}}$ *is given by*

$$b_{\mathbf{v}}(\mathbf{x}) = -\ln\left(\frac{1 - \|\mathbf{x}\|^2}{\|\mathbf{x} - \mathbf{v}\|^2}\right). \tag{11.146}$$

See Bridson and Haefliger [**BH99**, p. 273].

Just as with the distance function to a point, the Busemann function is **1-Lipschitz**. That is, for any Riemannian manifold (M^n, g) and geodesic ray γ, the Busemann function b_γ satisfies

$$|b_\gamma(x) - b_\gamma(y)| \leq d(x, y) \tag{11.147}$$

for all $x, y \in M^n$. This follows from the inequality $|d(\gamma(s), y) - d(\gamma(s), x)| \leq d(x, y)$ (see Exercise 11.12).

Note that by Rademacher's theorem (see Evans and Gariepy [**EG15**, §3.1]), b_γ is differentiable a.e.[10] We claim that

$$|\nabla b_\gamma| = 1 \tag{11.148}$$

at points wherever it is differentiable. To see this, let $x \in M^n$ and let $s \geq 0$. Let $\beta_s : [0, d(\gamma(s), x)] \to M^n$ be a unit-speed minimal geodesic with $\beta_s(0) = x$ and $\beta_s(d(\gamma(s), x)) = \gamma(s)$. Then, for $0 < \varepsilon < s$,

$$d(\beta_s(\varepsilon), x) = \varepsilon \quad \text{and} \quad d(\gamma(s), x) - d(\gamma(s), \beta_s(\varepsilon)) = \varepsilon.$$

Choose $s_i \to \infty$ so that β_{s_i} converges to a unit-speed ray $\beta_\infty : [0, \infty) \to M^n$ emanating from x. Since

$$\left(s_i - d(\gamma(s_i), \beta_{s_i}(\varepsilon))\right) - \left(s_i - d(\gamma(s_i), x)\right) = \varepsilon,$$

[10] Note that (11.147) implies that $b_\gamma \circ \mathbf{x}^{-1}$ is locally Lipschitz for any coordinate system $\mathbf{x}$.

by taking $i \to \infty$ we obtain

$$b_\gamma(\beta_\infty(\varepsilon)) - b_\gamma(x) = \varepsilon$$

and $d(\beta_\infty(\varepsilon), x) = \varepsilon$. Since this holds for all $\varepsilon > 0$, equation (11.148) follows.

We conclude that b_γ is a distance function in the sense of the following definition.

Definition 11.45. A Lipschitz function f is called a **distance function** if $|\nabla f| = 1$ wherever it is differentiable; in particular, $|\nabla f| = 1$ a.e.

A key fact is that Busemann functions are subharmonic on manifolds with nonnegative Ricci curvature.

Lemma 11.46. *If γ is a geodesic ray in a complete Riemannian manifold (M^n, g) with $\operatorname{Ric} \geq 0$, then $\Delta b_\gamma \geq 0$ in the barrier sense of Definition 11.25.*

Proof. Let $x \in M^n$. Choose a sequence $s_i \to \infty$ and let $\beta_i : [0, d(x, \gamma(s_i))] \to M^n$ be a unit-speed minimal geodesic joining x to $\gamma(s_i)$. Then $\beta_i'(0)$ subconverges to a unit vector V in T_xM. The unit-speed geodesic $\beta : [0, \infty) \to M^n$, defined by $\beta'(0) = V$, is a geodesic ray. For $t \geq 0$, define $f_t : M^n \to \mathbb{R}$ by

$$f_t(y) = t - d(y, \beta(t)) + b_\gamma(x). \tag{11.149}$$

Claim. For each $t \geq 0$, f_t is a lower barrier for b_γ at x.

Clearly, $f_t(x) = b_\gamma(x)$. We just need to show that $f_t(y) \leq b_\gamma(y)$ for $y \in M^n$. By the triangle inequality,

$$-d(y, \beta(t)) \leq s - d(y, \gamma(s)) - s + d(\beta(t), \gamma(s)).$$

Taking $s \to \infty$, we obtain

$$-d(y, \beta(t)) \leq b_\gamma(y) - b_\gamma(\beta(t)). \tag{11.150}$$

On the other hand, we have (note that $t \leq d(x, \gamma(s_i))$ for i large enough)

$$\begin{aligned} b_\gamma(x) &= \lim_{i\to\infty} \left(s_i - d(x, \gamma(s_i))\right) \\ &= \lim_{i\to\infty} \left(s_i - d(\beta_i(t), \gamma(s_i)) - d(x, \beta_i(t))\right) \\ &= b_\gamma(\beta(t)) - t. \end{aligned} \tag{11.151}$$

The claim follows directly from equations (11.150) and (11.151).

Now, by Theorem 11.27, we have that

$$\Delta f_t(x) \geq -\frac{n-1}{d(x, \beta(t))}$$

in the barrier sense. Hence, the lemma follows from taking $t \to \infty$. □

11.5.4. Hessians, splittings, and cones.

Lemma 11.47. *Let (M^n, g) be a complete Riemannian manifold. Suppose that there exists a function $u : M^n \to \mathbb{R}$ such that $\nabla^2 u = 0$ and $|\nabla u| = 1$ on M^n. Then there exists a complete Riemannian manifold (N^{n-1}, h) such that (M^n, g) is isometric to $(N^{n-1}, h) \times \mathbb{R}$.*

Proof. Let $p \in M^n$ and let $s_0 = u(p)$. Let $N^{n-1} := u^{-1}(s_0)$, with the induced metric $h = g|_{N^{n-1}}$. Since $\nabla u \neq 0$ everywhere on M^n, by the implicit function theorem we have that N^{n-1} is a smooth hypersurface and h is a smooth metric. Define $\varphi_t : M^n \to M^n$ to be the 1-parameter group of diffeomorphisms of M^n generated by the complete vector field ∇u. Since $\mathcal{L}_{\nabla u} g = 2\nabla^2 u = 0$, we have that the φ_t are isometries.

Let $i : N^{n-1} \to M^n$ be the inclusion map. Define $F : N^{n-1} \times \mathbb{R} \to M^n$ by $F(y, s) = \varphi_s(i(y))$. We claim that F is an isometry. We compute that

$$\frac{d}{dt} u(\varphi_t(x)) = \nabla u \cdot \frac{d}{dt} \varphi_t(x) = |\nabla u|^2(\varphi_t(x)) = 1.$$

Hence $u(\varphi_t(x)) = u(x) + t$. Furthermore, $\varphi_t : u^{-1}(t_0) = u^{-1}(t_0 + t)$ is an isometry with inverse φ_{-t}. So F maps $N^{n-1} \times \{t\}$ isometrically to $u^{-1}(s_0+t)$. Moreover, $dF\left(\frac{\partial}{\partial s}\right) = \nabla u$. Since ∇u is a unit vector field normal to each level set of u, we conclude that F is an isometry. Since $(N^{n-1}, h) \times \mathbb{R} = (M^n, g)$ is complete, we have that (N^{n-1}, h) is complete. ☐

Now we are ready to prove another fundamental consequence of Theorem 11.24, which is the following **Cheeger–Gromoll splitting theorem** (see [**CG72**]).

Theorem 11.48 (Cheeger and Gromoll). *Let (M^n, g) be a complete Riemannian manifold with $\mathrm{Ric}_g \geq 0$, and suppose that there is a geodesic line γ in M^n. Then (M^n, g) is isometric to $\left(N^{n-1}, h\right) \times \mathbb{R}$ with the product metric, where $\left(N^{n-1}, h\right)$ is a complete Riemannian manifold with $\mathrm{Ric}_h \geq 0$.*

Proof. Consider the two Busemann functions $b_\pm$ associated to the geodesic rays $\gamma_\pm : [0, \infty) \to M^n$ defined by $\gamma_\pm(s) = \gamma(\pm s)$ for $s \geq 0$. Since (M^n, g) has $\mathrm{Ric} \geq 0$, by Lemma 11.46 we have that $\Delta b_\pm \geq 0$ in the barrier sense. Hence

$$\Delta\left(b_+ + b_-\right) \geq 0.$$

Since $d\left(\gamma(s), \gamma(-s)\right) = 2s$, we have that $b_+ + b_- \leq 0$ on M^n. Moreover, $\left(b_+ + b_-\right)(x) = 0$ for $x \in \gamma$. Given $x \in M^n$ such that $\left(b_+ + b_-\right)(x) = 0$, by applying the mean value inequality (Proposition 11.39) on $B_r(x)$, where $r < \mathrm{inj}(x)$, we obtain that $b_+ + b_- = 0$ in $B_r(x)$. Thus we have the set of points in M^n where $b_+ + b_- = 0$ is open. Since this set is also closed and nonempty, we conclude that $b_+ + b_- \equiv 0$ on M^n. Hence $\Delta\left(b_+ + b_-\right) \equiv 0$.

Since $\Delta b_{\pm} \geq 0$, this implies that $\Delta b_{\pm} = 0$ in the barrier sense. Standard elliptic regularity theory implies that $b_{\pm}$ is smooth. Therefore $|\nabla b_{\pm}| \equiv 1$. Recall the fundamental Bochner formula (see Proposition 11.3)

$$\frac{1}{2}\Delta|\nabla u|^2 = \langle \nabla \Delta u, \nabla u\rangle + |\nabla^2 u|^2 + \mathrm{Ric}(\nabla u, \nabla u). \tag{11.152}$$

Thus $|\nabla^2 b_{\pm}|^2 + \mathrm{Ric}(\nabla b_{\pm}, \nabla b_{\pm}) = 0$. Since $\mathrm{Ric} \geq 0$, this implies that $\nabla^2 b_{\pm} = 0$ on M^n. We conclude from Lemma 11.47 that there is a complete Riemannian manifold (N^{n-1}, h) such that (M^n, g) is isometric to $(N^{n-1}, h) \times \mathbb{R}$. □

11.6. Harmonic functions on complete noncompact manifolds

In the panorama of geometric analysis on Riemannian manifolds, Yau's Liouville theorem for positive harmonic functions on complete Riemannian manifolds with nonnegative Ricci curvature is a seminal result. It is a crucial milestone, offering a remarkable example of a gradient estimate that brings about a global conclusion. This theorem is a testament to the power of the maximum principle. In this section we discuss the proof of Yau's Liouville theorem [**Yau75**].

11.6.1. The Liouville theorem on Euclidean space. Before delving into the nuances of Yau's Liouville theorem within the context of Riemannian manifolds, we first discuss the much simpler case of the Liouville theorem on Euclidean space, albeit the proof is by a different method. Let $u : \mathbb{R}^n \to \mathbb{R}$ be a positive harmonic function. Given any points $\mathbf{x}, \mathbf{y} \in \mathbb{R}^n$, consider the balls

$$B_{R-\|\mathbf{x}-\mathbf{y}\|}(\mathbf{x}) \subset B_R(\mathbf{y})$$

for $R > \|\mathbf{x} - \mathbf{y}\|$. See Figure 11.6.1. By the **mean value property** for harmonic functions (see Gilbarg and Trudinger [**GT01**, Theorem 2.1]), we have with $r := R - \|\mathbf{x} - \mathbf{y}\|$ that

$$u(\mathbf{x}) = \frac{\int_{B_r(\mathbf{x})} u\, d\mu_{\mathbb{E}}}{\operatorname{Vol} B_r(\mathbf{x})} \leq \frac{\int_{B_R(\mathbf{y})} u\, d\mu_{\mathbb{E}}}{\operatorname{Vol} B_r(\mathbf{x})} = \frac{\operatorname{Vol} B_R(\mathbf{y})}{\operatorname{Vol} B_r(\mathbf{x})} u(\mathbf{y}),$$

where the inequality in the display above is because u is positive. Since $\lim_{R\to\infty} \frac{\operatorname{Vol} B_R(\mathbf{y})}{\operatorname{Vol} B_r(\mathbf{x})} = 1$, we obtain that

$$u(\mathbf{x}) \leq u(\mathbf{y}).$$

Since $\mathbf{x}$ and $\mathbf{y}$ are arbitrary points, we conclude that u is a constant function.

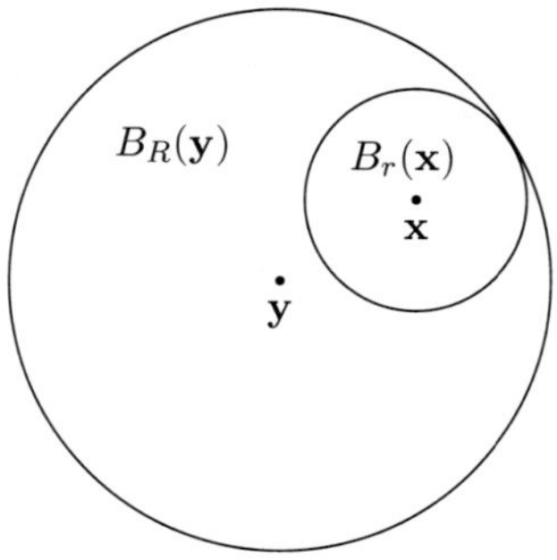

Figure 11.6.1. The Liouville theorem for positive harmonic functions on $\mathbb{R}^n$ follows from applying the mean value property on the balls $B_R(\mathbf{y})$ and $B_r(\mathbf{x}) := B_{R-\|\mathbf{x}-\mathbf{y}\|}(\mathbf{x})$ and taking $R \to \infty$.

11.6.2. Statement of Yau's theorem.

We have the following **Liouville-type theorem** for harmonic functions on manifolds with nonnegative Ricci curvature.

Theorem 11.49 (Yau [**Yau75**]). *If (M^n, g) is a complete noncompact Riemannian manifold with nonnegative Ricci curvature and u is a positive harmonic function, then u is constant.*

11.6.3. The elliptic gradient estimate.

Yau's proof of the Liouville theorem for Riemannian manifolds with nonnegative Ricci curvature is based on an elliptic gradient estimate, a.k.a. differential Harnack estimate. In particular, Theorem 11.49 is an immediate consequence of taking $\mathrm{R} \to \infty$ in the following (see Cheng and Yau [**CY75**]).

Theorem 11.50 (Differential Harnack estimate). *Let (M^n, g) be a complete Riemannian manifold with nonnegative Ricci curvature. Then there exists a constant C depending only on the dimension n such that if u is a positive harmonic function on a geodesic ball $B_{2\mathrm{R}}(p)$, where $p \in M^n$ and $\mathrm{R} \in (0, \infty)$, then*

$$\sup_{x \in B_{\mathrm{R}}(p)} |\nabla \ln u|(x) \leq \frac{C}{\mathrm{R}}. \tag{11.153}$$

Proof. The general idea of the proof of the gradient estimate (11.153) is to apply the elliptic maximum principle to the gradient quantity in (11.153). A difficulty that one encounters is related to the noncompactness of the manifold M^n, which requires the use of a 'cutoff function'. Recall the fundamental Bochner formula (11.21):

$$\frac{1}{2}\Delta|\nabla f|^2 = \langle \nabla\Delta f, \nabla f\rangle + |\nabla^2 f|^2 + \mathrm{Ric}(\nabla f, \nabla f). \tag{11.154}$$

Let u be a positive harmonic function on M^n. The positivity of u allows us to choose $f := \ln u$. Using that u is harmonic, we compute that

$$\Delta f = \operatorname{div}\left(\frac{\nabla u}{u}\right) = \frac{\Delta u}{u} - \frac{|\nabla u|^2}{u^2} = -|\nabla f|^2. \tag{11.155}$$

By applying this formula to (11.154) and using $\mathrm{Ric} \geq 0$, we obtain the elliptic inequality:

$$\frac{1}{2}\Delta|\nabla f|^2 \geq -\langle \nabla f, \nabla|\nabla f|^2\rangle + |\nabla^2 f|^2. \tag{11.156}$$

Note, since M^n is noncompact, we do not know that the maximum of $|\nabla f|^2$ is attained somewhere. We solve this problem by multiplying this gradient quantity by a cutoff function so that the resulting function has compact support. Then we have a function that attains its global maximum and we can apply the maximum principle.

Let $\phi : M^n \to \mathbb{R}$ be a nonnegative function with compact support, which we will choose below. We call ϕ a **cutoff function**. Define the nonnegative function

$$F := \frac{\phi}{2}|\nabla f|^2. \tag{11.157}$$

The $\frac{1}{2}$ factor is introduced to eliminate factors of 2 in the equations that we will derive below. We compute from (11.156) and using the product rule (11.230) on $\frac{\phi}{2}$ times $|\nabla f|^2$ that

$$\begin{aligned} \Delta F &= \langle \nabla \phi, \nabla|\nabla f|^2\rangle + \phi\Delta\left(\frac{1}{2}|\nabla f|^2\right) + \Delta\phi\frac{1}{2}|\nabla f|^2 \\ &\geq \langle \nabla\phi - \phi\nabla f, \nabla|\nabla f|^2\rangle + \phi|\nabla^2 f|^2 + \Delta\phi\frac{1}{2}|\nabla f|^2. \end{aligned} \tag{11.158}$$

In the following, we compute at any point where $\phi \neq 0$. By (5.108) and (11.155), we see that

$$|\nabla^2 f|^2 \geq \frac{1}{n}(\Delta f)^2 = \frac{1}{n}|\nabla f|^4 = \frac{4}{n\phi^2}F^2.$$

(This is a simple but crucial inequality. Note that this argument with u instead of f does not work as in that case one only obtains $|\nabla u|^2 \geq \frac{1}{n}(\Delta u)^2 = 0$, which is trivial.) Thus, (11.158) may be rewritten as

$$\begin{aligned} \Delta F &\geq -2\langle \tfrac{1}{\phi^2}\nabla\phi - \tfrac{1}{\phi}\nabla f, \nabla\phi\rangle F + 2\langle \tfrac{1}{\phi}\nabla\phi - \nabla f, \nabla F\rangle \\ &\quad + \frac{4}{n\phi}F^2 + \frac{1}{\phi}\Delta\phi F \\ &= \langle X, \nabla F\rangle + \frac{4}{n\phi}F^2 + \psi F, \end{aligned} \tag{11.159}$$

where $X := 2\big(\frac{1}{\phi}\nabla\phi - \nabla f\big)$ and where

$$\psi := \frac{1}{\phi}\Delta\phi - \frac{2}{\phi^2}|\nabla\phi|^2 + \frac{2}{\phi}\langle\nabla f, \nabla\phi\rangle . \tag{11.160}$$

Now, the idea is that the F squared term on the right-hand side of (11.159) with the positive coefficient $\frac{4}{n\phi}$ will enable us to bound F from above. We carry out this idea by the maximum principle as follows. Since the nonnegative cutoff function ϕ has compact support, the maximum of F is attained at some point p_0 in M^n. We have

$$\nabla F(p_0) = 0, \qquad \Delta F(p_0) \leq 0 .$$

Suppose that $F(p_0) > 0$, for otherwise we are done. Then the elliptic inequality (11.159) yields at p_0 that

$$0 \geq \frac{4}{n}F + \phi\psi = \frac{4}{n}F + \Delta\phi - \frac{2}{\phi}|\nabla\phi|^2 + 2\langle\nabla f, \nabla\phi\rangle .$$

Observe by the Peter–Paul inequality (i.e., $2ab \leq ca^2 + c^{-1}b^2$ provided $c > 0$) that

$$-2\langle\nabla f, \nabla\phi\rangle \leq 2|\nabla f||\nabla\phi| \leq \frac{\phi}{n}|\nabla f|^2 + \frac{n}{\phi}|\nabla\phi|^2 = \frac{2}{n}F + \frac{n}{\phi}|\nabla\phi|^2 .$$

Thus, we obtain at p_0 that

$$0 \geq \frac{2}{n}F + \Delta\phi - (n+2)\frac{|\nabla\phi|^2}{\phi} . \tag{11.161}$$

In order to obtain a 'small' upper bound for F, we see that the name of the game is to obtain small upper bounds for both $-\Delta\phi$ and $|\nabla\phi|^2$.

Now let R be a positive number. It is easy to define a smooth cutoff function $\phi : [0,\infty) \to [0,1]$ satisfying the conditions

$$\begin{aligned} &\phi(r) = 1 \text{ if } r \in [0, \mathrm{R}], && \phi(r) = 0 \text{ if } r \in [2\,\mathrm{R}, \infty), \\ &0 \leq -\,\phi^{-1/2}\phi' \leq C\,\mathrm{R}^{-1} \text{ where } \phi > 0, && |\phi''| \leq C\,\mathrm{R}^{-2} \end{aligned} \tag{11.162}$$

for some constant C. Note that $\phi' \leq 0$. Fix a point $p \in M^n$ and define, using redundant notation, the radial cutoff function $\phi : M^n \to [0,1]$ by

$$\phi(x) := \phi(d(x,p)) .$$

We see that ϕ has compact support inside $\overline{B}_{2\,\mathrm{R}}(p)$, ϕ is smooth outside the cut locus of p, and ϕ is Lipschitz continuous everywhere. See Figure 11.6.2.

Recall from the Laplacian comparison theorem (Theorem 11.24) for Riemannian manifolds with nonnegative Ricci curvature that, outside of p and the cut locus of p, i.e., where the function $x \mapsto d_p(x) := d(x,p)$ is smooth, we have

$$\Delta d_p \leq \frac{n-1}{d_p} . \tag{11.163}$$

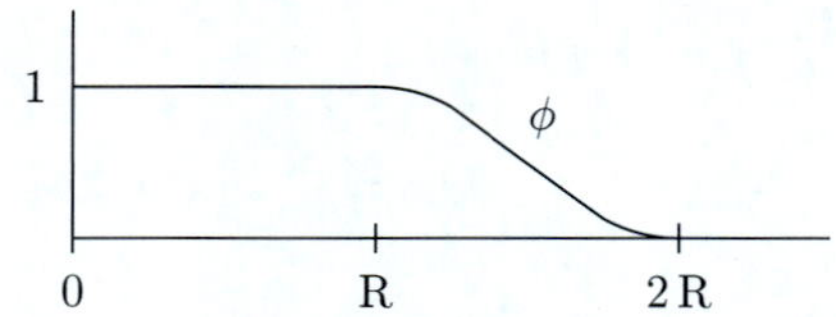

Figure 11.6.2. The cutoff function ϕ.

We also have $|\nabla d_p| = 1$ wherever d_p is smooth, so the cutoff function ϕ on M satisfies

$$\frac{1}{\phi}|\nabla \phi|^2 = \frac{(\phi')^2}{\phi} \leq C\,\mathrm{R}^{-2},$$

where the constant C may change from term to term in our discussion. We further compute using (11.163) that

$$\Delta\phi = \operatorname{div}(\phi'\nabla d_p) = \phi'\Delta d_p + \phi''|\nabla d_p|^2 \geq -C\,\mathrm{R}^{-2}, \tag{11.164}$$

where we also used that on the support of $\phi' \circ d_p$ we have $d_p \geq \mathrm{R}$. Since $2F = |\nabla f|^2$ in $B_\mathrm{R}(p)$, substituting these inequalities into (11.161) yields that *provided* p_0 *is not in the cut locus of* p, we have

$$\sup_{B_\mathrm{R}(p)} |\nabla f|^2 \leq 2F(p_0) \leq C\,\mathrm{R}^{-2}\,. \tag{11.165}$$

This is the desired estimate in this case. The proof in the case where p_0 is contained in the cut locus is discussed below. □

11.6.4. Dealing with the cut locus: Calabi's trick. Now we handle the case where the maximum point p_0 of F is a cut point of the basepoint p. Here, we can use **Calabi's trick** (see [**Cal67**]) and argue as follows. Let $\gamma : [0, L] \to M^n$, where $L := d\,(p, p_0)$, be a minimal geodesic joining p to p_0. Let $p_\varepsilon := \gamma\,(\varepsilon)$ for $0 < \varepsilon < \operatorname{inj}(p)$. With ϕ as above, define the new cutoff function $\phi_\varepsilon : M^n \to [0, 1]$ by

$$\phi_\varepsilon(x) := \phi(d_{p_\varepsilon}(x) + \varepsilon)\,.$$

Define

$$F_\varepsilon := \frac{\phi_\varepsilon}{2}|\nabla f|^2. \tag{11.166}$$

By the triangle inequality, for all $x \in M^n$ we have

$$d_{p_\varepsilon}(x) + \varepsilon = d(x, p_\varepsilon) + d(p_\varepsilon, p) \geq d(x, p) = d_p(x)\,,$$

with equality if $x = p_0$ since γ is minimal. Since $\phi : [0, \infty) \to [0, 1]$ is nonincreasing, this implies that

$$\phi_\varepsilon(x) = \phi(d_{p_\varepsilon}(x) + \varepsilon) \leq \phi(d_p(x)) = \phi(x) \qquad \text{for } x \in M^n,$$

with equality at $x = p_0$. Thus, $F_\varepsilon(x) \leq F(x)$ on M^n, with equality at $x = p_0$. This implies that the maximum point p_0 of F is also a maximum point of F_ε.

Now, p_0 is not in the cut locus of p_ε by the characterization of the cut locus in Lemma 7.81,[11] so that ϕ_ε is smooth at p_0. Thus, by applying the maximum principle to F_ε in exactly the same way we applied it to F, we obtain as in (11.161) that at p_0,

$$0 \geq \frac{2}{n} F_\varepsilon + \Delta\phi_\varepsilon - \frac{n+2}{\phi_\varepsilon} |\nabla\phi_\varepsilon|^2.$$

As before,

$$|\nabla\phi_\varepsilon|(x) = |\phi'(d_{p_\varepsilon}(x) + \varepsilon)| \leq C\,\mathrm{R}^{-1}(\phi_\varepsilon)^{1/2}(x),$$

and, similarly to (11.164),

$$\Delta\phi_\varepsilon(x) = \phi'(d_{p_\varepsilon}(x) + \varepsilon)\Delta d_{p_\varepsilon}(x) + \phi''(d_{p_\varepsilon}(x) + \varepsilon)|\nabla d_{p_\varepsilon}|^2(x) \geq -C\,\mathrm{R}^{-2}$$

since ε is small, especially compared to R. Therefore we conclude that

$$F_\varepsilon(p_0) \leq C\,\mathrm{R}^{-2}$$

for all $\varepsilon > 0$ sufficiently small, where the constant C is independent of ε. Inequality (11.165) and hence Theorem 11.50 now follow since $F_\varepsilon \to F$ as $\varepsilon \to 0$.

11.7. The heat equation

Given a Riemannian manifold (M^n, g), a function $u : M^n \times (\alpha, \omega) \to \mathbb{R}$ is a solution to the heat equation if

$$\frac{\partial u}{\partial t}(x,t) = \Delta u(x,t) \tag{11.167}$$

for all $x \in M^n$ and $t \in (\alpha, \omega)$. By considering $u(x,t)$ as the temperature at the point x and time t, one can derive the heat equation from the physical principle that the rate of change of thermal energy in a region is equal to the net flux of heat across the region's boundary.

11.7.1. The Euclidean heat kernel.

In the exploration of heat diffusion, we can first examine Euclidean space as a simpler example. The iconic solution to the heat equation is the **Euclidean heat kernel** on Euclidean space $\mathbb{R}^n$, which is defined by

$$K(\mathbf{x}, t; \mathbf{y}, s) = (4\pi(t-s))^{-\frac{n}{2}} \mathrm{e}^{-\frac{\|\mathbf{x}-\mathbf{y}\|^2}{4(t-s)}} \tag{11.168}$$

[11] The geodesic $-\gamma$ from p_0 to p is minimizing past p_ε, so (1) p_ε is not a conjugate point of p_0, which implies that p_0 is not a conjugate point of p_ε, and (2) there cannot be two distinct minimal geodesics from p_ε to p_0.

for $\mathbf{x}, \mathbf{y} \in \mathbb{R}^n$ and $s < t$. Let $\tau := t - s > 0$ and $\mathbf{z} := \mathbf{x} - \mathbf{y}$. Then

$$K(\mathbf{x}, t; \mathbf{y}, s) = (4\pi\tau)^{-\frac{n}{2}} \mathrm{e}^{-\frac{\|\mathbf{z}\|^2}{4\tau}} =: H(\mathbf{z}, \tau). \tag{11.169}$$

It is easy to compute that H is a solution to the heat equation:

$$\frac{\partial H}{\partial \tau}(\mathbf{z}, \tau) = \Delta_{\mathbb{E}} H(\mathbf{z}, \tau), \tag{11.170}$$

where $\Delta_{\mathbb{E}}$ denotes the Euclidean Laplacian. Observe the self-similarity (scaling) property

$$H(\mathbf{z}, \tau) = \tau^{-\frac{n}{2}} H\left(\frac{\mathbf{z}}{\sqrt{\tau}}, 1\right). \tag{11.171}$$

So the heat kernel at time τ looks like the conjugate heat kernel at time 1 at space scale $\sqrt{\tau}$ (**self-similar**).

The averaging of the heat equation spreads out the heat kernel. To see the averaging property of the heat equation in general, let $\mathbf{x}_0 \in \mathbb{R}^n$, $r_0 > 0$, and suppose that $u : B_{r_0}(\mathbf{x}_0) \to \mathbb{R}$ is a smooth function. Define the function $\overline{u} : [0, r_0) \to \mathbb{R}$ by

$$\overline{u}(r) := \frac{1}{n\omega_n r^{n-1}} \int_{S_r(\mathbf{x}_0)} u(\mathbf{x})\, d\sigma(\mathbf{x})$$

for $r \in (0, r_0)$ and $\overline{u}(0) := u(\mathbf{x}_0)$. That is, $\overline{u}(r)$ is equal to the average of u over the sphere $S_r(\mathbf{x}_0)$. Then $\overline{u}$ is smooth up to the boundary $\{0\}$ of $[0, r_0)$. One may compute that (we leave this as an exercise; hint: use Taylor's theorem)

$$\overline{u}'(0) = 0 \quad \text{and} \quad \overline{u}''(0) = \frac{1}{n} \Delta u(\mathbf{x}_0). \tag{11.172}$$

Thus, by Taylor's theorem, we have that

$$\begin{aligned} \overline{u}(r) &= \overline{u}(0) + \overline{u}'(0) r + \frac{1}{2} \overline{u}''(0) r^2 + \mathrm{O}(r^3) \\ &= u(\mathbf{x}_0) + \frac{1}{2n} \Delta u(\mathbf{x}_0) r^2 + \mathrm{O}(r^3). \end{aligned}$$

Now, if $u(\mathbf{x}, t)$ is a solution to the heat equation $\partial_t u = \Delta u$, then

$$u(\mathbf{x}_0, t_0 + \tau) = u(\mathbf{x}_0, t_0) + \tau \partial_t u(\mathbf{x}_0, t_0) + \mathrm{O}(\tau^2).$$

So, by letting $u(\mathbf{x}) = u(\mathbf{x}, t_0)$ and taking r so that $\frac{r^2}{2n} = \tau$, we obtain that u after a time lapse τ,

$$u(\mathbf{x}_0, t_0 + \tau) = \overline{u}(\sqrt{2n\tau}) + \mathrm{O}\left(\tau^{\frac{3}{2}}\right),$$

is approximately equal to the average of u on the sphere of radius $\sqrt{2n\tau}$.

Furthermore, we have the 'total heat conservation law'

$$\int_{\mathbb{R}^n} H(\mathbf{z},\tau)d\mu_{\mathbb{E}}(\mathbf{z}) = 1 \tag{11.173}$$

for all $\tau \in (0,\infty)$, and the 'point concentration phenomenon'

$$\lim_{\tau\to 0} H(\mathbf{z},\tau) = \delta_{\mathbf{0}}(\mathbf{z}), \tag{11.174}$$

where $\delta_{\mathbf{0}}$ is the Dirac delta function centered at $\mathbf{0}$, which is a distribution, and the limits in (11.174) are distributional (a.k.a. weak) limits. That is, for any smooth function φ with compact support on $\mathbb{R}^n$, we have[12]

$$\lim_{\tau\to 0}\int_{\mathbb{R}^n} H(\mathbf{z},\tau)\varphi(\mathbf{z})\,d\mu_{\mathbb{E}}(\mathbf{z}) = \varphi(\mathbf{0}).$$

See the famous book by Schwartz [**Sch66**] for the theory of distributions.

11.7.2. Elementary gradient estimate on a Riemannian manifold.

Let u be a solution to the heat equation

$$\frac{\partial u}{\partial t} = \Delta u \tag{11.175}$$

on a Riemannian manifold (M^n, g). Recall the fundamental Bochner formula (11.21)

$$\frac{1}{2}\Delta|\nabla u|^2 = \langle \nabla u, \nabla(\Delta u)\rangle + |\nabla^2 u|^2 + \mathrm{Ric}(\nabla u, \nabla u). \tag{11.176}$$

Using this, we compute that

$$\begin{aligned}\frac{\partial}{\partial t}|\nabla u|^2 &= 2\left\langle \nabla\left(\frac{\partial u}{\partial t}\right), \nabla u\right\rangle \\ &= 2\langle \nabla(\Delta u), \nabla u\rangle \\ &= \Delta|\nabla u|^2 - 2|\nabla^2 u|^2 - 2\,\mathrm{Ric}(\nabla u, \nabla u).\end{aligned} \tag{11.177}$$

In particular, if g has nonnegative Ricci curvature, then

$$\left(\frac{\partial}{\partial t} - \Delta\right)|\nabla u|^2 \le -2|\nabla^2 u|^2 \le 0. \tag{11.178}$$

Thus,

$$\left(\frac{\partial}{\partial t} - \Delta\right)|\nabla u| \le -\frac{|\nabla^2 u|^2}{|\nabla u|} + \frac{\left|\nabla|\nabla u|\right|^2}{|\nabla u|} \le 0, \tag{11.179}$$

where the last inequality is by Kato's inequality.

The parabolic maximum principle says the following (see Theorem 14.7 below for a generalization of this result to time-dependent metrics and its proof).

[12] This formula holds more generally for $\varphi \in L^1_{\mathrm{loc}}(\mathbb{R}^n)$ such that φ grows exponentially subquadratically; i.e., $\varphi(\mathbf{x}) = o(e^{c\|\mathbf{x}\|^2})$ for all $c > 0$ (see, e.g., Chavel [**Cha84**, Chapter VI, §2]).

Proposition 11.51. *Let (M^n, g) be a closed Riemannian manifold and let $v : M^n \times [0,T)$ be a subsolution to the heat-type equation*

$$\frac{\partial v}{\partial t} \leq \Delta v + \langle X, \nabla v \rangle, \tag{11.180}$$

where $X = X(t)$ is a (time-dependent) vector field on M^n. If $v(x,0) \leq C$ for all $x \in M^n$, then $v(x,t) \leq C$ for all $x \in M^n$ and $t \in [0,T)$.

The idea of the proof is that for any time t_0, at any spatial maximum point x_0, we have that $\Delta v(x_0, t_0) \leq 0$ and $\nabla v(x_0, t_0) = \vec{0}$; thus, $\frac{\partial v}{\partial t}(x_0, t_0) \leq 0$. This says that the subsolution v is infinitesimally nonincreasing at any spacetime point where it achieves a spatial maximum.

By (11.178) and the parabolic maximum principle, we have the following.

Proposition 11.52 (Elementary gradient estimate). *For a solution u to the heat equation on a closed Riemannian manifold with nonnegative Ricci curvature, if $|\nabla u| \leq C$ at $t = t_0$, then $|\nabla u| \leq C$ for $t \geq t_0$.*

This estimate is sharp if we allow noncompact manifolds, as is exhibited for example by considering a linear function on Euclidean space.

11.7.3. The heat kernel on a Riemannian manifold.

Let

$$\square^* := -\partial_t - \Delta \tag{11.181}$$

be the **conjugate heat operator**. For any C^2 functions ϕ, ψ on $M^n \times I$ with compact support, by integration by parts we have that

$$\iint_{M^n \times I} \square \phi \, \psi \, d\mu dt = \iint_{M^n \times I} \phi \, \square^* \psi \, d\mu dt. \tag{11.182}$$

We also observe that if $\square^* v = 0$, then

$$\frac{d}{dt} \int_{M^n} v(x,t) \, d\mu(x) = 0 \tag{11.183}$$

by the divergence theorem.

Let $K(x,t;y,s)$ denote the unique **minimal fundamental solution to the heat equation** on (M^n, g). By 'fundamental solution' we mean that

$$\square_{x,t} K(x,t;y,s) = 0, \quad \lim_{t \to s+} K(x,t;y,s) = \delta_y(x), \tag{11.184a}$$

$$\square^*_{y,s} K(x,t;y,s) = 0, \quad \lim_{s \to t-} K(x,t;y,s) = \delta_x(y), \tag{11.184b}$$

where δ_y is the Dirac delta function centered at y and the limits in (11.184) are distributional limits (this is defined exactly analogously to the Euclidean case). By 'minimal', we mean that for any positive fundamental solution H, $K \leq H$.

Note that (11.184a) implies (11.184b) and vice versa. Indeed, if (11.184a) holds, then for $s < r < t$ and $x, y \in M^n$, we have by integrating by parts that

$$\begin{aligned} 0 &= \int_{M^n} K(x,t;z,r)\,\Box_{z,r} K(z,r;y,s)\,d\mu(z) \\ &= \int_{M^n} \Box^*_{z,r} K(x,t;z,r)\,K(z,r;y,s)\,d\mu(z). \end{aligned}$$

By letting $r \to s_+$, we obtain $\Box^*_{y,s} K(x,t;y,s) = 0$, so that (11.184b) holds. One proves the converse similarly. By (11.184a), we have that $K(\cdot,\cdot;y,s)$ is a solution to the heat equation for (y,s) fixed. On the other hand, by (11.184b), we have that $K(x,t;\cdot,\cdot)$ is a solution to the conjugate heat equation for (x,t) fixed. We also call the minimal fundamental solution K the **(conjugate) heat kernel**.

Observe that by (11.183) and (11.184b), we have

$$\int_{M^n} K(x,t;y,s)\,d\mu(y) \equiv 1. \tag{11.185}$$

Let $h_t : M^n \to \mathbb{R}$, and let $h(y,s)$ be the solution to the conjugate heat equation with $h(y,t) = h_t(y)$. Then

$$h(y,s) = \int_{M^n} h_t(x) K(x,t;y,s)\,d\mu(x). \tag{11.186}$$

For the following result on noncompact manifolds that are not necessarily complete, see for example Chavel [**Cha84**, pp. 188–191] (cf. [**CCG⁺10**, Theorem 24.39]).

Proposition 11.53. *Without any assumption on g except smoothness, the existence and uniqueness of $K(x,t;y,s)$ holds when M^n is noncompact.*

Diffeomorphism invariance of K: If φ is a diffeomorphism of M^n, then

$$K_{\varphi^* g}(x,t;y,s) = K_g(\varphi(x),t;\varphi(y),s) =: (K\circ\varphi)(x,t;y,s). \tag{11.187}$$

11.7.4. The Li–Yau Harnack estimate. Having acquainted ourselves with some foundational properties of the heat equation on Riemannian manifolds, we now transition to a deeper exploration, focusing on one of the critical tools in geometric analysis — the Li–Yau Harnack estimate. This powerful estimate, both elegant and profound, brings forth essential insights into the behavior of solutions to the heat equation. Its significance in the landscape of geometric analysis is paramount, acting as a cornerstone for further, more complex investigations.

Recall from (11.169) that the Euclidean heat kernel is given by $H(\mathbf{z},\tau) = (4\pi\tau)^{-\frac{n}{2}} e^{-\frac{|\mathbf{z}|^2}{4\tau}}$. Its logarithm is

$$\ln H(\mathbf{z},\tau) = -\frac{n}{2}\ln(4\pi\tau) - \frac{|\mathbf{z}|^2}{4\tau}.$$

We compute that

$$\Delta_{\mathbb{E}} \ln H(\mathbf{z},\tau) = -\Delta_{\mathbb{E}}\left(\frac{|\mathbf{z}|^2}{4\tau}\right) = -\frac{n}{2\tau}.$$

Remarkably, for arbitrary positive solutions to the heat equation, not just heat kernels, we have the following differential Harnack estimate.

Theorem 11.54 (Li–Yau differential Harnack estimate)**.** *Let (M^n, g) be a complete Riemannian manifold with nonnegative Ricci curvature. Let $u : M^n \to \mathbb{R}$ be a positive solution to the heat equation $\partial_t u = \Delta u$. Then*

$$\partial_t \ln u - |\nabla \ln u|^2 = \Delta \ln u \geq -\frac{n}{2t}. \tag{11.188}$$

Proof. Define

$$Q := \Delta \ln u. \tag{11.189}$$

We compute that

$$\partial_t \ln u = \frac{1}{u}\partial_t u = \frac{1}{u}\Delta u = \Delta \ln u + |\nabla \ln u|^2. \tag{11.190}$$

Note that this formula implies the equality in (11.188). We compute that

$$\begin{aligned}\partial_t Q &= \partial_t(\Delta \ln u) = \Delta(\partial_t \ln u)\\ &= \Delta(\Delta \ln u + |\nabla \ln u|^2).\end{aligned}$$

By applying the fundamental Bochner formula (11.21) to this and using the nonnegativity of the Ricci curvature, we obtain

$$\begin{aligned}\partial_t Q &= \Delta Q + 2|\nabla^2 \ln u|^2 + 2\langle \nabla \ln u, \nabla Q\rangle + 2\operatorname{Ric}(\nabla \ln u, \nabla \ln u)\\ &\geq \Delta Q + 2\langle \nabla \ln u, \nabla Q\rangle + \frac{2}{n}Q^2.\end{aligned}$$

If M^n is closed, then the Li–Yau differential Harnack estimate follows from the parabolic maximum principle (Proposition 11.51). On the other hand, in the noncompact case one can use a cutoff function to prove the result; see, e.g., Schoen and Yau [**SY94**, §4.1] for an exposition. □

Corollary 11.55 (Path-integrated Harnack estimate)**.** *Let (M^n, g) be a complete Riemannian manifold with nonnegative Ricci curvature. If u is a* *positive solution to the heat equation on M^n, then for any $x_1, x_2 \in M^n$ and $0 < t_1 < t_2$,*

$$\frac{u(x_2,t_2)}{u(x_1,t_1)} \geq \left(\frac{t_2}{t_1}\right)^{-\frac{n}{2}} \exp\left(-\frac{1}{4}\frac{d(x_1,x_2)^2}{t_2-t_1}\right). \tag{11.191}$$

Proof. Let $\gamma : [t_1, t_2] \to M^n$ be a path joining x_1 to x_2. Then $t \mapsto (\gamma(t), t)$ is a spacetime path joining (x_1, t_1) to (x_2, t_2). Integrating the Li–Yau differential Harnack estimate (11.188) over this path and applying the fundamental theorem of calculus yields

$$\begin{aligned}
\ln\frac{u(x_2,t_2)}{u(x_1,t_1)} &= \int_{t_1}^{t_2} \frac{d}{dt}\big(\ln u(\gamma(t),t)\big)dt \\
&= \int_{t_1}^{t_2} \left(\frac{\partial}{\partial t}\ln u + \nabla \ln u \cdot \frac{d\gamma}{dt}\right) dt \\
&\geq \int_{t_1}^{t_2} \left(-\frac{n}{2t} + |\nabla \ln u|^2 + \nabla \ln u \cdot \frac{d\gamma}{dt}\right) dt \\
&\geq -\frac{n}{2}\ln\left(\frac{t_2}{t_1}\right) - \frac{1}{4}\int_{t_1}^{t_2} \left|\frac{d\gamma}{dt}(t)\right|^2 dt.
\end{aligned}$$

The result now follows from taking $\gamma(t)$ to be a constant-speed minimal geodesic, which always exists since the metric g is complete;[13] observe that in this case we have that $\left|\frac{d\gamma}{dt}(t)\right| \equiv \frac{d(x_1,x_2)}{t_2-t_1}$. □

The following consequence of the Li–Yau Harnack estimate was proved earlier by Cheeger and Yau [**CY81**] by a different method.

Corollary 11.56. *If (M^n, g) is a complete Riemannian manifold with nonnegative Ricci curvature, then the heat kernel $K(x,t;y,s)$ on M^n satisfies*

$$K(x,t;y,s) \geq (4\pi(t-s))^{-\frac{n}{2}} e^{-\frac{d(x,y)^2}{4(t-s)}}. \tag{11.192}$$

Proof. Without loss of generality, we may assume that $s = 0$. By taking $t_2 = t$, $t_1 \to 0$, $x_2 = x$, and $x_1 = y$ in (11.191), we obtain that

$$\begin{aligned}
K(x,t;y,0) &\geq t^{-\frac{n}{2}} \lim_{t_1\to 0} \exp\left(-\frac{1}{4}\frac{d(y,x)^2}{t-t_1}\right) t_1^{\frac{n}{2}} K(y,t_1;y,0) \\
&= (4\pi t)^{-\frac{n}{2}} \exp\left(-\frac{d(y,x)^2}{4t}\right)
\end{aligned}$$

since the heat kernel satisfies $\lim_{t_1\to 0} t_1^{\frac{n}{2}} K(y,t_1;y,0) = (4\pi)^{-\frac{n}{2}}$; this last limit follows from the asymptotics of the heat kernel (see, e.g., [**CCG**$^+$**10**, Theorem 23.17]). □

[13] If g is not complete, then the result also holds by taking a sequence of constant-speed paths whose lengths limit to $d(x_1, x_2)$.

11.8. Hodge theory

In this section, after a brief introduction to de Rham cohomology, we will discuss the Hodge theorem and some of its consequences. For proofs, see the books by Warner [**War83**, Chapter 6], Jost [**Jos17**, §3.4, §3.6] for Riemannian manifolds, and Griffiths and Harris [**GH94**, Chapter 0, §6] and Wells [**Wel08**, Chapter V, §5] for Kähler manifolds.

Throughout this section, (M^n, g) will denote a closed oriented Riemannian manifold.

11.8.1. De Rham cohomology and the de Rham theorem.

The exterior derivative $d_p : \Omega^p(M^n) \to \Omega^{p+1}(M^n)$ acting on differential forms (defined by (8.26)) gives rise to a 'complex' of differential operators:
(11.193)

$$0 \to \Omega^0(M^n) \xrightarrow{d_0} \Omega^1(M^n) \xrightarrow{d_1} \cdots \xrightarrow{d_{n-2}} \Omega^{n-1}(M^n) \xrightarrow{d_{n-1}} \Omega^n(M^n) \to 0,$$

where $d^2 := d_p \circ d_{p-1} = 0$ (this last condition characterizes **differential complexes** by definition). Hence $\mathrm{image}(d_{p-1}) \subset \ker(d_p)$ and we can define the **p-th de Rham cohomology group** as the quotient (real vector) space:

$$H^p_{\mathrm{dRh}}(M^n) := \frac{\ker(d_p)}{\mathrm{image}(d_{p-1})}. \tag{11.194}$$

We have the following.

Theorem 11.57 (De Rham). *If M^n is a closed manifold, then the p-th de Rham cohomology group is isomorphic to the p-th singular cohomology group with real coefficients:*

$$H^p_{\mathrm{dRh}}(M^n) \cong H^p(M^n; \mathbb{R}).$$

For treatments of de Rham cohomology, including the proof of the de Rham theorem, we refer the reader to Bott and Tu [**BT82**] and Lee [**Lee13**, Chapters 17 and 18]; for singular cohomology, see the references therein.

11.8.1.1. *The Poincaré lemma.* We say that a subset S of $\mathbb{R}^n$ is **starshaped** with respect to a point $\mathbf{x}_0 \in \mathbb{R}^n$ if for every $\mathbf{x} \in S$, we have that $\mathbf{x}_0 + t(\mathbf{x} - \mathbf{x}_0) \in S$ for all $t \in [0, 1]$.

The following says that the de Rham cohomology groups of starshaped open subsets of Euclidean spaces are zero.

Theorem 11.58 (Poincaré lemma). *Any closed differential form on an open starshaped subset U of $\mathbb{R}^n$ is exact.*

Proof. Without loss of generality, we may assume that U is starshaped with respect to the origin $\mathbf{0}$; that is, if $\mathbf{x} \in U$, then $t\mathbf{x} \in U$ for all $t \in [0, 1]$.

Define dilation maps $f_t : \mathbb{R}^n \to \mathbb{R}^n$ by $f_t(\mathbf{x}) = t\mathbf{x}$ for $\mathbf{x} \in \mathbb{R}^n$ and $t \in [0,1]$. Then f_t is a diffeomorphism for $t \in (0,1]$, $f_0(\mathbf{x}) \equiv \mathbf{0}$, and $f_1 = \mathrm{id}$. Moreover, $f_t(U) \subset U$ for $t \in [0,1]$. For $t \in (0,1)$, define the vector field V_t on $\mathbb{R}^n$ by

$$V_t(\mathbf{x}) := d(f_t)^{-1}\left(\frac{df_t}{dt}\right) = \frac{1}{t}\mathbf{x}.$$

Let α be a closed differential form on U, so that $d\alpha = 0$. By (6.93), we have that

$$\frac{d}{dt} f_t^* \alpha = f_t^*(\mathcal{L}_{V_t}\alpha).$$

Now, for any $0 < t_0 < t_1 < 1$, we calculate that

$$\begin{aligned} f_{t_1}^*\alpha - f_{t_0}^*\alpha &= \int_{t_0}^{t_1} \frac{d}{dt} f_t^* \alpha\, dt \\ &= \int_{t_0}^{t_1} f_t^*(\mathcal{L}_{V_t}\alpha)\, dt \\ &= \int_{t_0}^{t_1} f_t^*\big(\iota_{V_t}(d\alpha) + d(\iota_{V_t}\alpha)\big)\, dt \\ &= \int_{t_0}^{t_1} f_t^*\big(d(\iota_{V_t}\alpha)\big)\, dt, \end{aligned}$$

where the third equality is true by Cartan's magic formula (8.34) and where the fourth equality is true since α is closed. By applying the naturality property (8.30), we obtain

$$f_{t_1}^*\alpha - f_{t_0}^*\alpha = \int_{t_0}^{t_1} d\big(f_t^*(\iota_{V_t}\alpha)\big)\, dt = d\left(\int_{t_0}^{t_1} f_t^*(\iota_{V_t}\alpha)\, dt\right).$$

By taking the limits as $t_0 \to 0_+$ and $t_1 \to 1_-$ and using the facts that $f_0^*\alpha = 0$ and $f_1^*\alpha = \alpha$, we conclude that

$$\alpha = f_1^*\alpha - f_0^*\alpha = d\beta,$$

where

$$\beta := \int_0^1 f_t^*(\iota_{V_t}\alpha)\, dt$$

is a smooth differential form. This proves that α is exact. □

11.8.2. The pointwise inner product on $\Lambda^p T_x^* M$. Recall that the **pointwise inner product** on the vector space of p-forms $\Lambda^p T_x^* M$ at a point $x \in M^n$ is defined by

$$\langle \gamma, \eta \rangle_x := \sum_{i_1,\ldots,i_p=1}^{n} \gamma\left(e_{i_1},\ldots,e_{i_p}\right) \eta\left(e_{i_1},\ldots,e_{i_p}\right) \tag{11.195}$$

for $\gamma, \eta \in \Lambda^p T_x^* M$, where $\{e_i\}_{i=1}^n$ is an orthonormal frame at x. With respect to local coordinates $(\mathcal{U}, \{x^i\}_{i=1}^n)$, we have for $x \in \mathcal{U}$ that

$$\langle \gamma, \eta \rangle_x = \sum_{i_1,\ldots,i_p,j_1,\ldots,j_p=1}^{n} g^{i_1 j_1} \cdots g^{i_p j_p} \gamma_{i_1 \cdots i_p} \eta_{j_1 \cdots j_p}, \tag{11.196}$$

where the components of the forms are defined by

$$\gamma_{i_1 \cdots i_p} := \gamma(\partial_{i_1}, \ldots, \partial_{i_p}) \quad \text{and} \quad \eta_{j_1 \cdots j_p} := \eta(\partial_{j_1}, \ldots, \partial_{j_p}). \tag{11.197}$$

The **pointwise norm** of γ is defined by

$$|\gamma|_x := \sqrt{\langle \gamma, \gamma \rangle_x}. \tag{11.198}$$

Let $\{\omega^j\}_{j=1}^n$ denote the orthonormal coframe at x dual to $\{e_i\}_{i=1}^n$. We have

$$\left\langle \omega^{i_1} \wedge \cdots \wedge \omega^{i_p}, \omega^{j_1} \wedge \cdots \wedge \omega^{j_p} \right\rangle_x = \det\left((\delta_{i_k j_\ell})_{k,\ell=1}^p \right).$$

In particular, one can show from this that $\{\omega^{i_1} \wedge \cdots \wedge \omega^{i_p}\}_{1 \le i_1 < \cdots < i_p \le n}$ is an orthonormal basis for $\Lambda^p T_x^* M$.

11.8.3. The Hodge star operator.

The **Hodge star operator**

$$* : \Lambda^p T^* M \to \Lambda^{n-p} T^* M, \tag{11.199}$$

where $0 \le p \le n$, is well-defined by the property that

$$\langle \gamma, \eta \rangle_x \, d\mu(x) = (\gamma \wedge * \eta)(x) \tag{11.200}$$

for any $\gamma, \eta \in \Lambda^p T_x^* M$, $x \in M^n$. For example, we have

$$*\left(\omega^1 \wedge \cdots \wedge \omega^p\right) = \omega^{p+1} \wedge \cdots \wedge \omega^n$$

for a positively oriented orthonormal coframe $\left\{\omega^j\right\}_{j=1}^n$. Acting on p-forms, we have

$$*^2 := * \circ * = (-1)^{p(n-p)} \, \mathrm{id} : \Lambda^p T^* M \to \Lambda^p T^* M. \tag{11.201}$$

In particular, if p or $n-p$ is even, then $*^2 = \mathrm{id}$, whereas if p and $n-p$ are both odd, then $*^2 = -\mathrm{id}$.

If $n = 2m$ is even, then the Hodge star operator is **conformally invariant** in the middle dimension m: Denoting $*_g$ to be the Hodge star with respect to the metric g, we have for any function u on M^{2m} that

$$*_{e^u g} = *_g : \Lambda^m T^* M \to \Lambda^m T^* M. \tag{11.202}$$

This is because for m-forms γ and η on M^{2m}, we have

$$\langle \gamma, \eta \rangle_{e^u g} \, d\mu_{e^u g} = \langle \gamma, \eta \rangle_g \, d\mu_g,$$

where $\langle \cdot, \cdot \rangle_g$ denotes the pointwise inner product defined by (11.196) with respect to g.

11.8.4. The codifferential. The **codifferential** acting on p-forms

$$\delta : \Omega^p(M^n) \to \Omega^{p-1}(M^n)$$

is defined by

$$\delta\alpha = (-1)^{n(p+1)+1} * d * \alpha. \tag{11.203}$$

One verifies that δ is the L^2**-adjoint** operator of the differential d; i.e.,

$$\langle\langle d\beta, \alpha\rangle\rangle_{L^2} = \langle\langle \beta, \delta\alpha\rangle\rangle_{L^2} \tag{11.204}$$

for $\alpha \in \Omega^p(M^n)$ and $\beta \in \Omega^{p-1}(M^n)$, where $\langle\langle\cdot,\cdot\rangle\rangle_{L^2}$ is the L^2-inner product defined by (11.27). Indeed, we compute that

$$\begin{aligned}
\langle\langle \beta, \delta\alpha\rangle\rangle_{L^2} &= \int_{M^n} \beta \wedge *(\delta\alpha) \\
&= (-1)^{n(p+1)+1} \int_{M^n} \beta \wedge \left(*^2 \left(d * \alpha\right)\right) \\
&= (-1)^{n(p+1)+1}(-1)^{(p-1)(n-p+1)} \int_{M^n} \beta \wedge d(*\alpha) \\
&= (-1)^p \int_{M^n} \beta \wedge d(*\alpha)
\end{aligned}$$

since $d * \alpha$ being an $(n-p+1)$-form implies that $*^2 = (-1)^{(p-1)(n-p+1)}\,\mathrm{id}$. Now, by applying (8.28) and Stokes's theorem to the last line above, we obtain

$$\langle\langle \beta, \delta\alpha\rangle\rangle_{L^2} = \int_{M^n} d\beta \wedge (*\alpha) = \langle\langle d\beta, \alpha\rangle\rangle_{L^2}.$$

In terms of covariant derivatives, the codifferential δ can be expressed as

$$(\delta\alpha)\left(X_1, \ldots, X_{p-1}\right) = -\sum_{j=1}^{n} \left(\nabla_{e_j}\alpha\right)\left(e_j, X_1, \ldots, X_{p-1}\right), \tag{11.205}$$

where $\{e_j\}_{j=1}^n$ is an orthonormal frame. One can prove (11.205) by using (8.32), the definition of δ as the L^2-adjoint of d, and integration by parts. That is,

$$\delta\alpha = -\operatorname{div}(\alpha),$$

or in local coordinates (we leave this as an exercise)

$$(\delta\alpha)_{i_1\cdots i_{p-1}} = -\sum_{j,k=1}^{n} g^{jk}\nabla_j \alpha_{k\, i_1\cdots i_{p-1}}. \tag{11.206}$$

11.8.5. The Hodge Laplacian. The **Hodge Laplacian** acting on differential p-forms is defined by

$$\Delta_d := -(d\delta + \delta d) = -(d+\delta)^2. \tag{11.207}$$

We remark that in the literature, it is at least as common to define the Hodge Laplacian to be the negative of the definition that we adopt. The chosen sign convention ensures that the Bochner formulas directly relate the Hodge Laplacian to the rough Laplacian instead of minus the rough Laplacian.

Observe that Δ_d is a self-adjoint operator:

$$\begin{aligned} \langle\langle \Delta_d \alpha, \beta \rangle\rangle_{L^2} &= -\langle\langle d\delta\alpha + \delta d\alpha, \beta\rangle\rangle_{L^2} \\ &= -\langle\langle \delta\alpha, \delta\beta\rangle\rangle_{L^2} - \langle\langle d\alpha, d\beta\rangle\rangle_{L^2} \\ &= \langle\langle \alpha, \Delta_d \beta\rangle\rangle_{L^2}. \end{aligned} \tag{11.208}$$

Acting on functions, it is the same as the Laplacian defined in (8.86). We also have that the Hodge Laplacian commutes with the Hodge star operator:

$$*\Delta_d = \Delta_d * . \tag{11.209}$$

Indeed, we compute that acting on p-forms,

$$\begin{aligned} - * \Delta_d\alpha &= (-1)^{n(p+1)+1} * d * d * \alpha + (-1)^{n(p+2)+1} *^2 d * d\alpha \\ &= (-1)^{n(p+1)+1} * d * d * \alpha + (-1)^{np+1}(-1)^{p(n-p)} d * d\alpha \\ &= (-1)^{n(p+1)+1} * d * d(*\alpha) + (-1)^{np+1} d * d * (*\alpha) \\ &= -\Delta_d * \alpha, \end{aligned}$$

where we used

$$(-1)^{n(p+1)+1} = (-1)^{n(n-p+2)+1} \quad \text{and} \quad (-1)^{np+1} = (-1)^{n(n-p+1)+1}$$

to rewrite the first and second terms on the third line, respectively.

11.8.6. Harmonic forms. A differential form α is said to be **harmonic** if

$$\Delta_d \alpha = 0. \tag{11.210}$$

Since, by (11.208),

$$\langle\langle \Delta_d \alpha, \alpha\rangle\rangle_{L^2} = -\|\delta\alpha\|_{L^2} - \|d\alpha\|_{L^2}, \tag{11.211}$$

we have that α is harmonic if and only if

$$d\alpha = 0 \quad \text{and} \quad \delta\alpha = 0. \tag{11.212}$$

In particular, a function f is harmonic if and only if $df = 0$; that is, f is constant.

The **space of harmonic p-forms** is denoted by

$$\mathbf{H}^p(M^n) = \{\alpha \in \Omega^p(M^n) : \Delta_d\alpha = 0\}. \tag{11.213}$$

11.8.7. The Hodge decomposition theorem. The essence of the Hodge decomposition theorem, stated below, is understanding when, given $\gamma \in \Omega^p(M^n)$, the equation

$$\Delta_d \alpha = \gamma$$

has a solution $\alpha \in \Omega^p(M^n)$. If $\beta \in \mathbf{H}^p(M)$ is a harmonic p-form, then

$$\langle\langle \gamma, \beta \rangle\rangle_{L^2} = \langle\langle \Delta_d \alpha, \beta \rangle\rangle_{L^2} = \langle\langle \alpha, \Delta_d \beta \rangle\rangle_{L^2} = 0.$$

So a necessary condition to solve $\Delta_d \alpha = \gamma$ is $\langle\langle \gamma, \beta \rangle\rangle_{L^2} = 0$ for all $\beta \in \mathbf{H}^p(M^n)$. The converse of this is also true. In particular, we have the following, which is known as the **Hodge decomposition theorem**.

Theorem 11.59. *Let (M^n, g) be a closed Riemannian manifold.*

(1) *Given $\gamma \in \Omega^p(M^n)$, the equation*

$$\Delta_d \alpha = \gamma \tag{11.214}$$

has a solution $\alpha \in \Omega^p(M^n)$ if and only if

$$\langle\langle \gamma, \beta \rangle\rangle_{L^2} = 0, \tag{11.215}$$

for all $\beta \in \mathbf{H}^p(M^n)$.

(2) *Consequently, we have the following orthogonal decomposition (with respect to the L^2-inner product) of the space of p-forms:*

$$\begin{aligned} \Omega^p(M^n) &= \Delta_d\big(\Omega^p(M^n)\big) \oplus \mathbf{H}^p(M^n) \\ &= d\delta\big(\Omega^p(M^n)\big) \oplus \delta d\big(\Omega^p(M^n)\big) \oplus \mathbf{H}^p(M^n). \end{aligned} \tag{11.216}$$

Moreover, the space $\mathbf{H}^p(M^n)$ is finite dimensional.

Note that any two solutions to (11.214) differ by a harmonic p-form.

For proofs of Theorem 11.59, see Jost [**Jos17**], where in §3.4 a proof is given using the calculus of variations, Sobolev spaces, and regularity theory and where in §3.6 a proof is given using heat equation methods.

By the Hodge decomposition theorem, we can identify the p-th de Rham cohomology group with the space of harmonic p-forms.

Corollary 11.60. *Let (M^n, g) be a closed Riemannian manifold. In each de Rham cohomology class of differential forms, there exists a unique harmonic form representing the cohomology class. In particular, for each $0 \leq p \leq n$ the p-th de Rham cohomology group $H^p_{\mathrm{dRh}}(M^n)$ is isomorphic to the space of harmonic p-forms $\mathbf{H}^p(M^n)$.*

Proof. Let $[\alpha] \in H^p_{\mathrm{dRh}}(M^n)$ be a de Rham cohomology class, so that α is a closed p-form. By (11.216), we have that

$$\alpha = d\delta\beta + \delta d\gamma + \eta, \tag{11.217}$$

where η is a harmonic p-form and β and γ are p-forms. Since $d\alpha = 0$, we have that $d\delta d\gamma = 0$, which implies that $\delta d\gamma = 0$. Therefore η is a harmonic p-form in the same de Rham cohomology class as α. This proves the existence of a harmonic form in each de Rham cohomology class.

Now suppose that η and θ are harmonic p-forms in the same de Rham cohomology class. Then there exists a $(p-1)$-form σ such that $d\sigma = \eta - \theta$. Since η and θ are harmonic, we have that $\delta d\sigma = 0$. We conclude that

$$\|\eta - \theta\|_{L^2}^2 = \|d\sigma\|_{L^2}^2 = \langle\langle \sigma, \delta d\sigma \rangle\rangle_{L^2} = 0.$$

This proves the uniqueness of the harmonic form in each de Rham cohomology class. □

We have the following existence theorem for solutions to the Poisson equation on a closed Riemannian manifold.

Corollary 11.61. *Let $\phi : M^n \to \mathbb{R}$ be a function satisfying $\int_{M^n} \phi d\mu = 0$. Then there exists a function $f : M^n \to \mathbb{R}$ satisfying the Poisson equation*

$$\Delta f = \phi. \tag{11.218}$$

The function f is unique up to an additive constant.

Proof. By Lemma 11.7, we have that $\mathbf{H}^0(M^n)$ is the 1-dimensional vector space of constant functions on M^n. Hence, a function $\phi \in \Omega^0(M^n) = C^\infty(M^n)$ satisfies condition (11.215) if and only if $\int_{M^n} \phi d\mu = 0$. The existence of f now follows from Theorem 11.59(1).[14] The uniqueness of f up to an additive constant follows from the fact that any harmonic function on a closed manifold is constant. □

The statement in Corollary 11.60 that in each de Rham cohomology class there exists a harmonic form can be considered as a special case of the Hodge theorem (Theorem 11.59). The idea of a direct proof of this special case is to employ the methods of the calculus of variations and partial differential equations to prove that in each de Rham cohomology class $[\alpha] \in H^p_{\mathrm{dRh}}(M^n)$ there exists a minimizer of the energy functional $E : [\alpha] \to \mathbb{R}$ defined by

$$E(\beta) := \|\beta\|_{L^2}^2. \tag{11.219}$$

See Jost [**Jos17**] for the details of such a proof. Now, given a closed p-form η that is a minimizer of E, we can see that η is coclosed and hence harmonic

[14] We remark that, equivalently, by (11.216), for any smooth function ϕ on M^n there exists a smooth function f such that

$$\phi = -\delta df + \phi_{\mathrm{avg}} = \Delta f + \phi_{\mathrm{avg}}.$$

We can think of the constant function ϕ_{avg} as the harmonic representative of the 0-form ϕ.

as follows. Since η is a minimizer of E, it is a critical point of E, so that for any exact p-form $\gamma = d\theta$ we have that

$$0 = \frac{1}{2}\frac{d}{dt}\bigg|_{t=0} \|\eta + t\gamma\|_{L^2}^2 = \langle\langle \eta, \gamma\rangle\rangle_{L^2} = \langle\langle \eta, d\theta\rangle\rangle_{L^2} = \langle\langle \delta\eta, \theta\rangle\rangle_{L^2}.$$

Since θ is an arbitrary $(p-1)$-form, we conclude that $\delta\eta = 0$ as desired.

11.8.8. Commutator formulas for covariant differentiation. A useful tool for comparing different types of Laplacians acting on tensors and forms is the commutator formula for the covariant derivative. In local coordinates, the commutator formula for ∇ acting on tensors of type (p,q) is given by

$$\begin{aligned} \nabla_i \nabla_j \alpha_{k_1\cdots k_p}^{\ell_1\cdots\ell_q} - \nabla_j \nabla_i \alpha_{k_1\cdots k_p}^{\ell_1\cdots\ell_q} = &- \sum_{h=1}^{p}\sum_{m=1}^{n} R_{ijk_h}^m \alpha_{k_1\cdots k_{h-1}mk_{h+1}\cdots k_p}^{\ell_1\cdots\ell_q} \\ &+ \sum_{h=1}^{q}\sum_{m=1}^{n} R_{ijm}^{\ell_h} \alpha_{k_1\cdots k_p}^{\ell_1\cdots\ell_{h-1}m\ell_{h+1}\cdots\ell_q}. \end{aligned} \tag{11.220}$$

In particular, if α is a covariant p-tensor (e.g., a p-form), then

$$\nabla_i \nabla_j \alpha_{k_1\cdots k_p} - \nabla_j \nabla_i \alpha_{k_1\cdots k_p} = - \sum_{h=1}^{p}\sum_{m=1}^{n} R_{ijk_h}^m \alpha_{k_1\cdots k_{h-1}mk_{h+1}\cdots k_p}. \tag{11.221}$$

11.8.9. Bochner–Weitzenböck formulas for p-forms. In this subsection we discuss vanishing theorems for harmonic p-forms on Riemannian manifolds using the Bochner–Weitzenböck technique (a.k.a. the Bochner technique).

The **rough Laplacian** Δ acting on tensors is the trace of the second covariant derivative:

$$\Delta\alpha := \operatorname{trace}_{1,2}(\nabla^2\alpha). \tag{11.222}$$

In local coordinates, if α is a tensor of type (p,q), then

$$(\Delta\alpha)_{k_1\cdots k_p}^{\ell_1\cdots\ell_q} = g^{ij}\nabla_i\nabla_j\alpha_{k_1\cdots k_p}^{\ell_1\cdots\ell_q}. \tag{11.223}$$

11.8.9.1. *The Bochner–Weitzenböck formula for* 1*-forms.* We begin with the following **Bochner–Weitzenböck formula for 1-forms**.

Proposition 11.62. *For any Riemannian manifold* (M^n, g) *and* 1*-form* α, *we have*

$$\Delta_d\alpha - \Delta\alpha = -\operatorname{Ric}(\alpha), \tag{11.224}$$

where Δ_d *is the Hodge Laplacian,* Δ *is the rough Laplacian, and we consider the Ricci tensor as a (self-adjoint) linear map* $\operatorname{Ric} : T_x^*M \to T_x^*M$ *for each* $x \in M^n$.

Proof. Recall that the Hodge Laplacian is defined by $\Delta_d = -(d\delta + \delta d)$. We compute in local coordinates and using (8.33) that

$$\begin{aligned}(d\alpha)_{ij} &= \nabla_i \alpha_j - \nabla_j \alpha_i, \\ (\delta d\alpha)_j &= -g^{ik}\nabla_i (d\alpha)_{kj} = -g^{ik}\nabla_i (\nabla_k \alpha_j - \nabla_j \alpha_k)\end{aligned}$$

and

$$\begin{aligned}\delta\alpha &= -g^{ik}\nabla_i \alpha_k, \\ (d\delta\alpha)_j &= -g^{ik}\nabla_j \nabla_i \alpha_k,\end{aligned}$$

so that

$$\begin{aligned}(\Delta_d \alpha)_j &= g^{ik}\nabla_i (\nabla_k \alpha_j - \nabla_j \alpha_k) + g^{ik}\nabla_j \nabla_i \alpha_k \\ &= (\Delta\alpha)_j + g^{ik}(\nabla_j \nabla_i - \nabla_i \nabla_j)\alpha_k \\ &= (\Delta\alpha)_j - g^{kl} R_{jl} \alpha_k,\end{aligned}$$

where for the last equality we used (11.20). □

By applying (11.211) and integration by parts, we obtain the identity

$$\begin{aligned}\int_{M^n} \left(|d\alpha|^2 + |\delta\alpha|^2\right) d\mu &= -\int_{M^n} \langle \Delta_d \alpha, \alpha\rangle \, d\mu \\ &= \int_{M^n} |\nabla\alpha|^2 \, d\mu + \int_{M^n} \operatorname{Ric}(\alpha,\alpha)\, d\mu.\end{aligned} \tag{11.225}$$

In particular, if (M^n, g) satisfies $\operatorname{Ric} \geq 0$ and if α is a harmonic 1-form, then $\nabla\alpha \equiv 0$ and $\operatorname{Ric}(\alpha,\alpha) \equiv 0$ on M^n. If we strengthen the Ric assumption to $\operatorname{Ric} > 0$, then we conclude that $\alpha \equiv 0$ on M^n. That is:

Proposition 11.63. *If (M^n, g) is a closed Riemannian manifold with positive Ricci curvature, then any harmonic 1-form vanishes. In particular, $H^1_{\mathrm{dRh}}(M^n) = 0$.*

Observe that this proposition is also a consequence of Myers's theorem and the de Rham theorem since if a manifold has finite fundamental group, then $H^1(M^n; \mathbb{R}) = 0$.

11.8.9.2. *The Bochner–Weitzenböck formula for p-forms.* Let α be a p-form on M^n. In local coordinates, formula (8.32) for the exterior derivative says that

$$(d\alpha)_{i_0 i_1 \cdots i_p} = \sum_{\ell=0}^{p} (-1)^\ell \nabla_{i_\ell} \alpha_{i_0 i_1 \cdots \widehat{i_\ell} \cdots i_p}.$$

By this formula and (11.206), we compute that $-\delta(d\alpha)$ may be written as

$$
\begin{aligned}
-\delta(d\alpha)_{i_1\cdots i_p} &= \sum_{j,k=1}^{n} g^{jk}\nabla_j (d\alpha)_{k\,i_1\cdots i_p} \\
&= \sum_{j,k=1}^{n} \Big(g^{jk}\nabla_j\nabla_k\alpha_{i_1\cdots i_p} + \sum_{\ell=1}^{p} (-1)^{\ell} g^{jk}\nabla_j\nabla_{i_\ell}\alpha_{k\,i_1\cdots\widehat{i_\ell}\cdots i_p}\Big).
\end{aligned}
$$

We also compute that $-d(\delta\alpha)$ may be written as

$$
\begin{aligned}
-d(\delta\alpha)_{i_1\cdots i_p} &= -\sum_{j=1}^{p} (-1)^{j-1}\nabla_{i_j}(\delta\alpha)_{i_1\cdots\widehat{i_j}\cdots i_p} \\
&= \sum_{\ell=1}^{p}\sum_{j,k=1}^{n} (-1)^{\ell-1} g^{jk}\nabla_{i_\ell}\nabla_j\alpha_{k\,i_1\cdots\widehat{i_\ell}\cdots i_p}.
\end{aligned}
$$

Hence

$$
\begin{aligned}
(\Delta_d\alpha)_{i_1\cdots i_p} &= -\delta(d\alpha)_{i_1\cdots i_p} - d(\delta\alpha)_{i_1\cdots i_p} \\
&= (\Delta\alpha)_{i_1\cdots i_p} + \sum_{j,k=1}^{n}\sum_{\ell=1}^{p} (-1)^{\ell} g^{jk}\left(\nabla_j\nabla_{i_\ell} - \nabla_{i_\ell}\nabla_j\right)\alpha_{k\,i_1\cdots\widehat{i_\ell}\cdots i_p},
\end{aligned}
$$

where $\Delta\alpha$ is the rough Laplacian of α. So, by using the covariant derivative commutator formula (11.221), we obtain that

$$
\begin{aligned}
(\Delta_d\alpha - \Delta\alpha)_{i_1\cdots i_p} &= \sum_{\ell=1}^{p} (-1)^{\ell} R_{i_\ell}^{r}\alpha_{r\,i_1\cdots\widehat{i_\ell}\cdots i_p} \\
&\quad - \sum_{j,k=1}^{n}\sum_{\ell=1}^{p}\sum_{\substack{1\le m\le p\\ m\neq\ell}} (-1)^{\ell} g^{jk} R_{j i_\ell i_m}^{r}\alpha_{k\,i_1\cdots i_{m-1}\,r\,i_{m+1}\cdots\widehat{i_\ell}\cdots i_p}
\end{aligned}
\tag{11.226}
$$

since $g^{jk}R_{ji_\ell k}^{r} = -R_{i_\ell}^{r}$. We conclude the following.

Proposition 11.64. *If (M^n, g) is a Riemannian manifold and α is a p-form on M^n, then*

$$
\begin{aligned}
(\Delta_d\alpha)_{i_1\cdots i_p} &= (\Delta\alpha)_{i_1\cdots i_p} - \sum_{\ell=1}^{p} R_{i_\ell}^{r}\alpha_{i_1\cdots i_{\ell-1}\,r\,i_{\ell+1}\cdots i_p} \\
&\quad + \sum_{j,k=1}^{n}\sum_{\ell=1}^{p}\sum_{\substack{1\le m\le p\\ m\neq\ell}} g^{jk} R_{j i_\ell i_m}^{r}\alpha_{i_1\cdots i_{m-1}\,r\,i_{m+1}\cdots i_{\ell-1}\,k\,i_{\ell+1}\cdots i_p}.
\end{aligned}
\tag{11.227}
$$

For example, if $p = 1$, then $(\Delta_d\alpha - \Delta\alpha)_i = -R_i^r\alpha_r$. This agrees with (11.224).

We have the following result of Gallot and Meyer [**GM75**].

Theorem 11.65. *If* (M^n, g) *is a closed Riemannian manifold with positive curvature operator, then* $H^p(M^n; \mathbb{R}) = 0$ *for all* $0 < p < n$.

Proof. Consider any harmonic p-form α on (M^n, g), where $0 < p < n$. The idea of the proof is to integrate $\langle \alpha, \Delta_d \alpha \rangle$ and to apply the Bochner formula (11.227). One shows that $\|d\alpha\|_{L^2} + \|\delta\alpha\|_{L^2}$ is equal to the integral of an expression of the form $\mathrm{Rm} * \alpha * \alpha$, where $*$ denotes some product with contractions (not the Hodge star operator!) and where this expression is pointwise nonpositive and its integral is equal to zero if and only if $\alpha \equiv 0$. From this one concludes that $\alpha \equiv 0$ and hence $\mathbf{H}^p(M^n) = 0$ for all $0 < p < n$. The theorem follows from the Hodge decomposition theorem and the de Rham theorem. See [**GM75**] for details. □

Remark 11.66. Note that the case where $p = 1$ in the conclusion of the theorem follows from Proposition 11.63, under the weaker condition of positive Ricci curvature.

11.9. Notes and commentary

Important books on geometric analysis have been written by Aubin [**Aub82**], Schoen and Yau [**SY94**], Li [**Li12**], and Jost [**Jos17**]. These books discuss linear and nonlinear elliptic and parabolic equations in differential geometry.

The Bochner technique appears in [**Boc46**]. See also the book by Goldberg [**Gol98**] and the survey by Wu [**Wu17**].

Books on eigenvalues and eigenfunctions of the Laplacian on Riemannian manifolds have been written by Berger, Gauduchon, and Mazet [**BGM71**] and Chavel [**Cha84**].

The Lichnerowicz eigenvalue estimate (Theorem 11.10) appears in [**Lic77**] and the equality case is characterized by Obata [**Oba62**].

The Li and Yau eigenvalue estimate (Theorem 11.12) is in [**LY80**]; this was sharpened by Zhong and Yang [**ZY84**].

The Bishop–Gromov volume comparison theorem (Theorem 11.32) appears in Bishop and Crittenden [**BC01**]. The Bishop–Gromov relative volume comparison theorem (Theorem 11.34) appears in [**Gro07**].

The Milnor fundamental group growth estimate for manifolds with nonnegative Ricci curvature (Theorem 11.38) is in [**Mil68**]. Milnor conjectured that the fundamental group of any complete noncompact Riemannian manifold with nonnegative Ricci curvature is finitely generated. Brué, Naber, and Semola [**BNSa, BNSb**] proved that Milnor's conjecture is false in dimensions 6 and higher.

The splitting theorem by Cheeger and Gromoll (Theorem 11.48), which pertains to manifolds with nonnegative Ricci curvature that also admit a geodesic line, can be found in [**CG72**].

The Liouville theorem for harmonic functions on complete Riemannian manifolds with nonnegative curvature (Theorem 11.49) is proved by Yau in [**Yau75**]. The Cheng diameter rigidity theorem appears in [**Che75**]. The differential Harnack estimate for the heat equation is due to Li and Yau [**LY86**]. This inspired related estimates for geometric evolution equations and notably for the Ricci flow, Hamilton's matrix Harnack estimate [**Ham93**], and Perelman's conjugate heat kernel Harnack estimate [**Pera**, §9].

For a discussion of the asymptotics of the heat kernel on Riemannian manifolds, see Gilkey's book [**Gil95**]. For a detailed treatment of the heat kernel on Riemannian manifolds and estimates for it, see Grigor'yan's book [**Gri09**].

Manifolds with nonnegative Ricci curvature are somewhat rigid. The aforementioned Theorem 11.48 is an example of this. Much more generally, a fundamental direction in geometric analysis is the study of Ricci curvature bounds as pioneered by Cheeger and Colding [**CC97, CC00a, CC00b**] and also developed by Bamler, Jiang, Naber, and Tian. In particular, see also [**CCT02**], [**CN15**], [**Bam17**], and [**CJN21**]. For expositions of Cheeger–Colding theory, see the books by Cheeger [**Che01**] and by Bamler, Jiang, and the junior author [**BCJN**].

As opposed to nonnegative Ricci curvature, the condition of negative Ricci curvature does not appear to be geometrically significant. In particular, originally Gao [**Gao85**] and Gao and Yau [**GY86**] proved that on any closed 3-manifold there exists a metric with negative Ricci curvature. Later, Lohkamp [**Loh94, Loh95**] vastly generalized this by proving that for any closed smooth manifold M^n, any metric g_0 is C^0-close to a metric with negative Ricci curvature. That is, for any $\varepsilon > 0$, there exists a metric g on M^n with $\mathrm{Ric}_g < 0$ and $\|g - g_0\|_{C^0(g_0)} < \varepsilon$.

In contrast to the aforementioned Theorem 11.49, the behavior of positive harmonic functions on manifolds with negative curvature is very different than on manifolds with positive curvature; see Anderson and Schoen [**AS85**]. Harmonic maps between Riemannian manifolds, which are discussed in §12.8 below, generalize harmonic functions. The seminal paper on harmonic maps, which uses the heat flow method, is by Eells and Sampson [**ES64**]. For a book on harmonic maps, see Schoen and Yau [**SY97**].

A classic reference for the Hodge theorem (Theorem 11.59) is Hodge's book [**Hod41**]. For more modern references, see the books cited at the beginning of §11.8.

As compared to Theorem 11.65, much stronger results regarding manifolds with positive curvature, called sphere (or spherical space form) theorems, have been proved using the methods of partial differential equations applied to Riemannian geometry by Hamilton [**Ham82**, **Ham86**], Micallef and Moore [**MM88**], Böhm and Wilking [**BW08**], and Brendle and Schoen [**BS09**]. The result of Micallef and Moore uses the theory of harmonic maps from 2-spheres (including the work of Sacks and Uhlenbeck [**SU81**], which we state in §12.8.6 below) and a Bochner–Weitzenböck formula as parts of their proof and produces a topological sphere theorem. The rest of the aforementioned results use Hamilton's Ricci flow and yield differentiable sphere theorems.

Many important areas in differential geometry and geometric analysis are not discussed in this book, including complex and Kähler geometry, the Yang–Mills and the Seiberg–Witten equations (gauge theory), geometric measure theory, conformal geometry, and mathematical general relativity to name a few. For example, seminal papers in these areas include Yau [**Yau78**], Donaldson [**Don83**], Seiberg and Witten [**SW94**], Federer and Fleming [**FF60**], Schoen [**Sch84a**], and Christodoulou and Klainerman [**CK93**].

For books on complex and Kähler geometry, see those listed in §6.7. For books on gauge theory, see those listed in §6.6.4. Classic books on geometric measure theory are by Federer [**Fed69**] and Simon [**Sim83**]. For a selection of further topics in geometric analysis, see the book by Andrews, Langford, and the junior author [**ACL**].

For books on functional inequalities on manifolds, using either elliptic or heat kernel methods, see Bakry, Gentil, and Ledoux [**BGL14**] and Saloff-Coste [**SC02**]. Sobolev inequalities are also studied in detail in Hebey [**Heb99**].

11.10. Exercises

Exercise 11.1. *Let (M^n, g) be a Riemannian manifold. Prove that the Laplacian acting on functions may be expressed by*

$$\Delta f = \frac{1}{\sqrt{|g|}} \sum_{i,j=1}^{n} \frac{\partial}{\partial x^i}\left(\sqrt{|g|}\, g^{ij} \frac{\partial f}{\partial x^j}\right), \tag{11.228}$$

where $|g| := \det(g_{ij})$.
Hint: Use

$$\Delta f = \sum_{i,j=1}^{n} g^{ij}\left(\frac{\partial^2 f}{\partial x^i \partial x^j} - \Gamma_{ij}^k \frac{\partial f}{\partial x^k}\right). \tag{11.229}$$

Exercise 11.2. *Prove that if u and v are functions on a Riemannian manifold (M^n, g), then we have the following* **product rule for the Laplacian:**

$$\Delta(uv) = v\Delta u + u\Delta v + 2\langle \nabla u, \nabla v\rangle. \tag{11.230}$$

Observe that this formula generalizes (11.16).

Exercise 11.3. *Let (M^n, g) be a compact Riemannian manifold with boundary. Using the divergence theorem (Corollary* 8.9*), prove that for any smooth functions u and v on M^n,*

$$\int_{M^n} (u\Delta v - v\Delta u)\, d\mu = \int_{\partial M} \big(u\langle \nabla v, N\rangle - v\langle \nabla u, N\rangle\big) d\sigma, \tag{11.231}$$

where N is the unit outward normal to ∂M and where $d\sigma$ is the volume $(n-1)$-form of ∂M.

Exercise 11.4. *Let (M^n, g) be a closed Riemannian manifold. Prove that for any smooth function u on M^n,*

$$\int_{M^n} \big(|\nabla^2 u|^2 - (\Delta u)^2 + \mathrm{Ric}(\nabla u, \nabla u)\big) d\mu = 0. \tag{11.232}$$

Exercise 11.5 (Reilly's formula). *Let (M^n, g) be a compact Riemannian manifold with boundary. Prove that for any smooth function u on M^n,*

$$\begin{aligned}\int_{M^n} &\big((\Delta u)^2 - |\nabla^2 u|^2 - \mathrm{Ric}(\nabla u, \nabla u)\big) d\mu \\ &= \int_{\partial M} \big(\Delta_{\partial M} u + HN(u)\big) N(u)\, d\sigma \\ &\quad + \int_{\partial M} \big(-\langle \nabla_{\partial M} u, \nabla_{\partial M} N(u)\rangle + \mathrm{II}(\nabla_{\partial M} u, \nabla_{\partial M} u)\big)\, d\sigma,\end{aligned} \tag{11.233}$$

where H, II*, $\Delta_{\partial M}$, and $\nabla_{\partial M}$ are the mean curvature, second fundamental form, Laplacian, and gradient of the boundary ∂M with the induced metric, respectively.*

Exercise 11.6. *Let (M^n, g) be a compact Riemannian manifold with boundary. Let $u : M^n \to \mathbb{R}$ and $\lambda \in \mathbb{R}$ satisfy $\Delta u + \lambda u = 0$ on M^n.*

(1) *If $u = 0$ on ∂M, then we say that u and λ are* ***Dirichlet*** *eigenfunctions and eigenvalues of the Laplacian, respectively.*

(2) *If $N(u) = 0$ on ∂M, then we say that u and λ are* ***Neumann*** *eigenfunctions and eigenvalues of the Laplacian, respectively.*

Suppose that $\mathrm{Ric} \geq (n-1)Kg$*, where K is a positive real number. Prove in either case that if u is not identically zero, then $\lambda \geq nK$.*

Exercise 11.7. *Let (M^n, g) be a closed orientable Riemannian manifold. Let $P^{n-1} \subset M^n$ be a separating minimal hypersurface; i.e., it is closed, its mean curvature is zero, and its complement in M^n has two components.*

Suppose that Ric $\geq (n-1)Kg$, *where* K *is a positive real number. Prove that*

$$\lambda_1(P^{n-1}) \geq \frac{(n-1)K}{2}. \tag{11.234}$$

Hint: Let u *be an eigenfunction of the Laplacian on* P^{n-1}. *Let* M_1^n *and* M_2^n *be the closures of the components of* $M^n \setminus P^{n-1}$. *Let* $f_i : M_i^n \to \mathbb{R}$ *be the harmonic function on* M_i^n *with* $f_i|_{M_i^n} = u$, $i = 1, 2$. *Apply Reilly's formula.*

Exercise 11.8. *Prove that the function* $\Phi(r, s)$ *defined by* (11.35) *satisfies* $\frac{\partial \Phi}{\partial s}(r, s) \leq 0$.

Exercise 11.9. *Verify* (11.77).

Exercise 11.10. *Prove Lemma* 11.40.

Exercise 11.11. *Prove* (11.126).

Exercise 11.12. *Prove that the Busemann function is* 1*-Lipschitz, i.e.,* (11.147).

Exercise 11.13. *Fill in the details of the proof of Theorem* 11.38.

Exercise 11.14. *Let* $(M^n, g(t))$ *be a solution to the Ricci flow; that is,* $\partial_t g(t) = -2\operatorname{Ric}_{g(t)}$. *Prove that if* $u(t)$ *is a solution to the heat equation* $\partial_t u(t) = \Delta_{g(t)} u(t)$, *then*

$$\left(\frac{\partial}{\partial t} - \Delta_{g(t)}\right) |\nabla u(t)|^2_{g(t)} = -2|\nabla^2 u(t)|^2_{g(t)} \leq 0. \tag{11.235}$$

Exercise 11.15 (Diffeomorphism invariance of de Rham cohomology). *Prove that if closed smooth manifolds* M^n *and* N^n *are diffeomorphic, then the de Rham cohomology groups* $H^p_{\mathrm{dRh}}(M^n)$ *and* $H^p_{\mathrm{dRh}}(N^n)$ *are isomorphic.*

In fact, if M^n *and* N^n *are homotopy equivalent, then* $H^p_{\mathrm{dRh}}(M^n)$ *and* $H^p_{\mathrm{dRh}}(N^n)$ *are isomorphic (see Lee* [**Lee13**, Proposition 17.8]*).*

Exercise 11.16. *Prove the Poincaré lemma for smooth manifolds diffeomorphic to open balls.*

Exercise 11.17. *Show that the Hodge decomposition in* (11.216) *is orthogonal with respect to the* L^2*-inner product on* $\Omega^p(M^n)$.

Exercise 11.18. *Let* (M^n, g) *be a Riemannian manifold and let* $f : M^n \to \mathbb{R}$ *be a smooth function. Define Witten's deformation of the exterior derivative by*

$$d_t = \mathrm{e}^{-tf} d \circ \mathrm{e}^{tf} : \Omega^p(M^n) \to \Omega^{p+1}(M^n) \tag{11.236}$$

for $0 \leq p \leq n-1$; *that is,*

$$d_t\alpha = \mathrm{e}^{-tf} d(\mathrm{e}^{tf}\alpha) \quad \textit{for } \alpha \in \Omega^p(M^n).$$

Let $d_t^* : \Omega^{p+1}(M^n) \to \Omega^p$ *be the* L^2*-adjoint of* d_t*, so that (cf. (11.204))*

$$\langle\!\langle d_t\alpha, \beta\rangle\!\rangle_{L^2} = \langle\!\langle \alpha, d_t^*\beta\rangle\!\rangle_{L^2}$$

for $\alpha \in \Omega^p(M^n)$ *and* $\beta \in \Omega^{p+1}(M^n)$*. Show that*

$$d_t^* = \mathrm{e}^{tf} d^* \circ \mathrm{e}^{-tf}. \tag{11.237}$$

Exercise 11.19. *Define*

$$H_t^p(M^n) = \frac{\ker(d_t : \Omega^p \to \Omega^{p+1})}{\mathrm{image}(d_t : \Omega^{p-1} \to \Omega^p)}; \tag{11.238}$$

cf. (11.194). Show that the vector spaces $H_t^p(M^n)$ *and* $H_{\mathrm{dRh}}^p(M^n)$ *are isomorphic.*

Exercise 11.20. *Define the Witten–Hodge Laplacian by*

$$\Delta_t := d_t \circ d_t^* + d_t^* \circ d_t, \tag{11.239}$$

acting on p-forms, where $0 \le p \le n$*. Show that*

$$\Delta_t = \Delta_d + t(\mathcal{L}_{\nabla f} + \mathcal{L}_{\nabla f}^*) + t^2|\nabla f|^2, \tag{11.240}$$

where Δ_d *is the Hodge Laplacian,* $\mathcal{L}$ *is the Lie derivative, and* $\mathcal{L}_{\nabla f}^*$ *is the* L^2*-adjoint of* $\mathcal{L}_{\nabla f}$*.*

Exercise 11.21. *In the context of Euclidean space, let us consider a related operator. On* $\mathbb{R}^n$ *define the elliptic operator*

$$L = -\Delta + t^2\|\mathbf{x}\|^2;$$

that is, $Lf(\mathbf{x}) = -(\Delta f)(\mathbf{x}) + t^2\|\mathbf{x}\|^2 f(\mathbf{x})$*. Here,* $\Delta = \sum_{i=1}^n \frac{\partial^2}{\partial(x^i)^2}$*. Show that*

$$L\left(\mathrm{e}^{-t\frac{\|\mathbf{x}\|^2}{2}}\right) = nt\mathrm{e}^{-t\frac{\|\mathbf{x}\|^2}{2}}; \tag{11.241}$$

that is, $\mathrm{e}^{-t\frac{\|\mathbf{x}\|^2}{2}}$ *is an eigenvector of* L *with eigenvalue* nt*.*

Remark 11.67. In an influential paper, Witten [**Wit82**] gave a new proof of the Morse inequalities using an analysis of the Witten–Hodge Laplacian Δ_t associated to a Morse function f. Part of the idea of the proof is that, due to (11.240), the eigenfunctions of Δ_t are concentrated at the critical points of f for t large. See also the book by Zhang [**Zha01**].

Chapter 12

Elliptic Equations and the Geometry of Minimal Surfaces

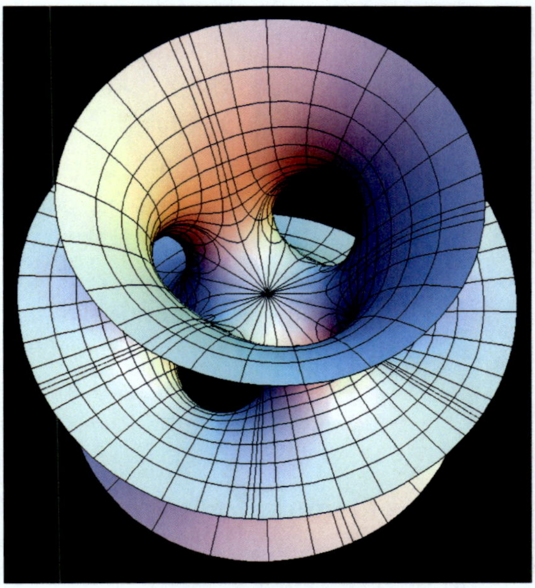

Figure 12.0.1. "Costa–Hoffman–Meeks Surface". Credit: 3DXM Virtual Mathematics Museum by The 3D-XplorMath Consortium. © Copyright 3DXM Consortium. `https://3d-xplormath.org/`

In this chapter, we focus on the minimal hypersurface equation, which is a nonlinear elliptic equation. This equation arises naturally in the variational setting, representing critical points for the area functional. Our discussion covers a small selection of topics regarding minimal hypersurfaces. In particular, we present the monotonicity and mean value formulas for minimal submanifolds, minimal graphs and the Bernstein theorem, the Simons identity, curvature estimates, the statement of the Plateau problem, minimal surfaces in 3-manifolds with positive scalar curvature, and an introduction to harmonic maps.

For further study, we refer the reader to the vast literature, which includes the books by Lawson [**Law80**] and Colding and Minicozzi [**CM11**] to name just a few. See also the notes and commentary section for additional references.

12.1. The definition of minimal submanifold

The study of minimal submanifolds is an important branch of differential geometry. In this introductory section, we provide the definition of a minimal submanifold.

Recall, by Definition 4.22, that the minimal surface equation for a hypersurface is the equation that says that the mean curvature function H is equal to zero. More generally, a submanifold M^n of Euclidean space $\mathbb{R}^N$ is **minimal** if its mean curvature vector $\vec{H}$, defined by (4.117), vanishes.

An immediate consequence of Proposition 5.2, which says that there exists a point at which all the principle curvatures are positive, is that there does not exist any closed minimal hypersurface of $\mathbb{R}^{n+1}$. This stems from the fact that, for any minimal hypersurface, the sum of the principal curvatures at each point is zero. Consequently, a nonpositive principal curvature must exist at each point, presenting a contradiction. Given this, our discussion will primarily center on noncompact minimal hypersurfaces without boundary or those that are compact with boundaries.

12.2. The monotonicity formula for minimal submanifolds

In this section, we prove a fundamental monotonicity formula for minimal submanifolds, with its generalization to the mean value formula.

12.2.1. The coarea formula. The coarea formula provides a relationship between the integrals of a function over a manifold and its integrals over level sets of another function. This formula is particularly useful in variational problems. We have the following **coarea formula** (see, e.g., Chavel [**Cha84**, pp. 85–86] and Lin and Yang [**LY02**, Theorem 4.2.1]).

Theorem 12.1 (Coarea formula). *Let (M^n, g) be a smooth compact Riemannian manifold with (possibly empty) boundary. If $f : M^n \to \mathbb{R}$ is a*

Lipschitz function and if $h : M^n \to \mathbb{R}$ is an integrable function, then

$$\int_{M^n} h\,|\nabla f|\,d\mu = \int_{-\infty}^{\infty}\left(\int_{\{f=t\}} h\,d\sigma\right) dt, \tag{12.1}$$

where $d\sigma$ is the induced Riemannian measure on $\{f = t\}$. In particular, by taking $h = |\nabla f|^{-1}$ we have that

$$\mathrm{Vol}(M^n) = \int_{-\infty}^{\infty}\left(\int_{\{f=t\}} |\nabla f|^{-1}\,d\sigma\right) dt. \tag{12.2}$$

To see the coarea formula intuitively, suppose that f is smooth with a nonzero gradient. We may assume that the level sets $\{f = t\}$, which are smooth hypersurfaces, are parametrized by maps $F_t : \Sigma^{n-1} \to \{f = t\}$ satisfying the flow $\partial_t F_t = \frac{1}{|\nabla f|} N_t$, where $N_t = \frac{\nabla f}{|\nabla f|}$, which is a unit vector field normal to $\{f = t\}$. In this case the coarea formula follows from applying the change of variables formula and Fubini's theorem, where we pull back the integral on the left-hand side of (12.1) by the map $F : \Sigma^{n-1} \times [f_{\min}, f_{\max}] \to M^n$ defined by $F(x,t) = F_t(x)$.

12.2.2. The monotonicity and mean value formulas for minimal submanifolds. Given a vector $V \in T_{\mathbf{x}}\mathbb{R}^N$, where $\mathbf{x}$ is in a submanifold M^n, let $V^\top$ and $V^\perp$ denote the tangential and normal components of V, respectively. We have the following monotonicity formula for minimal submanifolds in Euclidean space (see Simon [**Sim**] and Colding and Minicozzi [**CM11**, Proposition 1.12]).

Theorem 12.2. *Let M^n be a compact minimal submanifold with boundary ∂M in $\mathbb{R}^N$. For any point $\mathbf{x}_0 \in \mathbb{R}^N$ and for any $0 < t_1 < t_2$, we have*

$$\begin{aligned} &\frac{\mathrm{Area}\left(M^n \cap B_{t_2}(\mathbf{x}_0)\right)}{t_2^n} - \frac{\mathrm{Area}\left(M^n \cap B_{t_1}(\mathbf{x}_0)\right)}{t_1^n} \\ &\quad = \int_{M^n \cap (B_{t_2}(\mathbf{x}_0) \setminus B_{t_1}(\mathbf{x}_0))} \frac{\|(\mathbf{x}-\mathbf{x}_0)^\perp\|^2}{\|\mathbf{x}-\mathbf{x}_0\|^{n+2}}\,d\mu(\mathbf{x}), \end{aligned} \tag{12.3}$$

provided $\partial M \cap B_{t_2}(\mathbf{x}_0) = \emptyset$. In particular,

$$\frac{\mathrm{Area}\left(M^n \cap B_{t_2}(\mathbf{x}_0)\right)}{t_2^n} \geq \frac{\mathrm{Area}\left(M^n \cap B_{t_1}(\mathbf{x}_0)\right)}{t_1^n}, \tag{12.4}$$

with equality if and only if $(\mathbf{x} - \mathbf{x}_0)^\perp \equiv \mathbf{0}$ on $M^n \cap (B_{t_2}(\mathbf{x}_0) \setminus B_{t_1}(\mathbf{x}_0))$.

Proof. Without loss of generality, we may assume that $\mathbf{x}_0 = \mathbf{0}$ is the origin. Define $f : M^n \to \mathbb{R}$ by $f(\mathbf{x}) = \|\mathbf{x}\|$. Then

$$M^n \cap B_t(\mathbf{0}) = \{\mathbf{x} \in M^n \,|\, f(\mathbf{x}) \leq t\} =: \{f \leq t\}.$$

Recall from (5.63) that

$$\nabla(f^2) = 2\|\mathbf{x}^\top\|. \tag{12.5}$$

By (5.140) and since M^n is a minimal submanifold, so that $\vec{H} = \vec{0}$, we also have that

$$\Delta_{M^n}(f^2) = 2n.$$

Thus, using the divergence theorem, we calculate that

$$\begin{aligned} 2n \operatorname{Area}\left(M^n \cap B_t(\mathbf{0})\right) &= \int_{\{f \leq t\}} \Delta_{M^n}(f^2)\, d\mu \\ &= \int_{\{f=t\}} \langle \nabla(f^2), N_t \rangle\, d\sigma \\ &= \int_{\{f=t\}} \|\nabla(f^2)\|\, d\sigma \\ &= 2 \int_{\{f=t\}} \|\mathbf{x}^\top\|\, d\sigma(\mathbf{x}), \end{aligned} \tag{12.6}$$

since $N_t = \frac{\nabla(f^2)}{\|\nabla(f^2)\|}$ and where the last equality is by (12.5).

On the other hand, by the coarea formula (12.2), we have that

$$\begin{aligned} \operatorname{Area}\left(M^n \cap B_t(\mathbf{0})\right) &= \int_0^t \left(\int_{\{f=s\}} \|\nabla f\|^{-1} d\sigma \right) ds \\ &= \int_0^t \left(\int_{\{f=s\}} 2f \,\|\nabla(f^2)\|^{-1} d\sigma \right) ds \\ &= \int_0^t \left(\int_{\{f=s\}} \frac{\|\mathbf{x}\|}{\|\mathbf{x}^\top\|} d\sigma(\mathbf{x}) \right) ds. \end{aligned} \tag{12.7}$$

By differentiating this, we obtain

$$\begin{aligned} \frac{d}{dt} \frac{\operatorname{Area}\left(M^n \cap B_t(\mathbf{0})\right)}{t^n} &= -n \frac{\operatorname{Area}\left(M^n \cap B_t(\mathbf{0})\right)}{t^{n+1}} + \frac{1}{t^n} \int_{\{f=t\}} \frac{\|\mathbf{x}\|}{\|\mathbf{x}^\top\|} d\sigma(\mathbf{x}) \\ &= -\frac{1}{t^{n+1}} \int_{\{f=t\}} \|\mathbf{x}^\top\|\, d\sigma(\mathbf{x}) + \frac{1}{t^n} \int_{\{f=t\}} \frac{\|\mathbf{x}\|}{\|\mathbf{x}^\top\|} d\sigma(\mathbf{x}) \end{aligned}$$

where the last equality is by (12.6). Since $\|\mathbf{x}\| = t$ for $\mathbf{x} \in \{f = t\}$, we may rewrite this formula as

$$\frac{d}{dt}\frac{\text{Area}\left(M^n \cap B_t(\mathbf{0})\right)}{t^n} = \frac{1}{t^{n+1}}\int_{\{f=t\}} \frac{\|\mathbf{x}\|^2 - \|\mathbf{x}^\top\|^2}{\|\mathbf{x}^\top\|}\, d\sigma(\mathbf{x}) \tag{12.8}$$
$$= \int_{\{f=t\}} \frac{\|\mathbf{x}^\perp\|^2}{\|\mathbf{x}^\top\|\,\|\mathbf{x}\|^{n+1}}\, d\sigma(\mathbf{x})$$
$$= \int_{\{f=t\}} \frac{\|\mathbf{x}^\perp\|^2}{\|\mathbf{x}\|^{n+2}}\, \|\nabla f\|^{-1}\, d\sigma(\mathbf{x}).$$

By integrating this and using the coarea formula (12.1), we obtain

$$\left.\frac{\text{Area}\left(M^n \cap B_t(\mathbf{0})\right)}{t^n}\right|_{t=t_1}^{t_2} = \int_{t_1}^{t_2} \left(\int_{\{f=t\}} \frac{\|\mathbf{x}^\perp\|^2}{\|\mathbf{x}\|^{n+2}}\, \|\nabla f\|^{-1}\, d\sigma(\mathbf{x})\right) dt$$
$$= \int_{M^n \cap (B_{t_2}(\mathbf{0}) \setminus B_{t_1}(\mathbf{0}))} \frac{\|(\mathbf{x})^\perp\|^2}{\|\mathbf{x}\|^{n+2}}\, d\mu(\mathbf{x}).$$

This completes the proof of the monotonicity formula. $\square$

Definition 12.3. A subset C of $\mathbb{R}^N$ is a **cone** with vertex $\mathbf{x}_0$ if for every $\mathbf{x} \in C$ and $t \in [0, \infty)$ we have that $\mathbf{x}_0 + t(\mathbf{x} - \mathbf{x}_0) \in C$.

We say that a subset C of $B_r(\mathbf{x}_0) \subset \mathbb{R}^N$ is a **truncated cone of radius** r with vertex $\mathbf{x}_0$ if for every $\mathbf{x} \in C$ we have that $\mathbf{x}_0 + t(\mathbf{x} - \mathbf{x}_0) \in C$ provided $t \geq 0$ satisfies $t\|\mathbf{x} - \mathbf{x}_0\| < r$.

Corollary 12.4. *Suppose that* $\frac{\text{Area}\left(M^n \cap B_t(\mathbf{x}_0)\right)}{t^n}$ *is constant for* $t \in (0, t_2)$. *Then* $M^n \cap B_{t_2}(\mathbf{x}_0)$ *is an* n*-dimensional disk centered at* $\mathbf{x}_0$.

Proof. Without loss of generality, we may assume that $\mathbf{x}_0 = \mathbf{0}$. Then, by the monotonicity formula, we have that $\mathbf{x}^\perp \equiv \mathbf{0}$ for $\mathbf{x} \in M^n \cap B_{t_2}(\mathbf{x}_0)$; that is, $\mathbf{x} \in T_{\mathbf{x}}M$ (the position vector lies in its tangent space). This implies that $M^n \cap B_{t_2}(\mathbf{x}_0)$ is a truncated cone of radius t_2 with vertex $\mathbf{0}$. Since $\mathbf{0} \in M^n$ and since M^n is smooth, we conclude that $M^n \cap B_{t_2}(\mathbf{x}_0)$ is an n-dimensional disk of radius t_2 centered at $\mathbf{x}_0$. $\square$

12.2.3. The mean value formula for minimal submanifolds. The monotonicity formula (Theorem 12.2) generalizes to the following mean value formula for minimal submanifolds.

Theorem 12.5. *Let M^n be a compact minimal submanifold with boundary in $\mathbb{R}^N$ such that $\partial M \cap B_{t_2}(\mathbf{x}_0) = \emptyset$. Let $u : M^n \to \mathbb{R}$. For any point $\mathbf{x}_0 \in \mathbb{R}^N$ and for any $0 < t_1 < t_2$, we have*

(12.9)

$$\frac{\int_{M^n \cap B_{t_2}(\mathbf{x}_0)} u\, d\mu}{t_2^n} - \frac{\int_{M^n \cap B_{t_1}(\mathbf{x}_0)} u\, d\mu}{t_1^n}$$
$$= \int_{M^n \cap (B_{t_2}(\mathbf{x}_0) \setminus B_{t_1}(\mathbf{x}_0))} u \frac{\|(\mathbf{x} - \mathbf{x}_0)^{\perp}\|^2}{\|\mathbf{x} - \mathbf{x}_0\|^{n+2}}\, d\mu(\mathbf{x})$$
$$+ \frac{1}{2} \int_{t_1}^{t_2} \frac{1}{t^{n+1}} \left(\int_{M^n \cap B_t(\mathbf{x}_0)} (t^2 - \|\mathbf{x} - \mathbf{x}_0\|^2)\, \Delta_{M^n} u(\mathbf{x})\, d\mu(\mathbf{x}) \right) dt.$$

Proof. Without loss of generality, we may assume that $\mathbf{x}_0 = \mathbf{0}$. Let $f(\mathbf{x}) = \|\mathbf{x}\|$. Integrating by parts, we calculate for $0 < t < t_2$ that (cf. (12.6))

$$\begin{aligned}
2n \int_{M^n \cap B_t(\mathbf{0})} u\, d\mu &= \int_{\{f \le t\}} u \Delta_{M^n} (f^2 - t^2)\, d\mu \\
&= \int_{\{f \le t\}} (f^2 - t^2) \Delta_{M^n} u\, d\mu \\
&\quad + \int_{\{f = t\}} \langle u \nabla (f^2 - t^2) - (f^2 - t^2) \nabla u, N_t \rangle\, d\sigma \\
&= \int_{\{f \le t\}} (f^2 - t^2) \Delta_{M^n} u\, d\mu + \int_{\{f = t\}} u \|\nabla (f^2)\|\, d\sigma \\
&= \int_{M^n \cap B_t(\mathbf{0})} (\|\mathbf{x}\|^2 - t^2) \Delta_{M^n} u(\mathbf{x})\, d\mu(\mathbf{x}) \\
&\quad + 2 \int_{M^n \cap \partial B_t(\mathbf{0})} u \|\mathbf{x}^{\top}\|\, d\sigma(\mathbf{x}).
\end{aligned}$$

On the other hand, by the coarea formula (12.2), we have that (cf. (12.7))

$$\begin{aligned}
\int_{M^n \cap B_t(\mathbf{0})} u\, d\mu &= \int_0^t \left(\int_{\{f = s\}} u \|\nabla f\|^{-1}\, d\sigma \right) ds \\
&= \int_0^t \left(\int_{\{f = s\}} u \frac{\|\mathbf{x}\|}{\|\mathbf{x}^{\top}\|}\, d\sigma(\mathbf{x}) \right) ds.
\end{aligned}$$

Thus, we have that

$$\begin{aligned}
\frac{d}{dt}\frac{\int_{M^n\cap B_t(\mathbf{0})} u\,d\mu}{t^n} &= -n\frac{\int_{M^n\cap B_t(\mathbf{0})} u\,d\mu}{t^{n+1}} + \frac{1}{t^n}\int_{\{f=t\}} u\frac{\|\mathbf{x}\|}{\|\mathbf{x}^\top\|}\,d\sigma(\mathbf{x}) \\
&= -\frac{1}{2t^{n+1}}\int_{M^n\cap B_t(\mathbf{0})} (\|\mathbf{x}\|^2 - t^2)\Delta_{M^n}u(\mathbf{x})\,d\mu(\mathbf{x}) \\
&\quad - \frac{1}{t^{n+1}}\int_{\{f=t\}} u\|\mathbf{x}^\top\|\,d\sigma(\mathbf{x}) + \frac{1}{t^n}\int_{\{f=t\}} u\frac{\|\mathbf{x}\|}{\|\mathbf{x}^\top\|}\,d\sigma(\mathbf{x}) \\
&= -\frac{1}{2t^{n+1}}\int_{M^n\cap B_t(\mathbf{0})} (\|\mathbf{x}\|^2 - t^2)\,\Delta_{M^n}u(\mathbf{x})\,d\mu(\mathbf{x}) \\
&\quad + \frac{1}{t^{n+1}}\int_{\{f=t\}} u\frac{\|\mathbf{x}^\perp\|^2}{\|\mathbf{x}^\top\|}\,d\sigma(\mathbf{x}).
\end{aligned}$$

The mean value formula follows from integrating this formula over the interval $t_1 \leq t \leq t_2$. □

An immediate consequence of the mean value formula is the following mean value inequality for subharmonic functions on minimal submanifolds.

Corollary 12.6. *Let M^n be a compact minimal submanifold with boundary in $\mathbb{R}^N$ such that $\partial M \cap B_{t_2}(\mathbf{x}_0) = \emptyset$, where $\mathbf{x}_0 \in \mathbb{R}^N$. If $u : M^n \cap B_{t_2}(\mathbf{x}_0) \to \mathbb{R}$ is a nonnegative subharmonic function (i.e., $u \geq 0$ and $\Delta_{M^n} u \geq 0$ on $M^n \cap B_{t_2}(\mathbf{x}_0)$), then for any $0 < t_1 < t_2$, we have that*

$$\frac{\int_{M^n\cap B_{t_2}(\mathbf{x}_0)} u\,d\mu}{t_2^n} \geq \frac{\int_{M^n\cap B_{t_1}(\mathbf{x}_0)} u\,d\mu}{t_1^n}. \tag{12.10}$$

In particular, if $\mathbf{x}_0 \in M^n$, then

$$u(\mathbf{x}_0) \leq \frac{\int_{M^n\cap B_{t_2}(\mathbf{x}_0)} u\,d\mu}{\omega_n t_2^n}, \tag{12.11}$$

where the denominator on the right-hand side is the volume of the Euclidean n-ball of radius t_2.

We have the following generalization of Corollary 12.6 (see [**CM11**, Corollary 1.16]; we leave the proof as an exercise). This version is useful because the elliptic inequality in its hypothesis is suitable for applications, e.g., to curvature estimates.

Corollary 12.7. *Let M^n be a compact minimal submanifold with boundary in $\mathbb{R}^N$ such that $\partial M \cap B_{\bar{t}}(\mathbf{x}_0) = \emptyset$, where $\mathbf{x}_0 \in M^n$. If $u : M^n \cap B_{\bar{t}}(\mathbf{x}_0) \to \mathbb{R}$ is a nonnegative function satisfying*

$$\Delta_{M^n} u \geq -A\bar{t}^{-2}u,$$

where A is a real number, then

$$u(\mathbf{x}_0) \leq e^{\frac{A}{2}} \frac{\int_{B_{\bar{t}}(\mathbf{x}_0)\cap M^n} u\, d\mu}{\omega_n \bar{t}^n}. \tag{12.12}$$

12.3. Minimal graphs and the Bernstein theorem

A particular case of interest in the study of minimal surfaces is when they are represented as graphs over some domain in the Euclidean space. These are often referred to as 'minimal graphs'. Such graphs provide a deeper understanding of the geometric properties of minimal surfaces. In this section, we shall introduce the mathematical framework underlying minimal graphs and provide a detailed examination of their characteristics. One of the pivotal results in this context is the Bernstein theorem, which asserts that any entire minimal graph in $\mathbb{R}^3$ must be a plane.

12.3.1. Minimal graphs.

Definition 12.8. A **minimal graph** is a graphical hypersurface that is also a minimal hypersurface. A graph is **entire** if it is the graph of a function defined on all of $\mathbb{R}^n$.

Let $\mathcal{U}$ be an open subset of $\mathbb{R}^n$ and let $f : \mathcal{U} \to \mathbb{R}$ be a smooth function. Recall from (1.35) that the graph of f is the hypersurface

$$G^n := \{(\mathbf{x}, f(\mathbf{x})) \colon \mathbf{x} = (x^1, \dots, x^n) \in \mathcal{U}\}. \tag{12.13}$$

The graphical parametrization of G^n is the map $F : \mathcal{U} \to \mathbb{R}^{n+1}$ defined by

$$F(\mathbf{x}) := (\mathbf{x}, f(\mathbf{x})). \tag{12.14}$$

The components of a 2-tensor α on G^n are defined by

$$\alpha_{ij} := \alpha\left(\frac{\partial F}{\partial x^i}, \frac{\partial F}{\partial x^j}\right). \tag{12.15}$$

Recall from (4.113) that the components of the second fundamental form are

$$\mathrm{II}_{ij} = -\frac{1}{\sqrt{\|\nabla f\|^2 + 1}} \frac{\partial^2 f}{\partial x^i \partial x^j} \tag{12.16}$$

(where ∇ is the Euclidean gradient), that by (4.36) we have that the components of the first fundamental form are

$$\mathrm{I}_{ij} = \delta_{ij} + \frac{\partial f}{\partial x^i}\frac{\partial f}{\partial x^j}, \tag{12.17}$$

and that by (4.228) we have

$$\mathrm{I}^{ij} := (\mathrm{I}^{-1})_{ij} = \delta_{ij} - \frac{1}{\|\nabla f\|^2 + 1}\frac{\partial f}{\partial x^i}\frac{\partial f}{\partial x^j}. \tag{12.18}$$

Moreover, with $A(\mathbf{p}) = \sqrt{1 + \|\mathbf{p}\|^2}$, we have that

$$\mathrm{II}_{ik}\,\mathrm{I}^{kj} = -\partial_{x^i}\partial_{p^j} A(\nabla f). \tag{12.19}$$

Thus, the mean curvature of the graphical hypersurface G^n is given by

$$\begin{aligned} H &= \mathrm{I}^{ij}\,\mathrm{II}_{ij} \\ &= -\frac{\Delta f}{\sqrt{\|\nabla f\|^2+1}} + \frac{1}{2(\|\nabla f\|^2+1)^{3/2}} \sum_{i=1}^{n} \frac{\partial f}{\partial x^i}\frac{\partial \|\nabla f\|^2}{\partial x^i} \\ &= -\operatorname{div}\left(\frac{\nabla f}{\sqrt{\|\nabla f\|^2+1}}\right), \end{aligned} \tag{12.20}$$

where Δ is the Euclidean Laplacian and div is the Euclidean divergence. Thus, the **minimal hypersurface equation for the graph of a function** f (i.e., the equation for a minimal graph) is

$$\operatorname{div}\left(\frac{\nabla f}{\sqrt{\|\nabla f\|^2+1}}\right) = 0. \tag{12.21}$$

This is a **quasilinear** elliptic equation because the coefficients of the highest-(second-)order derivatives of the unknown function f depend on f and its first derivatives, but not on the second derivatives.

Recall that the upward pointing unit normal vector field to G^n is

$$N(\mathbf{x}) := N_{F(\mathbf{x})} = \left(-\frac{\nabla f}{\sqrt{\|\nabla f\|^2+1}}, \frac{1}{\sqrt{\|\nabla f\|^2+1}}\right). \tag{12.22}$$

So, if we define the vector field $\widetilde{N}$ on $\mathcal{U} \times \mathbb{R} \subset \mathbb{R}^{n+1}$ by extending N on G^n to be constant in the u^{n+1} variable, then we obtain

$$H = \operatorname{div}(\widetilde{N}). \tag{12.23}$$

In particular, G^n is a minimal graph if and only if $\operatorname{div}(\widetilde{N}) = 0$.

Now assume that $\partial\mathcal{U} := \overline{\mathcal{U}} \setminus \mathcal{U}$ is a smooth closed hypersurface of $\mathbb{R}^n$; a subset $\mathcal{U}$ for which this is true is defined to be a smooth bounded domain.

Lemma 12.9 (Minimizing property of minimal graphs). *Let G^n be a minimal graphical hypersurface with smooth boundary $\partial G = F(\partial\mathcal{U})$. Suppose that M^n is the graph of a function $g : \mathcal{U} \to \mathbb{R}$ such that $\partial M = \partial G$. Then*

$$\operatorname{Area}(M^n) \geq \operatorname{Area}(G^n), \tag{12.24}$$

where Area *denotes the n-dimensional volume of a hypersurface.*

Proof. For the purpose of integration, we define the 'signed' domain Ω in $\mathbb{R}^{n+1}$ which is 'bounded above' by M^n and 'bounded below' by G^n. Namely, to the regions of Ω where G^n is below M^n we assign a $+$ sign and to the regions of Ω where G^n is above M^n we assign a $-$ sign. Let N_M and

N_G denote the upward pointing unit normal vector fields to M^n and G^n, respectively. Then, since G^n is minimal and by the divergence theorem we have

$$
\begin{aligned}
0 &= \int_{\Omega} \operatorname{div}(\widetilde{N}) \\
&= \int_{M^n} \widetilde{N} \cdot N_M - \int_{G^n} \widetilde{N} \cdot N_G \\
&\leq \int_{M^n} 1 - \int_{G^n} 1 = \operatorname{Area}(M^n) - \operatorname{Area}(G^n),
\end{aligned}
\tag{12.25}
$$

where the last inequality holds since $\widetilde{N} \cdot N_M \leq 1$ on M^n and $\widetilde{N} = N_G$ on G^n. □

Remark 12.10. Note that

$$d(*\widetilde{N}^*) = \operatorname{div}(\widetilde{N})\, d\mu_{\mathbb{E}}, \tag{12.26}$$

where $\widetilde{N}^*$ is the 1-form dual to $\widetilde{N}$ on $\mathcal{U} \times \mathbb{R}$ and $*$ is the Hodge star operator.

12.3.2. The Bernstein theorem.

Bernstein [**Ber27**] proved the following fundamental result regarding the uniqueness of entire minimal graphs in 3-dimensional Euclidean space.

Theorem 12.11 (Bernstein). *Any entire minimal graph M^2 in $\mathbb{R}^3$ must be a plane.*

For the proof of the theorem, we closely follow the exposition in Colding and Minicozzi [**CM11**]. Recall that $2K = H^2 - |\,\mathrm{II}\,|^2$, so that for any minimal surface M^2 in $\mathbb{R}^3$ we have that

$$|\,\mathrm{II}\,|^2 d\mu = -2K\, d\mu = -2G^*(d\bar{\mu}), \tag{12.27}$$

where G is the Gauss map, $d\bar{\mu}$ is the area form of the standard metric $\bar{g}$ on S^2, and we used Lemma 5.5 for the last equality.

Now suppose that the minimal surface M^2 is a graph. Choose the unit normal vector field of M^2 to be the upward pointing normal field. We then have that the image of the Gauss map $G(M^2)$ lies in the upper hemisphere $S^2_+ := \{\mathbf{v} \in S^2 \colon \mathbf{v} \cdot \mathbf{e}_3 > 0\}$. Since S^2_+ is diffeomorphic to a disk, by the Poincaré lemma (see Theorem 11.58 and Exercise 11.16), there exists a bounded 1-form α on S^2_+ such that

$$d\bar{\mu} = d\alpha. \tag{12.28}$$

Let $\eta : M^2 \to \mathbb{R}$ be a nonnegative cutoff function (with compact support). We compute that

$$\begin{aligned}\int_{M^2} \eta^2 |\,\mathrm{II}\,|^2 d\mu &= -2\int_{M^2} \eta^2 G^*(d\bar{\mu}) = -2\int_{M^2} \eta^2 G^*(d\alpha)\\ &= -2\int_{M^2} \eta^2 d(G^*\alpha) = 4\int_{M^2} \eta\, d\eta \wedge G^*\alpha\\ &\leq 4\int_{M^2} \eta |\nabla\eta|\,|G^*\alpha|\, d\mu,\end{aligned}$$

where the last equality (on the second line) follows from Stokes's theorem (in the guise of integration by parts). Observe that

$$(G^*\alpha)(e_i) = \alpha(dG(e_i)) = \alpha(\mathrm{L}(e_i)),$$

where L is the Weingarten map. Choose $\{e_1, e_2\}$ to be an orthonormal basis of eigenvectors for L. We see that

$$|G^*\alpha|^2 = \sum_{i=1}^{2} |(G^*\alpha)(e_i)|^2 = \sum_{i=1}^{2} \big(\alpha(\mathrm{L}(e_i))\big)^2 = \sum_{i=1}^{2} \kappa_i^2 (\alpha(e_i))^2 \leq |\,\mathrm{II}\,|^2 |\alpha|^2.$$

More explicitly, for any $\mathbf{x} \in M^2$, we have

$$|G^*\alpha|(\mathbf{x}) \leq |\,\mathrm{II}\,|_g(\mathbf{x})\, |\alpha|_{\bar{g}}(G(\mathbf{x})).$$

Therefore, letting $C := \sup_{S^2_+} |\alpha|$, we have that

$$\begin{aligned}\int_{M^2} \eta^2 |\,\mathrm{II}\,|^2 d\mu &\leq 4\int_{M^2} \eta |\nabla\eta|\,|\,\mathrm{II}\,|\,|\alpha|\, d\mu\\ &\leq 4C\int_{M^2} \eta |\nabla\eta|\,|\,\mathrm{II}\,|\, d\mu\\ &\leq 4C\left(\int_{M^2} |\nabla\eta|^2 d\mu\right)^{1/2}\left(\int_{M^2} \eta^2 |\,\mathrm{II}\,|^2 d\mu\right)^{1/2}.\end{aligned}$$

We conclude that

$$\int_{M^2} \eta^2 |\,\mathrm{II}\,|^2 d\mu \leq 16C^2 \int_{M^2} |\nabla\eta|^2 d\mu. \tag{12.29}$$

For the sake of argument, instead of M^2 consider Euclidean space $\mathbb{R}^n$. A first attempt at a cutoff function η might be a radial function where given a radius R, $\eta(\mathbf{x}) := \eta(|\mathbf{x}|)$ and $\eta(r) = 1$ for $r \in [0, R]$, $\eta(r) = 0$ for $r \in [2R, \infty)$, and $0 \geq \eta'(r) \geq -C/R$. In this case we see that

$$\int_{\mathbb{R}^n} |\nabla\eta|^2(\mathbf{x})\, d\mu_{\mathbb{E}} \approx CR^{n-2},$$

where $d\mu_{\mathbb{E}}$ denotes the Euclidean volume form. In particular, if $n \geq 2$, then this quantity does not approach zero as $R \to \infty$.

But can we choose η so that we get that the Dirichlet integral of η decays to zero as $R \to \infty$, where $\eta(r) = 1$ for $r \in [0, R]$, $\eta'(r) \leq 0$, and $\eta(r) = 0$ for $r \in [a(R), \infty)$ for some function $a(R) > R$?

An answer is to let $a(R) = R^2$ and define $\eta(r) = 2 - \frac{\ln r}{\ln R}$ for $r \in [R, R^2]$. In this case we obtain

$$\int_{\mathbb{R}^n} |\nabla \eta|^2(\mathbf{x})\, d\mu_{\mathbb{E}} = n\omega_n \int_R^{R^2} \frac{1}{r^2 \ln^2 R} r^{n-1} dr.$$

In particular, if $n = 2$, then we obtain

$$\int_{\mathbb{R}^n} |\nabla \eta|^2(\mathbf{x})\, d\mu_{\mathbb{E}} = \frac{2\pi}{\ln R},$$

which has the desired property that it limits to 0 as $R \to \infty$.

We break up the integral over the annulus into a sum of integrals over annuli of fixed out-radii over in-radii as follows. Let $N := \lceil \ln R \rceil$ be the least integer $\geq \ln R$ and consider the annuli $A_i := B_{\mathrm{e}^i R}(\mathbf{0}) \setminus B_{\mathrm{e}^{i-1} R}(\mathbf{0})$ for $1 \leq i \leq N$. Again define η so that $\eta(r) = 2 - \frac{\ln r}{\ln R}$. We compute that

$$\begin{aligned}
\int_{M^2} |\nabla \eta|^2 &= \sum_{i=1}^{N} \int_{A_i} |\nabla \eta|^2 \\
&\leq \sum_{i=1}^{N} \int_{A_i} \frac{1}{r^2 \ln^2 R} \\
&\leq \sum_{i=1}^{N} C \mathrm{e}^{2i} R^2 \frac{1}{\mathrm{e}^{2(i-1)} R^2 \ln^2 R} \\
&= \frac{N C \mathrm{e}^2}{\ln^2 R} = \frac{\lceil \ln R \rceil C \mathrm{e}^2}{\ln^2 R}.
\end{aligned} \tag{12.30}$$

Applying this inequality to (12.29) yields

$$\int_{M^2} \eta^2 |\,\mathrm{II}\,|^2 d\mu \leq 16 C^2 \frac{\lceil \ln R \rceil C \mathrm{e}^2}{\ln^2 R}.$$

By taking $R \to \infty$, we obtain $\int_{M^2} |\,\mathrm{II}\,|^2 d\mu = 0$. The Bernstein theorem follows. □

In all dimensions, the following result of De Giorgi [**DG65**] and Fleming [**Fle62**] provides an equivalent formulation and avenue of attack for the Bernstein problem.

Theorem 12.12 (Fleming and De Giorgi). *Let $n \geq 2$. The property that there does not exist a nonplanar area-minimizing cone in $\mathbb{R}^n$ is equivalent to the property that any entire minimal graph M^n in $\mathbb{R}^{n+1}$ must be a hyperplane.*

Using this, the 2-dimensional Bernstein theorem was generalized by De Giorgi [**DG65**], Almgren [**Alm66**], and Simons [**Sim68**] to the following result.

Theorem 12.13. *Any entire minimal graph M^n in $\mathbb{R}^{n+1}$, where $n \leq 7$, must be a hyperplane.*

See the book by Colding and Minicozzi [**CM11**, Theorem 2.23] for a proof of Theorem 12.13.

The result is false for $n \geq 8$. This is because the cone

$$\{(\mathbf{x}, \mathbf{y}) \in \mathbb{R}^4 \times \mathbb{R}^4 \colon \|\mathbf{x}\|^2 = \|\mathbf{y}\|^2\} \tag{12.31}$$

is area-minimizing. This cone is an example of a Simons cone [**Sim68**].

Remark 12.14. Chodosh and Li [**CL**] proved that any complete, 2-sided, stable minimal immersed hypersurface in $\mathbb{R}^4$ must be a hyperplane.

12.3.3. Nitsche's proof of the Bernstein theorem. In this subsection we present J. C. C. Nitsche's proof of the Bernstein theorem (Theorem 12.11) (see [**Nit57**]). We are grateful to Yu Yuan for providing us his detailed notes on the proof, which we follow closely; any errors in the transcription are due to the authors.

The idea is, for 2-dimensional graphical hypersurfaces, to use the Heinz transformation to convert the minimal surface equation to the Monge–Ampère equation. Then, by the so-called Legendre–Lewy transformation we can convert the Monge–Ampère equation into the Laplace equation. Finally, we apply the Liouville theorem for harmonic functions to obtain the result. We now present how to carry out this proof.

Let $v : \mathbb{R}^n \to \mathbb{R}$ be a smooth function such that the gradient map $\nabla v : \mathbb{R}^n \to \mathbb{R}^n$, defined by

$$\nabla v(\mathbf{x}) := \nabla v_{\mathbf{x}},$$

is injective. Let

$$\mathcal{U} := \nabla v(\mathbb{R}^n) = \{\nabla v_{\mathbf{x}} : \mathbf{x} \in \mathbb{R}^n\}.$$

The **Legendre transformation** of v,

$$v^* : \mathcal{U} \to \mathbb{R},$$

is defined by

$$v^*(\mathbf{y}) := \langle \mathbf{y}, \mathbf{x} \rangle - v(\mathbf{x}), \tag{12.32}$$

where $\mathbf{x} := (\nabla v)^{-1}(\mathbf{y})$.

By differentiating the equation

$$v^*(\mathbf{y}) = \langle \mathbf{y}, (\nabla v)^{-1}(\mathbf{y})\rangle - v\big((\nabla v)^{-1}(\mathbf{y})\big)$$

in the direction of any vector $\mathbf{z}$, we obtain

$$\begin{aligned}(\nabla v^*)_{\mathbf{y}} \cdot \mathbf{z} &= \mathbf{z} \cdot (\nabla v)^{-1}(\mathbf{y}) + \mathbf{y} \cdot d\big((\nabla v)^{-1}\big)_{\mathbf{y}}(\mathbf{z}) \\ &\quad - (\nabla v)_{(\nabla v)^{-1}(\mathbf{y})} \cdot d\big((\nabla v)^{-1}\big)_{\mathbf{y}}(\mathbf{z}) \\ &= (\nabla v)^{-1}(\mathbf{y}) \cdot \mathbf{z}.\end{aligned}$$

That is,

$$\nabla v^* = (\nabla v)^{-1} : \mathcal{U} \to \mathbb{R}^n. \tag{12.33}$$

So the gradient of the Legendre transformation of a function is equal to the inverse of the gradient of the function, where the gradient is considered as a map. In terms of 'gradient graphs', we have that

$$(\mathbf{x}, \nabla v(\mathbf{x})) = (\nabla v^*(\mathbf{y}), \mathbf{y}) \in \mathbb{R}^n \times \mathbb{R}^n, \tag{12.34}$$

where $\mathbf{y} = \nabla v(\mathbf{x})$. By differentiating this equation, we obtain

$$\big[\nabla^2 v(\mathbf{x})\big]^{-1} = \big[\nabla^2 v^*(\mathbf{y})\big]. \tag{12.35}$$

Theorem 12.15 (Jörgens [**Joe54**]). *Let $u : \mathbb{R}^2 \to \mathbb{R}$ be a smooth strictly convex solution to the Monge–Ampère equation*

$$1 = \det(\nabla^2 u) = \begin{vmatrix} u_{11} & u_{12} \\ u_{21} & u_{22}\end{vmatrix}, \tag{12.36}$$

where we denote $u_i = \frac{\partial u}{\partial x^i}$ and $u_{ij} = \frac{\partial^2 u}{\partial x^i \partial x^j}$. Then u is a quadratic polynomial.

Proof. Given a solution $u : \mathbb{R}^2 \to \mathbb{R}$ to the Monge–Ampère equation (12.36), the **Legendre–Lewy transformation** of u is defined to be the Legendre transformation $v^* : \mathcal{U} \to \mathbb{R}$ of the function $v : \mathbb{R}^2 \to \mathbb{R}$ defined by

$$v(\mathbf{x}) := u(\mathbf{x}) + \frac{\|\mathbf{x}\|^2}{2}.$$

We have that

$$\nabla v_{\mathbf{x}} = \nabla u_{\mathbf{x}} + \mathbf{x}$$

and

$$\nabla v^* = (\nabla v)^{-1} = (\nabla u + \mathrm{id})^{-1}.$$

By differentiating this equation, we have that

$$d(\nabla v^*)_{\mathbf{y}} = \big(d(\nabla u + \mathrm{id})_{(\nabla u + \mathrm{id})^{-1}(\mathbf{y})}\big)^{-1} = \big((d\nabla u)_{\nabla v^*(\mathbf{y})} + \mathrm{id}\big)^{-1}$$

and hence

$$(d\nabla u)_{\nabla v^*(\mathbf{y})} = \big(d(\nabla v^*)_{\mathbf{y}}\big)^{-1} - \mathrm{id}\,. \tag{12.37}$$

Since u is a solution to the Monge–Ampère equation (12.36), we have that

$$1 = \det\big((d\nabla u)_{\nabla v^*(\mathbf{y})}\big) = \det\big(\big(d(\nabla v^*)_{\mathbf{y}}\big)^{-1} - \mathrm{id}\big). \tag{12.38}$$

Since $d(\nabla v^*)_{\mathbf{y}} = (\nabla^2 v^*)_{\mathbf{y}}$ is a symmetric 2×2-matrix, it has real eigenvalues, which we denote by μ_1 and μ_2. Since $(d\nabla u)_{\nabla v^*(\mathbf{y})}$ is positive-definite, equation (12.38) says that $0 < \mu_1, \mu_2 < 1$ and

$$1 = \begin{vmatrix} \mu_1^{-1} - 1 & 0 \\ 0 & \mu_2^{-1} - 1 \end{vmatrix} = 1 - \mu_1^{-1} - \mu_2^{-1} + \mu_1^{-1}\mu_2^{-1}.$$

In other words,

$$1 = \mu_1 + \mu_2 = \Delta v^* \tag{12.39}$$

on $\mathcal{U}$. This implies that

$$\Delta(\nabla^2 v^*) = 0. \tag{12.40}$$

Claim 12.16.

$$\mathcal{U} = \nabla v(\mathbb{R}^2) = \mathbb{R}^2. \tag{12.41}$$

Proof of Claim 12.16. We have that $d(\nabla v) = d(\nabla u) + \mathrm{id} > \mathrm{id}$, so that the map ∇v is an immersion. Let $\mathbf{x}_1, \mathbf{x}_2 \in \mathbb{R}^2$. We have that

$$\begin{aligned} \|\nabla v_{\mathbf{x}_1} - \nabla v_{\mathbf{x}_2}\|^2 &= \|\nabla u_{\mathbf{x}_1} - \nabla u_{\mathbf{x}_2} + \mathbf{x}_1 - \mathbf{x}_2\|^2 \\ &= \|\nabla u_{\mathbf{x}_1} - \nabla u_{\mathbf{x}_2}\|^2 + \|\mathbf{x}_1 - \mathbf{x}_2\|^2 + \langle \nabla u_{\mathbf{x}_1} - \nabla u_{\mathbf{x}_2}, \mathbf{x}_1 - \mathbf{x}_2\rangle \\ &\geq \|\mathbf{x}_1 - \mathbf{x}_2\|^2, \end{aligned}$$

since $\langle \nabla u_{\mathbf{x}_1} - \nabla u_{\mathbf{x}_2}, \mathbf{x}_1 - \mathbf{x}_2\rangle \geq 0$ by the convexity of u.

Thus the map ∇v is distance nondecreasing. Furthermore, this implies that ∇v is an injective immersion. By the smooth invariance of domain theorem, this implies that $\mathcal{U}$ is an open set and $\nabla v : \mathbb{R}^2 \to \mathcal{U}$ is a diffeomorphism. Moreover, by the distance nondecreasing property of ∇v, we have that $\mathcal{U}$ is also a closed set. This implies that $\mathcal{U} = \mathbb{R}^2$. □

Since $0 < \nabla^2 v^* < \mathrm{id}$ is an entire bounded matrix function by Claim 12.16 and since $\Delta(\nabla^2 v^*) = 0$, by the Liouville theorem for bounded entire harmonic functions on Euclidean space (see §11.6.1), we have that

$$\nabla^2 v^* = C, \tag{12.42}$$

where C is a constant matrix. By (12.37), we obtain that

$$\nabla^2 u =: C' \tag{12.43}$$

is also a constant matrix. Therefore $u : \mathbb{R}^2 \to \mathbb{R}$ is a quadratic function. This completes the proof of Jörgens's theorem. □

Let $F : \mathbb{R}^2 \to \mathbb{R}^3$, where

$$F(x^1, x^2) = \big(x^1, x^2, f(x^1, x^2)\big),$$

be the graphical parametrization of an entire graphical minimal surface. Let $g := F^* g_{\mathbb{E}}$ be the pulled-back metric on $\mathbb{R}^2$. Recall from (4.36) that

$$g_{ij} := g\,(\mathbf{e}_i, \mathbf{e}_j) = \delta_{ij} + f_i f_j \tag{12.44}$$

for $i, j = 1, 2$, and recall by (4.147) that

$$g^{ij} = (g^{-1})_{ij} = \delta_{ij} - \frac{f_i f_j}{\|\nabla f\|^2 + 1}. \tag{12.45}$$

By (5.69), $\Delta_g F = -HN$ is the mean curvature vector. Let $F := (F_1, F_2, F_3)$, where $F_\alpha : \mathbb{R}^2 \to \mathbb{R}$ for $\alpha = 1, 2, 3$. Since F parametrizes a minimal surface, we have that

$$\Delta_g F_1 = 0, \quad \Delta_g F_2 = 0, \quad \Delta_g F_3 = 0. \tag{12.46}$$

We have that

$$|g| := \det(g_{ij}) = \begin{vmatrix} 1 + f_1^2 & f_1 f_2 \\ f_1 f_2 & 1 + f_2^2 \end{vmatrix} = 1 + \|\nabla f\|^2.$$

Hence, by (11.228), (12.46), and we have for $\alpha = 1, 2$ that

$$0 = \sum_{i,j=1}^{2} \partial_i \left(\sqrt{|g|}\, g^{ij} (F_\alpha)_j\right) = \sum_{i=1}^{2} \partial_i \left(\sqrt{|g|}\, g^{i\alpha}\right) \tag{12.47}$$

where we used $(F_\alpha)_j = \delta_{\alpha j}$. In particular, by taking $\alpha = 1$, we obtain that

$$0 = \partial_1 \left(\frac{1 + f_2^2}{\sqrt{\|\nabla f\|^2 + 1}}\right) - \partial_2 \left(\frac{f_1 f_2}{\sqrt{\|\nabla f\|^2 + 1}}\right). \tag{12.48}$$

Similarly, by taking $\alpha = 2$, we obtain that

$$0 = -\partial_1 \left(\frac{f_1 f_2}{\sqrt{\|\nabla f\|^2 + 1}}\right) + \partial_2 \left(\frac{1 + f_1^2}{\sqrt{\|\nabla f\|^2 + 1}}\right). \tag{12.49}$$

By (12.48), (12.49), and the Poincaré lemma, there exist conjugate functions F_1^* and F_2^* on $\mathbb{R}^2$ such that

$$\nabla F_1^* = \left(\frac{f_1 f_2}{\sqrt{\|\nabla f\|^2 + 1}}, \frac{1 + f_2^2}{\sqrt{\|\nabla f\|^2 + 1}}\right), \tag{12.50a}$$

$$\nabla F_2^* = \left(\frac{1 + f_1^2}{\sqrt{\|\nabla f\|^2 + 1}}, \frac{f_1 f_2}{\sqrt{\|\nabla f\|^2 + 1}}\right). \tag{12.50b}$$

In particular,

$$\partial_1 F_1^* = \partial_2 F_2^*.$$

By this and the Poincaré lemma again, there exists a function $u : \mathbb{R}^2 \to \mathbb{R}$ such that

$$\nabla u = (F_2^*, F_1^*); \tag{12.51}$$

that is,

$$\partial_1 u = F_2^* \quad \text{and} \quad \partial_2 u = F_1^*. \tag{12.52}$$

This function u is called the **Heinz transformation** of f.

We calculate that

$$\begin{aligned} (\nabla^2 u) = \begin{pmatrix} \nabla \partial_1 u \\ \nabla \partial_2 u \end{pmatrix} = \begin{pmatrix} \nabla F_2^* \\ \nabla F_1^* \end{pmatrix} = \frac{1}{\sqrt{|g|}} \begin{pmatrix} g_{11} & g_{12} \\ g_{21} & g_{22} \end{pmatrix} \\ = \frac{1}{\sqrt{1 + \|\nabla f\|^2}} \begin{pmatrix} 1 + f_1^2 & f_1 f_2 \\ f_1 f_2 & 1 + f_2^2 \end{pmatrix}. \end{aligned} \tag{12.53}$$

Since g is positive-definite, we have that the entire function $u : \mathbb{R}^2 \to \mathbb{R}$ is strictly convex. By (12.53), we also have that u is a solution to the Monge–Ampère equation:

$$\det(\nabla^2 u) = 1.$$

By Jörgens's theorem (Theorem 12.15), we obtain that u is a quadratic polynomial. This implies that the matrix function $\nabla^2 u$ on $\mathbb{R}^2$ is constant.

Finally, taking the trace of (12.53) yields

$$C = \frac{1}{\sqrt{1 + \|\nabla f\|^2}} + \sqrt{1 + \|\nabla f\|^2},$$

where C is a constant. Hence, the function $\|\nabla f\|$ is constant. Consequently, (f_1, f_2) is a constant vector function. We conclude that f is a linear function. This completes Nitsche's proof of the Bernstein theorem (Theorem 12.11).

Remark 12.17. By (4.148), we have that the Christoffel symbols are given by

$$\Gamma_{ij}^k = \sum_{\ell=1}^{2} \left(\delta_{k\ell} - \frac{f_k f_\ell}{\|\nabla f\|^2 + 1} \right) f_{ij} f_\ell.$$

Therefore,

$$\Delta_g F_\alpha = \sum_{i,j=1}^{2} g^{ij} \left((F_\alpha)_{ij} - \sum_{k,\ell=1}^{2} \left(\delta_{k\ell} - \frac{f_k f_\ell}{\|\nabla f\|^2 + 1} \right) f_{ij} f_\ell (F_\alpha)_k \right). \tag{12.54}$$

If $\alpha = 1$ or $\alpha = 2$, then $F_\alpha(x^1, x^2) = x^\alpha$. Thus $(F_\alpha)_k = \delta_{\alpha k}$ and $(F_\alpha)_{ij} = 0$. Thus, in this case, (12.54) yields

$$\begin{aligned}\Delta_g F_\alpha &= -\sum_{i,j,\ell=1}^{2} g^{ij}\left(\delta_{\alpha\ell} - \frac{f_\alpha f_\ell}{\|\nabla f\|^2+1}\right) f_{ij} f_\ell \\ &= \frac{f_\alpha}{\|\nabla f\|^2+1}\sum_{i,j=1}^{2} g^{ij} f_{ij} \\ &= -\frac{f_\alpha}{\sqrt{\|\nabla f\|^2+1}} H \\ &= 0\end{aligned} \tag{12.55}$$

for $\alpha = 1, 2$, since $H = 0$. This gives another derivation of the first two formulas in (12.46).

12.3.4. Maximal space-like graphical hypersurfaces in Minkowski space. Recall that the **matrix determinant lemma** says that for any invertible matrix A and column vectors V and W, we have that

$$\det(A + VW^T) = (1 + W^T A^{-1} V)\det(A). \tag{12.56}$$

In particular, by taking $A = I$ and $V = W = \nabla f$, we have that

$$\det\left(I + \nabla f (\nabla f)^T\right) = 1 + \|\nabla f\|^2. \tag{12.57}$$

Let $\mathcal{U}$ be an open subset of $\mathbb{R}^n$ and let $f : \mathcal{U} \to \mathbb{R}$ be a smooth function. By (12.17) and (12.57), the area of the graph G^n of f in Euclidean space $\mathbb{R}^{n+1}$ is given by

$$\operatorname{Area}(G^n) = \int_{\mathcal{U}} \sqrt{1 + \|\nabla f\|^2}\, d\mu_{\mathbb{E}}. \tag{12.58}$$

The Euler–Lagrange equation of the convex area functional is the minimal surface equation (12.21).

Instead of a Euclidean ambient space, we may consider a Minkowskian ambient space. As in (7.120), let $\mathbb{E}^{n,1}$ denote Minkowski $(n+1)$-space, i.e., $\mathbb{R}^{n+1}$ with the Minkowski metric

$$g_{\mathbb{E}^{n,1}} = dx_1^2 + \cdots + dx_n^2 - dx_{n+1}^2. \tag{12.59}$$

We say that the graph $\widetilde{G}^n$ of f is **space-like** if $\|\nabla f\| < 1$. In this case, the area of $\widetilde{G}^n$ in Minkowski space $\mathbb{E}^{n,1}$ is given by

$$\operatorname{Area}\left(\widetilde{G}^n\right) = \int_{\mathcal{U}} \sqrt{1 - \|\nabla f\|^2}\, d\mu_{\mathbb{E}}. \tag{12.60}$$

The Euler–Lagrange equation of the concave area functional is the **maximal surface equation**:

$$\operatorname{div}\left(\frac{\nabla f}{\sqrt{1-\|\nabla f\|^2}}\right)=0. \tag{12.61}$$

Now consider the case where $\mathcal{U}=\mathbb{R}^2$. The **conjugate function** of f is given by

$$f^*(x^1,x^2):=\int^{(x^1,x^2)}\frac{1}{\sqrt{1-\|\nabla f\|^2}}\big(-f_2\,dx^1+f_1\,dx^2\big). \tag{12.62}$$

We have that

$$\nabla f^*=\frac{(-f_2,f_1)}{\sqrt{1-\|\nabla f\|^2}}$$

and so

$$\sqrt{1+\|\nabla f^*\|^2}=\frac{1}{\sqrt{1-\|\nabla f\|^2}}.$$

Then we deduce that

$$\operatorname{div}\left(\frac{\nabla f^*}{\sqrt{1+\|\nabla f^*\|^2}}\right)=\operatorname{div}(-f_2,f_1)=0\quad\text{on }\mathbb{R}^2. \tag{12.63}$$

Note that, by reversing the process above, the conjugate function of f^* is just f and satisfies (12.61).

By the Bernstein theorem (Theorem 12.11), ∇f^* is a constant vector and, in turn, so is ∇f. We have proven a Bernstein theorem for maximal space-like graphical hypersurfaces in $\mathbb{E}^{2,1}$.

The Bernstein theorem for maximal space-like graphical hypersurfaces in Minkowski space $\mathbb{E}^{n,1}$ was proved by Calabi [**Cal70**] for $n\le 4$ and by Cheng and Yau [**CY76**] in all dimensions.

12.4. The Simons identity

In this section, we discuss an important identity for the covariant derivatives of the second fundamental form discovered by Simons [**Sim68**]. This identity is crucial and has numerous applications in minimal surface theory.

Let M^n be a hypersurface in $\mathbb{R}^{n+1}$. Recall that, for each $\mathbf{x}\in M^n$, the Weingarten map $\mathrm{L}:T_{\mathbf{x}}M\to T_{\mathbf{x}}M$ is a self-adjoint linear map reflecting the infinitesimal 'bending' or rate of change of the unit normal vector field N.

The first and second covariant derivatives of the Weingarten map are defined via the product rule by

$$(\nabla_V\mathrm{L})(Y):=\nabla_V(\mathrm{L}(Y))-\mathrm{L}(\nabla_VY) \tag{12.64}$$

and (compare with (4.156))

$$(\nabla^2_{U,V}\,\mathrm{L})(Y) := \big(\nabla_U(\nabla_V\,\mathrm{L})\big)(Y) - (\nabla_{\nabla_U V}\,\mathrm{L})(Y), \tag{12.65}$$

respectively. By applying the product rule to this, we calculate that

$$\begin{aligned}(\nabla^2_{U,V}\,\mathrm{L})(Y) &= \nabla_U\big((\nabla_V\,\mathrm{L})(Y)\big) - (\nabla_V\,\mathrm{L})(\nabla_U Y) - (\nabla_{\nabla_U V}\,\mathrm{L})(Y)\\ &= \nabla_U\big(\nabla_V(\mathrm{L}(Y))\big) - \nabla_U\big(\mathrm{L}(\nabla_V Y)\big)\\ &\quad - (\nabla_V\,\mathrm{L})(\nabla_U Y) - (\nabla_{\nabla_U V}\,\mathrm{L})(Y)\\ &= \nabla_U\big(\nabla_V(\mathrm{L}(Y))\big) - (\nabla_U\,\mathrm{L})(\nabla_V Y)\\ &\quad - \mathrm{L}\,\big(\nabla_U(\nabla_V Y)\big) - (\nabla_V\,\mathrm{L})(\nabla_U Y) - (\nabla_{\nabla_U V}\,\mathrm{L})(Y).\end{aligned}$$

By applying the product rule to the term on the last line and rearranging terms, we obtain

$$\begin{aligned}(\nabla^2_{U,V}\mathrm{L})(Y) = \nabla_U\big(\nabla_V(\mathrm{L}(Y))\big) - \nabla_{\nabla_U V}(\mathrm{L}(Y)) - \mathrm{L}\big(\nabla_U(\nabla_V Y)\big)\\ + \mathrm{L}(\nabla_{\nabla_U V} Y) - (\nabla_U\mathrm{L})(\nabla_V Y) - (\nabla_V\mathrm{L})(\nabla_U Y).\end{aligned}$$

Note that the last line is symmetric in U and V. Thus,

$$\begin{aligned}(\nabla^2_{U,V}\mathrm{L})(Y) - (\nabla^2_{V,U}\mathrm{L})(Y) &= \nabla_U\big(\nabla_V(\mathrm{L}(Y))\big) - \nabla_{\nabla_U V}(\mathrm{L}(Y))\\ &\quad - \mathrm{L}\big(\nabla_U(\nabla_V Y)\big) + \mathrm{L}(\nabla_{\nabla_U V} Y)\\ &\quad - \nabla_V\big(\nabla_U(\mathrm{L}(Y))\big) + \nabla_{\nabla_V U}(\mathrm{L}(Y))\\ &\quad + \mathrm{L}\big(\nabla_V(\nabla_U Y)\big) - \mathrm{L}(\nabla_{\nabla_V U} Y)\\ &= \nabla_U\big(\nabla_V(\mathrm{L}(Y))\big) - \nabla_V\big(\nabla_U(\mathrm{L}(Y))\big)\\ &\quad - \nabla_{\nabla_U V}(\mathrm{L}(Y)) + \nabla_{\nabla_V U}(\mathrm{L}(Y))\\ &\quad - \mathrm{L}\big(\nabla_U(\nabla_V Y)\big) + \mathrm{L}\big(\nabla_V(\nabla_U Y)\big)\\ &\quad + \mathrm{L}(\nabla_{\nabla_U V} Y) - \mathrm{L}(\nabla_{\nabla_V U} Y).\end{aligned}$$

By the torsion-free property of the Levi-Civita connection, we obtain

$$\begin{aligned}&(\nabla^2_{U,V}\mathrm{L})(Y) - (\nabla^2_{V,U}\mathrm{L})(Y)\\ &\quad = \nabla_U\big(\nabla_V(\mathrm{L}(Y))\big) - \nabla_V\big(\nabla_U(\mathrm{L}(Y))\big) - \nabla_{[U,V]}(\mathrm{L}(Y))\\ &\qquad - \mathrm{L}\big(\nabla_U(\nabla_V Y) - \nabla_V(\nabla_U Y) - \nabla_{[U,V]} Y\big)\\ &\quad = \mathrm{Rm}(U,V)(\mathrm{L}(Y)) - \mathrm{L}(\mathrm{Rm}(U,V)Y).\end{aligned} \tag{12.66}$$

Now, substituting the Gauss equation (4.167) into this, we have

$$\begin{aligned}(\nabla^2_{U,V}\mathrm{L})(Y) - (\nabla^2_{V,U}\mathrm{L})(Y) &= \mathrm{II}(V,\mathrm{L}(Y))\,\mathrm{L}(U) - \mathrm{II}(U,\mathrm{L}(Y))\,\mathrm{L}(V)\\ &\quad - \mathrm{L}\,\big(\mathrm{II}(V,Y)\,\mathrm{L}(U) - \mathrm{II}(U,Y)\,\mathrm{L}(V)\big)\\ &= \mathrm{I}(V,\mathrm{L}^2(Y))\,\mathrm{L}(U) - \mathrm{I}(U,\mathrm{L}^2(Y))\,\mathrm{L}(V)\\ &\quad - \mathrm{II}(V,Y)\,\mathrm{L}^2(U) + \mathrm{II}(U,Y)\,\mathrm{L}^2(V).\end{aligned}$$

We define

$$\mathrm{II}^2(V,Y) := \mathrm{I}(V, \mathrm{L}^2(Y)). \tag{12.67}$$

We may then rewrite this equation as

$$\begin{aligned}(\nabla^2_{U,V}\mathrm{L})(Y) - (\nabla^2_{V,U}\mathrm{L})(Y) = {} & \mathrm{II}^2(V,Y)\,\mathrm{L}(U) - \mathrm{II}^2(U,Y)\,\mathrm{L}(V) \\ & - \mathrm{II}(V,Y)\,\mathrm{L}^2(U) + \mathrm{II}(U,Y)\,\mathrm{L}^2(V).\end{aligned}$$

Taking the inner product of this formula with a vector Z, we obtain the following commutator formula for the covariant derivatives of the second fundamental form:

(12.68)

$$\begin{aligned}(\nabla^2_{U,V})\,\mathrm{II}(Y,Z) - (\nabla^2_{V,U}\,\mathrm{II})(Y,Z) = {} & \mathrm{II}^2(V,Y)\,\mathrm{II}(U,Z) - \mathrm{II}^2(U,Y)\,\mathrm{II}(V,Z) \\ & - \mathrm{II}(V,Y)\,\mathrm{II}^2(U,Z) + \mathrm{II}(U,Y)\,\mathrm{II}^2(V,Z).\end{aligned}$$

On the other hand, by the Codazzi equation,

(12.69)

$$(\nabla^2_{U,V})\,\mathrm{II}(Y,Z) - (\nabla^2_{V,U}\,\mathrm{II})(Y,Z) = (\nabla^2_{U,Y})\,\mathrm{II}(V,Z) - (\nabla^2_{V,Z}\,\mathrm{II})(Y,U).$$

We conclude:

Proposition 12.18 (Simons's equation)**.**

$$\begin{aligned}(\nabla^2_{U,Y})\,\mathrm{II}(V,Z) - (\nabla^2_{V,Z}\,\mathrm{II})(U,Y) = {} & \mathrm{II}^2(V,Y)\,\mathrm{II}(U,Z) - \mathrm{II}^2(U,Y)\,\mathrm{II}(V,Z) \\ & - \mathrm{II}(V,Y)\,\mathrm{II}^2(U,Z) + \mathrm{II}(U,Y)\,\mathrm{II}^2(V,Z).\end{aligned}$$

We can now trace this over the U and Y components. Namely, let $\mathbf{x} \in M^n$ and let $\{e_i\}_{i=1}^n$ be an orthonormal basis for $T_{\mathbf{x}}M$. Taking $U = Y = e_i$ and summing over i yields

$$(\Delta\,\mathrm{II})(V,Z) - (\nabla^2 H)(V,Z) = -|\,\mathrm{II}\,|^2\,\mathrm{II}(V,Z) + H\,\mathrm{II}^2(V,Z). \tag{12.70}$$

That is, as a tensor equation,

$$\Delta\,\mathrm{II} - \nabla^2 H = -|\,\mathrm{II}\,|^2\,\mathrm{II} + H\,\mathrm{II}^2\,. \tag{12.71}$$

Note that if we trace again, then we obtain a vacuous equation.

There are numerous applications of Simons's identity in the study of minimal hypersurfaces; see the survey by Codá Marques [**CM14**]. We explore one such application in the next section.

12.5. Curvature estimate for stable minimal hypersurfaces

In this section, for stable minimal hypersurfaces of dimension at most 5 in Euclidean space, we will prove a curvature estimate in terms of the distance to the boundary.

12.5.1. An inequality for the covariant derivative of the second fundamental form. We start by proving a useful inequality for $\nabla \mathrm{II}$. Recall from Kato's inequality that

$$|\nabla \mathrm{II}|^2 \geq \left|\nabla|\mathrm{II}|\right|^2.$$

By taking advantage of the fact that the covariant derivative of the second fundamental form is fully symmetric, we can improve this as follows.

Lemma 12.19. *Let M^n be a minimal hypersurface in $\mathbb{R}^{n+1}$. At any point $\mathbf{x} \in M^n$ where II is not the zero tensor, we have that*

$$|\nabla \mathrm{II}|^2 \geq \frac{n+2}{n}\left|\nabla|\mathrm{II}|\right|^2.$$

Proof. Let $\mathbf{x} \in M^n$ and let $\{e_i\}_{i=1}^n$ be an orthonormal basis of eigenvectors of the Weingarten map L of M^n at $\mathbf{x}$, so that $\mathrm{I}_{ij} := \mathrm{I}(e_i, e_j) = \delta_{ij}$. Then $\mathrm{II}_{ij} := \mathrm{II}(e_i, e_j) = \kappa_i \delta_{ij}$, where the $\kappa_i = \mathrm{II}(e_i, e_i)$ are the principal curvatures of M^n at $\mathbf{x}$.

Firstly, using the Cauchy–Schwarz inequality, with respect to the frame above we calculate that

$$\begin{aligned}
|\mathrm{II}|^2\left|\nabla|\mathrm{II}|\right|^2 &= \frac{1}{4}\sum_{k=1}^n (\nabla_k |\mathrm{II}|^2)^2 \\
&= \sum_{k=1}^n \left(\sum_{i,j=1}^n \mathrm{II}_{ij}\, \nabla_k \mathrm{II}_{ij}\right)^2 \\
&= \sum_{k=1}^n \left(\sum_{i=1}^n \kappa_i \nabla_k \mathrm{II}_{ii}\right)^2 \\
&\leq \sum_{k=1}^n \left(\sum_{i=1}^n \kappa_i^2 \sum_{i=1}^n (\nabla_k \mathrm{II}_{ii})^2\right) \\
&= |\mathrm{II}|^2 \sum_{k,i=1}^n (\nabla_k \mathrm{II}_{ii})^2,
\end{aligned}$$

so that

$$\left|\nabla|\mathrm{II}|\right|^2 \leq \sum_{k,i=1}^n (\nabla_k \mathrm{II}_{ii})^2. \tag{12.72}$$

Secondly, using the minimal hypersurface equation and the algebraic inequality (5.108), we calculate that

$$
\begin{aligned}
\sum_{k,i=1}^{n} (\nabla_k \mathrm{II}_{ii})^2 &= \sum_{k\neq i}^{n} (\nabla_k \mathrm{II}_{ii})^2 + \sum_{i=1}^{n} (\nabla_i \mathrm{II}_{ii})^2 \\
&= \sum_{k\neq i}^{n} (\nabla_k \mathrm{II}_{ii})^2 + \sum_{i=1}^{n} \Big(-\sum_{j\neq i} \nabla_i \mathrm{II}_{jj} \Big)^2 \\
&\leq \sum_{k\neq i}^{n} (\nabla_k \mathrm{II}_{ii})^2 + \sum_{i=1}^{n} (n-1) \sum_{j\neq i} (\nabla_i \mathrm{II}_{jj})^2 \\
&= n \sum_{k\neq i}^{n} (\nabla_k \mathrm{II}_{ii})^2.
\end{aligned}
\tag{12.73}
$$

Finally, using the Codazzi equation, (12.73), and (12.72), we calculate that

$$
\begin{aligned}
|\nabla \mathrm{II}|^2 &= \sum_{i,j,k=1}^{n} (\nabla_k \mathrm{II}_{ij})^2 \\
&\geq \sum_{i,k=1}^{n} (\nabla_k \mathrm{II}_{ii})^2 + \sum_{i\neq j}^{n} (\nabla_i \mathrm{II}_{ij})^2 + \sum_{i\neq j}^{n} (\nabla_j \mathrm{II}_{ij})^2 \\
&= \sum_{i,k=1}^{n} (\nabla_k \mathrm{II}_{ii})^2 + \sum_{i\neq j}^{n} (\nabla_j \mathrm{II}_{ii})^2 + \sum_{i\neq j}^{n} (\nabla_i \mathrm{II}_{jj})^2 \\
&\geq \frac{n+2}{n} \sum_{i,k=1}^{n} (\nabla_k \mathrm{II}_{ii})^2 \\
&\geq \frac{n+2}{n} \big|\nabla |\mathrm{II}|\big|^2.
\end{aligned}
$$

This completes the proof of the lemma. □

12.5.2. Simons's inequalities. By Simons's identity (12.71), for a minimal hypersurface, we have that

$$\Delta \mathrm{II} = -|\mathrm{II}|^2 \mathrm{II}.$$

This implies that

$$
\begin{aligned}
\frac{1}{2}\Delta |\mathrm{II}|^2 &= \langle \Delta \mathrm{II}, \mathrm{II} \rangle + |\nabla \mathrm{II}|^2 \\
&= -|\mathrm{II}|^4 + |\nabla \mathrm{II}|^2 \\
&\geq -|\mathrm{II}|^4 + \frac{n+2}{n} \big|\nabla |\mathrm{II}|\big|^2,
\end{aligned}
\tag{12.74}
$$

where the last inequality is by Lemma 12.19. This, in turn, implies that

$$|\,\mathrm{II}\,|\Delta|\,\mathrm{II}\,| \geq -|\,\mathrm{II}\,|^4 + \frac{2}{n}\big|\nabla|\,\mathrm{II}\,|\big|^2. \tag{12.75}$$

We will call these inequalities, which we may view as elliptic inequalities, **Simons's inequalities**.

12.5.3. Consequences of the stability inequality.

Let M^n be a minimal submanifold in $\mathbb{R}^N$. We say that M^n is **stable** if for every normal variation of M^n with compact support in the interior of M^n, the second variation of area is nonnegative.

By the stability inequality (Proposition 12.26 below), under a normal variation of a minimal hypersurface M^n with speed f, we have that the second variation of the area of M^n is given by

$$\frac{d^2}{du^2}\bigg|_{u=0} \mathrm{Vol}_n(M^n) = \int_{M^n} (|\nabla f|^2 - |\,\mathrm{II}\,|^2 f^2)\,d\mu. \tag{12.76}$$

Hence, if M^n is a stable minimal hypersurface, then for each smooth function $f : M^n \to \mathbb{R}$ with compact support in the interior of M^n, we have that

$$\int_{M^n} |\,\mathrm{II}\,|^2 f^2 d\mu \leq \int_{M^n} |\nabla f|^2 d\mu. \tag{12.77}$$

This is the stability inequality for minimal hypersurfaces of Euclidean space.

Lemma 12.20. *If M^n is a stable minimal hypersurface in $\mathbb{R}^{n+1}$, then for any $2 \leq p < 2 + \sqrt{\frac{2}{n}}$ there exists a constant $C(n,p)$ such that*

$$\int_{M^n} |\,\mathrm{II}\,|^{2p} \eta^2 d\mu \leq C(n,p) \int_{M^n} |\,\mathrm{II}\,|^{2p-2} |\nabla \eta|^2 d\mu. \tag{12.78}$$

Hence,

$$\int_{M^n} |\,\mathrm{II}\,|^{2p} \eta^{2p} d\mu \leq 2^{2p} C(n,p)^p \int_{M^n} |\nabla \eta|^{2p} d\mu. \tag{12.79}$$

Proof. Taking $f = |\,\mathrm{II}\,|^{p-1}\eta$ in (12.77), where $\eta : M^n \to \mathbb{R}$ has compact support, yields

$$\begin{aligned}\int_{M^n} |\,\mathrm{II}\,|^{2p} \eta^2 d\mu \leq & \int_{M^n} (p-1)^2 |\,\mathrm{II}\,|^{2p-4} \big|\nabla|\,\mathrm{II}\,|\big|^2 \eta^2 d\mu \\ & + \int_{M^n} |\,\mathrm{II}\,|^{2p-2} |\nabla \eta|^2 d\mu \\ & + 2(p-1) \int_{M^n} |\,\mathrm{II}\,|^{2p-3} \langle \nabla|\,\mathrm{II}\,|, \nabla \eta\rangle \eta\, d\mu.\end{aligned} \tag{12.80}$$

On the other hand, multiplying Simons's inequality (12.75) by $|\,\mathrm{II}\,|^{2p-4}\eta^2$ and integrating yields

$$\int_{M^n}\left(-|\,\mathrm{II}\,|^{2p}\eta^2+\frac{2}{n}|\,\mathrm{II}\,|^{2p-4}\big|\nabla|\,\mathrm{II}\,|\big|^2\eta^2\right)d\mu\leq\int_{M^n}|\,\mathrm{II}\,|^{2p-3}\Delta|\,\mathrm{II}\,|\eta^2d\mu.$$

Integrating by parts on the right-hand side then yields

$$\int_{M^n}\left(\left(\frac{2}{n}+2p-3\right)|\,\mathrm{II}\,|^{2p-4}\big|\nabla|\,\mathrm{II}\,|\big|^2\eta^2+2|\,\mathrm{II}\,|^{2p-3}\langle\nabla|\,\mathrm{II}\,|,\nabla\eta\rangle\eta\right)d\mu \leq\int_{M^n}|\,\mathrm{II}\,|^{2p}\eta^2d\mu. \tag{12.81}$$

We may view this as a reverse Hölder inequality. By combining this with (12.80), we obtain

$$\begin{aligned}\int_{M^n}&\left(\frac{2}{n}+2p-3-(p-1)^2\right)|\,\mathrm{II}\,|^{2p-4}\big|\nabla|\,\mathrm{II}\,|\big|^2\eta^2d\mu\\&\leq\int_{M^n}|\,\mathrm{II}\,|^{2p-2}|\nabla\eta|^2d\mu+2(p-2)\int_{M^n}|\,\mathrm{II}\,|^{2p-3}\langle\nabla|\,\mathrm{II}\,|,\nabla\eta\rangle\eta\,d\mu.\end{aligned}$$

For this inequality to be useful, we need $\frac{2}{n}+2p-3-(p-1)^2>0$; that is, $-(p-2)^2+\frac{2}{n}>0$. Since $2\leq p<2+\sqrt{\frac{2}{n}}$, these inequalities hold. Thus, for such p and by the Peter–Paul inequality $2ab\leq ca^2+c^{-1}b^2$, we obtain

$$\int_{M^n}|\,\mathrm{II}\,|^{2p-4}\big|\nabla|\,\mathrm{II}\,|\big|^2\eta^2d\mu\leq C(n,p)\int_{M^n}|\,\mathrm{II}\,|^{2p-2}|\nabla\eta|^2d\mu. \tag{12.82}$$

By applying this to (12.80) and using the Peter–Paul inequality again, we conclude (12.78).

Now, replacing η by η^p in (12.78) and applying Hölder's inequality yields

$$\begin{aligned}\int_{M^n}|\,\mathrm{II}\,|^{2p}\eta^{2p}d\mu&\leq C(n,p)\int_{M^n}|\,\mathrm{II}\,|^{2p-2}|\nabla(\eta^p)|^2d\mu\\&\leq p^2C(n,p)\int_{M^n}|\,\mathrm{II}\,|^{2p-2}\eta^{2p-2}|\nabla\eta|^2d\mu\\&\leq p^2C(n,p)\left(\int_{M^n}|\,\mathrm{II}\,|^{2p}\eta^{2p}d\mu\right)^{\frac{p-1}{p}}\left(\int_{M^n}|\nabla\eta|^{2p}d\mu\right)^{\frac{1}{p}}.\end{aligned}$$

Finally, (12.79) follows easily. This completes the proof of the lemma. □

12.5.4. Statement and proof of the curvature estimate. Given a stable minimal hypersurface M^n with boundary in $\mathbb{R}^{n+1}$, its boundary could be arbitrarily 'jagged'. Thus, for any estimate of the second fundamental form of M^n in terms of the distance to the boundary, the bound must tend to infinity as the distance to the boundary tends to zero. It turns out that one may obtain a curvature estimate that grows inverse quadratically in terms of the distance to the boundary.

Theorem 12.21. *There exists a constant C with the following property. If M^n is a smooth compact oriented immersed stable minimal hypersurface in $\mathbb{R}^{n+1}$ with boundary ∂M, where $2 \leq n \leq 5$, then*

$$|\,\mathrm{II}\,|^2(\mathbf{x}) \leq \frac{C}{d(\mathbf{x}, \partial M)^2} \tag{12.83}$$

for $\mathbf{x} \in M^n$.

Proof. Suppose for a contradiction that the theorem is false. Then there exists a sequence of stable minimal hypersurfaces M_i^n as in the hypothesis and points $\mathbf{x}_i \in M_i^n$ such that

$$d(\mathbf{x}_i, \partial M_i)^2 |\,\mathrm{II}_i\,|^2(\mathbf{x}_i) = \max_{\mathbf{x} \in M_i^n} d(\mathbf{x}, \partial M_i)^2 |\,\mathrm{II}_i\,|^2(\mathbf{x}) \geq i^2, \tag{12.84}$$

where II_i denotes the second fundamental form of M_i^n. Let $d_i = d(\mathbf{x}_i, \partial M_i)$. Note that $M_i^n \cap B_{d_i}(\mathbf{x}_i)$ has compact closure in M_i^n. Define a radius

$$r_i := \frac{i}{2|\,\mathrm{II}_i\,|(\mathbf{x}_i)}.$$

From our assumption, we note that $r_i \leq \frac{d_i}{2}$, and this radius is large as compared to the curvature scale for i large (we think of II as the extrinsic curvature).

Let $\mathbf{x} \in M_i^n \cap B_{r_i}(\mathbf{x}_i)$. Observe that $d(\mathbf{x}, \partial M_i) \geq \frac{d_i}{2}$. Thus,

$$\begin{aligned} |\,\mathrm{II}_i\,|^2(\mathbf{x}) &= \frac{d(\mathbf{x}, \partial M_i)^2 |\,\mathrm{II}_i\,|^2(\mathbf{x})}{d(\mathbf{x}, \partial M_i)^2} \\ &\leq \frac{d_i^2 |\,\mathrm{II}_i\,|^2(\mathbf{x}_i)}{d(\mathbf{x}, \partial M_i)^2} \\ &\leq 4|\,\mathrm{II}_i\,|^2(\mathbf{x}_i). \end{aligned} \tag{12.85}$$

We dilate the hypersurfaces according to their curvature scales. Namely, let $k_i = |\,\mathrm{II}_i\,|(\mathbf{x}_i)$ and consider[1]

$$k_i\big((M_i^n \cap B_{r_i}(\mathbf{x}_i)) - \mathbf{x}_i\big) = \widetilde{M}_i^n \cap B_{\frac{i}{2}}(\mathbf{0}), \tag{12.86}$$

where $\widetilde{M}_i^n := k_i(M_i^n - \mathbf{x}_i)$, which is a rescaling and translation of M_i^n. Note that $\frac{i}{2} = k_i r_i \to \infty$ as $i \to \infty$. Let II_i denote the second fundamental form of $\widetilde{M}_i^n$ and let $\Omega_i := \widetilde{M}_i^n \cap B_{\frac{i}{2}}(\mathbf{0})$. We have for all $\mathbf{x} \in \Omega_i$ that

$$|\,\mathrm{II}_i\,|^2(\mathbf{x}) \leq 4|\,\mathrm{II}_i\,|^2(\mathbf{0}) = 4. \tag{12.87}$$

[1]For any subset S of $\mathbb{R}^N$, positive real number k, and point $\mathbf{x}_0 \in \mathbb{R}^N$, we define $k(S - \mathbf{x}_0) = \{k(\mathbf{x} - \mathbf{x}_0) : \mathbf{x} \in S\}$.

Since Ω_i is a minimal hypersurface, by (12.74) we have that

$$\frac{1}{2}\Delta|\,\mathrm{II}_i\,|^2 \geq -|\,\mathrm{II}_i\,|^4 + \frac{n+2}{n}\left|\nabla|\,\mathrm{II}_i\,|\right|^2 \geq -|\,\mathrm{II}_i\,|^4,$$

where the Laplacian Δ is with respect to the induced metric on Ω_i. Let $p > 1$ be a real number. We compute that

$$\begin{aligned}(12.88)\qquad \Delta|\,\mathrm{II}_i\,|^{2p} &= \Delta(|\,\mathrm{II}_i\,|^2)^p \\ &= p(|\,\mathrm{II}_i\,|^2)^{p-1}\Delta|\,\mathrm{II}_i\,|^2 + p(p-1)(|\,\mathrm{II}_i\,|^2)^{p-2}\left|\nabla|\,\mathrm{II}_i\,|^2\right|^2 \\ &\geq -2p|\,\mathrm{II}_i\,|^{2p+2}.\end{aligned}$$

Let $\eta_i : \mathbb{R}^{n+1} \to [0,1]$ be a smooth radial cutoff function with $\eta_i = 1$ on $B_{i/2}(\mathbf{0})$, $\eta_i = 0$ outside of $B_i(\mathbf{0})$, and $|\nabla\eta_i| \leq 4/i$. Let $d\mu_i$ denote the volume form of $\widetilde{M}_i^n$. Since $\widetilde{M}_i^n$ is a stable minimal hypersurface, by Lemma 12.20 we have for $i \geq 2$ that

$$\begin{aligned}\int_{\widetilde{M}_i^n\cap B_1(\mathbf{0})} |\,\mathrm{II}_i\,|^{2p} d\mu_i &\leq \int_{\Omega_i} |\,\mathrm{II}_i\,|^{2p} d\mu_i \\ &\leq \int_{\widetilde{M}_i^n\cap B_i(\mathbf{0})} |\,\mathrm{II}_i\,|^{2p}\eta_i^{2p} d\mu_i \\ &\leq 2^{2p}C(n,p)^p \int_{\widetilde{M}_i^n\cap B_i(\mathbf{0})} |\nabla\eta_i|^{2p} d\mu_i \\ &\leq \widetilde{C}(n,p)\, i^{-2p}\,\mathrm{Vol}_n\big(\widetilde{M}_i^n \cap B_i(\mathbf{0})\big)\end{aligned}$$

provided that $2 \leq p < 2 + \sqrt{\frac{2}{n}}$.

Now, by our hypothesis, we have that

$$\mathrm{Vol}_n\big(\widetilde{M}_i^n \cap B_i(\mathbf{0})\big) \leq A i^n,$$

so that

$$(12.89)\qquad \int_{\widetilde{M}_i^n\cap B_1(\mathbf{0})} |\,\mathrm{II}_i\,|^{2p} d\mu_i \leq A\widetilde{C}(n,p)\, i^{n-2p}.$$

Furthermore, by (12.88) and (12.87), we also have that

$$\Delta|\,\mathrm{II}_i\,|^{2p} \geq -2p|\,\mathrm{II}_i\,|^{2p+2} \geq -8p|\,\mathrm{II}_i\,|^{2p}$$

on Ω_i. Thus, by the mean value inequality (Corollary 12.7 with $\bar{t} = 1$ and $A = 8p$),

$$(12.90)\qquad 1 = |\,\mathrm{II}_i\,|^{2p}(\mathbf{0}) \leq \frac{\mathrm{e}^{4p}}{\omega_n} \int_{\widetilde{M}_i^n\cap B_1(\mathbf{0})} |\,\mathrm{II}_i\,|^{2p} d\mu_i.$$

Combining this with (12.89) yields

$$1 \leq \frac{e^{4p}}{\omega_n} A\widetilde{C}(n,p)\, i^{n-2p}.$$

Thus, we obtain the desired contradiction for i sufficiently large provided that $p > \frac{n}{2}$. Since we may choose p to be any real number in the interval $\left[2, 2+\sqrt{\frac{2}{n}}\right)$, we obtain a contradiction provided $2+\sqrt{\frac{2}{n}} > \frac{n}{2}$. It is easy to see that this inequality holds if $2 \leq n \leq 5$. □

12.6. Statement of Plateau's problem

Imagine a rigid wire looped into an arbitrary shape in space. If you were to dip this wire into a soapy solution and pull it out, a soap film would form inside the loop. Plateau's problem inquires: Does there always exist a soap film, or more precisely, a minimal surface, that spans any given wire configuration? In this section, we will present the mathematical formalization of this question, as addressed by Douglas and Radó.

Let $F : M^n \to \mathbb{R}^N$ be an immersed submanifold, also denoted by M^n. By Exercise 12.1, the mean curvature vector of M^n is equal to

$$\vec{H} = \Delta_g F,$$

where $g = F^* g_{\mathbb{E}}$ and $g_{\mathbb{E}}$ is the Euclidean metric on $\mathbb{R}^N$. Thus, M^n is a minimal submanifold of $\mathbb{R}^N$ if and only if $F : (M^n, g) \to (\mathbb{R}^N, g_{\mathbb{E}})$ is a harmonic map; i.e.,

$$\Delta_g F = \mathbf{0}.$$

While at first glance this may seem like a linear equation, it is important to note that the Laplacian is taken with respect to g, which itself depends on F. And so, the equation is inherently nonlinear. Indeed, hypersurfaces are locally hypersurfaces. By (12.21), the minimal hypersurface equation is a quasilinear nonlinear elliptic equation.

Let $(M^2, [g])$ be a Riemann surface, possibly with boundary. By Exercise 12.2 and the above, a conformal immersion $F : (M^2, [g]) \to (\mathbb{R}^N, g_{\mathbb{E}})$ is a minimal surface if and only if $F : (M^2, g) \to (\mathbb{R}^N, g_{\mathbb{E}})$ is a harmonic map.

A **simple closed curve** in $\mathbb{R}^N$ is an injective continuous map γ from the unit circle S^1 to $\mathbb{R}^N$. We also call the image C of γ a simple closed curve.

Let $\overline{B}^2$ denote the closed unit 2-dimensional ball, which has boundary S^1. The solution to the **Plateau problem** by Douglas [**Dou31**] and Radó [**Rad30**] states the following.

Theorem 12.22. *For any simple closed curve C in $\mathbb{R}^N$, $N \geq 3$, there exists a branched immersion $F : \overline{B}^2 \to \mathbb{R}^N$ such that:*

(1) *$F|_{S^1}$ parametrizes C.*

(2) *Outside a finite set of branch points in B^2, F is a smooth conformal and harmonic immersion and hence parametrizes a minimal surface.*

(3) *In the class of branched immersions, F has minimal area.*

When $N = 3$, Osserman [**Oss70**] proved that in the Douglas–Radó solution, branch points do not exist; i.e., F is a smooth minimal immersion on B^2.

See Harrison and Pugh [**HP16**] for a survey as well as a discussion of further developments on Plateau's problem.

12.7. Minimal surfaces in 3-manifolds with positive scalar curvature

In this section we outline the application of minimal surfaces and harmonic maps to understand the topology of closed 3-manifolds with positive scalar curvature.

Analogous to calculations of the first and second variations of the arc length of a path in Chapter 7, we may calculate the first and second variations of the $(n-1)$-volume of a hypersurface in M^n.

12.7.1. The first variation of the hypersurface volume formula.

Let Σ^{n-1} be an oriented smooth manifold and let $F : \Sigma^{n-1} \to M^n$ parametrize an immersed hypersurface in a Riemannian manifold $(M^n, \overline{g})$. Let $g := F^*(\overline{g})$ be the induced (pulled-back) metric on Σ^{n-1}. Let $(\mathcal{U}, \{x^i\}_{i=1}^{n-1})$ be local coordinates on Σ^{n-1}. The components of the metric g are defined by

$$g_{ij} := g\left(\frac{\partial}{\partial x^i}, \frac{\partial}{\partial x^i}\right) = \overline{g}\left(\frac{\partial F}{\partial x^i}, \frac{\partial F}{\partial x^i}\right). \tag{12.91}$$

The volume element of (Σ^{n-1}, g) is given by

$$d\mu_g = \sqrt{\det g_{ij}}\, dx^1 \wedge \cdots \wedge dx^{n-1}.$$

The $(n-1)$-dimensional volume of (Σ^{n-1}, g) is defined as

$$\mathrm{Vol}_{n-1}(g) := \int_{\Sigma^{n-1}} d\mu_g. \tag{12.92}$$

Let $F_u : \Sigma^{n-1} \to M^n$, $u \in I$, be a 1-parameter family of parametrized hypersurfaces in a Riemannian manifold $(M^n, \overline{g})$. Assume that this is a normal deformation in the sense that $F(\cdot, u) := F_u$ satisfies

$$\frac{\partial F}{\partial u}(x,u) = f(x)N(x,u), \tag{12.93}$$

where $f : \Sigma^{n-1} \to \mathbb{R}$ is a smooth function and $N(x,u)$ is a unit normal to $\Sigma_u := F_u(\Sigma^{n-1})$ at $F(x,u)$ depending smoothly on (x,u). The metric g_u on Σ^{n-1} induced by the parametrization F_u is given by

$$g_u(V,W) := g\big(dF_u(V), dF_u(W)\big) \tag{12.94}$$

for $V, W \in T\Sigma$.

The $(n-1)$-dimensional volume of Σ_u (equivalently, (Σ^{n-1}, g_u)) is defined by

$$\mathrm{Vol}_{n-1}(\Sigma_u) := \mathrm{Vol}(g_u) = \int_{\Sigma^{n-1}} d\mu_{g_u}. \tag{12.95}$$

Lemma 12.23. *The first variation of the $(n-1)$-dimensional volume of Σ_u is given by*

$$\frac{d}{du}\mathrm{Vol}_{n-1}(\Sigma_u) = \int_{\Sigma^{n-1}} fH_u d\mu_{g_u}, \tag{12.96}$$

where $H_u(x)$ is the mean curvature of Σ_u at $F_u(x)$.

The rest of this subsection is devoted to the proof of this lemma. Let $(\mathcal{U}, \{x^i\}_{i=1}^{n-1})$ be local coordinates on Σ^{n-1}. The components of the metric are defined by

$$g_{ij} := (g_u)_{ij} := g\left(\frac{\partial F_u}{\partial x^i}, \frac{\partial F_u}{\partial x^i}\right), \tag{12.97}$$

where we usually suppress the dependence on u for the notation g_{ij}. The components of the second fundamental form are

$$\mathrm{II}_{ij} := \mathrm{II}_u\left(\frac{\partial F_u}{\partial x^i}, \frac{\partial F_u}{\partial x^i}\right), \tag{12.98}$$

where $\mathrm{II}_u(V,W) = \langle \nabla_V N, W\rangle$ denotes the second fundamental form of Σ_u.

Let $V_0 \in T\Sigma$ and let $\gamma : (-\varepsilon, \varepsilon) \to \Sigma^{n-1}$ satisfy $\gamma'(0) = V_0$. We have the associated map $\Gamma : (-\varepsilon, \varepsilon) \times I \to M^n$ defined by

$$\Gamma(v,u) := F_u\big(\gamma(v)\big). \tag{12.99}$$

Define $V := \frac{\partial}{\partial v}\Gamma = dF_u(\gamma'(v))$ and $U := \frac{\partial}{\partial u}\Gamma$, which are vector fields along the map Γ. Observe that

$$V(0,u) = dF_u(\gamma'(0)) = dF_u(V_0) \tag{12.100}$$

and

$$U(v,u) = \frac{\partial}{\partial u}F_u(\gamma(v)) = f(\gamma(v))N(\gamma(v),u), \tag{12.101}$$

where the last equality is by (12.93).

Let ∇^Γ denote the covariant derivative along the map Γ. We compute that

$$\nabla^\Gamma_U\big(dF_u(V_0)\big) = \nabla^\Gamma_U V = \nabla^\Gamma_V U = \nabla^\Gamma_V(fN) = V_0(f)N + f\nabla^\Gamma_{dF_u(V_0)}N.$$

Using this, we compute that (the ∇^Γ for W_0 is defined similarly to ∇^Γ for V_0)

$$\begin{aligned}
\frac{\partial}{\partial u}g_u(V_0,W_0) &= \frac{\partial}{\partial u}g\big(dF_u(V_0),dF_u(W_0)\big)\\
&= g\big(\nabla^\Gamma_U\big(dF_u(V_0)\big),dF_u(W_0)\big) + g\big(dF_u(V_0),\nabla^\Gamma_U\big(dF_u(W_0)\big)\big)\\
&= g\big(V_0(f)N + f\nabla^\Gamma_{dF_u(V_0)}N, dF_u(W_0)\big)\\
&\quad + g\big(dF_u(V_0), W_0(f)N + f\nabla^\Gamma_{dF_u(W_0)}N\big)\\
&= f\,g(\nabla^\Gamma_{dF_u(V_0)}N, dF_u(W_0)) + f\,g(dF_u(V_0),\nabla^\Gamma_{dF_u(W_0)}N)\\
&= 2f\,\mathrm{II}_u(V_0,W_0).
\end{aligned}\tag{12.102}$$

In local coordinates, (12.102) says that

$$\frac{\partial}{\partial u}g_{ij} = 2f\,\mathrm{II}_{ij}\,. \tag{12.103}$$

Using (12.103), we compute that

$$\frac{\partial}{\partial u}d\mu_{g_u} = \frac{1}{2}g^{ij}\left(\frac{\partial}{\partial u}g_{ij}\right)d\mu_{g_u} = fH\,d\mu_{g_u}. \tag{12.104}$$

Lemma 12.23 follows from integrating this formula over Σ^{n-1}.

12.7.2. The second variation of the hypersurface volume formula.

We compute the second variation by taking the variation of the first variation. For this, we first compute the evolution of the unit normal vector field.

Lemma 12.24. *If a 1-parameter family F_u of parametrized hypersurfaces in a Riemannian manifold satisfies $\frac{\partial F}{\partial u} = fN$, then*

$$\nabla^\Gamma_U N_u = -dF_u(\nabla^{g_u}f), \tag{12.105}$$

where ∇^Γ is the covariant derivative along the map Γ defined by (12.99).

Proof. Let $V \in T\Sigma$. We compute that

$$\begin{aligned}\bar{g}\big(\nabla^{\Gamma}_{U} N_u, dF_u(V)\big) &= -\bar{g}\big(N_u, \nabla^{\Gamma}_{U}(dF_u(V))\big)\\ &= -\bar{g}\big(N_u, \nabla^{\Gamma}_{V}(fN_u)\big)\\ &= -\bar{g}\big(\nabla^{\bar{g}}(f\circ F_u), dF_u(V)\big)\\ &= -\bar{g}\big(dF_u(\nabla^{g_u} f), dF_u(V)\big).\end{aligned}$$

The lemma follows since this formula holds for all $V \in T\Sigma$. □

By examining (12.96), we see that we need to compute $\frac{\partial}{\partial u} H_u$.

Lemma 12.25. *Under the variation of the parametrization $\frac{\partial F_u}{\partial u} = fN_u$, we have that*

$$\frac{\partial}{\partial u} H_u = -\Delta_u f - |\operatorname{II}_u|^2 f - \operatorname{Ric}_{\bar{g}}(N,N) f, \tag{12.106}$$

where Δ_u denotes the Laplacian with respect to g_u.

Proof. We compute that the evolution of the second fundamental form of Σ_u is given by

$$\begin{aligned}\frac{\partial}{\partial u}\big(\operatorname{II}_u(V,W)\big) &= \frac{\partial}{\partial u}\bar{g}\big(\nabla^{\Gamma}_{V} N_u, dF_u(W)\big)\\ &= \bar{g}\big(\nabla^{\Gamma}_{U}(\nabla^{\Gamma}_{V} N_u), dF_u(W)\big) + \bar{g}\big(\nabla^{\Gamma}_{V} N_u, \nabla^{\Gamma}_{U}(dF_u(W))\big)\\ &= \operatorname{Rm}_{\bar{g}}\left(fN_u, dF_u(V), N_u, dF_u(W)\right)\\ &\quad + \bar{g}\big(\nabla^{\Gamma}_{V}(\nabla^{\Gamma}_{U} N_u), dF_u(W)\big) + \bar{g}\big(\nabla^{\Gamma}_{V} N_u, \nabla^{\Gamma}_{W}(fN_u)\big)\\ &= f\operatorname{Rm}_{\bar{g}}\left(N_u, dF_u(V), N_u, dF_u(W)\right)\\ &\quad - \nabla^{g_u}\nabla^{g_u} f(V,W) + f\operatorname{II}^2_u(V,W)\end{aligned} \tag{12.107}$$

for $V, W \in T\Sigma$, where for the last equality we used Lemma 12.24. Thus the evolution of the mean curvature of Σ_u is given by

$$\begin{aligned}\frac{\partial}{\partial u} H_u &= \frac{\partial}{\partial u}\operatorname{trace}_{g_u}(\operatorname{II}_u)\\ &= \operatorname{trace}_{g_u}(\partial_u \operatorname{II}_u) - \partial_u g_u \cdot \operatorname{II}_u\\ &= -\operatorname{Ric}_{\bar{g}}(N_u, N_u) - \Delta_{g_u} f + \operatorname{trace}_{g_u}(\operatorname{II}^2_u) - 2f\operatorname{II}_u\cdot\operatorname{II}_u .\end{aligned}$$

The lemma follows. □

Proposition 12.26. *Under the variation $\frac{\partial F_u}{\partial u} = fN_u$, we have that the second variation of the hypersurface volume is given by*

$$\frac{d^2}{du^2}\operatorname{Vol}_{n-1}(\Sigma_u) = \int_{\Sigma^{n-1}} \left(|\nabla f|^2_{g_u} + (H_u^2 - |\operatorname{II}_u|^2_{g_u} - \operatorname{Ric}_{\bar{g}}(N,N))f^2\right) d\mu_{g_u}. \tag{12.108}$$

Proof. Using Lemma 12.25, we compute that

$$\begin{aligned}\frac{d^2}{du^2}\operatorname{Vol}_{n-1}(\Sigma_u) &= \frac{d}{du}\int_{\Sigma^{n-1}} fH_u d\mu_{g_u} \\ &= \int_{\Sigma^{n-1}} f\frac{\partial H_u}{\partial u}d\mu_{g_u} + \int_{\Sigma^{n-1}} fH_u\frac{\partial}{\partial u}d\mu_{g_u} \\ &= \int_{\Sigma^{n-1}} f\big(-\Delta_u f - |\mathrm{II}_u|^2_{g_u} f - \operatorname{Ric}_{\overline{g}}(N,N)f + H_u^2 f\big)d\mu_{g_u}\end{aligned}$$

and the proposition follows from integrating by parts on the first term of the last line. □

As a special case, we have the 'parallel' variation given by $f \equiv 1$ (i.e., $\frac{\partial F_u}{\partial u} = N_u$), in which case we obtain

$$\frac{d^2}{du^2}\operatorname{Vol}_{n-1}(\Sigma_u) = \int_{\Sigma^{n-1}} \big(H_u^2 - |\mathrm{II}_u|^2_{g_u} - \operatorname{Ric}_{\overline{g}}(N,N)\big)d\mu_{g_u}. \tag{12.109}$$

Now recall from the double trace of the Gauss equation (8.133) that

$$R_{g_u} = R_{\overline{g}} - 2\operatorname{Ric}_{\overline{g}}(N,N) + H_u^2 - |\mathrm{II}_u|^2_{g_u}. \tag{12.110}$$

By applying (12.110) to Proposition 12.26, we obtain the following.

Proposition 12.27. *Under the parallel variation $\frac{\partial F_u}{\partial u} = N_u$, the second variation of the hypersurface volume may be expressed as*

$$\frac{d^2}{du^2}\operatorname{Vol}_{n-1}(\Sigma_u) = \frac{1}{2}\int_{\Sigma^{n-1}} \big(H_u^2 - |\mathrm{II}_u|^2_{g_u} + R_g - R_{\overline{g}}\big)d\mu_{g_u}. \tag{12.111}$$

Definition 12.28. We say that a closed minimal hypersurface Σ^{n-1} in a Riemannian manifold $(M^n, \overline{g})$ is **stable** if the second variation of the hypersurface volume of Σ^{n-1} is positive semidefinite. That is, under any variation $\frac{\partial F_u}{\partial u} = fN_u$, we have

$$\left.\frac{\partial^2}{\partial u^2}\right|_{u=0} \operatorname{Vol}_{n-1}(\Sigma_u) \geq 0.$$

In dimension 3 we can use the second variation formula to understand stable minimal surfaces in 3-manifolds with positive scalar curvature.

Theorem 12.29. *If Σ^2 is an orientable closed stable minimal surface in an orientable 3-manifold $(M^3, \overline{g})$ with positive scalar curvature, then Σ^2 is diffeomorphic to a 2-sphere.*

Proof. Since Σ^2 is orientable, we may let Σ_u, $u \in (-\varepsilon, \varepsilon)$, be a variation of Σ^2 with $\Sigma_0 = \Sigma^2$ and $f \equiv 1$. Let g, H, and II denote the induced metric, mean curvature, and second fundamental form, respectively, of Σ^2. Denote $\operatorname{Area} := \operatorname{Vol}_2$. Since Σ^2 is a closed stable minimal surface, so that $H \equiv 0$

and the second variation is nonnegative, and by Proposition 12.27, we have that

$$(12.112)\qquad 0 \leq \frac{d^2}{du^2}\bigg|_{u=0} \mathrm{Area}(\Sigma_u) = \frac{1}{2}\int_{\Sigma^2} \big(-|\mathrm{II}|_g^2 + R_g - R_{\overline{g}}\big)d\mu_g.$$

Since R_g is equal to twice the Gauss curvature of g, by the Gauss–Bonnet formula, we obtain

$$(12.113)\qquad 4\pi\chi(\Sigma^2) \geq \int_{\Sigma^2} \big(|\mathrm{II}|_g^2 + R_{\overline{g}}\big)d\mu_g.$$

Since $R_{\overline{g}} > 0$, this implies $\chi(\Sigma^2) > 0$. Since Σ^2 is orientable, we conclude that Σ^2 is diffeomorphic to S^2. □

Remark 12.30. From the proof of the theorem, we see that if Σ^{n-1}, with induced metric g, is an orientable closed stable minimal hypersurface in a Riemannian manifold $(M^n, \overline{g})$ with positive scalar curvature, then $\int_{\Sigma^{n-1}} R_g d\mu_g > 0$.

Theorem 12.29 is due to Schoen and Yau [**SY79b**]. They used this result to classify closed oriented 3-manifolds with positive scalar curvature up to homotopy; more precisely, they proved that any such Riemannian manifold is diffeomorphic to the connected sums of $S^2 \times S^1$'s and quotients of homotopy 3-spheres. The complete classification is now known due to the Perelman–Hamilton proof of Thurston's geometrization conjecture. Further seminal works on positive scalar curvature are by Schoen and Yau [**SY79a**] and Gromov and Lawson [**GL80b, GL80a**].

One of the ingredients of Schoen and Yau's proof is the following existence result.

Theorem 12.31. *Let $(M^3, \overline{g})$ be a closed orientable Riemannian 3-manifold. Suppose that Σ^2 is a closed orientable Riemannian surface and $f_0 : \Sigma^2 \to M^3$ is a continuous map such that the induced map on fundamental groups $(f_0)_* : \pi_1(\Sigma^2) \to \pi_1(M^3)$ is an injection. Then there exists a minimal immersion $f : \Sigma^2 \to M^3$ such that $f_* = (f_0)_*$ and the area of $f(\Sigma^2)$ is equal to the minimum of the area of all immersions f' with $f'_* = (f_0)_*$.*[2]

The idea of the proof of Theorem 12.31 is summarized in [**SY79b**, p. 127], where Schoen and Yau write:

> "We fix a conformal structure on Σ^2 and minimize the energy over all maps which have the same action as f_0 on $\pi_1(\Sigma^2)$. ... The extremal map is harmonic with respect to the above conformal structure. We then vary the conformal structure

[2]The regularity of the minimal surface, i.e., that there are no branch points, is due to Osserman [**Oss70**] and Gulliver [**Gul73**].

> over Teichmüller space. Using the injectivity of $(f_0)_*$ and a result of L. Keen and B. Randol, we prove the energy of the harmonic map achieves its minimum somewhere. This gives rise to the branched minimal immersion f mentioned above."

By combining Theorems 12.29 and 12.31, Schoen and Yau proved the following.

Theorem 12.32. *Let $(M^3, \overline{g})$ be a closed orientable Riemannian 3-manifold with positive scalar curvature. Then $\pi_1(M^3)$ does not contain a finitely generated noncyclic abelian subgroup nor does it contain a subgroup isomorphic to the fundamental group of a surface of genus at least 2.*

Consequently, M^3 is diffeomorphic to the connected sums of $S^2 \times S^1$'s and quotients of homotopy 3-spheres.

Manifolds with positive scalar curvature are significant in differential geometry due to their broad connections and implications. See Gromov [**Gro23**].

Given any closed Riemannian manifold (M^n, g), there exists a metric $\widetilde{g}$ conformal to g that has constant scalar curvature. This result, which provides a solution to the **Yamabe problem**, was conclusively proved by Schoen [**Sch84a**], building upon the foundational works of Yamabe, Trudinger, and Aubin.

12.8. Introduction to harmonic maps

Harmonic maps between Riemannian manifolds are a natural generalization of harmonic functions as well as geodesics. Maps from surfaces are related to minimal surfaces. As canonical geometric objects, harmonic maps are useful for understanding the geometry and topology of Riemannian manifolds. In this section we give a brief introduction to harmonic maps.

12.8.1. The map-Hessian and the map-Laplacian. Let $f : (M^n, g) \to (N^m, h)$ be a smooth map between Riemannian manifolds. Let g^* denote the dual metric on the fibers of the cotangent bundle T^*M defined by (8.51). Let ∇^{g^*} denote the induced connection on T^*M derived from the Levi-Civita connection ∇^g of g, acting on 1-forms by the usual covariant differentiation. The differential of f, $df : TM \to TN$, may be considered as a section of the vector bundle $E := T^*M \otimes f^*(TN)$ over M^n, where the vector bundle $\pi : f^*(TN) \to M^n$ is the pull-back bundle of TN by f.

Let $\nabla^{g,h} := \nabla^{g^*} \otimes \nabla^{f^*h}$ be the unique linear connection on E satisfying

$$\nabla^{g,h}(\alpha \otimes W) = (\nabla^{g^*}\alpha) \otimes W + \alpha \otimes \nabla^{f^*h} W \tag{12.114}$$

for $\alpha \in C^\infty(T^*M)$, $W \in C^\infty(f^*(TN))$. This connection is compatible with the bundle metric $g^* \otimes f^*h$ on E; that is,

$$V\langle\sigma,\tau\rangle_{g,h} = \langle\nabla_V^{g,h}\sigma,\tau\rangle_{g,h} + \langle\sigma,\nabla_V^{g,h}\tau\rangle_{g,h}$$

for any sections σ, τ in $C^\infty(E)$ and any $V \in TM$. Here, $\langle\cdot,\cdot\rangle_{g,h}$ denotes the inner product with respect to $g^* \otimes f^*h$. In particular, $\langle\sigma,\tau\rangle_{g,h} = (f^*h)(\sigma(e_i),\tau(e_i))$, where $\{e_i\}_{i=1}^n$ is an orthonormal frame with respect to g.

The **map-Hessian** of f is defined to be $\nabla^{g,h}df$, which is a section of the bundle $T^*M \otimes T^*M \otimes f^*TN$. This section is symmetric in the first two components; that is,

$$\nabla^{g,h}df(V,W) = \nabla^{g,h}df(W,V) \in f^*(T_{f(x)}N)$$

for $V, W \in T_xM$. The **map-Laplacian** of f is defined to be the trace of the map-Hessian with respect to g of $\nabla^{g,h}df$; that is,

$$\Delta_{g,h}f := \operatorname{trace}_g(\nabla^{g,h}df) = \sum_{i=1}^n \nabla^{g,h}df(e_i.e_i) \in C^\infty(f^*TN). \tag{12.115}$$

If N^m is the Euclidean line $\mathbb{R}$, then the map-Laplacian is equal to the usual Laplacian acting on functions.

12.8.2. Definition of harmonic map. We say that f is a **harmonic map** if $\Delta_{g,h}f = \vec{0}$ is the zero-section of f^*TN. It is natural to think of harmonic maps as canonical maps between Riemannian manifolds.

The **map energy** of a map f is defined by

$$E_{g,h}(f) := \int_{M^n} |df|_{g,h}^2\, d\mu_g = \int_{M^n} \operatorname{trace}_g(f^*h)d\mu_g. \tag{12.116}$$

The function $|df|_{g,h}^2$ is called the **map energy density** of f.

12.8.3. First variation of the map energy. The first variation formula for the map energy is given by the following.

Proposition 12.33. *Let $\{f_s\}_{s\in(-\varepsilon,\varepsilon)}$ be a smooth 1-parameter family of maps from (M^n,g) to (N^m,h). Denote $f := f_0$ and $V := \left.\frac{\partial f_s}{\partial s}\right|_{s=0} \in f^*TN$. Then the first variation of the map energy density is given by*

$$\left.\frac{\partial}{\partial s}\right|_{s=0} |df_s|_{g,h}^2 = 2\langle\nabla^{f^*h}V, df\rangle_{g,h}. \tag{12.117}$$

Furthermore, if V has compact support in M^n, then the first variation of the map energy is given by

$$\left.\frac{d}{ds}\right|_{s=0} E_{g,h}(f_s) = -2\int_{M^n} \langle V, \Delta_{g,h} f\rangle_{f^*h}\, d\mu_g. \tag{12.118}$$

Therefore, a map f is a critical point of the map energy if and only if f is a harmonic map.

Proof. One computes that (see Exercise 12.12)

$$\left.\frac{\partial}{\partial s}\right|_{s=0} (f_s^* h) = 2\big\langle \nabla^{f^*h} V, df\big\rangle_{f^*h}, \tag{12.119}$$

where the inner product $\langle\cdot,\cdot\rangle_{f^*h}$ contracts the last components of sections of $TM \otimes f^*TN$. By tracing (12.119), we obtain

$$\begin{aligned}\left.\frac{\partial}{\partial s}\right|_{s=0} |df_s|^2_{g,h} &= \operatorname{trace}_g\left(\left.\frac{\partial}{\partial s}\right|_{s=0} (f_s^* h)\right) \\ &= 2\big\langle \nabla^{f^*h} V, df\big\rangle_{g,h},\end{aligned} \tag{12.120}$$

which is (12.117).

Next, using (12.120) we calculate that

$$\begin{aligned}\left.\frac{d}{ds}\right|_{s=0} E_{g,h}(f_s) &= \int_{M^n} \left(\left.\frac{\partial}{\partial s}\right|_{s=0} |df_s|^2_{g,h}\right) d\mu_g \\ &= 2\int_{M^n} \big\langle \nabla^{f^*h} V, df\big\rangle_{g,h} d\mu_g \\ &= -2\int_{M^n} \big\langle V, \operatorname{trace}_g\big(\nabla^{g,h} df\big)\big\rangle_{f^*h} d\mu_g \\ &= -2\int_{M^n} \big\langle V, \Delta_{g,h} f\big\rangle_{f^*h} d\mu_g.\end{aligned}$$

To arrive at the third equality above, we used the divergence theorem and the formula

$$\operatorname{div}_g(\langle V, df\rangle_{f^*h}) = \big\langle \nabla^{f^*h} V, df\big\rangle_{g,h} + \big\langle V, \operatorname{trace}_g\big(\nabla^{g,h} df\big)\big\rangle_{f^*h}, \tag{12.121}$$

where $\langle V, df\rangle_{f^*h}$ contracts V with the second component of df using the metric f^*h on f^*TN. □

12.8.4. Second variation of the map energy. To compute the second variation of the map energy, we note the following.

Lemma 12.34. *Denote $V_s := \frac{\partial f_s}{\partial s}$ and $V := V_0$. Then*

$$\begin{aligned}\left.\frac{\partial}{\partial s}\right|_{s=0} \big\langle V_s, \Delta_{g,h} f_s\big\rangle_{f_s^*h} &= \big\langle V, \Delta_{g,h} V\big\rangle_{f^*h} + \operatorname{trace}_g^{2,3}\big(\operatorname{Rm}_h(V, df, df, V)\big) \\ &\quad + \big\langle \nabla_V^{f^*h} V, \Delta_{g,h} f\big\rangle_{f^*h} + \big\langle \left.\tfrac{\partial V_s}{\partial s}\right|_{s=0}, \Delta_{g,h} f\big\rangle_{f^*h},\end{aligned}$$

where

$$\operatorname{trace}_g^{2,3}\left(\mathrm{Rm}_h(V, df, df, V)\right)(x) := \sum_{i=1}^{n} \mathrm{Rm}_h\left(V, df(e_i), df(e_i), V\right)(f(x))$$

and $\{e_i\}_{i=1}^n$ *is an orthonormal frame with respect to* g *at* x.

By the first variation formula and integrating the formula in the lemma above, we obtain the second variation formula for the map energy (see Exercise 12.15 for the proof).

Proposition 12.35. *If the section* V *of* f^*TN *has compact support in* M^n, *then*

$$\begin{aligned}\left.\frac{d^2}{ds^2}\right|_{s=0} E_{g,h}(f_s) =& 2\int_{M^n}\left(\left|\nabla^{f^*h}V\right|_{g,h}^2 - \operatorname{trace}_g^{2,3}\left(\mathrm{Rm}_h(V, df, df, V)\right)\right) d\mu_g \\ & -2\int_{M^n}\left(\left\langle \nabla_V^{f^*h}V, \Delta_{g,h}f\right\rangle_{f^*h} + \left\langle \left.\tfrac{\partial V_s}{\partial s}\right|_{s=0}, \Delta_{g,h}f\right\rangle_{f^*h}\right) d\mu_g.\end{aligned}$$

In particular, if f *is a harmonic map, then*

(12.122)
$$\begin{aligned}\left.\frac{d^2}{ds^2}\right|_{s=0} E_{g,h}(f_s) &= 2\int_{M^n}\left(\left|\nabla^{f^*h}V\right|_{g,h}^2 - \operatorname{trace}_g^{2,3}\left(\mathrm{Rm}_h(V, df, df, V)\right)\right) d\mu_g \\ &=: I(V, V).\end{aligned}$$

The quadratic form I on the space of sections of f^*TN appearing on the right-hand side of (12.122) is called the **index form**.

We say that a harmonic map f is **weakly stable** if $I(V,V) \geq 0$ for all $V \in f^*TN$ with compact support in M^n. If (N^m, h) has nonpositive sectional curvature, then

$$I(V,V) \geq 2\int_{M^n}\left|\nabla^{f^*h}V\right|_{g,h}^2 d\mu_g \geq 0.$$

Thus, any harmonic map with its target being a Riemannian manifold with nonpositive sectional curvature is weakly stable.

12.8.5. The harmonic map heat flow. Eells and Sampson [**ES64**] proved the following existence theorem for harmonic maps into targets with nonpositive sectional curvatures.

Theorem 12.36. *Let* (M^n, g) *and* (N^m, h) *be closed Riemannian manifolds, where* (N^m, h) *has nonpositive sectional curvature. If* $f_0 : M^n \to N^m$ *is a smooth map, then there exists a smooth harmonic map* $f : M^n \to N^m$ *which is smoothly homotopic to* f_0.

The original proof of this result by Eells and Sampson, which is beyond the scope of this book, employed the **harmonic map heat flow**. Given closed Riemannian manifolds (M^n, g) and (N^m, h), we define a 1-parameter family of maps $f(t) : M^n \to N^m$ as a solution to the harmonic map heat flow if

$$\frac{\partial}{\partial t} f(t) = \Delta_{g,h} f(t). \tag{12.123}$$

As in the hypothesis of the theorem above, suppose that (N^m, h) has nonpositive sectional curvature and that $f_0 : M^n \to N^m$ is a smooth map. Eells and Sampson proved that there exists a smooth solution $f(t) : M^n \to N^m$ to the harmonic map heat flow with $f(0) = f_0$ and that it defined for all time $t \in [0, \infty)$. Furthermore, as $t \to \infty$, the map $f(t)$ converges in C^∞ to a smooth harmonic map f.

Eells and Sampson's result is seminal in the application of nonlinear heat-type equations in differential geometry. It is also one of the motivations for Hamilton's discovery of the Ricci flow (discussed for surfaces in Chapter 14).

For a variational (often referred to as 'elliptic') approach to the proof of the Eells–Sampson theorem, one can consult the work by Schoen [**Sch84b**]. See Jost [**Jos17**, Chapter 8] for a detailed exposition.

12.8.6. Sacks and Uhlenbeck.

In this subsection, we state the fundamental existence theorems of Sacks and Uhlenbeck [**SU81**] for harmonic maps of surfaces.

Theorem 12.37. *Let (N^m, h) be a closed Riemannian manifold. If N^m has nontrivial second homotopy group, i.e., $\pi_2(N^m) \neq 0$, then N^m contains at least one nonconstant immersed minimal 2-sphere.*

Theorem 12.38. *If (N^m, h) is a simply connected closed Riemannian manifold, then it contains at least one nonconstant immersed minimal 2-sphere.*

The proofs of these results are beyond the scope of this book. Describing the idea of Sacks and Uhlenbeck's proofs, Treibergs [**Tre**] wrote:

> "This is an example of a geometric variational problem in which a special minimizing sequence is chosen (by solving a related variational problem) that exhibits concentration, yet with sufficient regularity properties to show convergence to a smooth solution away from finitely many singular points. By conformally blowing up near the singular points, another minimizing sequence is constructed that converges to another solution, a 'bubble'."

Let $f : (M^2, g) \to (N^m, h)$ be a smooth map with domain a Riemannian surface. Given $x \in M^2$, let $\{e_1, e_2\}$ be a positively oriented orthonormal frame at x with respect to g. Then the area form of g at x is given by

$$d\mu_g(x) = e_1^* \wedge e_2^*.$$

The map energy density of f at x is given by

$$|df|_{g,h}^2 = |df(e_1)|_h^2 + |df(e_2)|_h^2.$$

On the other hand, consider the contravariant 2-form

$$df(e_1) \wedge df(e_2) \in T_{f(x)}N \otimes T_{f(x)}N.$$

We have that $|df(e_1) \wedge df(e_2)|_h$ is equal to the area of the parallelogram in $T_{f(x)}N$ spanned by $df(e_1)$ and $df(e_2)$. In particular, by the change of variables formula we can express the area of the immersed surface $f(M^2)$ in N^m as

$$\operatorname{Area}\left(f(M^2)\right) = \int_{M^2} |df(e_1) \wedge df(e_2)|_h \, d\mu_g,$$

where $\{e_1, e_2\}$ is a global (but not necessarily continuous) orthonormal frame field on M^2.

Now, by the Cauchy–Schwarz inequality,

$$2|df(e_1) \wedge df(e_2)|_h \leq |df(e_1)|_h^2 + |df(e_2)|_h^2 = |df|_{g,h}^2.$$

Indeed, if $\theta \in [0, \pi]$ denotes the angle between $df(e_1)$ and $df(e_2)$, then

$$\begin{aligned} 2|df(e_1) \wedge df(e_2)|_h &= 2|df(e_1)|_h |df(e_2)|_h \sin(\theta) \\ &\leq 2|df(e_1)|_h |df(e_2)|_h \\ &\leq |df(e_1)|_h^2 + |df(e_2)|_h^2, \end{aligned}$$

with equality if and only if $\theta = \frac{\pi}{2}$ and $|df(e_1)|_h = |df(e_2)|_h$. Therefore, we have that

$$E_{g,h}(f) \geq 2 \operatorname{Area}\left(f(M^2)\right), \tag{12.124}$$

with equality if and only if $f : (M^2, g) \to (N^m, h)$ is a weakly conformal map.

12.9. Notes and commentary

For some books on minimal surfaces, see Lawson [**Law80**], Giusti [**Giu84**], Colding and Minicozzi [**CM11**], and Dierkes, Hildebrandt, and Sauvigny [**DHS10b**].

The Simons identity is presented in [**Sim68**].

An important application of minimal surfaces is Schoen and Yau's proof of the positive mass theorem [**SY79c, SY81**]. Witten [**Wit81**] provided an alternative and distinct proof using an analysis of harmonic spinors. Parker and Taubes [**PT82**] later offered a rigorous treatment of Witten's approach.

In a series of papers, culminating in [**DLSS20**], De Lellis, Spadaro, and Spolaor proved that 2-dimensional area minimizing integral currents in Riemannian manifolds must have isolated singularities. Their works complete a program initiated by Almgren [**Alm00**] and Chang [**Cha88**].

The multiplicity one conjecture for min-max hypersurfaces in Riemannian manifolds with bumpy metrics of dimension at most 7 was solved by Zhou [**Zho20**]; see the references therein, including works by Marques and Neves, for related works.

Extending the notion of minimal surface are constant mean curvature (CMC) surfaces, which are immersed surfaces in $\mathbb{R}^3$ with H equal to a nonzero constant, which we may normalize to equal 1 for example. H. Hopf [**Hop83**] proved that if $F : S^2 \to \mathbb{R}^3$ is an immersion of a 2-sphere with constant mean curvature, then F is an embedding and $F(S^2)$ is a round 2-sphere. He then conjectured that one can generalize this result to include any closed orientable surface as the domain. Wente [**Wen86**] made the breakthrough of proving the existence of immersed CMC tori in $\mathbb{R}^3$. This provided a counterexample to H. Hopf's conjecture. CMC surfaces of all genera were proved to exist by Kapouleas [**Kap91**]. For a survey of CMC surfaces, see [**BKK22**].

For a survey of the regularity theory of maximal surfaces, which have applications to general relativity, see Bartnik [**Bar87**].

For applications of harmonic maps to rigidity and superrigidity, see Siu [**Siu80**], Corlette [**Cor92**], and Gromov and Schoen [**GS92**].

Siu and Yau's [**SY80**] resolution of the Frankel conjecture ([**Fra61**]) and Micallef and Moore's [**MM88**] work on positive isotropic curvature both utilize the Sacks–Uhlenbeck theorem. On the other hand, Mori [**Mor79**] provided a stronger result in algebraic geometry using completely different methodologies. More specifically, Siu and Yau classified closed Kähler manifolds with positive bisectional curvature. For the classification of closed Kähler manifolds with nonnegative bisectional curvature, see Mok [**Mok88**].

12.10. Exercises

Exercise 12.1. *Show that if $F : M^n \to \mathbb{R}^N$ is a smooth immersion, then its mean curvature vector is given by $\vec{H} = \Delta_g F$, where $g = F^* g_{\mathbb{E}}$ and $g_{\mathbb{E}}$ is the Euclidean metric on $\mathbb{R}^N$.*
Hint: See (5.69).

Exercise 12.2. *Let* (M^2, g) *be a Riemannian surface and let* $F : M^2 \to \mathbb{R}^N$ *is a smooth immersion. Show that*

$$\Delta_{e^{2u}g} F = e^{-2u} \Delta_g F. \tag{12.125}$$

Exercise 12.3. *Prove Corollary* 12.7.

12.10.1. Mean curvature flow. Let $F_t : M^n \to \mathbb{R}^{n+1}$ be a smooth family of embedded hypersurfaces. Recall, from (4.35) and (4.95), that the components of the first and second fundamental forms of the hypersurfaces $F_t(M^n)$ are given by

$$g_{ij}(t) = \left\langle \frac{\partial F_t}{\partial u^i}, \frac{\partial F_t}{\partial u^j} \right\rangle \quad \text{and} \quad \mathrm{II}_{ij}(t) = -\left\langle \frac{\partial^2 F_t}{\partial u^i \partial u^j}, N \right\rangle, \tag{12.126}$$

respectively.

Exercise 12.4. *Using* (12.126)*, show that if* $\frac{\partial F_t}{\partial t} = \phi_t N_t$*, where* $\phi_t : M^n \to \mathbb{R}$ *is a family of smooth functions, then*

$$\frac{\partial}{\partial t} N_t = -\nabla^{g_t} \phi_t, \tag{12.127}$$

$$\frac{\partial}{\partial t} g_{ij}(t) = 2\phi_t \, \mathrm{II}_{ij}(t), \tag{12.128}$$

and

$$\frac{\partial}{\partial t} \mathrm{II}_{ij}(t) = -\nabla_i \nabla_j \phi_t + \phi_t \, \mathrm{II}^2_{ij}(t), \tag{12.129}$$

where $\mathrm{II}^2_{ij} := \mathrm{II}_{ik} \, g^{k\ell} \, \mathrm{II}_{\ell j}$ *and* $\nabla_i \nabla_j \phi_t := (\nabla^2 \phi_t)(\partial_i, \partial_j)$.

Exercise 12.5. *The* ***mean curvature flow*** *is the case where* $f_t = -H_t$ *is minus the mean curvature. Show that under the mean curvature flow,*

$$\frac{\partial H_t}{\partial t} = \Delta_{g_t} H_t + |\, \mathrm{II}_t \,|^2_{g_t} H_t. \tag{12.130}$$

Exercise 12.6. *Consider a parametrized mean curvature flow* $F_t : M^n \to M_t := F_t(M^n) \subset \mathbb{R}^{n+1}$ *for* $t \in [0, T)$*, where* $\partial_t F_t = -H_t N_t$*. Using the parabolic maximum principle as given in Theorem* 14.7 *below, prove that if the initial hypersurface possesses nonnegative mean curvature, then the hypersurfaces at all subsequent times will also maintain a nonnegative mean curvature.*

Exercise 12.7. *Using* (12.129) *and the (trace of the) Simons identity* (12.71)*, show that under the mean curvature flow,*

$$\frac{\partial}{\partial t} \mathrm{II}_t = \Delta_{g_t} \mathrm{II}_t + |\, \mathrm{II}_t \,|^2_{g_t} \mathrm{II}_t - 2H_t \, \mathrm{II}^2_t \,. \tag{12.131}$$

Exercise 12.8. *Let* $F_t : M^n \to M_t := F_t(M^n) \subset \mathbb{R}^{n+1}$, *for* $t \in [0,T)$, *represent a parametrized mean curvature flow, so that* $\partial_t F_t = -H_t N_t$. *Suppose that* α_t *is a symmetric* 2*-tensor satisfying a heat-type equation of the form*

$$\partial_t \alpha_t = \Delta_{g_t} \alpha_t + \beta_t,$$

where β_t *is a polynomial in* α_t, g_t *and satisfies the* ***null-eigenvector condition***

$$\beta_t(V,V) = 0 \quad \text{whenever } \alpha_t(V,W) = 0 \text{ for all } W. \tag{12.132}$$

By Hamilton [**Ham82**, Theorem 9.1] (*cf. the exposition in* [**CK04**, Theorem 4.6]), *the parabolic tensor maximum principle states that if* $\alpha_0(V,V) \geq 0$ *for all tangent vectors* V, *then* $\alpha_t(V,V) \geq 0$ *for all* V *and for all* $t \in [0,T)$.

Prove that, under the mean curvature flow, if the initial hypersurface has a nonnegative second fundamental form, then the hypersurfaces at all positive times also have a nonnegative second fundamental form.

Exercise 12.9. *Let* $F_t : M^n \to M_t := F_t(M^n) \subset \mathbb{R}^{n+1}$ *for* $t \in [0,T)$ *be a solution to the mean curvature flow. Prove that if the initial hypersurface satisfies* $\mathrm{II}_0 \geq cH_0 g_0$ *for some positive constant* c, *meaning that* $\mathrm{II}_0 - cH_0 g_0$ *is positive semidefinite, then* $\mathrm{II}_t \geq cH_t g_t$ *for all* $t \in [0,T)$.

12.10.2. Stable minimal surfaces.

Exercise 12.10. *Let* Σ^2 *be an orientable closed stable minimal surface in a* 3*-manifold with nonnegative scalar curvature. Show that* Σ^2 *is diffeomorphic to either a* 2*-sphere or a* 2*-torus.*

12.10.3. Harmonic maps.

Exercise 12.11. *Let* $f : (M^n, g) \to (N^m, h)$ *be a smooth map. Let* $\{x^i\}_{i=1}^n$ *and* $\{y^\alpha\}_{\alpha=1}^m$ *be local coordinates on* M^n *and* N^m, *respectively. Let* $f^\gamma := y^\gamma \circ f$ *and let* $\Gamma(h)_{\alpha\beta}^\gamma$ *denote the Christoffel symbols of* h *with respect to* $\{y^\alpha\}_{\alpha=1}^m$. *Show that we may express the map-Laplacian in local coordinates as*

$$\begin{aligned}(\Delta_{g,h} f)^\gamma &= \Delta_g (f^\gamma) + g^{ij} (\Gamma(h)_{\alpha\beta}^\gamma \circ f) \frac{\partial f^\alpha}{\partial x^i} \frac{\partial f^\beta}{\partial x^j} \\ &= g^{ij} \left(\frac{\partial^2 f^\gamma}{\partial x^i \partial x^j} - \Gamma(g)_{ij}^k \frac{\partial f^\gamma}{\partial x^k} + (\Gamma(h)_{\alpha\beta}^\gamma \circ f) \frac{\partial f^\alpha}{\partial x^i} \frac{\partial f^\beta}{\partial x^j} \right).\end{aligned}$$

Exercise 12.12. *Prove* (12.119).

Exercise 12.13. *Show, using local coordinates, that the variation of the map-Laplacian is given by*

$$\left.\frac{\partial}{\partial s}\right|_{s=0} (\Delta_{g,h} f_s)^\gamma = \Delta_g(V^\gamma) + g^{ij}\left(\Gamma\left(h\right)_{\alpha\beta}^\gamma \circ f\right) \frac{\partial V^\alpha}{\partial x^i}\frac{\partial f^\beta}{\partial x^j} + g^{ij}\left(\Gamma\left(h\right)_{\alpha\beta}^\gamma \circ f\right)\frac{\partial f^\alpha}{\partial x^i}\frac{\partial V^\beta}{\partial x^j} + g^{ij}\left(\left(V^\delta \frac{\partial}{\partial y^\delta}\Gamma\left(h\right)_{\alpha\beta}^\gamma\right)\circ f\right)\frac{\partial f^\alpha}{\partial x^i}\frac{\partial f^\beta}{\partial x^j}. \tag{12.133}$$

Exercise 12.14. *Show, using local coordinates, that*

$$\left.\frac{\partial}{\partial s}\right|_{s=0} \langle V, \Delta_{g,h} f_s\rangle_{f_s^* h} = h_{\alpha\gamma} V^\alpha \left(\Delta_{g,h} V\right)^\gamma + \left(\mathrm{Rm}_h\right)_{\delta\alpha\beta\varepsilon} V^\delta g^{ij}\frac{\partial f^\alpha}{\partial x^i}\frac{\partial f^\beta}{\partial x^j} V^\varepsilon + h_{\alpha\gamma} V^\beta \left(\nabla_\beta^{f^* h} V\right)^\alpha \left(\Delta_{g,h} f\right)^\gamma . \tag{12.134}$$

Exercise 12.15. *Prove Proposition* 12.35.

Chapter 13

Geometric Flows of Curves in the Plane

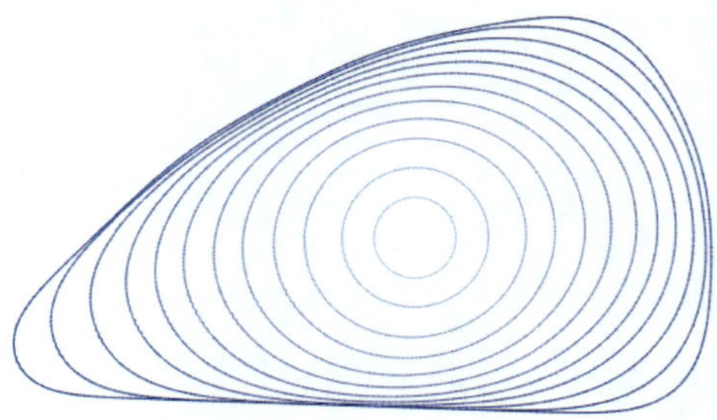

Figure 13.0.1. The curve shortening flow of a convex curve shrinking to a round point. Credit: By David Eppstein – Own work, CC0, `https://commons.wikimedia.org/w/index.php?curid=44763188`

In this chapter, we consider geometric flows of convex and embedded closed curves in the plane in the directions of their normal vector fields with speeds that are functions of their curvatures. The first geometric flow that we consider, the *inverse curve shortening flow*, is equivalent to the 1-dimensional heat equation $\frac{\partial u}{\partial t}(x,t) = \frac{\partial^2 u}{\partial x^2}(x,t)$. In our exposition, we discuss some basic properties of the 1-dimensional heat equation and their implications for the geometric flow. The second geometric flow that we consider is the *curve shortening flow* (CSF); this is a nonlinear parabolic equation and we will show that it shrinks embedded closed curves in the plane to round points.

We will explore, under the curve shortening flow, how geometric quantities evolve. This includes length, area, the isoperimetric ratio, and curvature and its derivatives. We prove monotonicity formulas of Gage, Hamilton, and Huisken. Using these monotonicity formulas, we discuss singularity analysis, which leads to the convergence theorems of Gage–Hamilton and Grayson.

13.1. Isoperimetric inequality in the plane

In the class of embedded closed curves in the plane $\mathbb{R}^2$, round circles are evidently candidates for being canonical. For many geometric flows of embedded closed plane curves, the goal is to establish global-in-time existence and convergence to round circles. This leads to the questions: How can we characterize round circles (besides having constant curvature) and how can we characterize when curves under a geometric flow become rounder as time increases. In this section, we discuss a fundamental way to characterize round circles among embedded closed plane curves.

Observe that for a circle of radius r, its length (circumference) is equal to $\mathrm{L} = 2\pi r$ and the area of the disk that it bounds is equal to $\mathrm{A} = \pi r^2$. Thus, for round circles, we have the length-area relation

$$\mathrm{L}^2 = 4\pi\,\mathrm{A}\,. \tag{13.1}$$

The isoperimetric inequality says the following:

Theorem 13.1. *For any smooth embedded closed curve in $\mathbb{R}^2$, we have*

$$\mathrm{L}^2 \geq 4\pi\,\mathrm{A}, \tag{13.2}$$

where L *is the* ***length*** *of the curve and* A *is the* ***area*** *of the region that it bounds. Equality in* (13.2) *holds if and only if the curve is a round circle.*

We now give a proof, which is due to E. Schmidt [**Sch39**], of the theorem. Let C be a closed curve in the plane bounding a compact region R. We have

$$dx \wedge dy = d(xdy) = -d(ydx). \tag{13.3}$$

Hence, by Stokes's theorem,

$$\mathrm{A} := \mathrm{Area}(R) = \int_D dx \wedge dy = \frac{1}{2}\int_C (xdy - ydx). \tag{13.4}$$

Parametrize C by a unit-speed path $\gamma : [0, \mathrm{L}] \to \mathbb{R}^2$ with the counterclockwise orientation. Let

$$\gamma(s) = (x(s), y(s)).$$

Then we have

$$\mathrm{A} = \mathrm{Area}(R) = \int_0^{\mathrm{L}} \left(x(s)\frac{dy}{ds}(s) - y(s)\frac{dx}{ds}(s) \right) ds. \tag{13.5}$$

By translating the curve C in the x-direction, we may assume without loss of generality that the smallest vertical slab that R is contained in is of the form

$$\Sigma := [-r, r] \times \mathbb{R},$$

where $r > 0$. Note that this uniquely defines r. We also assume that the parametrization γ is such that $\gamma(0)$ has its x-value equal to r. Choose $s_1 \in (0, \mathrm{L})$ such that the x-value of $\gamma(s_1)$ is equal to $-r$. We can now define, for the sake of comparison, a surjective map to the circle of radius r by

$$\overline{\gamma} : [0, \mathrm{L}] \to S^1(r), \tag{13.6}$$

where $x(\overline{\gamma}(s)) = x(\gamma(s))$ for $s \in [0, \mathrm{L}]$, $y(\overline{\gamma}(s)) \geq 0$ for $s \in [0, s_1]$, and $y(\overline{\gamma}(s)) \leq 0$ for $s \in [s_1, \mathrm{L}]$. That is, the 'comparison' parametrization $\overline{\gamma}$ of the round circle has the same x-values as the parametrization γ of the embedded closed curve, and we specify the intervals on which $\overline{\gamma}$ maps onto the upper and lower semicircles.

Define $\overline{y} : [0, \mathrm{L}] \to [-r, r]$ by

$$\overline{\gamma}(s) =: (x(s), \overline{y}(s)). \tag{13.7}$$

Then $\overline{y}(s) = \sqrt{r^2 - x(s)^2}$ for $s \in [0, s_1]$ and $\overline{y}(s) = -\sqrt{r^2 - x(s)^2}$ for $s \in [s_1, \mathrm{L}]$. Let $D(r)$ denote the disk of radius r, which is equal to the region bounded by $S^1(r)$. We have

$$\mathrm{Area}(R) = \int_0^{\mathrm{L}} x(s) \frac{dy}{ds}(s)\, ds \tag{13.8}$$

and

$$-\int_0^{\mathrm{L}} \overline{y}(s) \frac{dx}{ds}(s)\, ds = \mathrm{Area}(D(r)) = \pi r^2. \tag{13.9}$$

The reason why the first equality in the display above is true is that the integral depends only on the winding number of the map $\overline{\gamma}$ onto $S^1(r)$, which is equal to that of the usual counterclockwise parametrization of $S^1(r)$. Thus,

$$\mathrm{Area}(R) + \pi r^2 = \int_0^{\mathrm{L}} \left(x(s) \frac{dy}{ds}(s) - \overline{y}(s) \frac{dx}{ds}(s) \right) ds. \tag{13.10}$$

Claim 13.2.

$$\left(x \frac{dy}{ds} - \overline{y} \frac{dx}{ds} \right)^2 \leq (x^2 + \overline{y}^2) \left(\left(\frac{dx}{ds} \right)^2 + \left(\frac{dy}{ds} \right)^2 \right) = r^2. \tag{13.11}$$

Note that the equality in (13.11) is because the points on $S^1(r)$ have distance r to the origin and because γ is a unit-speed path.

By assuming the claim for now, we obtain

$$\operatorname{Area}(R) + \pi r^2 \leq \int_0^{\mathrm{L}} \left| x(s)\frac{dy}{ds}(s) - \overline{y}(s)\frac{dx}{ds}(s) \right| ds \leq \int_0^{\mathrm{L}} r\, ds = \mathrm{L}\, r. \tag{13.12}$$

Now recall that (a special case of the arithmetic-geometric mean inequality)

$$\sqrt{ab} \leq \frac{a+b}{2}. \tag{13.13}$$

Indeed, if we define $A = \sqrt{a}$ and $B = \sqrt{b}$, then this is equivalent to

$$0 \leq \frac{A^2 + B^2}{2} - AB = \frac{1}{2}(A-B)^2,$$

which is true. By (13.13) and (13.12), we have

$$\sqrt{\pi r^2 \operatorname{Area}(R)} \leq \frac{\operatorname{Area}(R) + \pi r^2}{2} \leq \frac{\mathrm{L}\, r}{2}. \tag{13.14}$$

This implies

$$\mathrm{L}^2 \geq 4\pi \operatorname{Area}(R),$$

which is the isoperimetric inequality.

Now we prove the claim, which follows immediately from the general inequality

$$(ac - bd)^2 \leq (a^2 + b^2)(c^2 + d^2) \tag{13.15}$$

for $a, b, c, d \in \mathbb{R}$. This inequality is true by the following straightforward calculation:

$$\begin{aligned} (a^2 + b^2)(c^2 + d^2) - (ac - bd)^2 &= a^2 d^2 + b^2 c^2 + 2abcd \\ &= (ad + bc)^2 \\ &\geq 0, \end{aligned} \tag{13.16}$$

which proves the claim. This completes the proof of the isoperimetric inequality.

Finally, we characterize the equality case for (13.2). Firstly, we have equality in (13.16) if and only if $a = 0$ or $d = 0$ and $b = 0$ or $c = 0$. Since we are considering smooth curves, via (13.11) this translates to equality holding in the isoperimetric inequality (13.2) if and only if

$$x(s)\frac{dx}{ds}(s) + \overline{y}(s)\frac{dy}{ds}(s) \equiv 0. \tag{13.17}$$

Let $x'(s) := \frac{dx}{ds}(s)$ and $y'(s) := \frac{dy}{ds}(s)$. Substituting this equation into (13.11), where we now assume equality therein, yields

$$\begin{aligned}
r^2 &= \big(x(s)y'(s) - \overline{y}(s)x'(s)\big)^2 \\
&= x^2(s)(y')^2(s) + \overline{y}^2(s)(x')^2(s) - 2x(s)x'(s)\overline{y}(s)y'(s) \\
&= x^2(s)(y')^2(s) + \overline{y}^2(s)(x')^2(s) + 2x^2(s)(x')^2(s) \\
&= x^2(s)\big((x')^2(s) + (y')^2(s)\big) + \big(x^2(s) + \overline{y}^2(s)\big)(x')^2(s) \\
&= x^2(s) + r^2(x')^2(s).
\end{aligned}$$

Thus, $x^2(s) = r^2 - r^2(x')^2(s) = r^2(y')^2(s)$, which implies that

$$x(s) = \pm r y'(s). \tag{13.18}$$

Furthermore, since we have pointwise equality in (13.11), we have that

$$\big|x(s)y'(s) - \overline{y}(s)x'(s)\big| \equiv r. \tag{13.19}$$

This and the positivity of the quantities in (13.10) imply that

$$x(s)y'(s) - \overline{y}(s)x'(s) \equiv r$$

and hence

$$\frac{1}{4\pi}\,\mathrm{L}^2 + \pi r^2 = \mathrm{Area}(R) + \pi r^2 = r\,\mathrm{L}\,. \tag{13.20}$$

Therefore, $\mathrm{L} = 2\pi r$, which in turn implies that $\mathrm{A} = \pi r^2$.

We can now repeat the same argument above, but instead we translate the curve C in the y-direction to put it in the smallest horizontal (as opposed to vertical) slab of the form $\Sigma' := \mathbb{R} \times [-r', r']$, where $r' > 0$. Since we are in the equality case of the isoperimetric inequality, we have $\mathrm{A} = \pi(r')^2$, which implies that $r' = r$. By the same argument as used above to obtain (13.18), we have that $y(s) = \pm r x'(s)$. Finally, we conclude that

$$x^2(s) + y^2(s) = r^2(y')^2(s) + r^2(x')^2(s) = r^2. \tag{13.21}$$

Thus, γ parametrizes a circle of radius r. This completes the proof of Theorem 13.1.

In all dimensions, the isoperimetric inequality says the following.

Theorem 13.3. *For any embedded closed hypersurface M^n bounding a compact domain Ω in $\mathbb{R}^{n+1}$, we have*

$$\mathrm{Vol}_n(M^n) \geq (n+1)\omega_{n+1}^{\frac{1}{n+1}} \,\mathrm{Vol}_{n+1}(\Omega)^{\frac{n}{n+1}}. \tag{13.22}$$

Some articles and books discussing the isoperimetric inequality, including its proof, are Osserman [**Oss78**], Bandle [**Ban80**], Chavel [**Cha01**], and Maggi [**Mag12**].

The Cartan–Hadamard conjecture posits that the isoperimetric inequality holds with the Euclidean constant in any simply connected complete Riemannian manifold with nonpositive sectional curvature. This conjecture has been confirmed for ambient dimensions 2, 3, and 4 by Weil [**Wei26**], Kleiner [**Kle92**], and Croke [**Cro84**], respectively.

For further developments on isoperimetric inequalities, especially for minimal hypersurfaces, see Brendle [**Bre23**].

13.2. The inverse curve shortening flow

In this section, we examine the inverse curve shortening flow of convex embedded plane curves. By demonstrating its equivalence to the linear heat equation on the circle, we explore the behavior of the flow's solutions. Specifically, we find that the curves expand towards infinity while simultaneously becoming rounder.

13.2.1. Definition of the inverse curve shortening flow. Let M^1 be a closed 1-dimensional manifold diffeomorphic to the circle. Suppose that $F_t : M^1 \to \mathbb{R}^2$ parametrizes an embedded strictly convex plane curve

$$M_t^1 := F_t(M^1)$$

for each $t \in [0, T)$, where F_t depends smoothly on t. By definition, $F(\cdot, t) := F_t$ is a parametric solution to the **inverse curve shortening flow** if it satisfies the equation

$$\frac{\partial F}{\partial t}(u,t) = \frac{1}{\kappa(u,t)} N_t(u), \tag{13.23}$$

where $\kappa(u,t) > 0$ denotes the curvature of M_t^1 at $F_t(u)$ and where $N_t(u)$ denotes the unit outward normal of M_t^1 at $F_t(u)$. Note that for fixed $u \in M^1$, $t \mapsto F(u,t)$ is an $\mathbb{R}^2$-valued function of one variable. Thus, for each $u \in M^1$ and $t \in [0, T)$, $\frac{\partial F}{\partial t}(u,t) \in \mathbb{R}^2$. Therefore, (13.23) makes sense as an equation since its right-hand side is also a vector in $\mathbb{R}^2$ for each $u \in M^1$ and $t \in [0, T)$.

13.2.2. Using the Gauss map parametrization. Let $G_t : M^1 \to S^1$ be the Gauss map defined by $G_t(u)$ equaling the unit outward normal to M_t^1 at $F_t(u)$. Here, $S^1 \subset \mathbb{R}^2$ is the unit circle and we identify S^1 with $\mathbb{R}/2\pi\mathbb{Z}$ by the map $(\cos(\theta), \sin(\theta)) \mapsto [\theta]$, where $[\theta] \in \mathbb{R}/2\pi\mathbb{Z}$ denotes the equivalence class of $\theta \in \mathbb{R}$. That is, we have

$$G_t(u) := N_t(u). \tag{13.24}$$

Since we are identifying S^1 with $\mathbb{R}/2\pi\mathbb{Z}$, we have that each G_t, $t \in [0, T)$, is uniquely determined by the property that

$$G_t(u) = \theta, \tag{13.25}$$

where θ is determined up to an additive multiple of 2π by

$$N_t(u) = (\cos(\theta), \sin(\theta)). \tag{13.26}$$

Since each closed curve M_t^1 is embedded and strictly convex, the Gauss map $G_t : M^1 \to S^1$ is a diffeomorphism. Therefore, the inverse of the Gauss map $G_t^{-1} : S^1 \to M^1$ is also a diffeomorphism. We have that

$$G_t^{-1}(\theta) = u, \tag{13.27}$$

where $u \in M^1$ is uniquely determined by (13.26).

Recall from (5.18) that the support function $\sigma_t : M^1 \to \mathbb{R}$ of M_t^1 is defined as the dot product of the position vector and its normal by

$$\sigma_t(u) := \sigma(u,t) := F(u,t) \cdot N_t(u) = F(u,t) \cdot G_t(u). \tag{13.28}$$

Example 13.4. *Let r_0 be a positive real number and define $F_t : S^1 = \mathbb{R}/2\pi\mathbb{Z} \to \mathbb{R}^2$ by*

$$F_t(u) = \big(r_0 e^t \cos(u), r_0 e^t \sin(u)\big).$$

Then

$$\frac{\partial F_t}{\partial t}(u,t) = \big(r_0 e^t \cos(u), r_0 e^t \sin(u)\big) = F_t(u).$$

On the other hand, since F_t parametrizes a circle of radius $r_0 e^t$, we have that

$$\kappa_t(u) = \frac{1}{r_0 e^t},$$

so that $\frac{1}{\kappa_t(u)} = r_0 e^t$. Since the outward unit normal is

$$N_t(u) = (\cos(u), \sin(u)),$$

we conclude that $\frac{\partial F_t}{\partial t}(u) = \frac{1}{\kappa_t(u)} N_t(u)$, so that F_t is a solution to the inverse curve shortening flow. Note that the support function is given by $\sigma_t(u) = r_0 e^t$.

We can use the Gauss parametrization to reparametrize our solution. Namely, define $\tilde{F} : S^1 \times [0,T) \to \mathbb{R}^2$ by

$$\tilde{F}(\theta,t) := F(G_t^{-1}(\theta),t). \tag{13.29}$$

Denote $\tilde{F}_t(\theta) := \tilde{F}(\theta,t)$. Observe that $\tilde{F}_t(S^1) = F_t(M^1) = M_t^1$, so that geometrically $\tilde{F}_t$ parametrizes the same family of closed curves as F_t. Let

$$\begin{aligned}
\tilde{N}(\theta,t) &:= N(G_t^{-1}(\theta),t) = (\cos(\theta), \sin(\theta)),\\
\tilde{\sigma}(\theta,t) &:= \sigma(G_t^{-1}(\theta),t) = \tilde{F}(\theta,t) \cdot \tilde{N}(\theta,t),\\
\tilde{\kappa}(\theta,t) &:= \kappa(G_t^{-1}(\theta),t)
\end{aligned}$$

be the reparametrized outward unit normal, the reparametrized support function, and the reparametrized curvature, respectively. Note that $\tilde{N}(\theta,t)$

is independent of t and that $\tilde{N}(\theta, t)$ is the unique unit vector making the angle θ with the vector $e_1 = (1, 0)$. Recall that

$$\widetilde{\kappa}(\theta, t) = \left(\frac{\partial^2 \tilde{\sigma}}{\partial \theta^2}(\theta, t) + \tilde{\sigma}(\theta, t)\right)^{-1}. \tag{13.30}$$

By differentiating (13.29) using the chain rule, we have that

$$\frac{\partial \tilde{F}}{\partial t}(\theta, t) = \frac{\partial F}{\partial t}(G_t^{-1}(\theta), t) + \frac{\partial F}{\partial u}(G_t^{-1}(\theta), t)\frac{\partial G_t^{-1}}{\partial \theta}(\theta, t). \tag{13.31}$$

Note that $\frac{\partial G_t^{-1}}{\partial \theta}(\theta, t)$ is interpreted as a scalar and that $\frac{\partial F}{\partial u}$ is a vector tangent to M_t^1 at $\widetilde{F}(\theta, t)$. For the purpose of computing the evolution of the support function, we write this equation slightly more abstractly as

$$\begin{aligned}\frac{\partial \tilde{F}}{\partial t}(\theta, t) &= \frac{\partial F}{\partial t}(G_t^{-1}(\theta), t) + V(\theta, t) \\ &= \frac{1}{\widetilde{\kappa}(\theta, t)}\widetilde{N}(\theta, t) + V(\theta, t),\end{aligned} \tag{13.32}$$

where $V(\theta, t)$ is tangential to M_t^1.

13.2.3. Evolution of the support function. We compute that the reparametrized support function satisfies

$$\begin{aligned}\frac{\partial \tilde{\sigma}}{\partial t}(\theta, t) &= \frac{\partial \widetilde{F}}{\partial t}(\theta, t) \cdot \widetilde{N}(\theta, t) + \widetilde{F}(\theta, t) \cdot \frac{\partial \widetilde{N}}{\partial t}(\theta, t) \\ &= \left(\frac{1}{\widetilde{\kappa}}(\theta, t)\widetilde{N}(\theta, t) + V(\theta, t)\right) \cdot \widetilde{N}(\theta, t) \\ &= \frac{1}{\widetilde{\kappa}(\theta, t)},\end{aligned}$$

where we used that $\frac{\partial \widetilde{N}}{\partial t}(\theta, t) = 0$ for the second equality and that $V(\theta, t) \cdot \widetilde{N}(\theta, t) = 0$ for the third equality. Therefore, by (13.30), we have that the reparametrized support function satisfies the linear heat-type equation:

$$\frac{\partial \tilde{\sigma}}{\partial t}(\theta, t) = \frac{\partial^2 \tilde{\sigma}}{\partial \theta^2}(\theta, t) + \tilde{\sigma}(\theta, t). \tag{13.33}$$

Let $\tilde{\sigma}(\theta, t)$, $\theta \in \mathbb{R}$, $t \in [0, T)$, be a solution to (13.33), so that it satisfies the periodicity condition $\tilde{\sigma}(\theta + 2\pi, t) = \tilde{\sigma}(\theta, t)$. Then, for any real number θ_0, both $\tilde{\sigma}(\theta_0 + \theta, t)$ and $\tilde{\sigma}(\theta_0 - \theta, t)$ are also solutions to (13.33) on S^1. In particular, observe that

$$\tilde{\sigma}_{a,b}(\theta, t) = a\cos(\theta) + b\sin(\theta) \tag{13.34}$$

is an explicit static solution for each $a, b \in \mathbb{R}$ and $t \in (-\infty, \infty)$.

13.2.4. Expanding round circles. We have the even simpler solutions

$$\tilde{\sigma}_{r_0}(\theta, t) := \tilde{\sigma}_{r_0}(t) := r_0 \mathrm{e}^t, \tag{13.35}$$

$t \in [0, \infty)$, since they are constant in space (the S^1 factor) and they solve the ordinary differential equation

$$\frac{d\tilde{\sigma}_{r_0}}{dt}(t) = \tilde{\sigma}_{r_0}(t) \tag{13.36}$$

with the initial condition $\tilde{\sigma}_{r_0}(0) = r_0$. The solution $\tilde{\sigma}_{r_0}(t)$ corresponds to an expanding circle centered at the origin with radius $r_0 \mathrm{e}^t$ at time t.

By the linearity of the heat-type equation (13.33), we have that

$$\tilde{\sigma}_{a,b,r_0}(\theta, t) := \tilde{\sigma}_{r_0}(\theta, t) + \tilde{\sigma}_{a,b}(\theta, t) = r_0 \mathrm{e}^t + a\cos(\theta) + b\sin(\theta) \tag{13.37}$$

is a solution to (13.33).

Now, suppose that we change the center of the circle to a point $(a, b) \in \mathbb{R}^2$. Then a circle of radius $r_0 \mathrm{e}^t$ centered at (a, b) is a solution to the inverse curve shortening flow. A parametrized solution $F : M^1 \times [0, T) = S^1 \times [0, T) \to \mathbb{R}^2$ to (13.23) is given by

$$F(u, t) = \big(a + r_0 \mathrm{e}^t \cos(u), b + r_0 \mathrm{e}^t \sin(u)\big). \tag{13.38}$$

Note that

$$N_t(u) = (\cos(u), \sin(u)). \tag{13.39}$$

Therefore, the Gauss map reparametrization is the same as the original parametrization: $\theta = u$ and $\widetilde{N}(\theta, t) = N_t(u)$. In other words,

$$\tilde{F}(\theta, t) = \big(a + r_0 \mathrm{e}^t \cos(\theta), b + r_0 \mathrm{e}^t \sin(\theta)\big). \tag{13.40}$$

Thus,

$$\begin{aligned} \tilde{\sigma}(\theta, t) &= \tilde{F}(\theta, t) \cdot \tilde{N}(\theta, t) \\ &= \big(a + r_0 \mathrm{e}^t \cos(\theta), b + r_0 \mathrm{e}^t \sin(\theta)\big) \cdot (\cos(\theta), \sin(\theta)) \\ &= r_0 \mathrm{e}^t + a\cos(\theta) + b\sin(\theta) \\ &= \tilde{\sigma}_{a,b,r_0}(\theta, t). \end{aligned} \tag{13.41}$$

Observe that

$$\big|\tilde{\sigma}_{a,b,r_0}(\theta, t) - \mathrm{e}^t\big| = \big|a\cos(\theta) + b\sin(\theta)\big| \leq \sqrt{a^2 + b^2} \tag{13.42}$$

and that

$$r_0 \mathrm{e}^t = \frac{1}{2\pi} \int_{S^1} \sigma_{a,b,r_0}(\theta, t)\, d\theta \tag{13.43}$$

is the average of $\sigma_{a,b,r_0}(\cdot, t)$.

13.2.5. Analyzing the general solution using Fourier series. Now we consider the general solution to (13.33). Let $u_0 : S^1 \to \mathbb{R}$ be a smooth function. Let

$$u_0(\theta) = r_0 + \sum_{k=1}^{\infty} \big(a_k \cos(k\theta) + b_k \sin(k\theta)\big) \tag{13.44}$$

be the Fourier series expansion of u_0. Define

$$u(\theta, t) = r_0 \mathrm{e}^t + \sum_{k=1}^{\infty} \big(a_k \cos(k\theta) + b_k \sin(k\theta)\big) \mathrm{e}^{(1-k^2)t}. \tag{13.45}$$

We see that $u(\cdot, 0) = u_0$. Differentiating the Fourier series term by term, we find that $u(\theta, t)$ is a solution to (13.33). Since this solution exists for all forward time $0 \leq t < \infty$, it is a **global solution**. This is also called an **eternal solution**.

Note that there exists a constant C_0 such that

$$\left| u(\theta, t) - r_0 \mathrm{e}^t \right| \leq C_0 \tag{13.46}$$

for all $(\theta, t) \in S^1 \times [0, \infty)$. We also obtain that there exists a constant C_1 such that

$$\left| \frac{\partial u}{\partial \theta}(\theta, t) \right| \leq C_1 \tag{13.47}$$

for all $(\theta, t) \in S^1 \times [0, \infty)$.

Now, let

$$r_{\text{in}}(t) := \min_{\theta \in S^1} |\widetilde{F}(\theta, t)| \quad \text{and} \quad r_{\text{out}}(t) := \max_{\theta \in S^1} |\widetilde{F}(\theta, t)| \tag{13.48}$$

be the **inradius** and **outradius** (a.k.a. **circumradius**) of M_t^1, respectively. Since $r_0 \mathrm{e}^t - C_0 \leq u(\theta, t) \leq r_0 \mathrm{e}^t + C_0$, we have

$$r_0 \mathrm{e}^t - C_0 \leq r_{\text{in}}(t) \leq r_{\text{out}}(t) \leq r_0 \mathrm{e}^t + C_0. \tag{13.49}$$

See Figure 13.2.1.

We in fact have that for each $k \geq 1$ there exists a constant C_k such that

$$\left| \frac{\partial^k u}{\partial \theta^k}(\theta, t) \right| \leq C_k \tag{13.50}$$

for all $(\theta, t) \in S^1 \times [0, \infty)$.

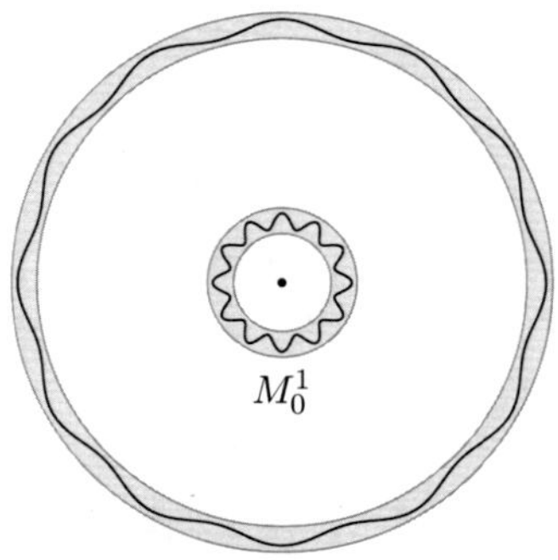

Figure 13.2.1. The curves M_0^1 and M_t^1 are in annuli of width $2C_0$, where C_0 is independent of t.

13.2.6. Convergence of the rescaled solutions to a round circle. It is natural to rescale the solution $u(\theta, t)$ so that it converges to the constant r_0, which corresponds to the circle of radius r_0. Namely, define

$$v(\theta, t) := \mathrm{e}^{-t} u(\theta, t). \tag{13.51}$$

Then, by (13.49), we have that

$$r_0 - C_0 \mathrm{e}^{-t} \leq \mathrm{e}^{-t} r_{\mathrm{in}}(t) \leq \mathrm{e}^{-t} r_{\mathrm{out}}(t) \leq r_0 + C_0 \mathrm{e}^{-t}. \tag{13.52}$$

13.3. Introduction to the curve shortening flow

The rest of this chapter is devoted to the curve shortening flow, which we define below. The aim is to give a proof of the Gage–Hamilton and Grayson result that the curve shortening flow shrinks embedded plane curves to round points. After giving the definition and examples of the curve shortening flow in this section, we calculate how geometric invariants of the curve evolve and we present various important monotonicity formulas in the subsequent sections. In the last part of this chapter, we show how the monotonicity formulas can be used to analyze singularities and prove the Gage–Hamilton and Grayson theorem.

13.3.1. Motivation for the curve shortening flow. John von Neumann [**vN52**], in a discussion of metal grains, wrote:

> "The diffusion-flow across a particular bubble-wall is ... proportional to the pressure difference between the two bubbles adjacent to this wall, multiplied by the length of the wall. The pressure difference of the two adjacent bubbles, at a given point P of a wall, is $2\gamma/R$, where γ is the surface tension of the liquid forming the froth and R is the radius of curvature of the wall P."

We see that $1/R$ is equal to the curvature κ of P.

Mullins [**Mul56**] wrote:

> "... a boundary of mean curvature H and free energy per unit area σ is urged toward its nearest center of curvature with pressure given by $p = H\sigma$. ... This, in turn, causes the boundary to move in the direction favored by the pressure with speed S given by $S = pM = H\sigma M$, ..."

Here, M is the constant of speed per unit pressure. So, up to a constant, the boundary surface is evolving by its mean curvature H in the normal direction $-N$.

The aforementioned two papers were historically at the beginning stage of the study of the **curve shortening flow** and **mean curvature flow**. The mean curvature flow was revived by the work on Brakke [**Bra78**], which was from the point of view of geometric measure theory.

The modern study of the curve shortening flow was instigated by Gage [**Gag83**] and Gage and Hamilton [**GH86**].

13.3.2. The curve shortening flow equation.

13.3.2.1. *Definition of the curve shortening flow.* The **curve shortening flow equation** for a 1-parameter family of parametrized closed curves $F_t : M^1 \to \mathbb{R}^2$, $t \in I$ an interval, is

$$\frac{\partial F}{\partial t}(u,t) = -\kappa(u,t)N(u,t) \tag{13.53}$$

for $u \in M^1$ and $t \in I$, where $F(u,t) := F_t(u)$ and $N_t = N(\cdot,t)$ is the outward unit normal vector field.

13.3.2.2. *The curve shortening flow as a heat-type equation.* Let s_t denote a choice of arc-length parameter at time t (s_t is determined up to an additive time-dependent constant), so that the arc-length element is ds_t, which is a 1-form along the path F_t. Recall from (4.3) that

$$\frac{\partial F_t}{\partial s_t} = T_t,$$

where T_t is the unit tangent vector field of the parametrization $F_t := F(\cdot,t)$. Since we have chosen N_t to be the outward unit normal vector field, by (4.7) we have that

$$\frac{\partial^2 F_t}{\partial s_t^2} = \frac{\partial T_t}{\partial s_t} = -\kappa_t N_t. \tag{13.54}$$

Thus, we may write the curve shortening flow as

$$\frac{\partial F_t}{\partial t} = \frac{\partial^2 F_t}{\partial s_t^2}. \tag{13.55}$$

This looks like the heat equation. The one caveat is that the space derivatives are with respect to the arc-length parameter s_t, which depends on t. In particular, if $(F_1)_t$ and $(F_2)_t$ are solutions to the curve shortening flow, then $(F_1)_t + (F_2)_t$ is not necessarily a solution to the curve shortening flow.

13.3.3. The statement of the Gage–Hamilton and Grayson theorem.

Theorem 13.5 (Gage–Hamilton and Grayson). *For any embedded closed curve in the plane, a unique solution to the curve shortening flow with this curve as the initial curve exists on a time interval $[0, T)$, where T is equal to the area enclosed by the initial curve divided by 2π and the curve shrinks to a round point as $t \to T$.*

We have that a solution exists for a short time; see Gage and Hamilton [**GH86**, §2]. So the main difficulty is to understand what possible singularities can form under the curve shortening flow (CSF) of embedded plane curves.

13.3.4. Examples of curve shortening flows. We look at two of the simplest examples of curve shortening flows, one of which is somewhat obvious and one of which is not so obvious. Both examples are important to the theory.

Example 13.6 (Shrinking circles). *Let r_0 be a positive real number and define $F_t : S^1 \to \mathbb{R}^2$ by*

$$F_t(u) = \left(\sqrt{r_0^2 - 2t}\cos(u), \sqrt{r_0^2 - 2t}\sin(u)\right), \tag{13.56}$$

$t \in [0, r_0^2/2)$. Observe that F_t parametrizes the circle of radius $\sqrt{r_0^2 - 2t}$. Therefore, the curvature is given by

$$\kappa_t(u) = \frac{1}{\sqrt{r_0^2 - 2t}}. \tag{13.57}$$

We calculate that

$$\frac{\partial F_t}{\partial t}(u) = \left(-\frac{1}{\sqrt{r_0^2 - 2t}}\cos(u), -\frac{1}{\sqrt{r_0^2 - 2t}}\sin(u)\right) = -\kappa_t N_t. \tag{13.58}$$

Thus, F_t is a solution to the CSF. Observe that we have that the solution is defined on the ancient time interval $(-\infty, r_0^2/2)$. Since the shape of a shrinking round circle remains unchanged, we call this solution a ***shrinking self-similar solution*** *of the CSF.*

A (parametrized) curve shortening flow $F_t : M^1 \to \mathbb{R}^2$, $t \in I$, is a **graphical curve shortening flow** provided that for each $t \in I$ the curve

$M^1_t = F_t(M^1)$ is the graph of a function $f_t : J \to \mathbb{R}$, where J is an interval. Thus, for each $t \in I$ the graphical parametrization $G_t : J \to \mathbb{R}^2$ of M^1_t defined by

$$G_t(x) = (x, f_t(x)) \tag{13.59}$$

is a reparametrization of the original parametrization $F_t : M^1 \to \mathbb{R}^2$ of M^1_t. Let $F_t =: (x_t, y_t)$. Choose N_t to be the downward pointing unit normal vector field to M^1_t. A choice of the unit tangent vector $T_t(u) := T(u,t)$ is given by

$$T_t(u) = \frac{\left(1, \frac{\partial y}{\partial x}\right)}{\sqrt{1 + \left(\frac{\partial y}{\partial x}\right)^2}} \tag{13.60}$$

as functions of (u,t). Thus,

$$N_t(u) = \frac{\left(\frac{\partial y}{\partial x}, -1\right)}{\sqrt{1 + \left(\frac{\partial y}{\partial x}\right)^2}}. \tag{13.61}$$

Note that $\{N_t, T_t\}$ is a positively oriented orthonormal frame field along M^1_t.

Since G_t is a reparametrization of F_t, we have that the vector fields $\frac{\partial G_t}{\partial t}$ and $\frac{\partial F_t}{\partial t}$ differ by a tangential vector field; that is,

$$\frac{\partial G}{\partial t}(x_t, t) = \frac{\partial F}{\partial t}(u,t) + a(u,t)\frac{\partial F}{\partial x_t}(u,t) \tag{13.62}$$

$$= -\kappa_t(u,t)N_t(u) + a(u,t)\frac{\partial F}{\partial x_t}(u,t) \tag{13.63}$$

for some function $a : M^1 \times I \to \mathbb{R}$. Since G_t is a graphical parametrization, the time-derivative of the x component of G_t is equal to zero. Hence, we have that

$$a = \kappa_t \frac{\frac{\partial y}{\partial x}}{\sqrt{1 + \left(\frac{\partial y}{\partial x}\right)^2}}. \tag{13.64}$$

We have that

$$\frac{\partial s_t}{\partial x} = \sqrt{1 + \left(\frac{\partial y}{\partial x}\right)^2}. \tag{13.65}$$

Using this, we compute that the curvature of a graph is given by

$$\kappa_t(u) = \frac{\partial N_t}{\partial s_t} \cdot T_t = \left(\frac{\partial s_t}{\partial x}\right)^{-1} \frac{\partial N_t}{\partial x} \cdot T_t = \frac{1}{\left(1+\left(\frac{\partial y}{\partial x}\right)^2\right)^{3/2}} \frac{\partial^2 y}{\partial x^2}. \tag{13.66}$$

In terms of the graphical parametrization, this says that

$$\kappa_t(x) = \frac{1}{\left(1+\left(\frac{\partial f_t}{\partial x}(x)\right)^2\right)^{3/2}} \frac{\partial^2 f_t}{\partial x^2}(x). \tag{13.67}$$

By applying this to (13.62), we obtain

$$\begin{aligned} \frac{\partial f_t}{\partial t} &= \kappa_t \left(1+\left(\frac{\partial f_t}{\partial x}\right)^2\right)^{-1/2} + \kappa_t \frac{\frac{\partial f_t}{\partial x}}{\sqrt{1+\left(\frac{\partial f_t}{\partial x}\right)^2}} \frac{\partial f_t}{\partial x} \\ &= \kappa_t \left(1+\left(\frac{\partial f_t}{\partial x}\right)^2\right)^{1/2} = \frac{1}{1+\left(\frac{\partial f_t}{\partial x}\right)^2} \frac{\partial^2 f_t}{\partial x^2}. \end{aligned} \tag{13.68}$$

We have shown that if $G_t(x) = (x, f_t(x))$ is a reparametrization of a curve shortening flow, then f_t satisfies

$$\frac{\partial f_t}{\partial t} = \frac{1}{1+\left(\frac{\partial f_t}{\partial x}\right)^2} \frac{\partial^2 f_t}{\partial x^2}. \tag{13.69}$$

Conversely, it can be easily shown that if $f_t : J \to \mathbb{R}$, $t \in [0,T)$, satisfies (13.69), then there exists a reparametrization F_t of the map $G_t(x) = (x, f_t(x))$ that satisfies the curve shortening flow with $F_0 = G_0$.

Example 13.7. *The* ***Grim Reaper*** *is the graphical curve shortening flow given by*

$$f_t(x) = \ln(\sec(x)) + t \tag{13.70}$$

for $x \in (-\frac{\pi}{2}, \frac{\pi}{2})$ *and* $t \in \mathbb{R}$. *See Figure* 13.3.1.

It is easy to see that

$$\frac{\partial f_t}{\partial t}(x,t) = 1, \quad \frac{\partial f_t}{\partial x}(x,t) = \tan(x), \quad \frac{\partial^2 f_t}{\partial x^2}(x,t) = \sec^2(x). \tag{13.71}$$

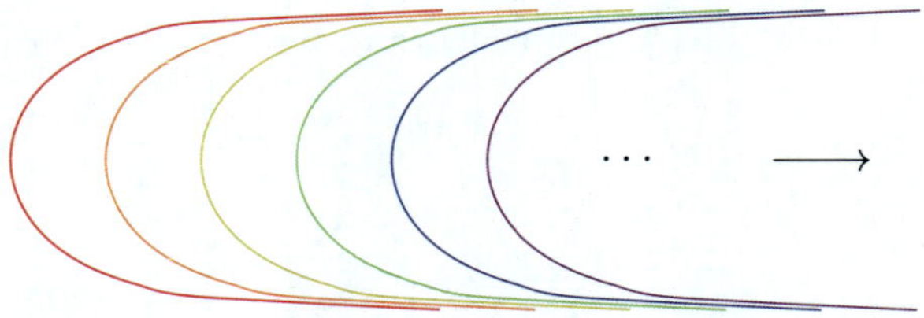

Figure 13.3.1. The Grim Reaper (rotated clockwise by 90 degrees) is a translating self-similar solution to the CSF.

From this we clearly see that f_t is a solution to the curve shortening flow since

$$\frac{1}{1+\left(\frac{\partial f_t}{\partial x}\right)^2}\frac{\partial^2 f_t}{\partial x^2} = \frac{1}{1+\tan^2(x)}\sec^2(x) = 1 = \frac{\partial f_t}{\partial t}; \tag{13.72}$$

that is, (13.68) *is satisfied.*

Observe that since the functions f_t are of the form $f_t(x) = g(x) + t$ for some function g, the curves $M_1^t := G_t(-\frac{\pi}{2}, \frac{\pi}{2})$ in the xy-plane evolve by translation in the y-direction with unit speed. Since the shapes and sizes of these curves M_1^t remain unchanged, the Grim Reaper is a ***translating self-similar solution*** *to the CSF.*

13.4. Evolution of geometric quantities under the curve shortening flow

It is natural to begin the qualitative study of the curve shortening flow by calculating how the various basic geometric quantities associated to the curve evolve. Our eventual goal is to control suitable geometric quantities well enough to establish the global existence and convergence of solutions.

13.4.1. Evolution of arc-length element. As before, let s_t be a choice of the arc-length parameter for each $t \in I$. We have that

$$\frac{\partial s_t}{\partial u} = \left\|\frac{\partial F}{\partial u}(\cdot, t)\right\| \tag{13.73}$$

is the speed of the parametrization and

$$ds_t = \frac{\partial s_t}{\partial u}du = \left\|\frac{\partial F}{\partial u}(\cdot, t)\right\| du \tag{13.74}$$

is the arc-length element. We compute that

$$\frac{\partial}{\partial t}\left\|\frac{\partial F_t}{\partial u}\right\| = \frac{1}{2\left\|\frac{\partial F_t}{\partial u}\right\|}\frac{\partial}{\partial t}\left(\left\|\frac{\partial F_t}{\partial u}\right\|^2\right) = \frac{1}{\left\|\frac{\partial F_t}{\partial u}\right\|}\frac{\partial^2 F_t}{\partial t\partial u}\cdot\frac{\partial F_t}{\partial u} = \frac{1}{\left\|\frac{\partial F_t}{\partial u}\right\|}\frac{\partial^2 F_t}{\partial u\partial t}\cdot\frac{\partial F_t}{\partial u}, \tag{13.75}$$

where in the last line we used Clairaut's theorem that mixed partial derivatives are equal. Now we can apply the curve shortening equation to this to obtain

$$\frac{\partial}{\partial t}\left\|\frac{\partial F_t}{\partial u}\right\| = \frac{1}{\left\|\frac{\partial F_t}{\partial u}\right\|}\frac{\partial}{\partial u}(-\kappa_t N_t)\cdot\frac{\partial F_t}{\partial u} = -\kappa_t\frac{1}{\left\|\frac{\partial F_t}{\partial u}\right\|}\frac{\partial N_t}{\partial u}\cdot\frac{\partial F_t}{\partial u} = -\kappa_t\frac{\partial N_t}{\partial s_t}\cdot\frac{\partial F_t}{\partial s_t}\left\|\frac{\partial F_t}{\partial u}\right\|, \tag{13.76}$$

where for the second equality we used that $N\cdot\frac{\partial F_t}{\partial u} = 0$ in reducing the product rule to one term.

Now recall from (4.7) that

$$\frac{\partial T_t}{\partial s_t} = -\kappa_t N_t. \tag{13.77}$$

The reason we have a minus sign on the right-hand side is that we are using the normal in the direction opposite to the one used in Chapter 4. We further compute that

$$0 = \frac{\partial}{\partial s_t}(N_t\cdot T_t) = \frac{\partial N_t}{\partial s_t}\cdot T_t + N_t\cdot\frac{\partial T_t}{\partial s_t},$$

so that

$$\frac{\partial N_t}{\partial s_t}\cdot T_t = -N_t\cdot\frac{\partial T_t}{\partial s_t}.$$

Since, on the other hand, $\frac{\partial N_t}{\partial s_t}\cdot N_t = 0$, we conclude that

$$\frac{\partial N_t}{\partial s_t} = \kappa_t T_t. \tag{13.78}$$

We may now plug this identity into (13.76) to obtain that the time-derivative of the speed is given by

$$\frac{\partial}{\partial t}\left\|\frac{\partial F_t}{\partial u}\right\| = -\kappa_t^2\left\|\frac{\partial F_t}{\partial u}\right\|. \tag{13.79}$$

This, in turn, by (13.74) immediately implies the following formula for the evolution of the arc-length element:

$$\frac{\partial}{\partial t} ds_t = -\kappa_t^2 ds_t. \tag{13.80}$$

Note that the arc-length element ds_t is an *invariant quantity* of the curve M_t^1; that is, it is independent of the parametrization F_t of M_t^1. On the other hand, the speed $\|\frac{\partial F_t}{\partial u}\|$ is not an invariant quantity as it certainly depends on the parametrization F_t. In general, we are interested in deriving equations for invariant quantities and understanding how they behave under the curve shortening flow.

13.4.2. Evolution of the length. Let L_t denote the length of the curve M_t^1. By integrating (13.80), we obtain that the evolution of the length L_t is given by

$$\frac{d\,\mathrm{L}_t}{dt} = -\int_{M_t^1} \kappa_t^2\, ds_t. \tag{13.81}$$

So we see that the length of a closed embedded plane curve is decreasing under the curve shortening flow, justifying the name of this equation!

13.4.3. Evolution of the enclosed area. Let A_t denote the area of the bounded region enclosed by M_t^1. The evolution of A_t is given by

$$\frac{d\,\mathrm{A}_t}{dt} = -\int_{M_t^1} \kappa_t\, ds_t = -2\pi, \tag{13.82}$$

where the second equality is by Theorem 4.9. Intuitively, we can see this formula as follows. Let $t \in I$ and imagine the curve at a slightly later time $t + \Delta t$. The curve $M_{t+\Delta t}^1$ looks like the 'normal graph' of the function κ_t over the curve M_t^1. Previously we have defined radial graphs over round circles in §1.3.3.3. Normal graphs are just generalizations of radial graphs. In any case, we have the following formula:

$$\mathrm{A_t} - \mathrm{A_{t+\Delta t}} = \Delta t \int_{M_t^1} \kappa_t ds_t + O((\Delta t)^2). \tag{13.83}$$

This implies (13.82). See §13.4.4 below for a rigorous proof of this formula.

As a consequence of the evolution of area formula (13.82), we have that if an embedded solution to the CSF exists on a time interval $[0, T)$, then

$$T \leq \frac{\mathrm{A}_0}{2\pi}. \tag{13.84}$$

13.4.4. Evolution of unit tangent and normal vectors. A more rigorous proof of (13.82) can be given as follows. Recall from (5.73) that

$$\mathrm{A}_t = \frac{1}{2}\int_{M_t^1} \sigma_t ds_t, \tag{13.85}$$

where $\sigma_t = F_t \cdot N_t$ is the support function of M_t^1. We have

$$\frac{\partial}{\partial t}\sigma_t = \frac{\partial F_t}{\partial t}\cdot N_t + F_t \cdot \frac{\partial N_t}{\partial t}. \tag{13.86}$$

Now,

$$\begin{aligned}
\frac{\partial}{\partial t}T_t &= \frac{\partial}{\partial t}\left(\frac{\partial F_t}{\partial s_t}\right) \\
&= \frac{\partial}{\partial t}\left(\left\|\frac{\partial F_t}{\partial u}\right\|^{-1}\frac{\partial F_t}{\partial u}\right) \\
&= \frac{\partial}{\partial t}\left(\left\|\frac{\partial F_t}{\partial u}\right\|^{-1}\right)\frac{\partial F_t}{\partial u} + \left\|\frac{\partial F_t}{\partial u}\right\|^{-1}\frac{\partial^2 F_t}{\partial u \partial t} \\
&= \kappa_t^2\left\|\frac{\partial F_t}{\partial u}\right\|^{-1}\frac{\partial F_t}{\partial u} + \left\|\frac{\partial F_t}{\partial u}\right\|^{-1}\frac{\partial}{\partial u}(-\kappa_t N_t) \\
&= \kappa_t^2 T_t - \frac{\partial \kappa_t}{\partial s_t}N_t - \kappa_t\frac{\partial N_t}{\partial s_t} \\
&= -\frac{\partial \kappa_t}{\partial s_t}N_t,
\end{aligned} \tag{13.87}$$

where we used Clairaut's theorem, (13.79), the curve shortening flow equation, and (13.78). This equation makes sense for the following reasons. Firstly, the T_t are unit vectors, so their time-derivatives must be perpendicular to themselves. Secondly, if we have a shrinking round circle, then the time-derivatives of T_t are zero, and this corresponds to the case where $\frac{\partial \kappa_t}{\partial s_t} = 0$.

We can use the calculation above and that N_t and T_t are orthonormal to derive that

$$\frac{\partial}{\partial t}N_t = \frac{\partial \kappa_t}{\partial s_t}T_t. \tag{13.88}$$

Now we can plug this equation into (13.86) to yield

$$\frac{\partial}{\partial t}\sigma_t = -\kappa_t + F_t \cdot T_t\frac{\partial \kappa_t}{\partial s_t}. \tag{13.89}$$

Using (13.89), we compute that

$$\frac{d}{dt}\mathrm{A}_t = \frac{1}{2}\int_{M_t^1} \frac{\partial \sigma_t}{\partial t} ds_t + \sigma_t \frac{\partial}{\partial t} ds_t \qquad (13.90)$$
$$= \frac{1}{2}\int_{M_t^1}\left(-\kappa_t + F_t \cdot T_t \frac{\partial \kappa_t}{\partial s_t} - \sigma_t \kappa_t^2\right) ds_t.$$

Observe that

$$\frac{\partial}{\partial s_t}(F_t \cdot T_t) = T_t \cdot T_t + F_t \cdot \frac{\partial T_t}{\partial s_t} = 1 - F_t \cdot N_t \kappa_t, \qquad (13.91)$$

where we used (13.77). By integrating by parts on the right-hand side of (13.90) and applying (13.91), we obtain

$$\frac{d}{dt}\mathrm{A}_t = \frac{1}{2}\int_{M_t^1}\left(-\kappa_t - \frac{\partial}{\partial s_t}(F_t \cdot T_t)\kappa_t - \sigma_t \kappa_t^2\right) ds_t = -\int_{M_t^1} \kappa_t ds_t.$$

This proves (13.82) again.

13.5. Preserving embeddedness

In this section, we demonstrate that as long as a solution to the (parametrized) curve shortening flow is smooth, it remains embedded.

Let $F_t : M^1 \to \mathbb{R}^2$, $t \in [0,T)$, be an embedded curve shortening flow, where M^1 is diffeomorphic to S^1, and let $M_t^1 := F_t(S^1)$. The **Euclidean distance squared** $f_t : M^1 \times M^1 \to \mathbb{R}$ between points on M_t^1 is defined by

$$f_t(u_1,u_2) := \|F_t(u_1) - F_t(u_2)\|^2 = \|V_t(u_1,u_2)\|^2, \qquad (13.92)$$

where the norm $\|\cdot\|$ is the Euclidean norm of $\mathbb{R}^2$ and where

$$V_t(u_1,u_2) := F_t(u_1) - F_t(u_2). \qquad (13.93)$$

Let s_1 and s_2 denote arc-length parameters on the first and second factors of $M^1 \times M^1$, respectively; both parameters also depend on t. The time-derivative of f_t is given by

$$\frac{\partial f_t}{\partial t}(u_1,u_2) = 2V_t(u_1,u_2)\cdot\left(\frac{\partial F_t}{\partial t}(u_1) - \frac{\partial F_t}{\partial t}(u_2)\right) \qquad (13.94)$$
$$= 2V_t(u_1,u_2)\cdot\left(\frac{\partial^2 F_t}{\partial s^2}(u_1) - \frac{\partial^2 F_t}{\partial s^2}(u_2)\right).$$

On the other hand, the spatial-derivative of f_t with respect to arc length on the first factor is given by

$$\frac{\partial f_t}{\partial s_1}(u_1,u_2) = 2V_t(u_1,u_2)\cdot\frac{\partial F_t}{\partial s}(u_1). \qquad (13.95)$$

The second derivative is

$$\frac{\partial^2 f_t}{\partial s_1^2}(u_1,u_2) = 2V_t(u_1,u_2)\cdot\frac{\partial^2 F_t}{\partial s^2}(u_1)+2 \tag{13.96}$$
$$= -2\kappa_t(u_1)V_t(u_1,u_2)\cdot N_t(u_1)+2,$$

since $\frac{\partial F_t}{\partial s}(u_1)\cdot\frac{\partial F_t}{\partial s}(u_1)=1$. Similarly, we have

$$\frac{\partial f_t}{\partial s_2}(u_1,u_2) = -2V_t(u_1,u_2)\cdot\frac{\partial F_t}{\partial s}(u_2) \tag{13.97}$$

and

$$\frac{\partial^2 f_t}{\partial s_2^2}(u_1,u_2) = -2V_t(u_1,u_2)\cdot\frac{\partial^2 F_t}{\partial s^2}(u_2)+2 \tag{13.98}$$
$$= 2\kappa_t(u_2)V_t(u_1,u_2)\cdot N_t(u_2)+2.$$

By putting the above equations together, we obtain

$$\frac{\partial f_t}{\partial t} = \frac{\partial^2 f_t}{\partial s_1^2}+\frac{\partial^2 f_t}{\partial s_2^2}-4. \tag{13.99}$$

This is a heat-type equation on $M^1\times M^1$.

Example 13.8. *Recall from Example* 13.6 *that the shrinking round circle solution to the CSF with the initial curve a circle of radius* r_0 *is given by*

$$F_t(u) = \big(r_t\cos(u), r_t\sin(u)\big) = r_t\mathrm{e}^{iu}, \tag{13.100}$$

where $r_t=\sqrt{r_0^2-2t}$ *and we used complex numbers. We have that*

$$f_t(u_1,u_2) = r_t^2\left|\mathrm{e}^{iu_1}-\mathrm{e}^{iu_2}\right|^2 \tag{13.101}$$
$$= r_t^2(2-\mathrm{e}^{i(u_1-u_2)}-\mathrm{e}^{i(u_2-u_1)})$$
$$= 2(r_0^2-2t)(1-\cos(u_1-u_2)).$$

Thus

$$\frac{\partial f_t}{\partial t}(u_1,u_2) = -4(1-\cos(u_1-u_2)),$$
$$\frac{\partial^2 f_t}{\partial s_1^2}(u_1,u_2) = \frac{1}{r_0^2-2t}\frac{\partial^2 f_t}{\partial u_1^2}(u_1,u_2) = 2\cos(u_1-u_2),$$
$$\frac{\partial^2 f_t}{\partial s_2^2}(u_1,u_2) = \frac{1}{r_0^2-2t}\frac{\partial^2 f_t}{\partial u_2^2}(u_1,u_2) = 2\cos(u_1-u_2).$$

From these equations we can rederive (13.99) *directly.*

Let

$$\ell(u_1,u_2) := \int_{u_1}^{u_2}\left\|\frac{\partial F}{\partial u}(u)\right\| du \tag{13.102}$$

be the arc length of the path F between u_1 and u_2. Given an immersion with a curvature bound, we have an associated **chord-arc inequality**.

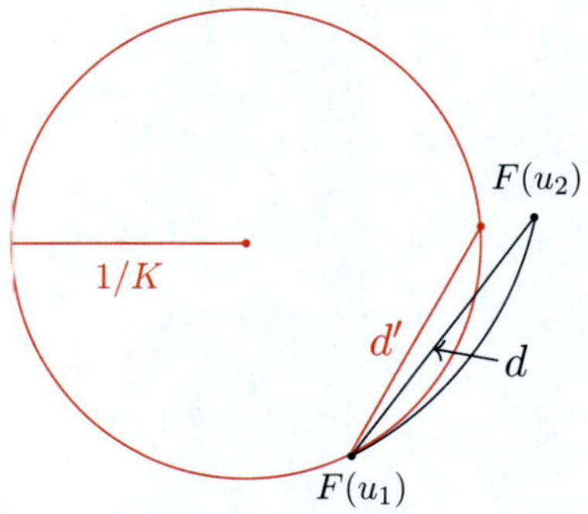

Figure 13.5.1. d' is the length of the chord of the circular arc of radius $1/K$ and length $\ell(u_1, u_2)$.

Lemma 13.9. *Let M^1 be a 1-manifold diffeomorphic to S^1 and let K be a positive real number. If $F : M^1 \to \mathbb{R}^2$ is an immersion with $|\kappa(u)| \leq K$ for all $u \in M^1$, then*

$$d(u_1, u_2) := \|F(u_1) - F(u_2)\| \geq \frac{2}{K} \sin\left(\frac{K\,\ell(u_1, u_2)}{2}\right) =: d'(u_1, u_2) \tag{13.103}$$

for all $u_1, u_2 \in M^1$ satisfying $\ell(u_1, u_2) \leq \frac{\pi}{K}$. See Figure 13.5.1.

Proof. Let $\ell := \ell(u_1, u_2)$ and suppose that $\ell \leq \frac{\pi}{K}$. Choose the arc-length parameter s on M^1 so that $s(u_1) = -\ell/2$ and $s(u_2) = \ell/2$. Let $\theta : [-\ell/2, \ell/2] \to \mathbb{R}$ denote the angle between $\frac{\partial F}{\partial s}(s)$ and $\frac{\partial F}{\partial s}(0)$. By the curvature bound, we have that

$$|\theta(s)| = \left| \int_0^s \frac{\partial \theta}{\partial s}(\bar{s})\, d\bar{s} \right| = \left| \int_0^s \kappa(\bar{s})\, d\bar{s} \right| \leq K\,|s|. \tag{13.104}$$

We compute, as functions of arc length, that

$$\begin{aligned} \|F(\ell/2) - F(-\ell/2)\| &\geq (F(\ell/2) - F(-\ell/2)) \cdot \frac{\partial F}{\partial s}(0) \\ &= \int_{-\ell/2}^{\ell/2} \frac{\partial F}{\partial s}(s) \cdot \frac{\partial F}{\partial s}(0)\, ds \\ &= \int_{-\ell/2}^{\ell/2} \cos(\theta(s))\, ds \\ &\geq \int_{-\ell/2}^{\ell/2} \cos(K\,|s|)\, ds \\ &= \frac{2}{K} \sin\left(\frac{K\ell}{2}\right), \end{aligned}$$

where we used the monotonicity of cosine on $\left[0, \frac{K\ell}{2}\right] \subset \left[0, \frac{\pi}{2}\right]$. □

As a consequence of (13.99) and Lemma 13.9, one can prove the following; cf. [**GH86**, Theorem 3.2.1].

Proposition 13.10. *Let $F_t : M^1 \to \mathbb{R}^2$, $t \in [0,T)$, be an immersed curve shortening flow, where F_0 is an embedding. Then F_t is an embedding for each $t \in [0,T)$.*

Proof. Observe by differentiating (13.92) that

$$\frac{\partial^2 f_t}{\partial s_1 \partial s_2}(u_1,u_2) + 2\frac{\partial F_t}{\partial s}(u_1) \cdot \frac{\partial F_t}{\partial s}(u_2) = 0. \tag{13.105}$$

By adding or subtracting this to the right-hand side of (13.99), we obtain

$$\begin{aligned}\frac{\partial f_t}{\partial t} &= \frac{\partial^2 f_t}{\partial s_1^2} + \frac{\partial^2 f_t}{\partial s_2^2} \pm 2\frac{\partial^2 f_t}{\partial s_1 \partial s_2} \pm 4\frac{\partial F_t}{\partial s}(u_1) \cdot \frac{\partial F_t}{\partial s}(u_2) - 4 \\ &= \left(\frac{\partial}{\partial s_1} \pm \frac{\partial}{\partial s_2}\right)^2 f_t - 2\left\|\frac{\partial F_t}{\partial s}(u_1) \mp \frac{\partial F_t}{\partial s}(u_2)\right\|^2\end{aligned} \tag{13.106}$$

where we used $\left\|\frac{\partial F_t}{\partial s}(u_1)\right\| = \left\|\frac{\partial F_t}{\partial s}(u_2)\right\| = 1$ for the second equality.

Now suppose, at some time t, that $(\overline{u}_1, \overline{u}_2)$ is a critical point of f_t. By (13.95) and (13.97), we then have that

$$V_t(\overline{u}_1,\overline{u}_2) \cdot \frac{\partial F_t}{\partial s}(\overline{u}_1) = V_t(\overline{u}_1,\overline{u}_2) \cdot \frac{\partial F_t}{\partial s}(\overline{u}_2) = 0, \tag{13.107}$$

where $V_t(\overline{u}_1,\overline{u}_2) = F_t(\overline{u}_1) - F_t(\overline{u}_2)$ as defined by (13.93).

Assume that $V_t(\overline{u}_1,\overline{u}_2) \neq \mathbf{0}$. Then we have that the unit tangent vectors $\frac{\partial F_t}{\partial s}(\overline{u}_1)$ and $\frac{\partial F_t}{\partial s}(\overline{u}_2)$ are each perpendicular to $V_t(\overline{u}_1,\overline{u}_2)$. This implies that[1]

$$N_t(\overline{u}_1) = \mp N_t(\overline{u}_2), \tag{13.108}$$

so that

$$\frac{\partial F_t}{\partial s}(\overline{u}_1) = \pm\frac{\partial F_t}{\partial s}(\overline{u}_2). \tag{13.109}$$

Applying this to (13.106) yields[2]

$$\frac{\partial f_t}{\partial t}(\overline{u}_1,\overline{u}_2) = \left(\left(\frac{\partial}{\partial s_1} \pm \frac{\partial}{\partial s_2}\right)^2 f_t\right)(\overline{u}_1,\overline{u}_2). \tag{13.110}$$

[1]That is, $N_t(\overline{u}_1) = -N_t(\overline{u}_2)$ or $N_t(\overline{u}_1) = N_t(\overline{u}_2)$.

[2]We have + in (13.110) if we have − in (13.108) and vice versa.

Now let $T' \in (0,T)$ and let $K := \max_{M^1\times[0,T']} |\kappa|$. By Lemma 13.9, we have that

$$\sqrt{f_t}(u_1,u_2) =: d_t(u_1,u_2) \geq \frac{2}{K}\sin\left(\frac{K\,\ell_t(u_1,u_2)}{2}\right) \tag{13.111}$$

for all $u_1, u_2 \in M^1$ and $t \in [0,T']$ satisfying $\ell_t(u_1,u_2) \leq \frac{\pi}{K}$.

For any $t \in [0,T']$ and $(\overline{u}_1,\overline{u}_2) \in M^1 \times M^1$ such that $F_t(\overline{u}_1) \neq F_t(\overline{u}_2)$ and f_t attains a local minimum at $(\overline{u}_1,\overline{u}_2)$ in $M^1 \times M^1$, by (13.110) we have that

$$\frac{\partial f_t}{\partial t}(\overline{u}_1,\overline{u}_2) \geq 0. \tag{13.112}$$

By the parabolic maximum principle, we have that the function

$$t \mapsto \min\left\{\min_{\ell_t(u_1,u_2)\geq\frac{\pi}{K}} f_t(u_1,u_2), \frac{4}{K^2}\right\} \tag{13.113}$$

is nondecreasing. From this and (13.111) we conclude that $F_t(u_1) \neq F_t(u_2)$ for all $u_1 \neq u_2$ and $t \in [0,T']$. The proposition follows. □

Since the curve remains embedded under the curve shortening flow, we have the following.

Corollary 13.11. *If $F_t : M^1 \to \mathbb{R}^2$ is a solution to the CSF where $M^1 \cong S^1$ and F_0 is an embedding, then the supremal time of existence T is finite; in fact, $T \leq \frac{\mathrm{A}_0}{2\pi}$, where A_0 is the area of the region that the initial curve encloses.*

Later, we will see that the maximal time of a curve shortening flow with initial area A_0 is equal to $T = \frac{\mathrm{A}_0}{2\pi}$; that is, $\mathrm{A}_t \to 0$ as $t \to T$, where A_t denotes the area of the region that the curve at time t encloses.

13.6. Gage's monotonicity formula

Recall that the isoperimetric inequality (Theorem 13.1) says that $\mathrm{L}^2 \geq 4\pi\,\mathrm{A}$; in other words, $\frac{\mathrm{L}^2}{\mathrm{A}} \geq 4\pi$, with equality if and only if the curve is a round circle. Thus, if we expect that the curve shortening flow deforms embedded convex curves to round circles, then we may hope that the isoperimetric ratio $\frac{\mathrm{L}^2}{\mathrm{A}}$ decreases under the flow. Gage's monotonicity formula says that this is indeed true! See [**Gag83**].

Proposition 13.12 (Monotonicity of the isoperimetric ratio under curve shortening flow)**.** *For any embedded closed convex curve evolving by the curve shortening flow, we have*

$$\frac{d}{dt}\left(\frac{\mathrm{L}_t^2}{\mathrm{A}_t}\right) \leq 0. \tag{13.114}$$

Proof. Using (13.81) and (13.82), we compute that

$$\frac{d}{dt}\ln\left(\frac{\mathrm{L}_t^2}{\mathrm{A}_t}\right) = \frac{2}{\mathrm{L}_t}\frac{d\,\mathrm{L}_t}{dt} - \frac{1}{\mathrm{A}_t}\frac{d\,\mathrm{A}_t}{dt} = -\frac{2}{\mathrm{L}_t}\int_{M_t^1}\kappa_t^2\,ds_t + \frac{2\pi}{\mathrm{A}_t}.$$

The quantity on the right-hand side is nonpositive by the following lemma. The proposition follows. □

Lemma 13.13 (Gage's inequality). *For any embedded closed convex curve M^1 in the plane,*

$$\int_{M^1}\kappa^2 ds \geq \frac{\pi\,\mathrm{L}}{\mathrm{A}}, \tag{13.115}$$

where L is the length of the curve and A is the area of the region it encloses.

Proof. The main idea is to show that there exists a choice of origin for $\mathbb{R}^2$ (by translating the curve) such that the support function σ, defined by (5.18), satisfies the inequality

$$\int_{M^1}\sigma^2 ds \leq \frac{\mathrm{L}\,\mathrm{A}}{\pi}. \tag{13.116}$$

We give a sketch of the proof of this in the next subsection.

Assuming this inequality, we deduce from the length identity (5.71) and the Cauchy–Schwarz inequality for integrals that

$$\mathrm{L}^2 = \left(\int_{M^1}\sigma\,\kappa\,ds\right)^2 \leq \int_{M^1}\sigma^2 ds\int_{M^1}\kappa^2 ds \leq \frac{\mathrm{L}\,\mathrm{A}}{\pi}\int_{M^1}\kappa^2 ds,$$

which yields inequality (13.115). □

13.6.1. Proof of Gage's inequality.

In this section, closely following Mike Gage's paper [**Gag83**], we present the proof of inequality (13.116), which in turn completes the proof of Lemma 13.13.

Define the Bonnesen functional $B : [r_{\mathrm{in}}, r_{\mathrm{out}}] \to \mathbb{R}$ by[3]

$$B(r) := -\pi r^2 + \mathrm{L}\,r - \mathrm{A}\,. \tag{13.117}$$

We have that $B(r) \geq 0$ for $r \in [r_{\mathrm{in}}, r_{\mathrm{out}}]$ (see Osserman [**Oss79**]).

[3]The function B has the following property. Let M^1 be a convex curve and let M_r^1 be the parallel curve to M^1 of distance r from M^1 and to the inside of M^1. Then the area A_r enclosed by M_r^1 satisfies $\mathrm{A}_r = \mathrm{A} - \mathrm{L}\,r + \pi r^2 + \mathrm{O}(r^3) = -B(r) + \mathrm{O}(r^3)$.

Firstly, suppose that M^1 is **centrally symmetric**; i.e., M^1 has the property that if $\mathbf{x} \in M^1$, then $-\mathbf{x} \in M^1$. In this case we have that $r_{\text{in}} \leq \sigma(s) \leq r_{\text{out}}$ and hence $B(\sigma(s)) \geq 0$.[4] Thus,

$$\begin{aligned} 0 &\leq \int_{M^1} B(\sigma(s))\, ds \\ &= -\pi \int_{M^1} \sigma(s)^2\, ds + \mathrm{L} \int_{M^1} \sigma(s)\, ds - \mathrm{A}\,\mathrm{L} \\ &= -\pi \int_{M^1} \sigma(s)^2\, ds + \mathrm{A}\,\mathrm{L}, \end{aligned}$$

where we used (13.85) for the last equality. We conclude that

$$\int_{M^1} \sigma(s)^2\, ds \leq \frac{\mathrm{L}\,\mathrm{A}}{\pi}$$

for centrally symmetric curves.

Secondly, to obtain this inequality for all convex curves one does a cut, flip, and paste symmetrization procedure to show that the inequality holds for all convex curves (see Gage [**Gag83**] for the details).[5]

13.7. The evolution of the curvature κ

We first discuss the arc length and angle reparametrizations of embedded and convex curves, respectively. Then we calculate the evolution of the curvature κ under the curve shortening flow with respect to the original parametrization. In the case of convex embedded curves, we also present the evolution of curvature under the angle parametrization.

13.7.1. Reparametrizations of the curve shortening flow.

Recall that for the *inverse curve shortening flow* of convex embedded plane curves, we considered the parametrization by the inverse of the Gauss map in §13.2.2. It is useful to do the same for the *curve shortening flow* in the case of convex embedded curves.

Firstly, we discuss the effect of reparametrizations on curve shortening flows (this discussion easily generalizes to arbitrary flows).

13.7.2. The arc-length reparametrization of the curve shortening flow.

Let $F_t : M^1 \to \mathbb{R}^2$, $t \in I$, be a curve shortening flow. Let $s_t : M^1 \to \mathbb{R}/\,\mathrm{L}_t\,\mathbb{Z}$ be reparametrizations by arc length for each $t \in I$. Let $f_t : M^1_t \to \mathbb{R}$, $t \in I$, be any 1-parameter family of functions. By calculating in a similar

[4] The inradius circle and the outradius circle can both be chosen to be centered at the origin.

[5] We would like to thank Mike Gage for explaining his proof to us.

fashion to that for (13.87), we obtain that the commutator of the derivatives $\frac{\partial}{\partial t}$ and $\frac{\partial}{\partial s_t}$ acting on f_t is given by

$$\begin{aligned}\frac{\partial}{\partial t}\frac{\partial}{\partial s_t}f_t - \frac{\partial}{\partial s_t}\frac{\partial}{\partial t}f_t &= \frac{\partial}{\partial t}\left(\left\|\frac{\partial F_t}{\partial u}\right\|^{-1}\frac{\partial f_t}{\partial u}\right) - \left\|\frac{\partial F_t}{\partial u}\right\|^{-1}\frac{\partial}{\partial u}\left(\frac{\partial f_t}{\partial t}\right)\\ &= \frac{\partial}{\partial t}\left(\left\|\frac{\partial F_t}{\partial u}\right\|^{-1}\right)\frac{\partial f_t}{\partial u}\\ &= \kappa_t^2\left\|\frac{\partial F_t}{\partial u}\right\|^{-1}\frac{\partial f_t}{\partial u}\\ &= \kappa_t^2\frac{\partial f_t}{\partial s_t}.\end{aligned}$$

That is, we have the following.

Lemma 13.14. *Under the curve shortening flow, we have*

$$\frac{\partial}{\partial t}\left(\frac{\partial f_t}{\partial s_t}\right) = \frac{\partial}{\partial s_t}\left(\frac{\partial f_t}{\partial t}\right) + \kappa_t^2\frac{\partial f_t}{\partial s_t} \tag{13.118}$$

for any 1*-parameter family of scalar or vector-valued functions* f_t*.*

For example, if $f_t = F_t$ (the formula applies equally well to vector-valued functions), we recover (13.87):

$$\begin{aligned}\frac{\partial}{\partial t}T_t &= \frac{\partial}{\partial t}\frac{\partial}{\partial s_t}F_t = \frac{\partial}{\partial s_t}\frac{\partial}{\partial t}F_t + \kappa_t^2\frac{\partial F_t}{\partial s_t}\\ &= \frac{\partial}{\partial s_t}(-\kappa_t N_t) + \kappa_t^2 T_t = -\frac{\partial \kappa_t}{\partial s_t}N_t.\end{aligned}$$

By applying Lemma 13.14 to the unit tangent vector field T_t, which is a vector-valued function, we obtain

$$\frac{\partial}{\partial t}(-\kappa_t N_t) = \frac{\partial}{\partial s_t}\left(-\frac{\partial \kappa_t}{\partial s_t}N_t\right) - \kappa_t^3 N_t. \tag{13.119}$$

That is,

$$-\frac{\partial \kappa_t}{\partial t}N_t - \kappa_t\frac{\partial \kappa_t}{\partial s_t}T_t = -\frac{\partial^2 \kappa_t}{\partial s_t^2}N_t - \frac{\partial \kappa_t}{\partial s_t}\kappa_t T_t - \kappa_t^3 N_t. \tag{13.120}$$

By taking the normal component of this equation, we obtain the evolution equation for the curvature:

Corollary 13.15. *Under the curve shortening flow, the curvature evolves by*

$$\frac{\partial \kappa_t}{\partial t} = \frac{\partial^2 \kappa_t}{\partial s_t^2} + \kappa_t^3. \tag{13.121}$$

13.7.3. A lower bound for the blow-up rate of the curvature. Define

$$\mathrm{K}(t) := \max_{u\in M^1} \kappa(u,t). \tag{13.122}$$

Then $\mathrm{K}(t)$ is a Lipschitz continuous function. Define

$$\frac{d^+\mathrm{K}}{dt}(t) := \limsup_{\Delta t\to 0_+} \frac{\mathrm{K}(t+\Delta t)-\mathrm{K}(t)}{\Delta t}, \tag{13.123}$$

which is the lim sup of forward difference quotients, and define

$$S_t := \{u\in M^1 : \kappa(u,t)=\mathrm{K}(t)\}. \tag{13.124}$$

Then (see, e.g., Lemma B.40 in [**CLN06**])

$$\frac{d^+\mathrm{K}}{dt}(t) = \sup_{u\in S_t} \frac{\partial\kappa}{\partial t}(u,t). \tag{13.125}$$

We then see that

$$\frac{d^+\mathrm{K}}{dt}(t) = \sup_{u\in S_t}\left(\frac{\partial^2\kappa}{\partial s^2}+\kappa^3\right)(u,t) \le \mathrm{K}(t)^3 \tag{13.126}$$

since $\frac{\partial^2\kappa}{\partial s^2}(u,t)\le 0$ for $u\in S_t$.

By the short-time existence theorem and Corollary 13.11, the supremal time of existence T of the CSF is positive and finite. By the long-time existence theorem (see Theorem 13.26 below), we have

$$\sup_{t\in[0,T))]} \mathrm{K}(t) = \infty.$$

Thus, there exists a sequence $t_i \to T$ such that $\mathrm{K}(t_i)\to\infty$. Now let $t\in[0,T)$. For all i such that $t_i > t$, we have

$$\mathrm{K}(t_i)^{-2} - \mathrm{K}(t)^{-2} = \int_t^{t_i} \frac{d\,\mathrm{K}^{-2}}{dt}(\bar{t})\,d\bar{t} \ge -2(t_i - T)$$

by (13.126). By taking $i\to\infty$, we obtain

$$-\mathrm{K}(t)^{-2} \ge -2(T-t),$$

so that we have the following lower bound for the maximum curvature:

$$\mathrm{K}(t) \ge \frac{1}{\sqrt{2(T-t)}}. \tag{13.127}$$

13.7.4. The angle reparametrization of the curve shortening flow. In the case of convex embedded plane curves, we have another parametrization, which we call the *angle reparametrization.* Namely, for each $t \in I$, define $\theta_t \in \mathbb{R}/2\pi\mathbb{Z}$ by

$$N_t(u) = \big(\cos(\theta_t(u)), \sin(\theta_t(u))\big). \tag{13.128}$$

Since $\theta_t \in \mathbb{R}/2\pi\mathbb{Z}$ is in one-to-one correspondence with $N_t \in S^1$, the angle parametrization is equivalent to the Gauss map parametrization.

Now, on one hand,

$$\frac{\partial}{\partial t}N_t(u) = \frac{\partial \theta_t}{\partial t}(u)\big(-\sin(\theta_t(u)), \cos(\theta_t(u))\big),$$

and on the other hand,

$$\frac{\partial}{\partial t}N_t(u) = \frac{\partial \kappa_t}{\partial s_t}(u)T_t(u).$$

Assume that F_t is a counterclockwise parametrization of M_t^1. We then conclude that

$$\frac{\partial \theta_t}{\partial t}(u) = \frac{\partial \kappa_t}{\partial s_t}(u). \tag{13.129}$$

Since M_t^1 is convex and embedded, we have that $u \mapsto \theta(u,t)$ is a diffeomorphism from M^1 to $\mathbb{R}/2\pi\mathbb{Z}$ for each t. Therefore, its inverse map $\theta \mapsto u(\theta, t)$ is also a diffeomorphism. Let

$$\widetilde{F}(\theta, t) := F(u(\theta, t), t) \tag{13.130}$$

be the angle parametrization of the curve shortening flow. By the chain rule,

$$\frac{\partial \widetilde{F}}{\partial t}(\theta, t) = \frac{\partial F}{\partial t}(u(\theta,t),t) + \frac{\partial F}{\partial u}(u(\theta,t),t)\frac{\partial u}{\partial t}(\theta, t). \tag{13.131}$$

Now,

$$\theta = \theta\big(u(\theta, t), t\big).$$

By taking the time-derivative of this, we obtain

$$\begin{aligned} 0 &= \frac{\partial \theta_t}{\partial t}(u(\theta,t)) + \frac{\partial \theta_t}{\partial u}(u(\theta,t))\frac{\partial u}{\partial t}(\theta, t) \\ &= \frac{\partial \kappa_t}{\partial s_t} + \kappa_t \left\| \frac{\partial F}{\partial u} \right\| \frac{\partial u}{\partial t}. \end{aligned}$$

Thus,

$$\frac{\partial u}{\partial t} = -\kappa_t^{-1} \left\| \frac{\partial F}{\partial u} \right\|^{-1} \frac{\partial \kappa_t}{\partial s_t} \tag{13.132}$$

and we obtain

$$\frac{\partial \widetilde{F}}{\partial t}(\theta, t) = \frac{\partial F}{\partial t}(u(\theta, t), t) - \kappa_t^{-1} \frac{\partial \kappa_t}{\partial s_t} T_t. \tag{13.133}$$

By a similar calculation, we have that for any family of functions $f_t = (\cdot, t) : M^1 \to \mathbb{R}$, $t \in I$,

$$\frac{\partial \widetilde{f}}{\partial t}(\theta, t) = \frac{\partial f}{\partial t}(u(\theta, t), t) - \kappa_t^{-1} \frac{\partial \kappa_t}{\partial s_t} \frac{\partial f}{\partial s_t}, \tag{13.134}$$

where $\widetilde{f}(\theta, t) := f(u(\theta, t), t)$.

Now let $\widetilde{\kappa}(\theta, t) := \kappa(u(\theta, t), t)$. We compute that

$$\begin{aligned} \frac{\partial \widetilde{\kappa}_t}{\partial t}(\theta, t) &= \frac{\partial \kappa_t}{\partial t}(u(\theta, t), t) - \kappa_t^{-1} \frac{\partial \kappa_t}{\partial s_t} \frac{\partial \kappa_t}{\partial s_t} \\ &= \frac{\partial^2 \widetilde{\kappa}_t}{\partial s_t^2} - \widetilde{\kappa}_t^{-1} \frac{\partial \widetilde{\kappa}_t}{\partial s_t} \frac{\partial \widetilde{\kappa}_t}{\partial s_t} + \widetilde{\kappa}_t^3. \end{aligned}$$

On the other hand, denoting $\kappa_t := \widetilde{\kappa}_t$ for simplicity, we have

$$\begin{aligned} \frac{\partial^2 \kappa_t}{\partial s_t^2} &= \kappa_t \frac{\partial}{\partial \theta_t} \left(\kappa_t \frac{\partial \kappa_t}{\partial \theta_t} \right) \\ &= \kappa_t^2 \frac{\partial^2 \kappa_t}{\partial \theta_t^2} + \kappa_t \left(\frac{\partial \kappa_t}{\partial \theta_t} \right)^2 \\ &= \kappa_t^2 \frac{\partial^2 \kappa_t}{\partial \theta_t^2} + \kappa_t^{-1} \frac{\partial \kappa_t}{\partial s_t} \frac{\partial \kappa_t}{\partial s_t}. \end{aligned} \tag{13.135}$$

Thus, we may rewrite (13.121) as

$$\frac{\partial \kappa_t}{\partial t} = \kappa_t^2 \left(\frac{\partial^2 \kappa_t}{\partial \theta_t^2} + \kappa_t \right), \tag{13.136}$$

where $\kappa_t := \widetilde{\kappa}_t$. We may write this more succinctly as

$$\partial_t \kappa = \kappa^2 (\partial_\theta^2 \kappa + \kappa). \tag{13.137}$$

For our original parametrization, we have that $\frac{\partial}{\partial t}$ is the partial derivative in the direction of t with respect to the variables u and t. Now let $\tau = \tau_t = t$ and let $\frac{\partial}{\partial \tau}$ denote the partial derivative in the direction of τ with respect to the variables $\theta = \theta_t$ and τ. In particular, by Clairaut's theorem, we have that

$$\frac{\partial}{\partial \tau} \frac{\partial}{\partial \theta} = \frac{\partial}{\partial \theta} \frac{\partial}{\partial \tau}. \tag{13.138}$$

We now compute the evolution of the curvature κ with respect to the variables θ and τ.

13.8. Hamilton's Harnack estimate

A key question in the search for monotone quantities is: What geometric invariants should we estimate? For geometric evolution equations, Hamilton discovered the principle of considering quantities that vanish on self-similar solutions, which in the case of the curve shortening flow include the Grim Reaper. In this section, we present Hamilton's Harnack estimate for the curve shortening flow, which is inspired by the Li–Yau differential Harnack estimate (discussed in §11.7.4).

Recall from (13.70) that the Grim Reaper at time $t = 0$ is the graph of the function

$$f_0(x) = \ln(\sec(x)). \tag{13.139}$$

We have that

$$\frac{\partial G_0}{\partial x}(x) = (1, \tan(x)). \tag{13.140}$$

Thus, a choice of the unit tangent vector field is given by

$$T(x) = T_{G_0(x)} = (\cos(x), \sin(x)). \tag{13.141}$$

By using the unit tangent vector field instead of the unit normal vector field to define the angle parameter θ in (13.128), we have that $\theta(x) = x$ is an angle parameter for the Grim Reaper.

By (13.67) and since $\frac{\partial f_0}{\partial x}(x) = \tan(x)$ and $\frac{\partial^2 f_0}{\partial x^2}(x) = \sec^2(x)$, we have that the curvature of the Grim Reaper is given by

$$\kappa_0(x) = \frac{1}{\left(1 + \left(\frac{\partial f_0}{\partial x}(x)\right)^2\right)^{3/2}} \frac{\partial^2 f_0}{\partial x^2}(x) = \cos(x). \tag{13.142}$$

We may observe that, of course,

$$\frac{\partial^2 \kappa_0}{\partial \theta^2} + \kappa_0 = \frac{\partial^2 \kappa_0}{\partial x^2} + \kappa_0 = 0.$$

Conversely, if we have a complete curve in the plane with positive curvature and satisfying $\frac{\partial^2 \kappa}{\partial \theta^2} + \kappa = 0$, then after a rescaling of the curve and a translation of the angle parameter we may assume that

$$\kappa(\theta) = \cos(\theta). \tag{13.143}$$

From this one can easily derive that the curve is a Grim Reaper.

13.8.1. Hamilton's Harnack quantity and its evolution. The Harnack quantity for the curve shortening flow is

$$Q := \frac{\partial}{\partial t}\ln\kappa = \frac{1}{\kappa}\frac{\partial\kappa}{\partial t} = \kappa\left(\frac{\partial^2\kappa}{\partial\theta^2} + \kappa\right). \tag{13.144}$$

By our calculation in the previous subsection, we have that Q vanishes on the Grim Reaper. By Hamilton's principle, Q is a good candidate for being a monotone quantity.

Lemma 13.16. *Under the curve shortening flow with respect to the angle parameter, the Harnack quantity evolves by*

$$\frac{\partial Q}{\partial t} = \kappa^2\frac{\partial^2 Q}{\partial\theta^2} + 2\kappa\frac{\partial\kappa}{\partial\theta}\frac{\partial Q}{\partial\theta} + 2Q^2. \tag{13.145}$$

Proof. We compute that

$$\begin{aligned}\partial_t Q &= \partial_t\big(\kappa(\partial_\theta^2\kappa + \kappa)\big)\\ &= (\partial_\theta^2\kappa + \kappa)\,\partial_t\kappa + \kappa(\partial_\theta^2\partial_t\kappa + \partial_t\kappa)\\ &= Q^2 + \kappa\big(\partial_\theta^2(\kappa Q) + \kappa Q\big).\end{aligned}$$

By applying the product rule for each of the two derivatives in the last line, we obtain

$$\partial_t Q = Q^2 + \kappa^2\partial_\theta^2 Q + \kappa Q\,\partial_\theta^2\kappa + 2\kappa\,\partial_\theta\kappa\,\partial_\theta Q + \kappa^2 Q.$$

The lemma now follows from combining terms. □

13.8.2. Hamilton's Harnack estimate and its implications. Using Lemma 13.16, we can now prove the Harnack estimate for the curve shortening flow. The ODE associated to the heat-type equation (13.145) is

$$\frac{dq}{dt}(t) = 2q(t)^2. \tag{13.146}$$

The solution to this ODE with the 'initial condition'

$$\lim_{t\to 0} q(t) = -\infty \tag{13.147}$$

is given by

$$q(t) = -\frac{1}{2t}. \tag{13.148}$$

Since Q is finite at time $t = 0$, we have that $Q(x,t) > q(t)$ for all $x \in \mathbb{R}/2\pi\mathbb{Z}$ and $t > 0$ sufficiency small. Thus, by the parabolic maximum principle, we have

$$\frac{\partial}{\partial t}\ln\kappa(\theta,t) = Q(\theta,t) \geq -\frac{1}{2t}. \tag{13.149}$$

In other words, for each $\theta \in \mathbb{R}/2\pi\mathbb{Z}$, we have that $t \mapsto \sqrt{t}\,\kappa(\theta,t)$ is nondecreasing. Note that, here, $\frac{\partial}{\partial t}$ is the partial derivative with θ fixed.

We can rewrite the Harnack quantity Q in terms of the arc-length parameter. Namely, by (13.135), we have

$$Q = \kappa^{-1}\frac{\partial^2 \kappa}{\partial s^2} - \kappa^{-2}\left(\frac{\partial \kappa}{\partial s}\right)^2 + \kappa^2 \tag{13.150}$$
$$= \frac{\partial^2}{\partial s^2}\ln\kappa + \kappa^2$$
$$= \frac{\partial}{\partial t}\ln\kappa - \left(\frac{\partial}{\partial s}\ln\kappa\right)^2.$$

Thus, Hamilton's differential Harnack estimate says that

$$\frac{\partial}{\partial t}\ln\kappa - \left(\frac{\partial}{\partial s}\ln\kappa\right)^2 + \frac{1}{2t} \geq 0. \tag{13.151}$$

Here, $\frac{\partial}{\partial t}$ is the partial derivative with u fixed.

13.9. Huisken's monotonicity formula

We discuss Huisken's powerful monotonicity formula for the curve shortening flow. As not to confuse notations, let t_0 denote the singularity time and as usual let T denote the unit tangent vector of the evolving curve. Define the function $\Phi : \mathbb{R}^2 \times (-\infty, t_0) \to \mathbb{R}$ by

$$\Phi(\mathbf{x}, t) := (4\pi(t_0 - t))^{-\frac{1}{2}} \mathrm{e}^{-\frac{\|\mathbf{x}\|^2}{4(t_0-t)}}. \tag{13.152}$$

It is easy to check that

$$\Psi := (4\pi(t_0 - t))^{-\frac{1}{2}}\Phi = (4\pi(t_0 - t))^{-1} \mathrm{e}^{-\frac{\|\mathbf{x}\|^2}{4(t_0-t)}} \tag{13.153}$$

is a solution to the conjugate heat equation:

$$\frac{\partial \Psi}{\partial t} + \Delta_{\mathbb{R}^2}\Psi = 0 \tag{13.154}$$

(we leave this as an exercise). One can show that as $t \to t_0$, the function $\Psi(\cdot, t)$ converges in the distributional sense to the Dirac delta function centered at $\mathbf{0}$. So we call Ψ the **conjugate heat kernel**.

Example 13.17. *Let $t_0 > 0$ and let M_t^1, $t \in [0, t_0)$, be the round shrinking circles solution to the curve shortening flow that shrinks to the origin $\mathbf{0}$ at time t_0. Then M_t^1 is given as the circle of radius $r(t) = \sqrt{2(t_0 - t)}$. Thus, for $\mathbf{x} \in M_t^1$, we have*

$$\Phi(\mathbf{x}, t) = (4\pi(t_0 - t))^{-1/2}\mathrm{e}^{-\frac{1}{2}}. \tag{13.155}$$

Thus,

$$\int_{M_t^1} \Phi \, ds_t = 2\pi\sqrt{2(t_0 - t)}(4\pi(t_0 - t))^{-1/2}\mathrm{e}^{-\frac{1}{2}} = \sqrt{2\pi/\mathrm{e}}, \tag{13.156}$$

which is a constant.

Huisken's monotonicity formula [**Hui90**] says the following.

Theorem 13.18. *If $F_t : M^1 \to \mathbb{R}^2$, $t \in [0, T)$, is a curve shortening flow, then*

$$\begin{aligned} &\frac{d}{dt}\int_{M^1} \Phi(F_t(u), t)\, ds_t(u) \\ &\quad = -\int_{M^1} \left(\kappa(u,t) - \frac{F_t(u)\cdot N_t(u)}{2(t_0 - t)}\right)^2 \Phi(F_t(u), t)\, ds_t(u). \end{aligned} \tag{13.157}$$

Proof. Let $\tau := t_0 - t$ and $ds := ds_t$. Using (13.80), we compute that

$$\begin{aligned} \frac{d}{dt}\int_{M^1} \Phi(F_t(u), t)\, ds_t(u) &= \int_{M^1} \left(\frac{\partial \Phi}{\partial t} - \Phi\kappa^2\right) ds \\ &= \int_{M^1} \left(\frac{1}{2\tau} - \frac{\|F\|^2}{4\tau^2} - \kappa^2\right) \Phi\, ds. \end{aligned} \tag{13.158}$$

On the other hand,

$$\frac{\partial \Phi}{\partial s} = -\frac{F\cdot T}{2\tau}\Phi$$

and

$$\begin{aligned} \frac{\partial^2 \Phi}{\partial s^2} &= -\Phi\frac{\partial}{\partial s}\left(\frac{F\cdot T}{2\tau}\right) - \frac{F\cdot T}{2\tau}\frac{\partial \Phi}{\partial s} \\ &= \left(\frac{(F\cdot T)^2}{4\tau^2} - \frac{1}{2\tau} + \frac{\kappa F\cdot N}{2\tau}\right)\Phi \end{aligned}$$

since $\frac{\partial T}{\partial s} = -\kappa N$. So, by the fundamental theorem of calculus, we have

$$0 = \int_{M^1} \frac{\partial^2 \Phi}{\partial s^2} ds = \int_{M^1} \left(\frac{(F\cdot T)^2}{4\tau^2} - \frac{1}{2\tau} + \frac{\kappa F\cdot N}{2\tau}\right)\Phi\, ds.$$

By applying this identity and the fact that

$$\|F\|^2 = (F\cdot T)^2 + (F\cdot N)^2$$

to (13.158), we obtain

$$\frac{d}{dt}\int_{M^1} \Phi\, ds = \int_{M^1} \left(-\frac{(F\cdot N)^2}{4\tau^2} - \kappa^2 + \frac{\kappa F\cdot N}{2\tau}\right)\Phi\, ds.$$

The theorem now follows from completing the square on the right-hand side. □

The theorem implies that **Huisken's entropy** (see [**Hui90**])

$$E(F) := \int_{M^1} \Phi \, ds \tag{13.159}$$

is a monotonically nonincreasing function of time under the curve shortening flow. Whenever one has a monotone quantity under a geometric flow, one can ask: For what solutions is Huisken's entropy constant in time? From (13.157), we see that this is the case if and only if

$$\kappa = \frac{F \cdot N}{2\tau}, \tag{13.160}$$

where $\tau = \tau(t) = 2(t_0 - t)$. Let us consider this equation for a specific time t, so that τ is just some positive real number. By multiplying the scalar equation (13.160) by the vector N, we obtain the equivalent equation

$$\kappa N = \frac{1}{2\tau} F^{\perp}, \tag{13.161}$$

where $F^{\perp}$ is the normal component of the vector F. That is, for the associated solution F_t to the curve shortening flow, we have

$$\frac{\partial F_t}{\partial t} = -\kappa_t N_t = -\frac{1}{2\tau(t)} F_t^{\perp}. \tag{13.162}$$

Such a solution is called a **shrinking self-similar solution** to the curve shortening flow.

We now explain the reason for the terminology 'self-similar solution'. Suppose that $\widetilde{F}_t = a(t) F_0$, where $a(t)$ is some function. Then $\{\widetilde{F}_t\}$ is a 1-parameter family of parametrized curves that is self-similar in the sense that the map $\widetilde{F}_t$ is similar to the map F_0; indeed, the curve $\widetilde{F}_t(M^1)$ is the constant multiple $a(t)$ of the curve $F_0(M^1)$. We compute that

$$\frac{\partial}{\partial t} \widetilde{F}_t = a'(t) F_0 = \frac{a'(t)}{a(t)} \widetilde{F}_t. \tag{13.163}$$

We choose $a(t)$ to satisfy $a(0) = 1$ and

$$\frac{a'(t)}{a(t)} = -\frac{1}{2\tau(t)} = -\frac{1}{2(t_0 - t)}. \tag{13.164}$$

The unique solution to this ODE initial-value problem is

$$a(t) = \left(1 - \frac{t}{t_0}\right)^{1/2}. \tag{13.165}$$

So our self-similar family of parametrizations is

$$\widetilde{F}_t = \left(1 - \frac{t}{t_0}\right)^{1/2} F_0 \tag{13.166}$$

and we have

$$\frac{\partial}{\partial t}\widetilde{F}_t = -\frac{1}{2\tau(t)}\widetilde{F}_t. \tag{13.167}$$

In comparison with (13.162), the right-hand sides of the equations differ by the tangential vector field $\frac{1}{2\tau(t)}F_t^\top := \frac{1}{2\tau(t)}(F_t \cdot T)T$. What this means is that since $\widetilde{F}(0) = F(0)$, the map F_t is a reparametrization of the map $\widetilde{F}_t$ for each t. That is, there exist self-diffeomorphisms $\phi_t : M^1 \to M^1$ such that

$$F_t = \widetilde{F}_t \circ \phi_t.$$

This implies that the image curves are equal; that is, $F_t(M^1) = \widetilde{F}_t(M^1)$. So the curve shortening flow F_t is geometrically indistinguishable from the 'self-similar' family of maps $\widetilde{F}_t$.

Let us now look at the shrinking self-similar equation in more detail. By (13.54) and (13.161), we have that

$$-\frac{d^2F}{ds^2} = \kappa N = \frac{1}{2\tau}(F - F^\top) = \frac{1}{2\tau}\left(F - \left(F \cdot \frac{dF}{ds}\right)\frac{dF}{ds}\right). \tag{13.168}$$

This is a second-order ODE system of two equations for $F(s)$. By the uniqueness theorem for ODEs, the solution to this ODE with the given initial data $F(0) =: \mathbf{v}_0$ and $\frac{dF}{ds}(0) =: \mathbf{v}_1$ is unique. Note that $\mathbf{v}_1$ is a unit vector. We also observe that by taking the dot product of equation (13.168) with N, we obtain

$$\kappa = \frac{1}{2\tau}F \cdot N. \tag{13.169}$$

Conversely, (13.169) implies that

$$\kappa N = \frac{1}{2\tau}(F \cdot N)N = \frac{1}{2\tau}\left(F - \left(F \cdot \frac{dF}{ds}\right)\frac{dF}{ds}\right).$$

Suppose that $F(0) = \mathbf{0}$ is the origin. In this case, we may define

$$\overline{F}(s) := s\mathbf{v}_1.$$

We then have that

$$\frac{d\overline{F}}{ds} = \mathbf{v}_1 \qquad \text{and} \qquad \frac{d^2\overline{F}}{ds^2} = \mathbf{0}.$$

From this it is easy to deduce that $\overline{F}(s)$ satisfies (13.168). By uniqueness, we have that $F(s) = \overline{F}(s)$. Summarizing, we have proved the following result.

Lemma 13.19. *If $F : M^1 \to \mathbb{R}^2$ is a shrinking self-similar solution to the curve shortening flow and if $\mathbf{0} \in F(M^1)$, then F parametrizes a straight line.*

13.10. Arzelà–Ascoli theorem and equicontinuous families of functions

A key basic result in real analysis is the Arzelà–Ascoli theorem, which we now consider in the context of geometric analysis.

Let (M, d) be a metric space. Recall that a family $\mathcal{F}$ of real-valued functions on M is **equicontinuous** if for any $\varepsilon > 0$ there exists $\delta > 0$ such that for all $\phi \in \mathcal{F}$ and all $x, y \in M$, if $d(x, y) < \delta$, then

$$|\phi(x) - \phi(y)| < \varepsilon.$$

Example 13.20. *Let (M^n, g) be a Riemannian manifold. Suppose that $\mathcal{F}$ is a family of functions on M^n that are uniformly Lipschitz; that is, there exists a positive constant C such that for all $\phi \in \mathcal{F}$ and all $x, y \in M^n$,*

$$|\phi(x) - \phi(y)| \leq C d(x, y). \tag{13.170}$$

Then $\mathcal{F}$ is an equicontinuous family. Indeed, given $\varepsilon > 0$, we may let $\delta = \frac{\varepsilon}{C}$ for the definition of equicontinuity.

In particular, if $\mathcal{F}$ is a family of differentiable functions on M^n that satisfy a uniform derivative bound, i.e., for all $\phi \in \mathcal{F}$ and $x \in M^n$,

$$\|d\phi_x\| \leq C, \tag{13.171}$$

then $\mathcal{F}$ is an equicontinuous family.

Now suppose that $\phi : M^n \to \mathbb{R}$ is twice differentiable. We have that its derivative is a real-valued function on the tangent bundle: $d\phi : TM \to \mathbb{R}$. Observe that if $\|\nabla d\phi\| \leq C$ on M^n, then we have that the derivative of the restricted function $d\phi : SM \to \mathbb{R}$ is bounded by C, where SM denotes the unit tangent bundle. Thus, if $\mathcal{F}$ is a family of twice-differentiable functions on M^n such that $\|\nabla d\phi\| \leq C$ for all $\phi \in \mathcal{F}$ for some constant C, then the family

$$\mathcal{G} := \{d\phi \colon \phi \in \mathcal{F}\} \tag{13.172}$$

of functions on SM is equicontinuous.

We have the following fundamental result in analysis; see, e.g., [**Rud76**, Theorem 7.25].

Theorem 13.21 (Arzelà and Ascoli)**.** *Suppose that (M, d) is a compact metric space. If $\{\phi_i\}$ is a uniformly bounded and equicontinuous sequence of real-valued functions on M, then there exists a subsequence $\{\phi_{i_j}\}$ that converges uniformly to a continuous function ϕ_∞ on M.*

We also have the following regarding the uniform convergence of derivatives; see, e.g., [**Rud76**, Theorem 7.17] for the 1-dimensional case.

Theorem 13.22. *Let (M^n, g) be a Riemannian manifold and let $\{\phi_i\}$ be a sequence of real-valued functions on M^n. Suppose that $\{\phi_i\}$ converges uniformly to a function ϕ_∞ and that $\{d\phi_i\}$ converges uniformly to a 1-form ψ_∞. Then ϕ_∞ is differentiable and $d\phi_\infty = \psi_\infty$; that is, $d\phi_i$ converges uniformly to $d\phi_\infty$.*

By combining the preceding theorem with Theorem 13.21, we obtain the following result.

Theorem 13.23. *Let (M^n, g) be a Riemannian manifold and let $\{\phi_i\}$ be a sequence of real-valued functions on M^n with the property that the functions and their first and second derivatives are uniformly bounded. Then there exists a subsequence $\{\phi_{i_j}\}$ such that $\{\phi_{i_j}\}$ converges uniformly to a continuously differentiable function ϕ_∞ on M^n and $\{d\phi_{i_j}\}$ converges uniformly to the function $d\phi_\infty$ on the unit tangent bundle SM.*

We remark that the subsequence $\{d\phi_{i_j}\}$ converging uniformly to the function $d\phi_\infty$ on SM implies that $\{d\phi_{i_j}\}$ converges uniformly to $d\phi_\infty$ as sections of the cotangent bundle T^*M, i.e., as maps from M to T^*M whose composition with the projection map $T^*M \to M^n$ is the identity map of M^n.

For higher derivatives, we have the following.

Theorem 13.24. *Let (M^n, g) be a Riemannian manifold and let $\{\phi_i\}$ be a sequence of real-valued functions on M^n with the property that there exists a constant C such that*

$$\|\nabla^m \phi_i\| \leq C \qquad \text{for } 0 \leq m \leq k+1,$$

where k is a nonnegative integer. Then there exists a subsequence $\{\phi_{i_j}\}$ such that $\{\phi_{i_j}\}$ converges uniformly to a C^k function ϕ_∞ on M^n and $\{\nabla^m \phi_{i_j}\}$ converges uniformly to the m-tensor $\nabla^m \phi_\infty$ (as sections of the tensor bundle $\bigotimes^m T^*M$) for $1 \leq m \leq k$.

The C^k**-norm** of a function ϕ is defined by

$$\|\phi\|_{C^k} = \sum_{m=0}^{k} \sup_{x \in M^n} \|\nabla^m \phi\|(x). \tag{13.173}$$

We may rephrase the conclusion of Theorem 13.24 as $\|\phi_{i_j} - \phi_\infty\|_{C^k} \to 0$.

We also recall the following result.

Lemma 13.25. *Let $\phi_t : X \to \mathbb{R}$, $t \in (0, T)$, where $T \in (0, \infty]$, be a family of functions on a set X with the property that*

$$|\partial_t \phi_t(x)| \leq \alpha(t) \tag{13.174}$$

for all $x \in X$ *and* $t \in [0,T)$, *where* $\alpha : [0,T) \to \mathbb{R}_+$ *is a function satisfying*

$$\int_0^T \alpha(t)\,dt < \infty. \tag{13.175}$$

Then there exists a function $\phi_T : X \to \mathbb{R}$ *such that* ϕ_t *converges uniformly to* ϕ_T *as* $t \to T$; *that is, for any* $\varepsilon > 0$, *there exists* $\delta > 0$ *such that for all* $x \in X$ *and* $t \in (T-\delta, T)$, *we have*

$$|\phi_t(x) - \phi_T(x)| < \varepsilon. \tag{13.176}$$

Proof. For any $0 \le t_1 < t_2 < T$ and $x \in X$, we have

$$|\phi_{t_1}(x) - \phi_{t_2}(x)| \le \int_{t_1}^{t_2} |\partial_t \phi_t(x)|\,dt \le \int_{t_1}^{t_2} \alpha(t)\,dt.$$

Now let $\varepsilon > 0$. By hypothesis, there exists $\delta > 0$ such that $\int_{t_1}^T \alpha(t)\,dt < \varepsilon$ provided $t_1 \ge T - \delta$. Thus, for any $T - \delta \le t_1 < t_2 < T$ and $x \in X$, we have

$$|\phi_{t_1}(x) - \phi_{t_2}(x)| < \varepsilon.$$

We leave it as an exercise to deduce the lemma. □

13.11. Singularity analysis

In this section, we begin the analysis of singularity formation for the curve shortening flow. Firstly, we discuss derivative of curvature estimates and the consequent long-time existence theorem which says that we can continue the solution as long as the curvature stays bounded. Complementarily, a solution is singular if its curvature is unbounded. We divide singular solutions into those whose curvature blow-up rate is comparable to the circle, called 'Type I', and those whose blow-up rate is faster, called 'Type II'. In the latter part of this section, we analyze the Type I case and show that in this case the CSF shrinks an embedded curve to a round point.

13.11.1. Derivative estimates and the long-time existence theorem. The long-time existence theorem for the curve shortening flow is as follows; see [**GH86**].

Theorem 13.26. *For any parametrized embedded closed curve* $F_o : S^1 \to \mathbb{R}^2$, *there exists a unique maximal solution to the curve shortening flow* $F_t : S^1 \to \mathbb{R}^2$, $t \in [0,T)$, *with* $F_0 = F_o$ *and* $T \in (0,\infty)$. *This solution has the properties that* F_t *parametrizes an embedded closed curve for each* $t \in [0,T)$ *and the curvature satisfies*

$$\sup_{S^1 \times [0,T)} \kappa(u,t) = \infty. \tag{13.177}$$

We will use this theorem as a black box, although we give the idea of its proof below. The essence of the proof of long-time existence is contained in the proof of Theorem 4.5 in [**GH86**], where they write:

> "Then κ has a limit as t goes to T which is C^∞ and we can extend the solution past T."

We remark that one of the ingredients of the proof of Theorem 13.26 comprises the Bernstein-type derivative of curvature estimates given as follows.

Proposition 13.27. *For each positive integer m, there exists a constant C_m depending only on m, τ, and K such that if $F_t : S^1 \to \mathbb{R}^2$, $t \in [0, \tau]$, is a curve shortening flow satisfying the curvature bound*

$$\sup_{(u,t)\in S^1\times[0,\tau]} |\kappa(u,t)| \leq K, \tag{13.178}$$

then

$$\left|\frac{\partial^m \kappa}{\partial s^m}\right|(u,t) \leq \frac{C_m}{t^{m/2}} \tag{13.179}$$

for all $u \in S^1$ and $t \in (0, \tau]$.

In particular, for $u \in S^1$ and $t \in \left[\frac{\tau}{2}, \tau\right]$,

$$\left|\frac{\partial^m \kappa}{\partial s^m}\right|(u,t) \leq \frac{2^{m/2} C_m}{\tau^{m/2}}. \tag{13.180}$$

Thus, for finite times t, a bound for the curvature yields a bound for the higher derivatives of the curvature of each order (for t not too small). To get a taste for the proof of the proposition, we provide the argument for the case of one derivative; i.e., $m = 1$. Using Corollary 13.15, we compute that

$$\frac{\partial}{\partial t}(\kappa^2) = \frac{\partial^2}{\partial s^2}(\kappa^2) - 2\left(\frac{\partial \kappa}{\partial s}\right)^2 + 2\kappa^4. \tag{13.181}$$

Using Lemma 13.14 in addition, we compute that

$$\begin{aligned} \frac{\partial}{\partial t}\left(\frac{\partial \kappa}{\partial s}\right) &= \frac{\partial}{\partial s}\left(\frac{\partial \kappa}{\partial t}\right) + \kappa^2 \frac{\partial \kappa}{\partial s} \\ &= \frac{\partial}{\partial s}\left(\frac{\partial^2 \kappa}{\partial s^2} + \kappa^3\right) + \kappa^2 \frac{\partial \kappa}{\partial s} \\ &= \frac{\partial^2}{\partial s^2}\left(\frac{\partial \kappa}{\partial s}\right) + 4\kappa^2 \frac{\partial \kappa}{\partial s}. \end{aligned} \tag{13.182}$$

Using this, we compute that

$$\frac{\partial}{\partial t}\left(\frac{\partial \kappa}{\partial s}\right)^2 = \frac{\partial^2}{\partial s^2}\left(\frac{\partial \kappa}{\partial s}\right)^2 - 2\left(\frac{\partial^2 \kappa}{\partial s^2}\right)^2 + 8\kappa^2\left(\frac{\partial \kappa}{\partial s}\right)^2. \tag{13.183}$$

Now we define the Bernstein-type gradient quantity

$$B := t\left(\frac{\partial \kappa}{\partial s}\right)^2 + b\kappa^2, \tag{13.184}$$

where $b \in \mathbb{R}_+$ is to be defined below. Using (13.181) and (13.183), we compute that

$$\begin{aligned}\frac{\partial B}{\partial t} &= \frac{\partial^2 B}{\partial s^2} + \left(\frac{\partial \kappa}{\partial s}\right)^2 - 2t\left(\frac{\partial^2 \kappa}{\partial s^2}\right)^2 + 8t\kappa^2\left(\frac{\partial \kappa}{\partial s}\right)^2 - 2b\left(\frac{\partial \kappa}{\partial s}\right)^2 + 2b\kappa^4 \\ &\leq \frac{\partial^2 B}{\partial s^2} + (-2b + 1 + 8t\kappa^2)\left(\frac{\partial \kappa}{\partial s}\right)^2 + 2b\kappa^4.\end{aligned}$$

Thus, choosing $b = \frac{1}{2}(1 + 8\tau K^2)$ and using that $|\kappa| \leq K$, we obtain that

$$\frac{\partial B}{\partial t} \leq \frac{\partial^2 B}{\partial s^2} + 2bK^4 \tag{13.185}$$

on $S^1 \times [0, \tau]$. Now, at $t = 0$ we have that $B \leq bK^2$. Therefore, by applying the maximum principle to (13.185), we obtain that

$$t\left(\frac{\partial \kappa}{\partial s}\right)^2 (u, t) \leq B(u, t) \leq bK^2 + 2bK^4\tau =: C_1^2 \tag{13.186}$$

for all $u \in S^1$ and $t \in [0, \tau]$. We conclude that

$$\left|\frac{\partial \kappa}{\partial s}\right|(u, t) \leq \frac{C_1}{\sqrt{t}} \tag{13.187}$$

for all $u \in S^1$ and $t \in (0, \tau]$. Clearly C_1 depends only on τ and K; thus, we have proved Proposition 13.27 in the case where $m = 1$.

We remark that if $\tau \leq \frac{2}{K^2}$, then $b \leq 9$ and by (13.186) we may take $C_1 = 7K$, so we obtain

$$\left|\frac{\partial \kappa}{\partial s}\right|(u, t) \leq \frac{7K}{\sqrt{t}} \tag{13.188}$$

for all $u \in S^1$ and $t \in (0, \tau]$. One can show that there exist universal constants C_m such that if $\tau \leq \frac{2}{K^2}$, then

$$\left|\frac{\partial^m \kappa}{\partial s^m}\right|(u, t) \leq \frac{C_m K}{t^{m/2}} \tag{13.189}$$

for all $u \in S^1$ and $t \in (0, \tau]$.

13.11.2. Singular solutions. As a consequence of (13.177), all solutions to the curve shortening flow on closed curves are **singular**. For each such solution, T is defined to be the **singularity time**.

By Example 13.6, if F_o parametrizes a circle of radius r_0, then $T := r_0^2/2$ is the singularity time. As $t \to T$, F_t parametrizes a circle whose radius tends to zero. See Figure 13.11.1.

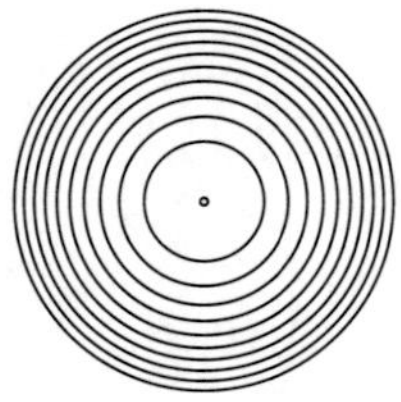

Figure 13.11.1. A shrinking circle. Shown are snapshots of the evolving circle taken at equal time steps. As the circle shrinks to a point, its speed increases to infinity.

13.11.2.1. *Type I singular solutions.* Observe that for the shrinking circle, by formula (13.57) its curvature satisfies $\kappa_t(u) \equiv \frac{1}{\sqrt{r_0^2-2t}}$. In other words,

$$\kappa_t(u) \equiv \frac{1}{\sqrt{2(T-t)}}. \tag{13.190}$$

In general, a singular solution $F_t : M^1 \to \mathbb{R}^2$, $t \in [0,T)$, to the curve shortening flow forms a **Type I singularity** (or is **Type I**) if there exists a constant C such that

$$|\kappa_t(u)| \leq \frac{C}{\sqrt{T-t}} \tag{13.191}$$

for all $u \in M^1$ and $t \in [0,T)$. Here, M^1 may be diffeomorphic to either a circle or a line. By (13.190), a shrinking circle forms a Type I singularity. So a Type I singular solution is one for which the 'curvature blow-up rate' is comparable to that of the shrinking round circle.

13.11.2.2. *Type II singular solutions.* If a singular solution is not Type I, then it is **Type II**. This case is characterized by

$$\sup_{M^1\times[0,T)} |\kappa_t(u)|\sqrt{T-t} = \infty. \tag{13.192}$$

Grayson [**Gra87**] originally proved that the curve shortening flow of embedded curves cannot form a Type II singularity. In fact, he showed that embedded curves eventually become convex. One can then invoke the Gage–Hamilton theorem to conclude that the curves shrink to round points. Such singularities are necessarily Type I.

On the other hand, Huisken's chord-arc estimate (see §13.12 below) can also be used to preclude the formation of Type II singularities.

We remark that for the curve shortening flow of immersed plane curves, Type II singularities can form. In fact, looking at certain solutions under a microscope yields the Grim Reaper translating self-similar solution as a limit.

13.11.3. Convergence of Type I singular solutions of the CSF. Let $F_t : M^1 \to \mathbb{R}^2$, $t \in [0,T)$, be a solution to the curve shortening flow. A point $\mathbf{y}_0$ in the plane is a **blow-up point** of the CSF if there exists $u_0 \in M^1$ such that

$$\lim_{t\to T} F_t(u_0) = \mathbf{y}_0 \quad \text{and} \quad \sup_{t\in[0,T)} |\kappa|(u_0,t) = \infty. \tag{13.193}$$

By translating our solution by a fixed vector, we will assume the following:

the blow-up point is the origin; i.e., $\mathbf{y}_0 = \mathbf{0}$.

It is easy to see that the parametrizations F_t converge uniformly to a map $F_T : M^1 \to \mathbb{R}^2$ as $t \to T$. This is because for any $u \in M^1$ and $0 \leq t_1 < t_2 < T$, we have

(13.194)

$$\|F_{t_2}(u) - F_{t_1}(u)\| = \left| \int_{t_1}^{t_2} |\kappa(u,t)|\, dt \right| \leq \int_{t_1}^{t_2} \frac{C}{\sqrt{T-t}} dt \leq 2C(t_2 - t_1)^{\frac{1}{2}}.$$

Later, we will see that $F_T(u) \equiv \mathbf{0}$.

Since $\mathbf{y}_0 = \mathbf{0}$ and by taking $t_2 \to T$ in (13.194) and calling $t_1 =: t$, we have that

$$\|F_t(u_0)\| = \|\mathbf{y}_0 - F_t(u_0)\| \leq 2C(T-t)^{1/2} \tag{13.195}$$

for all $t \in [0,T)$. Thinking of $F_t(u_0)$ as a 'basepoint' for the curve $F_t(S^1)$, we call this a **basepoint estimate**. This estimate is expected as qualitatively this is what a shrinking round circle satisfies, and in that case, the basepoint moves in a straight line with speed $\kappa(t) = \frac{1}{\sqrt{2(T-t)}}$, which is comparable to the Type I bound assumption.

To understand the singularity formation as time t approaches T, we rescale our singular solution in both space and time. Namely, define the new time variable and scaled parametrization by

$$\tilde{t}(t) := -\frac{1}{2}\ln(T-t), \tag{13.196a}$$

$$\widetilde{F}_{\tilde{t}} := (2(T-t))^{-1/2} F_t. \tag{13.196b}$$

The solution $\widetilde{F}_{\tilde{t}}$ is a **spacetime rescaling** of F_t. Observe that the new time parameter $\tilde{t}$ takes all values in the immortal time interval $[-\frac{1}{2}\ln T, \infty)$. We compute that

$$\begin{aligned}
(13.197) \qquad \frac{\partial \widetilde{F}_{\tilde{t}}}{\partial \tilde{t}} &= \left(\frac{\partial \tilde{t}}{\partial t}\right)^{-1} \left((2(T-t))^{-3/2} F_t + (2(T-t))^{-1/2} \frac{\partial F_t}{\partial t}\right) \\
&= (2(T-t))^{-1/2} F_t - (2(T-t))^{1/2} \kappa_t N_t \\
&= \widetilde{F}_{\tilde{t}} - \tilde{\kappa}_{\tilde{t}} \widetilde{N}_{\tilde{t}}.
\end{aligned}$$

This equation is defined to be the **normalized curve shortening flow**.

Next, we see what Huisken's monotonicity formula says about the normalized solution $\widetilde{F}_{\tilde{t}}$. Let

$$(13.198) \qquad \tau(t) := T - t.$$

Recall from Theorem 13.18 that if F_t is a curve shortening flow, then

$$\begin{aligned}
(13.199) \quad & \frac{d}{dt} \int_{M^1} (4\pi\tau(t))^{-1/2} \mathrm{e}^{-\frac{\|F_t(u)\|^2}{4\tau(t)}} \, ds_t(u) \\
&= -\int_{M^1} \left(\kappa(u,t) - \frac{F_t(u)\cdot N_t(u)}{2\tau(t)}\right)^2 (4\pi\tau(t))^{-1/2} \mathrm{e}^{-\frac{\|F_t(u)\|^2}{4\tau(t)}} \, ds_t(u).
\end{aligned}$$

The effect of rescaling on the quantities in the monotonicity formula is manifest as

$$\|\widetilde{F}_{\tilde{t}}\|^2 = \frac{\|F_t\|}{2\tau(t)}, \quad \tilde{\kappa}_{\tilde{t}} = \sqrt{2\tau(t)}\,\kappa_t, \quad \widetilde{N}_{\tilde{t}} = N_t, \quad d\tilde{s}_{\tilde{t}} = \frac{1}{\sqrt{2\tau(t)}} ds_t.$$

Thus,

$$\int_{M^1} (4\pi\tau(t))^{-\frac{1}{2}} \mathrm{e}^{-\frac{\|F_t\|^2}{4\tau(t)}} \, ds_t = \int_{M^1} (2\pi)^{-\frac{1}{2}} \exp\left(-\frac{\|\widetilde{F}_{\tilde{t}}\|^2}{2}\right) d\tilde{s}_{\tilde{t}}.$$

Similarly, we can express the right-hand side of (13.199) in terms of the normalized solution. We then see that

(13.200)

$$\frac{d}{d\tilde{t}} \int_{M^1} \exp\left(-\frac{\|\widetilde{F}_{\tilde{t}}\|^2}{2}\right) d\tilde{s}_{\tilde{t}} = -\int_{M^1} \left(\tilde{\kappa}_{\tilde{t}} - \widetilde{F}_{\tilde{t}} \cdot \widetilde{N}_{\tilde{t}}\right)^2 \exp\left(-\frac{\|\widetilde{F}_{\tilde{t}}\|^2}{2}\right) d\tilde{s}_{\tilde{t}}$$

for all $\tilde{t} \in [-\frac{1}{2}\ln T, \infty)$, where we factored out $(2\pi)^{-\frac{1}{2}}$ from both sides of the equation. By applying the fundamental theorem of calculus to this, we obtain

$$(13.201) \qquad \int_{-\frac{1}{2}\ln T}^{\infty} \int_{M^1} \left(\tilde{\kappa}_{\tilde{t}} - \widetilde{F}_{\tilde{t}} \cdot \widetilde{N}_{\tilde{t}}\right)^2 \exp\left(-\frac{\|\widetilde{F}_{\tilde{t}}\|^2}{2}\right) d\tilde{s}_{\tilde{t}} d\tilde{t} < \infty.$$

By (13.195), we have the basepoint bound for the normalized flow:

$$(13.202) \qquad \|\widetilde{F}_{\tilde{t}}(u_0)\| \leq \sqrt{2}C$$

for all $\tilde{t} \in [-\frac{1}{2}\ln T, \infty)$. This says that for each $\tilde{t} \in [-\frac{1}{2}\ln T, \infty)$, the curve $\widetilde{F}_{\tilde{t}}(M^1)$ intersects the closed ball $\overline{B}_{\sqrt{2}C}(\mathbf{0})$. Furthermore, by Proposition 13.27,[6] we have the Bernstein-type derivative estimates that for each $m \geq 0$ there exist constants C_m such that

$$\left|\frac{\partial^m \widetilde{\kappa}}{\partial s^m}(u, \tilde{t})\right| \leq C_m \tag{13.203}$$

for $u \in M^1$ and $\tilde{t} \in [1 - \frac{1}{2}\ln T, \infty)$.

Now let $\widetilde{\mathrm{L}}_{\tilde{t}}$ denote the length of $\widetilde{M}^1_{\tilde{t}} := \widetilde{F}_{\tilde{t}}(M^1)$. Let $S^1_{\tilde{t}} := \mathbb{R}/\widetilde{\mathrm{L}}_{\tilde{t}}\mathbb{Z}$ denote the circle of circumference $\widetilde{\mathrm{L}}_{\tilde{t}}$. Let

$$\widetilde{G}_{\tilde{t}} : S^1_{\tilde{t}} \to \mathbb{R}^2$$

denote the reparametrization by arc length of $\widetilde{F}_{\tilde{t}}$ such that the parameter 0 for $\widetilde{G}_{\tilde{t}}$ corresponds to the parameter u_0 for $\widetilde{F}_{\tilde{t}}$. The uniform derivatives of curvature bounds imply that there exist constants C'_m such that

$$\left\|\frac{\partial^m \widetilde{G}_{\tilde{t}}}{\partial s^m}(u, \tilde{t})\right\| \leq C'_m \tag{13.204}$$

for $u \in M^1$ and $\tilde{t} \in [1 - \frac{1}{2}\ln T, \infty)$. For example,

$$\left\|\frac{\partial^2 \widetilde{G}_{\tilde{t}}}{\partial s^2}\right\| = \| - \widetilde{\kappa}\widetilde{N}\| \leq |\widetilde{\kappa}| \leq C_0$$

and

$$\left\|\frac{\partial^3 \widetilde{G}_{\tilde{t}}}{\partial s^3}\right\| = \left\|-\frac{\partial \widetilde{\kappa}}{\partial s}\widetilde{N} - \widetilde{\kappa}\frac{\partial \widetilde{N}}{\partial s}\right\| \leq \left|\frac{\partial \widetilde{\kappa}}{\partial s}\right| + \widetilde{\kappa}^2 \leq C_1 + C_0^2.$$

Let $\{\tilde{t}_i\}$ be a sequence of times limiting to ∞. Then there exists a subsequence $\{\tilde{t}_i\}$ such that $\widetilde{\mathrm{L}}_{\tilde{t}_i}$ converges to some nonnegative extended real number $\widetilde{\mathrm{L}}_\infty$. Since the curvatures of the solution to the normalized CSF are uniformly bounded, we have that $\widetilde{\mathrm{L}}_\infty \in (0, \infty]$. By the basepoint bound (13.202), the derivative estimates in (13.204), and the Arzelà–Ascoli theorem (Theorem 13.23), we have that there exists a subsequence such that the maps $\widetilde{G}_{\tilde{t}_i} : S^1_{\tilde{t}_i} \to \mathbb{R}^2$ converge in each C^k-norm on compact subsets to a C^∞ complete immersion

$$\widetilde{G}_\infty : S^1_\infty := \mathbb{R}/\widetilde{\mathrm{L}}_\infty\mathbb{Z} \to \mathbb{R}^2. \tag{13.205}$$

The parametrized curve $\widetilde{G}_\infty$ is defined to be a **singularity model** of the curve shortening flow F_t since it is a limit of spacetime rescalings about a blow-up point of F_t.

[6] This proposition applies to the normalized curve shortening flow and we apply it with time step equal to 1.

Now we show that $\widetilde{G}_\infty$ is an embedding. By the same proof as for Proposition 13.10, we have that for the original solution F_t to the curve shortening flow:

(1) (cf. (13.111))

$$d_t(u_1,u_2) \geq \frac{2\sqrt{T-t}}{C}\sin\left(\frac{C\ell_t(u_1,u_2)}{2\sqrt{T-t}}\right) \tag{13.206}$$

for all $u_1, u_2 \in M^1$ and $t \in [0,T)$ satisfying $\ell_t(u_1,u_2) \leq \frac{\pi\sqrt{T-t}}{C}$;

(2) the function

$$\phi(t) := \min_{\ell_t(u_1,u_2)\geq\frac{\pi\sqrt{T-t}}{C}} d_t^2(u_1,u_2) \tag{13.207}$$

satisfies

$$\frac{d}{dt}\phi(t) \geq 0 \tag{13.208}$$

whenever $\phi(t) < \frac{4(T-t)}{C^2}$.

Thus, the solution $\widetilde{G}_{\tilde{t}}$ to the normalized curve shortening flow has the following properties:

(1) For all $u_1, u_2 \in M^1$ and $\tilde{t} \in [-\frac{1}{2}\ln T, \infty)$ satisfying $\tilde{\ell}_{\tilde{t}}(u_1,u_2) \leq \frac{\pi}{2C}$, we have

$$\tilde{d}_{\tilde{t}}(u_1,u_2) \geq \frac{1}{C}\sin\left(C\tilde{\ell}_{\tilde{t}}(u_1,u_2)\right). \tag{13.209}$$

(2) The function

$$\tilde{\phi}(\tilde{t}) := \min_{\tilde{\ell}_{\tilde{t}}(u_1,u_2)\geq\frac{\pi}{K}} \tilde{d}_{\tilde{t}}^2(u_1,u_2) \tag{13.210}$$

satisfies

$$\frac{d}{d\tilde{t}}\tilde{\phi}(\tilde{t}) \geq 0 \tag{13.211}$$

whenever $\tilde{\phi}(\tilde{t}) < \frac{\pi}{2C}$.

It follows from the above that the singularity model $\widetilde{G}_\infty$ is an embedding.

Now we recall from (13.201) that

$$\int_{-\frac{1}{2}\ln T}^{\infty}\int_{M^1}\left(\tilde{\kappa}_{\tilde{t}} - \widetilde{G}_{\tilde{t}}\cdot\widetilde{N}_{\tilde{t}}\right)^2\exp\left(-\frac{\|\widetilde{G}_{\tilde{t}}\|^2}{2}\right)d\tilde{s}_{\tilde{t}}d\tilde{t} < \infty. \tag{13.212}$$

Because the convergence of $\widetilde{G}_{\tilde{t}}$ to $\widetilde{G}_\infty$ as $\tilde{t}\to\infty$ is in each C^k on compact subsets, from this we conclude that $\widetilde{G}_\infty$ satisfies

$$\tilde{\kappa}_\infty - \widetilde{G}_\infty\cdot\widetilde{N}_\infty = 0; \tag{13.213}$$

cf. (13.169). That is, $\widetilde{G}_\infty$ is a complete embedded shrinking self-similar solution of the curve shortening flow.

We have the following result of Abresch and Langer [**AL86**]; see [**AL86**] for the proof.

Theorem 13.28. *Any complete embedded shrinking self-similar solution of the curve shortening flow is either a straight line through the origin or the circle of radius $\sqrt{2}$ centered at the origin.*

Corollary 13.29. *If F_t is a Type I singular solution to the CSF, then the singularity model $\widetilde{G}_\infty$ parametrizes either a straight line through the origin or the circle of radius $\sqrt{2}$ centered at the origin.*

The case of a straight line singularity model can be ruled out by a local regularity theorem; see White [**Whi05**]. Thus $\widetilde{G}_\infty$ must parametrize the circle of radius $\sqrt{2}$ centered at the origin, which implies that the original solution F_t parametrizes a curve that shrinks to a round point.

13.12. Huisken's chord-arc estimate

We now discuss Huisken's chord-arc estimate [**Hui98**].

Let $F_t : S^1 \to \mathbb{R}^2$, $t \in [0, T)$, be an embedded curve shortening flow, where $S^1 := \mathbb{R}/2\pi\mathbb{Z}$. Without loss of generality, we may assume that F_t parametrizes $M_t^1 := F_t(S^1)$ in the counterclockwise direction. That is, as $u \in \mathbb{R}$ increases, $F_t(\mathrm{e}^{iu})$ traverses M_t^1 in the counterclockwise direction. For simplicity, we denote the equivalence class in S^1 of $u \in \mathbb{R}$ redundantly by u itself instead of the usual $[u]$ notation. So u and $u + 2\pi k$, where $k \in \mathbb{Z}$, denote the same point in S^1.

Recall that the Euclidean distance between the points $F_t(u_1)$ and $F_t(u_2)$ on M_t^1 is defined by

$$d_t(u_1, u_2) := \|F_t(u_1) - F_t(u_2)\|, \tag{13.214}$$

where the norm $\|\cdot\|$ is the Euclidean norm of $\mathbb{R}^2$. This is the length of the straight *chord* between $F_t(u_1)$ and $F_t(u_2)$.

The **arc length** between points $F_t(u_1)$ and $F_t(u_2)$ on M_t^1 is defined by

$$\ell_t(u_1, u_2) := \int_{u_1}^{u_2} ds_t, \tag{13.215}$$

where this is the arc length along M_t^1 from $F_t(u_1)$ to $F_t(u_2)$ in the counterclockwise direction. That is, given $u_1 \in \mathbb{R}$, then u_2 is defined as $u_2 + 2\pi k$, where k is the smallest integer such that $u_2 + 2\pi k \geq u_1$.

Note that $d_t(u_1, u_2)$ and $\ell_t(u_1, u_2)$ are the extrinsic and intrinsic distances between $F_t(u_1)$ and $F_t(u_2)$, respectively.

Example 13.30 (Round circles). *For the curve shortening flow of closed embedded curves, shrinking round circles are model cases. Let $F_t : S^1 \to \mathbb{R}^2$ be the parametrization of the shrinking circle of radius $r_t := \sqrt{r_0^2 - 2t}$ defined by*

$$F_t(u) := \big(r_t \cos(u), r_t \sin(u)\big). \tag{13.216}$$

We have that $\mathrm{L}_t = 2\pi r_t$ is the length of $M_t^1 = F_t(S^1)$. Via the canonical projection $\mathbb{R} \to \mathbb{R}/2\pi\mathbb{Z} = S^1$, we may assume that $u \in \mathbb{R}$. Let $u_2 = u_1 + \theta$, where $\theta \in [0, 2\pi)$. Then $\ell_t(u_1, u_2) = r_t\theta$. By elementary trigonometry, we have that

$$d_t(u_1, u_2) = 2r_t \sin\left(\frac{\theta}{2}\right) \tag{13.217}$$

if $\theta \in [0, \pi]$. Therefore, provided $\theta \in [0, \pi]$, we have

$$\frac{\mathrm{L}_t}{d_t} \sin\left(\frac{\pi \ell_t}{\mathrm{L}_t}\right) = \frac{2\pi r_t}{2r_t \sin\left(\frac{\theta}{2}\right)} \sin\left(\frac{\pi r_t \theta}{2\pi r_t}\right) = \pi.$$

This, of course, is a constant independent of θ and t.

Let $\triangle := \{(u, u) : u \in S^1\}$ be the diagonal of $S^1 \times S^1$. In view of (13.217), we define for any embdedded curve shortening flow the function $\ell' : \big((S^1 \times S^1) \setminus \triangle\big) \times [0, T) \to \mathbb{R}$ by

$$\ell'_t(u_1, u_2) := \mathrm{L}_t \sin\left(\frac{\pi \ell_t(u_1, u_2)}{\mathrm{L}_t}\right), \tag{13.218}$$

where L_t is the length of M_t^1. Define the **chord-arc comparison function** $\Phi : [0, T) \to \mathbb{R}$ by

$$\Phi(t) := \sup_{(S^1 \times S^1)\setminus\triangle} \frac{\ell'_t}{d_t} = \sup_{(S^1 \times S^1)\setminus\triangle} \frac{\mathrm{L}_t}{d_t} \sin\left(\frac{\pi \ell_t}{\mathrm{L}_t}\right), \tag{13.219}$$

where ℓ_t is the arc length, d_t is the chord length, and L_t is the total arc length, all at time t. Note that the quantity Φ is scale-invariant since the quantities L, d, and ℓ all scale the same. See Figure 13.12.1.

Since the quantity $\frac{\ell'_t}{d_t} = \frac{\mathrm{L}_t}{d_t} \sin\left(\frac{\pi \ell_t}{\mathrm{L}_t}\right)$ is constant in the special case of shrinking round circles, Φ may be a good candidate for being a monotone quantity under the CSF. Indeed, this is the case. The proof of the following takes some effort.

Theorem 13.31 (Chord-arc monotonicity). *If $F_t : S^1 \to \mathbb{R}^2$, $t \in [0, T)$, is an embedded curve shortening flow, then the chord-arc comparison function*

$$t \mapsto \Phi(t)$$

defined by (13.219) is a nonincreasing function of t. If F_t is not a round shrinking circle, then $t \mapsto \Phi(t)$ is strictly decreasing.

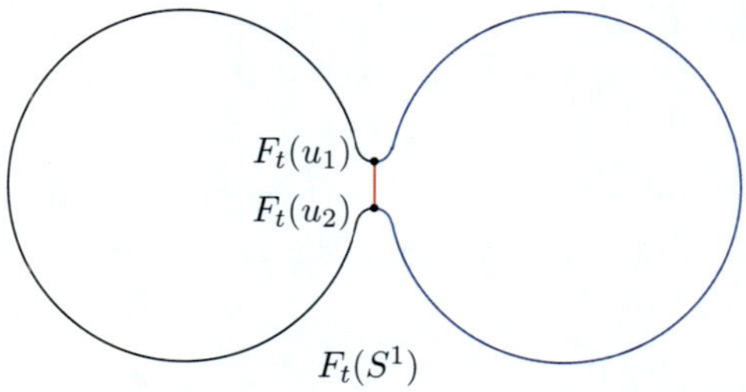

Figure 13.12.1. The red vertical segment is the 'chord' and the blue curve is one of the 'arcs', each from $F_t(u_1)$ to $F_t(u_2)$. For this embedded closed curve $F_t(S^1)$, Huisken's chord-arc quantity $\Phi(t)$ is large since d_t but not ℓ_t is small relative to L_t.

Proof. The original proof is by Huisken [**Hui98**], which we follow in conjunction with the exposition by Brendle [**Bre14**]. Suppose, for a contradiction, that the theorem were false. Then there would exist a real number α and times $t_1 < t_2$ such that

$$\Phi(t_1) < \alpha < \Phi(t_2). \tag{13.220}$$

In view of the definition of Φ, we define the function $Z : S^1 \times S^1 \times [0,T) \to \mathbb{R}$ by

$$Z_t(u_1,u_2) := Z(u_1,u_2,t) := -\mathrm{L}_t \sin\left(\frac{\pi\,\ell_t(u_1,u_2)}{\mathrm{L}_t}\right) + \alpha\, d_t(u_1,u_2), \tag{13.221}$$

where $d_t(u_1,u_2) = \|F_t(u_1) - F_t(u_2)\|$ and $\ell_t(u_1,u_2) = \int_{u_1}^{u_2} ds_t$ are as defined before.

Since $\Phi(t_1) < \alpha$, we have $Z_{t_1}(u_1,u_2) \geq 0$ for all $u_1,u_2 \in S^1$. Note also that for u_1 and u_2 such that $\ell_t(u_1,u_2)$ is small, we have that

$$Z_t(u_1,u_2) = (-\pi+\alpha)\ell_t(u_1,u_2) + \mathrm{O}\big(\ell_t(u_1,u_2)^2\big) \tag{13.222}$$

for any t. Since, e.g., $Z_{t_1} \geq 0$, we have $\alpha \geq \pi$. Clearly, we may assume that $\alpha > \pi$ and still have that (13.220) is satisfied. This then implies that Z_t is positive near, but not at, the diagonal $\triangle$ of $S^1 \times S^1$ for all $t \in [t_1,t_2]$.

By (13.220), for some $u_2 \neq v_2$ and for all $u_1 \neq v_1$, we have

$$Z_{t_1}(u_1,v_1) > 0 > Z_{t_2}(u_2,v_2). \tag{13.223}$$

By the properties above and since Z is a continuous function, there exists a first time $\bar{t} \in (t_1,t_2)$ such that there exist $\overline{u}_1,\overline{u}_2 \in (S^1 \times S^1) \setminus \triangle$ such that

$$Z_{\bar{t}}(\overline{u}_1,\overline{u}_2) = 0 \tag{13.224}$$

and

$$Z_{\bar{t}}(u_1,u_2) \geq 0 \quad \text{for all } (u_1,u_2) \in S^1 \times S^1. \tag{13.225}$$

Note that then $Z_t > 0$ on $(S^1 \times S^1) \setminus \Delta$ for $t \in [t_1, \bar{t})$. The idea now is to derive a contradiction by establishing a heat-type equation for Z at $(\overline{u}_1, \overline{u}_2, \bar{t})$.

In view of the invariance of the curve shortening flow under reparametrizations, we may reparametrize the solution to the curve shortening flow so that $F_{\bar{t}}$ satisfies

$$\left\| \frac{\partial F_{\bar{t}}}{\partial u_1}(\overline{u}_1) \right\| = 1 \quad \text{and} \quad \left\| \frac{\partial F_{\bar{t}}}{\partial u_2}(\overline{u}_2) \right\| = 1. \tag{13.226}$$

That is, $\frac{\partial F_{\bar{t}}}{\partial u_1}(\overline{u}_1)$ and $\frac{\partial F_{\bar{t}}}{\partial u_2}(\overline{u}_2)$ are each unit tangent vectors. Let

$$V_{\bar{t}}(\overline{u}_1, \overline{u}_2) := F_{\bar{t}}(\overline{u}_1) - F_{\bar{t}}(\overline{u}_2). \tag{13.227}$$

By the first derivative test, we have

$$\begin{aligned} 0 &= \frac{\partial Z_{\bar{t}}}{\partial u_1}(\overline{u}_1, \overline{u}_2) \\ &= \pi \cos\left(\frac{\pi \, \ell_{\bar{t}}(\overline{u}_1, \overline{u}_2)}{\mathrm{L}_{\bar{t}}} \right) + \alpha \frac{V_{\bar{t}}(\overline{u}_1, \overline{u}_2) \cdot \frac{\partial F_{\bar{t}}}{\partial u_1}(\overline{u}_1)}{d_{\bar{t}}(\overline{u}_1, \overline{u}_2)} \end{aligned} \tag{13.228}$$

since $\frac{\partial \ell_{\bar{t}}}{\partial u_1}(\overline{u}_1, \overline{u}_2) = -1$ by (13.226). Similarly, we have

$$\begin{aligned} 0 &= \frac{\partial Z_{\bar{t}}}{\partial u_2}(\overline{u}_1, \overline{u}_2) \\ &= -\pi \cos\left(\frac{\pi \, \ell_{\bar{t}}(\overline{u}_1, \overline{u}_2)}{\mathrm{L}_{\bar{t}}} \right) - \alpha \frac{V_{\bar{t}}(\overline{u}_1, \overline{u}_2) \cdot \frac{\partial F_{\bar{t}}}{\partial u_2}(\overline{u}_2)}{d_{\bar{t}}(\overline{u}_1, \overline{u}_2)} \end{aligned} \tag{13.229}$$

since $\frac{\partial \ell_{\bar{t}}}{\partial u_2}(\overline{u}_1, \overline{u}_2) = 1$. By adding the two equations together, we obtain

$$V_{\bar{t}}(\overline{u}_1, \overline{u}_2) \cdot \left(\frac{\partial F_{\bar{t}}}{\partial u_1}(\overline{u}_1) - \frac{\partial F_{\bar{t}}}{\partial u_2}(\overline{u}_2) \right) = 0. \tag{13.230}$$

We will find this identity useful.

Next, we consider the second derivatives of Z at $(\overline{u}_1, \overline{u}_2, \overline{t})$:

(13.231)

$$\begin{aligned}\frac{\partial^2 Z}{\partial u_1^2}(\overline{u}_1, \overline{u}_2, \overline{t}) &= \frac{\pi^2}{\mathrm{L}_{\overline{t}}} \sin\left(\frac{\pi \ell_{\overline{t}}(\overline{u}_1, \overline{u}_2)}{\mathrm{L}_{\overline{t}}}\right) + \alpha \frac{\left(\frac{\partial F}{\partial u_1} \cdot \frac{\partial F}{\partial u_1}\right)(\overline{u}_1, \overline{t})}{d(\overline{u}_1, \overline{u}_2, \overline{t})} \\ &\quad + \alpha \frac{V_{\overline{t}}(\overline{u}_1, \overline{u}_2) \cdot \frac{\partial^2 F}{\partial u_1^2}(\overline{u}_1, \overline{t})}{d(\overline{u}_1, \overline{u}_2, \overline{t})} - \alpha \frac{\left(V_{\overline{t}}(\overline{u}_1, \overline{u}_2) \cdot \frac{\partial F}{\partial u_1}(\overline{u}_1, \overline{t})\right)^2}{d(\overline{u}_1, \overline{u}_2, \overline{t})^3} \\ &= \frac{\pi^2}{\mathrm{L}_{\overline{t}}} \sin\left(\frac{\pi \ell_{\overline{t}}(\overline{u}_1, \overline{u}_2)}{\mathrm{L}_{\overline{t}}}\right) + \frac{\alpha}{d(\overline{u}_1, \overline{u}_2, \overline{t})} \\ &\quad - \alpha \frac{\kappa(\overline{u}_1, \overline{t}) V_{\overline{t}}(\overline{u}_1, \overline{u}_2) \cdot N(\overline{u}_1, \overline{t})}{d(\overline{u}_1, \overline{u}_2, \overline{t})} \\ &\quad - \alpha \frac{\left(V_{\overline{t}}(\overline{u}_1, \overline{u}_2) \cdot \frac{\partial F}{\partial u_1}(\overline{u}_1, \overline{t})\right)^2}{d(\overline{u}_1, \overline{u}_2, \overline{t})^3},\end{aligned}$$

where we used (13.226). Similarly, we compute that

$$\begin{aligned}\frac{\partial^2 Z}{\partial u_2^2}(\overline{u}_1, \overline{u}_2, \overline{t}) &= \frac{\pi^2}{\mathrm{L}_{\overline{t}}} \sin\left(\frac{\pi \ell_{\overline{t}}(\overline{u}_1, \overline{u}_2)}{\mathrm{L}_{\overline{t}}}\right) + \frac{\alpha}{d(\overline{u}_1, \overline{u}_2, \overline{t})} \\ &\quad + \alpha \frac{\kappa(\overline{u}_2, \overline{t}) V_{\overline{t}}(\overline{u}_1, \overline{u}_2) \cdot N(\overline{u}_2, \overline{t})}{d(\overline{u}_1, \overline{u}_2, \overline{t})} \\ &\quad - \alpha \frac{\left(V_{\overline{t}}(\overline{u}_1, \overline{u}_2) \cdot \frac{\partial F}{\partial u_2}(\overline{u}_2, \overline{t})\right)^2}{d(\overline{u}_1, \overline{u}_2, \overline{t})^3}.\end{aligned} \tag{13.232}$$

Typically, we might anticipate, regarding the expectation of deriving a heat-type equation for Z, that it is natural to consider the Laplacian $\frac{\partial^2 Z}{\partial u_1^2} + \frac{\partial^2 Z}{\partial u_2^2}$ on $S^1 \times S^1$ at each time $t \in [0, T)$. However, for our considerations, the operator on Z given by

$$\frac{\partial^2 Z}{\partial u_1^2} + \frac{\partial^2 Z}{\partial u_2^2} - 2\frac{\partial^2 Z}{\partial u_1 \partial u_2} = \left(\frac{\partial}{\partial u_1} - \frac{\partial}{\partial u_2}\right)^2 Z \tag{13.233}$$

turns out to be more useful. So we calculate

(13.234)

$$\begin{aligned}\frac{\partial^2 Z}{\partial u_1 \partial u_2}(\overline{u}_1, \overline{u}_2, \overline{t}) &= -\frac{\pi^2}{\mathrm{L}_{\overline{t}}} \sin\left(\frac{\pi \ell_{\overline{t}}(\overline{u}_1, \overline{u}_2)}{\mathrm{L}_{\overline{t}}}\right) - \alpha \frac{\frac{\partial F}{\partial u_1}(\overline{u}_1, \overline{t}) \cdot \frac{\partial F}{\partial u_2}(\overline{u}_2, \overline{t})}{d(\overline{u}_1, \overline{u}_2, \overline{t})} \\ &\quad + \alpha \frac{\left(V_{\overline{t}}(\overline{u}_1, \overline{u}_2) \cdot \frac{\partial F_{\overline{t}}}{\partial u_1}(\overline{u}_1)\right)\left(V_{\overline{t}}(\overline{u}_1, \overline{u}_2) \cdot \frac{\partial F_{\overline{t}}}{\partial u_2}(\overline{u}_2)\right)}{d_{\overline{t}}(\overline{u}_1, \overline{u}_2)^3}.\end{aligned}$$

Using the formulas above, while noting that the $\frac{\alpha}{d}$ terms cancel and using (13.230) to cancel more terms, we compute that

$$\begin{aligned}(13.235)\quad &\left(\frac{\partial^2 Z}{\partial u_1^2}+\frac{\partial^2 Z_t}{\partial u_2^2}-2\frac{\partial^2 Z_t}{\partial u_1\partial u_2}\right)(\overline{u}_1,\overline{u}_2,\overline{t})\\ &=\frac{4\pi^2}{\mathrm{L}_{\overline{t}}}\sin\left(\frac{\pi\,\ell_{\overline{t}}(\overline{u}_1,\overline{u}_2)}{\mathrm{L}_{\overline{t}}}\right)-\alpha\frac{\kappa(\overline{u}_1,\overline{t})V_{\overline{t}}(\overline{u}_1,\overline{u}_2)\cdot N(\overline{u}_1,\overline{t})}{d(\overline{u}_1,\overline{u}_2,\overline{t})}\\ &\quad+\alpha\frac{\kappa(\overline{u}_2,\overline{t})V_{\overline{t}}(\overline{u}_1,\overline{u}_2)\cdot N(\overline{u}_2,\overline{t})}{d(\overline{u}_1,\overline{u}_2,\overline{t})}.\end{aligned}$$

Next, we calculate the time-derivative of Z at $(\overline{u}_1,\overline{u}_2,\overline{t})$ as follows. Using

$$\frac{\partial d_t}{\partial t}(\overline{u}_1,\overline{u}_2)=-\frac{\kappa(\overline{u}_1,\overline{t})V_{\overline{t}}(\overline{u}_1,\overline{u}_2)\cdot N(\overline{u}_1,\overline{t})}{d(\overline{u}_1,\overline{u}_2,\overline{t})}+\frac{\kappa(\overline{u}_2,\overline{t})V_{\overline{t}}(\overline{u}_1,\overline{u}_2)\cdot N(\overline{u}_2,\overline{t})}{d(\overline{u}_1,\overline{u}_2,\overline{t})}$$

and

(13.236)

$$\frac{\partial Z}{\partial t}(\overline{u}_1,\overline{u}_2,\overline{t})=-\left(\frac{\partial}{\partial t}\bigg|_{t=\overline{t}}\mathrm{L}_t\sin\left(\frac{\pi\,\ell_t(\overline{u}_1,\overline{u}_2)}{\mathrm{L}_t}\right)\right)+\alpha\left.\frac{\partial d_t}{\partial t}(\overline{u}_1,\overline{u}_2)\right|_{t=\overline{t}},$$

we obtain

$$\begin{aligned}(13.237)\quad &\left(-\frac{\partial Z}{\partial t}+\frac{\partial^2 Z}{\partial u_1^2}+\frac{\partial^2 Z}{\partial u_2^2}-2\frac{\partial^2 Z}{\partial u_1\partial u_2}\right)(\overline{u}_1,\overline{u}_2,\overline{t})\\ &=\frac{4\pi^2}{\mathrm{L}_{\overline{t}}}\sin\left(\frac{\pi\,\ell_{\overline{t}}(\overline{u}_1,\overline{u}_2)}{\mathrm{L}_{\overline{t}}}\right)+\frac{\partial}{\partial t}\bigg|_{t=\overline{t}}\left(\mathrm{L}_t\sin\left(\frac{\pi\,\ell_t(\overline{u}_1,\overline{u}_2)}{\mathrm{L}_t}\right)\right).\end{aligned}$$

Now, by (13.224) and (13.225), we have that

$$(13.238)\qquad \frac{\partial Z_{\overline{t}}}{\partial t}(\overline{u}_1,\overline{u}_2)\le 0.$$

Also, by the second derivative test,

$$\left(\frac{\partial^2 Z}{\partial u_1^2}+\frac{\partial^2 Z}{\partial u_2^2}-2\frac{\partial^2 Z}{\partial u_1\partial u_2}\right)(\overline{u}_1,\overline{u}_2,\overline{t})=\left(\left(\frac{\partial}{\partial u_1}-\frac{\partial}{\partial u_2}\right)^2 Z\right)(\overline{u}_1,\overline{u}_2,\overline{t})\ge 0.$$

Therefore, we have the inequality

$$(13.239)\qquad 0\le\frac{4\pi^2}{\mathrm{L}_{\overline{t}}}\sin\left(\frac{\pi\,\ell_{\overline{t}}(\overline{u}_1,\overline{u}_2)}{\mathrm{L}_{\overline{t}}}\right)+\frac{\partial}{\partial t}\bigg|_{t=\overline{t}}\left(\mathrm{L}_t\sin\left(\frac{\pi\,\ell_t(\overline{u}_1,\overline{u}_2)}{\mathrm{L}_t}\right)\right).$$

To evaluate the time-derivative on the right-hand side, we first calculate and estimate the evolution equations for ℓ_t and L_t. We have

$$(13.240)\qquad \left.\frac{\partial\ell}{\partial t}(\overline{u}_1,\overline{u}_2,t)\right|_{t=\overline{t}}=-\int_{\overline{u}_1}^{\overline{u}_2}\kappa_{\overline{t}}(u)^2 ds_{\overline{t}}(u)$$

and

$$\frac{d\,\mathrm{L}}{dt}(\bar{t}) = -\int_{S^1} \kappa_{\bar{t}}^2 ds_{\bar{t}}. \tag{13.241}$$

On the other hand, by the Cauchy–Schwarz inequality for integrals,

$$\left(\int \kappa ds\right)^2 \leq \int \kappa^2 ds \int ds, \tag{13.242}$$

which is an equality if and only if κ is constant. By our hypothesis, $M_{\bar{t}}^1$ does not have constant curvature. Therefore,

$$\int_{S^1} \kappa_{\bar{t}}^2 ds_{\bar{t}} > \frac{1}{\mathrm{L}_{\bar{t}}}\left(\int_{S^1} \kappa_{\bar{t}} ds_{\bar{t}}\right)^2 = \frac{4\pi^2}{\mathrm{L}_{\bar{t}}}. \tag{13.243}$$

Thus, by (13.241), we have that

$$\frac{d\,\mathrm{L}}{dt}(\bar{t}) < -\frac{4\pi^2}{\mathrm{L}_{\bar{t}}}. \tag{13.244}$$

Similarly,

$$\int_{\overline{u}_1}^{\overline{u}_2} \kappa_{\bar{t}}(u)^2 ds_{\bar{t}}(u) \geq \frac{1}{\ell_{\bar{t}}(\overline{u}_1, \overline{u}_2)}\left(\int_{\overline{u}_1}^{\overline{u}_2} \kappa_{\bar{t}}(u)\, ds_{\bar{t}}(u)\right)^2. \tag{13.245}$$

We proceed to evaluate the right-hand side. Now,

$$\begin{aligned}\int_{\overline{u}_1}^{\overline{u}_2} \kappa_{\bar{t}}(u)\, ds_{\bar{t}}(u) &= \int_{\overline{u}_1}^{\overline{u}_2} d\theta_{\bar{t}}(u) = \angle\left(\frac{\partial F}{\partial u_1}(\overline{u}_1, \bar{t}), \frac{\partial F}{\partial u_2}(\overline{u}_2, \bar{t})\right) \\ &= \cos^{-1}\left(\frac{\partial F}{\partial u_1}(\overline{u}_1, \bar{t}) \cdot \frac{\partial F}{\partial u_2}(\overline{u}_2, \bar{t})\right),\end{aligned} \tag{13.246}$$

where $\angle$ denotes angle. Let

$$W_{\bar{t}}(\overline{u}_1, \overline{u}_2) := -\frac{V_{\bar{t}}(\overline{u}_1, \overline{u}_2)}{d_{\bar{t}}(\overline{u}_1, \overline{u}_2)} = \frac{F_{\bar{t}}(\overline{u}_2) - F_{\bar{t}}(\overline{u}_1)}{d_{\bar{t}}(\overline{u}_1, \overline{u}_2)}, \tag{13.247}$$

which is a unit vector. By (13.230) and (13.228), we have

$$W_{\bar{t}}(\overline{u}_1, \overline{u}_2) \cdot \frac{\partial F_{\bar{t}}}{\partial u_1}(\overline{u}_1) = W_{\bar{t}}(\overline{u}_1, \overline{u}_2) \cdot \frac{\partial F_{\bar{t}}}{\partial u_2}(\overline{u}_2) = \frac{1}{\alpha}\pi \cos\left(\frac{\pi \ell_{\bar{t}}(\overline{u}_1, \overline{u}_2)}{\mathrm{L}_{\bar{t}}}\right). \tag{13.248}$$

That is, the unit vector $W_{\bar{t}}(\overline{u}_1, \overline{u}_2)$ bisects the unit tangent vectors $\frac{\partial F_{\bar{t}}}{\partial u_1}(\overline{u}_1)$ and $\frac{\partial F_{\bar{t}}}{\partial u_2}(\overline{u}_2)$. Define

$$\begin{aligned}\xi &= \cos^{-1}\left(W_{\bar{t}}(\overline{u}_1, \overline{u}_2) \cdot \frac{\partial F_{\bar{t}}}{\partial u_1}(\overline{u}_1)\right) \\ &= \cos^{-1}\left(W_{\bar{t}}(\overline{u}_1, \overline{u}_2) \cdot \frac{\partial F_{\bar{t}}}{\partial u_2}(\overline{u}_2)\right).\end{aligned} \tag{13.249}$$

By (13.246), we then have

$$\int_{\overline{u}_1}^{\overline{u}_2} \kappa_{\bar{t}}(u)\,ds_{\bar{t}}(u) = \cos^{-1}\left(\frac{\partial F}{\partial u_1}(\overline{u}_1,\bar{t})\cdot\frac{\partial F}{\partial u_2}(\overline{u}_2,\bar{t})\right) = 2\xi, \tag{13.250}$$

where the last equality is by elementary trigonometry. Thus, by (13.240) and (13.245), we have

$$\frac{\partial \ell}{\partial t}(\overline{u}_1,\overline{u}_2,t)\bigg|_{t=\bar{t}} \le -\frac{1}{\ell_{\bar{t}}(\overline{u}_1,\overline{u}_2)}4\xi^2. \tag{13.251}$$

Therefore, by (13.239), we have

(13.252)

$$\begin{aligned}
0 &\le \frac{4\pi^2}{\mathrm{L}_{\bar{t}}}\sin\left(\frac{\pi\,\ell_{\bar{t}}(\overline{u}_1,\overline{u}_2)}{\mathrm{L}_{\bar{t}}}\right) + \frac{\partial}{\partial t}\bigg|_{t=\bar{t}}\left(\mathrm{L}_t\sin\left(\frac{\pi\,\ell_t(\overline{u}_1,\overline{u}_2)}{\mathrm{L}_t}\right)\right)\\
&= \left(\frac{4\pi^2}{\mathrm{L}_{\bar{t}}} + \frac{d\,\mathrm{L}}{dt}(\bar{t})\right)\sin\left(\frac{\pi\,\ell_{\bar{t}}(\overline{u}_1,\overline{u}_2)}{\mathrm{L}_{\bar{t}}}\right)\\
&\quad + \mathrm{L}_{\bar{t}}\cos\left(\frac{\pi\,\ell_{\bar{t}}(\overline{u}_1,\overline{u}_2)}{\mathrm{L}_{\bar{t}}}\right)\frac{\pi\ell_{\bar{t}}(\overline{u}_1,\overline{u}_2)}{\mathrm{L}_{\bar{t}}}\left(\frac{\frac{\partial\ell}{\partial t}(\overline{u}_1,\overline{u}_2,t)\big|_{t=\bar{t}}}{\ell_{\bar{t}}(\overline{u}_1,\overline{u}_2)} - \frac{\frac{d\mathrm{L}}{dt}(\bar{t})}{\mathrm{L}_{\bar{t}}}\right).
\end{aligned}$$

This says that

(13.253)

$$\begin{aligned}
0 &\le \frac{4\pi^2}{\mathrm{L}_{\bar{t}}}\sin\left(\frac{\pi\,\ell_{\bar{t}}(\overline{u}_1,\overline{u}_2)}{\mathrm{L}_{\bar{t}}}\right) + \pi\cos\left(\frac{\pi\,\ell_{\bar{t}}(\overline{u}_1,\overline{u}_2)}{\mathrm{L}_{\bar{t}}}\right)\frac{\partial\ell}{\partial t}(\overline{u}_1,\overline{u}_2,t)\bigg|_{t=\bar{t}}\\
&\quad + \left(\sin\left(\frac{\pi\,\ell_{\bar{t}}(\overline{u}_1,\overline{u}_2)}{\mathrm{L}_{\bar{t}}}\right) - \frac{\pi\ell_{\bar{t}}(\overline{u}_1,\overline{u}_2)}{\mathrm{L}_{\bar{t}}}\cos\left(\frac{\pi\,\ell_{\bar{t}}(\overline{u}_1,\overline{u}_2)}{\mathrm{L}_{\bar{t}}}\right)\right)\frac{d\,\mathrm{L}}{dt}(\bar{t}).
\end{aligned}$$

Since the coefficient in front of $\frac{d\mathrm{L}}{dt}(\bar{t})$ on the last line in the display above is negative, by (13.244) and (13.251) we obtain

$$\begin{aligned}
0 &< \frac{4\pi^2}{\mathrm{L}_{\bar{t}}}\sin\left(\frac{\pi\,\ell_{\bar{t}}(\overline{u}_1,\overline{u}_2)}{\mathrm{L}_{\bar{t}}}\right) - \pi\cos\left(\frac{\pi\,\ell_{\bar{t}}(\overline{u}_1,\overline{u}_2)}{\mathrm{L}_{\bar{t}}}\right)\frac{1}{\ell_{\bar{t}}(\overline{u}_1,\overline{u}_2)}4\xi^2\\
&\quad - \left(\sin\left(\frac{\pi\,\ell_{\bar{t}}(\overline{u}_1,\overline{u}_2)}{\mathrm{L}_{\bar{t}}}\right) - \frac{\pi\ell_{\bar{t}}(\overline{u}_1,\overline{u}_2)}{\mathrm{L}_{\bar{t}}}\cos\left(\frac{\pi\,\ell_{\bar{t}}(\overline{u}_1,\overline{u}_2)}{\mathrm{L}_{\bar{t}}}\right)\right)\frac{4\pi^2}{\mathrm{L}_{\bar{t}}}\\
&= \pi\cos\left(\frac{\pi\,\ell_{\bar{t}}(\overline{u}_1,\overline{u}_2)}{\mathrm{L}_{\bar{t}}}\right)\left(\frac{4\pi^3\ell_{\bar{t}}(\overline{u}_1,\overline{u}_2)}{\mathrm{L}_{\bar{t}}^2} - \frac{4\pi\xi^2}{\ell_{\bar{t}}(\overline{u}_1,\overline{u}_2)}\right).
\end{aligned} \tag{13.254}$$

Now we can obtain a contradiction. Namely, by (13.248) and (13.249), we have

(13.255)

$$\cos(\xi) = W_{\bar{t}}(\overline{u}_1,\overline{u}_2)\cdot\frac{\partial F_{\bar{t}}}{\partial u_1}(\overline{u}_1) = \frac{\pi}{\alpha}\cos\left(\frac{\pi\ell_{\bar{t}}(\overline{u}_1,\overline{u}_2)}{\mathrm{L}_{\bar{t}}}\right) \le \cos\left(\frac{\pi\ell_{\bar{t}}(\overline{u}_1,\overline{u}_2)}{\mathrm{L}_{\bar{t}}}\right),$$

where the last inequality is because $\alpha > \pi$. This, in turn, implies that $\xi \geq \frac{\pi \ell_{\bar{t}}(\overline{u}_1, \overline{u}_2)}{\mathrm{L}_{\bar{t}}}$, which is the desired contradiction. This completes the proof of the chord-arc monotonicity theorem. □

13.13. Convergence to a round point

Recall that we concluded §13.11 by showing that in the Type I case, the CSF shrinks embedded curves to round points. So, to prove the main convergence theorem, it suffices to show that an embedded curve cannot develop a Type II singularity.

13.13.1. Looking at the curve shortening flow under a microscope. In §13.11 we considered what can be characterized as a Type I rescaling about a blow-up point. We now discuss some other types of rescalings that are useful for understanding singularities.

Suppose that $F_t : M^1 \to \mathbb{R}^2$, $t \in [0, T)$, is a singular solution to the curve shortening flow. We may ask the following question:

> What does the geometry of the curve $M_t^1 = F_t(M^1)$ look like as t approaches the singularity time T?

Since F_t is a singular solution, the curvature κ of M_t^1 blows up as t approaches T. More precisely,

$$\sup_{M^1 \times [0,T)} |\kappa|(u, t) = \infty. \tag{13.256}$$

Thus, as the singularity forms, i.e., as $t \to T$, to see what is happening to the curve F_t, we want to rescale it so that the curvature $\kappa(t)$ becomes bounded.

13.13.1.1. *Parabolic rescaling at the curvature scale.* More specifically, suppose that $\{(u_i, t_i)\}$ is a sequence along which the curvatures blow up:

$$k_i := |\kappa|(u_i, t_i) \to \infty. \tag{13.257}$$

Note that this implies that $t_i \to T$.

Consider the curve $M_{t_i}^1$ and its parametrization $F_{t_i} = F(\cdot, t_i)$. We can translate in space this curve by defining

$$G_{t_i}(u) := G(u, t_i) := F(u, t_i) - F(u_i, t_i), \tag{13.258}$$

so that

$$G(u_i, t_i) = \mathbf{0} \in \mathbb{R}^2 \tag{13.259}$$

is the origin. We can next rescale the curve by defining

$$H_{t_i}(u) := H(u, t_i) := k_i\big(F(u, t_i) - F(u_i, t_i)\big). \tag{13.260}$$

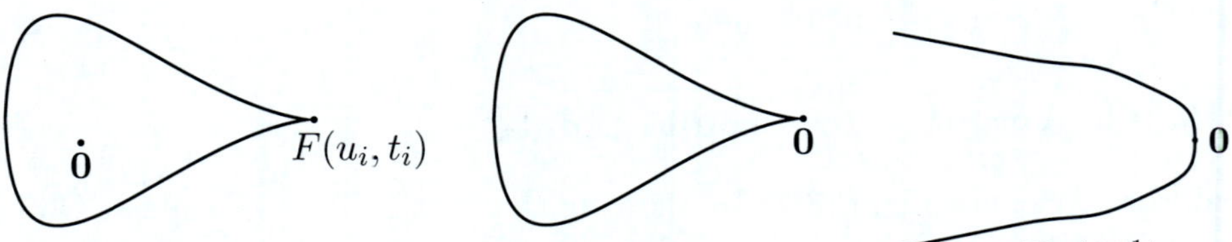

Figure 13.13.1. Looking at a hypothetical singular solution to the curve shortening flow under a microscope. G_{t_i} translates the curve so that $F(u_i, t_i)$ corresponds to the origin $\mathbf{0}$; see (13.259). H_{t_i} rescales the curve so that the curvature at $\mathbf{0}$ is equal to 1; see (13.261).

Here, we are **rescaling the solution at the curvature scale** at a sequence of times.

Let $\kappa_F := \kappa$ and let κ_H denote the curvatures of the curves parametrized by F and H, respectively. Because of the rescaling, we have that

$$\kappa_H(u, t_i) = k_i^{-1} \kappa_F(u, t_i);$$

in particular,

$$|\kappa_H|(u_i, t_i) = 1. \tag{13.261}$$

See Figure 13.13.1.

Example 13.32. *If the maps F_t parametrize shrinking round circles, then $H(\cdot, t_i)$ parametrizes a circle of radius 1 for each i. For example, if we choose $F_t(u) = \sqrt{r_0^2 - 2t}(\cos(u) - a, \sin(u) - b)$, where $r_0 > 0$ and $(a, b) \in \mathbb{R}^2$, which is a round circle shrinking to the point (a, b), then $H(\cdot, t_i)$ defined by (13.260) parametrizes a circle of radius 1 centered at the point $(-\cos(u_i), -\sin(u_i))$; this circle passes through the origin $\mathbf{0}$. Note that the parametrizations $H(\cdot, t_i)$ do not necessarily converge as $i \to \infty$. However, we may pass to a subsequence of i so that $u_i \to u_\infty \in \mathbb{R}/2\pi\mathbb{Z}$, that is, so that $(\cos(u_i), \sin(u_i))$ converges to $(\cos(u_\infty), \sin(u_\infty))$. In this case, the parametrizations $H(\cdot, t_i)$ converge to the standard parametrization of the circle of radius 1 centered at the point $(-\cos(u_\infty), -\sin(u_\infty))$. We leave it as an exercise for the reader to verify all of these assertions.*

Now, in general, each parametrized curve H_{t_i} is a microscopic snapshot of the curve shortening flow. That is, $H(\cdot, t_i)$ is the rescaling by k_i of the curve shortening flow F at the time t_i. So instead of seeing the whole movie (we are interested in how the movie ends), we only see blown-up movie stills at a sequence of times approaching the end. So we blow up the whole movie,

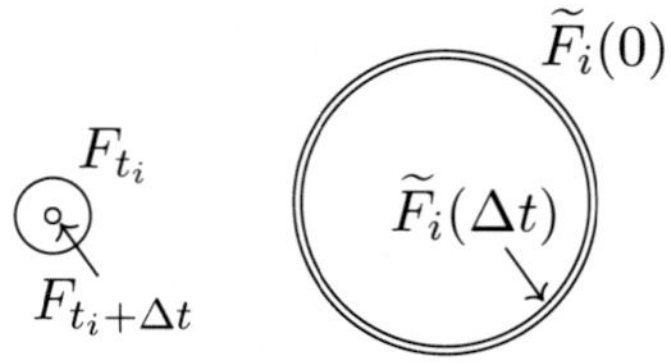

Figure 13.13.2. Left: The original shrinking circle F_t at a time t_i near the singularity time; the action is moving fast. Right: The shrinking circle $\widetilde{F}_i(t)$ after rescaling space and time; the action at $t = 0$ is happening at 'unit speed'.

space, time, and all. That is, we define the sequence of rescaled solutions to the CSF as follows:

$$J_i(u,t) := k_i F(u, t_i + k_i^{-2} t) \tag{13.262}$$

for all t satisfying $t_i + k_i^{-2} t \in [0, T)$, that is, for all $t \in [-k_i^2 - t_i, k_i^2(T - t_i))$. Observe that since $k_i \to \infty$ and $t_i \to T > 0$, we have that the bottom time endpoint $-k_i^2 t_i$ tends to $-\infty$. By the way we have defined J_i, we have

$$J_i(u,0) := k_i F(u, t_i). \tag{13.263}$$

So the snapshot time t_i for F becomes the time 0 for J_i. Finally, we translate the solution so that we are focusing our microscope at the origin, similarly to how we defined G in terms of F. Namely, we define the sequence of rescaled and translated solutions to the CSF as follows:

$$\widetilde{F}_i(u,t) := k_i \big(F(u, t_i + k_i^{-2} t) - F(u_i, t_i) \big). \tag{13.264}$$

This is a **parabolic rescaling at the curvature scale** of the solution to the CSF based at the sequence of points $\{(u_i, t_i)\}$. With this definition, we have

$$\widetilde{F}_i(u,0) = k_i \big(F(u, t_i) - F(u_i, t_i) \big) = H_{t_i}(u) \tag{13.265}$$

and

$$\widetilde{F}_i(u_i, 0) = \mathbf{0}. \tag{13.266}$$

See Figure 13.13.2.

13.13.2. Basepoint selection for parabolic rescalings. Assume that our solution to the CSF forms a Type II singularity, so that

$$\sup_{M^1 \times [0,T)} \sqrt{T - t}\, |\kappa|(u,t) = \infty.$$

First, choose any sequence $T_i \to T$. Observe that by the Type II singularity assumption, we have that

$$\lim_{i \to \infty} \max_{M^1 \times [0, T_i]} \sqrt{T - t}\, |\kappa|(u,t) = \infty. \tag{13.267}$$

Then choose $(u_i, t_i) \in M^1 \times [0, T_i]$ such that

$$\sqrt{T_i - t_i}\,|\kappa|(u_i, t_i) = \max_{M^1 \times [0,T_i]} \sqrt{T_i - t}\,|\kappa|(u, t), \tag{13.268}$$

where, by (13.267), this quantity tends to ∞ as $i \to \infty$. As in §13.13.1.1, let $k_i := |\kappa|(u_i, t_i)$ and consider the sequence of solutions to the CSF defined by (13.264):

$$\widetilde{F}_i(u, t) := k_i\big(F(u, t_i + k_i^{-2}t) - F(u_i, t_i)\big).$$

Recall the observation that $\widetilde{F}_i(u_i, 0) = \mathbf{0}$. By (13.268), we have that the curvatures $\widetilde{\kappa}_i$ of $\widetilde{F}_i$ satisfy

(13.269)

$$|\widetilde{\kappa}_i(u, t)| = k_i^{-1}\big|\kappa(u, t_i + k_i^{-2}t)\big| \leq \left(\frac{T_i - t_i}{T_i - (t_i + k_i^{-2}t)}\right)^{\frac{1}{2}} = \left(\frac{k_i^2(T_i - t_i)}{k_i^2(T_i - t_i) - t}\right)^{\frac{1}{2}}$$

for $t_i + k_i^{-2}t \in [0, T_i]$, that is, for $-k_i^2 t_i \leq t \leq k_i^2(T_i - t_i)$. By (13.267) and (13.268), we have that

$$k_i^2(T_i - t_i) \to \infty \quad \text{as } i \to \infty \tag{13.270}$$

and of course $k_i^2 t_i \to \infty$.

Note that the estimate in (13.269) is uniform in the following sense:

(1) For all $u \in M^1$ and $t \in (-\infty, 0]$, we have

$$|\widetilde{\kappa}_i(u, t)| \leq 1. \tag{13.271}$$

(2) For any $\varepsilon > 0$, there exists a positive integer i_0 such that

$$|\widetilde{\kappa}_i(u, t)| \leq 1 + \varepsilon \tag{13.272}$$

for all $u \in M^1$, $t \in [0, \varepsilon^{-1}]$, and $i \geq i_0$ (this is by (13.270)).

Lemma 13.33. *For each $m \geq 1$ there exist constants C_m such that*

$$\left|\frac{\partial^m \widetilde{\kappa}_i}{\partial s^m}(u, t)\right| \leq C_m \tag{13.273}$$

for $u \in M^1$ and $t \in \big(-\infty, \frac{1}{2}k_i^2(T_i - t_i)\big]$.

Proof. By (13.269) we have the uniform curvature estimate $|\widetilde{\kappa}_i(u, t)| \leq \sqrt{2}$ for $u \in M^1$ and $t \in \big(-\infty, \frac{1}{2}k_i^2(T_i - t_i)\big]$. Thus, the lemma follows from the Bernstein-type derivatives of curvature estimate (13.189). □

Since $\widetilde{F}_i(u_i, 0) = \mathbf{0}$, by passing to a subsequence, we may assume that the unit tangent vectors

$$\frac{\partial \widetilde{F}_i}{\partial s}(u_i, 0) \to \widetilde{T}_\infty$$

converge to some unit vector $\widetilde{T}_\infty$ in $\mathbb{R}^2$. This, and the uniform derivatives of curvature estimate, implies that a subsequence of the solutions $\widetilde{F}_i(t)$ converges smoothly to a smooth limit solution $\widetilde{F}_\infty(t)$ to the curve shortening flow with

$$|\widetilde{\kappa}_\infty(u_\infty, 0)| = 1 \quad \text{and} \quad \widetilde{F}_\infty(u_\infty, 0) = \mathbf{0}, \tag{13.274}$$

where $u_i \to u_\infty \in M^1_\infty$ and where M^1_∞ is the 1-manifold that is the domain of $\widetilde{F}_\infty(t)$. Moreover, since $k_i^2(T_i - t_i) \to \infty$, this limit solution exists on the time interval $(-\infty, \infty)$ and by (13.269) we have

$$|\widetilde{\kappa}_\infty(u, t)| \leq 1 \quad \text{for all } u \in M^1_\infty,\ t \in (-\infty, \infty). \tag{13.275}$$

By changing the direction of the unit normal field if necessary, we may assume that

$$\widetilde{\kappa}_\infty(u_\infty, 0) = 1.$$

Claim 13.34.

$$\widetilde{\kappa}_\infty(u, t) > 0 \tag{13.276}$$

for all $u \in M^1_\infty$ and $t \in (-\infty, \infty)$.

Proof of Claim 13.34. Firstly, for the solution $F_t : M^1 \to \mathbb{R}^2$ to the CSF, we have that for each $t \in (0, T)$ the zeroes of the function κ_t on M^1 are isolated, $\frac{\partial \kappa_t}{\partial s} \neq 0$ at each zero of κ_t, and

$$\begin{aligned} \frac{d}{dt} \int_{M^1} |\kappa_t| \, ds_t &= \frac{d}{dt} \int_{\{\kappa_t \geq 0\}} \kappa_t \, ds_t - \frac{d}{dt} \int_{\{\kappa_t \leq 0\}} \kappa_t \, ds_t \\ &= \frac{d}{dt} \int_{\{\kappa_t \geq 0\}} \frac{\partial^2 \kappa_t}{\partial s^2} ds - \frac{d}{dt} \int_{\{\kappa_t \leq 0\}} \frac{\partial^2 \kappa_t}{\partial s^2} ds \\ &= -2 \sum_{u:\, \kappa_t(u) = 0} \left| \frac{\partial \kappa_t}{\partial s} \right| (u), \end{aligned} \tag{13.277}$$

where for the last equality we used the fundamental theorem of calculus, that the normal derivative of κ_t at $\partial\{\kappa_t \geq 0\}$ points in the direction in which κ_t is nonincreasing, and that the normal derivative of κ_t at $\partial\{\kappa_t \leq 0\}$ points in the direction in which κ_t is nondecreasing. Hence,

$$\int_0^T \left(\sum_{u:\, \kappa_t(u) = 0} \left| \frac{\partial \kappa_t}{\partial s} \right| (u) \right) dt \leq \frac{1}{2} \int_{M^1} |\kappa_0| \, ds_0 =: C. \tag{13.278}$$

Secondly, consider the rescaled solution $\widetilde{F}_\infty : M^1_\infty \to \mathbb{R}^2$, $t \in (-\infty, \infty)$. For each $t \in (-\infty, \infty)$, the set $\{u \in M^1_\infty : \widetilde{\kappa}_\infty(u,t) = 0\}$ is discrete and for each u in this set we have $\frac{\partial \widetilde{\kappa}_\infty}{\partial s}(u,t) \neq 0$. From this and (13.278), we obtain

$$\int_{-\infty}^{\infty} \left(\sum_{u:\, \widetilde{\kappa}_\infty(u,t)=0} \left| \frac{\partial \widetilde{\kappa}_\infty}{\partial s} \right| (u,t) \right) dt = 0. \tag{13.279}$$

However, this contradicts that $\frac{\partial \widetilde{\kappa}_\infty}{\partial s}(u,t) \neq 0$ for each u such that $\widetilde{\kappa}_\infty(u,t) = 0$. Hence the function $\widetilde{\kappa}_\infty$ has no zeroes and we conclude that $\widetilde{\kappa}_\infty$ is positive everywhere on $M^1_\infty \times (-\infty, \infty)$. □

Now since $\widetilde{\kappa}_\infty > 0$, we can apply Hamilton's differential Harnack estimate (13.149) for solutions to the curve shortening flow which have positive curvature. In particular, for the solution $\widetilde{F}_\infty(t)$, $t \in (-\infty, \infty)$, we have that the Harnack quantity satisfies

$$\widetilde{Q}_\infty := \widetilde{\kappa}_\infty \left(\frac{\partial^2 \widetilde{\kappa}_\infty}{\partial \theta^2} + \widetilde{\kappa}_\infty \right) \geq -\frac{1}{2(t-\alpha)}, \tag{13.280}$$

where θ is the angle parameter, on $M^1_\infty \times (\alpha, \infty)$ and for any $\alpha \in (-\infty, \infty)$. By taking $\alpha \to -\infty$, we obtain

$$\widetilde{Q}_\infty = \widetilde{\kappa}_\infty \left(\frac{\partial^2 \widetilde{\kappa}_\infty}{\partial \theta^2} + \widetilde{\kappa}_\infty \right) \geq 0 \tag{13.281}$$

on $M^1_\infty \times (-\infty, \infty)$. Moreover, since $\widetilde{\kappa}_\infty$ attains its spacetime maximum at $(u_\infty, 0)$, we have equality in (13.281) at $(u_\infty, 0)$. By (13.145), we have that

$$\frac{\partial \widetilde{Q}_\infty}{\partial t} = \widetilde{\kappa}_\infty^2 \frac{\partial^2 \widetilde{Q}_\infty}{\partial \theta^2} + 2\widetilde{\kappa}_\infty \frac{\partial \widetilde{\kappa}_\infty}{\partial \theta} \frac{\partial \widetilde{Q}_\infty}{\partial \theta} + 2\widetilde{Q}_\infty^2. \tag{13.282}$$

Since $\widetilde{Q}_\infty$ satisfies (13.282), $\widetilde{Q}_\infty \geq 0$ on $M^1_\infty \times (-\infty, \infty)$, and $\widetilde{Q}_\infty(u_\infty, 0) = 0$, we may apply the *strong maximum principle* (see, e.g., [**CCG+08**, §12.4.1]) to conclude that

$$\widetilde{\kappa}_\infty \left(\frac{\partial^2 \widetilde{\kappa}_\infty}{\partial \theta^2} + \widetilde{\kappa}_\infty \right) = \widetilde{Q}_\infty \equiv 0 \tag{13.283}$$

on $M^1_\infty \times (-\infty, \infty)$. This implies that $\widetilde{F}_\infty$ parametrizes the Grim Reaper translating self-similar solution to the CSF. Finally, we obtain a contradiction to Huisken's chord-arc estimate as follows. For the original solution F_t of the CSF, for a sequence of times $t_i \to T$ we have that an arc of F_{t_i} is after rescaling ε-close in $C^{\varepsilon^{-1}}$-norm to the Grim Reaper truncated at length ε^{-1}. Take two points P_1 and P_2 on opposite sides of the Grim Reaper at distance $\varepsilon^{-1}/2$ from the basepoint of the Grim Reaper (where it attains its maximum curvature). This corresponds (approximately) to two points

$F_{t_i}(u_1)$ and $F_{t_i}(u_2)$ for which after rescaling the chord length $d_{t_i}(u_1, u_2) \approx \pi$ and the minimum of the two arc lengths satisfies $\ell_{t_i}(u_1, u_2) \approx \varepsilon^{-1} \gg \pi$. This implies that Huisken's chord-arc quantity $\Phi(t_i)$ is unbounded, which is a contradiction to his monotonicity formula. This completes the proof of Theorem 13.5.

13.14. Notes and commentary

As mentioned at the beginning of this chapter, the earliest papers on the curve shortening flow are by von Neumann [**vN52**] and Mullins [**Mul56**]. The modern study of the curve shortening flow is due to Gage [**Gag83**], Gage and Hamilton [**GH86**], and Grayson [**Gra87**].

Besides Grayson's original proof and Huisken's [**Hui98**] chord-arc proof, there are alternative proofs of the convergence to round points of the CSF of embedded plane curves. See Hamilton [**Ham95e**] for a proof based on an isoperimetric ratio monotonicity formula. See Andrews and Bryan [**AB11**] for a proof using the isoperimetric profile.

Huisken's proof of the chord-arc estimate uses a two-point function. For further developments on the application of the method of two-point functions in geometric analysis, see Brendle [**Bre14**]. In particular, this method has been used by Brendle to prove the Lawson conjecture [**Bre13a**] and by Andrews [**And12**] in his simpler proof of the Sheng and Wang [**SW09**] noncollapsing theorem for mean curvature flow.

The modern study of the mean curvature flow of submanifolds in Euclidean spaces began with Brakke [**Bra78**], who considered it from the point of view of geometric measure theory. The further development of the mean curvature flow began with Huisken [**Hui84**] and we refer the reader to the books by Ecker [**Eck04**] and by Andrews, Guenther, Langford, and the junior author [**ACGL20**] for an exposition of some of the developments in the mean curvature flow. Some important developments in the understanding of mean curvature flow singularity formation include the papers by Huisken [**Hui90**], Hamilton [**Ham95c**], Huisken and Sinestrari [**HS99**], White [**Whi03, Whi05**], Colding and Minicozzi [**CM12**], and Bamler and Kleiner [**BK**] to only name a few.

Ancient solutions of the mean curvature flow have been studied by Haslhofer [**Has15**], Brendle and Choi [**BC21**], and Bourni, Langford, and Lynch [**BLL23**]; see also the references therein.

The inverse mean curvature flow was used by Huisken and Ilmanen [**HI01**] to prove the Riemannian Penrose inequality; see Bray [**Bra01**] for an improvement using a completely different method.

13.15. Exercises

13.15.1. The curve shortening flow.

Exercise 13.1. *The* ***hairclip*** *is the graph of the time-dependent function* $y_t : \mathbb{R} \to \mathbb{R}$ *defined by*

$$y_t(x) = \ln\left(\cos x + \sqrt{e^{2t} + \cos^2 x}\right) - t \tag{13.284}$$

for $x, t \in \mathbb{R}$. *Observe that*

$$\lim_{t\to-\infty} (y_t(x) + t) = \ln(2\cos x),$$

which is the Grim Reaper translating self-similar solution. We also have that $\lim_{t\to\infty}(y_t(x) + t) = 0$, *which represents the* x*-axis. See Figure* 13.15.1.

Prove that the hairclip is an (eternal) solution to the curve shortening flow on the 'eternal' time interval $(-\infty, \infty)$.

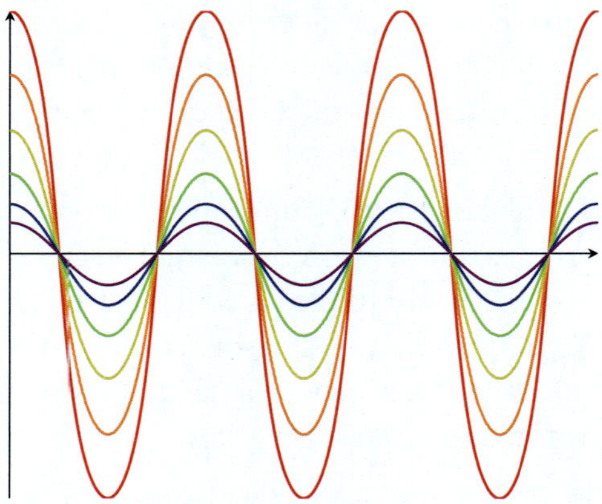

Figure 13.15.1. The hairclip solution to the curve shortening flow. Time flows forward, transitioning from red to orange, yellow, green, blue, purple, and finally limiting to the x-axis as $t \to \infty$.

Exercise 13.2. *For* $\theta \in S^1$ *and* $t < 0$, *define*

$$\kappa(\theta, t) = \sqrt{\cos 2\theta + \coth(-2t)}. \tag{13.285}$$

Prove that κ *satisfies the curvature equation for the CSF* (13.137); *i.e.,* $\partial_t \kappa = \kappa^2(\partial_\theta^2 \kappa + \kappa)$.

One can show that κ *is the curvature of a solution to the CSF on a closed curve. This solution is known as the* ***paperclip*** *ancient solution. See for example* [**ACGL20**, §4.4.2] *for details.*

Exercise 13.3. *Given a strictly convex embedded closed curve* M^1, *let* $\alpha := \partial_\theta(\kappa^2) = 2\kappa\partial_\theta\kappa$ *and define*

$$I := \int_{M^1} \left((\partial_\theta \alpha)^2 - 4\alpha^2\right) d\theta. \tag{13.286}$$

(1) *Prove that for the paperclip ancient solution,* $\partial_\theta^2 \alpha + 4\alpha \equiv 0$ *and hence* $I \equiv 0$.

(2) *Prove that for any closed convex solution to the CSF,*

$$\partial_t \alpha = \kappa^2 (\partial_\theta^2 \alpha + 4\alpha)$$

and

$$\frac{dI}{dt} = -2 \int_{M^1} \frac{(\partial_t \alpha)^2}{\kappa^2} d\theta \leq 0.$$

See Daskalopoulos, Hamilton, and Sesum [**DHS10a**] *for the classification of ancient convex solutions to the curve shortening flow of closed curves.*

Exercise 13.4. *Let*

$$Q := \frac{\partial^2}{\partial s^2} \ln \kappa + \kappa^2 = \frac{\partial}{\partial t} \ln \kappa - \left(\frac{\partial}{\partial s} \ln \kappa \right)^2 \tag{13.287}$$

be Hamilton's Harnack quantity for the CSF with positive curvature. Prove that under the CSF,

$$\frac{\partial Q}{\partial t} = \frac{\partial^2 Q}{\partial s^2} + 2 \frac{\partial \ln \kappa}{\partial s} \frac{\partial Q}{\partial s} + 2Q^2. \tag{13.288}$$

Exercise 13.5. *Consider an embedded convex solution to the CSF. Define* ***Hamilton's entropy*** *by*

$$\mathcal{N} := - \int_{M^1} \kappa \ln \kappa ds, \tag{13.289}$$

which is well-defined since $\kappa > 0$, *and define* $\mathcal{P} := -\frac{d\mathcal{N}}{dt}$. *Prove that*

$$\mathcal{P} = \int_{M^1} Q \kappa ds \tag{13.290}$$

and

$$\frac{d\mathcal{P}}{dt} = 2 \int_{M^1} Q^2 \kappa ds \geq \frac{1}{\pi} \mathcal{P}^2. \tag{13.291}$$

For further discussion of Hamilton's entropy under the CSF, see the expository [**ACGL20**, §4.4.3].

Exercise 13.6. *Prove that we have equality in the chord-arc inequality* (13.103) *if and only if* $F|_{[u_1,u_2]}$ *parametrizes an arc of a round circle of radius* $\frac{1}{K}$.

13.15.2. The mean curvature flow.

13.15.3. Mean curvature flow (continued from the previous chapter's exercises). Let $F_t : M^n \to \mathbb{R}^{n+1}$ be a smooth family of hypersurfaces satisfying the mean curvature flow; that is, $\frac{\partial F_t}{\partial t} = -H_t N_t$.

Exercise 13.7. *Using* (12.128), *prove that the induced volume form* $d\mu_t$ *satisfies*

$$\frac{\partial}{\partial t} d\mu_t = -H_t^2 d\mu_t. \tag{13.292}$$

Exercise 13.8. *Define*

$$\Phi(\mathbf{x}, t) := (4\pi(t_0 - t))^{-\frac{n}{2}} e^{-\frac{\|\mathbf{x}\|^2}{4(t_0 - t)}}. \tag{13.293}$$

Using (5.69) *and integration by parts, prove that*

$$\int_{M^n} H\langle x, N\rangle \Phi \, d\mu = \int_{M^n} \left(n - \frac{1}{2(t_0 - t)} \|\mathbf{x}^\top\|^2 \right) \Phi \, d\mu. \tag{13.294}$$

Exercise 13.9. *Define*

$$\Theta(t) = \int_{M^n} \Phi(\mathbf{x}, t) \, d\mu_t(\mathbf{x}). \tag{13.295}$$

Generalizing Theorem 13.18, *prove Huisken's monotonicity formula for the mean curvature flow:*

$$\frac{d}{dt}\Theta(t) = \int_{M^n} \left(H_t - \frac{\langle \mathbf{x}, N\rangle}{2(t_0 - t)} \right)^2 \Phi(t) \, d\mu_t. \tag{13.296}$$

Chapter 14

Uniformization of Surfaces via Heat Flow

Recall that the differential geometric version of the uniformization theorem (Theorem 8.27) says that for any Riemannian metric g_0 on a closed surface M^2, there exists a positive function v such that the conformally related metric vg_0 on M^2 has constant curvature. That is, by changing infinitesimal lengths but not infinitesimal angles associated to the metric, one can arrange the geometry so that the new metric is nice in the sense that it has constant curvature.

In this chapter, we consider Hamilton's heat flow approach to the proof of this result. Namely, we start with a Riemannian metric on a closed surface and we deform the metric in its conformal class by a heat-type equation, called *Hamilton's Ricci flow*, to a constant curvature metric. Figure 14.0.1 shows snapshots of a solution to the Ricci flow on a 2-sphere.

Some of techniques for studying the Ricci flow on surfaces, as for geometric flows in general, are that of pointwise and integral monotonicity formulas. Specifically, we will see how to estimate the curvature and its derivatives, including differential Harnack and entropy estimates. Additional ingredients include noncollapsing and self-similar solutions, called *Ricci solitons*. We will find it natural to enlarge the category of our discussion to 2-dimensional orbifolds. We include an appendix on the Kazdan–Warner identity.

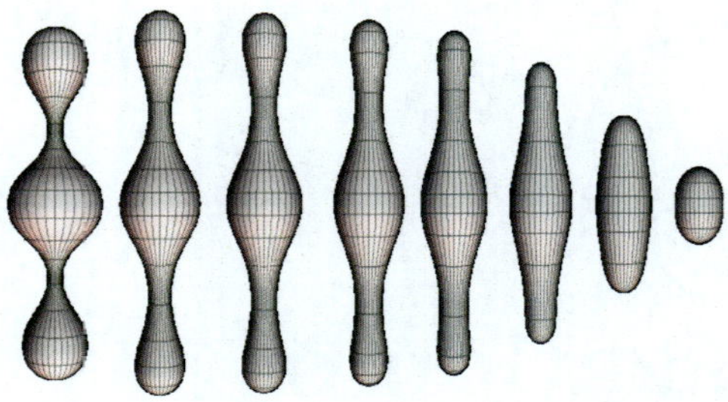

Figure 14.0.1. A rotationally symmetric solution to the Ricci flow on the 2-sphere. Metrics at later times are to the right. As the area decreases to zero, the metrics become rounder. Eventually the 2-sphere will shrink to a 'round' point. Credit: By CBM – Own work, Public Domain, `https://commons.wikimedia.org/w/index.php?curid=8867273`

14.1. Families of conformally equivalent metrics on surfaces

It turns out that the Ricci flow on surfaces preserves angles (as we will see below, this is true since $\mathrm{Ric} = \frac{1}{2}Rg$ is equal to a function times the metric when $n = 2$). So we consider families of metrics with this property.

Let M^2 be a closed oriented 2-dimensional manifold. Let g_0 be a Riemannian metric on M^2; see Figure 14.1.1. Let $u(t) : M^2 \to \mathbb{R}$, $t \in I$, where I is an interval, be a 1-parameter family of functions. Then

$$g(x,t) := \mathrm{e}^{2u(x,t)} g_0(x), \tag{14.1}$$

$t \in I$, is a 1-parameter family of metrics. By definition, each metric $g(t) = \mathrm{e}^{2u(t)} g_0$ is *conformal to* (or *conformally equivalent to*) g_0; that is, the infinitesimal angles defined by $g(t)$ are the same as those defined by g_0 (see §8.6). The function $\mathrm{e}^{2u(t)}$ is the **conformal factor**. For simplicity, we will also call u the conformal factor.

We now consider the variation of a 1-parameter family of conformal metrics. Let

$$v(x,t) := 2\frac{\partial u}{\partial t}(x,t). \tag{14.2}$$

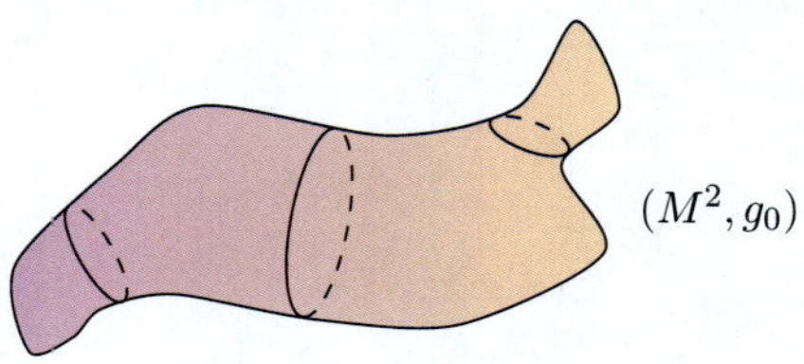

Figure 14.1.1. A Riemannian surface (M^2, g_0). The metric g_0 can be taken to be the initial condition for a Ricci flow on M^2.

Differentiating (14.1) yields the equivalent formula

$$\left(\frac{\partial}{\partial t}g\right)(t) = 2\frac{\partial u}{\partial t}(t)\mathrm{e}^{2u(t)}g_0 = v(t)\,g(t). \tag{14.3}$$

It is easy to see that the conformal deformation of the metric $g(t)$ equation

$$\frac{\partial}{\partial t}g(t) = v(t)\,g(t) \tag{14.4}$$

holds if and only if the conformal factors $u(t)$ satisfy

$$2\frac{\partial u}{\partial t}(t) = v(t). \tag{14.5}$$

Even though $g(t)$ is any 1-parameter family of conformally equivalent metrics, the $g(t)$ satisfying (14.4) is a **conformal deformation with velocity** $v(t)$.

Let $R(t) = 2K(t)$ denote the scalar curvature of $g(t)$, which is equal to twice the Gauss curvature of $g(t)$. By (8.98), we have that if $g(t) = \mathrm{e}^{2u(t)}g_0$, then

$$R(t) = \mathrm{e}^{-2u(t)}\big(R_0 - 2\Delta_0 u(t)\big) = \mathrm{e}^{-2u(t)}R_0 - 2\Delta_{g(t)}u(t), \tag{14.6}$$

where R_0 and Δ_0 denote the scalar curvature and Laplacian of g_0, respectively; the second equality follows from Lemma 8.22.

14.2. Variation of the curvature under a conformal variation of the metric

For families of Riemannian metrics, we can track their curvatures. In the case of a conformal variation of a metric on a surface, the variation of the scalar curvature satisfies a nice equation, which we derive in this section.

By differentiating (14.6), we calculate that if the metrics $g(t)$ on M^2 satisfy (14.4), i.e., $\partial_t g = vg$, then

$$\begin{aligned}\frac{\partial R}{\partial t}(t) &= -2\frac{\partial u}{\partial t}(t)\mathrm{e}^{-2u(t)}\big(R_0 - 2\Delta_0 u(t)\big) - 2\mathrm{e}^{-2u(t)}\Delta_0\left(\frac{\partial u}{\partial t}(t)\right)\\ &= -v(t)\,R(t) - \mathrm{e}^{-2u(t)}\Delta_0 v(t)\\ &= -v(t)\,R(t) - \Delta_{g(t)}v(t).\end{aligned}$$

Summarizing, we have proved the following.

Lemma 14.1. *If a 1-parameter family of Riemannian metrics $g(t)$, $t \in I$, on a 2-dimensional smooth manifold M^2 satisfies $\frac{\partial}{\partial t}g(t) = v(t)\,g(t)$, where $v(t) : M^2 \to \mathbb{R}$ for each $t \in I$, then their scalar curvatures satisfy the equation*

$$\frac{\partial R}{\partial t}(t) = -\Delta_{g(t)}v(t) - v(t)\,R(t). \tag{14.7}$$

If we take $v(t) = -R(t)$, then we obtain the Ricci flow on surfaces.

Corollary 14.2. *If a 1-parameter family of Riemannian metrics $g(t)$ on a 2-dimensional manifold M^2 satisfies the equation*

$$\frac{\partial}{\partial t} g(t) = -R(t)\, g(t), \tag{14.8}$$

defined to be the ***Ricci flow on surfaces****, then their scalar curvatures satisfy the equation*

$$\frac{\partial R}{\partial t}(t) = \Delta_{g(t)} R(t) + R(t)^2. \tag{14.9}$$

Equation (14.9) is a nonlinear heat-type equation and is also called a **reaction-diffusion equation**. On the right-hand side, the *diffusion term* is the Laplacian ΔR and the *reaction term* is the nonlinear function R^2. Without the reaction term, from (14.9) we obtain the heat equation, which smooths out the solution. Without the diffusion term, from (14.9) we obtain an ODE, which in this case is $\frac{\mathrm{dR}}{dt} = \mathrm{R}^2$, where R is a function of t.

Example 14.3 (Shrinking 2-sphere)**.** *Suppose that g_0 is the 2-sphere of radius ρ_0. Then its scalar curvature is $R_0 = \frac{2}{\rho_0^2}$. As we will see in Example* 14.4, *there exists a (unique) solution $g(t)$ to the Ricci flow satisfying the initial condition $g(0) = g_0$ which consists of round shrinking 2-spheres. Hence, for each t, $R(t)$ is a constant. Thus, $R(t)$ satisfies the ODE*

$$\frac{dR}{dt}(t) = R(t)^2.$$

Solving this ODE, we obtain

$$R(t) = \frac{1}{R_0^{-1} - t} = \frac{1}{\frac{\rho_0^2}{2} - t}. \tag{14.10}$$

Observe that this solution exists on the maximal time interval $\left[0, \frac{\rho_0^2}{2}\right)$; in fact, it can be defined on the ***ancient time interval*** $\left(-\infty, \frac{\rho_0^2}{2}\right)$. *As $t \to \frac{\rho_0^2}{2}$, we have that $R(t) \to \infty$ and the radius of the 2-sphere at time t tends to zero. See Figure* 14.2.1

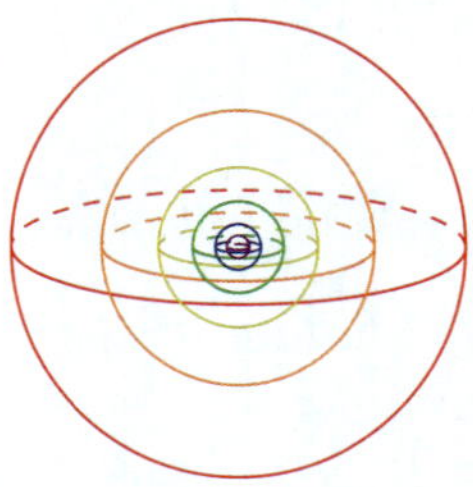

Figure 14.2.1. A constant curvature 2-sphere shrinking to a point under the Ricci flow.

14.3. The normalized Ricci flow equation on surfaces

As we have seen from the shrinking spheres in Example 14.3, the areas of the metrics are not preserved by the Ricci flow in general. The *normalized Ricci flow* rectifies this defect by scaling the metrics so that the area is constant in time.

14.3.1. Definition of the normalized Ricci flow equation.

Let $g(t)$ be a family of metrics on a closed oriented surface M^2. Let $r(t)$ denote the average scalar curvature of $g(t)$; that is,

$$r(t) := \frac{\int_{M^2} R(t)\, d\mu(t)}{\int_{M^2} d\mu(t)}, \tag{14.11}$$

where $d\mu(t) := d\mu_{g(t)}$ denotes the area form of $g(t)$. This is equal to the average of the function $R(t)$ on M^2 with respect to the area form $d\mu(t)$. Observe that

$$\int_{M^2} \big(R(t) - r(t)\big) d\mu(t) = 0. \tag{14.12}$$

Hamilton [**Ham88**] considered the following equation for $g(t)$:

$$\frac{\partial}{\partial t} g(t) = \big(r(t) - R(t)\big)\, g(t). \tag{14.13}$$

This equation, defined to be the **normalized Ricci flow on surfaces**, is equivalent to the equation

$$2\frac{\partial u}{\partial t}(t) = r(t) - R(t) \tag{14.14}$$

for the conformal factor $u(t)$ defined by $g(t) = \mathrm{e}^{2u(t)} g_0$.

As for any equation, the main questions are: Do solutions exist and how do they behave?

Firstly, geometrically, we will see below that the metrics $\bar{g}(\bar{t})$ of a normalized Ricci flow are just metric rescalings and time reparametrizations[1] of the metrics $g(t)$ of a Ricci flow:

$$\frac{\partial}{\partial t} g(t) = -R(t)\, g(t). \tag{14.15}$$

Observe that the metric is (conformally) shrinking at points where the curvature is positive and the metric is expanding at points where the curvature is negative. See Figure 14.3.1.

In the next subsection (see (14.19) below), we will prove that the area of $\widetilde{g}(\widetilde{t})$ is constant under the normalized Ricci flow. On the other hand (see

[1] This is why we denote the time parameter by $\bar{t}$ instead of t.

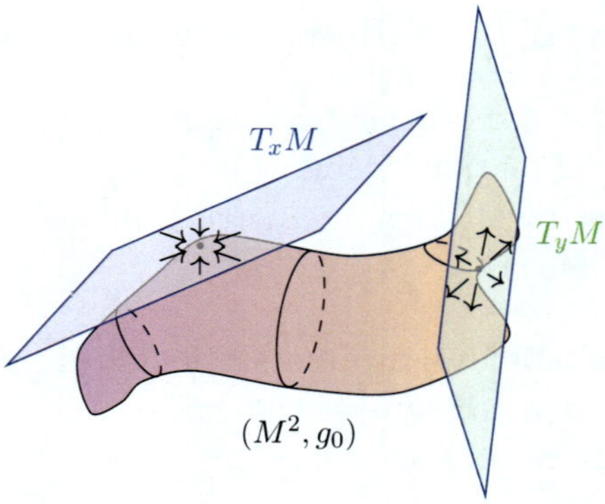

Figure 14.3.1. A Riemannian surface (M^2, g_0), where M^2 is diffeomorphic to the 2-sphere S^2. The arrows indicate that at points with positive curvature (such as x), the metric shrinks conformally under the Ricci flow and at points with negative curvature (such as at y), the metric expands conformally under the Ricci flow.

(14.23) below), under the Ricci flow the area of $g(t)$ is given by

$$\text{Area}(g(t)) = \text{Area}(g_0) - 4\pi\chi(M^2)t. \tag{14.16}$$

So:

(1) If $\chi(M^2) > 0$, then the area of $g(t)$ decreases at a constant rate.

(2) If $\chi(M^2) = 0$, then the area of $g(t)$ is constant.

(3) If $\chi(M^2) < 0$, then the area of $g(t)$ increases at a constant rate.

In particular, if M^2 is diffeomorphic to the 2-sphere S^2, then under the Ricci flow the area of $g(t)$ decreases at a constant rate until it limits to zero in a finite amount of time (provided one can show the solution exists as long as the area is positive).

Example 14.4 (Constant curvature solutions). *Suppose that (M^2, g_0) is a closed Riemannian surface with constant curvature $r_0 := R(g_0)$. Then:*

(1) *$g(t) \equiv g_0$, $t \in [0, \infty)$, is the unique maximal solution to the* normalized *Ricci flow with $g(0) = g_0$.*

(2) *$g(t) := (1 - r_0 t)g_0$, for all $t \geq 0$ satisfying $1 - r_0 t > 0$, is the unique maximal solution to the (*unnormalized*) Ricci flow with $g(0) = g_0$. Indeed, we check that*

$$\partial_t g(t) = -r_0 g_0 = -R(0)g(0) = -R(t)g(t).$$

If $r_0 \leq 0$, then this solution exists for all $t \in [0, \infty)$. On the other hand, if $r_0 > 0$, then this solution exists on the maximal time interval $[0, r_0^{-1})$; this agrees with Example 14.3.

14.3.2. Evolution of the area under the normalized and unnormalized Ricci flows. Now, suppose that we are given a solution $g(t)$ to the normalized Ricci flow on a closed oriented surface M^2. Suppose in addition that the time interval of existence is $I = [0, T)$, where $T \in (0, \infty]$, and that $g(0) = g_0$. (The last equality is equivalent to the conformal factor satisfying $u(0) = 0$.)

Let $\{\omega_0^1, \omega_0^2\}$ be a positively oriented orthonormal coframe field for g_0 defined on an open subset $\mathcal{U}$ of M^2. Then $\{\omega^1(t), \omega^2(t)\} := \{e^{u(t)}\omega_0^1, e^{u(t)}\omega_0^2\}$ is a positively oriented orthonormal coframe field for $g(t) = e^{2u}g_0$ on $\mathcal{U}$. Recall from (8.38) that the area form $d\mu(t) = d\mu_{g(t)}$ of $g(t)$ is given by

$$d\mu(t) = \omega^1(t) \wedge \omega^2(t) = e^{2u(t)}\omega_0^1 \wedge \omega_0^2 = e^{2u(t)} d\mu_{g_0}$$

on $\mathcal{U}$. Thus,

$$\frac{\partial}{\partial t} d\mu(t) = 2\frac{\partial u}{\partial t}(t) e^{2u(t)} d\mu_{g_0} = \big(r(t) - R(t)\big) d\mu(t).$$

Hence, on all of M^2 we have under the normalized Ricci flow that the area form of $g(t)$ evolves by

$$\frac{\partial}{\partial t} d\mu(t) = \big(r(t) - R(t)\big) d\mu(t). \tag{14.17}$$

Since $r(t)$ is the average of $R(t)$, we have

$$\begin{aligned} \frac{d}{dt}\operatorname{Area}(g(t)) &= \frac{d}{dt}\int_{M^2} d\mu(t) = \int_{M^2} \frac{\partial}{\partial t} d\mu(t) \\ &= \int_{M^2} \big(r(t) - R(t)\big) d\mu(t) \\ &= 0. \end{aligned} \tag{14.18}$$

Thus, under the normalized Ricci flow,

$$\operatorname{Area}(g(t)) \equiv \operatorname{Area}(g_0) \tag{14.19}$$

for all $t \in [0, T)$. As a consequence, by the Gauss–Bonnet formula, we have that the average scalar curvature

$$r(t) = \frac{\int_{M^2} R(t) d\mu(t)}{\int_{M^2} d\mu(t)} \equiv \frac{4\pi\chi(M^2)}{\operatorname{Area}(g_0)} \tag{14.20}$$

is a constant independent of t. So we denote $r := r(t)$.

On the other hand, under the (unnormalized) Ricci flow (14.13), we have similarly to (14.17) that

$$\frac{\partial}{\partial t} d\mu(t) = -R(t) d\mu(t). \tag{14.21}$$

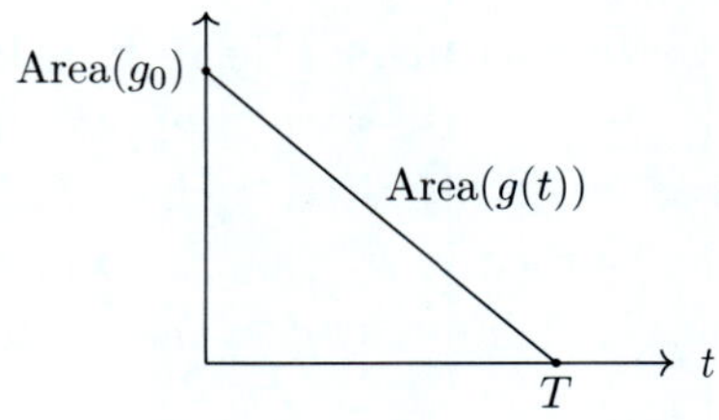

Figure 14.3.2. Area, as a function of time, of a closed surface with positive Euler characteristic under the (unnormalized) Ricci flow. The supremal time is $T = \frac{\mathrm{Area}(g_0)}{4\pi\chi(M^2)}$.

Therefore, under the Ricci flow we have (cf. (14.18))

$$\frac{d}{dt}\,\mathrm{Area}(g(t)) = -\int_{M^2} R(t)\,d\mu(t) = -4\pi\chi(M^2) = -r_0\,\mathrm{Area}(g_0), \tag{14.22}$$

where r_0 is the average scalar curvature at time zero. We conclude that under the Ricci flow,

$$\mathrm{Area}(g(t)) = \mathrm{Area}(g_0) - 4\pi\chi(M^2)t = \mathrm{Area}(g_0)(1 - r_0 t). \tag{14.23}$$

See Figure 14.3.2.

14.3.3. The relation between the unnormalized and normalized Ricci flows. In this subsection, we show that the unnormalized and normalized Ricci flows are related by a change in time parameter and by homothetic rescalings, depending on time, of the metrics. It is in this sense that solutions to the two flows with the same initial conditions are geometrically comparable: The shapes, but not the sizes, of the metrics are the same for the two flows.

Let $(M^2, g(t))$ be a solution of the Ricci flow. Define space and time rescaled metrics by

$$\bar{g}(\bar{t}) := \frac{1}{1 - r_0 t} g(t), \tag{14.24}$$

where

$$\bar{t}(t) := \int_0^t \frac{1}{1 - r_0 \tau} d\tau = -\frac{1}{r_0}\ln(1 - r_0 t). \tag{14.25}$$

By (14.23), we have that

$$\mathrm{Area}\left(\bar{g}(\bar{t})\right) \equiv \mathrm{Area}(g_0). \tag{14.26}$$

We have

$$\frac{d\bar{t}}{dt}(t) = \frac{1}{1 - r_0 t}.$$

Using this, we compute that

$$\begin{aligned}\frac{\partial}{\partial \bar{t}}\bar{g}(\bar{t}) &= \frac{1}{d\bar{t}/dt}\left(\frac{1}{1-r_0 t}g(t)\right)\\ &= (1-r_0 t)\frac{\partial}{\partial t}\left(\frac{1}{1-r_0 t}g(t)\right)\\ &= \frac{\partial}{\partial t}g(t) + r_0\frac{1}{1-r_0 t}g(t)\\ &= -R(t)g(t) + r_0\bar{g}(\bar{t})\\ &= \big(r_0 - \bar{R}(\bar{t})\big)\bar{g}(\bar{t}).\end{aligned}$$

Thus, $\bar{g}(\bar{t})$ is a solution to the normalized Ricci flow with $\bar{g}(0) = g_0$.

Conversely, suppose that $\bar{g}(\bar{t})$ is a solution to the normalized Ricci flow with $\bar{g}(0) = g_0$. By reversing the discussion above, we have that if $t(\bar{t}) := \frac{1}{r_0}(1 - \mathrm{e}^{-r_0\bar{t}})$ and $g(t) := \mathrm{e}^{-r_0\bar{t}}\bar{g}(\bar{t})$, then $g(t)$ is a solution to the (unnormalized) Ricci flow with $g(0) = g_0$.

14.4. Short-time existence of the normalized Ricci flow

In order to use the Ricci flow, we need to first establish the short-time existence of solutions given an initial metric. The Ricci flow in dimension 2 is a conformal deformation and it reduces to a scalar equation which is of strictly parabolic type. This allows us to apply general theory to establish short-time existence.

By (14.14) and (14.6), we have that the function $u(x,t)$ satisfies

$$\frac{\partial u}{\partial t}(t) = \mathrm{e}^{-2u(t)}\Delta_0 u(t) - \mathrm{e}^{-2u(t)}\frac{R_0}{2} + \frac{r}{2}. \tag{14.27}$$

This is a heat-type equation for u. Technically, it has the fancy name of a *quasilinear second-order parabolic partial differential equation.* In any case, there is a well-developed theory of such equations and in particular we have the following well-known result. The proof of this result is beyond the scope of this book. See, e.g., Friedman [**Fri64**] and Taylor [**Tay97**, Chapter 15] for the methods to prove such a result; another reference is Ladyženskaja, Solonnikov, and Ural′ceva [**LSU68**].

Lemma 14.5. *Given any function $u_0 : M^2 \to \mathbb{R}$, there exist $T \in (0,\infty]$ and a unique family of functions $u(t)$, $t \in [0,T)$, that satisfy the heat-type equation* (14.27) *with the initial condition $u(0) = u_0$. Moreover, the function $u : M^2 \times [0,T) \to \mathbb{R}$, defined by $u(t)(x) := u(x,t)$, is smooth.*

By taking $u_0 = 0$, i.e., the zero function, and by the equivalence of equations (14.27) and (14.13), we have the following.

Corollary 14.6 (Short-time existence and uniqueness). *For any closed Riemannian surface* (M^2, g_0), *there exists* $T \in (0, \infty]$ *and a unique family of metrics* $g(t)$, $t \in [0, T)$, *that satisfy the normalized Ricci flow* (14.13) *with the initial condition* $g(0) = g_0$. *Moreover,* $g(x, t)$ *is smooth in space and time.*

We take T to be the supremal time of existence. (In other words, $[0, T)$ is the maximal time interval of future existence.) That is, by definition no continuation of the solution exists beyond time T. Later, we shall show that the supremal time of existence T of the normalized Ricci flow on surfaces is equal to ∞.

14.5. A lower bound for the curvature under the normalized Ricci flow

An important tool for studying heat-type equations is the parabolic maximum principle, which we introduce and apply in this section to study the behavior of the scalar curvatures of solutions to Ricci flow. We have seen the statement of the parabolic maximum principle for 1-space and 1-time-dimensional heat-type equations in the previous chapter on the curve shortening flow. In this section, we will give the statement and proof in more generality.

By Lemma 14.1, since $g(t)$ is a conformal deformation with velocity $v(t) = r - R(t)$, we have that scalar curvature satisfies the following evolution equation under the normalized Ricci flow:

$$\frac{\partial R}{\partial t}(t) = \Delta_{g(t)} R(t) + R(t)^2 - r\,R(t). \tag{14.28}$$

Using that r is constant in time, we may rewrite this formula as

$$\frac{\partial}{\partial t}(R(t) - r) = \Delta_{g(t)}(R(t) - r) + (R(t) - r)^2 + r\,(R(t) - r). \tag{14.29}$$

In particular, by dropping from the right-hand side the *square term*, which is nonnegative, we obtain

$$\frac{\partial}{\partial t}(R(t) - r) \geq \Delta_{g(t)}(R(t) - r) + r\,(R(t) - r). \tag{14.30}$$

This, in turn, implies that

$$\frac{\partial}{\partial t}\big(\mathrm{e}^{-rt}(R(t) - r)\big) \geq \Delta_{g(t)}\big(\mathrm{e}^{-rt}(R(t) - r)\big). \tag{14.31}$$

14.5.1. The parabolic maximum principle on manifolds. In general, if $w(t) : M^n \to \mathbb{R}$, $t \in [0, T)$, are functions satisfying

$$\frac{\partial w}{\partial t}(x, t) \geq \Delta_{g(t)} w(x, t) + \langle X, \nabla w \rangle (x, t), \tag{14.32}$$

where $g(t)$, $t \in [0,T)$, is a 1-parameter family of Riemannian metrics on M^n and where $X := X(t)$ is a time-dependent vector field on M^n, then w is defined to be a **supersolution to a heat-type equation**. So $\mathrm{e}^{-rt}(R(t) - r)$ is a supersolution to the heat equation (with normalized Ricci flow background) by (14.31).

The following is fundamentally important to estimating solutions to second-order parabolic partial differential equations. It has a wide range of applications and is 'unreasonably effective'. Cf. Proposition 11.51.

Theorem 14.7 (Parabolic minimum principle for supersolutions to the heat equation). *If $w : M^n \times [0,T) \to \mathbb{R}$, where M^n is compact, satisfies* (14.32), *and if $w(x,0) \geq -C$ for all $x \in M^n$, where C is some constant, then*

$$w(x,t) \geq -C \quad \text{for all } x \in M^n,\ t \in [0,T). \tag{14.33}$$

Proof. The proof relies on the first and second derivative tests from calculus. The trick to implement this is to introduce a so-called fudge factor. To this end, let $\epsilon > 0$ and define

$$w_\epsilon(x,t) := w(x,t) + \epsilon t + \epsilon.$$

By (14.32), we have

$$\frac{\partial w_\epsilon}{\partial t}(x,t) \geq \Delta w_\epsilon(x,t) + \langle X, \nabla w_\epsilon \rangle(x,t) + \epsilon. \tag{14.34}$$

By hypothesis, $w_\epsilon(x,0) \geq -C + \epsilon$ for all $x \in M^n$.

Suppose for a contradiction that the function w_ϵ were less than $-C$ somewhere in $M^n \times [0,T)$. Then there would exist a first time $t_0 \in (0,T)$ such that

$$w_\epsilon(x_0,t_0) = -C \quad \text{for some } x_0 \in M^n. \tag{14.35}$$

This is a rather intuitive result, true since w_ϵ is continuous and M^n is compact, which we will prove in the remark right after this proof.

By the choice of t_0, we have that $w_\epsilon(x,t) \geq -C$ for all $(x,t) \in M^n \times [0,t_0]$. By the first derivative test, since w_ϵ on $M^n \times [0,t_0]$ attains its minimum at (x_0,t_0), we have

$$\frac{\partial w_\epsilon}{\partial t}(x_0,t_0) \leq 0,$$
$$\nabla w_\epsilon(x_0,t_0) = \vec{0};$$

see Figure 14.5.1. By the second derivative test (11.2), we have that

$$(\nabla^2 w_\epsilon)_{(x_0,t_0)} \geq 0$$

is positive semidefinite. In particular, by tracing this, we obtain

$$(\Delta w_\epsilon)(x_0,t_0) \geq 0;$$

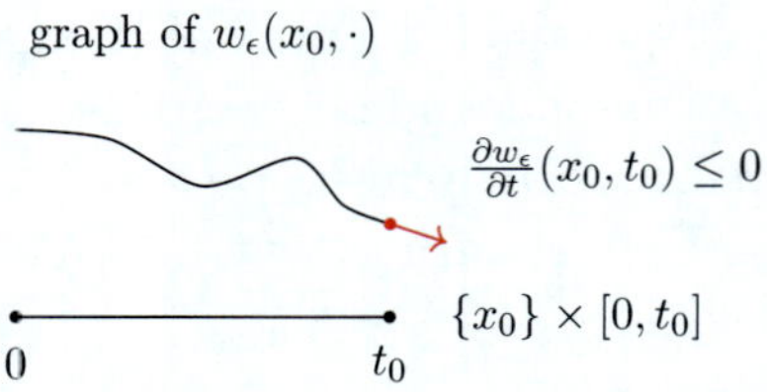

Figure 14.5.1. The first derivative test: At the minimum point (x_0, t_0) we have $\frac{\partial w_\epsilon}{\partial t} \leq 0$.

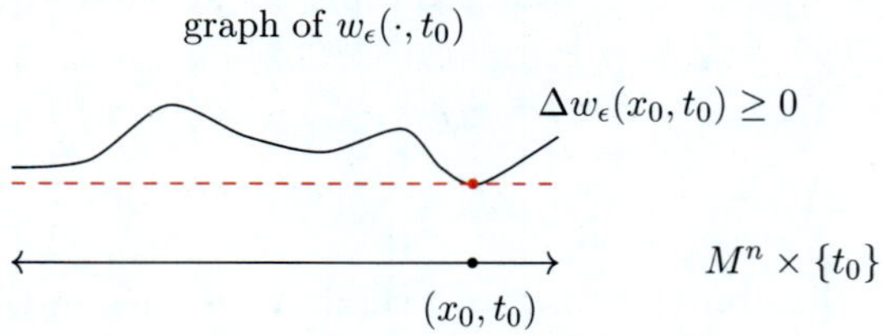

Figure 14.5.2. The second derivative test: At the minimum point (x_0, t_0) we have $\Delta w_\epsilon \geq 0$.

see Figure 14.5.2. By applying the first and second derivative tests to (14.34), we obtain

$$0 \geq \frac{\partial w_\epsilon}{\partial t}(x_0, t_0) \geq (\Delta w_\epsilon)(x_0, t_0) + \langle X, \nabla w_\epsilon \rangle (x_0, t_0) + \epsilon \geq \epsilon.$$

This is a contradiction since $\epsilon > 0$. Therefore, $w_\epsilon \geq -C$ on all of $M^n \times [0, T)$. By taking $\epsilon \to 0$, we conclude that $w \geq -C$ on all of $M^n \times [0, T)$. ☐

Remark 14.8. We give a proof of (14.35). Let

$$t_0 := \sup \left\{ \bar{t} \in [0, T) \colon w_\epsilon > -C \text{ on } M^n \times [0, \bar{t}] \right\}.$$

Firstly, since $w_\epsilon(\cdot, 0) \geq -C + \epsilon$ on M^n and since w_ϵ is continuous, we have that $t_0 > 0$. Secondly, since $w_\epsilon < -C$ somewhere in $M^n \times [0, T)$, we have $t_0 < T$. Thirdly, by the definition of t_0, we have $w_\epsilon(\cdot, t_0) \geq -C$ on M^n.

Suppose for a contradiction that $w_\epsilon(\cdot, t_0) > -C$ on all of M^n. Since M^n is compact, this would imply that $w_\epsilon(\cdot, t_0) \geq -C + \delta$ on M^n for some constant $\delta > 0$. Since w_ϵ is continuous and since M^n is compact, there would exist $\eta > 0$ such that $w_\epsilon \geq -C$ on $M^n \times [t_0 + \eta]$. This is a contradiction to the definition of t_0.[2] We conclude that $w_\epsilon(\cdot, t_0) = -C$ somewhere on M^n.

[2] A proof by contradiction of this: If no such η exists, then there would exist a sequence (x_i, t_i) with $x_i \in M^n$ and $t_i \searrow t_0$ such that $w_\epsilon(x_i, t_i) \leq -C + \frac{1}{i}$. Since M^n is compact, we may pass to a subsequence so that $x_i \to x_\infty \in M^n$. By the continuity of w_ϵ, we have $w_\epsilon(x_\infty, t_0) = \lim_{i \to \infty} w_\epsilon(x_i, t_i) \leq -C$, which is a contradiction.

14.5.2. Applying the maximum principle to bound the scalar curvature from below.

By applying the parabolic minimum principle (Theorem 14.7) to (14.31), we have that if $R_0 - r \geq -C$ (such a C always exists since M^2 is compact), then

$$\mathrm{e}^{-rt}(R(t) - r) \geq -C. \tag{14.36}$$

That is, under the normalized Ricci flow on surfaces, we have the estimate

$$R(t) - r \geq -C\mathrm{e}^{rt}. \tag{14.37}$$

This estimate is particularly effective when $r < 0$. This is because in this case, we have a lower bound for $\min_{x \in M^2}(R(x,t) - r)$ that is exponentially decaying in time. By the Gauss–Bonnet formula, the condition that $r < 0$ is equivalent to the topological condition that $\chi(M^2) < 0$; that is, the genus of M^2 is $\mathbf{g} := \mathbf{g}(M^2) := 1 - \frac{1}{2}\chi(M^2) > 1$.

Lemma 14.9 (Parabolic maximum principle for subsolutions of the heat equation). *If $w : M^n \times [0,T) \to \mathbb{R}$, where M^n is compact, satisfies*

$$\frac{\partial w}{\partial t}(x,t) \leq \Delta_{g(t)} w(x,t) + \langle X, \nabla w \rangle (x,t), \tag{14.38}$$

where $X = X(t)$ is a time-dependent vector field, and if $w(x,0) \leq C$ for all $x \in M^n$, where C is some constant, then

$$w(x,t) \leq C \quad \text{for all } x \in M^n,\ t \in [0,T). \tag{14.39}$$

For the proof of Lemma 14.9, see Exercise 14.1. We will find the following slight generalization useful.

Lemma 14.10 (Parabolic maximum principles for linear heat-type equations). *Let $g(t)$ be a 1-parameter family of Riemannian metrics on a closed manifold M^n and let $X = X(t)$ be a 1-parameter family of vector fields on M^n.*

(1) *If $w : M^n \times [0,T) \to \mathbb{R}$ satisfies*

$$\frac{\partial w}{\partial t}(x,t) \geq \Delta_{g(t)} w(x,t) + \langle X, \nabla w \rangle (x,t) + cw(x,t) \tag{14.40}$$

and if $w(x,0) \geq -C$ for all $x \in M^n$, where c and C are constants, then

$$w(x,t) \geq -C\mathrm{e}^{ct} \quad \text{for all } x \in M^n,\ t \in [0,T). \tag{14.41}$$

(2) *Similarly, if*

$$\frac{\partial w}{\partial t}(x,t) \leq \Delta_{g(t)} w(x,t) + \langle X, \nabla w \rangle (x,t) + cw(x,t) \tag{14.42}$$

and if $w(\cdot,0) \leq C$, then

$$w(x,t) \leq C\mathrm{e}^{ct}. \tag{14.43}$$

For the proof of Lemma 14.10, see Exercise 14.2.

14.6. Estimating the curvature from above under the normalized Ricci flow

In this section, we describe an elegant technique to obtain a time-dependent upper bound for the scalar curvature of a solution to the normalized Ricci flow on surfaces, where the quality of the bound depends on the sign of the Euler characteristic.

14.6.1. The difficulty in obtaining an upper bound for the curvature. Unlike the case of a lower bound, an effective *upper* bound for $R(x,t) - r$ under the normalized Ricci flow on surfaces is not as obvious. Indeed, let

$$\overline{R}(x,t) := R(x,t) - r$$

be the scalar curvature minus its average. Then (14.29) is the reaction-diffusion equation

$$\left(\frac{\partial}{\partial t} - \Delta\right)\overline{R} = \overline{R}^2 + r\overline{R}. \tag{14.44}$$

The associated ODE to the PDE (14.44) is obtained by dropping the Laplacian term; this yields the equation

$$\frac{d}{dt}\mathrm{S} = \mathrm{S}^2 + r\mathrm{S}, \tag{14.45}$$

where $\mathrm{S} := \mathrm{S}(t)$. The solution to this ODE with initial data $\mathrm{S}(0) = \mathrm{S}_0$ is given by

$$\mathrm{S}(t) = \frac{r}{\left(1 + \frac{r}{\mathrm{S}_0}\right)\mathrm{e}^{-rt} - 1} \tag{14.46}$$

provided $\mathrm{S}_0 \neq 0$ and $r \neq 0$ (the solution with $S_0 = 0$ is given by $S(t) \equiv 0$; if $r = 0$ and $\mathrm{S}_0 \neq 0$, then the solution is given by $\mathrm{S}(t) = \left(\frac{1}{\mathrm{S}_0} - t\right)^{-1}$). Observe that if $\mathrm{S}_0 > 0$ when $r \geq 0$ or $\mathrm{S}_0 > -r$ when $r < 0$, then

$$\mathrm{S}(t) \to \infty \quad \text{as } t \to \overline{T},$$

where $\overline{T} := \frac{1}{r}\ln\left(1 + \frac{r}{\mathrm{S}_0}\right)$ if $r \neq 0$ and $\overline{T} := \frac{1}{\mathrm{S}_0}$ if $r = 0$. That is, we have finite-time blow-up of the solutions to the ODE (14.45) with large enough initial data.

The statement of the parabolic maximum principle for reaction-diffusion equations with nonlinear reaction terms is as follows.

Lemma 14.11. *Suppose that $g(t)$, $t \in [0,T)$, is a smooth 1-parameter family of Riemannian metrics on a closed differentiable manifold M^n. Let $u : M^n \times [0,T) \to \mathbb{R}$ be a subsolution to*

$$\frac{\partial u}{\partial t}(x,t) \leq \Delta_{g(t)}u(x,t) + \langle X, \nabla u\rangle(x,t) + F(u(x,t)), \tag{14.47}$$

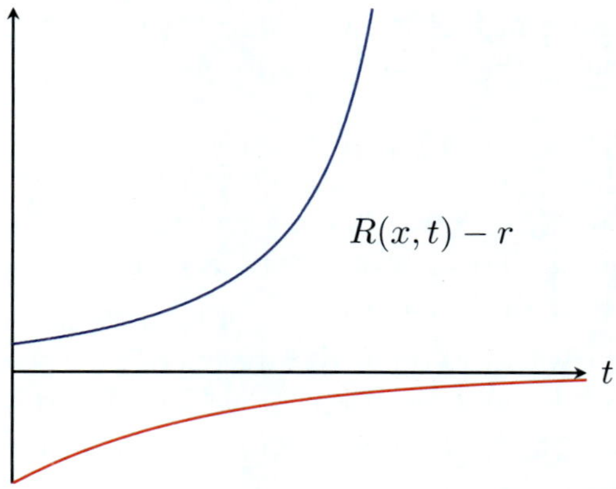

Figure 14.6.1. The lower bound (14.37) for $R(x,t)-r$ is represented by the red curve. When $r \geq 0$, the blue curve represents the upper bound given by the solution (14.46) to the associated ODE, which blows up as $t \to \overline{T}$.

where $F : \mathbb{R} \to \mathbb{R}$ *is some smooth one-variable function and where* $X = X(t)$ *is a time-dependent vector field on* M^n*. Let* $U_0 \in \mathbb{R}$ *satisfy* $U_0 \geq \max_{M^n} u(\cdot, 0)$*. Let* $U(t)$*,* $t \in T'$*, be the solution to the associated ODE*

$$\frac{dU}{dt}(t) = F(U(t)), \quad U(0) = U_0. \tag{14.48}$$

Then we have that

$$u(x,t) \leq U(t) \tag{14.49}$$

for all $x \in M^n$ *and* $t \in [0, \min\{T, T'\})$.

As a consequence of this parabolic maximum principle, by choosing $\mathrm{S}_0 := \max_{M^2} \overline{R}(\cdot, 0)$, we obtain the following upper estimate for the scalar curvature:

$$R(x,t) - r \leq \mathrm{S}(t) \tag{14.50}$$

for all $x \in M^2$ and $t \in [0, \min\{T, T'\})$. See Figure 14.6.1. Unfortunately, $T' < \infty$ provided g_0 does not have constant curvature (which means $\mathrm{S}_0 > 0$, at least when $r \geq 0$), so we cannot get an upper bound for all time for R. We need another method.

14.6.2. A key tool: The potential function. Necessity is the mother of invention. A simple, but not obvious, method to obtain an effective upper bound for $R(x,t) - r$ proceeds as follows. We carry this out in a few steps. Firstly, by definition,

$$\int_{M^2} \overline{R}(x,t) \, d\mu(x,t) = 0$$

for each time t. Because of this, by Corollary 11.61 there exists a function $f(t) : M^2 \to \mathbb{R}$ satisfying the **Poisson-type equation**

$$\Delta_{g(t)} f(t) = \overline{R}(t) \tag{14.51}$$

on M^2. Note that each $f(t)$ is determined up to an additive constant. This is because any harmonic function on M^2 is a constant (see Lemma 11.7). We call $f(t)$ the **potential function**.

Recall that the curvature is defined in terms of the second derivatives of the metric. On the other hand, from (14.6) we saw that the scalar curvature of a conformally related metric may be expressed in terms of the Laplacian of the conformal factor. So, by analogy, the consideration of the potential function seems to be a reasonable thing to do. Let us now see if it helps.

14.6.3. Estimates for the potential function and its derivatives.

Because we are in dimension 2, using Lemma 8.22 we can calculate that

$$\begin{aligned} \frac{\partial}{\partial t}\big(\Delta_{g(t)} f(t)\big) &= \frac{\partial}{\partial t}\big(\mathrm{e}^{-2u(t)} \Delta_{g_0} f(t)\big) \\ &= -2\frac{\partial u}{\partial t}(t) \mathrm{e}^{-2u(t)} \Delta_{g_0} f(t) + \mathrm{e}^{-2u(t)} \Delta_{g_0} \left(\frac{\partial f}{\partial t}\right) \\ &= \overline{R}(t)\, \Delta_{g(t)} f(t) + \Delta_{g(t)} \left(\frac{\partial f}{\partial t}\right). \end{aligned} \tag{14.52}$$

Thus, by taking the time-derivative of (14.51), we obtain

$$\overline{R}(t)\, \Delta_{g(t)} f(t) + \Delta_{g(t)} \left(\frac{\partial f}{\partial t}\right) = \Delta \overline{R} + \overline{R}^2 + r\overline{R}.$$

In view of (14.51), we can rewrite this equation as

$$\Delta_{g(t)} \left(\frac{\partial f}{\partial t}\right) = \Delta\big(\Delta_{g(t)} f(t)\big) + r\Delta_{g(t)} f(t).$$

Again, since any harmonic function on M is a constant, this implies that there exist constants $C(t)$ such that

$$\frac{\partial f}{\partial t}(t) = \Delta_{g(t)} f(t) + rf(t) + C(t).$$

In the definition of $f(t)$ we can choose $f(t)$ so that these constants $C(t)$ are identically zero, that is, so that

$$\frac{\partial f}{\partial t}(t) = \Delta_{g(t)} f(t) + rf(t). \tag{14.53}$$

For simplicity, we write this equation as follows.

Lemma 14.12. *Under the normalized Ricci flow on a closed surface, the potential function f satisfies*

$$\left(\frac{\partial}{\partial t} - \Delta\right) f = rf. \tag{14.54}$$

If, given a family of metrics $g(t)$, we consider the equation

$$\left(\frac{\partial}{\partial t} - \Delta_{g(t)}\right) w(t) = rw(t), \tag{14.55}$$

then we have an equation that is linear in w. It is in this sense that (14.53) is a linear heat-type equation. On the other hand, $f(t)$ itself does not depend linearly on $g(t)$.

It is useful to consider the gradient of f. Let $g(t)^* = \langle \cdot, \cdot \rangle$ denote the inner product on T^*M dual to the metric $g(t)$. Then $\partial_t g(t)^* = \overline{R} g(t)^*$. So we compute that

$$\begin{aligned} \frac{\partial}{\partial t} |\nabla f(t)|^2_{g(t)} &= \frac{\partial}{\partial t} \left(g(t)^* \left(df(t), df(t) \right) \right) \\ &= \overline{R} g(t)^* \left(df(t), df(t) \right) + 2 \left\langle \partial_t \left(df(t) \right), df(t) \right\rangle. \end{aligned} \tag{14.56}$$

Now,

$$\begin{aligned} \partial_t \left(df(t) \right) &= d\left(\partial_t f(t) \right) = d(\Delta f + rf) \\ &= \Delta df - \operatorname{Ric}(df) + rdf \\ &= \Delta df - \frac{1}{2} R df + rdf, \end{aligned} \tag{14.57}$$

where $\operatorname{Ric} : T^*M \to T^*M$ in the second line, where we used Lemma 11.2 to obtain the second equality, and where we used that $\operatorname{Ric} = \frac{1}{2} Rg$ from $n = 2$ in the third line. Thus, by applying (14.57) to (14.56), we have that

$$\begin{aligned} \frac{\partial}{\partial t} |\nabla f(t)|^2_{g(t)} &= \overline{R} |\nabla f(t)|^2_{g(t)} + 2 \langle \Delta df, df \rangle \\ &\quad - R |\nabla f(t)|^2_{g(t)} + 2r |\nabla f(t)|^2_{g(t)} \\ &= 2 \langle \Delta df, df \rangle + r |\nabla f(t)|^2_{g(t)} \\ &= \Delta_{g(t)} |\nabla f(t)|^2_{g(t)} - 2 |\nabla^2 f(t)|^2_{g(t)} + r |\nabla f(t)|^2_{g(t)}. \end{aligned} \tag{14.58}$$

For simplicity, we write this equation as follows.

Lemma 14.13. *Under the normalized Ricci flow on a closed surface, the norm squared of the gradient of the potential function satisfies*

$$\left(\frac{\partial}{\partial t} - \Delta\right) |\nabla f|^2 = -2 |\nabla^2 f|^2 + r |\nabla f|^2. \tag{14.59}$$

We remark that the general formula for computing the heat operator applied to $|\nabla v|^2$ for a function $v = v(x,t)$ is given by (14.77) below. The 'good' Hessian norm squared term is common to such calculations.

From the point of view of bounding the quantity by the parabolic maximum principle, the term $-2|\nabla^2 f|^2$ is a good term. In fact, by (5.108) we have

$$|\nabla^2 f|^2 \geq \frac{1}{2}\left(\operatorname{trace}_g(\nabla^2 f)\right)^2 = \frac{1}{2}(\Delta f)^2 = \overline{R}^2. \tag{14.60}$$

Because of this good term, the heat-type equation (14.59) for $|\nabla f|^2$ is useful for controlling the bad term $\overline{R}^2$ on the right-hand side of the equation (14.44) for $\overline{R}$. So we consider the sum

$$h := \overline{R} + |\nabla f|^2. \tag{14.61}$$

By (14.59) and (14.44), we have the following important equation:

$$\begin{aligned}\left(\frac{\partial}{\partial t} - \Delta\right) h &= \overline{R}^2 + r\overline{R} - 2|\nabla^2 f|^2 + r|\nabla f|^2 \\ &= -2\left|-\frac{1}{2}\overline{R}g + \nabla^2 f\right|^2 + rh.\end{aligned} \tag{14.62}$$

To see the last equality, we calculate using $|g|^2 = n = 2$ and $\langle \nabla^2 f, g\rangle = \Delta f$ that

$$\left|-\frac{1}{2}\overline{R}g + \nabla^2 f\right|^2 = \frac{1}{2}\overline{R}^2 + |\nabla^2 f|^2 - \overline{R}\Delta f = |\nabla^2 f|^2 - \frac{1}{2}\overline{R}^2.$$

Consequently, we obtain the following.

Lemma 14.14. *Under the normalized Ricci flow on a closed surface, the quantity $h = \overline{R} + |\nabla f|^2$ satisfies*

$$\left(\frac{\partial}{\partial t} - \Delta\right) h \leq rh. \tag{14.63}$$

By applying the parabolic maximum principle (Exercise 14.2, (2) of Lemma 14.10), we have

$$h(x,t) \leq C\mathrm{e}^{rt}, \tag{14.64}$$

where $C := \max_{y\in M^2} h(y,0)$. In particular, since $\overline{R} \leq h$, we have

$$\overline{R}(x,t) \leq C\mathrm{e}^{rt}. \tag{14.65}$$

To wit, in order to estimate the curvature $\overline{R}$, we estimated the larger quantity h since it satisfies a better heat-type equation.

On the other hand, by (14.37) we have

$$\overline{R}(x,t) \geq -C\mathrm{e}^{rt} \tag{14.66}$$

for some constant C. Thus, we have the following.

Lemma 14.15 (Curvature estimate under the normalized Ricci flow). *Under the normalized Ricci flow on a closed surface, there exists a constant C depending only on the initial metric g_0 such that*

$$|\overline{R}|(x,t) \leq C\mathrm{e}^{rt} \tag{14.67}$$

for all $x \in M^2$ and $t \in [0,T)$, where T is the supremal time of future existence of $g(t)$. In particular, if the genus $\mathbf{g} > 1$, or equivalently $\chi(M^2) < 0$, so that $r < 0$, we have the exponential decay of $|\overline{R}|$.

14.7. Uniform convergence of the metric $g(t)$ as t approaches the supremal time, when $\chi(M^2) < 0$

We now assume that $\chi(M^2) < 0$; that is, $r < 0$. Under this condition, we show that the exponential decay estimate in Lemma 14.15 is sufficient to prove the uniform convergence of $g(t)$ as $t \to T$, where $T \in (0,\infty]$ is the supremal time of future existence. Later we will show that $T = \infty$ and that the limit metric has constant curvature.

As in (14.1), define $u(t) : M^2 \to \mathbb{R}$, $t \in [0,T)$, by

$$g(t) =: \mathrm{e}^{2u(t)} g_0.$$

Then, by (14.14), the conformal factor u satisfies

$$\frac{\partial u}{\partial t} = -\frac{1}{2}\overline{R}. \tag{14.68}$$

Integrating this, we see that for each $x \in M^2$ and $t_1 < t_2$,

$$u(x,t_1) - u(x,t_2) = \frac{1}{2}\int_{t_1}^{t_2} \overline{R}(x,t)dt.$$

Hence, using $r < 0$, we compute that

$$|u(x,t_1) - u(x,t_2)| \leq \frac{1}{2}\int_{t_1}^{t_2} |\overline{R}|(x,t)dt \leq C\int_{t_1}^{t_2} \mathrm{e}^{rt}dt \leq \frac{C}{|r|}\mathrm{e}^{rt_1} \tag{14.69}$$

for some constant C. Note that C is independent of $x \in M^2$ and $t_2 \in (t_1, T)$.

As a consequence of (14.69), we have the following:

(1) There exists a constant C such that

$$|u|(x,t) \leq C \tag{14.70}$$

for all $x \in M^2$ and $t \in [0,T)$. To see this, we simply let $t_1 = 0$ and $t_2 = t$ in (14.69). This yields the result for $C := \sup_{x\in M^2} |u|(x,0) + \frac{C}{|r|}$.

(2) For each $x \in M^2$, the limit

$$\lim_{t\to T} u(x,t) =: u_T(x) \tag{14.71}$$

exists. This statement is true even if $T = \infty$.

The proof of (2) is as follows. Choose any sequence $t_i \to T$. We have for any $i < j$ that

$$|u(x,t_i) - u(x,t_j)| \leq \frac{C}{|r|}(\mathrm{e}^{rt_i} - \mathrm{e}^{rt_j}) \leq \frac{C}{|r|}(\mathrm{e}^{rt_i} - \mathrm{e}^{rT}), \tag{14.72}$$

where $\mathrm{e}^{rT} := 0$ if $T = \infty$. This shows that $\{u(x,t_i)\}_{i=1}^{\infty}$ is a Cauchy sequence. Since every Cauchy sequence of real numbers converges, we have that

$$\lim_{i\to\infty} u(x,t_i) =: u_T(x)$$

exists for each $x \in M^2$. Now, for any $t \in (t_i, T)$ and $x \in M^2$, we have

$$|u(x,t_i) - u(x,t)| \leq \frac{C}{|r|}(\mathrm{e}^{rt_i} - \mathrm{e}^{rT}). \tag{14.73}$$

This implies that the convergence

$$\lim_{t\to T} u(x,t) =: u_T(x)$$

is *uniform*. By definition, this means that for any $\epsilon > 0$, there exists $t_\epsilon < T$ such that for all $x \in M^2$ and $t \in (t_\epsilon, T)$ we have

$$|u(x,t) - u_T(x)| < \epsilon.$$

Note that we have not yet established any regularity properties of u_T such as continuity or higher differentiability. This will be a goal of the following sections.

In any case, as a consequence of (14.70) in (1), we have

$$\mathrm{e}^{-2C} g_0 \leq g(t) \leq \mathrm{e}^{2C} g_0 \tag{14.74}$$

for all $t \in [0,T)$. In general, given two metrics g and g', we say that $g \leq g'$ if $g' - g$ is a positive semidefinite symmetric 2-tensor. Hence, if α is any k-tensor, then

$$\mathrm{e}^{-kC} |\alpha|_{g_0} \leq |\alpha|_{g(t)} \leq \mathrm{e}^{kC} |\alpha|_{g_0} \tag{14.75}$$

for all $t \in [0, T)$. As a consequence of (2) and (1), we have that

$$\lim_{t \to T} |g(t) - g_T|_{g_0} = 0, \tag{14.76}$$

where

$$g_T := \mathrm{e}^{2u_T} g_0.$$

14.8. Estimating derivatives of the curvature

Similarly to the previous chapter on the curve shortening flow, in view of the Arzelà–Ascoli theorem, we need to estimate the derivatives of the curvature of our solution to the normalized Ricci flow. We will accomplish this via the parabolic maximum principle.

14.8.1. Estimating the gradient of the curvature. In general, for a time-dependent function $v(t)$ and under the normalized Ricci flow on surfaces, using the same method as that to obtain (14.56) and (14.58), we compute that

(14.77)

$$\left(\frac{\partial}{\partial t} - \Delta\right) |\nabla v(t)|^2_{g(t)} = -2|\nabla^2 v|^2 - r|\nabla v|^2 + 2d\left(\left(\frac{\partial}{\partial t} - \Delta\right) v\right) \cdot dv.$$

By applying this formula to $v(t) = R(t)$, we obtain

$$\begin{aligned}\left(\frac{\partial}{\partial t} - \Delta\right)|\nabla R|^2 &= -2|\nabla^2 R|^2 - r|\nabla R|^2 + 2d(R^2) \cdot dR - 2r dR \cdot dR \\ &= -2|\nabla^2 R|^2 + 4R|\nabla R|^2 - 3r|\nabla R|^2.\end{aligned}$$

We rewrite this as

$$\left(\frac{\partial}{\partial t} - \Delta\right) |\nabla R|^2 = -2|\nabla^2 R|^2 + 4\overline{R}|\nabla R|^2 + r|\nabla R|^2. \tag{14.78}$$

Assume that $\chi(M^2) < 0$. Since $\overline{R} \leq Ce^{rt}$ and $r < 0$, there exists $t_0 < \infty$ such that $\overline{R}(x, t) \leq -\frac{1}{8}r$ for all $t \geq t_0$ and $x \in M^2$. We then obtain for $t \geq t_0$ that

$$\left(\frac{\partial}{\partial t} - \Delta\right) |\nabla R|^2 \leq -2|\nabla^2 R|^2 + \frac{r}{2}|\nabla R|^2 \leq \frac{r}{2}|\nabla R|^2. \tag{14.79}$$

Hence, by the parabolic maximum principle (Lemma 14.10), we have the following.

Lemma 14.16. *Under the normalized Ricci flow on a closed surface M^2 with $\chi(M^2) < 0$, there exists a constant C depending only on the initial metric g_0 such that*

$$|\nabla R|^2(x, t) \leq C\mathrm{e}^{\frac{r}{2}t}, \tag{14.80}$$

where the norm is with respect to $g(t)$.

14.8.2. Estimating the higher derivatives of curvature.

For the higher-order derivatives of R, one can prove the following bounds.

Lemma 14.17 (Higher derivatives of curvature estimate). *Under the normalized Ricci flow on a closed surface M^2 with $\chi(M^2) < 0$ and for each positive integer k, there exists a positive constants C_k depending only on the initial metric g_0 and k such that*

$$|\nabla^k R|^2(x,t) \leq C_k \mathrm{e}^{\frac{r}{2}t} \tag{14.81}$$

for all $x \in M^2$ and $t \in [0,T)$, where ∇^k denotes the k-th covariant derivative, so that $\nabla^k R$ is a covariant k-tensor.

Given the estimates of the curvature and its derivatives from the previous section, we sketch the proof of the second derivative estimate, i.e., the case where $k = 2$. Details of the proof for all $k \geq 1$ are given in Chapter 5 of [**CK04**]. By [**CK04**, Lemma 5.25], we have (see also Exercise 14.23)

$$\begin{aligned} \frac{\partial}{\partial t}|\nabla^2 R|^2 = \Delta|\nabla^2 R|^2 - 2|\nabla^3 R|^2 + (2R - 4r)\,|\nabla^2 R|^2 \\ + 2R\,(\Delta R)^2 + 2\langle \nabla R, \nabla\,|\nabla R|^2\rangle. \end{aligned} \tag{14.82}$$

Now let

$$\varphi := |\nabla^2 R|^2 - 3r|\nabla R|^2.$$

Then there exists a constant C depending only on $g(0)$ such that (see the proof of Corollary 5.26 in [**CK04**])

$$\frac{\partial \varphi}{\partial t} \leq \Delta\varphi + \frac{2r}{3}\varphi + C\mathrm{e}^{rt}.$$

In particular, for any point (x,t) such that $\varphi(x,t) \geq -\frac{6C}{r}\mathrm{e}^{rt} = \frac{6C}{|r|}\mathrm{e}^{rt}$, we have

$$\frac{\partial \varphi}{\partial t}(x,t) \leq \Delta\varphi(x,t) + \frac{r}{2}\varphi(x,t).$$

By (a slight variant of) the parabolic maximum principle, we conclude that

$$|\nabla^2 R|^2 \leq \varphi \leq C\mathrm{e}^{\frac{r}{2}t} \tag{14.83}$$

for some constant C depending only on $g(0)$.

14.9. Long-time existence and convergence when $\chi(M^2) < 0$

Given the estimates of the curvature and its derivatives from the previous section, we are now in position to prove the long-time existence and convergence to constant negative curvature of the normalized Ricci flow with any initial metric on a surface with $\chi(M^2) < 0$.

14.9.1. Convergence of the metrics in each derivative-norm to a smooth metric. We now proceed to prove that $g_T = \lim_{t\to T} g(t)$ is a C^∞ Riemannian metric on M^2. We start by estimating the first spatial derivative of u. We have

$$\frac{\partial}{\partial t} du(x,t) = -\frac{1}{2} dR(x,t) \tag{14.84}$$

as an equation for 1-forms. Thus,

$$du(x,t_1) - du(x,t_2) = \frac{1}{2}\int_{t_1}^{t_2} dR(x,t)dt \in T_x^* M.$$

For the right-hand side, the integration of the vector-valued function $t \mapsto dR(x,t)$ from $[0,T)$ to T_x^*M is defined in the usual way. Taking norms, we obtain the estimate

$$\begin{aligned} |du(x,t_1) - du(x,t_2)|_{g_0} &\le \frac{1}{2}\int_{t_1}^{t_2} |dR(x,t)|_{g_0} dt \\ &\le C\int_{t_1}^{t_2} \mathrm{e}^{\frac{r}{4}t} dt \\ &\le \frac{4C}{|r|}\left(\mathrm{e}^{\frac{r}{4}t_1} - \mathrm{e}^{\frac{r}{4}t_2}\right). \end{aligned}$$

From this, it follows that the limit

$$\lim_{t\to T} du(x,t) =: v_T(x) \tag{14.85}$$

exists and that the convergence is uniform. By Theorem 13.22, u_T is differentiable and $du_T = v_T$. In fact, by Theorem 13.23 we have that, for all $k \ge 1$, u_T is k-times differentiable and $\nabla^k u(t)$ converges uniformly to $\nabla^k u_T$ in the bundle of k-tensors $\bigotimes^k T^*M$. Hence, $g_T = \mathrm{e}^{2u_T} g_0$ is a C^∞-metric.

Now, if $T < \infty$, then we may continue the solution and there exist $\epsilon > 0$ and metrics $g(t)$, $t \in [T, T+\epsilon)$, solving the normalized Ricci flow (14.13) with $g(T) = g_T$. As such the two families of metrics $\{g(t)\}_{t\in[0,T)}$ and $\{g(t)\}_{t\in[T,T+\epsilon)}$ combine to form a solution to (14.13) on the time interval $[0, T+\epsilon)$ with $g(0) = g_0$. This contradicts T being the maximal time. Hence, we conclude that $T = \infty$.

Now that we know that $T = \infty$, we have shown above that g_∞ is a C^∞-metric on M^2. Furthermore, since $\nabla^k u(t)$ converges uniformly to $\nabla^k u_\infty$ as $t \to \infty$, we have that $R(t)$ converges uniformly to $R(g_\infty)$. By the estimate (14.67), we conclude that $R(g_\infty) \equiv r$. That is, g_∞ is a constant negative scalar curvature r-metric. In summary, we have proved that for any initial metric on a surface of *genus greater than one* (i.e., *negative Euler characteristic*), the normalized Ricci flow exists for all positive time and converges to a constant negative curvature metric as time approaches infinity. This proves Theorem 14.18 below in the case where the genus $\mathbf{g} > 1$; i.e., the Euler characteristic of M^2 is negative.

14.10. Statement of the general convergence theorem

Using similar techniques, one can prove that for any initial metric g_0 on a closed oriented surface with *zero Euler characteristic*, i.e., on a *torus*, a unique solution to the normalized Ricci flow exists for all $t \in [0, \infty)$ and that $g(t)$ converges to a C^∞-metric g_∞ as $t \to \infty$, where the curvature of g_∞ is identically zero. For details in this case, the reader may consult the original paper [**Ham88**] or Chapter 5 of the expository book [**CK04**].

The statement of the global existence and convergence result for *all* closed surfaces is as follows.

Theorem 14.18 (Uniformization theorem by Ricci flow). *Let (M^2, g_0) be a closed oriented Riemannian surface. Then there exists a solution $g(t)$ to the normalized Ricci flow for all time $t \in [0, \infty)$ with $g(0) = g_0$. As $t \to \infty$, $g(t)$ converges in each C^k-norm to a C^∞-metric g_∞ with constant scalar curvature equal to $\frac{4\pi\chi(M^2)}{\mathrm{Area}(g_0)}$.*

In the next section we consider the proof of this theorem in the special case where the Euler characteristic χ of M^2 is positive, having covered the case where $\chi(M^2) < 0$ above (references containing the $\chi(M^2) = 0$ case are given above).

14.11. The Ricci flow on the 2-sphere

In this section and the next, we present the essential details of the proof of the convergence of the Ricci flow on closed surfaces with positive Euler characteristic. Since we assume that our surface is oriented, this means that our surface is diffeomorphic to the 2-sphere S^2.

14.11.1. Using monotone quantities to find more monotone quantities. Recall from (14.62) that, under the normalized Ricci flow on any closed surface M^2, the quantity $h = \overline{R} + |\nabla f|^2$, where $\overline{R} = R - r$, satisfies the evolution equation

$$\left(\frac{\partial}{\partial t} - \Delta\right) h = -2\left|-\frac{1}{2}\overline{R}g + \nabla^2 f\right|^2 + rh \leq rh. \tag{14.86}$$

This implies that

$$\left(\frac{\partial}{\partial t} - \Delta\right) (\mathrm{e}^{-rt}h) = -2\mathrm{e}^{-rt}\left|-\frac{1}{2}\overline{R}g + \nabla^2 f\right|^2 \leq 0.$$

So we have that $e^{-rt}h$ is a monotone quantity in the sense that it is a subsolution to the heat equation and hence its spatial maximum is a nonincreasing function of time.

14.11.1.1. *The trace-free part of the Hessian of the potential function.* Motivated by Hamilton's idea that quantities that arise in the evolution equations of monotone quantities may also behave nicely under the normalized Ricci flow, one considers the symmetric 2-tensor

$$\beta := -\frac{1}{2}\overline{R}g + \nabla^2 f, \tag{14.87}$$

which by (14.86) has the property that

$$\left(\frac{\partial}{\partial t} - \Delta\right) h = -2\,|\beta|^2 + rh.$$

We also note that β is trace-free; that is,

$$\operatorname{trace}_g(\beta) = -\overline{R} + \Delta f = 0. \tag{14.88}$$

14.11.1.2. *Characterizing when the trace-free part of the Hessian of* f *vanishes.* Observe that $\overline{R}$ vanishes if and only if $R \equiv r$; that is, g has constant curvature. Note also that if $\overline{R}$ vanishes, then f is constant, so that then β also vanishes. Conversely, we have the following.

Lemma 14.19. *If β vanishes for some closed oriented Riemannian surface (M^2, g), then g has constant curvature.*

Proof. Suppose that $\beta = 0$. Then

$$\mathcal{L}_{\nabla f} g = 2\nabla^2 f = \overline{R}g. \tag{14.89}$$

Case 1: $\chi \leq 0$. Here we have that $r \leq 0$, which is a condition we will use to our advantage. Taking the divergence of (14.89), we have

$$\begin{aligned} dR &= \operatorname{div}(\overline{R}g) \\ &= 2\operatorname{div}(\nabla^2 f) \\ &= 2d(\Delta f) + 2\operatorname{Ric}(df) \\ &= 2dR + Rdf. \end{aligned}$$

Therefore,

$$dR + Rdf = 0. \tag{14.90}$$

Taking a second divergence yields

$$\begin{aligned} 0 &= \Delta R + dR \cdot df + R\Delta f \\ &= \Delta R + dR \cdot df + R\overline{R} \\ &= \Delta\overline{R} + d\overline{R} \cdot df + \overline{R}^2 + r\overline{R}. \end{aligned}$$

Since $\overline{R}$ is a smooth function, and hence is continuous, and since M^2 is compact, there exists a point $x_0 \in M^2$ at which $\overline{R}$ attains its minimum: $\overline{R}(x_0) = \min_{x \in M^2} \overline{R}(x)$. We have

$$\Delta \overline{R}(x_0) \geq 0, \quad d\overline{R}(x_0) = \vec{0}.$$

Therefore,

$$\overline{R}(x_0)^2 + r\overline{R}(x_0) \leq 0.$$

Since $r \leq 0$, if $\overline{R}(x_0) < 0$, then $\overline{R}(x_0)^2 > 0$ and $r\overline{R}(x_0) \geq 0$ and thus we have a contradiction. Therefore, $\overline{R}(x_0) \geq 0$. Finally, since $\int_{M^2} \overline{R} d\mu = 0$, we conclude that $\overline{R} \equiv 0$ on all of M^2.

Case 2: $\chi > 0$. In this case, by the classification of surfaces (Theorem 8.25), we have that M^2 is diffeomorphic to the 2-sphere S^2.

By (14.89) and (14.164), we have that ∇f is a conformal vector field. Hence we may apply the Kazdan–Warner identity, i.e., Theorem 14.29 below, to obtain

$$0 = \int_{M^2} \langle \nabla_g R, \nabla f \rangle_g \, d\mu_g.$$

Integrating by parts, we obtain

$$0 = -\int_{M^2} R \Delta_g f \, d\mu_g = -\int_{M^2} \overline{R} \Delta_g f \, d\mu_g = -\int_{M^2} \overline{R}^2 \, d\mu_g.$$

We again conclude that $\overline{R} \equiv 0$ on M^2. □

14.11.1.3. *The evolution of the trace-free part of the Hessian of f and its norm.* Since we know the evolution equations for $\overline{R}$, g, and f, we can compute the evolution of $\beta = -\frac{1}{2}\overline{R}g + \nabla^2 f$. One catch is that we also have to calculate the evolution of the Hessian operator ∇^2 since it depends on $g(t)$. In any case, one arrives at the following formula (see Exercise 14.24):

$$\left(\frac{\partial}{\partial t} - \Delta\right) \beta = (r - 2R)\beta. \tag{14.91}$$

We also refer the reader to [**CK04**] for an exposition of the details of this calculation of Hamilton.

Let $\partial_t = \frac{\partial}{\partial t}$. In general, for any symmetric 2-tensor $\gamma(t)$, under the normalized Ricci flow on surfaces we have

$$\begin{aligned}
(\partial_t - \Delta) |\gamma(t)|^2_{g(t)} &= (\partial_t - \Delta) \left(g^{ik}(t) g^{j\ell}(t) \gamma_{ij}(t) \gamma_{k\ell}(t) \right) \\
&= -2 g^{ik}(t) g^{j\ell}(t) g^{pq}(t) \nabla_p \gamma_{ij}(t) \nabla_q \gamma_{k\ell}(t) \\
&\quad + 2\partial_t g^{ik}(t) g^{j\ell}(t) \gamma_{ij}(t) \gamma_{k\ell}(t) \\
&\quad + 2 g^{ik}(t) g^{j\ell}(t) (\partial_t - \Delta) \gamma_{ij}(t) \gamma_{k\ell}(t) \\
&= -2|\nabla \gamma|^2 + 2\overline{R}|\gamma|^2 + 2 (\partial_t - \Delta) \gamma \cdot \gamma,
\end{aligned} \tag{14.92}$$

where we used $\partial_t g^{ik}(t) = -g^{ir}(t)g^{is}(t)\partial_t g_{rs}(t)$. Hence, we obtain from (14.91) that

$$\left(\frac{\partial}{\partial t} - \Delta\right) |\beta|^2 = -2|\nabla\beta|^2 - 2R|\beta|^2. \tag{14.93}$$

14.11.1.4. *For any metric on the 2-sphere, a conformally equivalent metric has positive curvature.* Now assume that M^2 is diffeomorphic to the 2-sphere. Let g_0 be a Riemannian metric on M^2. Recall from (8.98) that if $g_1 = \mathrm{e}^{2u} g_0$, then

$$R_1 = \mathrm{e}^{-2u}(R_0 - 2\Delta_0 u).$$

Let r be the average scalar curvature of g_0. Recall by Corollary 11.61, which is a consequence of the Hodge theorem, that since $\int_M (R_0 - r) d\mu_0 = 0$, there exists a function $u : M^2 \to \mathbb{R}$ satisfying the Poisson equation

$$2\Delta_0 u = R_0 - r.$$

For this choice of u, we have

$$R_1 = \mathrm{e}^{-2u} r > 0.$$

To summarize, for any metric on the 2-sphere, by a conformal change we may assume that it has positive scalar curvature.

14.11.1.5. *A uniform lower bound for the scalar curvature.* We consider the normalized Ricci flow $g(t)$ starting from the metric g_1 with positive scalar curvature. Using the techniques in §14.9 (which are for the case where $\chi(M^2) < 0$), one can show for the case where $\chi(M^2) > 0$ that $g(t)$ exists for all time $t \in [0, \infty)$. By the parabolic maximum principle applied to the equation (14.28), we have that $R(t) > 0$ for all $t \geq 0$. Hamilton proved the following *a priori estimate*.

Proposition 14.20. *Under the normalized Ricci flow on a closed surface with positive curvature, there exists a constant c such that $R(x,t) \geq c > 0$ for all $x \in M^2$ and $t \in [0, \infty)$.*

We will finish the proof of this proposition in §14.12.6.4 below.

The point of the proposition is that the positive lower bound for the scalar curvature is *uniform*. That is, the proposition precludes the scalar curvature from decaying to zero as time tends to infinity. Another important significance of this estimate is that by (14.93) it implies that

$$\left(\frac{\partial}{\partial t} - \Delta\right) |\beta|^2 \leq -2c|\beta|^2.$$

Hence, by the parabolic maximum principle (Exercise 14.2, (2) of Lemma 14.10), there exists a constant C such that

$$|\beta|^2(x,t) \leq C\mathrm{e}^{-2ct} \tag{14.94}$$

for all $x \in M^2$ and $t \in [0, \infty)$, where $c > 0$. Since c is positive, we have that $|\beta|(\cdot, t)$ decays exponentially in t.

14.11.2. The modified Ricci flow. In Riemannian geometry, isometric metrics are considered to be geometrically the same. So we first discuss the effect of pulling back by diffeomorphisms on a 1-parameter family of metrics. Let $\varphi_t : M^2 \to M^2$, $t \in [0, \infty)$, be a 1-parameter family of diffeomorphisms. Let V_t be the **1-parameter family of vector fields generated by** φ_t; that is, by definition,

$$\frac{\partial}{\partial t}\varphi_t(x) =: V_t(\varphi_t(x)) = (V_t \circ \varphi_t)(x).$$

Let $g(t)$, $t \in [0, \infty)$, be a solution to the normalized Ricci flow on M^2. The 1-parameter family of pull-back metrics

$$\widetilde{g}(t) := \varphi_t^* g(t) \tag{14.95}$$

is by definition given by (see §6.5.1)

$$\widetilde{g}(t)(V, W) = g(t)\big(d\varphi_t(V), d\varphi_t(W)\big). \tag{14.96}$$

Also by definition, $\widetilde{g}(t)$ is isometric to $g(t)$. Thus, geometrically, the family $\widetilde{g}(t)$ is indistinguishable from the solution $g(t)$. The difference is that we will be able to first show that $\widetilde{g}(t)$ converges as $t \to \infty$.

Using the product rule and the consequence of the definition of the Lie derivative (6.93), we compute that

$$\begin{aligned} \frac{\partial}{\partial t}\widetilde{g}(t) &= \frac{\partial}{\partial t}(\varphi_t^* g(t)) \\ &= \varphi_t^*\left(\frac{\partial}{\partial t} g(t)\right) + \mathcal{L}_{d(\varphi_t^{-1})(\partial_t \varphi_t)}\widetilde{g}(t) \\ &= -\overline{R}_{\widetilde{g}(t)}\widetilde{g}(t) + \mathcal{L}_{d(\varphi_t^{-1})(V_t)}\widetilde{g}(t), \end{aligned} \tag{14.97}$$

where we used that $\varphi_t^*(\overline{R}_{g(t)} g(t)) = \overline{R}_{\widetilde{g}(t)}\widetilde{g}(t)$.

Define

$$\widetilde{f}(t) := f(t) \circ \varphi_t. \tag{14.98}$$

From now on, we choose the diffeomorphisms φ_t to be defined by

$$V_t = \nabla f(t) \quad \text{and} \quad \varphi_0 = \mathrm{id}_{M^2}, \tag{14.99}$$

so that

$$\frac{\partial}{\partial t}\varphi_t(x) = (\nabla f(t) \circ \varphi_t)(x). \tag{14.100}$$

Let $\widetilde{\nabla}$ denote the gradient with respect to $\widetilde{g}(t)$. By (14.97), we then have

$$\begin{aligned}\text{(14.101)}\qquad \frac{\partial}{\partial t}\widetilde{g}(t) &= -\overline{R}_{\widetilde{g}(t)}\widetilde{g}(t) + \mathcal{L}_{d(\varphi_t^{-1})(\nabla f(t))}\widetilde{g}(t)\\ &= -\overline{R}_{\widetilde{g}(t)}\widetilde{g}(t) + \mathcal{L}_{\widetilde{\nabla}\widetilde{f}(t)}\widetilde{g}(t)\\ &= -\overline{R}_{\widetilde{g}(t)}\widetilde{g}(t) + 2\widetilde{\nabla}^2\widetilde{f}(t)\\ &=: 2\widetilde{\beta}(t),\end{aligned}$$

where to obtain the second equality we used that

$$\text{(14.102)}\qquad \widetilde{\nabla}\widetilde{f}(t) = \nabla_{\varphi_t^* g(t)}(f(t)\circ\varphi_t) = \varphi_t^*(\nabla_{g(t)}f(t)) = d(\varphi_t^{-1})(\nabla f(t)).$$

We calculate that

$$\Delta_{\widetilde{g}(t)}\widetilde{f}(t) = \Delta_{\varphi_t^* g(t)}(f\circ\varphi_t) = (\Delta_{g(t)}f(t))\circ\varphi_t = (R_{g(t)} - r)\circ\varphi_t.$$

Therefore,

$$\text{(14.103)}\qquad \Delta_{\widetilde{g}(t)}\widetilde{f}(t) = \overline{R}_{\widetilde{g}(t)} := R_{\widetilde{g}(t)} - r$$

since $r = r\circ\varphi_t$ follows from r being constant and since, by $\widetilde{g}(t) = \varphi_t^* g(t)$, we have $R_{\widetilde{g}(t)} = R_{g(t)}\circ\varphi_t$. Equation (14.103) is analogous to (14.51). Namely, $\widetilde{f}(t)$ is the potential function for $\widetilde{g}(t)$.

Observe that

$$\text{(14.104)}\qquad \operatorname{trace}_{\widetilde{g}(t)}\left(\widetilde{\beta}(t)\right) = \operatorname{trace}_{\widetilde{g}(t)}\left(\frac{\partial}{\partial t}\widetilde{g}(t)\right) = -2\overline{R}_{\widetilde{g}(t)} + 2\widetilde{\Delta}\widetilde{f}(t) = 0.$$

Therefore, the area form of $\widetilde{g}(t)$ is independent of time under the modified Ricci flow:

$$\text{(14.105)}\qquad \frac{\partial}{\partial t}d\mu_{\widetilde{g}(t)} = 0.$$

This implies that $d\mu_{\widetilde{g}(t)}$ and $\operatorname{Area}(\widetilde{g}(t))$ are constant, the latter of which we already know from the area of $g(t)$ being constant and $\widetilde{g}(t) = \varphi_t^* g(t)$.

We now calculate the evolution of the potential function for $\widetilde{g}(t)$:

$$\begin{aligned}\frac{\partial\widetilde{f}}{\partial t}(t) &= \frac{\partial}{\partial t}(f(t)\circ\varphi_t)\\ &= \frac{\partial f}{\partial t}(t)\circ\varphi_t + df(t)\left(\frac{\partial}{\partial t}\varphi_t\right)\\ &= \left(\Delta_{g(t)}f(t) + rf(t)\right)\circ\varphi_t + df(t)\left(\nabla_{g(t)}f(t)\circ\varphi_t\right)\\ &= \Delta_{\varphi_t^* g(t)}(f(t)\circ\varphi_t) + rf(t)\circ\varphi_t + |\nabla_{g(t)}f(t)|^2\circ\varphi_t.\end{aligned}$$

That is,

$$\text{(14.106)}\qquad \frac{\partial\widetilde{f}}{\partial t}(t) = \Delta_{\widetilde{g}(t)}\widetilde{f}(t) + |\widetilde{\nabla}\widetilde{f}(t)|^2_{\widetilde{g}(t)} + r\widetilde{f}(t).$$

This is analogous to (14.53), except that we have a gradient term. Observe that this gradient term may be rewritten as $|\widetilde{\nabla}\widetilde{f}(t)|^2_{\widetilde{g}(t)} = \mathcal{L}_{\widetilde{\nabla}\widetilde{f}(t)}\widetilde{f}$.

14.11.3. Convergence to constant curvature for the normalized Ricci flow on the 2-sphere. We can now describe the amazing proof of Hamilton. The long-time existence of the solution of the normalized Ricci flow on the 2-sphere holds for the following reasons. Firstly, by (14.67), we have that $|R(x,t) - r| \leq Ce^{rt}$ for all $x \in M^2$ and $t \in [0,T)$ (Proposition 14.20 gives a much better lower bound for the scalar curvature). Secondly, by using this and similarly to Lemma 14.17, we can obtain time-dependent estimates for all derivatives of the curvature. Thirdly, similarly to §14.9, we can deduce from this that a unique solution $g(t)$ to the normalized Ricci flow on the 2-sphere exists for all time $t \in [0,\infty)$.

Now recall from (14.94) and (14.87) that

$$|\beta(t)|^2_{g(t)}(x,t) = \left|-\frac{1}{2}\overline{R}g + \nabla^2 f\right|^2_{g(t)}(x,t) \leq C\mathrm{e}^{-2ct}. \tag{14.107}$$

On the other hand, by (14.101),

$$2\widetilde{\beta}(t) = -\overline{R}_{\widetilde{g}(t)}\widetilde{g}(t) + 2\widetilde{\nabla}^2\widetilde{f}(t) = \varphi_t^*\left(-\overline{R}_{g(t)}g(t) + 2\nabla^2 f(t)\right).$$

Therefore,

$$\left|\frac{\partial}{\partial t}\widetilde{g}(t)\right|^2_{\widetilde{g}(t)}(x,t) = \left|-\overline{R}_{\widetilde{g}(t)}\widetilde{g}(t) + 2\widetilde{\nabla}^2\widetilde{f}(t)\right|^2_{\widetilde{g}(t)}(x,t) \leq C\mathrm{e}^{-2ct}. \tag{14.108}$$

One can show, analogously to Lemma 14.17, that for each positive integer k there exists a constant C_k such that

$$\left|\widetilde{\nabla}^k\widetilde{\beta}(t)\right|_{\widetilde{g}(t)} \leq C_k. \tag{14.109}$$

Similarly to §14.9, we can deduce from this that the solution $\widetilde{g}(t)$ to the modified Ricci flow exists for all time $t \in [0,\infty)$ and that the metrics $\widetilde{g}(t)$ converge as $t \to \infty$ to a smooth Riemannian metric $\widetilde{g}_\infty$.[3] Furthermore, this metric satisfies

$$\widetilde{\beta}_\infty := -\frac{1}{2}\overline{R}_{\widetilde{g}_\infty}\widetilde{g}_\infty + \widetilde{\nabla}^2\widetilde{f}_\infty = 0, \tag{14.110}$$

where $\widetilde{f}_\infty$ satisfies

$$\Delta_{\widetilde{g}_\infty}\widetilde{f}_\infty = R_{\widetilde{g}_\infty} - r. \tag{14.111}$$

By Lemma 14.19 we obtain that

$$R_{\widetilde{g}_\infty} \equiv r. \tag{14.112}$$

That is, since the metric $\widetilde{g}_\infty$ satisfies (14.110), it also satisfies (14.112).

[3]We leave this to the reader; see also Exercise 14.25.

Moreover, since the convergence of $\widetilde{g}(t)$ to $\widetilde{g}_\infty$ is exponentially fast in each C^k-norm, we have that $R_{\widetilde{g}(t)}$ converges to r exponentially fast under the modified Ricci flow. We also have that $|\widetilde{\nabla}^k R_{\widetilde{g}(t)}|_{\widetilde{g}(t)}$ decays exponentially to 0 as $t \to \infty$ for each positive integer k. Since the solution $g(t)$ satisfies $R_{\widetilde{g}(t)} = R_{g(t)} \circ \varphi_t$ and $\widetilde{\nabla}^k R_{\widetilde{g}(t)} = \varphi_t^* \nabla^k_{g(t)} R_{g(t)}$, we have that $R_{g(t)}$ converges to r exponentially fast and each $|\widetilde{\nabla}^k R_{\widetilde{g}(t)}|$ decays exponentially to 0 as $t \to \infty$. Therefore, the solution $g(t)$ to the normalized Ricci flow converges exponentially fast in each C^k-norm to a smooth Riemannian metric g_∞. Since $R_{g(t)}$ converges to r, we conclude that $R_{g_\infty} \equiv r$. This proves the convergence of the Ricci flow $g(t)$ on a 2-sphere with $R_{g_0} > 0$ to a constant positive curvature metric. It remains to prove Proposition 14.20.

14.12. The entropy and Harnack estimates

In this section, we discuss the entropy and Harnack estimates for solutions on S^2 that are used in the proof of the key estimate in Proposition 14.20, which says that the scalar curvature under the normalized Ricci flow is uniformly bounded from below by a positive constant.

14.12.1. The general idea of entropy. The idea of *entropy* is important in thermodynamics, statistical mechanics, cosmology, information theory, probability theory, and partial differential equations.

Let n be a positive integer and suppose that $\mathbf{p} := \{p_1, \ldots, p_n\}$ is a (discrete) probability distribution of a set of n elements; that is, $\sum_{i=1}^n p_i = 1$. Then the **entropy** of this probability distribution is defined to be equal to

$$N(\mathbf{p}) := -\sum_{i=1}^{n} p_i \ln(p_i). \tag{14.113}$$

14.12.2. Entropy for the heat equation. Let (M^n, g) be a closed Riemannian manifold and let $f : M^n \to \mathbb{R}$ be a positive function with $\int_{M^n} f d\mu = 1$. The **relative entropy** of the probability distribution $f d\mu$ is defined as

$$N(f) := -\int_{M^n} f \ln(f)\, d\mu. \tag{14.114}$$

Now suppose that $f(t) : M^n \to \mathbb{R}$ is a solution to the **heat equation**

$$\frac{\partial f}{\partial t} = \Delta f. \tag{14.115}$$

We compute that

$$\begin{aligned}\frac{dN}{dt} &= -\int_{M^n}\left(\ln(f)\frac{\partial f}{\partial t} + f\frac{\partial}{\partial t}\ln(f)\right)d\mu \\ &= -\int_{M^n}(\ln(f)\Delta f + \Delta f)\,d\mu = \int_{M^n}\frac{|\nabla f|^2}{f}d\mu \\ &\geq 0,\end{aligned}$$

where we integrated by parts and used the divergence theorem. Thus, the entropy of a solution to the heat equation is a nondecreasing function.

14.12.3. Entropy in comparison to Lebesgue p-norms. For any real number $p > 1$, we have

$$\int_{M^n} f\ln f\,d\mu \leq 2\left(\int_{M^n}|f-1|^p d\mu\right)^{1/p} + \frac{2}{p-1}\int_{M^n}|f-1|^p d\mu. \tag{14.116}$$

Now recall that the **L^p-norm** of a function $f : M^n \to \mathbb{R}$ is defined by

$$\|f\|_p := \left(\int_{M^n}|f|^p\,d\mu\right)^{1/p}. \tag{14.117}$$

Hölder's inequality says that for any $\alpha, \beta \in [1,\infty]$ with $\frac{1}{\alpha}+\frac{1}{\beta}=1$,[4] we have

$$\|fg\|_1 \leq \|f\|_\alpha \|g\|_\beta \tag{14.118}$$

for any functions f and g.

Now suppose that $\int_{M^n} d\mu = 1$ and let $q > p > 1$. By Hölder's inequality with $\alpha = \frac{q}{p}$, we have

$$\begin{aligned}(\|f\|_p)^p &= \||f|^p\|_1 = \||f|^p\cdot 1\|_1 \leq \||f|^p\|_\alpha \|1\|_\beta \\ &= \||f|^p\|_{\frac{q}{p}} = \left(\int_{M^n}|f|^q d\mu\right)^{p/q} = \|f\|_q^p.\end{aligned}$$

Hence,

$$\|f\|_p \leq \|f\|_q. \tag{14.119}$$

So, for $q > p > 1$, the L^q-norm is 'stronger' than the L^p-norm in the sense that $\|f\|_q \leq C$ tells you more than $\|f\|_p \leq C$.

Now, by (14.116), the entropy satisfies

$$\int_{M^n} f\ln f\,d\mu \leq 2\|f-1\|_p + \frac{2}{p-1}\|f-1\|_p^p. \tag{14.120}$$

So, the L^p-distance $\|f-1\|_p$ between f and the constant function 1 controls the entropy of f.

[4]We use the convention that $\frac{1}{\infty} := 0$.

We will now see that the idea of entropy is useful in Ricci flow.

14.12.4. Hamilton's entropy estimate. Let (M^2, g) be a closed Riemannian surface. If g has positive curvature, then we can define **Hamilton's surface entropy** by

$$N(g) := \int_{M^2} R \ln R \, d\mu. \tag{14.121}$$

(This is the opposite of the usual sign convention for entropy, so we want to show that Hamilton's entropy decreases.)

Let $(M^2, g(t))$ be a Ricci flow on a closed surface with positive curvature. The **surface entropy monotonicity formula** is

$$\begin{aligned} \frac{d}{dt}N(g(t)) &= -2\int_{M^2} \left| -\frac{1}{2}\overline{R}g + \nabla^2 f \right|^2 d\mu - \int_{M^2} \frac{|\nabla R + R\nabla f|^2}{R} d\mu \\ &= -2\int_{M^2} |\beta|^2 d\mu - 4\int_{M^2} \frac{|\operatorname{div}(\beta)|^2}{R} d\mu \\ &\leq 0. \end{aligned} \tag{14.122}$$

This implies Hamilton's result that his surface entropy is monotonically nonincreasing.

How did we obtain this monotonicity formula? The second equality in (14.122) follows from the definition of β and the calculations

$$\operatorname{div}(\beta) = -\frac{1}{2}\nabla R + \operatorname{div}(\nabla^2 f)$$

and

$$\begin{aligned} \operatorname{div}(\nabla^2 f) &= \sum_{i=1} \nabla^3 f(e_i, \cdot, e_i) \\ &= \sum_{i=1} \nabla^3 f(\cdot, e_i, e_i) + \operatorname{Ric}(\nabla f) \\ &= \nabla(\Delta f) + \frac{1}{2}R\nabla f \\ &= \nabla R + \frac{1}{2}R\nabla f. \end{aligned}$$

The first equality in (14.122) follows from the formula

$$\frac{d}{dt}N(g(t)) = -\int_{M^2} \frac{|\nabla R|^2}{R} d\mu + \int_{M^2} \overline{R}^2 d\mu \tag{14.123}$$

and an integration by parts. To see (14.123), we calculate as follows. Recall by (14.17) that

$$\frac{\partial}{\partial t} d\mu = -\overline{R} d\mu.$$

By combining this with (14.28), we obtain

$$\frac{\partial}{\partial t}(Rd\mu) = \frac{\partial R}{\partial t}d\mu + R\frac{\partial}{\partial t}d\mu = \Delta R d\mu. \tag{14.124}$$

Note that a consistency check for this formula is that as a consequence we have

$$\frac{d}{dt}\int_{M^2} Rd\mu = \int_{M^2} \Delta R d\mu = 0,$$

where the last equality is true by the divergence theorem. Indeed, we already know this from the Gauss–Bonnet formula.

Now, using (14.28) and (14.124), we calculate that

$$\begin{aligned}\frac{d}{dt}N &= \frac{d}{dt}\int_{M^2} \ln R R d\mu \\ &= \int_{M^2} \frac{\partial}{\partial t}(\ln R) R d\mu + \int_{M^2} \ln R \frac{\partial}{\partial t}(Rd\mu) \\ &= \int_{M^2} \frac{1}{R}(\Delta R + R\overline{R}) R d\mu + \int_{M^2} \ln R \Delta R d\mu.\end{aligned}$$

We can now integrate by parts to obtain

$$\frac{d}{dt}N = -\int_{M^2} \frac{|\nabla R|^2}{R} d\mu + \int_{M^2} \overline{R}^2 d\mu,$$

where we also used that $\int_{M^2} R\overline{R} d\mu = \int_{M^2} \overline{R}^2 d\mu$. See, e.g., [**CK04**] for an exposition of the details of how to carry out the integration by parts to obtain (14.122) from (14.123).

14.12.5. Hamilton's Harnack estimate.

In the study of the Ricci flow on surfaces, β is a natural quantity. Recall from the previous subsection that

$$2\operatorname{div}(\beta) = \nabla R + R\nabla f. \tag{14.125}$$

By taking a second divergence, we obtain

$$\begin{aligned}2\operatorname{div}^2(\beta) &= \operatorname{div}(\nabla R + R\nabla f) \\ &= \Delta R + \langle \nabla R, \nabla f\rangle + R\overline{R},\end{aligned} \tag{14.126}$$

where we used that $\Delta f = \overline{R}$. Now (14.125) implies that

$$-2\frac{\nabla R}{R}\cdot \operatorname{div}(\beta) = -\frac{|\nabla R|^2}{R} - \langle \nabla R, \nabla f\rangle.$$

Therefore,

$$Q := \frac{2}{R}\operatorname{div}^2(\beta) - 2\frac{\nabla R}{R^2}\cdot\operatorname{div}(\beta) = \frac{\Delta R}{R} - \frac{|\nabla R|^2}{R^2} + \overline{R} = \Delta \ln R + R - r. \tag{14.127}$$

The quantity Q is called **Hamilton's Harnack quantity**. As we will see in the next section, Q vanishes on self-similar solutions to the Ricci flow, called *Ricci solitons* (as we will see, β vanishes on Ricci solitons). This is one motivation for considering Q as a natural quantity for which to compute the evolution equation.

One can show the estimate

$$Q(x,t) \geq -\frac{Cre^{rt}}{Ce^{rt}-1} =: q(t), \tag{14.128}$$

where $C > 1$ is a constant depending only on g_0. Note that the function $q(t)$ is increasing. In particular, we have that if $t \geq 1$, then

$$Q(x,t) \geq -\frac{Cr}{C-1} =: -C' \tag{14.129}$$

for all $x \in M^2$. This is the **Harnack estimate** for the Ricci flow on surfaces.

The proof of (14.128) is to derive the following heat-type inequality:

$$\frac{\partial Q}{\partial t} \geq \Delta Q + 2\langle \nabla \ln R, \nabla Q\rangle + Q^2 + rQ. \tag{14.130}$$

By taking $C := \frac{q_0}{q_0+r} > 1$, where $q_0 := \min Q(\cdot,0)$, we have that $q(t)$ satisfies the ODE $\frac{dq}{dt} = q^2 + rq$ with $q(0) = \frac{Cr}{C-1} = q_0$. Now, applying the parabolic maximum principle to (14.130) yields the Harnack estimate (14.128).

Let us see why the Harnack estimate for the Ricci flow on surfaces is useful.

Using (14.28), we calculate that

$$\frac{\partial}{\partial t}\ln R = \frac{1}{R}\frac{\partial R}{\partial t} = \frac{1}{R}(\Delta R + R^2 - rR) = \Delta \ln R + |\nabla \ln R|^2 + R - r. \tag{14.131}$$

Therefore, the Harnack quantity Q defined by (14.127) may be re-expressed as the spacetime gradient quantity

$$Q = \frac{\partial}{\partial t}\ln R - |\nabla \ln R|^2. \tag{14.132}$$

Thus, the Harnack estimate (14.129) says that

$$\frac{\partial}{\partial t}\ln R - |\nabla \ln R|^2 \geq -C' \tag{14.133}$$

for some constant C', provided $t \geq 1$.

In order to compare the curvatures of the solution at two different points (x_1, t_1) and (x_2, t_2) in spacetime, we will integrate the differential expression Q along paths in spacetime. For this purpose, let

$$\gamma : [t_1, t_2] \to M^2$$

be a path with $\gamma(t_1) = x_1$ and $\gamma(t_2) = x_2$. Consider the associated spacetime path

$$\widetilde{\gamma} : [t_1, t_2] \to M^2 \times [t_1, t_2]$$

defined by

$$\widetilde{\gamma}(t) := (\gamma(t), t). \tag{14.134}$$

Observe that $\widetilde{\gamma}(t_1) = (x_1, t_1)$ and $\widetilde{\gamma}(t_2) = (x_2, t_2)$.

We now apply the fundamental theorem of calculus to the one-variable function $\ln R$ along $\widetilde{\gamma}$ to obtain

$$\begin{aligned} \ln R(x_2, t_2) &- \ln R(x_1, t_1) \\ &= \int_{t_1}^{t_2} \frac{d}{dt}(\ln R(\gamma(t), t))\, dt \\ &= \int_{t_1}^{t_2} \left(\nabla \ln R(\gamma(t), t) \cdot \gamma'(t) + \frac{\partial \ln R}{\partial t}(\gamma(t), t)\right) dt, \end{aligned} \tag{14.135}$$

where the dot product $\cdot$ denotes the inner product with respect to the metric $g(t)$, also denoted by $\langle \cdot, \cdot \rangle$. By applying the Harnack estimate (14.133) to this, we obtain

$$\begin{aligned} \ln \frac{R(x_2, t_2)}{R(x_1, t_1)} &\geq \int_{t_1}^{t_2} \left(\nabla \ln R(\gamma(t), t) \cdot \gamma'(t) + |\nabla \ln R|^2(\gamma(t), t) - C'\right) dt \\ &\geq -\int_{t_1}^{t_2} \frac{1}{4}|\gamma'(t)|^2_{g(t)} dt - C'(t_2 - t_1), \end{aligned}$$

where to obtain the last inequality we used the elementary (Peter–Paul) inequality $-ab + b^2 \geq -\frac{1}{4}a^2$ and that

$$\nabla \ln R(\gamma(t), t) \cdot \gamma'(t) \geq -|\nabla \ln R(\gamma(t), t)|_{g(t)} |\gamma'(t)|_{g(t)}.$$

We have proved the following proposition.

Proposition 14.21. *Let $(M^2, g(t))$ be a solution to the normalized Ricci flow on surfaces with positive curvature. Let $x_1, x_2 \in M^2$ and $t_1 < t_2$. Then for any path $\gamma : [t_1, t_2] \to M^2$ with $\gamma(t_1) = x_1$ and $\gamma(t_2) = x_2$, we have*

$$\frac{R(x_2,t_2)}{R(x_1,t_1)} \geq \mathrm{e}^{-C'(t_2-t_1)} \exp\left(-\int_{t_1}^{t_2} \frac{1}{4}|\gamma'(t)|^2_{g(t)} dt \right). \tag{14.136}$$

To get the best estimate from (14.136), on the right-hand side we should take the supremum over all such paths γ. Since, in general, we cannot compute the supremum, we will be satisfied with a rough lower estimate of the right-hand side which indeed will suffice for our purposes.

14.12.6. The uniform estimate for the scalar curvature.

We now proceed to obtain a uniform estimate for the scalar curvature R under the normalized Ricci flow.

14.12.6.1. *Uniform equivalence of the metrics on short-time intervals.* Let t_1 be any positive time. Let $x_1 \in M^2$ be a point at which $R(\cdot, t_1)$ attains its maximum. Let

$$K_1 := \max_{M^2} R(\cdot, t_1) = R(x_1, t_1). \tag{14.137}$$

We have

$$\frac{d}{dt} R_{\max}(t) \leq (R_{\max}(t))^2 - r R_{\max}(t) \leq (R_{\max}(t))^2.$$

The solution to the ODE $\frac{dk}{dt} = k^2$ with the initial condition $k(t_1) = K_1$ is given by

$$k(t) = \frac{1}{\frac{1}{K_1} + t_1 - t}$$

for $t \in [t_1, t_1 + \frac{1}{K_1}]$. Therefore, by the parabolic maximum principle, we have that

$$R(x,t) \leq 2K_1 \tag{14.138}$$

for all $x \in M^2$ and $t \in [t_1, t_1 + \frac{1}{2K_1}]$.

Let $t_2 := t_1 + \frac{1}{2K_1}$. Let $\bar{t} \in [t_1, t_2]$. We have for any $x \in M^2$,

$$\begin{aligned} g(x,t_2) &= \exp\left(\int_{\bar{t}}^{t_2} (r - R(x,t))\, dt \right) g(x,\bar{t}) \\ &\geq \exp\left(\int_{\bar{t}}^{t_2} (r - 2K_1)\, dt \right) g(x,\bar{t}) \\ &\geq \exp\left(\frac{r}{2K_1} - 1 \right) g(x,\bar{t}) \\ &\geq \mathrm{e}^{-1} g(x,\bar{t}) \end{aligned}$$

for all $\bar{t} \in [t_1, t_2]$. That is, we have the following.

Lemma 14.22. *For any normalized Ricci flow on a 2-sphere with positive curvature, we have*

$$g(x,t) \le \mathrm{e}\, g(x,t_2) \tag{14.139}$$

for all $x \in M^2$ and $t \in [t_1, t_2]$, where $t_2 = t_1 + \frac{1}{2K_1}$ and $K_1 = \max_{M^2} R(\cdot, t_1)$.

14.12.6.2. *Smoothing property of the curvature function.* Let x_2 be a point in M^2 and let $t_2 = t_1 + \frac{1}{2K_1}$, where (x_1, t_1) is as in the previous subsection so that $R(x_1, t_1) = \max_{M^2} R(\cdot, t_1) = K_1$. Let $\gamma : [t_1, t_2] \to M^2$ be a constant-speed minimal geodesic with respect to the metric $g(t_2)$ joining the point x_1 to the point x_2. Then

$$|\gamma'(t)|_{g(t_2)} = \frac{d_{g(t_2)}(x_1, x_2)}{t_2 - t_1}.$$

Further assume that $K \ge 1$. By Proposition 14.21, we have

$$\begin{aligned} \frac{R(x_2,t_2)}{R(x_1,t_1)} &\ge \mathrm{e}^{-C'(t_2-t_1)} \exp\left(-\int_{t_1}^{t_2} \frac{1}{4}|\gamma'(t)|^2_{g(t)} dt\right) \\ &\ge \mathrm{e}^{-\frac{C'}{2K}} \exp\left(-\mathrm{e}\int_{t_1}^{t_2} \frac{1}{4}|\gamma'(t)|^2_{g(t_2)} dt\right) \\ &= \mathrm{e}^{-\frac{C'}{2K}} \exp\left(-\frac{\mathrm{e}}{4}\frac{d^2_{g(t_2)}(x_1,x_2)}{t_2-t_1}\right). \end{aligned} \tag{14.140}$$

Recall that $t_2 - t_1 = \frac{1}{2K_1}$. Thus, if we assume that $d_{g(t_2)}(x_1, x_2) \le \frac{1}{\sqrt{K_1}}$, then

$$\frac{R(x_2,t_2)}{R(x_1,t_1)} \ge \mathrm{e}^{-\frac{C'}{2K_1}} \mathrm{e}^{-\frac{\mathrm{e}}{2}} \ge \mathrm{e}^{-\frac{C'+\mathrm{e}}{2}} \tag{14.141}$$

where the last inequality is true since $K_1 \ge 1$. Since $R(x_1, t_1) = K_1$, we obtain the following.

Lemma 14.23.

$$R(x_2, t_2) \ge \mathrm{e}^{-\frac{C'+\mathrm{e}}{2}} K_1 \tag{14.142}$$

for all $x_2 \in B^{g(t_2)}_{1/\sqrt{K_1}}(x_1)$ and where $t_2 = t_1 + \frac{1}{2K_1}$.

This lemma reflects the smoothing property of the curvature function. Namely, if the curvature is large at a point (x_1, t_1), then the curvature is large in a small ball centered at that point at a slightly later time.

14.12.6.3. *Combining the entropy and differential Harnack estimates.* We are now in a position to combine the entropy and differential Harnack estimates to obtain the uniform bound for the scalar curvature.

Lemma 14.24. *There exists a universal constant $c > 0$ such that*

$$\operatorname{Area}\left(B^{g(t_2)}_{1/\sqrt{K_1}}(x_1)\right) \geq \frac{c}{K_1}, \tag{14.143}$$

where $K_1 = R(x_1, t_1) = \max_{M^2} R(\cdot, t_1)$. That is, with respect to the metric $g(t_2)$, the ball of radius $\rho := \frac{1}{\sqrt{K_1}}$ centered at $x_1 \in M^2$ has area at least $c\rho^2$.

Proof. By Theorem 7.94(3) applied to $g(t_1)$, we have that $\operatorname{inj}(g(t_1)) \geq \frac{\pi}{\sqrt{K_1}}$. This and the curvature bound $R(t_1) \leq K_1$ implies the area estimate at time t_1. The lemma now follows from $g(t_2) \geq \mathrm{e}^{-1} g(t_1)$ by (14.139). □

Now recall that the monotonicity of the surface entropy says that there exists a constant (depending only on the initial metric g_0) such that

$$N(g(t)) = \int_{M^2} R \ln R \, d\mu \leq C$$

for all $t \in [0, \infty)$. On the other hand, recall the elementary inequality that for any $u \in (0, \infty)$, $u \ln u \geq -\frac{1}{\mathrm{e}}$. Thus we have that, where $B^{t_2} := B^{g(t_2)}$,

$$\begin{aligned} \int_{B^{t_2}_\rho(x_1)} R \ln R \, d\mu(t_2) &= \int_{M^2} R \ln R \, d\mu(t_2) - \int_{M^2 \setminus B^{t_2}_\rho(x_1)} R \ln R \, d\mu(t_2) \\ &\leq C + \frac{\operatorname{Area}(g(t_2))}{\mathrm{e}}, \end{aligned}$$

where the right-hand side is constant depending only on g_0. By applying Lemmas 14.23 and 14.24, we obtain that

$$\begin{aligned} C + \frac{\operatorname{Area}(g_0)}{\mathrm{e}} &\geq \operatorname{Area}(B^{t_2}_\rho(x_1)) \mathrm{e}^{-\frac{C' + \mathrm{e}}{2}} K_1 \ln\left(\mathrm{e}^{-\frac{C' + \mathrm{e}}{2}} K_1\right) \\ &\geq c \mathrm{e}^{-\frac{C' + \mathrm{e}}{2}} \ln\left(\mathrm{e}^{-\frac{C' + \mathrm{e}}{2}} K_1\right). \end{aligned}$$

This implies that for any time t_1, $\max_{t_1} R(\cdot, t_1) \leq K_1$ is bounded by a constant depending only on g_0. Since t_1 is arbitrary, this implies that the scalar curvature of the solution to the normalized Ricci flow is uniformly bounded.

14.12.6.4. *The uniform positive lower bound for the scalar curvature.* Since the metrics $g(t)$, $t \in [0, \infty)$, all have positive and uniformly bounded curvature and constant area, we have that the diameters of $g(t)$ are uniformly bounded from above (see, e.g., Corollary 5.52 in [**CK04**]). We claim that we can thus use the Harnack estimate again to obtain a uniform positive lower bound for the scalar curvatures of $g(t)$. This will complete the proof of Proposition 14.20 and hence also of Theorem 14.18 in the $\chi > 0$ case.

Proof of the lower bound. Let C be such that $R(x, t) \leq C$ for all $x \in M^2$ and $t \in [0, \infty)$ and $\operatorname{diam}(g(t)) \leq C$ for all $t \in [0, \infty)$. Let (x_2, t_2) be a point with $t_2 \geq 1$. Let $t_1 := t_2 - 1$ and let x_1 be a point at which

$$R(x_1, t_1) = r;$$

such a point always exists since r is equal to the average of R at time t_1. By the same argument as to obtain (14.140), we have

$$\frac{R(x_2, t_2)}{R(x_1, t_1)} \geq \mathrm{e}^{-\frac{C'}{2C}} \exp\left(-\frac{\mathrm{e}}{4}\frac{d^2_{g(t_2)}(x_1, x_2)}{t_2 - t_1}\right).$$

By applying the uniform diameter bound to this inequality, we obtain

$$R(x_2, t_2) \geq r\mathrm{e}^{-\frac{C'}{2C} - \frac{\mathrm{e}C^2}{4}}. \tag{14.144}$$

This is the desired uniform positive lower bound for the scalar curvature.

Remark 14.25. If the initial metric on a 2-sphere does not have positive curvature, then one can modify Hamilton's arguments to obtain the convergence of the solution; regarding this, one of the key issues is an injectivity radius estimate. See [**CK04**, Chapter 5].

14.13. Ricci solitons

In this section, we further discuss a notion of canonical Riemannian metric which appears implicitly in Hamilton's proof of the convergence of the Ricci flow on surfaces. This notion has become fundamental in the study of both Ricci flow in all dimensions and Riemannian geometry.

One may consider the possibility of a normalized Ricci flow on a surface M^2 that just moves by diffeomorphisms. That is, one may consider the possibility that the solution is of the form $g(t) = \phi_t^* g_0$ for some 1-parameter family of diffeomorphisms of M^2. Recall that isometric metrics are geometrically the same. Thus, such a solution is geometrically a fixed point of the normalized Ricci flow. One can think about this abstractly. That is, let $\mathfrak{Met}$ denote the set of Riemannian metrics on M^2. Let $\mathfrak{Diff}$ denote the group of self-diffeomorphisms of M^2. The group $\mathfrak{Diff}$ acts on the set $\mathfrak{Met}$ by pull-back: We have

$$\sigma : \mathfrak{Diff} \times \mathfrak{Met} \to \mathfrak{Met} \tag{14.145}$$

defined by

$$\sigma(\phi, g) := \phi^*(g). \tag{14.146}$$

The quotient space $\mathfrak{Met}/\mathfrak{Diff}$ is the set of isometry classes of Riemannian metrics on M^2. A Ricci flow $g(t)$, $t \in I$, may equivalently be considered as the path $\gamma : I \to \mathfrak{Met}$ defined by $\gamma(t) := g(t)$. Let $\pi : \mathfrak{Met} \to \mathfrak{Met}/\mathfrak{Diff}$ be the canonical projection map. Then

$$\pi \circ \gamma : I \to \mathfrak{Met}/\mathfrak{Diff}$$

maps t to the isometry class of $g(t)$. We see that a Ricci flow $g(t)$ evolves by diffeomorphisms if and only if the associated path $\pi \circ \gamma$ is constant.

14.13.1. Shrinking and steady Ricci solitons. It has been believed that constant curvature Riemannian metrics are the most natural metrics. Both the uniformization theorem and the Ricci flow version of its proof support this belief. The Ricci flow proof actually first proves convergence of the modified flow to what is called a shrinking Ricci soliton, which we now define.

Definition 14.26. A Riemannian surface (M^2, g) and a function f on M^2 is defined to be a **shrinking Ricci soliton** if[5]

$$\overline{R}g := (R - r)g = 2\nabla^2 f. \tag{14.147}$$

Observe that by tracing (14.147), we have that

$$\Delta f = R - r. \tag{14.148}$$

For this reason, we define f to be the **potential function**.

By (7.29), the shrinking Ricci soliton equation (14.147) says that

$$(R - r)g = \mathcal{L}_{\nabla f} g. \tag{14.149}$$

We claim that this equation is an infinitesimal version of the condition that a solution $g(t)$ to the normalized Ricci flow is of the form $g(t) = \phi_t^* g_0$ for some 1-parameter family of diffeomorphisms $\{\phi_t\}_{t\in\mathbb{R}}$. To see this, we compute that

$$(R_{g(t)} - r)g(t) = \frac{\partial}{\partial t} g(t) = \frac{\partial}{\partial t}(\phi_t^* g_0) = \mathcal{L}_{d(\phi_t^{-1})(\frac{\partial}{\partial t}\phi_t)} g(t).$$

Hence, if

$$d(\phi_t^{-1})\left(\frac{\partial}{\partial t}\phi_t\right) = \nabla_{g(t)} f(t) \tag{14.150}$$

for some function $f(t) : M^2 \to \mathbb{R}$, then we obtain

$$(R_{g(t)} - r)g(t) = \mathcal{L}_{\nabla f(t)} g(t). \tag{14.151}$$

Thus, in this case the Riemannian surface $(M^2, g(t))$ with $f(t)$ is a shrinking Ricci soliton. For such solutions to the *normalized* Ricci flow, the metric $g(t)$ is geometrically independent of time and moving only by the pull-back by diffeomorphisms. The reason why we call it a *shrinking* Ricci soliton is as follows. Define

$$\widetilde{g}(\tilde{t}) := \mathrm{e}^{-r_0 t} g(t) = \mathrm{e}^{-r_0 t} \phi_t^* g_0, \tag{14.152}$$

where $\tilde{t}(t) := \frac{1}{r_0}(1 - \mathrm{e}^{-r_0 t})$. By the discussion at the end of §14.3.3, we have that $\widetilde{g}(\tilde{t})$ is a solution to the unnormalized Ricci flow. These rescaled

[5]In the literature, this is typically called a **shrinking gradient Ricci soliton**.

metrics satisfy the Ricci flow and evolve by diffeomorphisms and scalings. Since $r_0 > 0$ and since $t(\tilde{t}) = -\frac{1}{r_0}\ln(1 - r_0\tilde{t})$ is an increasing function, we have that the metrics $\widetilde{g}(\tilde{t})$ are shrinking forward in time. This justifies the moniker 'shrinking Ricci soliton'.

In the previous section, we proved (albeit omitting some key details) that any solution $\widetilde{g}(t)$ to the modified Ricci flow on S^2 converges to a smooth metric $\widetilde{g}_\infty$ which satisfies equation (14.110); that is,

$$\overline{R}_{\widetilde{g}_\infty}\widetilde{g}_\infty = 2\nabla^2_{\widetilde{g}_\infty}\widetilde{f}_\infty.$$

Thus, we proved that for a flow that is geometrically the same as the normalized Ricci flow (i.e., the solutions of the two equations differ by the pull-back by diffeomorphisms), the solutions converge to shrinking Ricci solitons. We then used the Kazdan–Warner identity to prove that any shrinking Ricci soliton on S^2 must have constant curvature. So we proved that the solution to the *modified* Ricci flow converges to a constant curvature metric. Moreover, we can conclude the same for the *normalized* Ricci flow because of the exponential rate of convergence to constant curvature, including the derivatives of curvature decaying exponentially to 0. That is, the solutions to the normalized Ricci flow converge to constant curvature metrics. This completes the proof of the differential geometric version of the uniformization theorem.

The discussion above begs the following question: Are there Ricci solitons that are not constant curvature metrics (so that the potential functions are constants)?

We first consider **steady Ricci solitons**. These are Riemannian surfaces (M^2, g), together with functions $f : M^2 \to \mathbb{R}$, that satisfy the equation (cf. (14.147))

$$Rg = 2\nabla^2 f. \tag{14.153}$$

As in the shrinking case, we define f to be the potential function.

14.13.2. Cigar soliton. An iconic example of a *steady* Ricci soliton is the 2-dimensional **cigar soliton**. Its underlying manifold is the plane $\mathbb{R}^2$. Its Riemannian metric is defined by

$$g_\Sigma(x^1, x^2) := \frac{4g_\mathbb{E}}{1 + (x^1)^2 + (x^2)^2}, \tag{14.154}$$

where $g_\mathbb{E} = dx^1 \otimes dx^1 + dx^2 \otimes dx^2$ is the Euclidean metric and its potential function is defined by

$$f_\Sigma\left(x^1, x^2\right) := -\ln\left(1 + (x^1)^2 + (x^2)^2\right). \tag{14.155}$$

See Figure 14.13.1. The reason for the factor of 4 in (14.154) is so that the maximum scalar curvature of g_Σ will be equal to 1.

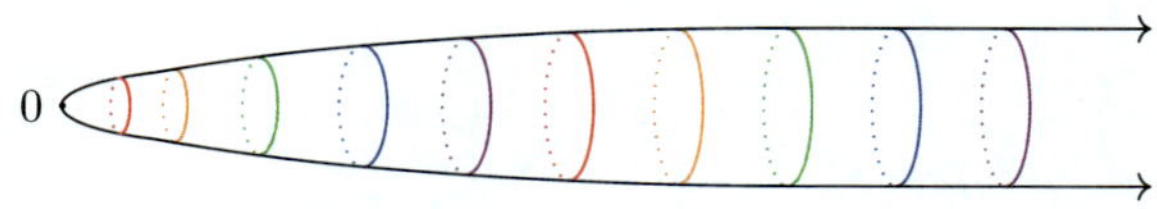

Figure 14.13.1. The cigar soliton metric g_Σ on $\mathbb{R}^2$.

The (exterior) derivative of the potential function f_Σ is given by

$$df_\Sigma = -\frac{1}{1+(x^1)^2+(x^2)^2}(2x^1\,dx^1 + 2x^2\,dx^2). \tag{14.156}$$

Thus, the gradient, with respect to g_Σ, of the potential function is given by

$$\nabla_{g_\Sigma} f_\Sigma\,(x^1,x^2) = \left(-\frac{x^1}{2}, -\frac{x^2}{2}\right). \tag{14.157}$$

Recall from (8.98) that if $\widetilde{g} = \mathrm{e}^{2u}g$, then

$$K_{\widetilde{g}} = \mathrm{e}^{-2u}(K_g - \Delta_g u).$$

Using this, we compute that the Gauss curvature of g_Σ is equal to

$$\begin{aligned} K_\Sigma &= -\frac{1+(x^1)^2+(x^2)^2}{4}\Delta_{\mathbb{E}}\left(\frac{1}{2}\ln\frac{4}{1+(x^1)^2+(x^2)^2}\right) \\ &= \frac{1}{2(1+(x^1)^2+(x^2)^2)}. \end{aligned} \tag{14.158}$$

On the Riemannian surface $(\mathbb{R}^2, g_\Sigma)$ we have the global orthonormal frame field defined by

$$e_1 := \sqrt{1+(x^1)^2+(x^2)^2}\frac{\partial}{\partial x^1}, \qquad e_2 := \sqrt{1+(x^1)^2+(x^2)^2}\frac{\partial}{\partial x^2}.$$

The Hessian of f_Σ with respect to this frame field is given by

$$\nabla^2 f_\Sigma(e_i,e_j) = e_i(e_j(f_\Sigma)) - \sum_{k=1}^{2}\omega_j^k(e_i)e_k(f_\Sigma). \tag{14.159}$$

We have that $(e_{\mathbb{E}})_1 = \frac{\partial}{\partial x^1}$, $(e_{\mathbb{E}})_2 = \frac{\partial}{\partial x^2}$ is a global orthonormal frame field for the Euclidean metric $g_{\mathbb{E}}$. Its dual orthonormal coframe field is given by $(\omega_{\mathbb{E}})^1 = dx^1$, $(\omega_{\mathbb{E}})^2 = dx^2$. By (8.95) and since the connection 1-forms of the Euclidean metric $(\omega_{\mathbb{E}})_j^i = 0$, we have that the connection 1-forms ω_j^i of g_Σ with respect to the orthonormal frame e_1, e_2 are given by

$$\begin{aligned} \omega_j^k &= \frac{\partial}{\partial x^j}\left(\frac{1}{2}\ln\frac{4}{1+(x^1)^2+(x^2)^2}\right)dx^k \\ &\quad - \frac{\partial}{\partial x^k}\left(\frac{1}{2}\ln\frac{4}{1+(x^1)^2+(x^2)^2}\right)dx^j \\ &= \frac{x^k dx^j - x^j dx^k}{1+(x^1)^2+(x^2)^2}. \end{aligned}$$

Thus,

$$\omega_j^k(e_i) = \frac{x^k\delta_{ij} - x^j\delta_{ik}}{\sqrt{1+(x^1)^2+(x^2)^2}},$$

where δ_{ij} is the Kronecker delta symbol. We have

$$e_j(f_\Sigma) = -\frac{2x^j}{\sqrt{1+(x^1)^2+(x^2)^2}}$$

and

$$e_i(e_j(f_\Sigma)) = -2\delta_{ij} + \frac{2x^i x^j}{1+(x^1)^2+(x^2)^2}.$$

Moreover,

$$\begin{aligned}-\sum_{k=1}^{2}\omega_j^k(e_i)e_k(f_\Sigma) &= \sum_{k=1}^{2}\frac{x^k\delta_{ij} - x^j\delta_{ik}}{\sqrt{1+(x^1)^2+(x^2)^2}}\frac{2x^k}{\sqrt{1+(x^1)^2+(x^2)^2}}\\ &= \frac{2((x^1)^2+(x^2)^2)\delta_{ij} - 2x^i x^j}{1+(x^1)^2+(x^2)^2}.\end{aligned}$$

Hence, by (14.159) and by summing the last two displays, we obtain

$$\nabla^2 f_\Sigma(e_i,e_j) = -\frac{2\delta_{ij}}{\sqrt{1+(x^1)^2+(x^2)^2}} = -K_\Sigma g_\Sigma(e_i,e_j).$$

This proves that the cigar soliton $(\mathbb{R}^2, g_\Sigma, f_\Sigma)$ is a steady Ricci soliton.

14.13.3. Shrinking Ricci solitons on orbifolds. Hamilton [**Ham88**, §10] proved that any shrinking Ricci soliton on a bad orbifold must be rotationally symmetric. He also proved that the soliton is unique up to scaling and diffeomorphisms.

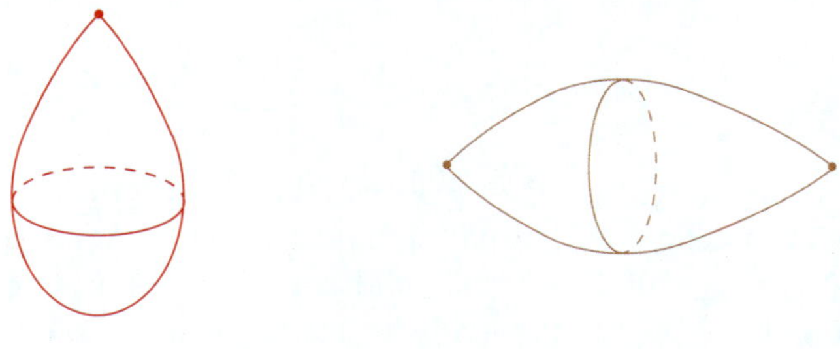

Figure 14.13.2. The teardrop (left) and football (right) shrinking Ricci solitons.

The proof (see Case 2 in the proof of Lemma 14.19) of the fact that on S^2 the only shrinking Ricci solitons are the constant curvature metrics uses the Kazdan–Warner identity and hence uses the uniformization theorem. Chen, Lu, and Tian [**CLT06**] proved this result without using the uniformization theorem.

14.14. Uniformization of 2-dimensional orbifolds

Both the uniformization theorem and the global existence and convergence of the Ricci flow on surfaces extend to 2-dimensional orbifolds. In this section, we survey a result in this direction.

Firstly, we remark that the Hodge decomposition theorem (Theorem 11.59) extends to orbifolds; see Baily [**Bai56**]. In particular, we have the following consequence in dimension 2 (cf. Corollary 11.61).

Proposition 14.27. *Let $(\mathcal{O}^2, g)$ be a closed Riemannian orbifold with isolated singularities. If $\phi : \mathcal{O}^2 \to \mathbb{R}$ is a function satisfying $\int_{\mathcal{O}^2} \phi \, d\mu = 0$, then there exists a function $f : \mathcal{O}^2 \to \mathbb{R}$ satisfying the Poisson equation*

$$\Delta f = \phi. \tag{14.160}$$

Thus, for any closed Riemannian orbifold $(\mathcal{O}^2, g)$ with isolated singularities, there exists a function $f : \mathcal{O}^2 \to \mathbb{R}$ satisfying

$$\Delta f = R - r, \tag{14.161}$$

where r is the average of the scalar curvature R.

By the works of Wu [**Wu91, CW91**], we have the following.

Theorem 14.28 (Uniformization of 2-dimensional orbifolds)**.** *Let $(\mathcal{O}^2, g_0)$ be a 2-dimensional closed oriented Riemannian orbifold. Then there exists a solution $g(t)$ to the modified Ricci flow for all time $t \in [0, \infty)$ with $g(0) = g_0$. As $t \to \infty$, $g(t)$ converges in each C^k-norm to a C^∞-metric g_∞. There exists a function f_∞ on $\mathcal{O}^2$ such that $(\mathcal{O}^2, g_\infty)$ together with f_∞ is a shrinking Ricci soliton. That is,*

$$(R_{g_\infty} - r) g_\infty = 2 \nabla^2_{g_\infty} f_\infty. \tag{14.162}$$

(Note that $\Delta_{g_\infty} f_\infty = R_{g_\infty} - r$.) For a good orbifold, both the normalized and modified Ricci flows converge to constant curvature metrics.

For *good* orbifolds, this result is originally due to Hamilton [**Ham88**]. Any closed orbifold that admits a constant curvature metric must be a good orbifold. Therefore, any shrinking Ricci soliton on a bad closed orbifold must be nontrivial; that is, its potential and curvature functions are not constant.

14.15. Appendix: The Kazdan–Warner identity

Let (M^n, g) be a Riemannian manifold and let Y be a vector field on M^n. We say that Y is a **conformal Killing vector field** if

$$\mathcal{L}_Y g = \frac{2}{n} \operatorname{div}(Y) g. \tag{14.163}$$

This condition on Y is equivalent to the condition that there exists some function $\phi : M^n \to \mathbb{R}$ such that

$$\mathcal{L}_Y g = \phi g. \tag{14.164}$$

Indeed, if this latter condition holds, then by tracing this equation, we obtain

$$2\operatorname{div}(Y) = g^{ij}(\nabla_i Y_j + \nabla_j Y_i) = n\phi,$$

which, together with (14.164), implies (14.163). Note that any Killing vector field is necessarily a conformal Killing vector field.

The following result is called the **Kazdan–Warner identity** [**KW74**]. This result can be used in the classification of Ricci solitons on the 2-sphere; see Lemma 14.19.

Theorem 14.29 (Kazdan and Warner). *For any conformal Killing vector field Y on a Riemannian 2-sphere $(\mathbb{S}^2, g)$, i.e., $\mathcal{L}_Y g = \operatorname{div}(Y) g$, we have*

$$\int_{\mathbb{S}^2} \langle \nabla_g R, Y \rangle_g \, d\mu_g = 0. \tag{14.165}$$

Proof. By the uniformization theorem (Theorem 8.19), there exists a function u on $\mathbb{S}^2$ such that $g = \mathrm{e}^{2u}\bar{g}$, where $\bar{g}$ has constant Gauss curvature $K_{\bar{g}} \equiv 1$. So $\Delta_{\bar{g}} u + \mathrm{e}^{2u} K_g = 1$. Let ℓ be the restriction of a linear function on $\mathbb{R}^3$ to $\mathbb{S}^2$ (considering $\mathbb{S}^2 \subset \mathbb{R}^3$ as the unit sphere), so that $\Delta_{\bar{g}}\ell + 2\ell = 0$. (The eigenspace associated to the eigenvalue 2 of $-\Delta_{\bar{g}}$ consists of the linear functions and the next smallest eigenvalue is 6.) In the following, ∇, Δ, and $\langle \cdot, \cdot \rangle$ are all with respect to $\bar{g}$. We compute, integrating by parts twice while using $\nabla^2 \ell + \ell \bar{g} = 0$ and $\Delta_{\bar{g}}\ell + 2\ell = 0$, that

$$\begin{aligned}
\int_{\mathbb{S}^2} \langle \nabla u, \nabla \ell \rangle \, \Delta u \, dA_{\bar{g}} &= -\int_{\mathbb{S}^2} (\langle \nabla^2 \ell, \nabla u \otimes \nabla u \rangle + \langle \nabla^2 u, \nabla u \otimes \nabla \ell \rangle) \, dA_{\bar{g}} \\
&= \int_{\mathbb{S}^2} \left(\ell |\nabla u|^2 - \frac{1}{2} \langle \nabla |\nabla u|^2, \nabla \ell \rangle \right) dA_{\bar{g}} \\
&= 0.
\end{aligned}$$

Integrating by parts and again using $\Delta_{\bar{g}}\ell = -2\ell$, we obtain

$$\begin{aligned}
\int_{\mathbb{S}^2} \langle \nabla u, \nabla \ell \rangle \, \mathrm{e}^{2u} K_g \, dA_{\bar{g}} &= \frac{1}{2} \int_{\mathbb{S}^2} \langle \nabla \mathrm{e}^{2u}, \nabla \ell \rangle \, K_g \, dA_{\bar{g}} \\
&= \int_{\mathbb{S}^2} \left(\ell K_g - \frac{1}{2} \langle \nabla K_g, \nabla \ell \rangle \right) \mathrm{e}^{2u} dA_{\bar{g}}.
\end{aligned}$$

Hence

$$\int_{\mathbb{S}^2}\left(\ell K_g-\frac{1}{2}\langle\nabla K_g,\nabla\ell\rangle\right)e^{2u}dA_{\bar{g}}=\int_{\mathbb{S}^2}\langle\nabla u,\nabla\ell\rangle\left(e^{2u}K_g+\Delta u\right)dA_{\bar{g}}$$
$$=\int_{\mathbb{S}^2}\langle\nabla u,\nabla\ell\rangle\,dA_{\bar{g}}$$
$$=-\int_{\mathbb{S}^2}\ell\Delta u\,dA_{\bar{g}}$$
$$=\int_{\mathbb{S}^2}\ell(e^{2u}K_g-1)\,dA_{\bar{g}}.$$

Since $\int_{\mathbb{S}^2}\ell dA_{\bar{g}}=0$, we conclude that

$$\int_{\mathbb{S}^2}\langle\nabla_g K_g,\nabla_{\bar{g}}\ell\rangle_g\,dA_g=\int_{\mathbb{S}^2}\langle\nabla_{\bar{g}}K_g,\nabla_{\bar{g}}\ell\rangle_{\bar{g}}\,e^{2u}dA_{\bar{g}}=0. \tag{14.166}$$

The group of conformal transformations of $\mathbb{S}^2$ is the Möbius group, which has complex dimension 3. Hence, the vector space of conformal Killing vector fields of g (which is the same as that of $\bar{g}$) is 6 real dimensional, of which 3 dimensions are spanned by Killing vector fields of $\bar{g}$ and the other 3 dimensions are spanned by the $\nabla_{\bar{g}}\ell$. If X is a Killing vector field of $\bar{g}$, then

$$\int_{\mathbb{S}^2}\langle\nabla_g K_g,X\rangle_g\,dA_g=\int_{\mathbb{S}^2}\langle\nabla_{\bar{g}}K_g,X\rangle_{\bar{g}}\,e^{2u}dA_{\bar{g}}$$
$$=-2\int_{\mathbb{S}^2}\langle\nabla_{\bar{g}}u,X\rangle_{\bar{g}}\,K_g e^{2u}dA_{\bar{g}}$$
$$=2\int_{\mathbb{S}^2}\langle\nabla_{\bar{g}}u,X\rangle_{\bar{g}}\,(\Delta_{\bar{g}}u-1)\,dA_{\bar{g}}$$
$$=-2\int_{\mathbb{S}^2}\left(\langle\nabla^2_{\bar{g}}u,\nabla_{\bar{g}}u\otimes X\rangle_{\bar{g}}+\langle\nabla_{\bar{g}}X,\nabla_{\bar{g}}u\otimes\nabla_{\bar{g}}u\rangle_{\bar{g}}\right)dA_{\bar{g}}$$
$$=-\int_{\mathbb{S}^2}\langle\nabla_{\bar{g}}|\nabla_{\bar{g}}u|^2,X\rangle_{\bar{g}}\,dA_{\bar{g}}$$
$$=0$$

by integrating by parts and by using $\nabla^{\bar{g}}_i X_j+\nabla^{\bar{g}}_j X_i=0$ and $\operatorname{div}_{\bar{g}}X=0$. The theorem follows. □

14.16. Notes and commentary

There are several different proofs of the convergence of the normalized Ricci flow on a topological 2-sphere to a round (i.e., constant positive curvature) metric. See Bartz, Struwe, and Ye [**BSY94**] for a proof based on the Aleksandrov reflection method, which can be viewed as the 2-dimensional version of Ye's result [**Ye94**] for the Yamabe flow on locally conformally flat manifolds in higher dimensions. See Hamilton [**Ham95d**] for a proof based on an isoperimetric ratio monotonicity formula. See Andrews and Bryan [**AB10**] for a proof using the isoperimetric profile, which also provides a direct proof of the uniform curvature estimate.

The Kazdan–Warner inequality is in [**KW74**].

The 3-dimensional Ricci flow has been used by Perelman and Hamilton to prove Thurston's geometrization conjecture, which subsumes the Poincaré conjecture. There is a vast literature on Hamilton's Ricci flow. Seminal works are by Hamilton [**Ham82, Ham86, Ham88, Ham93, Ham95a, Ham95b, Ham97, Ham99**] and Perelman [**Pera, Perb, Perc**]. Expositions of Hamilton and Perelman's work include Kleiner and Lott [**KL08**] and Morgan and Tian [**MT07, MT14**].

Further important developments in dimension 3 include the works of Bamler and Kleiner [**BK23**], Brendle [**Bre13b, Bre20**], Brendle, Daskalopoulos, and Sesum [**BDS21**], and Kleiner and Lott [**KL17, KL20**].

Differentiable spherical space form theorems were proved by Böhm and Wilking [**BW08**] and Brendle and Schoen [**BS09, BS08**].

High-dimensional compactness, regularity, and singularity analysis have been developed by Bamler [**Bam21, Bama, Bam23, Bamb**].

The Kähler–Ricci flow is an active area of research. See the survey by Song and Weinkove [**SW13**]. For the Kähler–Ricci flow approach to the analytic minimal model program, see Song and Tian [**ST17**].

There are books by Topping [**Top06**], Andrews and Hopper [**AH11**], and Brendle [**Bre10**] and the series of books by Chu, Glickenstein, Guenther, Isenberg, Ivey, Knopf, Lu, Luo, Ni, and the junior author [**CK04, CLN06, CCG⁺07, CCG⁺08, CCG⁺10, CCG⁺15**].

For books on Ricci solitons, see [**Cho23**] by the junior author and [**CKM**] by Kotschwar, Munteanu, and the junior author.

For books on the discrete Ricci flow and discrete conformal geometry, see Zeng and Gu [**ZG13**] and Glickenstein, Luo, and the junior author [**CGL**].

14.17. Exercises

Exercise 14.1. *Prove Lemma* 14.9.

Exercise 14.2. *Prove Lemma* 14.10.

Exercise 14.3. *Show that the cigar soliton metric, defined by* (14.154), *may be expressed (except at the origin* $\mathbf{0}$*) by a change of coordinates as*

$$g_\Sigma = ds^2 + \tanh^2(s)\, d\theta^2, \tag{14.167}$$

for $s \in (0, \infty)$ *and* $\theta \in \mathbb{R}/2\pi\mathbb{Z}$.

Exercise 14.4. *Prove that the Gauss curvature of the cigar soliton metric is given by*

$$K_\Sigma = 2\,\mathrm{sech}^2(s). \tag{14.168}$$

Exercise 14.5. *Prove that for a Ricci flow $g(t)$ on a closed surface M^2 with positive curvature, the evolution of Hamilton's entropy $N(g(t))$ defined by* (14.121) *is given by*

$$\frac{d}{dt}N(g(t)) = -\int_{M^2} \frac{|\nabla R|^2}{R}d\mu + \int_{M^2}(R-r)^2 d\mu. \tag{14.169}$$

Exercise 14.6. *Using* (14.169) *and using the fundamental Bochner formula* (11.21), *i.e.,*

$$\frac{1}{2}\Delta|\nabla f|^2 = \langle \nabla f, \nabla\Delta f\rangle + |\nabla^2 f|^2 + \mathrm{Ric}(\nabla f, \nabla f),$$

prove the entropy formula (14.122).

14.17.1. Exercises on Ricci solitons on surfaces.

The general form of the Ricci soliton equation on a surface is given by

$$(R-\lambda)g = 2\nabla^2 f, \tag{14.170}$$

where $\lambda \in \mathbb{R}$ (compare with (14.147)).

We say that the Ricci soliton is **shrinking**, **steady**, or **expanding** if $\lambda > 0$, $\lambda = 0$, or $\lambda < 0$, respectively.

Exercise 14.7. *Prove that for any Ricci soliton on a surface,*

$$\nabla R = R\nabla f. \tag{14.171}$$

Conclude that $R = Ce^f$ for some constant C.
Hint: Take the divergence of (14.170). *When considering the consequences of* (14.171), *first assume that you are at a point where $R \neq 0$.*

Exercise 14.8. *Prove that for any Ricci soliton on a surface,*

$$\Delta R = -\langle \nabla R, \nabla f\rangle + \lambda R - R^2. \tag{14.172}$$

Exercise 14.9. *Prove that for any Ricci soliton on a surface,*

$$R + |\nabla f|^2 - \lambda f = C \tag{14.173}$$

for some constant C.

Exercise 14.10. *Prove that for any Ricci soliton on a surface,*

$$\mathcal{L}_{J(\nabla f)}g = 0;$$

that is, $J(\nabla f)$ is a Killing vector field. Here, $J: TM \to TM$ is the almost complex structure (see (10.103)*).*

14.17.2. Linear Harnack estimate.

Exercise 14.11. *Let $\alpha > 0$ and let $\left(M^2, g(t)\right)$ be a solution to the α-**Ricci flow**:*

$$\frac{\partial}{\partial t} g = -\alpha R g \tag{14.174}$$

on a closed surface with $R > 0$. Couple this solution to a positive solution u to the linear heat-type equation:

$$\frac{\partial}{\partial t} u = \Delta u + \alpha R u. \tag{14.175}$$

Define the linear Harnack quantity $P := \Delta \ln u + \alpha R$. Prove that

$$\begin{aligned}\frac{\partial P}{\partial t} &= \Delta P + 2\left|P_{ij}\right|^2 - \alpha R P + 2\nabla P \cdot \nabla \ln u \\ &\quad + R\left|\nabla \ln u - \alpha \nabla \ln R\right|^2 + \alpha R\left(\frac{\partial}{\partial t} \ln R - \alpha\left|\nabla \ln R\right|^2\right),\end{aligned}$$

where $P_{ij} := \nabla_i \nabla_j \ln u + \frac{1}{2}\alpha R g_{ij}$. Deduce, using the parabolic maximum principle, that

$$\frac{\partial}{\partial t} \ln u - |\nabla \ln u|^2 + \frac{1}{t} = \Delta \ln u + \alpha R + \frac{1}{t} \geq 0. \tag{14.176}$$

*We call this a **linear Harnack estimate**. By taking $\alpha = 1$ and $u = R$, we see that this generalizes Hamilton's Harnack estimate on surfaces.*

14.17.3. Exercises on the Ricci flow on a 2-sphere with initial curvature changing sign.

Let $(M^2, g(t))$ be a solution to the normalized Ricci flow on a surface M^2 diffeomorphic to S^2. Let $s = s(t)$ be a solution to the ODE $\frac{d}{dt}s = s^2 - rs$ with $s(0) < \min_{M^2} R(\cdot, 0)$. By the maximum principle, we have that $R(x,t) - s(t) > 0$ for all $x \in M^2$ and $t \in [0, \infty)$.

14.17.3.1. *Hamilton's surface entropy.* Define the **modified Hamilton's surface entropy** by

$$N(g(t)) := \int_{M^2} \left(R(x,t) - s(t)\right) \ln\left(R(x,t) - s(t)\right) d\mu(x,t). \tag{14.177}$$

Exercise 14.12. *Show that under the normalized Ricci flow on surfaces,*

$$\frac{dN}{dt} = \int_{M^2}\left(-\frac{|\nabla R|^2}{R - s} + (R - s)\big(R - r + s + s\ln(R - s)\big)\right) d\mu. \tag{14.178}$$

Exercise 14.13. *Show that one can rewrite the evolution of the entropy as*

$$\begin{aligned}(14.179)\qquad \frac{dN}{dt} &= -\int_{M^2} \frac{|\nabla R + (R-s)\nabla f|^2}{R-s} d\mu - 2\int_{M^2} |\beta|^2 d\mu \\ &\quad - s\int_{M^2} \Big(|\nabla f|^2 + s - r - (R-s)\ln(R-s)\Big) d\mu,\end{aligned}$$

where $\beta = -\frac{1}{2}\overline{R}g + \nabla^2 f$.

Exercise 14.14. *Show that under the normalized Ricci flow on* $M^2 \cong S^2$,

$$(14.180)\qquad \frac{d}{dt}\int_{M^2} f\, d\mu = \int_{M^2} (rf + |\nabla f|^2) d\mu.$$

Use this to prove that there exists a constant C *such that*

$$(14.181)\qquad \int_0^\infty \mathrm{e}^{-rt} \int_{M^2} |\nabla f|^2 \, d\mu dt \le C.$$

Exercise 14.15. *Prove that*

$$\begin{aligned}\frac{dN}{dt} &\le -s\int_{M^2} \Big(|\nabla f|^2 + s - r - (R-s)\ln(R-s)\Big) d\mu \\ &\le C\mathrm{e}^{-rt}\int_{M^2} |\nabla f|^2 \, d\mu + C\mathrm{e}^{-rt}|N|.\end{aligned}$$

Conclude that there exists a constant C *such that under the normalized Ricci flow,*

$$(14.182)\qquad N(g(t)) \le C$$

for all $t \in [0,\infty)$.

14.17.3.2. *Hamilton's Harnack estimate for curvature changing sign.* Define $L := \ln(R-s)$ and define the **modified Hamilton's Harnack quantity** by

$$(14.183)\qquad Q := \Delta L + R - r = \frac{\partial L}{\partial t} - |\nabla L|^2 - s.$$

Exercise 14.16. *Prove that* Q *satisfies*

$$\begin{aligned}(14.184)\qquad \frac{\partial Q}{\partial t} &= \Delta Q + 2\langle \nabla Q, \nabla L\rangle + 2\left|\nabla^2 L + \frac{1}{2}(R-r)g\right|^2 \\ &\quad + s|\nabla L|^2 + (r-s)Q + s(R-r).\end{aligned}$$

Exercise 14.17. *Further modify Hamilton's Harnack quantity by defining $P := Q + sL$. Show that there exists a constant C depending only on g_0 such that*

$$\frac{\partial P}{\partial t} \geq \Delta P + 2\langle \nabla P, \nabla L\rangle + \frac{1}{2}\left(P^2 - C^2\right). \tag{14.185}$$

Hint: Show that there exists a constant C such that

$$\frac{\partial}{\partial t}(sL) \geq \Delta(sL) + 2\langle \nabla(sL), \nabla L\rangle - s|\nabla L|^2 - C. \tag{14.186}$$

Exercise 14.18. *For the Ricci flow on $M^2 \cong S^2$, show that for any initial metric g_0 there exists a constant $C > 0$ depending only on g_0 such that for any $x_1, x_2 \in M^2$ and $0 \leq t_1 < t_2$,*

$$\frac{R(x_2, t_2) - s(t_2)}{R(x_1, t_1) - s(t_1)} \geq e^{-\frac{A}{4} - C(t_2 - t_1)}, \tag{14.187}$$

where $A := \inf_\gamma \int_{t_1}^{t_2} \left|\frac{d\gamma}{dt}(t)\right|^2_{g(t)} dt$ and the infimum is taken over all paths $\gamma : [t_1, t_2] \to M^2$ satisfying $\gamma(t_1) = x_1$ and $\gamma(t_2) = x_2$.

14.17.4. Conformal Killing vector fields.

Exercise 14.19. *Let Y be a conformal Killing vector field on a Riemannian manifold (M^n, g). By taking the divergence of (14.163), show that*

$$\Delta Y + \mathrm{Ric}(Y) + \frac{n-2}{n}\nabla(\mathrm{div}(Y)) = 0. \tag{14.188}$$

Derive the integral identity

$$\int_{M^n}\left(|\nabla Y|^2 - \mathrm{Ric}(Y, Y) + \frac{n-2}{n}|\,\mathrm{div}(Y)|^2\right) d\mu = 0. \tag{14.189}$$

Conclude that if (M^n, g) has negative Ricci curvature, then $Y \equiv 0$.

Exercise 14.20. *Let Y be a conformal Killing vector field on a Riemannian manifold (M^n, g). Show that*

$$\frac{1}{n}\int_{M^n} R\,\mathrm{div}(Y)\,d\mu = -\int_{M^n}\langle \mathrm{div}(\mathrm{Ric}), Y\rangle\, d\mu. \tag{14.190}$$

By combining this with the general identity

$$\int_{M^n}\langle \mathrm{div}(\mathrm{Ric}), Y\rangle\, d\mu = -\frac{1}{2}\int_{M^n} R\,\mathrm{div}(Y)\,d\mu, \tag{14.191}$$

conclude that if $n \geq 3$, then

$$\int_{M^n}\langle \nabla R, Y\rangle\, d\mu = -\int_{M^n} R\,\mathrm{div}(Y)\,d\mu = 0. \tag{14.192}$$

This identity, due to Bourguignon and Ezin [**BE87**], *is the higher-dimensional version of the Kazdan–Warner identity.*

14.17.5. Tensor calculations.

Exercise 14.21. *Prove the following commutator identity of the Hessian and Laplacian:*

(14.193)
$$\nabla_i \nabla_j \Delta R = \Delta \nabla_i \nabla_j R - 2R\nabla_i \nabla_j R + \left(R\Delta R + \frac{1}{2} |\nabla R|^2 \right) g_{ij} - \nabla_i R \nabla_j R.$$

Hint: The Ricci identity for a surface is

$$\nabla \Delta = \Delta \nabla - \frac{1}{2} R \nabla. \tag{14.194}$$

Exercise 14.22. *Prove that under the normalized Ricci flow on surfaces,*

$$\frac{\partial}{\partial t} \nabla_i \nabla_j R = \Delta \nabla_i \nabla_j R + R\Delta R g_{ij} + 2\nabla_i R \nabla_j R - r\nabla_i \nabla_j R. \tag{14.195}$$

Hint: Use the formula

$$\frac{\partial}{\partial t} \Gamma_{ij}^k = -\frac{1}{2} \left(\nabla_i R \delta_j^k + \nabla_j R \delta_i^k - \nabla^k R g_{ij} \right). \tag{14.196}$$

Exercise 14.23. *Using the previous two exercises, prove equation* (14.82) *for the evolution of* $|\nabla^2 R|^2$ *under the normalized Ricci flow on surfaces.*

Exercise 14.24. *Prove* (14.91).

Exercise 14.25. *Let* g_t, $t \in [0, T]$, *be a smooth* 1*-parameter family of Riemannian metrics on* M^n. *Let* $p \in M^n$.

(1) *Prove that for all* $V \in T_p M \setminus \{\vec{0}\}$,

$$\ln \frac{g_T(V, V)}{g_0(V, V)} = \int_0^T \left(\frac{\partial}{\partial t} g_t \right) (U_t, U_t)\, dt \leq \int_0^T \left| \frac{\partial}{\partial t} g_t \right|_{g_t} dt, \tag{14.197}$$

where $U_t := \frac{V}{\sqrt{g_t(V,V)}}$.

(2) *Prove that if* $\left| \frac{\partial}{\partial t} g_t \right|_{g_t} (p, t) \leq C$ *on* $M^n \times [0, T]$, *then for all* $p \in M^n$ *and* $V \in T_p M$,

$$\mathrm{e}^{-CT} g_0(V, V) \leq g_T(V, V) \leq \mathrm{e}^{CT} g_0(V, V). \tag{14.198}$$

This extends the discussion in §14.7 *to families of metrics that are not necessarily conformal. In particular, this applies to the discussion in* §14.11.3.

Bibliography

[AB10] Ben Andrews and Paul Bryan, *Curvature bounds by isoperimetric comparison for normalized Ricci flow on the two-sphere*, Calc. Var. Partial Differential Equations **39** (2010), no. 3-4, 419–428, DOI 10.1007/s00526-010-0315-5. MR2729306

[AB11] Ben Andrews and Paul Bryan, *Curvature bound for curve shortening flow via distance comparison and a direct proof of Grayson's theorem*, J. Reine Angew. Math. **653** (2011), 179–187, DOI 10.1515/CRELLE.2011.026. MR2794630

[Abi81] William Abikoff, *The uniformization theorem*, Amer. Math. Monthly **88** (1981), no. 8, 574–592, DOI 10.2307/2320507. MR628026

[ACGL20] Ben Andrews, Bennett Chow, Christine Guenther, and Mat Langford, *Extrinsic geometric flows*, Graduate Studies in Mathematics, vol. 206, American Mathematical Society, Providence, RI, [2020] ©2020, DOI 10.1090/gsm/206. MR4249616

[ACL] Ben Andrews, Bennett Chow, and Mat Langford, *Geometric analysis*, in preparation.

[AH74] P. Alexandroff and H. Hopf, *Topologie. I*, Berichtigter Reprint, Die Grundlehren der mathematischen Wissenschaften, Band 45, Springer-Verlag, Berlin-New York, 1974. MR345087

[AH11] Ben Andrews and Christopher Hopper, *The Ricci flow in Riemannian geometry: A complete proof of the differentiable 1/4-pinching sphere theorem*, Lecture Notes in Mathematics, vol. 2011, Springer, Heidelberg, 2011, DOI 10.1007/978-3-642-16286-2. MR2760593

[Ahl10] Lars V. Ahlfors, *Conformal invariants: Topics in geometric function theory*, reprint of the 1973 original, with a foreword by Peter Duren, F. W. Gehring, and Brad Osgood, AMS Chelsea Publishing, Providence, RI, 2010, DOI 10.1090/chel/371. MR2730573

[AL86] U. Abresch and J. Langer, *The normalized curve shortening flow and homothetic solutions*, J. Differential Geom. **23** (1986), no. 2, 175–196. MR845704

[Ale24] J. W. Alexander, *An example of a simply connected surface bounding a region which is not simply connected*, Proc Natl Acad Sci U S A. **10** (1924), no. 1, 8–10.

[Ale48] A. D. Aleksandrov, *Vnutrennyaya Geometriya Vypuklyh Poverhnosteĭ* (Russian), Intrinsic Geometry of Convex Surfaces], OGIZ, Moscow-Leningrad, 1948. MR29518

[Ale68] D. V. Alekseevskiĭ, *Riemannian spaces with unusual holonomy groups* (Russian), Funkcional. Anal. i Priložen. **2** (1968), no. 2, 1–10. MR231313

[All40] Carl B. Allendoerfer, *The Euler number of a Riemann manifold*, Amer. J. Math. **62** (1940), 243–248, DOI 10.2307/2371450. MR2251

[Alm66] F. J. Almgren Jr., *Some interior regularity theorems for minimal surfaces and an extension of Bernstein's theorem*, Ann. of Math. (2) **84** (1966), 277–292, DOI 10.2307/1970520. MR200816

[Alm00] Frederick J. Almgren Jr., *Almgren's big regularity paper: Q-valued functions minimizing Dirichlet's integral and the regularity of area-minimizing rectifiable currents up to codimension 2*, with a preface by Jean E. Taylor and Vladimir Scheffer, World Scientific Monograph Series in Mathematics, vol. 1, World Scientific Publishing Co., Inc., River Edge, NJ, 2000. MR1777737

[And12] Ben Andrews, *Noncollapsing in mean-convex mean curvature flow*, Geom. Topol. **16** (2012), no. 3, 1413–1418, DOI 10.2140/gt.2012.16.1413. MR2967056

[Arn06] Vladimir I. Arnold, *Ordinary differential equations*, translated from the Russian by Roger Cooke, second printing of the 1992 edition, Universitext, Springer-Verlag, Berlin, 2006. MR2242407

[AS53] W. Ambrose and I. M. Singer, *A theorem on holonomy*, Trans. Amer. Math. Soc. **75** (1953), 428–443, DOI 10.2307/1990721. MR63739

[AS85] Michael T. Anderson and Richard Schoen, *Positive harmonic functions on complete manifolds of negative curvature*, Ann. of Math. (2) **121** (1985), no. 3, 429–461, DOI 10.2307/1971181. MR794369

[Aub82] Thierry Aubin, *Nonlinear analysis on manifolds. Monge-Ampère equations*, Grundlehren der mathematischen Wissenschaften [Fundamental Principles of Mathematical Sciences], vol. 252, Springer-Verlag, New York, 1982, DOI 10.1007/978-1-4612-5734-9. MR681859

[AW43] Carl B. Allendoerfer and André Weil, *The Gauss-Bonnet theorem for Riemannian polyhedra*, Trans. Amer. Math. Soc. **53** (1943), 101–129, DOI 10.2307/1990134. MR7627

[Bac21] Rudolf Bach, *Zur Weylschen Relativitätstheorie und der Weylschen Erweiterung des Krümmungstensorbegriffs* (German), Math. Z. **9** (1921), no. 1-2, 110–135, DOI 10.1007/BF01378338. MR1544454

[Bai56] Walter L. Baily Jr., *The decomposition theorem for V-manifolds*, Amer. J. Math. **78** (1956), 862–888, DOI 10.2307/2372472. MR100103

[Bama] Richard H. Bamler, *Entropy and heat kernel bounds on a Ricci flow background*, arXiv:2008.07093v3.

[Bamb] Richard H. Bamler, *Structure theory of non-collapsed limits of Ricci flows*, arXiv:2009.03243v2.

[Bam17] Richard H. Bamler, *Structure theory of singular spaces*, J. Funct. Anal. **272** (2017), no. 6, 2504–2627, DOI 10.1016/j.jfa.2016.10.020. MR3603307

[Bam21] Richard H. Bamler, *Recent developments in Ricci flows*, Notices Amer. Math. Soc. **68** (2021), no. 9, 1486–1498, DOI 10.1090/noti2343. MR4323820

[Bam23] Richard H. Bamler, *Compactness theory of the space of super Ricci flows*, Invent. Math. **233** (2023), no. 3, 1121–1277, DOI 10.1007/s00222-023-01196-3. MR4623543

[Ban80] Catherine Bandle, *Isoperimetric inequalities and applications*, Monographs and Studies in Mathematics, vol. 7, Pitman (Advanced Publishing Program), Boston, Mass.-London, 1980. MR572958

[Bar87] Robert Bartnik, *Maximal surfaces and general relativity*, Miniconference on geometry and partial differential equations, 2 (Canberra, 1986), Proc. Centre Math. Anal. Austral. Nat. Univ., vol. 12, Austral. Nat. Univ., Canberra, 1987, pp. 24–49. MR924425

[BBI01] Dmitri Burago, Yuri Burago, and Sergei Ivanov, *A course in metric geometry*, Graduate Studies in Mathematics, vol. 33, American Mathematical Society, Providence, RI, 2001, DOI 10.1090/gsm/033. MR1835418

[BC01] Richard L. Bishop and Richard J. Crittenden, *Geometry of manifolds*, reprint of the 1964 original, AMS Chelsea Publishing, Providence, RI, 2001, DOI 10.1090/chel/344. MR1852066

[BC21] Simon Brendle and Kyeongsu Choi, *Uniqueness of convex ancient solutions to mean curvature flow in higher dimensions*, Geom. Topol. **25** (2021), no. 5, 2195–2234, DOI 10.2140/gt.2021.25.2195. MR4310889

[BCG96] Gérard Besson, Gilles Courtois, and Sylvestre Gallot, *Minimal entropy and Mostow's rigidity theorems*, Ergodic Theory Dynam. Systems **16** (1996), no. 4, 623–649, DOI 10.1017/S0143385700009019. MR1406425

[BCJN] Richard Bamler, Bennett Chow, Wenshuai Jiang, and Aaron Naber, *Lectures on Ricci curvature*, in preparation.

[BDS21] Simon Brendle, Panagiota Daskalopoulos, and Natasa Sesum, *Uniqueness of compact ancient solutions to three-dimensional Ricci flow*, Invent. Math. **226** (2021), no. 2, 579–651, DOI 10.1007/s00222-021-01054-0. MR4323639

[BE87] Jean-Pierre Bourguignon and Jean-Pierre Ezin, *Scalar curvature functions in a conformal class of metrics and conformal transformations*, Trans. Amer. Math. Soc. **301** (1987), no. 2, 723–736, DOI 10.2307/2000667. MR882712

[Ber27] Serge Bernstein, *Über ein geometrisches Theorem und seine Anwendung auf die partiellen Differentialgleichungen vom elliptischen Typus* (German), Math. Z. **26** (1927), no. 1, 551–558, DOI 10.1007/BF01475472. MR1544873

[Ber55] Marcel Berger, *Sur les groupes d'holonomie homogène des variétés à connexion affine et des variétés riemanniennes* (French), Bull. Soc. Math. France **83** (1955), 279–330. MR79806

[Bes78] Arthur L. Besse, *Manifolds all of whose geodesics are closed*, with appendices by D. B. A. Epstein, J.-P. Bourguignon, L. Bérard-Bergery, M. Berger and J. L. Kazdan, Ergebnisse der Mathematik und ihrer Grenzgebiete [Results in Mathematics and Related Areas], vol. 93, Springer-Verlag, Berlin-New York, 1978. MR496885

[Bes08] Arthur L. Besse, *Einstein manifolds*, reprint of the 1987 edition, Classics in Mathematics, Springer-Verlag, Berlin, 2008. MR2371700

[BG72] Robert B. Brown and Alfred Gray, *Riemannian manifolds with holonomy group* Spin *(9)*, Differential geometry (in honor of Kentaro Yano), Kinokuniya Book Store, Tokyo, 1972, pp. 41–59. MR328817

[BGL14] Dominique Bakry, Ivan Gentil, and Michel Ledoux, *Analysis and geometry of Markov diffusion operators*, Grundlehren der mathematischen Wissenschaften [Fundamental Principles of Mathematical Sciences], vol. 348, Springer, Cham, 2014, DOI 10.1007/978-3-319-00227-9. MR3155209

[BGM71] Marcel Berger, Paul Gauduchon, and Edmond Mazet, *Le spectre d'une variété riemannienne* (French), Lecture Notes in Mathematics, Vol. 194, Springer-Verlag, Berlin-New York, 1971. MR282313

[BGV04] Nicole Berline, Ezra Getzler, and Michèle Vergne, *Heat kernels and Dirac operators*, corrected reprint of the 1992 original, Grundlehren Text Editions, Springer-Verlag, Berlin, 2004. MR2273508

[BH99] Martin R. Bridson and André Haefliger, *Metric spaces of non-positive curvature*, Grundlehren der mathematischen Wissenschaften [Fundamental Principles of Mathematical Sciences], vol. 319, Springer-Verlag, Berlin, 1999, DOI 10.1007/978-3-662-12494-9. MR1744486

[Bie11] Ludwig Bieberbach, *Über die Bewegungsgruppen der Euklidischen Räume* (German), Math. Ann. **70** (1911), no. 3, 297–336, DOI 10.1007/BF01564500. MR1511623

[Bie12] Ludwig Bieberbach, *Über die Bewegungsgruppen der Euklidischen Räume (Zweite Abhandlung.) Die Gruppen mit einem endlichen Fundamentalbereich* (German), Math. Ann. **72** (1912), no. 3, 400–412, DOI 10.1007/BF01456724. MR1511704

[BK] Richard H. Bamler and Bruce Kleiner, *On the multiplicity one conjecture for mean curvature flows of surfaces*, arXiv:2312.02106.

[BK23] Richard H. Bamler and Bruce Kleiner, *Ricci flow and diffeomorphism groups of 3-manifolds*, J. Amer. Math. Soc. **36** (2023), no. 2, 563–589, DOI 10.1090/jams/1003. MR4536904

[BKK22] Christine Breiner, Nikolaos Kapouleas, and Stephen Kleene, *Conservation laws and gluing constructions for constant mean curvature (hyper)surfaces*, Notices Amer. Math. Soc. **69** (2022), no. 5, 762–773, DOI 10.1090/noti2473. MR4415896

[BLL23] Theodora Bourni, Mat Langford, and Stephen Lynch, *Collapsing and noncollapsing in convex ancient mean curvature flow*, J. Reine Angew. Math. **801** (2023), 273–305, DOI 10.1515/crelle-2023-0045. MR4621884

[BM58] R. Bott and J. Milnor, *On the parallelizability of the spheres*, Bull. Amer. Math. Soc. **64** (1958), 87–89, DOI 10.1090/S0002-9904-1958-10166-4. MR102804

[BNSa] Elia Brué, Aaron Naber, and Daniele Semola, *Fundamental groups and the Milnor conjecture*, arXiv:2303.15347.

[BNSb] Elia Brué, Aaron Naber, and Daniele Semola, *Six-dimensional counterexample to the Milnor conjecture*, arXiv:2311.12155.

[Boc46] S. Bochner, *Vector fields and Ricci curvature*, Bull. Amer. Math. Soc. **52** (1946), 776–797, DOI 10.1090/S0002-9904-1946-08647-4. MR18022

[Bon48] Pierre Ossian Bonnet, *Mémoire sur la théorie générale des surfaces*, J. de l'Ecole Polytechnique **19** (1848), 1–146.

[Bot59] Raoul Bott, *The stable homotopy of the classical groups*, Ann. of Math. (2) **70** (1959), 313–337, DOI 10.2307/1970106. MR110104

[Bra78] Kenneth A. Brakke, *The motion of a surface by its mean curvature*, Mathematical Notes, vol. 20, Princeton University Press, Princeton, NJ, 1978. MR485012

[Bra01] Hubert L. Bray, *Proof of the Riemannian Penrose inequality using the positive mass theorem*, J. Differential Geom. **59** (2001), no. 2, 177–267. MR1908823

[Bre10] Simon Brendle, *Ricci flow and the sphere theorem*, Graduate Studies in Mathematics, vol. 111, American Mathematical Society, Providence, RI, 2010, DOI 10.1090/gsm/111. MR2583938

[Bre13a] Simon Brendle, *Embedded minimal tori in S^3 and the Lawson conjecture*, Acta Math. **211** (2013), no. 2, 177–190, DOI 10.1007/s11511-013-0101-2. MR3143888

[Bre13b] Simon Brendle, *Rotational symmetry of self-similar solutions to the Ricci flow*, Invent. Math. **194** (2013), no. 3, 731–764, DOI 10.1007/s00222-013-0457-0. MR3127066

[Bre14] Simon Brendle, *Two-point functions and their applications in geometry*, Bull. Amer. Math. Soc. (N.S.) **51** (2014), no. 4, 581–596, DOI 10.1090/S0273-0979-2014-01461-2. MR3237760

[Bre20] Simon Brendle, *Ancient solutions to the Ricci flow in dimension 3*, Acta Math. **225** (2020), no. 1, 1–102, DOI 10.4310/acta.2020.v225.n1.a1. MR4176064

[Bre23] Simon Brendle, *Minimal hypersurfaces and geometric inequalities*, Ann. Fac. Sci. Toulouse Math. (6) **32** (2023), no. 1, 179–201. MR4574744

[Bro11] L. E. J. Brouwer, *Über Abbildung von Mannigfaltigkeiten* (German), Math. Ann. **71** (1911), no. 1, 97–115, DOI 10.1007/BF01456931. MR1511644

[Bro69] William Browder, *The Kervaire invariant of framed manifolds and its generalization*, Ann. of Math. (2) **90** (1969), 157–186, DOI 10.2307/1970686. MR251736

[Bro72] William Browder, *Surgery on simply-connected manifolds*, Ergebnisse der Mathematik und ihrer Grenzgebiete [Results in Mathematics and Related Areas], Band 65, Springer-Verlag, New York-Heidelberg, 1972. MR358813

[BS08] Simon Brendle and Richard M. Schoen, *Classification of manifolds with weakly* 1/4-*pinched curvatures*, Acta Math. **200** (2008), no. 1, 1–13, DOI 10.1007/s11511-008-0022-7. MR2386107

[BS09] Simon Brendle and Richard Schoen, *Manifolds with* 1/4-*pinched curvature are space forms*, J. Amer. Math. Soc. **22** (2009), no. 1, 287–307, DOI 10.1090/S0894-0347-08-00613-9. MR2449060

[BSY94] J. Bartz, M. Struwe, and R. Ye, *A new approach to the Ricci flow on* S^2, Ann. Scuola Norm. Sup. Pisa Cl. Sci. (4) **21** (1994), no. 3, 475–482. MR1310637

[BT82] Raoul Bott and Loring W. Tu, *Differential forms in algebraic topology*, Graduate Texts in Mathematics, vol. 82, Springer-Verlag, New York-Berlin, 1982. MR658304

[BW08] Christoph Böhm and Burkhard Wilking, *Manifolds with positive curvature operators are space forms*, Ann. of Math. (2) **167** (2008), no. 3, 1079–1097, DOI 10.4007/annals.2008.167.1079. MR2415394

[Cal67] Eugenio Calabi, *On Ricci curvature and geodesics*, Duke Math. J. **34** (1967), 667–676. MR216429

[Cal70] Eugenio Calabi, *Examples of Bernstein problems for some nonlinear equations*, Global Analysis (Proc. Sympos. Pure Math., Vols. XIV, XV, XVI, Berkeley, Calif., 1968), Proc. Sympos. Pure Math., XIV-XVI, Amer. Math. Soc., Providence, RI, 1970, pp. 223–230. MR264210

[Car51] Élie Cartan, *Leçons sur la géométrie des espaces de Riemann* (French), 2nd ed., Gauthier-Villars, Paris, 1951. MR44878

[CC97] Jeff Cheeger and Tobias H. Colding, *On the structure of spaces with Ricci curvature bounded below. I*, J. Differential Geom. **46** (1997), no. 3, 406–480. MR1484888

[CC00a] Jeff Cheeger and Tobias H. Colding, *On the structure of spaces with Ricci curvature bounded below. II*, J. Differential Geom. **54** (2000), no. 1, 13–35. MR1815410

[CC00b] Jeff Cheeger and Tobias H. Colding, *On the structure of spaces with Ricci curvature bounded below. III*, J. Differential Geom. **54** (2000), no. 1, 37–74. MR1815411

[CCG+07] Bennett Chow, Sun-Chin Chu, David Glickenstein, Christine Guenther, James Isenberg, Tom Ivey, Dan Knopf, Peng Lu, Feng Luo, and Lei Ni, *The Ricci flow: techniques and applications. Part I. Geometric aspects*, Mathematical Surveys and Monographs, vol. 135, American Mathematical Society, Providence, RI, 2007, DOI 10.1090/surv/135. MR2302600

[CCG+08] Bennett Chow, Sun-Chin Chu, David Glickenstein, Christine Guenther, James Isenberg, Tom Ivey, Dan Knopf, Peng Lu, Feng Luo, and Lei Ni, *The Ricci flow: techniques and applications. Part II. Analytic aspects*, Mathematical Surveys and Monographs, vol. 144, American Mathematical Society, Providence, RI, 2008, DOI 10.1090/surv/144. MR2365237

[CCG+10] Bennett Chow, Sun-Chin Chu, David Glickenstein, Christine Guenther, James Isenberg, Tom Ivey, Dan Knopf, Peng Lu, Feng Luo, and Lei Ni, *The Ricci flow: techniques and applications. Part III. Geometric-analytic aspects*, Mathematical Surveys and Monographs, vol. 163, American Mathematical Society, Providence, RI, 2010, DOI 10.1090/surv/163. MR2604955

[CCG+15] Bennett Chow, Sun-Chin Chu, David Glickenstein, Christine Guenther, James Isenberg, Tom Ivey, Dan Knopf, Peng Lu, Feng Luo, and Lei Ni, *The Ricci flow: techniques and applications. Part IV. Long-time solutions and related topics*, Mathematical Surveys and Monographs, vol. 206, American Mathematical Society, Providence, RI, 2015, DOI 10.1090/surv/206. MR3409114

[CCT02] J. Cheeger, T. H. Colding, and G. Tian, *On the singularities of spaces with bounded Ricci curvature*, Geom. Funct. Anal. **12** (2002), no. 5, 873–914, DOI 10.1007/PL00012649. MR1937830

[CE08] Jeff Cheeger and David G. Ebin, *Comparison theorems in Riemannian geometry*, revised reprint of the 1975 original, AMS Chelsea Publishing, Providence, RI, 2008, DOI 10.1090/chel/365. MR2394158

[CG72] Jeff Cheeger and Detlef Gromoll, *The splitting theorem for manifolds of nonnegative Ricci curvature*, J. Differential Geometry **6** (1971/72), 119–128. MR303460

[CGL] Bennett Chow, David Glickenstein, and Feng Luo, *Discrete conformal geometry*, in preparation.

[CGT82] Jeff Cheeger, Mikhail Gromov, and Michael Taylor, *Finite propagation speed, kernel estimates for functions of the Laplace operator, and the geometry of complete Riemannian manifolds*, J. Differential Geometry **17** (1982), no. 1, 15–53. MR658471

[Cha84] Isaac Chavel, *Eigenvalues in Riemannian geometry*, including a chapter by Burton Randol, with an appendix by Jozef Dodziuk, Pure and Applied Mathematics, vol. 115, Academic Press, Inc., Orlando, FL, 1984. MR768584

[Cha88] Sheldon Xu-Dong Chang, *Two-dimensional area minimizing integral currents are classical minimal surfaces*, J. Amer. Math. Soc. **1** (1988), no. 4, 699–778, DOI 10.2307/1990991. MR946554

[Cha01] Isaac Chavel, *Isoperimetric inequalities: Differential geometric and analytic perspectives*, Cambridge Tracts in Mathematics, vol. 145, Cambridge University Press, Cambridge, 2001. MR1849187

[Cha06] Isaac Chavel, *Riemannian geometry: A modern introduction*, 2nd ed., Cambridge Studies in Advanced Mathematics, vol. 98, Cambridge University Press, Cambridge, 2006, DOI 10.1017/CBO9780511616822. MR2229062

[Che44] Shiing-shen Chern, *A simple intrinsic proof of the Gauss-Bonnet formula for closed Riemannian manifolds*, Ann. of Math. (2) **45** (1944), 747–752, DOI 10.2307/1969302. MR11027

[Che51] Shiing-shen Chern, *Topics in differential geometry*, Institute for Advanced Study (IAS), Princeton, NJ, 1951. MR90080

[Che55] Shiing-shen Chern, *An elementary proof of the existence of isothermal parameters on a surface*, Proc. Amer. Math. Soc. **6** (1955), 771–782, DOI 10.2307/2032933. MR74856

[Che70] Jeff Cheeger, *Finiteness theorems for Riemannian manifolds*, Amer. J. Math. **92** (1970), 61–74, DOI 10.2307/2373498. MR263092

[Che75] Shiu Yuen Cheng, *Eigenvalue comparison theorems and its geometric applications*, Math. Z. **143** (1975), no. 3, 289–297, DOI 10.1007/BF01214381. MR378001

[Che95] Shiing-shen Chern, *Complex manifolds without potential theory (with an appendix on the geometry of characteristic classes)*, 2nd ed., Universitext, Springer-Verlag, New York, 1995. MR2244174

[Che01] Jeff Cheeger, *Degeneration of Riemannian metrics under Ricci curvature bounds* Lezioni Fermiane. [Fermi Lectures], Scuola Normale Superiore, Pisa, 2001. MR2006642

[CHK00] Daryl Cooper, Craig D. Hodgson, and Steven P. Kerckhoff, *Three-dimensional orbifolds and cone-manifolds*, with a postface by Sadayoshi Kojima, MSJ Memoirs, vol. 5, Mathematical Society of Japan, Tokyo, 2000. MR1778789

[Cho23] Bennett Chow, *Ricci solitons in low dimensions*, Graduate Studies in Mathematics, vol. 235, American Mathematical Society, Providence, RI, [2023] ©2023, DOI 10.1090/gsm/235. MR4624811

[CJN21] Jeff Cheeger, Wenshuai Jiang, and Aaron Naber, *Rectifiability of singular sets of noncollapsed limit spaces with Ricci curvature bounded below*, Ann. of Math. (2) **193** (2021), no. 2, 407–538, DOI 10.4007/annals.2021.193.2.2. MR4226910

[CK93] Demetrios Christodoulou and Sergiu Klainerman, *The global nonlinear stability of the Minkowski space*, Princeton Mathematical Series, vol. 41, Princeton University Press, Princeton, NJ, 1993. MR1316662

[CK04] Bennett Chow and Dan Knopf, *The Ricci flow: an introduction*, Mathematical Surveys and Monographs, vol. 110, American Mathematical Society, Providence, RI, 2004, DOI 10.1090/surv/110. MR2061425

[CKM] Bennett Chow, Brett Kotschwar, and Ovidiu Munteanu, *Ricci solitons in dimension four*, submitted.

[CL] Otis Chodosh and Chao Li, *Stable minimal hypersurfaces in* $\mathbb{R}^4$, Acto Math., to appear.

[CLN06] Bennett Chow, Peng Lu, and Lei Ni, *Hamilton's Ricci flow*, Graduate Studies in Mathematics, vol. 77, American Mathematical Society, Providence, RI; Science Press Beijing, New York, 2006, DOI 10.1090/gsm/077. MR2274812

[CLT06] Xiuxiong Chen, Peng Lu, and Gang Tian, *A note on uniformization of Riemann surfaces by Ricci flow*, Proc. Amer. Math. Soc. **134** (2006), no. 11, 3391–3393, DOI 10.1090/S0002-9939-06-08360-2. MR2231924

[CM11] Tobias Holck Colding and William P. Minicozzi II, *A course in minimal surfaces*, Graduate Studies in Mathematics, vol. 121, American Mathematical Society, Providence, RI, 2011, DOI 10.1090/gsm/121. MR2780140

[CM12] Tobias H. Colding and William P. Minicozzi II, *Generic mean curvature flow I: generic singularities*, Ann. of Math. (2) **175** (2012), no. 2, 755–833, DOI 10.4007/annals.2012.175.2.7. MR2993752

[CM14] Fernando Codá Marques, *Minimal surfaces: variational theory and applications*, Proceedings of the International Congress of Mathematicians—Seoul 2014. Vol. 1, Kyung Moon Sa, Seoul, 2014, pp. 283–310. MR3728473

[CN15] Jeff Cheeger and Aaron Naber, *Regularity of Einstein manifolds and the codimension 4 conjecture*, Ann. of Math. (2) **182** (2015), no. 3, 1093–1165, DOI 10.4007/annals.2015.182.3.5. MR3418535

[Cod69] Delfino Codazzi, *Sulle coordinate curvilinee d'una superficie dello spazio*, Ann. Mat. Pura Appl. **2** (1868–1869), 101–119.

[Cor92] Kevin Corlette, *Archimedean superrigidity and hyperbolic geometry*, Ann. of Math. (2) **135** (1992), no. 1, 165–182, DOI 10.2307/2946567. MR1147961

[Cro84] Christopher B. Croke, *A sharp four-dimensional isoperimetric inequality*, Comment. Math. Helv. **59** (1984), no. 2, 187–192, DOI 10.1007/BF02566344. MR749103

[CV27] Stefan E. Cohn-Vossen, *Zwei sätze über die starrheit der eiflächen*, Nach. Gesellschaft Wiss. Götinger, Math Phys. Kl. (1927), 125–134.

[CV36] Stefan E. Cohn-Vossen, *Bending of surfaces in the large*, Uspekhi Mat. Nauk **1** (1936), 33–76.

[CW91] Bennett Chow and Lang-Fang Wu, *The Ricci flow on compact 2-orbifolds with curvature negative somewhere*, Comm. Pure Appl. Math. **44** (1991), no. 3, 275–286, DOI 10.1002/cpa.3160440302. MR1090433

[CY75] S. Y. Cheng and S. T. Yau, *Differential equations on Riemannian manifolds and their geometric applications*, Comm. Pure Appl. Math. **28** (1975), no. 3, 333–354, DOI 10.1002/cpa.3160280303. MR385749

[CY76] Shiu Yuen Cheng and Shing Tung Yau, *Maximal space-like hypersurfaces in the Lorentz-Minkowski spaces*, Ann. of Math. (2) **104** (1976), no. 3, 407–419, DOI 10.2307/1970963. MR431061

[CY81] Jeff Cheeger and Shing Tung Yau, *A lower bound for the heat kernel*, Comm. Pure Appl. Math. **34** (1981), no. 4, 465–480, DOI 10.1002/cpa.3160340404. MR615626

[dC92] Manfredo Perdigão do Carmo, *Riemannian geometry*, Mathematics: Theory & Applications, translated from the second Portuguese edition by Francis Flaherty, Birkhäuser Boston, Inc., Boston, MA, 1992, DOI 10.1007/978-1-4757-2201-7. MR1138207

[dC16] Manfredo P. do Carmo, *Differential geometry of curves & surfaces*, revised & updated second edition of [MR0394451], Dover Publications, Inc., Mineola, NY, 2016. MR3837152

[DG65] Ennio De Giorgi, *Una estensione del teorema di Bernstein* (Italian), Ann. Scuola Norm. Sup. Pisa Cl. Sci. (3) **19** (1965), 79–85. MR178385

[DHS10a] Panagiota Daskalopoulos, Richard Hamilton, and Natasa Sesum, *Classification of compact ancient solutions to the curve shortening flow*, J. Differential Geom. **84** (2010), no. 3, 455–464. MR2669361

[DHS10b] Ulrich Dierkes, Stefan Hildebrandt, and Friedrich Sauvigny, *Minimal surfaces*, 2nd ed., with assistance and contributions by A. Küster and R. Jakob, Grundlehren der mathematischen Wissenschaften [Fundamental Principles of Mathematical Sciences], vol. 339, Springer, Heidelberg, 2010, DOI 10.1007/978-3-642-11698-8. MR2566897

[DK90] S. K. Donaldson and P. B. Kronheimer, *The geometry of four-manifolds*, Oxford Science Publications, Oxford Mathematical Monographs, The Clarendon Press, Oxford University Press, New York, 1990. MR1079726

[DLSS20] Camillo De Lellis, Emanuele Spadaro, and Luca Spolaor, *Regularity theory for 2-dimensional almost minimal currents III: Blowup*, J. Differential Geom. **116** (2020), no. 1, 125–185, DOI 10.4310/jdg/1599271254. MR4146358

[Don83] S. K. Donaldson, *An application of gauge theory to four-dimensional topology*, J. Differential Geom. **18** (1983), no. 2, 279–315. MR710056

[Don87] S. K. Donaldson, *The orientation of Yang-Mills moduli spaces and 4-manifold topology*, J. Differential Geom. **26** (1987), no. 3, 397–428. MR910015

[Dou31] Jesse Douglas, *Solution of the problem of Plateau*, Trans. Amer. Math. Soc. **33** (1931), no. 1, 263–321, DOI 10.2307/1989472. MR1501590

[dSG16] Henri Paul de Saint-Gervais, *Uniformization of Riemann surfaces: Revisiting a hundred-year-old theorem*, translated from the 2010 French original [MR2768303] by Robert G. Burns; The name of Henri Paul de Saint-Gervais covers a group composed of fifteen mathematicians: Aurélien Alvarez, Christophe Bavard, François Béguin, Nicolas Bergeron, Maxime Bourrigan, Bertrand Deroin, Sorin Dumitrescu, Charles Frances, Étienne Ghys, Antonin Guilloux, Frank Loray, Patrick Popescu-Pampu, Pierre Py, Bruno Sévennec and Jean-Claude Sikorav, Heritage of European Mathematics, European Mathematical Society (EMS), Zürich, 2016, DOI 10.4171/145. MR3494804

[Eck04] Klaus Ecker, *Regularity theory for mean curvature flow*, Progress in Nonlinear Differential Equations and their Applications, vol. 57, Birkhäuser Boston, Inc., Boston, MA, 2004, DOI 10.1007/978-0-8176-8210-1. MR2024995

[EG15] Lawrence C. Evans and Ronald F. Gariepy, *Measure theory and fine properties of functions*, revised edition, Textbooks in Mathematics, CRC Press, Boca Raton, FL, 2015. MR3409135

[Eng89] Ryszard Engelking, *General topology*, 2nd ed., translated from the Polish by the author, Sigma Series in Pure Mathematics, vol. 6, Heldermann Verlag, Berlin, 1989. MR1039321

[ES64] James Eells Jr. and J. H. Sampson, *Harmonic mappings of Riemannian manifolds*, Amer. J. Math. **86** (1964), 109–160, DOI 10.2307/2373037. MR164306

[Esc87] J.-H. Eschenburg, *Comparison theorems and hypersurfaces*, Manuscripta Math. **59** (1987), no. 3, 295–323, DOI 10.1007/BF01174796. MR909847

[Eva10] Lawrence C. Evans, *Partial differential equations*, 2nd ed., Graduate Studies in Mathematics, vol. 19, American Mathematical Society, Providence, RI, 2010, DOI 10.1090/gsm/019. MR2597943

[Fed69] Herbert Federer, *Geometric measure theory*, Die Grundlehren der mathematischen Wissenschaften, Band 153, Springer-Verlag New York, Inc., New York, 1969. MR257325

[FF60] Herbert Federer and Wendell H. Fleming, *Normal and integral currents*, Ann. of Math. (2) **72** (1960), 458–520, DOI 10.2307/1970227. MR123260

[FK92] H. M. Farkas and I. Kra, *Riemann surfaces*, 2nd ed., Graduate Texts in Mathematics, vol. 71, Springer-Verlag, New York, 1992, DOI 10.1007/978-1-4612-2034-3. MR1139765

[Fle62] Wendell H. Fleming, *On the oriented Plateau problem*, Rend. Circ. Mat. Palermo (2) **11** (1962), 69–90, DOI 10.1007/BF02849427. MR157263

[Fra61] Theodore Frankel, *Manifolds with positive curvature*, Pacific J. Math. **11** (1961), 165–174. MR123272

[Fre82] Michael Hartley Freedman, *The topology of four-dimensional manifolds*, J. Differential Geometry **17** (1982), no. 3, 357–453. MR679066

[Fri64] Avner Friedman, *Partial differential equations of parabolic type*, Prentice-Hall, Inc., Englewood Cliffs, NJ, 1964. MR181836

[FU91] Daniel S. Freed and Karen K. Uhlenbeck, *Instantons and four-manifolds*, 2nd ed., Mathematical Sciences Research Institute Publications, vol. 1, Springer-Verlag, New York, 1991, DOI 10.1007/978-1-4613-9703-8. MR1081321

[G89a] Matthias Günther, *On the perturbation problem associated to isometric embeddings of Riemannian manifolds*, Ann. Global Anal. Geom. **7** (1989), no. 1, 69–77, DOI 10.1007/BF00137403. MR1029846

[G89b] Matthias Günther, *Zum Einbettungssatz von J. Nash* (German), Math. Nachr. **144** (1989), 165–187, DOI 10.1002/mana.19891440113. MR1037168

[G91] Matthias Günther, *Isometric embeddings of Riemannian manifolds*, Proceedings of the International Congress of Mathematicians, Vol. I, II (Kyoto, 1990), Math. Soc. Japan, Tokyo, 1991, pp. 1137–1143. MR1159298

[Gag83] Michael E. Gage, *An isoperimetric inequality with applications to curve shortening*, Duke Math. J. **50** (1983), no. 4, 1225–1229, DOI 10.1215/S0012-7094-83-05052-4. MR726325

[Gao85] L. Zhiyong Gao, *The construction of negatively Ricci curved manifolds*, Math. Ann. **271** (1985), no. 2, 185–208, DOI 10.1007/BF01455986. MR783551

[Gau65] Carl Friedrich Gauss, *General investigations of curved surfaces*, translated from the Latin and German by Adam Hiltebeitel and James Morehead, Raven Press, Hewlett, N.Y., 1965. MR182537

[GH86] M. Gage and R. S. Hamilton, *The heat equation shrinking convex plane curves*, J. Differential Geom. **23** (1986), no. 1, 69–96. MR840401

[GH94] Phillip Griffiths and Joseph Harris, *Principles of algebraic geometry*, reprint of the 1978 original, Wiley Classics Library, John Wiley & Sons, Inc., New York, 1994, DOI 10.1002/9781118032527. MR1288523

[GH19] Victor Guillemin and Peter Haine, *Differential forms*, World Scientific Publishing Co. Pte. Ltd., Hackensack, NJ, 2019, DOI 10.1142/11058. MR3931306

[GHL04] Sylvestre Gallot, Dominique Hulin, and Jacques Lafontaine, *Riemannian geometry*, 3rd ed., Universitext, Springer-Verlag, Berlin, 2004, DOI 10.1007/978-3-642-18855-8. MR2088027

[Gil95] Peter B. Gilkey, *Invariance theory, the heat equation, and the Atiyah-Singer index theorem*, 2nd ed., Studies in Advanced Mathematics, CRC Press, Boca Raton, FL, 1995. MR1396308

[Giu84] Enrico Giusti, *Minimal surfaces and functions of bounded variation*, Monographs in Mathematics, vol. 80, Birkhäuser Verlag, Basel, 1984, DOI 10.1007/978-1-4684-9486-0. MR775682

[GL80a] Mikhael Gromov and H. Blaine Lawson Jr., *The classification of simply connected manifolds of positive scalar curvature*, Ann. of Math. (2) **111** (1980), no. 3, 423–434, DOI 10.2307/1971103. MR577131

[GL80b] Mikhael Gromov and H. Blaine Lawson Jr., *Spin and scalar curvature in the presence of a fundamental group. I*, Ann. of Math. (2) **111** (1980), no. 2, 209–230, DOI 10.2307/1971198. MR569070

[GL94] Pengfei Guan and Yan Yan Li, *The Weyl problem with nonnegative Gauss curvature*, J. Differential Geom. **39** (1994), no. 2, 331–342. MR1267893

[GM75] S. Gallot and D. Meyer, *Opérateur de courbure et laplacien des formes différentielles d'une variété riemannienne* (French), J. Math. Pures Appl. (9) **54** (1975), no. 3, 259–284. MR454884

[Gol98] Samuel I. Goldberg, *Curvature and homology*, revised reprint of the 1970 edition, Dover Publications, Inc., Mineola, NY, 1998. MR1635338

[GP10] Victor Guillemin and Alan Pollack, *Differential topology*, reprint of the 1974 original, AMS Chelsea Publishing, Providence, RI, 2010, DOI 10.1090/chel/370. MR2680546

[Gra87] Matthew A. Grayson, *The heat equation shrinks embedded plane curves to round points*, J. Differential Geom. **26** (1987), no. 2, 285–314. MR906392

[Gra94] Jeremy Gray, *On the history of the Riemann mapping theorem*, Rend. Circ. Mat. Palermo (2) Suppl. **34** (1994), 47–94. MR1295591

[Gra04] Alfred Gray, *Tubes*, 2nd ed., with a preface by Vicente Miquel, Progress in Mathematics, vol. 221, Birkhäuser Verlag, Basel, 2004, DOI 10.1007/978-3-0348-7966-8. MR2024928

[Gri09] Alexander Grigor'yan, *Heat kernel and analysis on manifolds*, AMS/IP Studies in Advanced Mathematics, vol. 47, American Mathematical Society, Providence, RI; International Press, Boston, MA, 2009, DOI 10.1090/amsip/047. MR2569498

[Gro70] M. L. Gromov, *Isometric imbeddings and immersions* (Russian), Dokl. Akad. Nauk SSSR **192** (1970), 1206–1209. MR275456

[Gro07] Misha Gromov, *Metric structures for Riemannian and non-Riemannian spaces*, based on the 1981 French original, with appendices by M. Katz, P. Pansu, and S. Semmes, translated from the French by Sean Michael Bates, reprint of the 2001 English edition, Modern Birkhäuser Classics, Birkhäuser Boston, Inc., Boston, MA, 2007. MR2307192

[Gro23] Misha Gromov, *Four lectures on scalar curvature*, Perspectives in scalar curvature. Vol. 1, World Sci. Publ., Hackensack, NJ, [2023] ©2023, pp. 1–514. MR4577903

[GS92] Mikhail Gromov and Richard Schoen, *Harmonic maps into singular spaces and p-adic superrigidity for lattices in groups of rank one*, Inst. Hautes Études Sci. Publ. Math. **76** (1992), 165–246. MR1215595

[GS99] Robert E. Gompf and András I. Stipsicz, *4-manifolds and Kirby calculus*, Graduate Studies in Mathematics, vol. 20, American Mathematical Society, Providence, RI, 1999, DOI 10.1090/gsm/020. MR1707327

[GT01] David Gilbarg and Neil S. Trudinger, *Elliptic partial differential equations of second order*, reprint of the 1998 edition, Classics in Mathematics, Springer-Verlag, Berlin, 2001. MR1814364

[Gul73] Robert D. Gulliver II, *Regularity of minimizing surfaces of prescribed mean curvature*, Ann. of Math. (2) **97** (1973), 275–305, DOI 10.2307/1970848. MR317188

[GW88] R. E. Greene and H. Wu, *Lipschitz convergence of Riemannian manifolds*, Pacific J. Math. **131** (1988), no. 1, 119–141. MR917868

[GY86] L. Zhiyong Gao and S.-T. Yau, *The existence of negatively Ricci curved metrics on three-manifolds*, Invent. Math. **85** (1986), no. 3, 637–652, DOI 10.1007/BF01390331. MR848687

[Had97] Jacques Hadamard, *Sur certaines propriétés des trajectories en dynamique*, Journal de mathématiques pures et appliquées 5e série **3** (1897), 331–387.

[Had98] Jacques Hadamard, *Les surfaces à courbure opposée et leurs lignes géodésiques*, Journal de mathématiques pures et appliquées 5e série **4** (1898), 27–73.

[Ham82] Richard S. Hamilton, *Three-manifolds with positive Ricci curvature*, J. Differential Geometry **17** (1982), no. 2, 255–306. MR664497

[Ham86] Richard S. Hamilton, *Four-manifolds with positive curvature operator*, J. Differential Geom. **24** (1986), no. 2, 153–179. MR862046

[Ham88] Richard S. Hamilton, *The Ricci flow on surfaces*, Mathematics and general relativity (Santa Cruz, CA, 1986), Contemp. Math., vol. 71, Amer. Math. Soc., Providence, RI, 1988, pp. 237–262, DOI 10.1090/conm/071/954419. MR954419

[Ham93] Richard S. Hamilton, *The Harnack estimate for the Ricci flow*, J. Differential Geom. **37** (1993), no. 1, 225–243. MR1198607

[Ham95a] Richard S. Hamilton, *A compactness property for solutions of the Ricci flow*, Amer. J. Math. **117** (1995), no. 3, 545–572, DOI 10.2307/2375080. MR1333936

[Ham95b] Richard S. Hamilton, *The formation of singularities in the Ricci flow*, Surveys in differential geometry, Vol. II (Cambridge, MA, 1993), Int. Press, Cambridge, MA, 1995, pp. 7–136. MR1375255

[Ham95c] Richard S. Hamilton, *Harnack estimate for the mean curvature flow*, J. Differential Geom. **41** (1995), no. 1, 215–226. MR1316556

[Ham95d] Richard S. Hamilton, *An isoperimetric estimate for the Ricci flow on the two-sphere*, Modern methods in complex analysis (Princeton, NJ, 1992), Ann. of Math. Stud., vol. 137, Princeton Univ. Press, Princeton, NJ, 1995, pp. 191–200, DOI 10.1080/09502389500490321. MR1369139

[Ham95e] Richard S. Hamilton, *Isoperimetric estimates for the curve shrinking flow in the plane*, Modern methods in complex analysis (Princeton, NJ, 1992), Ann. of Math. Stud., vol. 137, Princeton Univ. Press, Princeton, NJ, 1995, pp. 201–222, DOI 10.1016/1053-8127(94)00130-3. MR1369140

[Ham97] Richard S. Hamilton, *Four-manifolds with positive isotropic curvature*, Comm. Anal. Geom. **5** (1997), no. 1, 1–92, DOI 10.4310/CAG.1997.v5.n1.a1. MR1456308

[Ham99] Richard S. Hamilton, *Non-singular solutions of the Ricci flow on three-manifolds*, Comm. Anal. Geom. **7** (1999), no. 4, 695–729, DOI 10.4310/CAG.1999.v7.n4.a2. MR1714939

[Has15] Robert Haslhofer, *Uniqueness of the bowl soliton*, Geom. Topol. **19** (2015), no. 4, 2393–2406, DOI 10.2140/gt.2015.19.2393. MR3375531

[Hat] Allen Hatcher, *Vector bundles and K-theory*, Version 2.2, November 2017. Available at `https://pi.math.cornell.edu/ hatcher/VBKT/VB.pdf`.

[Hat02] Allen Hatcher, *Algebraic topology*, Cambridge University Press, Cambridge, 2002. MR1867354

[Heb99] Emmanuel Hebey, *Nonlinear analysis on manifolds: Sobolev spaces and inequalities*, Courant Lecture Notes in Mathematics, vol. 5, New York University, Courant Institute of Mathematical Sciences, New York; American Mathematical Society, Providence, RI, 1999. MR1688256

[Hei59] Erhard Heinz, *On elliptic Monge-Ampère equations and Weyl's embedding problem*, J. Analyse Math. **7** (1959), 1–52, DOI 10.1007/BF02787679. MR111943

[Hei62] Erhard Heinz, *On Weyl's embedding problem*, J. Math. Mech. **11** (1962), 421–454. MR139127

[Hei01] Juha Heinonen, *Lectures on analysis on metric spaces*, Universitext, Springer-Verlag, New York, 2001, DOI 10.1007/978-1-4613-0131-8. MR1800917

[Her43] G. Herglotz, *Über die Starrheit der Eiflächen* (German), Abh. Math. Sem. Hansischen Univ. **15** (1943), 127–129, DOI 10.1007/BF02941079. MR14714

[HH06] Qing Han and Jia-Xing Hong, *Isometric embedding of Riemannian manifolds in Euclidean spaces*, Mathematical Surveys and Monographs, vol. 130, American Mathematical Society, Providence, RI, 2006, DOI 10.1090/surv/130. MR2261749

[HHR16] M. A. Hill, M. J. Hopkins, and D. C. Ravenel, *On the nonexistence of elements of Kervaire invariant one*, Ann. of Math. (2) **184** (2016), no. 1, 1–262, DOI 10.4007/annals.2016.184.1.1. MR3505179

[HI01] Gerhard Huisken and Tom Ilmanen, *The inverse mean curvature flow and the Riemannian Penrose inequality*, J. Differential Geom. **59** (2001), no. 3, 353–437. MR1916951

[Hic65] Noel J. Hicks, *Notes on differential geometry*, Van Nostrand Mathematical Studies, No. 3, D. Van Nostrand Co., Inc., Princeton, N.J.-Toronto-London, 1965. MR179691

[Hir95] Friedrich Hirzebruch, *Topological methods in algebraic geometry*, translated from the German and Appendix One by R. L. E. Schwarzenberger, with a preface to the third English edition by the author and Schwarzenberger, Appendix Two by A. Borel, reprint of the 1978 edition, Classics in Mathematics, Springer-Verlag, Berlin, 1995. MR1335917

[Hit74] Nigel Hitchin, *Compact four-dimensional Einstein manifolds*, J. Differential Geometry **9** (1974), 435–441. MR350657

[HK71] Kenneth Hoffman and Ray Kunze, *Linear algebra*, 2nd ed., Prentice-Hall, Inc., Englewood Cliffs, NJ, 1971. MR276251

[Hod41] W. V. D. Hodge, *The Theory and Applications of Harmonic Integrals*, Cambridge University Press, Cambridge; The Macmillan Company, New York, 1941. MR3947

[Hop83] Heinz Hopf, *Differential geometry in the large*, notes taken by Peter Lax and John Gray, with a preface by S. S. Chern, Lecture Notes in Mathematics, vol. 1000, Springer-Verlag, Berlin, 1983, DOI 10.1007/978-3-662-21563-0. MR707850

[HP16] Jenny Harrison and Harrison Pugh, *Plateau's problem*, Open problems in mathematics, Springer, [Cham], 2016, pp. 273–302. MR3526938

[HR31] H. Hopf and W. Rinow, *Ueber den Begriff der vollständigen differentialgeometrischen Fläche* (German), Comment. Math. Helv. **3** (1931), no. 1, 209–225, DOI 10.1007/BF01601813. MR1509435

[HS99] Gerhard Huisken and Carlo Sinestrari, *Convexity estimates for mean curvature flow and singularities of mean convex surfaces*, Acta Math. **183** (1999), no. 1, 45–70, DOI 10.1007/BF02392946. MR1719551

[Hui84] Gerhard Huisken, *Flow by mean curvature of convex surfaces into spheres*, J. Differential Geom. **20** (1984), no. 1, 237–266. MR772132

[Hui85] Gerhard Huisken, *Ricci deformation of the metric on a Riemannian manifold*, J. Differential Geom. **21** (1985), no. 1, 47–62. MR806701

[Hui90] Gerhard Huisken, *Asymptotic behavior for singularities of the mean curvature flow*, J. Differential Geom. **31** (1990), no. 1, 285–299. MR1030675

[Hui98] Gerhard Huisken, *A distance comparison principle for evolving curves*, Asian J. Math. **2** (1998), no. 1, 127–133, DOI 10.4310/AJM.1998.v2.n1.a2. MR1656553

[Hus94] Dale Husemoller, *Fibre bundles*, 3rd ed., Graduate Texts in Mathematics, vol. 20, Springer-Verlag, New York, 1994, DOI 10.1007/978-1-4757-2261-1. MR1249482

[Huy05] Daniel Huybrechts, *Complex geometry: An introduction*, Universitext, Springer-Verlag, Berlin, 2005. MR2093043

[IL16] Thomas A. Ivey and Joseph M. Landsberg, *Cartan for beginners: Differential geometry via moving frames and exterior differential systems*, second edition of [MR2003610], Graduate Studies in Mathematics, vol. 175, American Mathematical Society, Providence, RI, 2016, DOI 10.1090/gsm/175. MR3586335

[Joe54] Konrad Jörgens, *Über die Lösungen der Differentialgleichung $rt - s^2 = 1$* (German), Math. Ann. **127** (1954), 130–134, DOI 10.1007/BF01361114. MR62326

[Jos06] Jürgen Jost, *Compact Riemann surfaces: An introduction to contemporary mathematics*, 3rd ed., Universitext, Springer-Verlag, Berlin, 2006, DOI 10.1007/978-3-540-33067-7. MR2247485

[Jos17] Jürgen Jost, *Riemannian geometry and geometric analysis*, 7th ed., Universitext, Springer, Cham, 2017, DOI 10.1007/978-3-319-61860-9. MR3726907

[Joy00] Dominic D. Joyce, *Compact manifolds with special holonomy*, Oxford Mathematical Monographs, Oxford University Press, Oxford, 2000. MR1787733

[Joy07] Dominic D. Joyce, *Riemannian holonomy groups and calibrated geometry*, Oxford Graduate Texts in Mathematics, vol. 12, Oxford University Press, Oxford, 2007. MR2292510

[Kap91] Nicolaos Kapouleas, *Compact constant mean curvature surfaces in Euclidean three-space*, J. Differential Geom. **33** (1991), no. 3, 683–715. MR1100207

[Kap01] Michael Kapovich, *Hyperbolic manifolds and discrete groups*, Progress in Mathematics, vol. 183, Birkhäuser Boston, Inc., Boston, MA, 2001. MR1792613

[Ker58] Michel A. Kervaire, *Non-parallelizability of the n-sphere for $n > 7$*, Proc. Natl. Acad. Sci. USA **44** (1958), no. 3, 280–283, DOI 10.1073/pnas.44.3.280. MR3075371

[Kir] Rob Kirby, *Markov's theorem on the nonrecognizablility of 4-manifolds: an exposition*, preprint. Available at `https://celebratio.org/Scharlemann_M/article/785/`.

[Kir89] Robion C. Kirby, *The topology of 4-manifolds*, Lecture Notes in Mathematics, vol. 1374, Springer-Verlag, Berlin, 1989, DOI 10.1007/BFb0089031. MR1001966

[KL08] Bruce Kleiner and John Lott, *Notes on Perelman's papers*, Geom. Topol. **12** (2008), no. 5, 2587–2855, DOI 10.2140/gt.2008.12.2587. MR2460872

[KL17] Bruce Kleiner and John Lott, *Singular Ricci flows I*, Acta Math. **219** (2017), no. 1, 65–134, DOI 10.4310/ACTA.2017.v219.n1.a4. MR3765659

[KL20] Bruce Kleiner and John Lott, *Singular Ricci flows II*, Geometric analysis—in honor of Gang Tian's 60th birthday, Progr. Math., vol. 333, Birkhäuser/Springer, Cham, [2020] ©2020, pp. 137–155, DOI 10.1007/978-3-030-34953-0_8. MR4181000

[Kle92] Bruce Kleiner, *An isoperimetric comparison theorem*, Invent. Math. **108** (1992), no. 1, 37–47, DOI 10.1007/BF02100598. MR1156385

[Kli61] Wilhelm Klingenberg, *Über Riemannsche Mannigfaltigkeiten mit positiver Krümmung* (German), Comment. Math. Helv. **35** (1961), 47–54, DOI 10.1007/BF02567004. MR139120

[KM63] Michel A. Kervaire and John W. Milnor, *Groups of homotopy spheres. I*, Ann. of Math. (2) **77** (1963), 504–537, DOI 10.2307/1970128. MR148075

[KN96a] Shoshichi Kobayashi and Katsumi Nomizu, *Foundations of differential geometry. Vol. I*, reprint of the 1963 original, A Wiley-Interscience Publication, Wiley Classics Library, John Wiley & Sons, Inc., New York, 1996. MR1393940

[KN96b] Shoshichi Kobayashi and Katsumi Nomizu, *Foundations of differential geometry. Vol. II*, reprint of the 1969 original, A Wiley-Interscience Publication, Wiley Classics Library, John Wiley & Sons, Inc., New York, 1996. MR1393941

[KS77] Robion C. Kirby and Laurence C. Siebenmann, *Foundational essays on topological manifolds, smoothings, and triangulations*, with notes by John Milnor and Michael Atiyah, Annals of Mathematics Studies, No. 88, Princeton University Press, Princeton, NJ; University of Tokyo Press, Tokyo, 1977. MR645390

[KW74] Jerry L. Kazdan and F. W. Warner, *Curvature functions for compact 2-manifolds*, Ann. of Math. (2) **99** (1974), 14–47, DOI 10.2307/1971012. MR343205

[Law80] H. Blaine Lawson Jr., *Lectures on minimal submanifolds. Vol. I*, 2nd ed., Mathematics Lecture Series, vol. 9, Publish or Perish, Inc., Wilmington, DE, 1980. MR576752

[Lee13] John M. Lee, *Introduction to smooth manifolds*, 2nd ed., Graduate Texts in Mathematics, vol. 218, Springer, New York, 2013. MR2954043

[Lee18] John M. Lee, *Introduction to Riemannian manifolds*, second edition of [MR1468735], Graduate Texts in Mathematics, vol. 176, Springer, Cham, 2018. MR3887684

[Lew38] Hans Lewy, *On the existence of a closed convex surface realizing a given riemannian metric*, Proceedings of the National Academy of Sciences, U.S.A. **24** (1938), no. 2, 104–106.

[Li93] Peter Li, *Lecture notes on geometric analysis*, Lecture Notes Series, vol. 6, Seoul National University, Research Institute of Mathematics, Global Analysis Research Center, Seoul, 1993. MR1320504

[Li12] Peter Li, *Geometric analysis*, Cambridge Studies in Advanced Mathematics, vol. 134, Cambridge University Press, Cambridge, 2012, DOI 10.1017/CBO9781139105798. MR2962229

[Lic63] André Lichnerowicz, *Spineurs harmoniques* (French), C. R. Acad. Sci. Paris **257** (1963), 7–9. MR156292

[Lic77] André Lichnerowicz, *Geometry of groups of transformations*, translated from the French and edited by Michael Cole, Noordhoff International Publishing, Leiden, 1977. MR438364

[LM89] H. Blaine Lawson Jr. and Marie-Louise Michelsohn, *Spin geometry*, Princeton Mathematical Series, vol. 38, Princeton University Press, Princeton, NJ, 1989. MR1031992

[Loh94] Joachim Lohkamp, *Metrics of negative Ricci curvature*, Ann. of Math. (2) **140** (1994), no. 3, 655–683, DOI 10.2307/2118620. MR1307899

[Loh95] Joachim Lohkamp, *Curvature h-principles*, Ann. of Math. (2) **142** (1995), no. 3, 457–498, DOI 10.2307/2118552. MR1356779

[LSU68] O. A. Ladyženskaja, V. A. Solonnikov, and N. N. Ural′ceva, *Linear and quasilinear equations of parabolic type* (Russian), translated from the Russian by S. Smith, Translations of Mathematical Monographs, Vol. 23, American Mathematical Society, Providence, RI, 1968. MR241822

[LY80] Peter Li and Shing Tung Yau, *Estimates of eigenvalues of a compact Riemannian manifold*, Geometry of the Laplace operator (Proc. Sympos. Pure Math., Univ. Hawaii, Honolulu, Hawaii, 1979), Proc. Sympos. Pure Math., XXXVI, Amer. Math. Soc., Providence, RI, 1980, pp. 205–239. MR573435

[LY82] Peter Li and Shing Tung Yau, *A new conformal invariant and its applications to the Willmore conjecture and the first eigenvalue of compact surfaces*, Invent. Math. **69** (1982), no. 2, 269–291, DOI 10.1007/BF01399507. MR674407

[LY86] Peter Li and Shing-Tung Yau, *On the parabolic kernel of the Schrödinger operator*, Acta Math. **156** (1986), no. 3-4, 153–201, DOI 10.1007/BF02399203. MR834612

[LY02] Fanghua Lin and Xiaoping Yang, *Geometric measure theory—an introduction*, Advanced Mathematics (Beijing/Boston), vol. 1, Science Press Beijing, Beijing; International Press, Boston, MA, 2002. MR2030862

[Mac71] Saunders Mac Lane, *Categories for the working mathematician*, Graduate Texts in Mathematics, Vol. 5, Springer-Verlag, New York-Berlin, 1971. MR354798

[Mag12] Francesco Maggi, *Sets of finite perimeter and geometric variational problems: An introduction to geometric measure theory*, Cambridge Studies in Advanced Mathematics, vol. 135, Cambridge University Press, Cambridge, 2012, DOI 10.1017/CBO9781139108133. MR2976521

[Mai56] Gaspare Mainardi, *Su la teoria generale delle superficie*, Giornale Dell' Istituto Lombardo **9** (1856), 385–404.

[Man] Ciprian Manolescu, *Four-dimensional topology*, preprint. Available at `https://web.stanford.edu/~cm5/4D.pdf`.

[Mar58] A. Markov, *The insolubility of the problem of homeomorphy* (Russian), Dokl. Akad. Nauk SSSR **121** (1958), 218–220. MR97793

[Mar60] A. A. Markov, *Insolubility of the problem of homeomorphy* (Russian), Proc. Internat. Congress Math. 1958, Cambridge Univ. Press, New York, 1960, pp. 300–306. MR115160

[Mas77] William S. Massey, *Algebraic topology: an introduction*, reprint of the 1967 edition, Graduate Texts in Mathematics, Vol. 56, Springer-Verlag, New York-Heidelberg, 1977. MR448331

[Mat82] Yukio Matsumoto, *On the bounding genus of homology* 3*-spheres*, J. Fac. Sci. Univ. Tokyo Sect. IA Math. **29** (1982), no. 2, 287–318. MR672065

[Mic80] Peter W. Michor, *Manifolds of differentiable mappings*, Shiva Mathematics Series, vol. 3, Shiva Publishing Ltd., Nantwich, 1980. MR583436

[Mil56] John Milnor, *On manifolds homeomorphic to the* 7*-sphere*, Ann. of Math. (2) **64** (1956), 399–405, DOI 10.2307/1969983. MR82103

[Mil58] John Milnor, *On simply connected* 4*-manifolds*, Symposium internacional de topología algebraica International symposium on algebraic topology, Universidad Nacional Autónoma de México and UNESCO, México, 1958, pp. 122–128. MR103472

[Mil59] John W. Milnor, *Sommes de variétes différentiables et structures différentiables des sphères* (French), Bull. Soc. Math. France **87** (1959), 439–444. MR117744

[Mil63] J. Milnor, *Morse theory*, based on lecture notes by M. Spivak and R. Wells, Annals of Mathematics Studies, No. 51, Princeton University Press, Princeton, NJ, 1963. MR163331

[Mil65] John Milnor, *Lectures on the h-cobordism theorem*, notes by L. Siebenmann and J. Sondow, Princeton University Press, Princeton, NJ, 1965. MR190942

[Mil68] J. Milnor, *A note on curvature and fundamental group*, J. Differential Geometry **2** (1968), 1–7. MR232311

[Mil76] John Milnor, *Curvatures of left invariant metrics on Lie groups*, Advances in Math. **21** (1976), no. 3, 293–329, DOI 10.1016/S0001-8708(76)80002-3. MR425012

[Mil78] John Milnor, *Analytic proofs of the "hairy ball theorem" and the Brouwer fixed-point theorem*, Amer. Math. Monthly **85** (1978), no. 7, 521–524, DOI 10.2307/2320860. MR505523

[Mil97] John W. Milnor, *Topology from the differentiable viewpoint*, based on notes by David W. Weaver, revised reprint of the 1965 original, Princeton Landmarks in Mathematics, Princeton University Press, Princeton, NJ, 1997. MR1487640

[MK06] James Morrow and Kunihiko Kodaira, *Complex manifolds*, reprint of the 1971 edition with errata, AMS Chelsea Publishing, Providence, RI, 2006, DOI 10.1090/chel/355. MR2214741

[MM88] Mario J. Micallef and John Douglas Moore, *Minimal two-spheres and the topology of manifolds with positive curvature on totally isotropic two-planes*, Ann. of Math. (2) **127** (1988), no. 1, 199–227, DOI 10.2307/1971420. MR924677

[MN14] Fernando C. Marques and André Neves, *Min-max theory and the Willmore conjecture*, Ann. of Math. (2) **179** (2014), no. 2, 683–782, DOI 10.4007/annals.2014.179.2.6. MR3152944

[Moi52] Edwin E. Moise, *Affine structures in* 3*-manifolds. V. The triangulation theorem and Hauptvermutung*, Ann. of Math. (2) **56** (1952), 96–114, DOI 10.2307/1969769. MR48805

[Moi77] Edwin E. Moise, *Geometric topology in dimensions* 2 *and* 3, Graduate Texts in Mathematics, Vol. 47, Springer-Verlag, New York-Heidelberg, 1977. MR488059

[Mok88] Ngaiming Mok, *The uniformization theorem for compact Kähler manifolds of nonnegative holomorphic bisectional curvature*, J. Differential Geom. **27** (1988), no. 2, 179–214. MR925119

[Moo01] John Douglas Moore, *Lectures on Seiberg-Witten invariants*, 2nd ed., Lecture Notes in Mathematics, vol. 1629, Springer-Verlag, Berlin, 2001. MR1830497

[Mor79] Shigefumi Mori, *Projective manifolds with ample tangent bundles*, Ann. of Math. (2) **110** (1979), no. 3, 593–606, DOI 10.2307/1971241. MR554387

[Mor96] John W. Morgan, *The Seiberg-Witten equations and applications to the topology of smooth four-manifolds*, Mathematical Notes, vol. 44, Princeton University Press, Princeton, NJ, 1996. MR1367507

[Mos68] G. D. Mostow, *Quasi-conformal mappings in n-space and the rigidity of hyperbolic space forms*, Inst. Hautes Études Sci. Publ. Math. **34** (1968), 53–104. MR236383

[MS74] John W. Milnor and James D. Stasheff, *Characteristic classes*, Annals of Mathematics Studies, No. 76, Princeton University Press, Princeton, NJ; University of Tokyo Press, Tokyo, 1974. MR440554

[MT07] John Morgan and Gang Tian, *Ricci flow and the Poincaré conjecture*, Clay Mathematics Monographs, vol. 3, American Mathematical Society, Providence, RI; Clay Mathematics Institute, Cambridge, MA, 2007. MR2334563

[MT14] John Morgan and Gang Tian, *The geometrization conjecture*, Clay Mathematics Monographs, vol. 5, American Mathematical Society, Providence, RI; Clay Mathematics Institute, Cambridge, MA, 2014. MR3186136

[Mul56] W. W. Mullins, *Two-dimensional motion of idealized grain boundaries*, J. Appl. Phys. **27** (1956), 900–904. MR78836

[Mun56] James Raymond Munkres, *Some applications of triangulation theorems*, ProQuest LLC, Ann Arbor, MI, 1956. Thesis (Ph.D.)–University of Michigan. MR2612418

[Mun00] James R. Munkres, *Topology*, second edition of [MR0464128], Prentice Hall, Inc., Upper Saddle River, NJ, 2000. MR3728284

[Mye41] S. B. Myers, *Riemannian manifolds with positive mean curvature*, Duke Math. J. **8** (1941), 401–404. MR4518

[Nas56] John Nash, *The imbedding problem for Riemannian manifolds*, Ann. of Math. (2) **63** (1956), 20–63, DOI 10.2307/1969989. MR75639

[Nir53] Louis Nirenberg, *The Weyl and Minkowski problems in differential geometry in the large*, Comm. Pure Appl. Math. **6** (1953), 337–394, DOI 10.1002/cpa.3160060303 MR58265

[Nit57] Johannes C. C. Nitsche, *Elementary proof of Bernstein's theorem on minimal surfaces*, Ann. of Math. (2) **66** (1957), 543–544, DOI 10.2307/1969907. MR90833

[Oba62] Morio Obata, *Certain conditions for a Riemannian manifold to be isometric with a sphere*, J. Math. Soc. Japan **14** (1962), 333–340, DOI 10.2969/jmsj/01430333. MR142086

[Oss70] Robert Osserman, *A proof of the regularity everywhere of the classical solution to Plateau's problem*, Ann. of Math. (2) **91** (1970), 550–569, DOI 10.2307/1970637. MR266070

[Oss78] Robert Osserman, *The isoperimetric inequality*, Bull. Amer. Math. Soc. **84** (1978), no. 6, 1182–1238, DOI 10.1090/S0002-9904-1978-14553-4. MR500557

[Oss79] Robert Osserman, *Bonnesen-style isoperimetric inequalities*, Amer. Math. Monthly **86** (1979), no. 1, 1–29, DOI 10.2307/2320297. MR519520

[Pera] Grisha Perelman, *The entropy formula for the Ricci flow and its geometric applications*, arXiv:math.DG/0211159.

[Perb] Grisha Perelman, *Ricci flow with surgery on three-manifolds*, arXiv:math.DG/0303109.

[Perc] Grisha Perelman, *Finite extinction time for the solutions to the Ricci flow on certain three-manifolds*, arXiv:math.DG/0307245.

[Pet87] Stefan Peters, *Convergence of Riemannian manifolds*, Compositio Math. **62** (1987), no. 1, 3–16. MR892147

[Pet16] Peter Petersen, *Riemannian geometry*, 3rd ed., Graduate Texts in Mathematics, vol. 171, Springer, Cham, 2016, DOI 10.1007/978-3-319-26654-1. MR3469435

[Pog52] A. V. Pogorelov, *Regularity of a convex surface with given Gaussian curvature* (Russian), Mat. Sbornik N.S. **31(73)** (1952), 88–103. MR52807

[Poi85] Henri Poincaré, *Sur les courbes définies par les équations différentielles*, Journal de mathématiques pures et appliquées **4** (1885), 167–244.

[PT82] Thomas Parker and Clifford Henry Taubes, *On Witten's proof of the positive energy theorem*, Comm. Math. Phys. **84** (1982), no. 2, 223–238. MR661134

[Rad30] Tibor Radó, *On Plateau's problem*, Ann. of Math. (2) **31** (1930), no. 3, 457–469, DOI 10.2307/1968237. MR1502955

[Ran81] Andrew Ranicki, *Exact sequences in the algebraic theory of surgery*, Mathematical Notes, vol. 26, Princeton University Press, Princeton, NJ; University of Tokyo Press, Tokyo, 1981. MR620795

[Rat19] John G. Ratcliffe, *Foundations of hyperbolic manifolds*, third edition of [MR1299730], Graduate Texts in Mathematics, vol. 149, Springer, Cham, [2019] ©2019, DOI 10.1007/978-3-030-31597-9. MR4221225

[Rau51] H. E. Rauch, *A contribution to differential geometry in the large*, Ann. of Math. (2) **54** (1951), 38–55, DOI 10.2307/1969309. MR42765

[Ray45] John William Strutt Rayleigh Baron, *The Theory of Sound*, 2nd ed., Dover Publications, New York, 1945. MR16009

[Ric63] Ian Richards, *On the classification of noncompact surfaces*, Trans. Amer. Math. Soc. **106** (1963), 259–269, DOI 10.2307/1993768. MR143186

[Rie13] Bernhard Riemann, *Bernhard Riemann "Über die Hypothesen, welche der Geometrie zu Grunde liegen"* (German), historical and mathematical commentary by Jürgen Jost, Klassische Texte der Wissenschaft. [Classical Texts of Science], Springer Spektrum, 2013, DOI 10.1007/978-3-642-35121-1. MR3525305

[Roe98] John Roe, *Elliptic operators, topology and asymptotic methods*, 2nd ed., Pitman Research Notes in Mathematics Series, vol. 395, Longman, Harlow, 1998. MR1670907

[Roh52] V. A. Rohlin, *New results in the theory of four-dimensional manifolds* (Russian), Doklady Akad. Nauk SSSR (N.S.) **84** (1952), 221–224. MR52101

[Ros97] Steven Rosenberg, *The Laplacian on a Riemannian manifold: An introduction to analysis on manifolds*, London Mathematical Society Student Texts, vol. 31, Cambridge University Press, Cambridge, 1997, DOI 10.1017/CBO9780511623783. MR1462892

[RSN90] Frigyes Riesz and Béla Sz.-Nagy, *Functional analysis*, translated from the second French edition by Leo F. Boron, reprint of the 1955 original, Dover Books on Advanced Mathematics, Dover Publications, Inc., New York, 1990. MR1068530

[Rud76] Walter Rudin, *Principles of mathematical analysis*, 3rd ed., International Series in Pure and Applied Mathematics, McGraw-Hill Book Co., New York-Auckland-Düsseldorf, 1976. MR385023

[Sak96] Takashi Sakai, *Riemannian geometry*, translated from the 1992 Japanese original by the author, Translations of Mathematical Monographs, vol. 149, American Mathematical Society, Providence, RI, 1996, DOI 10.1090/mmono/149. MR1390760

[Sal89] Simon Salamon, *Riemannian geometry and holonomy groups*, Pitman Research Notes in Mathematics Series, vol. 201, Longman Scientific & Technical, Harlow; copublished in the United States with John Wiley & Sons, Inc., New York, 1989. MR1004008

[Sas58] Shigeo Sasaki, *On the differential geometry of tangent bundles of Riemannian manifolds*, Tohoku Math. J. (2) **10** (1958), 338–354, DOI 10.2748/tmj/1178244668. MR112152

[SC02] Laurent Saloff-Coste, *Aspects of Sobolev-type inequalities*, London Mathematical Society Lecture Note Series, vol. 289, Cambridge University Press, Cambridge, 2002. MR1872526

[Sch35] H. F. Scherk, *Bemerkungen über die kleinste Fläche innerhalb gegebener Grenzen* (German), J. Reine Angew. Math. **13** (1835), 185–208, DOI 10.1515/crll.1835.13.185. MR1578041

[Sch39] Erhard Schmidt, *Über das isoperimetrische Problem im Raum von n Dimensionen* (German), Math. Z. **44** (1939), no. 1, 689–788, DOI 10.1007/BF01210681. MR1545795

[Sch66] Laurent Schwartz, *Théorie des distributions* (French), nouvelle édition, entiérement corrigée, refondue et augmentée, Publications de l'Institut de Mathématique de l'Université de Strasbourg, IX-X, Hermann, Paris, 1966. MR209834

[Sch84a] Richard Schoen, *Conformal deformation of a Riemannian metric to constant scalar curvature*, J. Differential Geom. **20** (1984), no. 2, 479–495. MR788292

[Sch84b] Richard M. Schoen, *Analytic aspects of the harmonic map problem*, Seminar on nonlinear partial differential equations (Berkeley, Calif., 1983), Math. Sci. Res. Inst. Publ., vol. 2, Springer, New York, 1984, pp. 321–358, DOI 10.1007/978-1-4612-1110-5_17. MR765241

[Sch14a] Wilhelm Schlag, *A course in complex analysis and Riemann surfaces*, Graduate Studies in Mathematics, vol. 154, American Mathematical Society, Providence, RI, 2014, DOI 10.1090/gsm/154. MR3186310

[Sch14b] Rolf Schneider, *Convex bodies: the Brunn-Minkowski theory*, Second expanded edition, Encyclopedia of Mathematics and its Applications, vol. 151, Cambridge University Press, Cambridge, 2014. MR3155183

[Shi83] Katsuhiro Shiohama, *A sphere theorem for manifolds of positive Ricci curvature*, Trans. Amer. Math. Soc. **275** (1983), no. 2, 811–819, DOI 10.2307/1999055. MR682734

[Sie69] C. L. Siegel, *Topics in complex function theory. Vol. I: Elliptic functions and uniformization theory*, translated from the original German by A. Shenitzer and D. Solitar, Interscience Tracts in Pure and Applied Mathematics, No. 25, Wiley-Interscience [A division of John Wiley & Sons, Inc.], New York-London-Sydney, 1969. MR257326

[Sim] Leon Simon, *Introduction to geometric measure theory*, 2014. Available at `https://web.stanford.edu/class/math285/ts-gmt.pdf`.

[Sim62] James Simons, *On the transitivity of holonomy systems*, Ann. of Math. (2) **76** (1962), 213–234, DOI 10.2307/1970273. MR148010

[Sim68] James Simons, *Minimal varieties in riemannian manifolds*, Ann. of Math. (2) **88** (1968), 62–105, DOI 10.2307/1970556. MR233295

[Sim83] Leon Simon, *Lectures on geometric measure theory*, Proceedings of the Centre for Mathematical Analysis, Australian National University, vol. 3, Australian National University, Centre for Mathematical Analysis, Canberra, 1983. MR756417

[Sim93] Leon Simon, *Existence of surfaces minimizing the Willmore functional*, Comm. Anal. Geom. **1** (1993), no. 2, 281–326, DOI 10.4310/CAG.1993.v1.n2.a4. MR1243525

[Siu80] Yum Tong Siu, *The complex-analyticity of harmonic maps and the strong rigidity of compact Kähler manifolds*, Ann. of Math. (2) **112** (1980), no. 1, 73–111, DOI 10.2307/1971321. MR584075

[Spi65] Michael Spivak, *Calculus on manifolds. A modern approach to classical theorems of advanced calculus*, W. A. Benjamin, Inc., New York-Amsterdam, 1965. MR209411

[Spi79a] Michael Spivak, *A comprehensive introduction to differential geometry. Vol. I*, 2nd ed., Publish or Perish, Inc., Wilmington, DE, 1979. MR532830

[Spi79b] Michael Spivak, *A comprehensive introduction to differential geometry. Vol. II*, 2nd ed., Publish or Perish, Inc., Wilmington, DE, 1979. MR532831

[Spi79c] Michael Spivak, *A comprehensive introduction to differential geometry. Vol. III*, 2nd ed., Publish or Perish, Inc., Wilmington, DE, 1979. MR532832

[Spi79d] Michael Spivak, *A comprehensive introduction to differential geometry. Vol. IV*, 2nd ed., Publish or Perish, Inc., Wilmington, DE, 1979. MR532833

[Spi79e] Michael Spivak, *A comprehensive introduction to differential geometry. Vol. V*, 2nd ed., Publish or Perish, Inc., Wilmington, DE, 1979. MR532834

[Spr57] George Springer, *Introduction to Riemann surfaces*, Addison-Wesley Publishing Co., Inc., Reading, MA, 1957. MR92855

[ST17] Jian Song and Gang Tian, *The Kähler-Ricci flow through singularities*, Invent. Math. **207** (2017), no. 2, 519–595, DOI 10.1007/s00222-016-0674-4. MR3595934

[Sto36] J. J. Stoker, *Über die Gestalt der positiv gekrümmten offenen Flächen im dreidimensionalen Raume* (German), Compositio Math. **3** (1936), 55–88. MR1556933

[SU81] J. Sacks and K. Uhlenbeck, *The existence of minimal immersions of* 2*-spheres*, Ann. of Math. (2) **113** (1981), no. 1, 1–24, DOI 10.2307/1971131. MR604040

[SW94] N. Seiberg and E. Witten, *Electric-magnetic duality, monopole condensation, and confinement in $N = 2$ supersymmetric Yang-Mills theory*, Nuclear Phys. B **426** (1994), no. 1, 19–52, DOI 10.1016/0550-3213(94)90124-4. MR1293681

[SW09] Weimin Sheng and Xu-Jia Wang, *Singularity profile in the mean curvature flow*, Methods Appl. Anal. **16** (2009), no. 2, 139–155, DOI 10.4310/MAA.2009.v16.n2.a1. MR2563745

[SW13] Jian Song and Ben Weinkove, *An introduction to the Kähler-Ricci flow*, An introduction to the Kähler-Ricci flow, Lecture Notes in Math., vol. 2086, Springer, Cham, 2013, pp. 89–188, DOI 10.1007/978-3-319-00819-6_3. MR3185333

[SY79a] R. Schoen and S. T. Yau, *On the structure of manifolds with positive scalar curvature*, Manuscripta Math. **28** (1979), no. 1-3, 159–183, DOI 10.1007/BF01647970. MR535700

[SY79b] R. Schoen and Shing Tung Yau, *Existence of incompressible minimal surfaces and the topology of three-dimensional manifolds with nonnegative scalar curvature*, Ann. of Math. (2) **110** (1979), no. 1, 127–142, DOI 10.2307/1971247. MR541332

[SY79c] Richard Schoen and Shing Tung Yau, *On the proof of the positive mass conjecture in general relativity*, Comm. Math. Phys. **65** (1979), no. 1, 45–76. MR526976

[SY80] Yum Tong Siu and Shing Tung Yau, *Compact Kähler manifolds of positive bisectional curvature*, Invent. Math. **59** (1980), no. 2, 189–204, DOI 10.1007/BF01390043. MR577360

[SY81] Richard Schoen and Shing Tung Yau, *Proof of the positive mass theorem. II*, Comm. Math. Phys. **79** (1981), no. 2, 231–260. MR612249

[SY94] R. Schoen and S.-T. Yau, *Lectures on differential geometry*, lecture notes prepared by Wei Yue Ding, Kung Ching Chang [Gong Qing Zhang], Jia Qing Zhong, and Yi Chao Xu, translated from the Chinese by Ding and S. Y. Cheng, with a preface translated from the Chinese by Kaising Tso, Conference Proceedings and Lecture Notes in Geometry and Topology, I, International Press, Cambridge, MA, 1994. MR1333601

[SY97] R. Schoen and S. T. Yau, *Lectures on harmonic maps*, Conference Proceedings and Lecture Notes in Geometry and Topology, II, International Press, Cambridge, MA, 1997. MR1474501

[Syn36] John Synge, *On the connectivity of spaces of positive curvature*, Quarterly Journal of Mathematics **7** (1936), 316–320.

[Tau82] Clifford Henry Taubes, *Self-dual Yang-Mills connections on non-self-dual* 4*-manifolds*, J. Differential Geometry **17** (1982), no. 1, 139–170. MR658473

[Tay97] Michael E. Taylor, *Partial differential equations. III. Nonlinear equations*, corrected reprint of the 1996 original, Applied Mathematical Sciences, vol. 117, Springer-Verlag, New York, 1997. MR1477408

[Tho69] John A. Thorpe, *Some remarks on the Gauss-Bonnet integral*, J. Math. Mech. **18** (1969), 779–786. MR256307

[Thu22] William P. Thurston, *The geometry and topology of three-manifolds. Vol. IV*, edited and with a preface by Steven P. Kerckhoff and a chapter by J. W. Milnor, American Mathematical Society, Providence, RI, [2022] ©2022. MR4554426

[Top06] Peter Topping, *Lectures on the Ricci flow*, London Mathematical Society Lecture Note Series, vol. 325, Cambridge University Press, Cambridge, 2006, DOI 10.1017/CBO9780511721465. MR2265040

[Tre] Andrejs Treibergs, *Some variational problems in geometry, II*, Lecture Notes VIGRE Minicourse on Nonconvex Variational Problems and Applications, University of Utah, Salt Lake City, May 23, 2005. Available at `https://www.math.utah.edu/vigre/minicourses/nvpa/AT3.pdf`.

[Uhl82a] Karen K. Uhlenbeck, *Connections with L^p bounds on curvature*, Comm. Math. Phys. **83** (1982), no. 1, 31–42. MR648356

[Uhl82b] Karen K. Uhlenbeck, *Removable singularities in Yang-Mills fields*, Comm. Math. Phys. **83** (1982), no. 1, 11–29. MR648355

[vN52] John von Neumann, *Discussion: Shape of metal grains*, Metal Interfaces, American Society for Testing Materials, Cleveland, 1952, p. 108.

[Wal70] C. T. C. Wall, *Surgery on compact manifolds*, London Mathematical Society Monographs, No. 1, Academic Press, London-New York, 1970. MR431216

[Wal04] Gerard Walschap, *Metric structures in differential geometry*, Graduate Texts in Mathematics, vol. 224, Springer-Verlag, New York, 2004, DOI 10.1007/978-0-387-21826-7. MR2045823

[War83] Frank W. Warner, *Foundations of differentiable manifolds and Lie groups*, corrected reprint of the 1971 edition, Graduate Texts in Mathematics, vol. 94, Springer-Verlag, New York-Berlin, 1983. MR722297

[Wei23] Roland Weitzenböck, *Invariantentheorie*, P. Noordhoff, Groningen, 1923.

[Wei26] André Weil, *Sur les surfaces a courbure negative (on surfaces of negative curvature)*, C. R. Acad. Sci. Paris **182** (1926), 1069–1071.

[Wel08] Raymond O. Wells Jr., *Differential analysis on complex manifolds*, 3rd ed., with a new appendix by Oscar Garcia-Prada, Graduate Texts in Mathematics, vol. 65, Springer, New York, 2008, DOI 10.1007/978-0-387-73892-5. MR2359489

[Wen86] Henry C. Wente, *Counterexample to a conjecture of H. Hopf*, Pacific J. Math. **121** (1986), no. 1, 193–243. MR815044

[Wey16] Hermann Weyl, *Über die Bestimmheit einer geschlossenen konvex Fläche durch ihr Linienelement*, Vierteljahresschrift der nat.-Forsch. Ges. Zürich **61** (1916), 40–72.

[Wey97] Hermann Weyl, *The classical groups: Their invariants and representations*, 15th printing, Princeton Paperbacks, Princeton Landmarks in Mathematics, Princeton University Press, Princeton, NJ, 1997. MR1488158

[Whi49] J. H. C. Whitehead, *On simply connected, 4-dimensional polyhedra*, Comment. Math. Helv. **22** (1949), 48–92, DOI 10.1007/BF02568048. MR29171

[Whi78] George W. Whitehead, *Elements of homotopy theory*, Graduate Texts in Mathematics, vol. 61, Springer-Verlag, New York-Berlin, 1978. MR516508

[Whi03] Brian White, *The nature of singularities in mean curvature flow of mean-convex sets*, J. Amer. Math. Soc. **16** (2003), no. 1, 123–138, DOI 10.1090/S0894-0347-02-00406-X. MR1937202

[Whi05] Brian White, *A local regularity theorem for mean curvature flow*, Ann. of Math. (2) **161** (2005), no. 3, 1487–1519, DOI 10.4007/annals.2005.161.1487. MR2180405

[Wil65] T. J. Willmore, *Note on embedded surfaces* (English, with Romanian and Russian summaries), An. Şti. Univ. "Al. I. Cuza" Iaşi Secţ. I a Mat. (N.S.) **11B** (1965), 493–496. MR202066

[Wit81] Edward Witten, *A new proof of the positive energy theorem*, Comm. Math. Phys. **80** (1981), no. 3, 381–402. MR626707

[Wit82] Edward Witten, *Supersymmetry and Morse theory*, J. Differential Geometry **17** (1982), no. 4, 661–692 (1983). MR683171

[Wol11] Joseph A. Wolf, *Spaces of constant curvature*, 6th ed., AMS Chelsea Publishing, Providence, RI, 2011, DOI 10.1090/chel/372. MR2742530

[Wu91] Lang-Fang Wu, *The Ricci flow on 2-orbifolds with positive curvature*, J. Differential Geom. **33** (1991), no. 2, 575–596. MR1094470

[Wu17] Hung-Hsi Wu, *The Bochner technique in differential geometry*, new edition of the 1988 original [MR1079031], expanded version of [MR0714349], CTM. Classical Topics in Mathematics, vol. 6, Higher Education Press, Beijing, 2017. MR3838345

[WX17] Guozhen Wang and Zhouli Xu, *The triviality of the 61-stem in the stable homotopy groups of spheres*, Ann. of Math. (2) **186** (2017), no. 2, 501–580, DOI 10.4007/annals.2017.186.2.3. MR3702672

[Yau75] Shing Tung Yau, *Harmonic functions on complete Riemannian manifolds*, Comm. Pure Appl. Math. **28** (1975), 201–228, DOI 10.1002/cpa.3160280203. MR431040

[Yau76] Shing Tung Yau, *Some function-theoretic properties of complete Riemannian manifold and their applications to geometry*, Indiana Univ. Math. J. **25** (1976), no. 7, 659–670, DOI 10.1512/iumj.1976.25.25051. MR417452

[Yau78] Shing Tung Yau, *On the Ricci curvature of a compact Kähler manifold and the complex Monge-Ampère equation. I*, Comm. Pure Appl. Math. **31** (1978), no. 3, 339–411, DOI 10.1002/cpa.3160310304. MR480350

[Ye94] Rugang Ye, *Global existence and convergence of Yamabe flow*, J. Differential Geom. **39** (1994), no. 1, 35–50. MR1258912

[ZG13] Wei Zeng and Xianfeng David Gu, *Ricci flow for shape analysis and surface registration: Theories, algorithms and applications*, SpringerBriefs in Mathematics, Springer, New York, 2013, DOI 10.1007/978-1-4614-8781-4. MR3136003

[Zha01] Weiping Zhang, *Lectures on Chern-Weil theory and Witten deformations*, Nankai Tracts in Mathematics, vol. 4, World Scientific Publishing Co., Inc., River Edge, NJ, 2001, DOI 10.1142/9789812386588. MR1864735

[Zhe00] Fangyang Zheng, *Complex differential geometry*, AMS/IP Studies in Advanced Mathematics, vol. 18, American Mathematical Society, Providence, RI; International Press, Boston, MA, 2000, DOI 10.1090/amsip/018. MR1777835

[Zho20] Xin Zhou, *On the multiplicity one conjecture in min-max theory*, Ann. of Math. (2) **192** (2020), no. 3, 767–820, DOI 10.4007/annals.2020.192.3.3. MR4172621

[ZY84] Jia Qing Zhong and Hong Cang Yang, *On the estimate of the first eigenvalue of a compact Riemannian manifold*, Sci. Sinica Ser. A **27** (1984), no. 12, 1265–1273. MR794292

Index

SELECTED PUBLISHED TITLES IN THIS SERIES

245 **Bennett Chow and Yutze Chow,** Lectures on Differential Geometry, 2024
244 **John M. Lee,** Introduction to Complex Manifolds, 2024
242 **Jayadev S. Athreya and Howard Masur,** Translation Surfaces, 2024
241 **Thorsten Theobald,** Real Algebraic Geometry and Optimization, 2024
240 **Nam Q. Le,** Analysis of Monge–Ampère Equations, 2024
238 **Julio González-Díaz, Ignacio García-Jurado, and M. Gloria Fiestras-Janeiro,** An Introductory Course on Mathematical Game Theory and Applications, Second Edition, 2023
237 **Michael Levitin, Dan Mangoubi, and Iosif Polterovich,** Topics in Spectral Geometry, 2023
235 **Bennett Chow,** Ricci Solitons in Low Dimensions, 2023
234 **Andrea Ferretti,** Homological Methods in Commutative Algebra, 2023
233 **Andrea Ferretti,** Commutative Algebra, 2023
232 **Harry Dym,** Linear Algebra in Action, Third Edition, 2023
231 **Luís Barreira and Yakov Pesin,** Introduction to Smooth Ergodic Theory, Second Edition, 2023
230 **Barbara Kaltenbacher and William Rundell,** Inverse Problems for Fractional Partial Differential Equations, 2023
229 **Giovanni Leoni,** A First Course in Fractional Sobolev Spaces, 2023
228 **Henk Bruin,** Topological and Ergodic Theory of Symbolic Dynamics, 2022
227 **William M. Goldman,** Geometric Structures on Manifolds, 2022
226 **Milivoje Lukić,** A First Course in Spectral Theory, 2022
225 **Jacob Bedrossian and Vlad Vicol,** The Mathematical Analysis of the Incompressible Euler and Navier-Stokes Equations, 2022
224 **Ben Krause,** Discrete Analogues in Harmonic Analysis, 2022
223 **Volodymyr Nekrashevych,** Groups and Topological Dynamics, 2022
222 **Michael Artin,** Algebraic Geometry, 2022
221 **David Damanik and Jake Fillman,** One-Dimensional Ergodic Schrödinger Operators, 2022
220 **Isaac Goldbring,** Ultrafilters Throughout Mathematics, 2022
219 **Michael Joswig,** Essentials of Tropical Combinatorics, 2021
218 **Riccardo Benedetti,** Lectures on Differential Topology, 2021
217 **Marius Crainic, Rui Loja Fernandes, and Ioan Mărcuţ,** Lectures on Poisson Geometry, 2021
216 **Brian Osserman,** A Concise Introduction to Algebraic Varieties, 2021
215 **Tai-Ping Liu,** Shock Waves, 2021
214 **Ioannis Karatzas and Constantinos Kardaras,** Portfolio Theory and Arbitrage, 2021
213 **Hung Vinh Tran,** Hamilton–Jacobi Equations, 2021
212 **Marcelo Viana and José M. Espinar,** Differential Equations, 2021
211 **Mateusz Michałek and Bernd Sturmfels,** Invitation to Nonlinear Algebra, 2021
210 **Bruce E. Sagan,** Combinatorics: The Art of Counting, 2020
209 **Jessica S. Purcell,** Hyperbolic Knot Theory, 2020
208 **Vicente Muñoz, Ángel González-Prieto, and Juan Ángel Rojo,** Geometry and Topology of Manifolds, 2020
207 **Dmitry N. Kozlov,** Organized Collapse: An Introduction to Discrete Morse Theory, 2020

For a complete list of titles in this series, visit the
AMS Bookstore at **www.ams.org/bookstore/gsmseries/**.